MW00439514

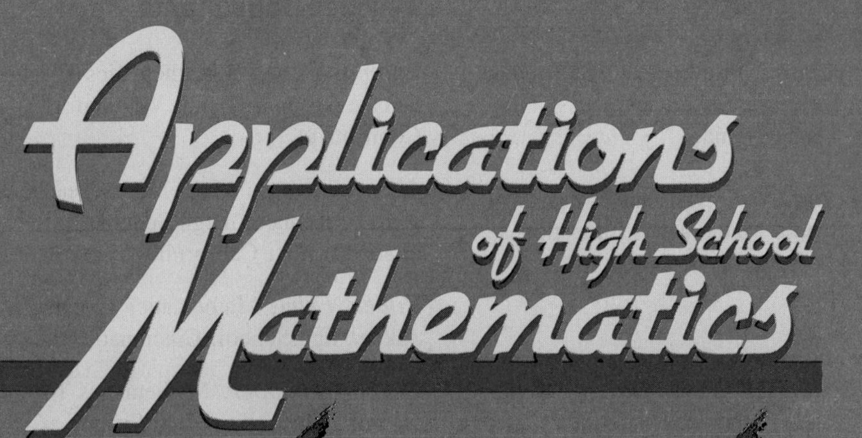

Martin P. Cohen
Gerald H. Elgarten
Francis J. Gardella
Wendy S. Lewis
Joanne E. Meldon
Marvin S. Weingarden

TEACHER CONSULTANTS
Thomas B. Day
Michael Fatur
Robert Friel
Barbara Skipper
Francis W. Stanley

Houghton Mifflin Company BOSTON
Atlanta Dallas Geneva, Illinois Palo Alto Princeton Toronto

Authors

Martin P. Cohen, Professor of Mathematics Education, University of Pittsburgh, Pittsburgh, Pennsylvania.

Gerald H. Elgarten, Assistant Professor of Secondary School Mathematics Education, City College of the City University of New York.

Francis J. Gardella, Supervisor of Mathematics and Computer Studies, East Brunswick Public Schools, East Brunswick, New Jersey.

Wendy S. Lewis, Mathematics Teacher, Royal High School, Simi Valley, California.

Joanne E. Meldon, Mathematics Teacher, Taylor Allderdice High School, Pittsburgh, Pennsylvania.

Marvin S. Weingarden, Supervisor of Secondary Mathematics, Detroit Public Schools, Detroit, Michigan.

Teacher Consultants

Thomas B. Day, Mathematics Teacher, Choate Rosemary Hall, Wallingford, Connecticut.

Michael Fatur, Mathematics Teacher, John F. Kennedy High School, Granada Hills, California.

Robert Friel, Mathematics Teacher, Aiken High School, Cincinnati, Ohio.

Barbara Skipper, Mathematics Teacher, Douglas MacArthur High School, North East Independent School District, San Antonio, Texas.

Francis W. Stanley, Mathematics Teacher, Shelby Junior High School, Shelby, North Carolina.

Printed in U.S.A.

ISBN: 0-395-59125-2

DEFGHIJ-VH-9654

contents

Unit 1 Using Whole Numbers and Decimals

Chapter 1 Whole Number and Decimal Concepts 1

Chapter 2
Adding and Subtracting Whole Numbers and Decimals 19

Chapter 3 Multiplying Whole Numbers 39

Chapter 4 Dividing Whole Numbers 59

Chapter 5 Multiplying and Dividing Decimals 77

Chapter 6 Metric Measurement 97

Unit 2 Using Fractions

Chapter 7 Number Theory and Fraction Concepts 121

Chapter 8 Multiplying and Dividing Fractions 143

Chapter 9 Adding and Subtracting Fractions 165

Chapter 10 Fractions and Decimals 185

Chapter 11 U.S. Customary Measurement 203

Unit 3 Using Ratio, Proportion, and Percent

Chapter 12 Ratio and Proportion 227

Chapter 13 Percent 245

Chapter 14 More Percent 265

Chapter 15 Using Percent 291

Unit 4 Statistics and Probability

Chapter 16 Graphing Data 317

Chapter 17 Interpreting Data 345

Chapter 18 Probability 373

Unit 5 Algebra

Chapter 19 Rational Numbers 401

Chapter 23 Surface Area and Volume 513

Chapter 24 Triangles 543

Consumer Applications

Income

Credit and Loans

Taxes

Housing

Automobiles/Transportation

Investing Money

Banking

Shopping

Food and Nutrition

Insurance

Statistics

Careers

Competitors in sports events usually wear numbers. The numbers make it easy for officials to keep track of the contestants.

CHAPTER 1

WHOLE NUMBER AND DECIMAL CONCEPTS

1-1 PLACE VALUE

There are ten digits in our number system. The **place** of a digit in a number determines the **value** of the digit. A **place value chart** shows the place of each digit in a number.

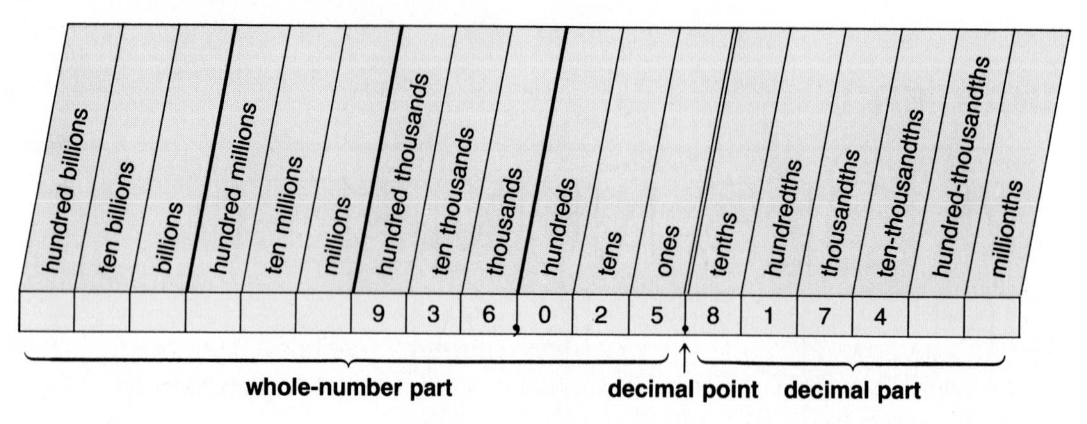

Commas separate the whole-number part into groups of three digits, called **periods.** You use the names of the periods to read a number.

46,207,690,433 → forty-six *billion,* two hundred seven *million,* six hundred ninety *thousand,* four hundred thirty-three

To read the decimal part of a number, read the digits after the decimal point as you would a whole number. Then read the place of the last digit. If there are digits greater than zero in the whole-number part, read the decimal point as *and.*

354.72 → three hundred fifty-four *and* seventy-two *hundredths*

You can write a number either in *words* or as a *numeral.*

example 1

Write each number in words: **a.** 4,603,259 **b.** 231.807 **c.** 0.2145

solution

	word form	**short word form**
a.	four million, six hundred three thousand, two hundred fifty-nine	4 million, 603 thousand, 259
b.	two hundred thirty-one and eight hundred seven thousandths	231 and 807 thousandths
c.	two thousand one hundred forty-five ten-thousandths	2145 ten-thousandths

your turn

Write the word form of each number.

1. 58,593
2. 81,000,818,100
3. 429.6
4. 0.0235

Write the short word form of each number.

5. 872,301,246
6. 1,700,000,017
7. 0.92
8. 3000.003

example 2

Write the numeral form of each number.

a. five million, eighty-one thousand, nine

b. seven thousand, forty-two and thirty-one thousandths

solution

a. Write the short word form. Leave a blank for each empty place. Put a zero in each blank.

5 million, _81 thousand, __9

5 million, 081 thousand, 009

The **numeral form** is 5,081,009.

b. Write the whole-number part. Determine the number of decimal places. Write the decimal part with the last digit in the last place. Put a zero in each empty decimal place.

7042._31

7042.031

The **numeral form** is 7042.031.

your turn

Write the numeral form of each number.

9. 28 million, 990 thousand
10. 157 thousand, 64 and 3 tenths
11. forty-nine billion, eight hundred three million, two hundred seventy-one
12. five thousand thirteen ten-thousandths

practice exercises

practice for example 1 (pages 2–3)

Write the word form of each number.

1. 14,612
2. 6,000,606,066
3. 0.23
4. 0.0708
5. 0.0017
6. 77.25
7. 4250.1
8. 987.339

Write the short word form of each number.

9. 7525
10. 16,000,182
11. 0.7
12. 0.348
13. 0.8401
14. 11.9
15. 494.49
16. 2380.0073

practice for example 2 (page 3)

Write the numeral form of each number.

17. 934 thousand, 276

18. five thousand, four hundred sixty

19. seven billion, three hundred

20. 801 ten-thousandths

21. ten and ninety-two thousandths

22. twenty and eighteen hundredths

mixed practice (pages 2–3)

Give the place of the underlined digit in each number.

23. $4{,}2\underline{9}6{,}451{,}004$ **24.** $904{,}\underline{8}12{,}366$ **25.** $0.4\underline{3}38$ **26.** $115.072\underline{7}5$

27. Eighty-one and seventy-five hundredths inches is the record height for the women's high jump. Write the numeral form of this number.

28. The estimated population of Japan in a recent year was 120,246,000. Write the short word form of this number.

29. The speed of light is about 186,281.7 mi/s (miles per second). Write the word form of this number.

30. The width of a small bandage is about three hundred seventy-five thousandths of an inch. Write the numeral form of this number.

review exercises

Find each answer.

1. 7 + 9 **2.** 14 − 6 **3.** 6 × 7 **4.** 30 ÷ 5 **5.** 8 − 4 **6.** 11 + 8

7. 54 ÷ 6 **8.** 5 × 10 **9.** 16 + 5 **10.** 20 − 7 **11.** 56 ÷ 8 **12.** 8 × 9

calculator corner

Calculators differ in the number of digits they display. They also differ in the way that they display zeros.

Write the answer as it appears in the calculator display.

1. Enter .25 and then press [=].

2. Enter 2.7500 and then press [=].

3. Enter 00463 and then press [=].

4. Enter 702.012 and then press [=].

5. Enter 9 as many times as possible. How many times does 9 appear? Write the numeral form of the number that is in the display.

6. Enter . and 9 as many times as possible. Does the 0 before the decimal point remain on the display? How many times does 9 appear?

1-2 WRITING GREATER NUMBERS

Greater numbers can be written in a short word form that includes decimals. The chart below shows some greater numbers in that form.

Registered Vehicles in U.S.

cars	127.9 million
motorcycles	5.5 million
trucks	38.1 million

example 1

Write the numeral form of the number of registered cars.

solution

The number is given as 127.9 million. Since there are 6 whole-number places to the right of the millions' place, move the decimal point 6 places to the right. Annex zeros as needed.

127.9 million
127.900000
127,900,000

your turn

Write the numeral form of each number.

1. 38.1 million
2. 7.25 billion
3. 460.9 thousand
4. 8.504 million

example 2

Complete.

a. 20,470,000 = 20.47 __?__

b. 89,200 = __?__ thousand

solution

a. Write the name of the greatest period in which there are nonzero digits.
20,470,000 = 20.47 million

b. Place the decimal point to the right of the digits in the period that is named.
89,200 = 89.2 thousand

your turn

Complete. Replace each __?__ with *thousand, million,* or *billion.*

5. 238,400 = 238.4 __?__
6. 15,650,000,000 = 15.65 __?__

Complete. Replace each __?__ with the correct decimal.

7. 4,942,000,000 = __?__ billion
8. 9,300,000 = __?__ million

practice exercises

practice for example 1 (page 5)

Write the numeral form of each number.

1. 4.3 million
2. 6.8 billion
3. 45.9 thousand
4. 37.9 thousand
5. 9.125 billion
6. 11.04 million
7. 700.16 million
8. 234.162 billion

practice for example 2 (page 5)

Complete. Replace each ___?___ with *thousand, million,* or *billion*.

9. 306,400,000,000 = 306.4 ___?___
10. 8,125,000 = 8.125 ___?___

Complete. Replace each ___?___ with the correct decimal.

11. 14,150,000 = ___?___ million
12. 9700 = ___?___ thousand

mixed practice (page 5)

Tell whether each statement is *true* or *false*.

13. 97,870,000,000 = 97.87 billion
14. 248,100,000 = 248.1 billion
15. 2.615 million = 2,650,000
16. 8.45 thousand = 8450
17. A telephone company reports that customers use the Yellow Pages 3.9 billion times per year. Write the numeral form of this number.
18. A well-known magazine sells 10.55 million copies each month. Write the numeral form of this number.
19. Film distributors collected $464,900,000 from the top ten movie rentals in a recent year. Write this amount of money in the short word form that includes decimals.
20. A national magazine reported that the highest-rated television show earns $2,550,000 per episode from advertising. Write this amount in the short word form that includes decimals.
21. In a recent year the Baltimore public library system had a total of 1.902 million volumes and a budget of $11.8 million. Write the numeral form of each number.

review exercises

Write the numeral form of each number.

1. 45 million, 12 thousand, 768
2. 16 and 856 ten-thousandths
3. two hundred eighty thousand, ten
4. four hundred one thousandths

1-3 ROUNDING WHOLE NUMBERS AND DECIMALS

To **round** a number to a given place, use the following method.

1. Circle the digit in the place to which you are rounding.
2. If the digit to the right of the circle is *less than 5,* copy the circled digit. If the digit to the right of the circle is *5 or more,* add 1 to the circled digit.
3. To the right of the circle:
   a. Drop all digits in the decimal places.
   b. Replace all digits in the whole-number places with zeros.

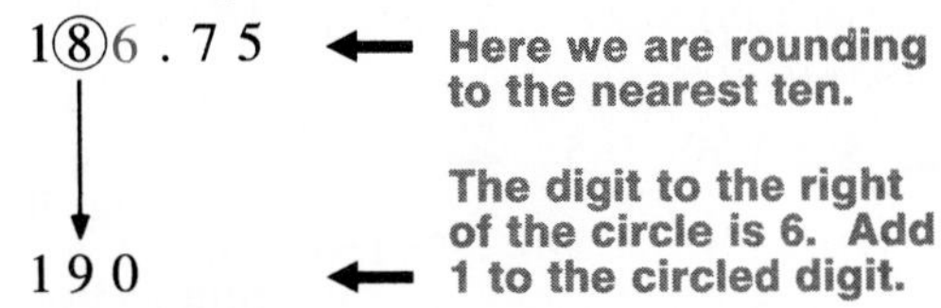

example 1

a. Round 0.283 to the nearest hundredth. **b.** Round 39,713 to the nearest thousand.

solution

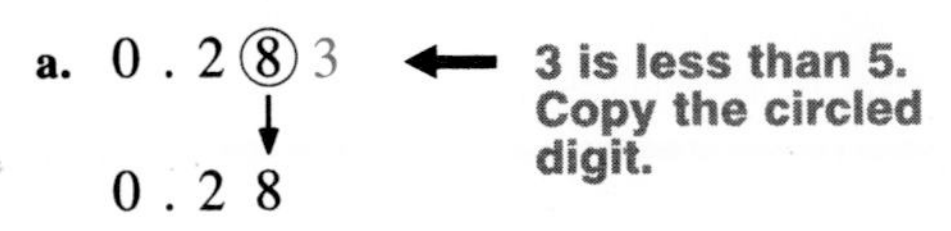

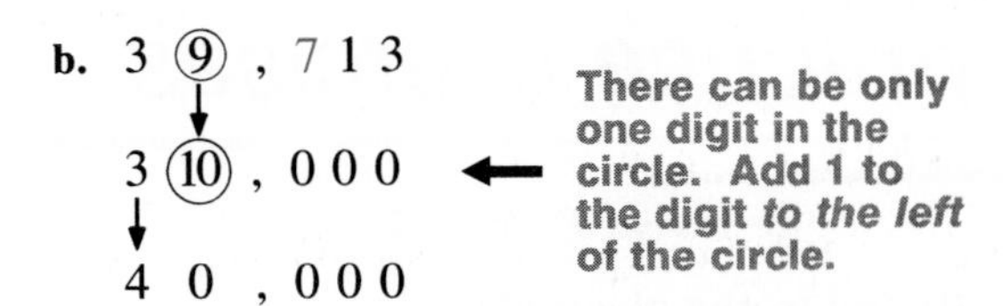

your turn

Round to the given place.

1. 15.84; nearest tenth
2. 45.21; nearest whole number
3. 0.0982; nearest hundredth
4. 731.05; nearest thousand

The **leading digit** of a number is its first *nonzero* digit.

example 2

Round to the place of the leading digit.

a. ⑥9 1 . 5 ← Round to the nearest hundred.

7 0 0

b. 0 . 0②5 ← Round to the nearest hundredth.

0 . 0 3

your turn

Round to the place of the leading digit.

5. 449
6. 9546
7. 0.707
8. 0.0883

Sometimes you may need to round an amount of money to the *nearest dollar* or to the *nearest cent.*

example 3

a. Round $436.95 to the nearest dollar.

b. Round $1.423 to the nearest cent.

solution

a. $ 4 3 ⑥ . 9 5 ← **Rounding to the nearest *dollar* is the same as rounding to the nearest *whole number.***

↓

$ 4 3 7

b. $ 1 . 4 ② 3 ← **Rounding to the nearest *cent* is the same as rounding to the nearest *hundredth.***

↓

$ 1 . 4 2

your turn

Round to the given place.

9. $15.3333; nearest dollar

10. $212.374; nearest cent

11. $0.97; nearest dollar

12. $109.999; nearest cent

practice exercises

practice for example 1 (page 7)

Round to the given place.

1. 803,803; nearest thousand

2. 25,963; nearest hundred

3. 5842.4; nearest whole number

4. 7.532; nearest ten

5. 0.395; nearest hundredth

6. 0.0491; nearest thousandth

7. 915.726; nearest tenth

8. 861.2033; nearest hundredth

practice for example 2 (page 7)

Round to the place of the leading digit.

9. 37 **10.** 6641 **11.** 517.35 **12.** 97,084

13. 9.81 **14.** 2.299 **15.** 0.0938 **16.** 0.0044

practice for example 3 (page 8)

Round to the nearest dollar.

17. $6.32 **18.** $307.09 **19.** $500.84 **20.** $3209.73

Round to the nearest cent.

21. $2.129 **22.** $3.1147 **23.** $798.091 **24.** $49.995

mixed practice (pages 7–8)

Round to the place of the underlined digit.

25. $4\underline{3}5.2$ **26.** $71\underline{8},643$ **27.** $16.47\underline{9}8$ **28.** $25,\underline{9}61$

29. $\underline{5}4,917,003$ **30.** $30.\underline{7}2$ **31.** $\underline{0}.89$ **32.** $0.02\underline{8}2$

33. $0.6\underline{9}5$ **34.** $\underline{9}73$ **35.** $\$19\underline{3}.67$ **36.** $\$4\underline{4}.38$

37. $\$6.0\underline{8}33$ **38.** $\$355.93\underline{7}5$ **39.** $72.5\underline{5}$ **40.** $8\underline{3}.125$

41. Write four different amounts that round to $38.

42. Write four different amounts that round to $9.50.

43. A calculator display shows 9090.909. Round this number to the nearest whole number.

44. A calculator display shows 3.1415927. Round this number to the nearest hundredth.

45. Jaime McSwain earned $324.49 last week. Round this amount to the nearest dollar.

46. The distance around Earth at the equator is about 24,902.4 mi (miles). Round this number to the nearest thousand miles.

review exercises

Write the numeral form of each number.

1. 4.5 million **2.** 3.8 thousand

3. 16.97 billion **4.** 8.302 million

5. 9.921 billion **6.** 27.44 million

7. 243.2 thousand **8.** 561.6 billion

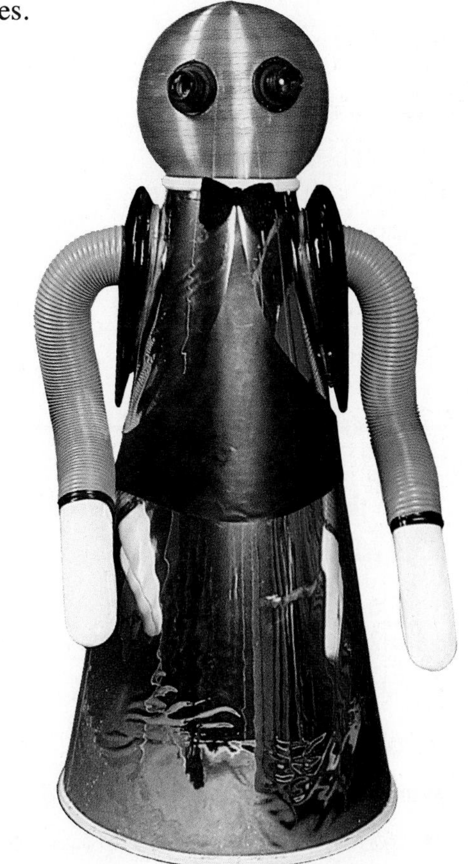

mental math

Often newspaper headlines and stories use *rounded* numbers instead of *exact* numbers.

Tell whether each number is more likely *rounded* or *exact*.

1. Local Company Hires 295 Employees
2. 62 Home Robots Sold at Computer Show
3. SCHOOL DEFICIT—1.2 Million Dollars
4. 55,000 Watch as Cardinals Defeat Giants
5. Voters Elect 20 New Council Members

1-4 COMPARING WHOLE NUMBERS AND DECIMALS

To **compare** two numbers, use the following method.

1. If one number has more *whole-number* places, then it is the greater number.
2. If the numbers have the same number of whole-number places, then compare digits from left to right. Find the first place in which the digits differ. The number with the greater digit in that place is the greater number.
3. Insert the correct symbol between the numbers.

   $>$ *is greater than* $<$ *is less than* $=$ *is equal to*

same

3 7 . 6 [8]
3 7 . 6 [3] 1

$8 > 3$

So $37.68 > 37.631$.

example 1

Compare. Replace each __?__ with $>$, $<$, or $=$.

a. 85,432 __?__ 85,423 **b.** 0.68 __?__ 0.684

solution

a. 8 5 4 [3] 2
8 5 4 [2] 3

$3 > 2$ So $85{,}432 > 85{,}423$.

b. 0 . 6 8 [0] ← **Write 0.68 as 0.680.**
0 . 6 8 [4]

$0 < 4$ So $0.68 < 0.684$.

your turn

Compare. Replace each __?__ with $>$, $<$, or $=$.

1. 5439 __?__ 5409 **2.** 3.040 __?__ 3.04 **3.** 0.0033 __?__ 0.03

example 2

Write the numbers in order from least to greatest: 8.762; 8.627; 8.67

solution

8 . [7] 6 2
8 . [6] 2 7
8 . [6] 7

$7 > 6$, so 8.762 is the *greatest* number.

8 . 6 [2] 7
8 . 6 [7]

$2 < 7$, so 8.627 is the *least* number.

The numbers in order from least to greatest: 8.627; 8.67; 8.762

your turn

Write each set of numbers in order from least to greatest.

4. 6571; 65,710; 65,791 **5.** 0.892; 0.8192; 0.8 **6.** \$14.44; \$14.54; \$14.50

practice exercises

practice for example 1 (page 10)

Compare. Replace each ? with >, <, or =.

1. 7893 ? 7889
2. $52,130 ? $5213
3. 0.48 ? 0.049
4. 0.06 ? 0.060
5. 7.443 ? 7.461
6. 7.52 ? 75.2

practice for example 2 (page 10)

Write each set of numbers in order from least to greatest.

7. $2583; $2385; $285
8. 264,915; 264,159; 264,905
9. 0.001; 0.0001; 0.01
10. 0.134; 1.34; 13.4
11. 5.067; 7.605; 7.506
12. $9.60; $9.66; $9.06

mixed practice (page 10)

Tell whether each statement is *true* or *false*.

13. $20{,}305 > 20{,}035$
14. $724 < 7240$
15. $0.045 = 0.45$
16. $\$34 = \$34.00$
17. $\$88.80 < \$88.08$
18. $19.02 > 19.020$
19. 2.5 million $<$ 2.55 million
20. 35.01 billion $>$ 35.1 billion
21. $18{,}253 < 18{,}523 < 18{,}325$
22. $515{,}035 < 518{,}932 < 518{,}942$
23. $0.309 > 0.0399 > 0.039$
24. $\$.48 > \$.45 > \$.46$
25. Sean earns $25,850 per year. Each year Lisa earns $24,890 and Janine earns $25,890. Who earns the most money per year?
26. France's population is about 55.4 million, and Egypt's is about 50.5 million. Which country has the greater population?
27. From 1964 to 1984, the winners of the women's Olympic pentathlon had these point totals: 5246; 5098; 4081; 4745; 5083; 5469. Write the point totals in order from least to greatest.
28. During a five-year period, the American League batting champions had these batting averages: 0.332; 0.361; 0.343; 0.368; 0.357. Write the batting averages in order from least to greatest.

review exercises

Write the word form of each number.

1. 26.1
2. 759
3. 1844
4. 92.17
5. 206,570,313
6. 1,000,985
7. 402.039
8. 6000.0008

SKILL REVIEW

Write the word form of each number. 1-1

1. 10,833
2. 56.56
3. 0.202
4. 0.0062
5. 481,620,470
6. 7,000,931,309
7. 3097.18
8. 123.514

Write the numeral form of each number.

9. two hundred ninety-eight thousand
10. thirty-six and one hundredth
11. The maximum running speed of a quarter horse is 47 and 5 tenths miles per hour. Write the numeral form of this number.

Write the numeral form of each number. 1-2

12. 5.9 million
13. 6.1 million
14. 18.43 thousand
15. 85.5 billion
16. 72.301 billion
17. 290.7 thousand

Complete. Replace each __?__ with *thousand, million,* or *billion.*

18. 1,120,000 = 1.12 __?__
19. 758,900 = 758.9 __?__
20. 4,600,000,000 = 4.6 __?__
21. 24,300,000 = 24.3 __?__
22. In a recent year manufacturers shipped 167.1 million record albums, worth $1.281 billion. Write the numeral form of each number.

Round to the given place. 1-3

23. 656,192; nearest thousand
24. 0.7938; nearest thousandth
25. $9.995; nearest dollar
26. $26.264; nearest cent

Round to the place of the leading digit.

27. 826,529
28. 99.77
29. 0.0016
30. 3.0765
31. The area of Lake Huron is about 59,596 km^2 (square kilometers). Round this number to the place of the leading digit.

Compare. Replace each __?__ with >, <, or =. 1-4

32. 8933 __?__ 8925
33. 1.0085 __?__ 1.083
34. 68.549 __?__ 68.54
35. $105.99 __?__ $109.55
36. 0.681 __?__ 0.6810
37. 536.2 __?__ 536.25

Write each set of numbers in order from least to greatest.

38. $73.94; $794.39; $79.43
39. 0.338; 0.0388; 0.0038
40. In four Olympic games, the winning times for men's downhill skiing were 111.43 s (seconds), 105.59 s, 105.5 s, and 105.72 s. Write these numbers in order from least to greatest.

1-5 WRITING CHECKS

A checking account is a convenient way to manage your money. You can pay for most items by writing a check. The amount of the check should not be greater than the amount you have in your checking account. A sample check is shown below.

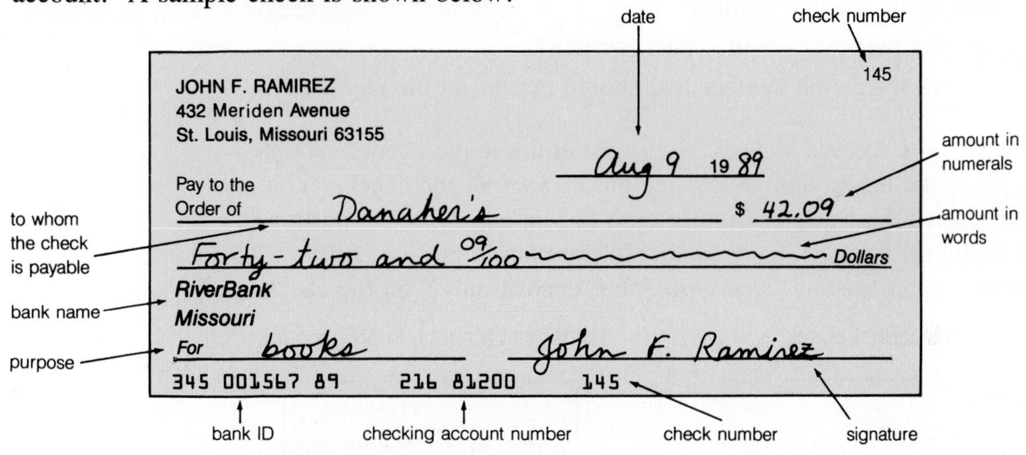

example

Use the check at the right.

a. To whom is the check payable?

b. What is the date of the check?

c. What is the amount of the check in words?

JOHN F. RAMIREZ
432 Meriden Avenue
St. Louis, Missouri 63155
146
Pay to the Order of West Insurance, Inc. Oct. 19 19 89 $ 601.47
Six hundred one and 47/100 Dollars
RiverBank
Missouri
For car insurance John F. Ramirez
345 001567 89 216 81200 146

solution

a. West Insurance, Inc. **b.** October 19, 1989 **c.** Six hundred one and $\frac{47}{100}$ dollars

exercises

Write the numeral form of each amount.

1. Six and $\frac{60}{100}$ dollars

2. Forty-nine and $\frac{94}{100}$ dollars

3. One hundred eighteen and $\frac{53}{100}$ dollars

4. Four thousand twelve and $\frac{00}{100}$ dollars

Write each amount in words as it would appear on a check.

5. \$8.75 **6.** \$308.50 **7.** \$215.06 **8.** \$1625

Use the check at the right.

JOHN F. RAMIREZ
432 Meriden Avenue
St. Louis, Missouri 63155

147

Oct. 23 19 89

Pay to the Order of F. W.'s Catalog Store $ 156.82

Dollars

RiverBank
Missouri

For clothes John F. Ramirez

345 001567 89 216 81200 147

9. Who wrote the check?
10. To whom is the check payable?
11. In what bank does John Ramirez have a checking account?
12. What is the account number?
13. For what purpose was the check written?
14. Write in words the number that should appear on the *Dollars* line.

After you receive a check, you must **endorse** the check. When you get to the bank, sign your name on the *back* of the check. You can either receive the entire amount in cash or you can deposit a portion of it in a savings or checking account. If you plan to put the *entire* amount in your account, also write "For deposit only" on the check.

To cash a check:

John F. Ramirez

To deposit a check in an account:

John F. Ramirez
For deposit only

15. You receive a paycheck and want to deposit it in your account. Make a sketch to show how you would endorse the check.
16. You receive a check and want to cash it. Make a sketch to show how you would endorse the check.

Find the answers to the following questions. You may want to talk to a local bank employee.

17. Is a check valid if its purpose has not been indicated?
18. Is a check valid if the front of the check has not been signed?
19. Sometimes when writing a check, a person may accidentally write two different amounts of money. Is the numeral form or the word form accepted as the correct amount?

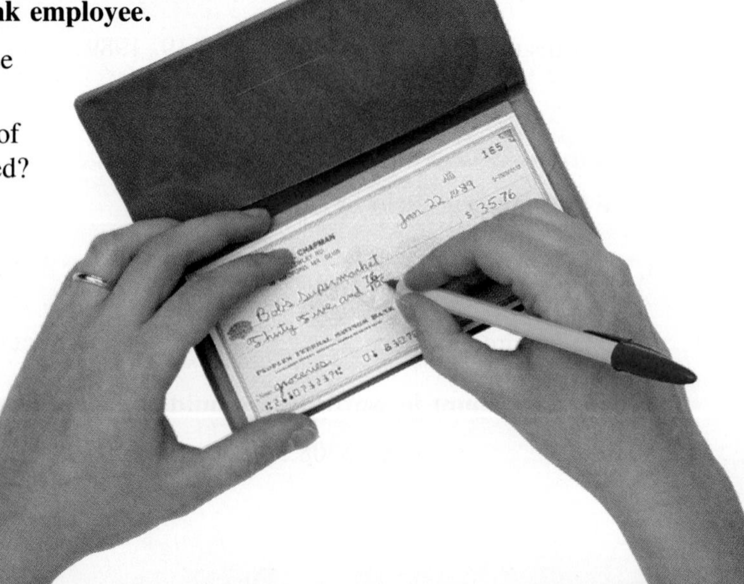

1-6 READING METERS

Electric companies use meters to measure the number of kilowatt-hours (kW·h) of electricity their customers use.

You read the dials on an electric meter from left to right. Pointers alternate in the directions in which they move. The pointer on the dial farthest right moves in a clockwise direction. If a pointer is between two numbers, read the *lesser* number. If the pointer is between 9 and 0, think of the 0 as 10 and read 9.

example

Write the number of kilowatt-hours shown on the meter.

solution

The number of kilowatt-hours is 7709.

exercises

Write the number of kilowatt-hours shown on each meter.

1.

2.

3.

4.

5.

6.

7.

8.

9. During one month, the Carsons used 1390 kW·h of electricity. They used 1345 kW·h during the next month. During which month did they use more electricity? Explain.

10. Sometimes you may have to read your own meter. The meter reader from the electric company may leave a form for you to complete. The form shows a sketch of a meter with no pointers. Make a sketch of an electric meter that has a reading of 4078.

A car has a type of meter called an **odometer.** It measures the number of miles that a car has been driven. The digit farthest right is in the *tenths'* place. For example, the odometer at the right shows that a car has been driven 40,572.8 mi (miles).

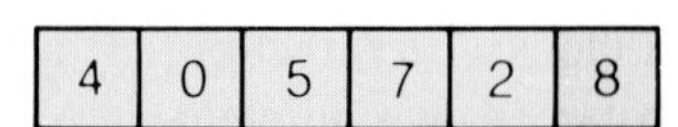

Write the number of miles shown on each odometer.

11. 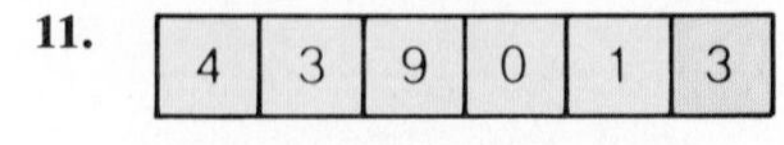

12. | 1 | 6 | 2 | 7 | 8 | 5 |

13. 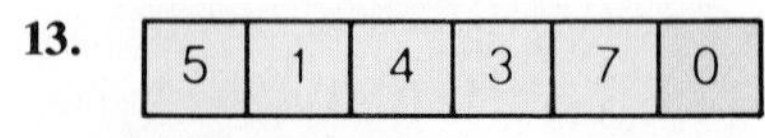

14. | 6 | 7 | 5 | 2 | 2 | 4 |

15. | 0 | 8 | 3 | 5 | 4 | 2 |

16. 

17. 

18. 

Used cars *A* and *B* are for sale. Their odometers are shown below. In each exercise, which car has been driven more miles?

19. *A:* | 8 | 4 | 1 | 5 | 9 | 7 |

B: | 8 | 0 | 9 | 9 | 8 | 1 |

20. *A:* | 3 | 9 | 0 | 6 | 4 | 6 |

B: | 3 | 9 | 0 | 6 | 6 | 4 |

21. When a car whose odometer is similar to the ones shown above has been driven exactly 100,000 mi, what does the odometer show?

A *palindrome* is a number that reads the same backward and forward. For example, 4114 is a palindrome. Find all 4-digit palindromes that can appear on the display of a digital clock similar to the one shown.

CHAPTER REVIEW

vocabulary vo·cab·u·lar·y

Choose the correct word or phrase to complete each sentence.

1. The symbol < means (*is less than, is greater than*).
2. The leading digit of a number is its (*first, first nonzero*) digit.

skills

Write the numeral form of each number.

3. 54 million, 280 thousand, 8
4. 4 thousand, 101 and 3 tenths
5. sixty-five thousandths
6. 6.75 billion
7. four hundred fifty-three billion, two hundred thirteen thousand, eighty-nine
8. National parks in the United States cover 47,157,330 acres. Write the word form of this number.

Complete. Replace each _?_ with the correct decimal.

9. 12,760,000 = _?_ million
10. 345,800,000,000 = _?_ billion
11. The population of Mexico is about 81.7 million. Write the numeral form of this number.

Round to the place of the underlined digit.

12. $0.5\underline{7}32$
13. $\underline{2}4.986$
14. $9{,}0\underline{9}7{,}411$
15. $\$631.5\underline{6}5$
16. Write four different numbers that round to 3.058.

Tell whether each statement is *true* or *false*.

17. 18.520 = 18.52
18. $62,640 < $6264
19. 8.077 > 8.74
20. Write the numbers in order from least to greatest: 7.42; 7.2; 7.4; 7; 7.3; 7.32

Write the numeral form of each amount.

21. Two hundred nine and $\frac{35}{100}$ dollars
22. Seventy-three and $\frac{00}{100}$ dollars
23. Write the number of kilowatt-hours shown on the meter.

24. Write the number of miles shown on the odometer.

0	7	3	4	0	8

CHAPTER TEST

Write the word form of each number.

1. 82,601 2. 79,000,893 3. 140.6 4. 0.0735 **1-1**

5. The planet Venus orbits the sun in six hundred fifteen thousandths of an Earth year. Write the numeral form of this number.

Write the numeral form of each number.

6. 185.4 thousand 7. 63.06 million 8. 7.901 million 9. 2.5 billion **1-2**

Complete. Replace each _?_ with *thousand, million,* or *billion.*

10. 39,410,000,000 = 39.41 _?_ 11. 269,900,000 = 269.9 _?_

12. The surface area of the world's oceans is about 139.5 million mi^2 (square miles). Write the numeral form of this number.

Round to the given place.

13. 5,666,093; nearest ten thousand 14. 7262.6304; nearest tenth **1-3**

15. $648.07; nearest dollar 16. $80.495; nearest cent

Round to the place of the leading digit.

17. 84,608 18. 0.0012 19. $26.89 20. 99.743

21. There are 29,589 students enrolled at a leading university. Round this number to the nearest hundred.

Compare. Replace each _?_ with >, <, or =.

22. 107,412 _?_ 107,421 23. $9087 _?_ $908 24. $45.00 _?_ $45 **1-4**

25. 6.4 _?_ 6.40 26. 15.302 _?_ 15.203 27. 3.256 _?_ 3.3

28. Four of the fastest times recorded for running an indoor mile are 231.2 s (seconds), 232.28 s, 232.56 s, and 229.78 s. Write these numbers in order from least to greatest.

Write each amount in words as it would appear on a check.

29. $54.82 30. $17.60 31. $2075.04 32. $910 **1-5**

33. Write the number of kilowatt-hours shown on the meter. 34. Write the number of miles shown on the odometer. **1-6**

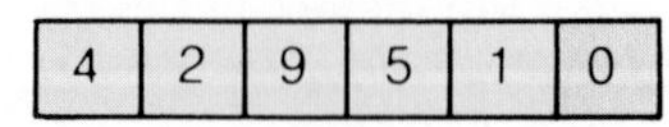

Making change, balancing your checkbook—in fact, most of your personal finances—require skill in addition and subtraction.

CHAPTER 2

ADDING AND SUBTRACTING WHOLE NUMBERS AND DECIMALS

2-1 ADDING WHOLE NUMBERS AND DECIMALS

To add two or more numbers, use the following method.

1. Line up the addends vertically. Make sure that any decimal points are aligned.
2. Add the digits in each column. If the sum is 10 or more, rename.
3. Continue column-by-column from right to left.

$$\begin{array}{r} \scriptstyle 1\,1\,1 \\ 1984 \\ +\ 218 \\ \hline 2202 \end{array} \quad \begin{array}{l} \leftarrow \text{addends} \\ \leftarrow \text{sum} \end{array}$$

example 1

Find each sum.

a. 437 + 698 **b.** \$1653 + \$84 + \$821

solution

a.
$$\begin{array}{r} \scriptstyle 1\,1\,1 \\ 437 \\ +\ 698 \\ \hline 1135 \end{array}$$
← **Rewrite the addition in vertical form.**

b.
$$\begin{array}{r} \scriptstyle 1\,1 \\ \$\,1653 \\ 84 \\ +\ \ 821 \\ \hline \$\,2558 \end{array}$$

your turn

Find each sum.

1. 327 + 867 **2.** 461 + 47 + 389 **3.** \$1579 + \$27 + \$842

example 2

Find each sum.

a. 187.92 + 4.35 **b.** \$1.64 + \$19 + \$7.10

solution

a.
$$\begin{array}{r} \scriptstyle 1\,1 \\ 187.92 \\ +\ \ \ 4.35 \\ \hline 192.27 \end{array}$$
← **Align the decimal points.**

b.
$$\begin{array}{r} \scriptstyle 1 \\ \$\ \ 1.64 \\ 19.00 \\ +\ \ 7.10 \\ \hline \$\,27.74 \end{array}$$
← **Annex zeros so that each number has the same number of decimal places.**

your turn

Find each sum.

4. 51.71 + 122.86 **5.** 0.39 + 8.4 + 17.2 **6.** \$11.07 + \$29 + \$17.69

You can add numbers in any order. This is the **commutative property of addition.**

$$18 + 7 = 7 + 18$$
$$25 = 25$$

When you add three numbers, you may group with parentheses as shown below. This is the **associative property of addition.**

$$(8 + 6) + 9 = 8 + (6 + 9)$$
$$14 + 9 = 8 + 15$$
$$23 = 23$$

You can use these properties to add numbers *in any order* and *in any groups of two*. You should group numbers whose sum is easier to find.

example 3

Use the properties of addition to find each sum mentally.

a. $18 + 6 + 2$ **b.** $6.8 + 12.9 + 4.2$

solution

a. $18 + 6 + 2 = 18 + 2 + 6$
$= (18 + 2) + 6$
$= 20 + 6$
$= 26$

b. $6.8 + 12.9 + 4.2 = 12.9 + 6.8 + 4.2$
$= 12.9 + (6.8 + 4.2)$
$= 12.9 + 11$
$= 23.9$

your turn

Use the properties of addition to find each sum mentally.

7. $14 + 63 + 17$ **8.** $32 + 46 + 28$ **9.** $9.6 + 2.5 + 3.5 + 1.3$

practice exercises

practice for example 1 (page 20)

Find each sum.

1. \$43 + \$182 + \$6 **2.** $895 + 72 + 1046$ **3.** $10{,}978 + 401 + 89$
4. $19 + 98{,}070 + 2353$ **5.** \$18 + \$256 + \$4092 **6.** \$2936 + \$908 + \$9

practice for example 2 (page 20)

Find each sum.

7. $0.64 + 5.23 + 1.62$ **8.** $23.6 + 0.816 + 18$ **9.** $0.0092 + 1.08 + 2.7$
10. \$19.87 + \$62.12 **11.** \$3.10 + \$26.34 + \$9 **12.** \$42 + \$17.65 + \$69.57

practice for example 3 (page 21)

Use the properties of addition to find each sum mentally.

13. 11 + 15 + 19 **14.** 21 + 19 + 17 **15.** 49 + 67 + 23

16. 36 + 14 + 83 + 22 **17.** 5.8 + 3.7 + 4.2 **18.** 6.5 + 2.7 + 5.5 + 4.3

mixed practice (pages 20–21)

Find each sum.

19. 304 + 921 **20.** 6395 + 6283 **21.** 214.3 + 179.8

22. $32.70 + $3.90 **23.** 25.97 + 165 + 23.9 **24.** $1.89 + $21.62 + $97

Use the properties of addition to find each sum mentally.

25. 25 + 13 + 37 + 5 **26.** 16 + 52 + 14 + 18 **27.** 78 + 52 + 23 + 97

28. 9.1 + 2.1 + 1.9 + 3.9 **29.** 6.2 + 7.9 + 3.8 + 7 **30.** 2.8 + 1.4 + 0.2 + 9.6

Name the property shown in each step. Then find each sum.

31. 225 + (38 + 75) = (225 + 38) + 75
= (38 + 225) + 75
= 38 + (225 + 75)

32. 535 + (486 + 65) + 14 = 535 + (65 + 486) + 14
= (535 + 65) + (486 + 14)

Use the chart shown at the right. A calculator may be helpful.

33. What is the total number of cars sold in 1975?

34. What is the total number of cars sold in 1980?

35. What is the total number of domestic cars sold in the three years?

36. What is the total number of exports sold in the three years?

SALES OF CARS MANUFACTURED IN THE UNITED STATES

Year	Domestic	Exports
1975	6,073,000	640,000
1980	5,840,000	560,000
1985	7,337,000	665,000

37. Were there more cars sold in the year 1980 or 1985?

38. There were 1,190,000 more domestic cars sold in 1984 than in 1980. What is the total number of domestic cars sold in 1984?

review exercises

Find each difference.

1. 5 − 2 **2.** 9 − 4 **3.** 7 − 1 **4.** 4 − 0 **5.** 16 − 6

6. 51 − 4 **7.** 36 − 23 **8.** 92 − 62 **9.** 63 − 45 **10.** 80 − 19

2-2 SUBTRACTING WHOLE NUMBERS AND DECIMALS

To subtract two numbers, use the following method.

1. Line up the numbers vertically. Make sure that any decimal points are aligned.
2. Subtract the digits in each column. Rename when necessary.
3. Continue column-by-column from right to left.

```
      8 13
  1 8,9 3 7
-   3,6 9 0
  1 5,2 4 7  ← difference
```

Since addition and subtraction are **inverse operations,** you can check a difference by adding.

```
Check:   1 5,2 4 7
       +   3,6 9 0
         1 8,9 3 7 ✓
```

example 1

Find each difference.

a. 694 − 362

b. 27,465 − 8936

solution

```
a.   6 9 4   ← No renaming
    −3 6 2     is needed.
     3 3 2
```

```
     1 16 14 5 15
b.   2 7,4 6 5   ← Rename twice in
   −   8,9 3 6     the thousands'
     1 8,5 2 9     place.
```

your turn

Find each difference.

1. 299 − 98 **2.** $127 − $36 **3.** 3423 − 2157 **4.** $8437 − $948

example 2

Find each difference.

a. 29.87 − 4.62

b. 9 − 2.07

solution

```
a.  2 9.8 7   ← Align the
   −  4.6 2     decimal points.
    2 5.2 5
```

```
    8 9 10
b.  9.0 0   ← Annex zeros so
   −2.0 7     that each number
    6.9 3     has the same number
              of decimal places.
```

your turn

Find each difference.

5. 18.36 − 1.25 **6.** 16.5 − 3.67 **7.** 702.6 − 498.7 **8.** $85 − $19.53

practice exercises

practice for example 1 (page 23)

Find each difference.

1. 8843 − 6721
2. 31,593 − 11,562
3. $4726 − $655
4. $6214 − $2382
5. 94,203 − 5120
6. 89,620 − 27,949

practice for example 2 (page 23)

Find each difference.

7. 23.34 − 1.32
8. $480.92 − $60.71
9. 12.9 − 9.86
10. 723.7 − 201.96
11. $4228 − $326.25
12. 134.3 − 25.277

mixed practice (page 23)

Find each difference.

13. 997 − 486
14. 825 − 597
15. $39.84 − $7.97
16. $46.88 − $15.63
17. $2094 − $1625
18. 28,763 − 495
19. 634 − 295.49
20. 4.32 − 0.29
21. 568.23 − 87.083

22. Cary traveled 23.4 mi (miles) by car and 25.6 mi by bus. How much farther did Cary travel by bus than by car?

23. The Transamerica Pyramid in San Francisco is 853 ft (feet) tall. The Sears Tower in Chicago is 1454 ft tall. How much taller is the Sears Tower than the Transamerica Pyramid?

24. Betty spent $29.95 for a sweater. She gave the cashier a $50 bill. How much change did Betty receive?

25. Ned bought four items costing $3.91, $1.25, $3.75, and $2.99. If he gave the cashier $12, how much change did Ned receive?

review exercises

Round to the place of the leading digit.

1. 28
2. 64.7
3. 3.38
4. 9398
5. 10,821
6. 46,117

Round to the place of the underlined digit.

7. $0.\underline{1}65$
8. $0.5\underline{0}5$
9. $7.11\underline{8}63$
10. $\$6\underline{3}.49$
11. $\underline{4}092$
12. $2\underline{4}1{,}900$

2-3 ESTIMATING SUMS AND DIFFERENCES

When you don't need an exact answer to a problem, you can **estimate.**

example 1

Estimate by rounding.

a.
```
 3110 →  3000
  981 →  1000   ← Round 981 to the same place as the other addends.
+4978 → +5000
        about 9000
```

b.
```
 $79.21 →  $80
− 18.49 → − 20
          about $60
```

your turn

Estimate by rounding.

1. 212 + 97 + 434
2. $785.30 − $369.40
3. $1.93 + $2.90 + $7.23
4. 42,995 − 18,672

Another way to estimate is by adjusting the sum of the *front-end digits*.

example 2

Estimate by adjusting the sum of the front-end digits.

a. 786 + 323 + 291

b. $13.69 + $23.06 + $13.37 + $9.85

solution

Add the front-end digits. Adjust the sum.

a.
```
 786 \
 323 /  about 100
+291 —  about 100
1200  +  200
   about 1400
```

b.
```
 $13.69 \
  23.06 —>  about $10
  13.37 /
+  9.85 —   about $10
 $40   +   $20
   about $60
```

your turn

Estimate by adjusting the sum of the front-end digits.

5. 31.37 + 21.09 + 17.74
6. 4326 + 1639 + 5978
7. $8352 + $9649 + $2546 + $1461
8. $129.54 + $99.05 + $269.47

You can use front-end digits to estimate a difference. If the front-end difference is zero, subtract the digits in the next place. If the front-end digits are not in the same place, subtract using as many digits as needed.

example 3

Estimate by using the front-end digits.

a. 64.7 − 28.5 **b.** 4783 − 4119 **c.** $56.21 − $3.48

solution

a.

```
  64.7
 −28.5
 about 40
```

b.

```
  4783
 −4119
 about 600
```

c.

```
  $56.21
 −  3.48
 about $53
```

your turn

Estimate by using the front-end digits.

9. 82.4 − 33.9 **10.** $14,743 − $13,629 **11.** 45.87 − 4.59

practice exercises

practice for example 1 (page 25)

Estimate by rounding.

1. 2104 + 3975
2. 859 + 4286
3. $2.16 + $1.43 + $3.79
4. 82.69 + 21.59 + 9.97
5. $724.46 − $388.72
6. 4754 − 2693

practice for example 2 (page 25)

Estimate by adjusting the sum of the front-end digits.

7. 5106 + 1239 + 4728
8. 6001 + 5634 + 2403
9. 55.6 + 64.9 + 69.2
10. $12.63 + $7.38 + $29.92
11. 193.21 + 245.49 + 62.14
12. $7.69 + $5.23 + $.76 + $1.35

practice for example 3 (page 26)

Estimate by using the front-end digits.

13. 4124 − 2394
14. $4.86 − $3.48
15. 8472 − 8321
16. 62,573 − 61,802
17. $18.49 − $7.55
18. $42,362 − $2411

mixed practice (pages 25–26)

Estimate.

19. \$46.09 + \$23.49 + \$11.29
20. \$16.21 + \$43.27 + \$10.11
21. 428.04 − 148.49
22. 78,261 − 25,934
23. 257.28 + 146.42 + 799.09
24. 12,964 + 26,871 + 49,435
25. \$9843 − \$9190
26. \$26.42 − \$21.27
27. 2848.91 − 467.26
28. 6.8142 − 0.8041
29. 582 + 612 + 412 + 509 + 407
30. \$1.55 + \$6.64 + \$3.37 + \$2.45
31. List the letters of *all* the differences for which 200 is a good estimate.
    a. 658 − 463 **b.** 604 − 42 **c.** 1648 − 1405
32. List the letters of *all* the differences for which 3000 is a good estimate.
    a. 85,089 − 50,000 **b.** 123,947 − 120,721 **c.** 17,894 − 14,951
33. There were 64,822 people at a football game. When it started raining, about 27,000 people went home. About how many people stayed at the game?
34. Bradford wanted to buy two albums for \$7.49 each and three tape cassettes for \$5.79 each. Estimate to determine if \$30 is enough to pay for these items.

review exercises

Find each sum or difference.

1. 681 + 78 + 1300
2. \$6 + \$1291 + \$920
3. \$1005 − \$361
4. 34,216 − 7088
5. 25.9 + 0.14 + 3.018
6. \$55 + \$23.49 + \$920
7. 12.043 − 0.2345
8. 924.39 − 16.82
9. \$69 − \$4.56

mental math

To subtract \$7.50 − \$3.90 mentally, *add on* from \$3.90 to \$7.50.

\$3.90 + \$.10 = \$4.00 \$4.00 + \$3.00 = \$7.00 \$7.00 + \$.50 = \$7.50

Since \$.10 + \$3.00 + \$.50 = \$3.60, \$7.50 − \$3.90 = \$3.60.

Subtract mentally.

1. \$5.40 − \$1.80
2. \$12.40 − \$5.90
3. \$23.90 − \$16.70
4. \$7.25 − \$3.55
5. \$9.45 − \$2.75
6. \$2.65 − \$1.35

PROBLEM SOLVING STRATEGY

2-4 IDENTIFYING TOO MUCH OR TOO LITTLE INFORMATION

To solve word problems, you need to carry out these four steps.

Four-Step Method

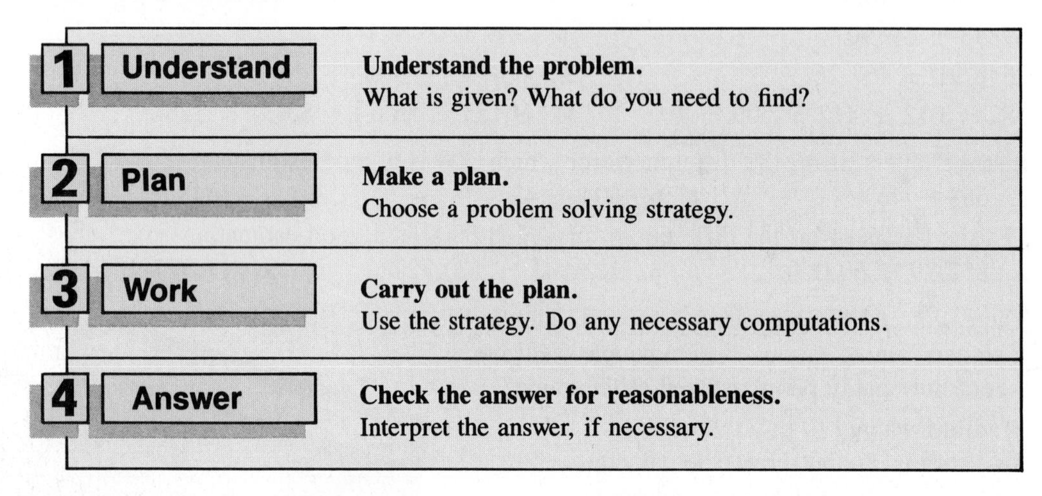

Sometimes the facts given are more than you need to solve the problem. Other times necessary facts may be missing.

example

If possible, solve. If it is not possible to solve, tell what additional information is needed.

Perry bought a gallon of milk and a dozen bagels.
The total cost was $4.08.

a. If Perry paid with a $20 bill, what was his change?

b. How much did a dozen bagels cost?

solution

a. **Step 1** Given: total cost of $4.08
a gallon of milk and a dozen bagels ← **This information is not needed here.**
Find: Perry's change from a $20 bill

Step 2 *Subtract* $4.08 from $20.

Step 3 $20 − $4.08 = $15.92

Step 4 *Estimate to check for reasonableness:* $20 − $4 = $16

Perry received $15.92 change.

b. **Step 1** Given: total cost of $4.08
a gallon of milk and a dozen bagels
Find: cost of a dozen bagels

Step 2 *Subtract* the cost of milk from the total cost.

You cannot solve the problem. To find how much a dozen bagels cost, you need to know the *cost* of the milk.

problems

If possible, solve. If it is not possible to solve, tell what additional information is needed.

1. Sue sold pennants at a football game. Each pennant cost $4.50. She had 3027 at the beginning of the game and 1886 at the end. How many pennants did Sue sell?
2. Paul bought tickets to a play. There was a $2.50 service charge on each ticket. If his total cost was $187.50, how many tickets did he purchase?
3. If gasoline costs $1.19 per gallon, how much will Tom have to pay to fill the tank of his car?
4. Lisa spent 3 h (hours) repairing a car. She used $45.87 in parts, charged $68 for labor, and added $2.29 for tax. Find the total repair bill.
5. The Choy family drove on a cross-country trip. When they returned, the odometer on their car showed 23,641.7. How many miles did the Choys drive?
6. Liz bought two ice hockey sticks for $12.95 each and a pair of ice hockey skates for $45.99. Her purchases totaled $71.89. How much change did she receive if she paid with a $100 bill?
7. This year Andre played in 20 basketball games. He averaged 32.25 points per game. Last year he played in 25 games and averaged 28.5 points per game. How much greater is his average this year?
8. One day a gas station sold 253.6 gal (gallons) of regular gas, 467.4 gal of unleaded gas, and 353.6 gal of super gas. How many gallons of each type of gas did the station have left?

review exercises

Compare. Replace each ? with >, <, or =.

1. 9230 ? 9320
2. 62.1 ? 6.21
3. 0.004 ? 0.0004
4. 17.5 ? 17.50
5. 316 ? 3016
6. 59.8 ? 59.79

SKILL REVIEW

Find each sum.

2-1

1. 49 + 6124
2. \$87 + \$32 + \$404
3. 20,144 + 38 + 560
4. \$16.25 + \$9.16
5. 7.514 + 27 + 12.09
6. 103 + 12.50 + 6.98

Use the properties of addition to find each sum mentally.

7. 17 + 18 + 12
8. 23 + 55 + 45
9. 4.1 + 2.5 + 7.9
10. 6.6 + 2.3 + 1.4
11. 31 + 26 + 14 + 19
12. 3.3 + 4 + 5.4 + 1.7

Find each difference.

2-2

13. 894 − 661
14. 43,620 − 19,473
15. 97.83 − 6.7
16. 61.3 − 8.56
17. \$13.16 − \$9.18
18. \$60 − \$19.53
19. A crowd of 48,151 fans attended a rock concert in Denver. The rock band traveled to Seattle and performed for 36,976 fans. How many more fans attended the concert in Denver than in Seattle?

Estimate by rounding.

2-3

20. 2458 + 6520
21. \$812 + \$68 + \$290
22. 7.20 + 2.67 + 4.99
23. \$794.65 − \$309.41
24. 86,093 − 57,124
25. \$61.21 − \$38.92

Estimate by using the front-end digits.

26. 8.16 + 6.94 + 3.8
27. 735 + 296 + 562
28. 29.7 + 16.5 + 43.9
29. 512.4 − 220.1
30. \$34.21 − \$3.19
31. \$6913 − \$6246
32. Kaitlin has items costing \$3.13, \$2.09, \$1.80, \$5.65, and \$5.30 in her grocery cart. Estimate to determine if she can buy all these items with a \$20 bill.

If possible, solve. If it is not possible to solve, tell what additional information is needed.

2-4

33. Marge bought a purse on sale for \$14 and a necklace on sale for \$7.50. How much did she save at the sale?
34. Kent sold 99 tickets on Tuesday and 67 tickets on Wednesday. How many tickets did Kent sell in those two days?
35. The population of Hamilton is 26,540. This is 87 more than last year. Sommer has 13,216 residents. How many more people live in Hamilton than in Sommer?
36. Tami bought 10 lb (pounds) of potatoes for \$3.88, four cucumbers for \$1, and a lemon for \$.34. How much did she spend?

APPLICATION

2-5 MAKING CHANGE

Many cash registers automatically calculate the amount of change due a customer.

example

Roxanne receives \$5 from a customer for a purchase totaling \$3.66. The cash register indicates that the customer should receive \$1.34 in change. Describe how Roxanne should make the given amount of change using the fewest coins and bills.

solution

Roxanne thinks:

$$\$1.34 = \$1.00 + \$.25 + \$.05 + \$.04$$

She gives the customer a dollar bill, a quarter, a nickel, and four pennies.

exercises

Describe how to make the given amount of change using the fewest coins and bills.

1. \$1.27 2. \$2.38 3. \$5.06 4. \$8.43 5. \$12.15 6. \$26.46

Describe how you would make the given amount of change using the fewest coins and bills.

7. Your cash register indicates change due of \$7.48.
8. Your cash register indicates change due of \$11.07.
9. Susan pays for a purchase with a \$10 bill. She receives \$3.32 in change. You have no nickels in your cash drawer.
10. Marvin pays for a purchase with a \$50 bill. He receives \$5.63 in change. You have no dimes in your cash drawer.
11. Juan pays for a purchase with a \$20 bill. He receives \$8.33 in change. You have no quarters in your cash drawer.
12. You have extra nickels in your cash drawer. Describe how you would make \$.39 in change using as many nickels as possible.
13. Todd spends \$3.18 on notebook paper. He has a \$5 bill and six pennies. How should Todd pay the salesperson in order to receive no pennies in change?

2-6 CHECKING ACCOUNTS

You use a **check register** to record the withdrawals, deposits, and service charges of your checking account. The **balance** is the amount of money in your account. If you make a deposit or receive interest, the amount is *added* to your balance. The amounts for checks and service charges are *subtracted* from your balance. Cash withdrawals made using an electronic teller or banking machine are also subtracted.

example

Here is Marilyn Folsom's check register.

a. To whom did she write check number 539?

b. What was her balance before she wrote check number 539?

c. What was her balance after she wrote check number 541?

RECORD ALL CHARGES OR CREDITS THAT AFFECT YOUR ACCOUNT

NUMBER	DATE	DESCRIPTION OF TRANSACTION	PAYMENT/DEBIT (-)		✓ T	FEE (IF ANY) (-)	DEPOSIT/CREDIT (+)		BALANCE	
									1278	31
539	2/18	Tremont Garage	178	89					178 1099	89 42
540	2/23	The Music Box	19	45					19 1079	45 97
	2/28	deposit					250	00	250 1329	00 97
541	2/28	Rent	750	00						
542	3/1	Granny's Grocery	52	63						
	3/1	Service charge				3.18				
	3/3	Electronic Withdrawal	50	00						

solution

a. Tremont Garage

b. $1278.31

c.
$$\begin{array}{r} \$1329.97 \\ -\ \ 750.00 \\ \hline \$\ \ 579.97 \end{array}$$

The balance was $579.97.

exercises

Use the check register at the right.

1. What was the balance after the deposit on July 8?
2. What was the amount of the check to Tony's Tailor Shop?
3. What was the balance after the electronic withdrawal on July 15?
4. Do you add or subtract the service charge?
5. Why doesn't the electronic withdrawal have a number?

RECORD ALL CHARGES OR CREDITS THAT AFFECT YOUR ACCOUNT

NUMBER	DATE	DESCRIPTION OF TRANSACTION	PAYMENT/DEBIT (−)	✓ T	FEE (IF ANY) (−)	DEPOSIT/ CREDIT (+)	BALANCE
							863 86
627	7/6	Malden Recreation	40 00				40 00 823 86
	7/8	deposit				375 00	375 00 1198 86
628	7/9	Tony's Tailor Shop	12 50				12 50 1186 36
629	7/12	Bicycle Peddler	52 50				52 50 1133 86
	7/15	service charge			4.20		4 20 1129 66
	7/15	electronic withdrawal	50 00				50 00 1079 66
630	7/19	Parkside School	156 00				156 00 923 66
	7/20	deposit				285 72	285 72 1209 38
631	7/26	Wyoming Wallpaper	97 83				97 83 1111 55
	7/28	electronic withdrawal	27 00				27 00 1084 55

Find the new balance after each withdrawal or deposit.

6. balance $356.89
   deposit $123.45
   new balance ?

7. balance $234.12
   check $ 93.56
   new balance ?

8. On October 1, Erna opened a checking account by depositing her entire paycheck of $845.39. She wrote a check for $120.76 on October 14. Erna made a deposit of $45.05 on October 20. What is the new balance in Erna's account?
9. Carl wrote a check for $578.23 on March 17. He then wrote another check for $32.78. What is the least amount of money Carl should have in his checking account before he writes these checks?
10. Lou Young's checking account had a balance of $324.21. He wrote a check for $23.90. Two days later Lou made a deposit of $87.23. What is the new balance in Lou's account?
11. Jean's checking account had a balance of $2357.90 on Tuesday. She then wrote a check for $381.47, a check for $239.01, and a check for $78.39. What is the new balance in Jean's account?

Today is the 27th day of a month. Ten days later, it will be the eighth day of the next month. What are the two months?

APPLICATION

2-7 RECONCILING AN ACCOUNT

If you have a checking account, you receive a monthly **bank statement** that lists all the transactions that took place in your account. Some banks return the checks you wrote. These are called **canceled checks.**

To **reconcile** your checking account, you must show that the bank statement and your check register agree. To do this, you compare the amount of each deposit and each check in your check register with those listed on the statement. Checks and deposits that are listed in your check register but are not credited on your statement are called **outstanding.**

The bank assumes that the balance on the statement is correct. You must notify the bank of any error.

example

Nikolai Dubetsky's check register showed a balance of $565.64. His bank statement showed an ending balance of $533.95. He had outstanding deposits of $121.34 and $84.11. His outstanding checks were for $60.36, $23.61, $78.76, and $10.21. The statement showed a service charge of $1.50 and interest of $2.32. Reconcile his account.

solution

Many banks provide a form like the one shown below to help reconcile an account. Complete the form as follows.

- List the outstanding checks and find their total.
- List the deposits not credited and find their total.
- Calculate Total 1 and Total 2 as shown below.

MONTHLY STATEMENT OF ACCOUNT
USE THIS AREA BELOW TO RECONCILE YOUR ACCOUNT

OUTSTANDING CHECKS		
	60	36
	23	61
	78	76
	10	21
TOTAL	172	94

BANK BALANCE SHOWN ON STATEMENT	$533.95
ADD (+) DEPOSITS NOT CREDITED IN THIS STATEMENT (IF ANY)	
$121.34	
$ 84.11	
$	
TOTAL DEPOSITS	$205.45
SUBTOTAL	$739.40
SUBTRACT (−) TOTAL OUTSTANDING CHECKS	$172.94
TOTAL 1	$566.46

REGISTER BALANCE	$565.64
LESS SERVICE CHARGE	$ 1.50
SUBTOTAL	$564.14
PLUS INTEREST	$ 2.32
TOTAL 2	$566.46
TOTAL 1 SHOULD EQUAL TOTAL 2	

Total 1 = Total 2 = $566.46. The account is reconciled.

exercises

Reconcile each account. A calculator may be helpful.

1. Mel Leebo's check register showed a balance of $271.44. His bank statement showed an ending balance of $302.67. He had an outstanding check for $32.40. The statement showed a service charge of $2.35 and interest of $1.18.
2. While reconciling her account, Ellen noticed that check 808 for $13.62 had been recorded in her register as $13.26. Her check register showed a balance of $217.44. Her bank statement showed an ending balance of $105.14. She had an outstanding deposit of $188.78, and outstanding checks for $56.39 and $20. The statement showed interest of $.45.
3. While reconciling her account, Tia discovered that she had failed to record check 85 for $12.06 and check 86 for $14.92. Her check register showed a balance of $672.52. Her bank statement showed a balance of $647.14. She had no outstanding checks or deposits. The statement showed a $1.30 service charge and interest of $2.90.
4. Rheta's check register showed a balance of $351.64. Her bank statement showed an ending balance of $339.27. She had outstanding deposits of $33.25 and $66.75. The outstanding checks were for $21.85, $8.32, $42.92, and $16.28. The statement showed a service charge of $3.25 and interest of $1.51.

calculator corner

Using a calculator to find a new register balance can save you time.

example Find the new register balance.
register balance: $168.93 interest: $2.39 service charge: $6.73

solution Enter 168.93 + 2.39 − 6.73 = .
The new register balance is $164.59.

Use a calculator to find the new register balance.

	REGISTER BALANCE	INTEREST	SERVICE CHARGE	NEW REGISTER BALANCE
1.	$349.87	$1.33	$1.20	?
2.	$2997.54	$13.11	$2.76	?
3.	$904.32	$3.95	$3.21	?
4.	$1284.89	$7.12	$6.75	?

TAXI DRIVER

Jo Fleet is a taxi driver. She works every night from 6:00 P.M. to 6:00 A.M. Jo must keep a record of her fares, mileage, and tips. For pay, Jo receives a fixed amount per shift and all her tips.

example

Jo had \$80 when she started her shift. Her first fare totaled \$19.25 and she received a \$4.75 tip. How much money did she have after this?

solution

$\$80 + \$19.25 + \$4.75 = \$104$ Jo had \$104.

exercises

Solve. A calculator may be helpful.

1. Mr. Fialli's fare was \$38.55. He paid with a \$50 bill. If Mr. Fialli told Jo to keep \$10 as a tip, how much change did Mr. Fialli receive?
2. During her shift, Jo spent \$8.72 for her supper, \$1.90 for peanuts, and \$14.30 for gasoline. How much money did she spend in all?
3. Sheila Wright gave Jo a \$20 bill for an \$8.75 fare. If Jo received a \$1.80 tip, describe how to make the change due using the fewest coins and bills.
4. Jo received \$21.35, \$19.20, \$24.85, \$17.25, and \$13.15 in fares. With each fare she received a tip of about \$3. About how much money did she receive in all?

CHAPTER REVIEW

vocabulary vo·cab·u·lar·y

Choose the correct word to complete each sentence.

1. The fact that 196 + 38 = 38 + 196 is an example of the (*associative, commutative*) property of addition.
2. A check that is listed in your register, but not on your statement, is called (*canceled, outstanding*).

skills

Find each sum or difference.

3. 8961 + 24 + 647
4. $9014 − $77
5. 5.3 + 1.4 + 0.7 + 3.6
6. $68 − $15.37
7. 143.24 − 91.5
8. 17,346 + 48 + 6003
9. 60,912 − 19,382
10. $6.40 + $55 + $.76
11. 35 + 12 + 68 + 45

Estimate by rounding.

12. 7813 − 2212
13. $35.40 − $17.78
14. $5.34 + $2.01 + $2.69

Estimate by using the front-end digits.

15. 274 + 311 + 119
16. 862.5 + 148.7
17. $6821 − $6468

If possible, solve. If it is not possible to solve, tell what additional information is needed.

18. Pat's Pet Shop now has 7 cats, 10 dogs, and 8 birds for sale. Last week Pat sold 2 dogs and 1 cat. How many pets are for sale?
19. The senior class held a talent show. They charged $1 admission and spent $24.50 on expenses. How much money did they raise?
20. When Lori bought a skirt, she received $6.68 in change. Describe how you would make the change using the fewest coins and bills.
21. Jay opened a checking account by depositing his entire paycheck of $932.76. He then wrote checks for $16.29, $32.50, and $6.44. Find the new balance.
22. Nancy's bank statement showed a balance of $5304.65. Her check register showed a balance of $4458.46. She had outstanding checks for $34.48, $730.20, and $68.45. The statement showed a $6.50 service charge for personalized checks and $19.56 in interest. Reconcile her account.

CHAPTER TEST

Find each sum.

1. \$9 + \$352 + \$65	**2.** 9.2 + 1.07 + 88.6	**2-1**
3. 21 + 26 + 19 + 44	**4.** 3.5 + 13.8 + 4.2 + 10.5	

Find each difference.

5. 24,663 − 421	**6.** \$150.80 − \$14.23	**2-2**
7. 9 − 6.37	**8.** 5172 − 4088	

Estimate by rounding.

9. 311 + 552 + 94	**10.** \$54.65 − \$13.22	**2-3**
11. 457 − 275	**12.** 8.7 + 4.32 + 9.17	

Estimate by using the front-end digits.

13. \$52.17 − \$33.64	**14.** 9.89 + 3.02 + 6.11	
15. 7972 − 7527	**16.** \$15.28 + \$32.23 + \$2.75	

If possible, solve. If it is not possible to solve, tell what additional information is needed.

2-4

17. Gerry used a \$25 gift certificate to help purchase a stereo on sale for \$375. Find the original price of the stereo.

18. Jan rode her bicycle four days this week at an average rate of 13 mi/h (miles per hour). Her distances were 14 mi, 23 mi, 26 mi, and 17 mi. Find her total mileage for the week.

Solve.

2-5

19. A cash register indicated change due of \$4.27. Describe how to make the given amount of change using the fewest coins and bills.

20. Describe how you would make \$15.44 in change using the fewest coins and bills.

2-6

21. Peter's checking account had a balance of \$1062.41. He made a deposit of \$124.65 and an electronic withdrawal of \$60. Find the new balance.

22. Joyce's checking account had a balance of \$674.08. On Friday, she wrote checks for \$6.11 and \$58.45. She also deposited her paycheck of \$253.90. Find the new balance.

2-7

23. Lena's check register showed a balance of \$640.19. Her bank statement showed a balance of \$1184.19. She has a \$6 service charge and an outstanding check for \$550. Reconcile her account.

The orderly arrangement of these pencils makes it possible to find the total number quickly by using multiplication.

CHAPTER 3

MULTIPLYING WHOLE NUMBERS

3-1 ESTIMATING PRODUCTS

When you multiply two numbers, the result is called the **product.** The numbers that you multiply are called **factors.**

$$\underbrace{6 \times 36}_{\text{factors}} = \underset{\text{product}}{216}$$

Sometimes you may only need to estimate to find a **range** of reasonable numbers for a product. To find the range, use the following method.

1. To find the *lower* end of the range, multiply using the leading digits.
2. To find the *upper* end of the range, round up any multi-digit factors and multiply.

example 1

Estimate by finding a range of reasonable numbers.

a. 6×36

b. $51 \times \$48$

solution

a. lower: $6 \times 30 = 180$
upper: $6 \times 40 = 240$
The product 6×36 is *between 180 and 240.*

b. lower: $50 \times \$40 = \$2000$
upper: $60 \times \$50 = \$3000$
The product $51 \times \$48$ is *between \$2000 and \$3000.*

your turn

Estimate by finding a range of reasonable numbers.

1. 3×84 **2.** 4×228 **3.** $91 \times \$37$ **4.** $53 \times \$293$

Sometimes you need a better estimate. Instead of a range, use a *single* number as the estimate.

example 2

Estimate by rounding to the place of the leading digit.

a. 6×31

b. $15 \times \$54$

solution

a.
$$\begin{array}{r} 31 \\ \times 6 \\ \hline \end{array} \rightarrow \begin{array}{r} 30 \\ \times 6 \\ \hline \text{about } 180 \end{array}$$

b.
$$\begin{array}{r} \$54 \\ \times 15 \\ \hline \end{array} \rightarrow \begin{array}{r} \$50 \\ \times 20 \\ \hline \text{about } \$1000 \end{array}$$

your turn

Estimate by rounding to the place of the leading digit.

5. 8 × 88 **6.** 49 × $37 **7.** $20 × 63 **8.** 38 × 245

practice exercises

practice for example 1 (page 40)

Estimate by finding a range of reasonable numbers.

1. 9 × 355 **2.** 8 × 661 **3.** 38 × 94 **4.** 47 × 52
5. 27 × 417 **6.** $64 × 73 **7.** $81 × 83 **8.** 45 × 540

practice for example 2 (pages 40–41)

Estimate by rounding to the place of the leading digit.

9. 3 × 76 **10.** 7 × $48 **11.** 67 × $69 **12.** 36 × 12
13. $56 × 142 **14.** 71 × 807 **15.** 82 × 931 **16.** $18 × 337

mixed practice (pages 40–41)

Estimate to tell if each answer is *reasonable* or *not reasonable*. If an answer is not reasonable, give an estimate of the product.

17. $25 × 42 $\stackrel{?}{=}$ $1050 **18.** $45 × 63 $\stackrel{?}{=}$ $284 **19.** 54 × 22 $\stackrel{?}{=}$ 1188
20. 85 × 614 $\stackrel{?}{=}$ 5219 **21.** 75 × 712 $\stackrel{?}{=}$ 35,000 **22.** 40 × 312 $\stackrel{?}{=}$ 12,480

23. Joe Johnston types 53 words per minute. Find a range of numbers to describe about how many words he types in 35 minutes.

24. Joy Gold studies 45 min daily. Find a range of numbers to describe about how many minutes she studies in six days.

25. At Ridge Vale High School, 285 students each contributed $3 to buy flags for classrooms. About how much money did the students contribute in all?

26. Mr. Ianocci grades 135 quizzes each week. About how many quizzes does Mr. Ianocci grade in 40 weeks?

review exercises

Find each product.

1. 8 × 6 **2.** 4 × 5 **3.** 3 × 9 **4.** 8 × 8 **5.** 6 × 9
6. 9 × 8 **7.** 3 × 8 **8.** 7 × 9 **9.** 6 × 7 **10.** 2 × 8

3-2 MULTIPLYING WHOLE NUMBERS

To multiply two whole numbers, use the following method.

1. Multiply one number by each digit of the other number, beginning with the ones' digit. Rename if necessary.
2. Add any partial products.

$$\begin{array}{r} {}^{1\,3} \\ 627 \\ \times 5 \\ \hline 3135 \end{array}$$

example 1

Find each product: **a.** 49 × 201 **b.** 249 × \$281

solution

a.

$$\begin{array}{r} 201 \\ \times 49 \\ \hline 1809 \\ 8040 \\ \hline 9849 \end{array}$$

1809 ← 9 × 201 partial products
8040 ← 40 × 201 partial products

b.

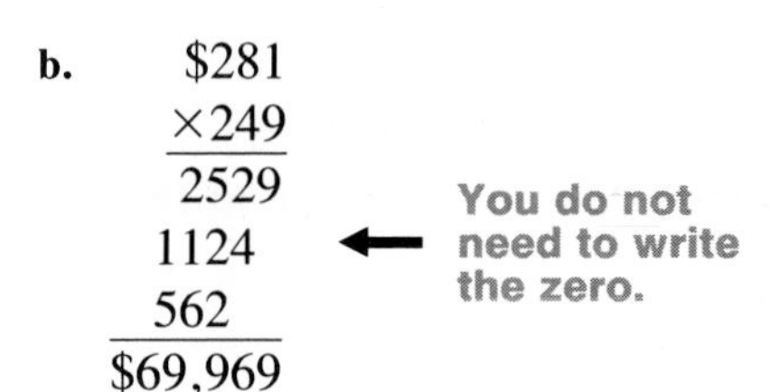

your turn

Find each product.

1. 6 × 352
2. 57 × 401
3. 604 × \$518
4. 123 × 634

To check the reasonableness of your answer, estimate by rounding factors to the place of the leading digit.

example 2

Find the product 88 × 481. Check the reasonableness of your answer.

solution

ESTIMATION

$$\begin{array}{r} 481 \\ \times 88 \\ \hline 3848 \\ 3848 \\ \hline 42{,}328 \end{array}$$

Estimate: 88 × 481 → 90 × 500 } about 45,000

The estimate is 45,000. The answer 42,328 is reasonable.

your turn

Find each product. Check the reasonableness of your answer.

5. 69 × 870
6. 42 × \$315
7. 374 × 844
8. 151 × 526

practice exercises

practice for example 1 (page 42)

Find each product.

1. 8 × 42	**2.** 37 × 9	**3.** $608 × 94	**4.** 31 × $705
5. 355 × 17	**6.** 29 × 665	**7.** 625 × 433	**8.** 781 × 214
9. 706 × 442	**10.** 676 × 802	**11.** $974 × 823	**12.** $575 × 329

practice for example 2 (page 42)

Find each product. Check the reasonableness of your answer.

13. 2 × 479	**14.** 7 × $543	**15.** 6 × 642	**16.** 4 × 171
17. $78 × 824	**18.** 54 × 863	**19.** 76 × 727	**20.** 39 × 845
21. 204 × 552	**22.** 298 × 308	**23.** $442 × 833	**24.** 622 × 117

mixed practice (page 42)

Find each product. Check the reasonableness of your answer.

25. 6 × 561	**26.** 8 × 338	**27.** 72 × 47	**28.** 61 × 43
29. 38 × 82	**30.** 59 × 19	**31.** 16 × 567	**32.** 74 × 392
33. 45 × 972	**34.** 34 × 623	**35.** 304 × 758	**36.** 703 × 561

Choose the letter of the best estimate for each product.

37. 75 × 38	**a.** 3200	**b.** 320	**c.** 32,000
38. 92 × 54	**a.** 45,000	**b.** 450	**c.** 4500
39. 478 × 502	**a.** 25,000	**b.** 250,000	**c.** 200,000
40. 63 × 161	**a.** 6000	**b.** 12,000	**c.** 600
41. 384 × 77	**a.** 3000	**b.** 3200	**c.** 32,000
42. 729 × 439	**a.** 280,000	**b.** 28,000	**c.** 30,000

43. The Bay High School auditorium has 26 rows of seats. Each row has 75 seats. Can the auditorium seat 2500 people? Explain.

44. Last semester Mr. Shore gave nine quizzes to each of his 109 students. How many quizzes did he give in all?

45. Sharon McCormack works at a restaurant for two hours each day after school. She earns $5 per hour. After working 15 days, has she earned enough money to buy a camera that costs $179? Explain.

46. The drama club ordered supplies for a benefit dinner. Boxes of forks contain 50 forks each. If the club ordered 65 boxes, were there enough forks for 3000 people? Explain.

review exercises

Use the addition properties to find each sum mentally.

1. $5 + 16 + 4 + 25$
2. $31 + 27 + 19 + 13$
3. $12 + 38 + 6 + 24$
4. $2.3 + 1.4 + 6.6 + 5.7$
5. $1.5 + 2.1 + 6.5 + 3.9$
6. $48 + 22 + 8 + 2$
7. $39 + 47 + 21 + 53$
8. $2.6 + 3.7 + 3.3 + 8.4$
9. $3.5 + 6.4 + 1.5 + 5.6$

Estimate.

10. \$47.16 − \$25.99
11. 35 × 319
12. \$1.40 + \$5.57 + \$9.98
13. 9 × \$134
14. 11,602 + 9851
15. 82,137 − 73,301
16. 283 + 419 + 594
17. \$5016 + \$4880
18. 16 × 64
19. \$71.88 − \$54.12
20. 21 × 380
21. \$298.15 − \$93.33

Find each answer.

22. \$55 + \$.08 + \$3.99
23. 9 × 77
24. 39.7 − 13.28
25. 1.4072 − 0.198
26. 3.91 + 0.067 + 24.3
27. 14 × \$45
28. \$100 − \$69.88
29. 306 × 512
30. \$3016 + \$29 + \$407
31. 25 × \$144
32. 370 − 82.1
33. \$84.75 + \$98.46

calculator corner

Before multiplying, you should decide whether it is faster to use mental math or a calculator.

example 1 20 × 400

solution Multiply mentally: 2 × 4 = 8, so 20 × 400 = 8000.

example 2 23 × 486

solution Enter 23 × 486 =. The calculator shows 11,178.

Tell whether you would find each product using *mental math* or a *calculator*. Then find the product.

1. 43 × 100
2. 700 × 32
3. 6000 × 200
4. 48 × 101
5. 78 × 6358
6. 286 × 736
7. 712 × 200
8. 400 × 400
9. 775 × 4962
10. 340 × 987
11. 30 × 2000
12. 4 × 820,576
13. 300 × 5000
14. 38 × 47,617
15. 2 × 61,875
16. 600 × 900
17. 770 × 1000
18. 1259 × 772

3-3 MULTIPLICATION PROPERTIES

Multiplication has properties similar to the properties of addition.

You can multiply two numbers in any order. This is the **commutative property of multiplication.**

$$9 \times 20 = 20 \times 9$$
$$180 = 180$$

When you multiply three numbers, you may group them as shown below. This illustrates the **associative property of multiplication.**

$$(2 \times 4) \times 5 = 2 \times (4 \times 5)$$
$$8 \times 5 = 2 \times 20$$
$$40 = 40$$

You can use these properties to multiply numbers *in any order* and *in any groups of two.*

example 1

Use the properties of multiplication to find each product.

a. 359×717

b. $50 \times 8 \times 2$

solution

a.

```
    359
   ×717
   2513
   359
 2513
257,403
```

← **Changing the order to multiply by 717 may be easier because 717 has repeated digits.**

b. $50 \times 8 \times 2$ (50 and 2 grouped: 100)

← **Group numbers whose product is easy to find.**

$$8 \times 100 = 800$$

your turn

Use the properties of multiplication to find each product.

1. 637×242 **2.** 762×355 **3.** $25 \times 8 \times 4$ **4.** $5 \times 5 \times 2 \times 2$

The **distributive property** allows you to multiply numbers inside parentheses by a factor outside the parentheses as shown below.

$$4 \times (6 + 2) = (4 \times 6) + (4 \times 2)$$
$$= 24 + 8$$
$$= 32$$

Using the distributive property can make some multiplications easier. You write one factor as a sum or a difference.

example 2

Use the distributive property to find each product.

a. 5×107 **b.** 8×28

solution

a. $5 \times 107 = 5 \times (100 + 7)$ ← **Write 107 as 100 + 7.**

$= (5 \times 100) + (5 \times 7)$

$= 500 + 35$

$= 535$

b. $8 \times 28 = 8 \times (30 - 2)$ ← **Write 28 as 30 − 2.**

$= (8 \times 30) - (8 \times 2)$

$= 240 - 16$

$= 224$

your turn

Use the distributive property to find each product.

5. 6×72 **6.** 7×310 **7.** 5×997 **8.** 9×195

practice exercises

practice for example 1 (page 45)

Use the properties of multiplication to find each product.

1. 15×66 **2.** 493×11 **3.** $5 \times 16 \times 20$ **4.** $25 \times 42 \times 4$

5. 125×313 **6.** $5 \times 95 \times 2$ **7.** $4 \times 84 \times 25$ **8.** 438×626

9. $50 \times 59 \times 2$ **10.** 231×555 **11.** 816×444 **12.** $5 \times 37 \times 2$

practice for example 2 (page 46)

Use the distributive property to find each product.

13. 4×52 **14.** 9×73 **15.** 8×29 **16.** 7×47

17. 6×699 **18.** 5×812 **19.** 9×914 **20.** 8×698

mixed practice (pages 45–46)

Use the properties of multiplication to find each product.

21. 62×88 **22.** 4×998 **23.** 7×990 **24.** 47×55

25. 3×512 **26.** $25 \times 16 \times 4$ **27.** $5 \times 81 \times 20$ **28.** 4×713

29. The Paramount movie theater has 25 rows containing 45 seats each. During a one-day showing of a film, owners sold tickets for every seat of every showing. They showed the film four times. How many tickets did the owners sell?

30. Bob Fox stores his baseball cards in six shoe boxes with 210 baseball cards in each box. How many baseball cards does he have?

31. Jan Keenan jogs three miles daily. How many miles does she jog in August?

32. Sue Kim buys five cases of dog food. Each case has 24 cans. Each can holds 14 oz (ounces). How many cans of dog food does Sue buy?

review exercises

Find each product.

1. 44×200
2. 600×37
3. 900×2000
4. 500×5000
5. 805×84
6. 309×99
7. 26×706
8. 63×401
9. 514×322
10. 679×183
11. 464×316
12. 923×911
13. 300×85
14. 75×400
15. 6000×700
16. 4000×800

mental math

Properties of multiplication often help you multiply mentally.

$2 \times 42 \times 50$ (2 and 50 → 100)
$100 \times 42 = 4200$

$$5 \times 17 = 5 \times (10 + 7) = (5 \times 10) + (5 \times 7) = 50 + 35 = 85$$

$$6 \times 99 = 6 \times (100 - 1) = (6 \times 100) - (6 \times 1) = 600 - 6 = 594$$

Use the multiplication properties to find each product mentally.

1. $2 \times 18 \times 50$
2. $25 \times 7 \times 4$
3. $500 \times 17 \times 2$
4. $4 \times 8 \times 25 \times 5$
5. 4×99
6. 7×199
7. 5×399
8. 15×999
9. 7×32
10. 6×12
11. 4×23
12. 8×15

3-4 EXPONENTS

A multiplication in which all the factors are the same number, such as $4 \times 4 \times 4 \times 4 \times 4$, can be written in **exponential form.**

$$\underbrace{4 \times 4 \times 4 \times 4 \times 4}_{\textbf{5 factors}} = 4^5 \leftarrow \textbf{exponent}$$

$\uparrow$ **base**

The **exponent** 5 indicates that the **base** 4 is used as a factor five times. You read 4^5 as *four to the fifth power*. You read 4^2 as *four to the second power* or *four squared*. You read 4^3 as *four to the third power* or *four cubed*.

Any base to the first power is equal to that base, as $3^1 = 3$. Any *nonzero* base to the zero power is equal to 1, as $5^0 = 1$. The number 1 to any power equals 1, as $1^9 = 1$.

example 1

Find each answer.

a. 6^3 **b.** 15^1 **c.** 10^0

solution

a. $6^3 = 6 \times 6 \times 6$
$= 216$ ← **To use a calculator, enter 6 y^x 3 =.**

b. $15^1 = 15$

c. $10^0 = 1$

your turn

Find each answer.

1. 3^2 **2.** 2^4 **3.** 5^3 **4.** 5^4 **5.** 8^0 **6.** 9^1

example 2

Find each answer.

a. 4×5^3 **b.** $2^2 \times 3^0 \times 50$

solution

a. $4 \times 5^3 = 4 \times 5 \times 5 \times 5$
$= 500$

b. $2^2 \times 3^0 \times 50 = 4 \times 1 \times 50$
$= 200$

your turn

Find each answer.

7. 3×4^4 **8.** $2^5 \times 3^2$ **9.** $4^2 \times 5^0 \times 6$ **10.** $2^3 \times 4^1 \times 50$

practice exercises

practice for example 1 (page 48)

Find each answer.

1. 10^4 **2.** 1^7 **3.** 7^2 **4.** 8^2 **5.** 14^1
6. 11^0 **7.** 1^6 **8.** 91^1 **9.** 56^0 **10.** 3^6

practice for example 2 (page 48)

Find each answer.

11. $3^3 \times 5$ **12.** $6^2 \times 9$ **13.** $3^4 \times 6^3$ **14.** $4^3 \times 5^0$
15. $5 \times 6^3 \times 2$ **16.** $100 \times 8^2 \times 4$ **17.** $3^1 \times 6^2 \times 9^2$ **18.** $10 \times 4^3 \times 2^2$

mixed practice (page 48)

Find each answer.

19. 9^2 **20.** 7^3 **21.** 2×6^2 **22.** $3^3 \times 3^2 \times 8$
23. 2^5 **24.** 7×5^4 **25.** $7^2 \times 12^0 \times 5$ **26.** 44^1
27. 21^1 **28.** 2^0 **29.** $2^3 \times 15$ **30.** $8^2 \times 7 \times 2^2$
31. 10^0 **32.** 10^1 **33.** 10^2 **34.** 10^3

35. Based on your answers to Exercises 31–34, state a short cut for writing the numeral form of ten to a power.

36. The planet Neptune is approximately 3×10^9 miles from the sun. Write the numeral form of this number.

37. Which score on a test would be better, 100^0 or 99^1? Explain.

review exercises

Estimate.

1. 391 + 477 **2.** \$16.38 + \$23.55 **3.** 9320 − 2416 **4.** \$105 − \$26.88
5. 51 × 309 **6.** 82 × 681 **7.** 36 × 947 **8.** 77 × 394

March 29, 1987, can be written 3/29/87. This is called a *product date* because 3 × 29 = 87. April 20, 1980, is also a product date because 4 × 20 = 80. Which dates in 1990 are product dates?

SKILL REVIEW

Choose the letter of the *best* estimate for each product. **3-1**

1. 5×49	**a.** 2000	**b.** 200	**c.** 250	**d.** 25
2. 62×58	**a.** 3000	**b.** 360	**c.** 300	**d.** 3600
3. 37×84	**a.** 2700	**b.** 3200	**c.** 2400	**d.** 3600
4. $9 \times \$73$	**a.** \$730	**b.** \$630	**c.** \$700	**d.** \$720

5. During the busiest hour of the day, 185 planes depart from the local airport. Find a range of reasonable numbers to describe the total number of planes that depart during that hour in three days.

6. Sarah Belson swims 22 laps daily in the local pool. Find a range of reasonable numbers to describe about how many laps she swims in 28 days.

7. Cy Gore sold 49 concert tickets costing \$8 each. About how much money did Cy collect in ticket sales?

Find each product. Check the reasonableness of your answer. **3-2**

8. $\$26 \times 61$ **9.** 34×107 **10.** $\$39 \times 457$
11. $\$85 \times 362$ **12.** 691×718 **13.** $77 \times \$15$
14. 803×57 **15.** 594×229 **16.** 416×549
17. 6545×37 **18.** 829×45 **19.** 337×247

Use the properties of multiplication to find each product. **3-3**

20. 6×89 **21.** 3×78 **22.** 271×454
23. 7×707 **24.** $5 \times 4 \times 20$ **25.** 823×616
26. 5×804 **27.** $25 \times 9 \times 4$ **28.** $50 \times 18 \times 2$
29. $4 \times 91 \times 25$ **30.** 8×699 **31.** 4×391

32. Mark Ippolito bought a pair of pants for \$13 and a sweatshirt for \$21. He also bought a watch for \$21 and a shirt for \$13. How much did Mark spend altogether?

Find each answer. **3-4**

33. 2^3 **34.** 3^5 **35.** 10^1 **36.** 15^0 **37.** 7^4 **38.** 2^6

39. 9×4^2 **40.** $5^3 \times 6^2$ **41.** $10 \times 3^2 \times 7$
42. $8^1 \times 3^3 \times 2^3$ **43.** $1^5 \times 3^2$ **44.** $6^2 \times 1^7$
45. 5×3^4 **46.** 8×2^5 **47.** $7^2 \times 2 \times 3^3$
48. $5^2 \times 10 \times 2^3$ **49.** $2^3 \times 11^0 \times 6^3$ **50.** $9^1 \times 9^0 \times 5^3$

APPLICATION

3-5 REBATES AND COUPONS

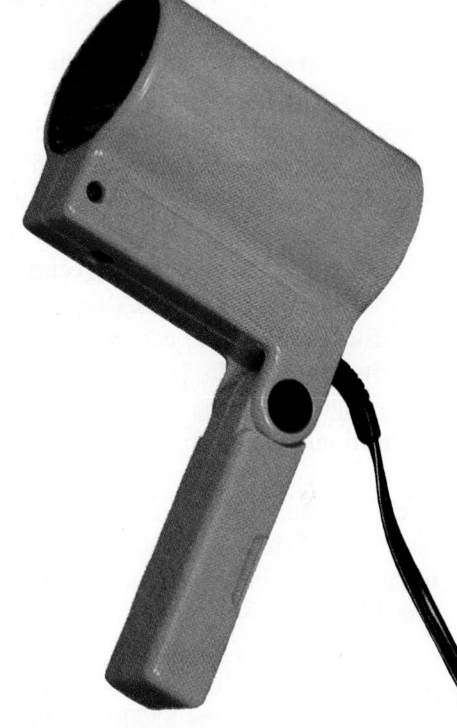

Some manufacturers offer mail-in *rebates* to consumers. A **rebate** is a refund of part of the purchase price. The final cost of an item purchased with a rebate is the price of the item minus the amount of the rebate.

example

Mario Rosselli buys a hair dryer for $44.98. The manufacturer offers a $7 rebate. Find the final cost of the dryer.

solution

Subtract:

	$44.98	← purchase price
	− 7.00	← rebate
	$37.98	← final cost

The final cost of the dryer is $37.98.

exercises

1. Warren Sampson buys a car stereo for $129.88. The manufacturer offers a factory rebate of $30. Find the final cost of the stereo.
2. Ronna Weiss buys a pair of jeans for $21.99 and mails in a rebate coupon for $3. Find the final cost of the jeans.
3. Sam Lowe mails in a $12 rebate coupon. His final cost for a parka is $100. What is the purchase price of the parka?
4. Alana Korda mails in a $6 rebate coupon. Her final cost for a phone is $54. Find the purchase price of the phone.

Consumers also use manufacturers' coupons to buy items at reduced prices. A coupon is like an instant rebate. To find the price of an item purchased with a coupon, subtract the value of the coupon from the price of the item. Some stores double the value of each coupon.

5. Fred Parker buys apple juice costing $1.58. He presents a coupon for 30¢. Find the final cost of the juice.
6. Sorota's Market doubles the value of each coupon. Sam Grogan presents a 25¢ coupon when he buys toothpaste that costs $1.39. Find the final cost of the toothpaste.
7. Find out which local stores, if any, double the value of coupons.
8. Find out where you can obtain coupons.

APPLICATION

3-6 FEDERAL INCOME TAX

Jonathan Powers is completing his federal income tax return. He needs both a **1040A tax form,** which is shown on pages 53–54, and his **W-2 form.** The W-2 form is a wage and tax statement that shows the amount of pay withheld for federal, local, and state taxes and for social security.

In Step 1 of the 1040A form, Jonathan fills in his name, address, and social security number. In Step 2, he checks his filing status. On line 5e, he claims three **exemptions.** Jonathan uses a calculator to add his salary, wages, tips, and taxable interest income. He enters this total income on line 10. He has no adjustments to his income, so he enters the same figure on line 12.

On line 14d, he takes a standard deduction of \$3760. Then he multiplies \$1900 × 3 and enters \$5700 on line 16. From his adjusted gross income, he subtracts \$3760 and \$5700. He enters \$12,142.65 on line 17 as taxable income.

The difference between total tax and any credits or payments already made determines whether there is a **refund** or **balance due.** If there is a refund, you send your tax form to the Internal Revenue Service and receive a check for the amount of the refund. If there is a balance due, you must submit a check or money order with your tax form.

example

Jonathan has a total tax amount due of \$1699. He has already had \$1585.20 withheld for taxes. Tell whether there is a refund or a balance due. Then find the exact amount.

solution

Subtract:		
	\$1699.00	← total tax due (line 20)
	− 1585.20	← total payments already made (line 22)
	\$113.80	

Jonathan has a balance due of \$113.80.

Form **1040A**

Department of the Treasury—Internal Revenue Service

U.S. Individual Income Tax Return (0) 19 –

OMB No. 1545-0085

Step 1
Label
Use IRS label. Otherwise, please print or type.

Your first name and initial (if joint return, also give spouse's name and initial): Jonathan T. and Samantha R. Last name: Powers

Present home address (number and street). (If you have a P.O. Box, see page 9 of the instructions.): 35112 Oak Street

City, town or post office, state, and ZIP code: Dallas, TX 75204

Your social security no. 123 45 6789

Spouse's social security no. 987 65 4321

For Privacy Act and Paperwork Reduction Act Notice, see page 31.

Presidential Election Campaign Fund

Do you want $1 to go to this fund? . . . . ☐ Yes ☐ No

If joint return, does your spouse want $1 to go to this fund?. ☐ Yes ☐ No

Note: *Checking "Yes" will not change your tax or reduce your refund.*

Step 2
Check your filing status
(Check only one)

1 ☐ Single (See if you can use Form 1040EZ.)
2 ☑ Married filing joint return (even if only one had income)
3 ☐ Married filing separate return. Enter spouse's social security number above and spouse's full name here.
4 ☐ Head of household (with qualifying person). If the qualifying person is your child but not your dependent, enter this child's name here.

Step 3
Figure your exemptions
(See page 12 of instructions.)

Caution: If you can be claimed as a dependent on another person's tax return (such as your parents' return), do not check box 5a. But be sure to check the box on line 14b on page 2.

5a ☑ **Yourself** **5b** ☑ **Spouse**

No. of boxes checked on 5a and 5b: 2

c Dependents: 1. Name (first, initial, and last name)	2. Check if under age 5	3. If age 5 or over, dependent's social security number	4. Relationship	5. No. of months lived in your home in 19 –
Paul Powers	✓		son	12

No. of children on 5c who lived with you: 1

No. of children on 5c who didn't live with you due to divorce or separation:

No. of parents listed on 5c:

No. of other dependents listed on 5c:

If more than 7 dependents, attach statement.

Attach Copy B of Form(s) W-2 here.

d If your child didn't live with you but is claimed as your dependent under a pre-1985 agreement, check here . . . . ▶ ☐

e Total number of exemptions claimed. (Also complete line 16.)

Add numbers entered on lines above: 3

Step 4
Figure your total income

Attach check or money order here.

6	Wages, salaries, tips, etc. This should be shown in Box 10 of your W-2 form(s). (Attach Form(s) W-2.)	6	21,364 49
7a	**Taxable** interest income (see page 17). (If over $400, also complete and attach Schedule 1, Part II.)	7a	238 16
b	**Tax-exempt** interest income (see page 17). (DO NOT include on line 7a.) 7b		
8	Dividends. (If over $400, also complete and attach Schedule 1, Part III.)	8	0 00
9	Unemployment compensation (insurance) from Form(s) 1099-G.	9	0 00
10	Add lines 6, 7a, 8, and 9. Enter the total. This is your **total income**.	▶ 10	21,602 65

Step 5
Figure your adjusted gross income

11a	Your IRA deduction from applicable Worksheet. New rules for IRAs begin on page 18. 11a		
b	Spouse's IRA deduction from applicable Worksheet. New rules for IRAs begin on page 18. 11b		
c	Add lines 11a and 11b. Enter the total. These are your **total adjustments.**	11c	0 00
12	Subtract line 11c from line 10. Enter the result. This is your **adjusted gross income**. (If this line is less than $15,432 and a child lived with you, see "Earned Income Credit" (line 21b) on page 27 of instructions.)	▶ 12	21,602 65

Form **1040A** (19 –)

Step 6
Figure your standard deduction,

13 Enter the amount from line 12. — 13: 21,602 | 65

14a Check if: ☐ **You** were 65 or over ☐ Blind; ☐ **Spouse** was 65 or over ☐ Blind — **Enter number of boxes checked ▶ 14a** ☐

b If you can be claimed as a dependent on another person's return (such as your parents' return), check here . . . . . . ▶ 14b ☐

c If you are married filing separately and your spouse files Form 1040 and itemizes deductions, check here . . . . ▶ 14c ☐

d Standard deduction. If you checked a box on line 14a, b, or c, see page 22 for amount to enter on line 14d. If no box is checked, enter amount shown below for your filing status.

Filing status from page 1:
- Single or Head of household, enter $2,540
- Married filing joint return, enter $3,760
- Married filing separate return, enter $1,880

14d: 3760 | 00

Exemption amount, and

15 Subtract line 14d from line 13. Enter the result. — 15: 17,842 | 65

16 Multiply $1,900 by the total number of exemptions claimed on line 5e. Or, figure your exemption amount from the chart on page 24 of the instructions. — 16: 5700 | 00

Taxable income

17 Subtract line 16 from line 15. Enter the result. This is your **taxable income.** ▶ 17: 12,142 | 65

If You Want IRS To Figure Your Tax, See Page 24 of the Instructions.

Step 7
Figure your tax, credits, and payments (including advance EIC payments)

Caution: If you are under age 14 and have more than $1,000 of investment income, see page 24 of the instructions and check here . . . . ▶ ☐

18 Find the tax on the amount on line 17. Check if from: ☑ Tax Table (pages 32–37); or ☐ Form 8615, Computation of Tax for Children Under Age 14 Who Have Investment Income of More Than $1,000. — 18: 1699 | 00

19 Credit for child and dependent care expenses. Complete and attach Schedule 1, Part I. — 19: 0 | 00

20 Subtract line 19 from line 18. Enter the result. (If line 19 is more than line 18, enter -0- on line 20.) This is your **total tax.** ▶ 20: 1699 | 00

21a Total Federal income tax withheld. This should be shown in Box 9 of your W-2 form(s). (If line 6 is more than $43,800, see page 26.) — 21a: 1585 | 20

b Earned income credit, from the worksheet on page 28 of the instructions. Also see page 27. — 21b:

22 Add lines 21a and 21b. Enter the total. These are your **total payments.** ▶ 22: 1585 | 20

Step 8
Figure your refund or amount you owe

23 If line 22 is larger than line 20, subtract line 20 from line 22. Enter the result. This is the **amount of your refund.** — 23:

24 If line 20 is larger than line 22, subtract line 22 from line 20. Enter the result. This is the **amount you owe.** Attach check or money order for full amount payable to "Internal Revenue Service." Write your social security number, daytime phone number, and "19– Form 1040A" on it. — 24: 113 | 80

Step 9
Sign your return

Under penalties of perjury, I declare that I have examined this return and accompanying schedules and statements, and to the best of my knowledge and belief, they are true, correct, and complete. Declaration of preparer (other than the taxpayer) is based on all information of which the preparer has any knowledge.

Your signature	Date	Your occupation
X Jonathan T. Powers	2/10/–	
Spouse's signature (if joint return, both must sign)	**Date**	**Spouse's occupation**
X Samantha R. Powers	2/10/–	

Paid preparer's use only

Preparer's signature	Date	Preparer's social security no.
X		
Firm's name (or yours if self-employed)		Employer identification no.
Address and ZIP code		Check if self-employed ☐

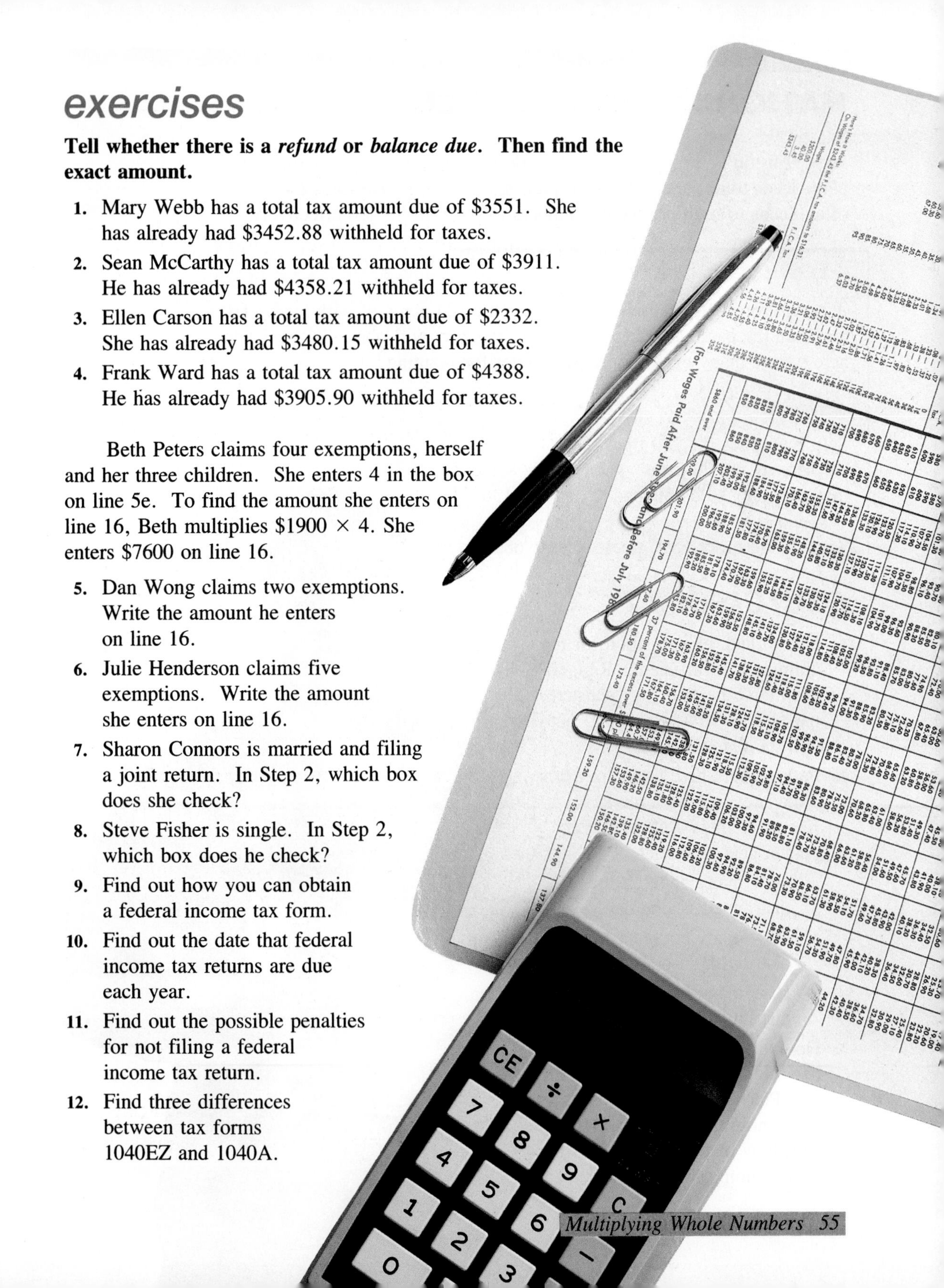

exercises

Tell whether there is a *refund* or *balance due*. Then find the exact amount.

1. Mary Webb has a total tax amount due of $3551. She has already had $3452.88 withheld for taxes.
2. Sean McCarthy has a total tax amount due of $3911. He has already had $4358.21 withheld for taxes.
3. Ellen Carson has a total tax amount due of $2332. She has already had $3480.15 withheld for taxes.
4. Frank Ward has a total tax amount due of $4388. He has already had $3905.90 withheld for taxes.

Beth Peters claims four exemptions, herself and her three children. She enters 4 in the box on line 5e. To find the amount she enters on line 16, Beth multiplies $1900 × 4. She enters $7600 on line 16.

5. Dan Wong claims two exemptions. Write the amount he enters on line 16.
6. Julie Henderson claims five exemptions. Write the amount she enters on line 16.
7. Sharon Connors is married and filing a joint return. In Step 2, which box does she check?
8. Steve Fisher is single. In Step 2, which box does he check?
9. Find out how you can obtain a federal income tax form.
10. Find out the date that federal income tax returns are due each year.
11. Find out the possible penalties for not filing a federal income tax return.
12. Find three differences between tax forms 1040EZ and 1040A.

CAREER

NATIONAL PARK RANGER

Sam Winston is a forest ranger at York National Park. He patrols the park, leads tours, and helps with park maintenance. He finds that he often needs to multiply whole numbers to perform his job effectively.

example

A recent forest fire caused the destruction of 24 acres of park land. To maintain the park, the rangers and park workers will plant 284 trees per acre. How many trees should Sam order for planting?

solution

Sam multiplies the 284 trees per acre by the 24 acres that are going to be planted.

$$\begin{array}{r} 284 \\ \times 24 \\ \hline 1136 \\ 568 \\ \hline 6816 \end{array}$$

Sam should order 6816 trees.

exercises

1. Due to lack of rain, food for the animals is insufficient. Rangers will drop three bales of hay per acre. The park has 6776 acres. How many bales will the rangers need?
2. Park rangers give tours six times a day, seven days a week. Each tour has 60 people. Each person gets a booklet.
   a. How many booklets are distributed in one day?
   b. How many booklets are distributed in one week?
   c. How many booklets are distributed in July?
3. On average, there are four people per tent site and six people per trailer site. The Pines Campground has 178 tent sites and 61 trailer sites. If each site is full, how many people are at the Pines Campground?

CHAPTER REVIEW

vocabulary vo·cab·u·lar·y

Choose the correct word to complete each sentence.

1. You may multiply numbers in any order. This is the (*commutative, associative*) property of multiplication.
2. In 5^3, 3 is the (*base, exponent*).

skills

Choose the letter of the way to arrive at the best estimate of each product.

		a.	b.	c.	d.
3.	789 × 22	800 × 20	800 × 10	700 × 20	700 × 10
4.	48 × 91	40 × 90	40 × 100	50 × 100	50 × 90
5.	32 × 57	30 × 50	40 × 50	30 × 60	40 × 60
6.	5 × 319	5 × 300	6 × 300	5 × 400	6 × 400
7.	84 × 39	80 × 30	90 × 30	90 × 40	80 × 40
8.	604 × 72	600 × 80	600 × 70	700 × 80	700 × 70
9.	362 × 741	300 × 700	400 × 700	300 × 800	400 × 800
10.	439 × 272	500 × 300	400 × 200	400 × 300	500 × 200

Find each product.

11. 8 × $493
12. $625 × 5
13. 136 × 694
14. 471 × 358
15. 48 × 925
16. 57 × 804
17. 4 × 96
18. 4 × 3 × 25 × 10
19. 392 × 777
20. 8 × 57
21. 7 × 196
22. 50 × 7 × 2 × 6
23. 9 × 599
24. 641 × 999
25. 10 × 12 × 3 × 10
26. 12 × 62
27. 25 × 6 × 8 × 4
28. 11 × 81
29. 5×6^2
30. 12×9^3
31. $3^5 \times 2^4$
32. $6^2 \times 4^3$
33. $2^2 \times 4^2 \times 25^0$
34. $5^2 \times 2^3 \times 10^0$
35. $9^1 \times 4^4$
36. $8^1 \times 6^4$
37. $6^3 \times 5^3 \times 1^2$
38. Betty Compton buys a hair dryer for $29.98. She mails in a rebate coupon for $3. Find the final cost of the hair dryer.
39. Val St. Pierre buys Shiny shampoo costing $1.59 and presents a coupon for 25¢. The store doubles the value of the coupon. Find the final cost of the shampoo.
40. Samantha Whitten claims five exemptions. Use the form on pages 53–54. Write the amount she enters on line 16.

CHAPTER TEST

Estimate by rounding to the place of the leading digit.

1. 19×74	**2.** 43×56	**3.** $\$6 \times 17$	**4.** $5 \times \$67$	**3-1**
5. 695×88	**6.** $\$32 \times 558$	**7.** 51×651	**8.** 406×49	

9. Miss Wilkins ordered calculators for 28 students in her class. If calculators cost $19 each, about how much did the order cost?

Find each product. Check the reasonableness of your answer.

10. 7×654	**11.** $51 \times \$49$	**12.** 687×322	**13.** 4×896	**3-2**
14. $17 \times \$65$	**15.** 608×81	**16.** 903×32	**17.** 926×483	
18. 36×73	**19.** 58×89	**20.** 102×806	**21.** 509×704	

22. To raise funds for charity, the senior class sold 402 boxes of granola bars. There were 24 bars in each box. How many granola bars did the seniors sell?

23. Tamika Jones works three hours a day after school. Her hourly pay is $6. How much does she earn in one day?

Use the properties of multiplication to find each product.

24. $2 \times 64 \times 5$	**25.** 162×636	**26.** 6×392	**27.** 3×598	**3-3**
28. 7×407	**29.** $25 \times 3 \times 4$	**30.** 249×818	**31.** 4×810	

32. Mr. Ferris orders seven make-up brushes for each student in his beauty school. There are 210 students in the school. How many brushes does he order?

33. John Kostas had 100 posters printed to advertise a school play. He displayed three posters in each of 32 stores. How many of the 100 posters did he display in these stores?

Find each answer.

34. 3^4	**35.** 6^2	**36.** 4^3	**37.** 8^3	**3-4**
38. $5^2 \times 7^1 \times 2^0$	**39.** $2^2 \times 50^0 \times 9^3$	**40.** $2^4 \times 4$	**41.** $3^2 \times 5^4$	

42. Carol Regent buys dishes and mails in a $3 rebate coupon. Her final cost is $39.95. What is the purchase price of the dishes? **3-5**

43. Barry Lymon buys a carton of orange juice costing $1.69 and presents a coupon for 20¢. Find the final cost of the juice.

44. Marilyn Southworth owes $3520.40 in taxes. She had $4025.80 in taxes withheld from her paychecks. Tell whether there is a refund or balance due. Then find the exact amount. **3-6**

In travel, times of arrival and departure are very important. Days are divided into smaller units that help to record times exactly.

CHAPTER 4

DIVIDING WHOLE NUMBERS

4-1 ESTIMATING QUOTIENTS

The example at the right shows the parts of a division. However, there are times that an estimated answer to a division is all that you need. To estimate a quotient you use numbers that divide easily, or **compatible numbers.**

$$\begin{array}{r} 32 \\ 6\overline{)192} \\ \underline{18\downarrow} \\ 12 \\ \underline{12} \\ 0 \end{array}$$

divisor → 6; 32 ← **quotient**; 192 ← **dividend**; 0 ← **remainder**

example 1

Estimate using compatible numbers.

a. $8\overline{)1703}$

about 200
$8\overline{)1600}$ ← **1600 is a number close to 1703 that is easily divided by 8.**

b. $39\overline{)30,751}$

about 800
$40\overline{)32,000}$ ← **Round 39 to 40. 40 divides both 28,000 and 32,000 easily, but 32,000 will give a closer estimate.**

your turn

Estimate using compatible numbers.

1. $5\overline{)1897}$ **2.** 47,264 ÷ 7 **3.** $73\overline{)65,492}$ **4.** 370,195 ÷ 587

If the divisor is not rounded, you can determine whether your estimate is greater than or less than the actual quotient by examining how you changed the dividend. If your estimate is greater than the actual quotient, it is an **overestimate.** If your estimate is less than the actual quotient, it is an **underestimate.**

example 2

Estimate each quotient. Determine whether the estimate is an *overestimate* or an *underestimate*.

a. 139 ÷ 3 **b.** 21,083 ÷ 90

solution

a. $3\overline{)139}$ → about 50 $3\overline{)150}$ → The dividend is increased. 50 is *greater than* the actual quotient. 50 is an *overestimate*.

b. $90\overline{)21,083}$ → about 200 $90\overline{)18,000}$ → The dividend is decreased. 200 is *less than* the actual quotient. 200 is an *underestimate*.

your turn

Estimate each quotient. Determine whether the estimate is an *overestimate* or an *underestimate*.

5. $4\overline{)2501}$ **6.** 6619 ÷ 80 **7.** $500\overline{)19,444}$ **8.** 40,735 ÷ 60

practice exercises

practice for example 1 (page 60)

Estimate using compatible numbers.

1. $3\overline{)113}$ **2.** $7\overline{)4051}$ **3.** 3642 ÷ 52 **4.** 27,103 ÷ 38

5. $42\overline{)20,227}$ **6.** $31\overline{)369,293}$ **7.** 450,713 ÷ 493 **8.** 495,027 ÷ 641

practice for example 2 (pages 60–61)

Estimate each quotient. Determine whether the estimate is an *overestimate* or an *underestimate*.

9. $9\overline{)2507}$ **10.** $6\overline{)3845}$ **11.** 37,414 ÷ 4 **12.** 53,641 ÷ 7

13. $40\overline{)49,625}$ **14.** $60\overline{)28,371}$ **15.** 88,722 ÷ 900 **16.** 553,322 ÷ 600

mixed practice (pages 60–61)

Estimate using compatible numbers.

17. $31\overline{)26,995}$ **18.** 4093 ÷ 7 **19.** 6381 ÷ 82 **20.** $48\overline{)53,942}$

21. $423\overline{)78,283}$ **22.** $88\overline{)90,037}$ **23.** 4302 ÷ 9 **24.** 250,921 ÷ 394

25. Sally worked for nine weeks last summer. If she earned $2525 for the summer, about how much did she earn each week?

26. The history class trip to Boston will cost $5800. If there are 28 students, about how much will each student have to pay?

27. Larry Chin will earn $34,900 this year. There are 52 weeks in a year. About how much will Larry earn each week?

review exercises

Round to the place of the leading digit.

1. 16 **2.** 34 **3.** 55 **4.** 98 **5.** 108

6. 281 **7.** 351 **8.** 349 **9.** 965 **10.** 3603

11. 5477 **12.** 6095 **13.** 10,965 **14.** 14,991 **15.** 15,037

4-2 DIVIDING WHOLE NUMBERS

To divide one whole number by another, use the following method.

Start at the left.

1. *Divide* in the greatest place possible.
2. *Multiply*.
3. *Subtract*.
4. *Bring down* the next digit or digits.

Repeat steps 1 to 4 until the remainder is less than the divisor.

example 1

Find the quotient. Check the reasonableness of your answer.

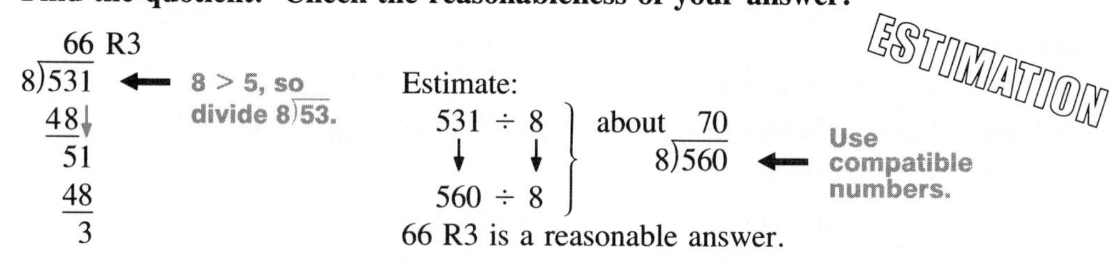

your turn

Find each quotient. Check the reasonableness of your answer.

1. $3\overline{)293}$ **2.** $4\overline{)853}$ **3.** $9892 \div 9$ **4.** $29{,}015 \div 5$

If the divisor has more than one digit, round the divisor to the place of its leading digit. This is your **trial divisor.**

After dividing, you should check your answer. Multiplication and division are *inverse operations*. To check your answer, you multiply the quotient by the divisor and then add the remainder.

example 2

Find the quotient.

Round to 20. → $17\overline{)5206}$ ← Try 2.

```
      2
17)5206
   34
   18
```

18 > 17, so 2 is not enough. Revise up.

```
    306 R4   ← Try 3.
17)5206
   51↓↓
    106
    102
      4
```

Check:

```
   306
   ×17
  2142
  306
  5202
+    4
  5206 ✓
```

your turn

Find each quotient.

5. $18\overline{)1035}$ **6.** $22\overline{)2356}$ **7.** $47\overline{)2486}$ **8.** $25\overline{)9810}$

9. $49{,}140 \div 44$ **10.** $73{,}695 \div 51$ **11.** $96{,}384 \div 32$ **12.** $90{,}562 \div 94$

practice exercises

practice for example 1 (page 62)

Find each quotient. Check the reasonableness of your answer.

1. $4\overline{)396}$ **2.** $5\overline{)733}$ **3.** $8\overline{)984}$ **4.** $9\overline{)890}$

5. $7\overline{)1718}$ **6.** $4\overline{)1461}$ **7.** $3\overline{)3825}$ **8.** $7\overline{)8593}$

9. $11{,}351 \div 6$ **10.** $16{,}335 \div 8$ **11.** $18{,}088 \div 9$ **12.** $26{,}565 \div 4$

13. $27{,}655 \div 5$ **14.** $24{,}962 \div 2$ **15.** $30{,}308 \div 6$ **16.** $71{,}849 \div 7$

practice for example 2 (pages 62–63)

Find each quotient.

17. $13\overline{)4165}$ **18.** $18\overline{)5616}$ **19.** $59\overline{)6491}$ **20.** $43\overline{)7149}$

21. $31\overline{)10{,}971}$ **22.** $33\overline{)67{,}405}$ **23.** $61\overline{)49{,}868}$ **24.** $78\overline{)92{,}320}$

25. $52{,}692 \div 25$ **26.** $37{,}989 \div 39$ **27.** $82{,}420 \div 20$ **28.** $90{,}240 \div 68$

29. $45{,}144 \div 76$ **30.** $67{,}493 \div 53$ **31.** $73{,}896 \div 77$ **32.** $56{,}015 \div 85$

mixed practice (pages 62–63)

Find each quotient. Check the reasonableness of your answer.

33. $9\overline{)7615}$ **34.** $6\overline{)3486}$ **35.** $2047 \div 5$ **36.** $9654 \div 8$

37. $16\overline{)10{,}645}$ **38.** $19\overline{)27{,}064}$ **39.** $24\overline{)18{,}792}$ **40.** $26\overline{)18{,}308}$

41. $34{,}842 \div 32$ **42.** $57{,}586 \div 67$ **43.** $37{,}720 \div 46$ **44.** $80{,}576 \div 92$

45. $234\overline{)49{,}945}$ **46.** $302\overline{)124{,}424}$ **47.** $281{,}261 \div 387$ **48.** $369{,}085 \div 455$

The following exercises were done using a calculator. Estimate to tell if each answer is reasonable. Then correct each unreasonable answer by finding the quotient.

49. $19{,}092 \div 74$ [258] **50.** $19{,}992 \div 98$ [204]

51. $27{,}720 \div 28$ [99] **52.** $23{,}460 \div 46$ [51]

53. $72{,}812 \div 218$ [334] **54.** $86{,}632 \div 476$ [182]

Solve.

55. Hal types 45 words per minute. How long will it take him to type an 1800-word report?
56. A car storage lot has 20,776 spaces. There are 98 rows in the lot. Each row has the same number of spaces. How many spaces are in each row?
57. Juanita earns $342 each week. She works 38 h (hours) each week. Juanita works 50 weeks each year. How much does she earn each hour?
58. The Milk Company packs 24 milk cartons in a case. It takes 5 min (minutes) to pack a case. How many cases does the company need to pack 7536 cartons?

review exercises

Find each answer.

1. $64 + 1027 + 786$
2. $33\overline{)8469}$
3. 60×4000
4. 5^3
5. $827.46 - 89.37$
6. $2862 \div 3$
7. $2.64 + 5.09 + 26$
8. $3^3 \times 4^2$
9. 526×209
10. $981 - 52.43$
11. $1.25 + 84.7 + 19$
12. $76\overline{)9728}$
13. $72{,}706 - 4828$
14. 815×34
15. $2^4 \times 3^2 \times 6^1$

mental math

You can use **short division** to divide 4189 by 6. Divide, multiply, and subtract mentally. Repeat as often as needed.

$$\overset{6}{6\overline{)4_{5}189}} \longrightarrow \overset{69}{6\overline{)4_{5}1_{5}89}} \longrightarrow \overset{698\ R1}{6\overline{)4_{5}1_{5}8_{4}9}}$$

Divide using short division.

1. $5\overline{)263}$
2. $2\overline{)573}$
3. $8\overline{)6325}$
4. $3\overline{)4287}$
5. $7\overline{)9340}$
6. $873 \div 9$
7. $702 \div 6$
8. $732 \div 4$
9. $8103 \div 8$
10. $7923 \div 5$

4-3 ORDER OF OPERATIONS

You must perform the operations of multiplication, division, addition, and subtraction in a specific order. The order of these operations is shown in the chart below.

Order of Operations

1. First do all work inside any parentheses.
2. Then find the value of each power.
3. Then do all multiplications and divisions in order from left to right.
4. Then do all additions and subtractions in order from left to right.

example 1

Find each answer: **a.** $72 \div 3 - 2 \times 8 + 4$ **b.** $12 + 18 \times 3^2$

solution

a.
$72 \div 3 - 2 \times 8 + 4$ ← **Multiply and divide in order.**
$24 - 2 \times 8 + 4$
$24 - 16 + 4$ ← **Add and subtract in order.**
$8 + 4$
12

b.
$12 + 18 \times 3^2$
$12 + 18 \times 9$ ← **$3^2 = 9$**
$12 + 162$ ← **Multiply, then add.**
174

your turn

Find each answer.

1. $5 + 7 \times 3 - 2$ **2.** $27 \div 3 \times 2 - 5$ **3.** $84 - 4^3 + 5 \times 3$

example 2

Find each answer: **a.** $94 - (3 + 7) - (11 \times 6)$ **b.** $(4^2 + 40) \div 2$

solution

a.
$94 - (3 + 7) - (11 \times 6)$ ← **Work inside the parentheses first.**
$94 - 10 - 66$
$84 - 66$
18

b.
$(4^2 + 40) \div 2$
$(16 + 40) \div 2$ ← **$4^2 = 16$**
$56 \div 2$
28

your turn

Find each answer.

4. $(14 + 16) \div 5 - 2$ **5.** $15 \div 3 + (2 \times 5)$ **6.** $24 \times 5 - (8^2 + 16)$

Division may be shown by a fraction bar. The bar acts like parentheses in the order of operations.

$$\frac{6 + 39}{18 - 3}$$ ← **fraction bar**

example 3

Find the answer.

$$\frac{6 + 39}{18 - 3} = \frac{45}{15} = 3$$ ← **Work above the fraction bar, then below the fraction bar. Then divide.**

your turn

Find each answer.

7. $\frac{3 \times 12}{4 + 5}$ **8.** $\frac{14 + 56}{7 \times 5}$ **9.** $\frac{3 + 2 \times 9}{11 - 4}$ **10.** $\frac{1025 \div 5}{21 + 2 \times 10}$

practice exercises

practice for example 1 *(page 65)*

Find each answer.

1. $15 \div 3 \times 6 - 14$ **2.** $100 - 6 \times 4 \times 3$ **3.** $78 + 2 \times 9 \div 3 - 21$

4. $85 - 3^3 \div 9$ **5.** $4^2 \times 20 + 52 \div 4$ **6.** $2^3 + 3^2 \times 12 - 41$

practice for example 2 *(page 65)*

Find each answer.

7. $84 \div 4 - (23 - 2)$ **8.** $(70 - 10) \div (8 + 52)$ **9.** $3 \times 40 \div (8 + 4)$

10. $10^2 \div (12 - 2)$ **11.** $4 + 2^3 \times (14 \div 7)$ **12.** $2 \times (8^2 - 10) - 50$

practice for example 3 *(page 66)*

Find each answer.

13. $\frac{5 + 23}{10 - 8}$ **14.** $\frac{3 \times 24}{18 \div 6}$ **15.** $\frac{6 + 9 \times 12}{96 \div 16}$ **16.** $\frac{112 - 56}{6 \div 2 + 5}$

mixed practice *(pages 65–66)*

Find each answer.

17. $14 \times 29 - 87$ **18.** $(39 + 7^2) \div (35 - 27)$ **19.** $39 - 2 \times 17 + 27$

20. $\frac{3 \times 7 - 5}{8 - 4}$ **21.** $\frac{5 \times 12}{12 - 3 \times 2}$ **22.** $\frac{19 + 45}{16 \div 4 + 4}$

23. $7 + 3 \times (8 + 11)$ **24.** $13 + 3^4 \div 9 - 17$ **25.** $7 \times (2^4 - 3) \div 13$

Replace each ___?___ with +, −, ×, or ÷ to make a true statement.

26. 24 ___?___ 9 ___?___ 7 = 8
27. (36 ___?___ 4) ___?___ 8 = 5
28. 4 ___?___ 7 ___?___ 8 = 224
29. 64 ___?___ 4 ___?___ 16 = 256
30. 72 ___?___ 12 ___?___ 2 = 12
31. (84 ___?___ 16) ___?___ 5 = 20
32. Tim works every day after school and earns $18 per day. On weekends, he earns $42 per day. How much does Tim earn in a week?
33. Lisa bought three cans of paint for $13 each and a brush for $6. How much was her total purchase?
34. Sandy worked 38 h (hours) the first week of June, 40 h the second week, and 36 h the third week. If Sandy worked a total of 156 h in June, how many hours did she work the fourth week?

review exercises

Use the properties of multiplication to find each product.

1. $20 \times 16 \times 5$
2. $5 \times 9 \times 8 \times 2$
3. $4 \times (50 + 18)$
4. 8×87

Estimate.

5. 27×44
6. 57×92
7. 211×68
8. 435×652
9. $7288 \div 8$
10. $2890 \div 47$
11. $72{,}033 \div 71$
12. $11{,}987 \div 378$

calculator corner

Some calculators have parentheses keys. Other calculators do not, and you can use memory keys instead.

example Find the answer: $50 \div (3 \times 5 - 10)$

solution If the calculator has parentheses keys, enter

50 ÷ [(] 3 × 5 − 10 [)] =.

The result is 10.

If the calculator does not have parentheses keys, enter

3 × 5 − 10 = [M+] . ← **Puts 5 into memory.**

Then enter 50 ÷ [MR] =. ← **Divides 50 by 5.**

The result is 10.

Use a calculator to find each answer.

1. $80 \div (11 - 6)$
2. $63 \div (12 - 9)$
3. $95 \div (3 \times 3 - 4)$
4. $88 \div (7 \times 3 - 10)$
5. $72 - (34 + 5)$
6. $51 - (9 \times 3)$
7. $30 - (4 + 6 + 3)$
8. $63 - (7 \times 5 - 11)$

PROBLEM SOLVING STRATEGY

4-4 CHOOSING THE CORRECT OPERATION

1 Understand
2 Plan
3 Work
4 Answer

An important part of using a four-step method to solve a problem is choosing the correct operation. Sometimes you need more than one operation. In such cases, you must decide not only which operations to use, but also in which order to use them.

example

Lesley bought a pair of pliers for \$10.99, five wrenches for \$3 each, and a hammer for \$9.59.

a. How much did Lesley pay for the wrenches?

b. Lesley gave the clerk \$40 to pay for her total purchase. How much change did she receive?

solution

a. Step 1 Given: bought 5 wrenches
\$3 for one wrench
Find: the cost of the wrenches

Step 2 *Multiply* \$3 by 5. ← **Only one operation is needed.**

Step 3 $\$3 \times 5 = \$15$

Step 4 *Check* against the wording of the problem:
If 5 wrenches cost \$15, then each wrench must cost $\$15 \div 5$, or \$3. √

Lesley paid \$15 for the wrenches.

b. Step 1 Given: bought pliers for \$10.99, wrenches for \$15, and a hammer for \$9.59
gave the clerk \$40
Find: how much change Lesley received

Step 2 First: *Add* \$10.99, \$15, and \$9.59 to find the total cost.
Second: *Subtract* the total cost from \$40. ← **Two operations are needed.**

Step 3 $\$10.99 + \$15 + \$9.59 = \$35.58$
$\$40 - \$35.58 = \$4.42$

Step 4 *Check:* $\$4.42 + (\$10.99 + \$15 + \$9.59) = \$40$ √

Lesley received \$4.42 change.

problems

1. Wanda belongs to a frequent flyer plan. She needs 25,000 mi (miles) to get a free airline ticket. By May Wanda had flown 14,358 mi. In June she flew 1537 mi. In July she flew 1259 mi.
   a. How many miles total did Wanda fly in June and July?
   b. How many more miles must Wanda fly to get a free ticket?
2. Maria bought a pair of sneakers for $24.95, two T-shirts for $7.75 each, and a pair of jeans for $22.50.
   a. How much did Maria pay for the T-shirts?
   b. Maria gave the clerk $80. How much change did she receive?
3. Barbara earns $12 per hour. Last week she worked 37 h (hours). How much did Barbara earn last week?
4. In 14 basketball games, Levon scored a total of 252 points. How many points per game is this?
5. In April Lorenzo drove 984 mi. In May he drove 1109 mi. In June he drove 1097 mi. How many miles in all did Lorenzo drive in the three months?
6. The Ozawas' grocery bill was $89.83. Mr. Ozawa gave the clerk five $20 bills. How much change did Mr. Ozawa receive?
7. Tony bought a shirt for $14.95, a pair of pants for $22.50, and a sweater for $27.85. How much did Tony pay for his purchase?
8. The distance from Rockville to Colton is 432 mi. Ellen left Rockville to go to Colton and drove at 55 mi/h (miles per hour) for 5 h. How much farther does Ellen have to drive to get to Colton?
9. Ralph received a tax refund check for $153.47. He deposited $55.25 in his checking account, $45.50 in his savings account, and took the rest in cash. How much did Ralph receive in cash?
10. Laura wants to buy eighteen wood beams at $3 each, twelve pieces of wallboard at $15 each, and two boxes of nails at $4 per box. Laura has $225. Is this enough money to buy what she wants?

review exercises

If possible, solve. If it is not possible to solve, tell what additional information is needed.

1. Roger drove toward Boynton for 4 h at 48 mi/h. It is 307 mi to Boynton. How far did Roger drive in four hours?
2. Serena bought a dress for $64.45 and a pair of shoes for $48.75. How much change did she receive?

SKILL REVIEW

Estimate using compatible numbers. **4-1**

1. $7\overline{)437}$	**2.** $4\overline{)3949}$	**3.** $2603 \div 6$
4. $61{,}471 \div 9$	**5.** $32\overline{)2807}$	**6.** $71\overline{)22{,}223}$
7. $30{,}827 \div 58$	**8.** $38{,}753 \div 47$	**9.** $61\overline{)665{,}322}$
10. $791\overline{)72{,}322}$	**11.** $184{,}058 \div 924$	**12.** $489{,}424 \div 413$

Estimate each quotient. Determine whether the estimate is an *overestimate* or an *underestimate*.

13. $5\overline{)322}$	**14.** $8\overline{)7781}$	**15.** $6202 \div 9$
16. $40\overline{)3719}$	**17.** $60\overline{)68{,}125}$	**18.** $27{,}102 \div 700$

Find each quotient. **4-2**

19. $8\overline{)856}$	**20.** $3\overline{)375}$	**21.** $4\overline{)6682}$
22. $3479 \div 6$	**23.** $21{,}849 \div 7$	**24.** $12{,}045 \div 9$
25. $78\overline{)5939}$	**26.** $46\overline{)8556}$	**27.** $24\overline{)4834}$
28. $23{,}994 \div 93$	**29.** $86{,}894 \div 62$	**30.** $34{,}604 \div 71$

Find each answer. **4-3**

31. $59 - 48 \div 3$	**32.** $74 - 5 \times 9$	**33.** $3 + 24 \div 2^3$
34. $6^2 \div 3 \times 5 - 43$	**35.** $18 \times 2 - 30 \div 6$	**36.** $42 \times 2 \div 4 + 15$
37. $(21 + 11) \div (12 - 8)$	**38.** $14 \times (3^2 - 2) + 15$	**39.** $86 - (2^2 + 7) \times 3$
40. $\frac{5 \times 9 - 10}{10 - 3}$	**41.** $\frac{64 \div 4 + 8}{42 \div 7}$	**42.** $\frac{66 - 2}{24 \div 12 \times 2}$

Solve. **4-4**

43. The Bilt-Rite Construction Company bought seven building lots for a total of $92,575. How much did each lot cost?

44. The athletic department ordered 144 baseballs. If each baseball costs $6, how much did the entire order cost?

45. At the sports banquet, there were 14 tables with eight people seated at each table. There were 15 people seated at the head table. How many people were at the sports banquet?

46. The students at Bedrock High School ordered a total of 1208 yearbooks. The freshman class ordered 197 yearbooks, the sophomore class ordered 281, and the junior class ordered 318. The rest of the yearbooks were ordered by the senior class. How many yearbooks did the senior class order?

APPLICATION

4-5 ELAPSED TIME

Elapsed time is the amount of time that passes between the start and the end of an event.

example

The longest professional baseball game ever played began at 8:00 P.M. one April night. It was stopped at 4:07 the next morning. When the game was resumed in June, the Pawtucket Red Sox beat the Rochester Red Wings in 18 min (minutes). Find the total playing time for the game.

solution

Think of a clock and count forward:

8:00 to 12:00 → 4 h (hours)
12:00 to 4:00 → 4 h
4:00 to 4:07 → 7 min
8 h 7 min

Add the minutes for the continued game:

  8 h 7 min
+ 18 min
8 h 25 min

The total playing time was 8 h 25 min.

exercises

Find the elapsed time.

1. 7:00 P.M. to 12:00 midnight
2. 2:00 P.M. to 9:00 P.M.
3. 11:16 P.M. to 11:50 P.M.
4. 2:32 P.M. to 3:18 P.M.
5. 4:05 A.M. to 3:30 P.M.
6. 8:20 A.M. to 1:12 P.M.
7. 12:00 noon to 2:11 A.M.
8. 9:33 A.M. to 11:13 P.M.
9. 8:15 A.M. to 5:02 P.M.
10. 5:47 A.M. to 3:54 P.M.
11. Susan made a phone call at 6:32 P.M. and hung up at 6:51 P.M. Find the elapsed time.
12. The Escadas left home at 12:20 P.M. They planned on arriving at their grandmother's house in about six hours. At what time do they expect to arrive at their grandmother's house?
13. Brian is making muffins that take 20 min to bake. He puts them in the oven at 10:18 A.M. At what time should the muffins be done?
14. Al's bus leaves at 7:42 A.M. Al takes 7 min to walk to the bus stop. By what time should Al leave his house to catch the bus?
15. Find the elapsed time between the start and end of your school day.

APPLICATION

4-6 READING A TIMETABLE

Lauren takes the train from Wellesley Square to South Station at 6:38 A.M. She checks the train schedule and finds that she should reach South Station at 7:10 A.M.

Follow these steps to read a train schedule.

1. Find the departure point.
2. Read down the column to find the departure time.
3. Read across the row to the column headed by the destination.
4. Read the arrival time at the place where the row and column meet.

Monday Thru Friday—Except Holidays

	Leave Framingham	West Natick	Natick	Wellesley Square	Wellesley Hills	Wellesley Farms	Auburndale	West Newton	Newtonville	Back Bay	Arrive South Station
A.M.	6 25	6 29	6 34	6 38	6 42	6 45	....	....	....	7 03	7 10
	7 00	7 04	7 09	7 15	7 19	7 22	7 26	7 29	7 32	7 45	7 50
	7 30	7 34	7 39	7 44	7 48	7 51	7 55	7 58	8 01	8 15	8 20
	8 00	8 04	8 09	8 15	8 19	8 23	....	....	8 31	8 42	8 47
	10 00	10 04	10 08	10 12	10 15	10 18	10 22	10 25	10 28	10 39	10 44
P.M.	12 00	12 04	12 08	12 12	12 15	12 18	12 22	12 25	12 28	12 39	12 44
	3 35	3 39	3 43	3 47	3 50	3 53	....	....	....	4 10	4 15
	5 30	....	....	5 39	....	....	....	....	....	6 07	6 12
	7 20	....	....	....	....	....	....	....	....	7 55	8 00
	8 20	....	....	....	....	....	....	....	....	8 55	9 00
	10 25	10 29	10 33	10 37	10 40	10 43	10 47	10 49	10 51	11 04	11 09
	12 25	12 29	12 33	12 37	12 40	12 43	12 47	12 49	12 51	1 04	1 09

example

Bob boards the train in West Natick at 7:34 A.M. At what time should the train reach Back Bay?

solution

Read down the West Natick column to 7:34 A.M. Read across to the Back Bay column. The train should arrive in Back Bay at 8:15 A.M.

exercises

Use the train schedule on page 72.

1. Joan boards the train in Auburndale at 7:26 A.M. At what time should the train reach South Station?
2. Porter takes the 8:00 A.M. train from Framingham. At what time should the train reach Newtonville?
3. Larry takes the 12:08 P.M. train from Natick. At what time should the train arrive in West Newton?
4. Peggy boards the train in Auburndale at 12:22 P.M. At what time should the train arrive in Back Bay?
5. Jeremy is planning to take a morning train from Wellesley Square to South Station. He wants to arrive in South Station at 7:50 A.M. By what time should he be at Wellesley Square?
6. Robin Nussbaum takes the 8:15 A.M. train from Wellesley Square every day. On Thursday the train was 17 min (minutes) late arriving in South Station. At what time did the train arrive?
7. There was a 24-min delay on the 10:08 A.M. train leaving Natick. What should be the new arrival time in Newtonville?
8. It takes Sue Tung 15 min to walk from her house to the Natick train station. What is the latest she can leave and still be on time for the 8:09 A.M. train from Natick?
9. Brad Donahue has a job interview at 9:00 A.M. He wants to arrive at least 10 min early. It will take him about 15 min to walk from Back Bay to the interview. What is the latest train he can take from Wellesley Hills?
10. Benjamin Estavez has a doctor's appointment at 4:30 P.M. The doctor's office is a 5-min walk from Back Bay. Benjamin takes the 3:35 P.M. train from Framingham. How early does he arrive for his appointment?
11. How long is the ride to Newtonville if you take the 10:18 A.M. train from Wellesley Farms?
12. When Simon arrived at the West Natick station, the time was 7:50 A.M. How long did he have to wait for the next train?

Jan has 10 coins in her purse. The value of the coins is 59¢. One of the coins is a quarter. What coins does Jan have in her purse?

4-7 UNITS OF TIME

60 seconds (s)	=	1 minute (min)
60 min	=	1 hour (h)
24 h	=	1 day
7 days	=	1 week
4 weeks (approximately)	=	1 month
365 days	=	1 year
52 weeks (approximately)	=	1 year
12 months	=	1 year
10 years	=	1 decade
100 years	=	1 century

The table at the right shows the relationships of some common units of time. To change from one unit to another, use the following method.

To change to a *smaller* unit, *multiply*.
To change to a *larger* unit, *divide*.

example

Complete: **a.** 17 years = _?_ weeks **b.** 760 h = _?_ days _?_ h

solution

a. A week is a smaller unit than a year, so *multiply* by 52.

$17 \times 52 = 884$ ← **A calculator may be helpful.**

17 years = 884 weeks

b. A day is a larger unit than an hour, so *divide* by 24.

$760 \div 24 = 31$ R16

760 h = 31 days 16 h ← **The remainder is given in hours.**

exercises

Complete. A calculator may be helpful.

1. 78 years = _?_ months
2. 152 weeks = _?_ months
3. 31 days = _?_ h = _?_ min
4. 86,400 s = _?_ min = _?_ h
5. 843 s = _?_ min _?_ s
6. 136 days = _?_ weeks _?_ days
7. 7 weeks 5 days = _?_ days
8. 15 days 8 h = _?_ h

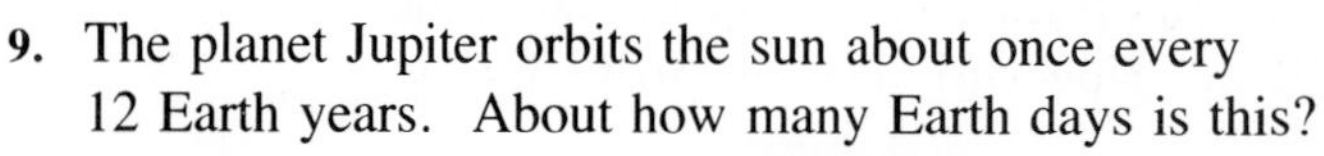

9. The planet Jupiter orbits the sun about once every 12 Earth years. About how many Earth days is this?
10. The Magna Charta was signed in England in 1215. About how many centuries ago was this?
11. The women's world record for a marathon is 2 h 26 min 6 s. About how many minutes is this?
12. Alexander Graham Bell invented the telephone in 1876. About how many decades ago was this?
13. The first manned moon landing was on July 20, 1969. How many months ago was this?

CHAPTER REVIEW

vocabulary vo·cab·u·lar·y

Choose the correct word to complete each sentence.

1. You use compatible numbers to get an (*exact, estimated*) answer to a division.
2. In the division 345 ÷ 10, 10 is the (*dividend, divisor*).

skills

Estimate using compatible numbers.

3. $8\overline{)704}$
4. $6\overline{)6731}$
5. 3408 ÷ 47
6. $92\overline{)43{,}620}$
7. $39\overline{)152{,}652}$
8. 141,065 ÷ 719

Estimate each quotient. Determine whether the estimate is an *overestimate* or an *underestimate*.

9. $4\overline{)332}$
10. $7\overline{)4827}$
11. 2682 ÷ 9
12. $80\overline{)2610}$
13. $50\overline{)51{,}799}$
14. 28,136 ÷ 600

Find each quotient.

15. $4\overline{)1439}$
16. $7\overline{)4067}$
17. 15,124 ÷ 4
18. $53\overline{)48{,}111}$
19. $63\overline{)106{,}602}$
20. 311,022 ÷ 42

Find each answer.

21. $47 - 2 \times 5 + 14$
22. $36 + 2^3 \div 2 - 10$
23. $(28 - 6) \times 8 - 5$
24. $9^2 - (9 \times 4)$
25. $\dfrac{88 - 44 \div 2}{11 - 5}$
26. $\dfrac{75 \div 3}{2 \times 10 + 5}$

Solve.

27. Sally bought four tires for her car. Each tire costs $49. She also bought a new battery for $57 and four shock absorbers costing $11 each. How much did Sally pay altogether?
28. Leon starts work at 7:45 A.M. He eats lunch at 12:30 P.M. How long does Leon work before lunch?
29. Use the train schedule on page 72. Cathy takes the 7:55 A.M. train from Auburndale. At what time should the train arrive at Back Bay?
30. The first successful expedition to the North Pole was in 1909. About how many decades ago was this?

CHAPTER TEST

Estimate using compatible numbers. **4-1**

1. $6\overline{)493}$
2. $9\overline{)8832}$
3. $78,854 \div 4$
4. $9006 \div 32$
5. $53\overline{)1455}$
6. $78\overline{)64,091}$
7. $677\overline{)47,454}$
8. $421\overline{)377,241}$
9. $417,349 \div 684$

Estimate each quotient. Determine whether the estimate is an *overestimate* or an *underestimate*.

10. $6\overline{)311}$
11. $5\overline{)2617}$
12. $3172 \div 4$
13. $30\overline{)29,516}$
14. $700\overline{)61,729}$
15. $191,158 \div 90$

Find each quotient. **4-2**

16. $6\overline{)3552}$
17. $8\overline{)3185}$
18. $2951 \div 3$
19. $38\overline{)20,837}$
20. $64\overline{)19,456}$
21. $60,164 \div 29$

Find each answer. **4-3**

22. $39 \div 3 \times 5 - 4$
23. $42 - 3 \times 4 + 19$
24. $54 + 2^4 \div 4 \times 3$
25. $58 + (4^3 \div 8) \div 2$
26. $(14 \times 3) - 12 \div 3$
27. $36 \div (6 - 2) + 15$
28. $\dfrac{11 + 24}{13 - 6}$
29. $\dfrac{81 \div 3 + 6}{44 \div 4}$
30. $\dfrac{28 \div 7}{19 - 3 \times 5}$

Solve.

31. In 11 football games, Stan rushed for 1298 yd (yards). How many yards per game is this? **4-4**
32. Wendy bought a tennis racquet for \$64 and four cans of tennis balls for \$3 per can. She gave the clerk \$80. How much change did she receive?
33. Jet Airlines flight 604 leaves New York at 11:20 A.M. and arrives in Atlanta at 1:33 P.M. Find the elapsed time. **4-5**
34. Use the train schedule on page 72. Julia boards the train in Natick at 12:08 P.M. At what time should the train reach South Station? **4-6**

Complete. **4-7**

35. 14 years = _?_ days
36. 112 days = _?_ weeks
37. 2908 s = _?_ min _?_ s
38. 34 h 17 min = _?_ min
39. The Olympic record for the men's 50,000-meter walk is 3 h 47 min 26 s. About how many minutes is this?

Utility workers usually receive an hourly rate of pay. Skill in multiplying decimals can help in computing regular and overtime earnings.

CHAPTER 5

MULTIPLYING AND DIVIDING DECIMALS

5-1 MULTIPLYING WITH DECIMALS

To multiply with decimals, use the following method.

1. Multiply as with whole numbers.
2. Count the number of decimal places in each factor.
3. Insert a decimal point in the product so that it has the same number of decimal places as the *total* number of decimal places in the factors.

example 1

Find each product: a. $\begin{array}{r} 22.5 \\ \times 5 \\ \hline \end{array}$ b. 15×1.78

solution

a. $\begin{array}{r} 22.5 \\ \times 5 \\ \hline 112.5 \end{array}$ ← **1 place** ← **+0 places** ← **1 place**

b. $\begin{array}{r} 1.78 \\ \times 15 \\ \hline 8\ 90 \\ 17\ 8 \\ \hline 26.70 \end{array} = 26.7$ ← **You may write the product as 26.7.**

your turn

Find each product.

1. $\begin{array}{r} 6.343 \\ \times 7 \\ \hline \end{array}$
2. 123.8×19
3. $\begin{array}{r} 3.05 \\ \times 26 \\ \hline \end{array}$
4. 8.4×5

example 2

Find each product: a. $\begin{array}{r} 3.25 \\ \times 0.61 \\ \hline \end{array}$ b. 0.4×0.07

solution

a. $\begin{array}{r} 3.25 \\ \times 0.61 \\ \hline 325 \\ 1\ 950 \\ \hline 1.9825 \end{array}$ ← **2 places** ← **+2 places** ← **4 places**

b. $\begin{array}{r} 0.07 \\ \times 0.4 \\ \hline 0.028 \end{array}$ ← **Insert one zero as a placeholder in the product.**

your turn

Find each product.

5. $\begin{array}{r} 6.7 \\ \times 4.5 \\ \hline \end{array}$
6. 0.2×0.4
7. $\begin{array}{r} 0.18 \\ \times 0.04 \\ \hline \end{array}$
8. 0.6×0.05

example 3

Find each product: **a.** $2 \times \$3.80$ **b.** $0.14 \times \$3.56$

solution

a.
$$\begin{array}{r} \$3.80 \\ \times 2 \\ \hline \$7.60 \end{array}$$
← Keep the zero in the product to show the number of cents.

b.
$$\begin{array}{r} \$3.56 \\ \times 0.14 \\ \hline 1424 \\ 356 \\ \hline \$.4984 \approx \$.50 \end{array}$$
← Round the product to the nearest cent. (The symbol ≈ means *is approximately equal to.*)

your turn

Find each product. If necessary, round to the nearest cent.

9. $\$4.32 \times 8$ **10.** $\$1.35 \times 4$ **11.** $\$.29 \times 7.7$ **12.** $\$.63 \times 0.92$

practice exercises

practice for example 1 (page 78)

Find each product.

1. $\begin{array}{r} 2.3 \\ \times 7 \\ \hline \end{array}$ **2.** $\begin{array}{r} 5.4 \\ \times 18 \\ \hline \end{array}$ **3.** $\begin{array}{r} 0.35 \\ \times 34 \\ \hline \end{array}$ **4.** $\begin{array}{r} 4.76 \\ \times 5 \\ \hline \end{array}$

5. 2×72.7 **6.** 5.31×9 **7.** 7.5×64 **8.** 84×0.25

9. 7.45×8 **10.** 35×0.68 **11.** 8×0.123 **12.** 0.974×6

practice for example 2 (page 78)

Find each product.

13. $\begin{array}{r} 3.4 \\ \times 6.6 \\ \hline \end{array}$ **14.** $\begin{array}{r} 0.87 \\ \times 0.52 \\ \hline \end{array}$ **15.** $\begin{array}{r} 0.03 \\ \times 0.4 \\ \hline \end{array}$ **16.** $\begin{array}{r} 0.006 \\ \times 0.7 \\ \hline \end{array}$

17. 0.2×0.04 **18.** 0.3×0.2 **19.** 0.629×0.4 **20.** 8.53×0.6

21. 0.4×0.05 **22.** 1.5×0.02 **23.** 71.4×0.7 **24.** 0.359×0.8

practice for example 3 (page 79)

Find each product. If necessary, round to the nearest cent.

25. $\begin{array}{r} \$22.64 \\ \times 6 \\ \hline \end{array}$ **26.** $\begin{array}{r} \$28.56 \\ \times 5 \\ \hline \end{array}$ **27.** $\begin{array}{r} \$24.23 \\ \times 0.51 \\ \hline \end{array}$ **28.** $\begin{array}{r} \$7.39 \\ \times 0.12 \\ \hline \end{array}$

29. $335 \times \$.24$ **30.** $147 \times \$.05$ **31.** $0.73 \times \$.96$ **32.** $0.8 \times \$3.96$

33. $64 \times \$1.35$ **34.** $\$2.98 \times 25$ **35.** $\$12.47 \times 0.9$ **36.** $0.27 \times \$5.65$

mixed practice (pages 78–79)

Find each product. If necessary, round to the nearest cent.

37. $\begin{array}{r} 4.483 \\ \times 6 \\ \hline \end{array}$ **38.** $\begin{array}{r} 87 \\ \times 2.3 \\ \hline \end{array}$ **39.** $\begin{array}{r} \$62.89 \\ \times 0.3 \\ \hline \end{array}$ **40.** $\begin{array}{r} \$6.34 \\ \times 7 \\ \hline \end{array}$

41. 0.3×0.3 **42.** $0.8 \times \$54.17$ **43.** 2×0.531 **44.** 5.1×56

45. $\$16.35 \times 6$ **46.** $0.14 \times \$6.84$ **47.** 5.8×4.7 **48.** 0.4×6.339

49. 6.3×9.1 **50.** 0.2×8.547 **51.** $0.52 \times \$3.79$ **52.** $\$.20 \times 0.5$

53. Find each product: **a.** 10×0.39 **b.** 100×0.39 **c.** 1000×0.39

54. Find each product: **a.** 0.1×1.8 **b.** 0.01×1.8 **c.** 0.001×1.8

55. Using your answers to Exercise 53, state a method for multiplying by 10, 100, and 1000.

56. Using your answers to Exercise 54, state a method for multiplying by 0.1, 0.01, and 0.001.

57. Ronald Peyton bought three sweaters for $12.99 each. How much was his total purchase?

58. Suzanne Pacello bought two pairs of sunglasses that sold for $10.25 each. She gave the store clerk $30. How much change did she receive?

review exercises

Round to the place of the leading digit.

1. 2.413 **2.** 152 **3.** 977 **4.** 0.0023

5. 5.638 **6.** 0.84 **7.** 11,618 **8.** 49.41

mental math

To multiply with money mentally, it may help to think of the dollars and the cents separately.

$3 \times \$2.60 = (3 \times \$2) + (3 \times \$.60) = \$6 + \$1.80 = \$7.80$

Multiply mentally.

1. $6 \times \$1.20$ **2.** $2 \times \$4.40$ **3.** $4 \times \$10.06$ **4.** $8 \times \$20.05$

5. $7 \times \$3.50$ **6.** $9 \times \$9.80$ **7.** $3 \times \$10.90$ **8.** $6 \times \$11.70$

5-2 ESTIMATING DECIMAL PRODUCTS

You can estimate a decimal product by rounding each factor to the place of the leading digit.

example 1

Estimate by rounding to the place of the leading digit.

a. 4.5 × $28.70 b. 0.62 × 7.42 c. 0.93 × 0.691

solution

a. $28.70 → $30; ×4.5 → ×5; about $150

b. 7.42 → 7; ×0.62 → ×0.6; about 4.2

c. 0.691 → 0.7; ×0.93 → ×0.9; about 0.63

your turn

Estimate by rounding to the place of the leading digit.

1. 3.8 × $49.91 2. 0.76 × 9.492 3. 0.33 × 0.54 4. 3.2 × $.97

Sometimes you can determine whether your estimate is an **overestimate** or an **underestimate.**

example 2

Estimate each product. Determine whether the estimate is an *overestimate*, an *underestimate*, or if you *can't tell*.

a. 8.7 × $5.81 b. 0.52 × 731 c. 0.82 × 3.995

solution

a. $5.81 → $6; ×8.7 → ×9; about $54

Both factors are rounded *up*, so $54 is an *overestimate*.

b. 731 → 700; ×0.52 → ×0.5; about 350

Both factors are rounded *down*, so 350 is an *underestimate*.

c. 3.995 → 4; ×0.82 → ×0.8; about 3.2

One factor is rounded *up*. The other is rounded *down*. You *can't tell*.

your turn

Estimate each product. Determine whether the estimate is an *overestimate*, an *underestimate*, or if you *can't tell*.

5. $2.70 × 26.5 6. 14.23 × 51.68 7. 0.71 × 8390 8. 0.47 × 4.1

practice exercises

practice for example 1 ***(page 81)***

Estimate by rounding to the place of the leading digit.

1. 4.36×12.517 **2.** 8.84×52.239 **3.** 0.67×7.51 **4.** 0.79×2.19
5. 0.32×0.86 **6.** 0.88×0.456 **7.** 0.9753×2.48 **8.** $0.991 \times \$8.49$

practice for example 2 ***(page 81)***

Estimate each product. Determine whether the estimate is an *overestimate*, an *underestimate*, or if you *can't tell.*

9. 4202×0.91 **10.** 0.42×734 **11.** 0.64×0.381 **12.** 58.35×11.99
13. 0.28×28.061 **14.** $0.75 \times \$26.44$ **15.** 0.48×61.492 **16.** $0.65 \times \$83.39$

mixed practice ***(page 81)***

Estimate each product.

17. 9.47×72.88 **18.** $0.55 \times \$6.75$ **19.** 0.833×0.41 **20.** $0.29 \times \$71.02$
21. 0.31×392 **22.** 7.943×20.95 **23.** 0.552×0.57 **24.** 676×0.818

25. A pound of yams costs $.49. About how much do 2.3 lb (pounds) of yams cost?

26. All Seasons Catering Service sells a picnic lunch for $8.98. About how much would 32 lunches cost?

27. Allison Givens bought a dress for $41.95, a blouse for $23.50, and slacks for $25.50. About how much was the total cost?

28. A case of 48 cans of soup costs $21.60. Jeff Gertman ordered 8 cases of soup. About how much did the order cost?

review exercises

Find each quotient.

1. $6\overline{)5812}$ **2.** $4\overline{)13,227}$ **3.** $7287 \div 21$ **4.** $60,109 \div 47$

Show how you can cut a circle into 11 pieces with only 4 straight cuts.

5-3 DIVIDING WITH DECIMALS

To divide a decimal by a nonzero whole number, use the following method.

1. Place the decimal point in the quotient directly above the decimal point in the dividend.
2. Divide as with whole numbers.

example 1

Find each quotient: **a.** $7\overline{)\$1.61}$ **b.** $5\overline{)1.2}$ **c.** $9.549 \div 9$

solution

a. $\begin{array}{r} \$.23 \\ 7\overline{)\$1.61} \\ \underline{1\ 4} \\ 21 \\ \underline{21} \\ 0 \end{array}$

b. $\begin{array}{r} 0.24 \\ 5\overline{)1.20} \\ \underline{1\ 0} \\ 20 \\ \underline{20} \\ 0 \end{array}$ ← **Annex zeros as needed in the dividend.**

c. $\begin{array}{r} 1.061 \\ 9\overline{)9.549} \\ \underline{9} \\ 54 \\ \underline{54} \\ 9 \\ \underline{9} \\ 0 \end{array}$ ← **Insert one zero as a placeholder in the quotient.**

your turn

Find each quotient.

1. $5\overline{)\$11.65}$ **2.** $4\overline{)13.8}$ **3.** $44 \div 16$ **4.** $8.056 \div 8$

To divide by a decimal, first multiply the divisor and the dividend by a *power* of 10 (10, 100, 1000, and so on) to make the divisor a whole number. Then divide.

example 2

Find each quotient: **a.** $0.5\overline{)0.15}$ **b.** $1.5 \div 0.25$

solution

a. $\begin{array}{r} 0.3 \\ 0.5\overline{)0.1\ 5} \\ \underline{1\ 5} \\ 0 \end{array}$ ← **$0.5 \times 10 = 5$, so move *both* decimal points *one* place to the right.**

b. $\begin{array}{r} 6 \\ 0.25\overline{)1.50} \\ \underline{1\ 50} \\ 0 \end{array}$ ← **$0.25 \times 100 = 25$, so move *both* decimal points *two* places to the right.**

your turn

Find each quotient.

5. $0.4\overline{)0.56}$ **6.** $0.16\overline{)4.8}$ **7.** $8.1 \div 0.009$ **8.** $24 \div 0.12$

Sometimes it is not desirable to find an *exact* quotient. In such cases, you *round* the quotient to an appropriate place. To round a quotient to a given place, you carry the division out to one additional decimal place.

example 3

Find each quotient to the indicated place.

a. $6\overline{)40.3}$; nearest tenth

b. $0.04 \div 0.24$; nearest hundredth

solution

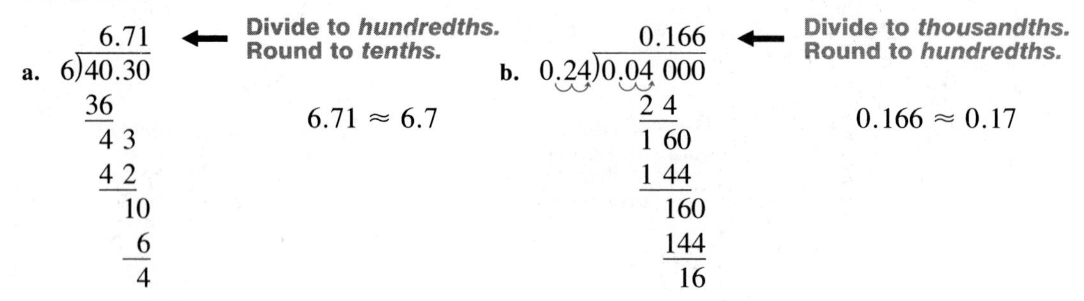

your turn

Find each quotient to the indicated place.

9. $7\overline{)17.7}$; nearest tenth

10. $3.3\overline{)2.566}$; nearest tenth

11. $0.3615 \div 0.42$; nearest hundredth

12. $0.007 \div 0.23$; nearest hundredth

practice exercises

practice for example 1 (page 83)

Find each quotient.

1. $3\overline{)\$4.32}$	**2.** $5\overline{)1.95}$	**3.** $13\overline{)29.12}$	**4.** $14\overline{)\$179.20}$
5. $2.04 \div 8$	**6.** $26.3 \div 5$	**7.** $124 \div 5$	**8.** $23 \div 4$
9. $0.377 \div 13$	**10.** $0.056 \div 8$	**11.** $60.24 \div 12$	**12.** $42.018 \div 6$
13. $17\overline{)34.85}$	**14.** $13\overline{)1.794}$	**15.** $25\overline{)30.2}$	**16.** $40\overline{)4.14}$

practice for example 2 (page 83)

Find each quotient.

17. $0.5\overline{)1.45}$	**18.** $0.3\overline{)5.655}$	**19.** $0.7\overline{)90.3}$	**20.** $0.8\overline{)54.4}$
21. $0.806 \div 0.13$	**22.** $0.2112 \div 0.24$	**23.** $615 \div 0.41$	**24.** $2176 \div 0.68$
25. $535.5 \div 8.5$	**26.** $502.2 \div 27.9$	**27.** $0.0544 \div 0.002$	**28.** $1.088 \div 0.004$
29. $0.65\overline{)59.8}$	**30.** $0.44\overline{)97.02}$	**31.** $0.16\overline{)0.36}$	**32.** $0.28\overline{)1.05}$

practice for example 3 (page 84)

Find each quotient to the indicated place.

33. $9\overline{)7.12}$; nearest tenth

34. $0.6\overline{)0.14}$; nearest tenth

35. $0.815 \div 2.4$; nearest hundredth

36. $0.283 \div 0.53$; nearest hundredth

37. $0.13\overline{)0.54}$; nearest hundredth

38. $0.11\overline{)0.245}$; nearest hundredth

39. $0.36 \div 3.7$; nearest thousandth

40. $0.12 \div 4.3$; nearest thousandth

mixed practice (pages 83–84)

Find each quotient. If necessary, round to the nearest thousandth.

41. $0.7\overline{)15.89}$ **42.** $22\overline{)14.35}$ **43.** $4.368 \div 9.1$ **44.** $\$18.54 \div 9$

45. $8\overline{)10.8}$ **46.** $0.12\overline{)6}$ **47.** $\$15.60 \div 13$ **48.** $0.945 \div 21$

49. $3.53 \div 3.5$ **50.** $49 \div 0.07$ **51.** $0.4\overline{)1.628}$ **52.** $3.6\overline{)7.218}$

53. $126.68 \div 4$ **54.** $0.343 \div 0.7$ **55.** $0.7\overline{)0.486}$ **56.** $20\overline{)6.3}$

57. Find each quotient: **a.** $10\overline{)1.5}$ **b.** $100\overline{)1.5}$ **c.** $1000\overline{)1.5}$

58. Find each quotient: **a.** $10\overline{)0.7}$ **b.** $100\overline{)0.7}$ **c.** $1000\overline{)0.7}$

59. Using your answers to Exercises 57 and 58, state a method for dividing a decimal by 10, 100, or 1000.

60. Diana Tompkins bought 15 lb (pounds) of swordfish for \$63.75. How much did one pound cost?

61. Juan Estevez paid \$20 for 25 rolls of tape. How much did one roll of tape cost?

62. Sue Tang paid \$17.97 for salmon. Salmon costs \$5.99/lb (per pound). How many pounds did she buy?

review exercises

Estimate by rounding to the place of the leading digit.

1. $2821 + 3417$ **2.** $155 + 118 + 546$ **3.** $32.68 - 14.2$ **4.** $772 - 459$

5. $\$19 \times 54$ **6.** 62×651 **7.** 0.56×0.941 **8.** $\$42.40 \times 0.75$

Estimate using compatible numbers.

9. $6\overline{)172}$ **10.** $5\overline{)2181}$ **11.** $843 \div 27$ **12.** $5524 \div 82$

13. $64{,}386 \div 70$ **14.** $9\overline{)1112}$ **15.** $5\overline{)4100}$ **16.** $25{,}041 \div 392$

5-4 ESTIMATING DECIMAL QUOTIENTS

You may want to estimate a quotient rather than find the exact amount. To estimate, round the divisor to the place of the leading digit. Then divide using compatible numbers, that is, numbers that divide easily.

example 1

Estimate using compatible numbers: **a.** $234.61 ÷ $7.99 **b.** 5.37)3187.2

solution

a. $7.99)$234.61

about 30
$8)$240 ← **Round $7.99 to $8, then use a compatible dividend.**

b. 5.37)3187.2

about 600
5)3000

your turn

Estimate using compatible numbers.

1. $6.60)$290.10 **2.** $9.39)$81.80 **3.** 4173.9 ÷ 7.9 **4.** 0.239 ÷ 4.16

example 2

Estimate using compatible numbers: **a.** 6253 ÷ 0.8 **b.** 0.64)0.4419

solution

a. 0.8)6253

about 8000
0.8)6400.0

b. 0.64)0.4419

about 0.7
0.6)0.4 2 ← **Round 0.64 to 0.6, then use a compatible dividend.**

your turn

Estimate using compatible numbers.

5. 0.3)593 **6.** 0.43)330.1 **7.** 0.336 ÷ 0.7 **8.** 0.715 ÷ 0.87

practice exercises

practice for example 1 *(page 86)*

Estimate using compatible numbers.

1. $6.50)$47.99 **2.** $4.25)$82.52 **3.** 8.6)89.92 **4.** 3.1)298.6
5. 3515.4 ÷ 6.79 **6.** 8927.5 ÷ 3.03 **7.** 0.41 ÷ 8.3 **8.** 0.32 ÷ 4.98

practice for example 2 (page 86)

Estimate using compatible numbers.

9. $0.8\overline{)555}$ 10. $0.72\overline{)2.15}$ 11. $0.9\overline{)0.531}$ 12. $0.62\overline{)0.548}$
13. $1757 \div 0.93$ 14. $2.53 \div 0.47$ 15. $0.121 \div 0.37$ 16. $0.0274 \div 0.7$

mixed practice (page 86)

Estimate.

17. $\$.52\overline{)\$197.98}$ 18. $4.4\overline{)11.53}$ 19. $0.48\overline{)0.213}$ 20. $\$2.72\overline{)\$89.10}$
21. $421.09 \div 6.83$ 22. $0.0638 \div 0.8$ 23. $359.1 \div 0.6$ 24. $252.38 \div 4.91$

25. John Ito bought a 9.1-oz (ounce) bag of pistachios for \$2.49. About how much did one ounce cost?
26. Sarah Hall bought a 4.25-oz bottle of shampoo for \$2.19. About how much did one ounce cost?
27. Darren Miscall paid \$11.95 for 3.8 lb (pounds) of veal. About how much did one pound of veal cost?
28. Perfume bottles contain 50.5 mL (milliliters) each. About how many bottles can a company fill with 3994.75 mL of perfume?

review exercises

Find each answer.

1. $19\overline{)1178}$ 2. $\$602 - \$16.49$ 3. $48{,}172 + 9066$ 4. 3.4×0.08
5. $72.39 - 3.648$ 6. $0.282 \div 4.7$ 7. $\$43 \times 380$ 8. $5.4 + 10.831$

calculator corner

Entering data incorrectly is a common mistake. If you estimate, you will be able to tell if the answer displayed is reasonable.

example a. 11.14×0.39
Answer displayed: 4.3446

b. $39.36 \div 4.8$
Answer displayed: 82

solution a. Estimate: $10 \times 0.4 = 4$
The answer is reasonable.

b. Estimate: $40 \div 5 = 8$
The answer is unreasonable.

Estimate first. Then use a calculator to find each product or quotient.

1. 8.3×9.9 2. 17.2×100.31 3. $63.918 \div 8.04$ 4. $11.97 \div 12.6$
5. $10.015\overline{)28.042}$ 6. 0.79×0.79 7. 0.203×20.3 8. $0.022\overline{)839.3}$

SKILL REVIEW

Find each product. If necessary, round to the nearest cent. **5-1**

1. $\begin{array}{r} 22.8 \\ \times 4 \\ \hline \end{array}$
2. $\begin{array}{r} 4.631 \\ \times 12 \\ \hline \end{array}$
3. $\begin{array}{r} 2.36 \\ \times 15 \\ \hline \end{array}$
4. $\begin{array}{r} \$4.44 \\ \times 0.22 \\ \hline \end{array}$
5. 7.9×3.6
6. $34 \times \$4.96$
7. 0.3×0.07
8. 0.009×0.2
9. 0.81×25.23
10. $85 \times \$7.92$
11. $0.5 \times \$1.57$
12. 4.05×0.8
13. 13.6×5
14. Joan Borden bought three video cassettes for $7.89 each. What was the total cost of her purchase?
15. The Skofield Art Gallery recently sold 1000 tickets to an exhibition of oil paintings. If each ticket cost $2.50, how much money did the gallery collect from ticket sales?

Estimate to place the decimal point in the product. **5-2**

16. $65.2 \times 26.9 =$ [175388]
17. $75.1 \times 46.6 =$ [349966]
18. $0.07 \times 0.44 =$ [000308]
19. $0.29 \times 0.53 =$ [0001537]
20. $8.157 \times 0.28 =$ [2283960]
21. $4.46 \times 0.752 =$ [3353920]
22. $0.71 \times 0.45 =$ [0031950]
23. $0.6 \times 0.79 =$ [000474]
24. Diana Reese bought 3.85 lb (pounds) of fish costing $6.99 per pound. About how much did she pay for the fish?

Find each quotient. If necessary, round to the nearest thousandth. **5-3**

25. $7\overline{)59.22}$
26. $6\overline{)20.4}$
27. $15.4 \div 0.04$
28. $1.002 \div 1.3$
29. $93\overline{)5.208}$
30. $7\overline{)4}$
31. $0.271 \div 0.03$
32. $6318 \div 0.9$
33. $0.127 \div 0.12$
34. Erik Jenson pays $5.76 to develop a roll of film. There are 24 pictures on the roll. How much does each picture cost to develop?
35. A 50-lb block of cheese costs a grocer $109.50. How much does the cheese cost per pound?

Estimate using compatible numbers. **5-4**

36. $371.6 \div 0.91$
37. $\$31.10 \div \$2.72$
38. $0.418 \div 5.82$
39. $6225 \div 0.7$
40. $\$2.87\overline{)\$29.81}$
41. $\$2.40\overline{)\$39.75}$
42. $0.6\overline{)0.531}$
43. $4.38\overline{)294.2}$
44. $0.5\overline{)0.516}$
45. $7.27\overline{)2088.4}$
46. $0.41\overline{)157.8}$
47. $\$6.71\overline{)\$135.80}$
48. $0.361 \div 5.94$
49. $0.319 \div 0.8$
50. $253 \div 0.49$

APPLICATION

5-5 ANNUAL SALARY

Employers often choose to pay an annual salary on a weekly or a monthly basis.

example

John and Suzanne each have an annual salary of $24,270. John is paid weekly, and Suzanne is paid monthly. Find John's weekly salary and Suzanne's monthly salary.

solution

Since there are 52 weeks in a year, divide the annual salary by 52 to find the weekly salary.

$\$24{,}270 \div 52 \approx \$466.73$

John's weekly salary is $466.73.

Since there are 12 months in a year, divide the annual salary by 12 to find the monthly salary.

$\$24{,}270 \div 12 = \$2022.50$

Suzanne's monthly salary is $2022.50.

exercises

Solve. If necessary, round to the nearest cent. A calculator may be helpful.

1. Susan Bing earns $12,900 annually. Find her weekly salary.
2. Peter Korbut earns $11,410 annually. Find his monthly salary.
3. Kenneth Frye earns $650.10 monthly. How much money does he earn annually?
4. Jeanne Dunn earns $798.40 monthly. Find her annual salary.
5. Chuck Henley earns $258.40 weekly. How much money does he earn annually?
6. Sarah Feltz earns $532.12 weekly. Carmen Ianelli earns $2088.95 monthly. Who earns the greater annual salary?
7. Louis Shore has a weekly salary of $220.47. Joshua Hairston has a monthly salary of $976.50. Who earns the greater annual salary?
8. Explain how to determine annual salary if you know monthly salary.
9. Explain how to determine annual salary if you know weekly salary.

APPLICATION

5-6 HOURLY PAY, OVERTIME, AND PIECEWORK

Some employers pay their employees a certain amount for each hour worked. If you know the **hourly rate** of pay and the number of hours worked in a week, you can find an employee's weekly earnings.

An employee who works more than 40 h in a week may receive **overtime pay.** Overtime pay is usually *time-and-a-half,* or 1.5 times the hourly rate. On holidays, pay is usually *double time,* or twice the hourly rate. The sum of regular and overtime earnings is called **gross pay.**

example

Mai Wong earns $5.50/h (per hour) for working a 40-h week in a bank. She earns time-and-a-half for overtime. Find her gross pay if she works 44 h one week.

solution

$$\begin{aligned}\text{regular earnings} &= \text{regular hourly rate} \times \text{number of regular hours}\\ &= \$5.50 \times 40\\ &= \$220\end{aligned}$$

$$\begin{aligned}\text{overtime earnings} &= \text{overtime rate} \times \text{number of overtime hours}\\ &= (\$5.50 \times 1.5) \times (44 - 40)\\ &= \$8.25 \times 4\\ &= \$33\end{aligned}$$

$$\begin{aligned}\text{gross pay} &= \text{regular earnings} + \text{overtime earnings}\\ &= \$220 + \$33\\ &= \$253\end{aligned}$$

Mai's gross pay is $253.

exercises

Find the gross pay. Assume time-and-a-half for any hours over 40.

1. 36 h at $13.72/h
2. 42 h at $6.50/h
3. 25 h at $8.55/h
4. 45 h at $13/h
5. 32 h at $10.25/h
6. 40 h at $12.44/h
7. Katie Chong works 8 h on a holiday and 32 h during the week. She earns $11/h and double time on holidays. Find her gross pay.
8. Debby White works 22 h each week as a cashier. Her hourly rate is $4.15. Find her weekly pay.
9. Josh Worth earns $7.80/h and double time on holidays. He works 35 h during the week and 4 h on a holiday. Find his gross pay.

10. Mary Bristol earns \$14.50/h. One week she works 38 h. The next week Mary works 44 h. Assume time-and-a-half for any hours over 40. How much more does Mary earn the second week?
11. Faye Jones earned \$78.75 picking fruit. She earned \$5.25/h. How many hours did she work?

Sometimes an employee earns money for each item produced. This is called **piecework.** People often earn piecework pay for services such as typing and sewing. To find the amount earned for an entire job, multiply the amount earned for one item by the number of items.

Sometimes an employer pays an hourly rate plus a bonus for any pieces produced over a set quota. Employers can reject pieces that an employee produces if the pieces are of poor quality. Employees do not earn pay for rejected pieces.

12. Kris Laronge is a typist. She charges \$1.55 for each page she types. How much does she earn for typing a 52-page report?
13. Josh Howe earns \$9.50 for mowing a small lawn and \$14.50 for a large lawn. How much does he earn for mowing three small lawns?
14. John Silver charges \$19.95 for cleaning a medium-sized rug and \$41.25 for cleaning a large rug. How much does he earn for cleaning one medium-sized rug and two large rugs?
15. Natalie Chetwin receives an hourly wage of \$5.50 plus a bonus of \$.10 for each item over her daily quota of 250 items. One day she worked 8 h and produced 320 items. Find her pay that day.
16. David Broz earns \$1.50 for each item produced. One day he makes 25 items, but his employer rejects two items. Find David's pay for his work that day.
17. Last month Susan Ruff earned \$36 sewing buttons on shirts. She sewed 1200 buttons. How much did she earn for each button?
18. Ron Pacello earned \$225 for typing 150 pages. How much did he charge for typing one page?
19. Read the classified advertisements in your local paper. Find a job that pays by piecework. Bring the advertisement to class.

APPLICATION

5-7 NET PAY

Employers make **deductions** for federal income tax and social security tax (F.I.C.A.). Employers may also deduct state income tax, union dues, and credit union payments. **Net pay,** or take-home pay, is the pay remaining after an employer makes all deductions from gross pay.

Employers issue an earnings statement with each paycheck. Employees can use earnings statements to check the accuracy of the amounts withheld.

example

YEAR-TO-DATE		CURRENT PAY PERIOD		
		EARNINGS	DEDUCTIONS	
TAXABLE EARNINGS	9200.00	GROSS EARN. 400.00		
FED. INC. TAX	1380.00		FED. INC. TAX	60.00
STATE TAX	460.00		STATE TAX	20.00
UNION DUES	64.40		INSURANCE	15.20
F.I.C.A.	690.92		F.I.C.A.	30.04
MISC.	348.80		CREDIT UNION	11.20
NET PAY	6255.88	NET PAY ?	TOTAL DEDUCTIONS	?
MORGAN, JOHN **week ending** 6/8				

Use the CURRENT PAY PERIOD information to find the following.

a. total deductions **b.** net pay **c.** annual payment for state tax

solution

a. Add the deductions shown on the statement. A calculator may be helpful.
$60.00 + $20.00 + $15.20 + $30.04 + $11.20 = $136.44
The total deductions are $136.44

b. Gross pay − total deductions = net pay
$400.00 − $136.44 = $263.56
The net pay is $263.56.

c. The statement shows that John pays $20.00 *weekly* for state tax.
$20.00 × 52 = $1040.00
John's annual payment for state tax is $1040.00.

exercises

Find the net pay and annual payment for state tax for the following weekly salaries. A calculator may be helpful.

	GROSS PAY	FED. INC. TAX	F.I.C.A.	STATE TAX	INS.
1.	$254.00	$ 38.10	$19.08	$12.70	$1.52
2.	$332.00	$ 49.80	$24.93	none	$3.71
3.	$396.00	$ 66.26	$29.74	$19.80	$2.38
4.	$549.25	$109.16	$41.25	$27.46	none
5.	$608.40	$125.73	$45.69	$30.42	$3.62
6.	$871.50	$199.40	$64.99	$43.58	$5.23

7. Peter Anderson's weekly pay is $332. His deductions are $74.71 for federal income tax, $24.93 for F.I.C.A., $16.60 for state tax, and $3.71 for insurance. Find Peter's net pay.

8. Yolanda Brown earns $396 weekly. Her deductions are $109.71 for federal income tax, $29.74 for F.I.C.A., $19.80 for state tax, and $2.38 for insurance. Find Yolanda's net pay.

9. Andre Markson works 38 h each week. He earns $14 per hour. His deductions are $120.98 for federal income tax, $39.95 for F.I.C.A., and $31.92 for state tax. Find Andre's net pay.

10. Jennifer Hamperson earns $11 per hour and time-and-a-half for any hours over 40 in a given week. Last week she worked 45 h. Her deductions were $117.95 for federal income tax and $39.24 for F.I.C.A. Find Jennifer's net pay.

F.I.C.A. stands for Federal Insurance Contributions Act. Money deducted for F.I.C.A. provides retirement income, survivors' benefits, and medical benefits for those who qualify. The F.I.C.A. deduction is on income up to $45,000. You can use the following formula to find the annual F.I.C.A. deduction taken.

$$\text{F.I.C.A. deduction} = 0.0751 \times \text{gross pay}$$

11. Dawn Keller's annual gross pay is $43,800. Find her annual F.I.C.A. deduction.

12. Mindy Falcom's annual gross pay is $25,000. Find her annual F.I.C.A. deduction.

13. Jeff Saxon's weekly deduction for F.I.C.A. is $34.38. Find his annual deduction.

14. Gordon Yanoff's employer deducts $266.16 annually for F.I.C.A. What is the monthly deduction?

CAREER

CAR RENTAL AGENCY MANAGER

Sondra Schur manages the West Car Rental Agency. She is responsible for the condition of each car and for the finances of the agency. Sondra determines the rates charged for each car rental.

example

Sondra charges $32.95 per day for the rental of a compact car. How much does she charge for a three-day rental?

solution

Sondra multiplies the daily rate by the number of days of rental.

$$3 \times \$32.95 = \$98.85$$

Sondra charges $98.85 for a three-day rental of a compact car.

exercises

1. The West Car Rental Agency charges $36.99 per day for the rental of a mid-size car. How much does the agency charge for a four-day rental of a mid-size car?
2. Sam's Rental Agency charges a daily rate of $26.99 for a compact car and $31.95 for a mid-size car. How much more does it cost to rent a mid-size car than a compact car for three days?
3. Ted Rome needs to rent a car for five days. Sharp Rent-a-Car charges $25.99 per day and $119.90 per week. Will it cost Ted less to rent by the day or by the week? Explain.
4. Bea Danner needs to rent a car for one day to drive 250 mi (miles). ABC Agency rents cars for $24.99 per day, including 150 free miles and $.15 for each additional mile. XYZ Agency rents cars for $19.95 per day, with 100 free miles and $.30 for each additional mile. Which agency offers the less costly rental? Explain.

CHAPTER REVIEW

vocabulary vo·cab·u·lar·y

Choose the correct word or phrase to complete each sentence.

1. The symbol ≈ means (*is approximately equal to, is not equal to*).
2. Pay after an employer makes deductions is called (*net, gross*) pay.

skills

Find each product. If necessary, round to the nearest cent.

3. 8 × 3.218
4. 5.33 × 0.26
5. 7 × $9.50
6. 1.1 × $2.85
7. 0.03 × 0.6
8. 9.95 × 4
9. 0.25 × $2.67
10. 8 × $34.85
11. 2.7 × 0.647

Estimate by rounding to the place of the leading digit.

12. 6.7 × $17.50
13. 2.3 × $44.77
14. 0.34 × 4.17
15. 0.85 × 6.75
16. 0.65 × 0.78
17. 0.51 × 0.32
18. $54.89 × 7.1
19. 2.41 × 0.33
20. 0.46 × 0.49
21. $9.78 × 3.6
22. 8.66 × 0.87
23. 0.77 × 0.29

Find each quotient. If necessary, round to the nearest thousandth.

24. $12\overline{)13}$
25. $0.9\overline{)0.728}$
26. 1.008 ÷ 0.03
27. 66.5 ÷ 7
28. $2.36\overline{)542.8}$
29. $0.18\overline{)216}$
30. 0.023 ÷ 0.36
31. 94.6 ÷ 0.5

Estimate using compatible numbers.

32. $5.3\overline{)298.2}$
33. $\$6.20\overline{)\$479.69}$
34. $0.8\overline{)0.462}$
35. $0.87\overline{)0.374}$
36. $\$3.89\overline{)\$19.89}$
37. $0.23\overline{)59.31}$
38. 0.252 ÷ 0.43
39. 0.615 ÷ 5.8
40. 12.188 ÷ 4.29
41. 0.423 ÷ 2.1
42. 0.58 ÷ 8.2
43. 0.315 ÷ 5.3

Solve.

44. Lorraine Paris works as a salesperson in a clothing store. Her annual salary is $15,060. Find her monthly salary.
45. Tom Walters earns $45.50 for cleaning one house. He cleans 15 houses each week. How much does he earn weekly?
46. Judy Frost works as an accountant. She earns $771.50 per week. Her deductions are $163.02 for federal income tax, $57.94 for F.I.C.A., $35.74 for state tax, and $5.23 for insurance. Find Judy's total deductions and net pay.

CHAPTER TEST

Find each product. If necessary, round to the nearest cent.

1. $1.45 ×5	2. 2.598 ×63	3. 3.55 ×18	4. 0.32 ×7.4	**5-1**
5. 9.1 × 4.5	6. 0.7 × 0.005	7. 3 × $3.90	8. $9.33 × 0.66	

9. Janice Thomas bought three skirts for $15.99 each. What was the total cost of the skirts?
10. Jack Bryson jogged 3.2 km (kilometers) every day for 15 days. What was the total distance he jogged in the 15 days?

Estimate each product. Determine if the estimate is an *overestimate*, an *underestimate*, or if you *can't tell*.

11. 5.37 × 14.652	12. $74.15 × 2.23	13. $56.32 × 4.91	**5-2**
14. 0.79 × 0.16	15. 4703 × 0.64	16. 0.92 × 5520	

17. At Elmo's Store pears cost $.69 per pound. About how much do 5.8 lb (pounds) cost?

Find each quotient.

18. 12 ÷ 8	19. 12.018 ÷ 6	20. $1.6\overline{)0.816}$	21. $0.32\overline{)224}$	**5-3**

Find each quotient to the indicated place.

22. $3\overline{)12.76}$; nearest tenth	23. $0.12\overline{)0.017}$; nearest tenth
24. 0.982 ÷ 0.6; nearest hundredth	25. 2.25 ÷ 0.17; nearest hundredth

Estimate using compatible numbers.

26. $9.7\overline{)1051.5}$	27. $0.9\overline{)824}$	28. $6.15\overline{)0.595}$	**5-4**
29. 0.476 ÷ 0.7	30. 789.2 ÷ 0.39	31. $419.83 ÷ $7.95	

32. The Choral Society collected $1215.50 by selling tickets for the school concert. Tickets cost $2.75 each. About how many tickets did the Choral Society sell?

33. Zach Lewis earns $525.25 weekly. His neighbor Johnny Black earns $2150.77 monthly. Who earns the greater annual salary? **5-5**

34. Joanne Majors earns $6.50/h working a 35-h week as a bank teller. How much does she earn for the week? **5-6**

35. Leroy Hughes earns $578 per week. His deductions are $91.16 for federal income tax, $43.41 for F.I.C.A., and $16.74 for state tax. Find Leroy's total deductions and net pay. **5-7**

Cookbooks and package labels that list the nutrients in various foods often state the quantities in grams and milligrams.

CHAPTER 6

METRIC MEASUREMENT

6-1 METRIC UNITS OF LENGTH

In the metric system, the basic unit of length is the **meter** (m). The length of one long step is about one meter.

You measure long distances in **kilometers** (km). The length of twelve city blocks is about one kilometer. You measure short lengths in **centimeters** (cm) and very short lengths in **millimeters** (mm). A dime is about two centimeters wide and one millimeter thick.

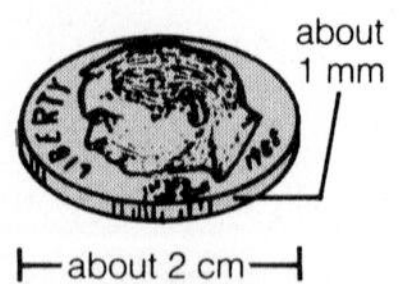

example 1

Select the most reasonable unit. Choose mm, cm, m, or km.

a. The distance from Los Angeles to New York is about 3944 __?__.

b. The length of a sheet of notebook paper is about 28 __?__.

solution

a. The distance is long. Choose km.

b. The length is short. Choose cm.

your turn

Select the most reasonable unit. Choose mm, cm, m, or km.

1. The height of a tree is about 5.5 __?__.

2. The length of a car key is about 57 __?__.

You can use a metric ruler to measure lengths in centimeters and millimeters. On the ruler, each centimeter is divided into ten equal parts. Each part is one millimeter long, so 1 cm = 10 mm and 1 mm = 0.1 cm.

example 2

Measure the line segment in: a. millimeters b. centimeters

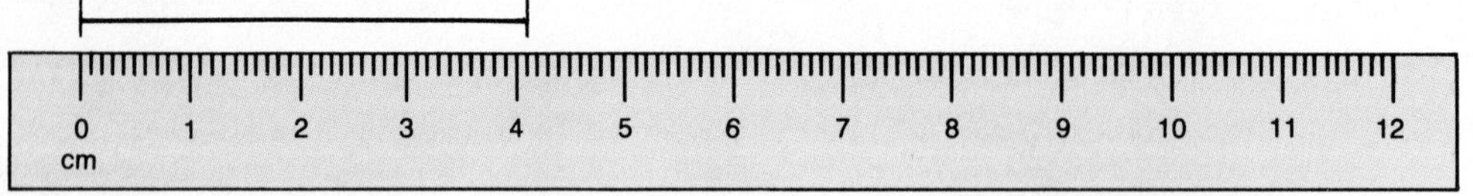

Place the ruler so that the leftmost mark lines up with one end of the line segment. Read the mark nearest the other end of the line segment.

a. The length is 41 mm.

b. The length is 4.1 cm.

your turn

Measure each line segment in: a. millimeters b. centimeters

3. ├──────────┤

4. ├──────────────────┤

The chart below shows how other metric units of length are related to the meter.

millimeter mm	centimeter cm	decimeter dm	meter m	dekameter dam	hectometer hm	kilometer km
1 mm = 0.001 m	1 cm = 0.01 m	1 dm = 0.1 m	1 m	1 dam = 10 m	1 hm = 100 m	1 km = 1000 m

The most commonly used units are the millimeter, centimeter, meter, and kilometer.

1 cm = 10 mm   1 m = 100 cm   1 m = 1000 mm   1 km = 1000 m

To change from one unit to another, use these rules.

To change from a larger unit to a smaller unit, you *multiply*.

To change from a smaller unit to a larger unit, you *divide*.

example 3

Complete: **a.** 6.2 m = _?_ cm   **b.** 183.6 m = _?_ km

solution

a. 1 m = 100 cm, so 6.2 m = 620 cm ← ***Think:*** **larger to smaller, so *multiply*.**

(× 100) (× 100)

b. 1000 m = 1 km, so 183.6 m = 0.1836 km ← ***Think:*** **smaller to larger, so *divide*.**

(÷ 1000) (÷ 1000)

your turn

Complete.

5. 8.3 cm = _?_ mm   **6.** 6000 mm = _?_ m   **7.** 117.2 m = _?_ km

practice exercises

practice for example 1 (page 98)

Select the most reasonable unit. Choose mm, cm, m, or km.

1. The distance from Boston to Dallas is about 2482 _?_.

2. The length of a pair of scissors is about 20.4 _?_.

3. The height of the Washington Monument is about 166.5 _?_.

4. The thickness of a nickel is about 2 _?_.

practice for example 2 (page 98)

Measure each line segment in: **a. millimeters** **b. centimeters**

5. ├──────────────────┤

6. ├──────────┤

practice for example 3 (page 99)

Complete.

7. 1365 mm = _?_ m **8.** 8200 m = _?_ km **9.** 3.6 m = _?_ cm

10. 500 km = _?_ m **11.** 232 cm = _?_ mm **12.** 78.3 mm = _?_ cm

mixed practice (pages 98–99)

Select the most reasonable measure.

13. length of a golf club: 105 mm 105 cm 105 m

14. distance from Chicago to Miami: 1911 cm 1911 m 1911 km

15. length of a table top: 1.6 mm 1.6 cm 1.6 m

16. length of a railroad car: 12.6 cm 12.6 m 12.6 km

Choose the letters of all measures that are equal to the given measure.

17. 13 m **a.** 0.013 km **b.** 0.13 km **c.** 1300 cm **d.** 13,000 cm

18. 2.35 km **a.** 235 m **b.** 2350 m **c.** 23.5 cm **d.** 0.235 cm

19. 6200 mm **a.** 620 cm **b.** 62 cm **c.** 6.2 m **d.** 0.62 m

20. 265 cm **a.** 26.5 mm **b.** 2650 mm **c.** 0.265 m **d.** 2.65 m

Draw a line segment of the given length.

21. 9 cm **22.** 13 cm **23.** 49 mm **24.** 81 mm

25. A mug is 9.1 cm high. How many millimeters high is the mug?

26. The Brooklyn Bridge is 0.486 km long. The Golden Gate Bridge is 1.28 km long. How many meters longer is the Golden Gate Bridge?

review exercises

Find each product or quotient.

1. 24×1000 **2.** 810×100 **3.** 1.27×1000 **4.** 0.06×10

5. $16 \div 1000$ **6.** $324 \div 100$ **7.** $3.71 \div 10$ **8.** $413.2 \div 1000$

mental math

To add 149 + 236 mentally, think of 236 as 200 + 30 + 6

149 + 200 = 349; 349 + 30 = 379; 379 + 6 = 385

Add mentally.

1. 87 + 14 **2.** 135 + 425 **3.** 129 + 345 **4.** 753 + 164

5. 3.6 + 4.7 **6.** 6.5 + 3.9 **7.** 9.6 + 3.5 **8.** 13.8 + 9.9

6-2 METRIC UNITS OF CAPACITY

The **liter** (L) is the basic unit of liquid capacity in the metric system. A can of motor oil holds about one liter. Small amounts of liquid are measured in **milliliters** (mL). An eyedropper holds about one milliliter.

1 L = 1000 mL

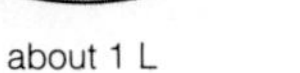
about 1 L

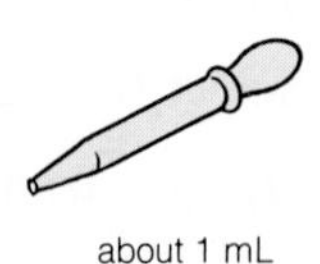
about 1 mL

example 1

Select the more reasonable unit. Choose mL or L.

a. A large pot holds about 8.5 _?_. **b.** A test tube holds about 50 _?_.

solution

a. The capacity is large. Choose L. **b.** The capacity is small. Choose mL.

your turn

Select the more reasonable unit. Choose mL or L.

1. A fish tank holds about 152 _?_. **2.** A glass holds about 240 _?_.

The chart below shows how other metric units of capacity are related to the liter. The asterisks (*) indicate the most commonly used units.

*milliliter	centiliter	deciliter	*liter	dekaliter	hectoliter	kiloliter
mL	cL	dL	L	daL	hL	kL
1 mL = 0.001 L	1 cL = 0.01 L	1 dL = 0.1 L	1 L	1 daL = 10 L	1 hL = 100 L	1 kL = 1000 L

example 2

Complete: **a.** 5 L = _?_ mL **b.** 2400 mL = _?_ L

solution

a. 1 L = 1000 mL, so 5 L = 5000 mL ← ***Think:* larger to smaller, so *multiply*.**

(× 1000) (× 1000)

b. 1000 mL = 1 L, so 2400 mL = 2.4 L ← ***Think:* smaller to larger, so *divide*.**

(÷ 1000) (÷ 1000)

your turn

Complete.

3. 3.7 L = _?_ mL **4.** 0.08 L = _?_ mL **5.** 7610 mL = _?_ L **6.** 468 mL = _?_ L

practice exercises

practice for example 1 (page 101)

Select the more reasonable unit. Choose mL or L.

1. A teakettle holds about 2.6 __?__.
2. A gas tank of a car holds about 60.8 __?__.
3. A teaspoon holds about 5 __?__.
4. A small can of juice holds about 180 __?__.

practice for example 2 (page 101)

Complete.

5. 2.85 L = __?__ mL
6. 0.631 L = __?__ mL
7. 0.047 L = __?__ mL
8. 4950 mL = __?__ L
9. 789 mL = __?__ L
10. 26,258 mL = __?__ L

mixed practice (page 101)

Compare. Replace each __?__ with >, <, or =.

11. 500 mL __?__ 0.5 L
12. 0.003 L __?__ 3 mL
13. 0.05 L __?__ 5 mL
14. 42 mL __?__ 0.42 L
15. 450 mL __?__ 4.5 L
16. 1.26 L __?__ 126 mL

Select the most reasonable measure.

17. capacity of a soup spoon:
    0.15 mL   15 mL   150 mL
18. capacity of a washing machine:
    4 L   40 L   400 L
19. capacity of a large can of paint:
    3.8 L   38 L   380 L
20. capacity of a bottle of liquid soap:
    5 mL   500 mL   5000 mL
21. How many milliliters of juice are in a half-liter bottle?
22. A case of spring water contains 24 bottles. Each bottle holds 330 mL. How many liters of spring water does a case contain?
23. One glass holds 250 mL. How many liters of water are needed to fill twenty of these glasses?
24. Denise bought 15 bottles of shampoo. Each bottle holds 207 mL. How many liters of shampoo did Denise buy?

review exercises

Find each answer.

1. $70 - 3 \times 4 + 3^3$
2. $18 \div 2 + 5^2 \times 10$
3. $3 + 5 \times 3 \times 6$
4. $(2^4 \times 5) \div (13 - 5)$
5. $(46 + 11) \div (21 - 18)$
6. $6 \times (14 - 8) \div 4$

6-3 METRIC UNITS OF MASS

In the metric system, the basic unit of mass is the **gram** (g). The mass of a dollar bill is about one gram.

You measure the mass of heavy objects in **kilograms** (kg). The mass of a steam iron is about one kilogram. You measure the mass of very light objects in **milligrams** (mg). The mass of a pin is about 100 mg.

1 kg = 1000 g   1 g = 1000 mg

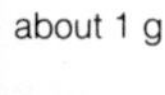

about 1 g

about 1 kg

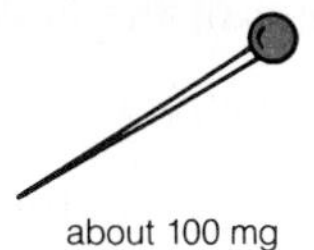

about 100 mg

example 1

Select the most reasonable unit. Choose mg, g, or kg.

a. The mass of a car is about 1400 _?_. **b.** The mass of a thumbtack is about 150 _?_.

solution

a. The object is heavy. Choose kg. **b.** The object is very light. Choose mg.

your turn

Select the most reasonable unit. Choose mg, g, or kg.

1. The mass of a chair is about 15 _?_. **2.** The mass of a tennis ball is about 57.6 _?_.

The chart below shows how other metric units of mass are related to the gram. The asterisks (*) indicate the most commonly used units.

*milligram mg	centigram cg	decigram dg	*gram g	dekagram dag	hectogram hg	*kilogram kg
1 mg = 0.001 g	1 cg = 0.01 g	1 dg = 0.1 g	1 g	1 dag = 10 g	1 hg = 100 g	1 kg = 1000 g

example 2

Complete: **a.** 3.6 kg = _?_ g **b.** 108 mg = _?_ g

solution

a. 1 kg = 1000 g, so 3.6 kg = 3600 g ← ***Think:*** **larger to smaller, so *multiply.***

(× 1000) (× 1000)

b. 1000 mg = 1 g, so 108 mg = 0.108 g ← ***Think:*** **smaller to larger, so *divide.***

(÷ 1000) (÷ 1000)

your turn

Complete: **3.** 0.5 g = _?_ mg **4.** 375 mg = _?_ g **5.** 6295 g = _?_ kg

Length, capacity, and mass are different types of measurement. You should be able to choose the appropriate unit of measurement.

example 3

Choose the letter of the appropriate unit for measuring the given object.

length of the Golden Gate Bridge **a.** L **b.** m **c.** g

solution

You must find the *length* of an object. Length is measured in meters. Choose **b.**

your turn

Choose the letter of the appropriate unit for measuring the given object.

6. capacity of a bucket **a.** L **b.** m **c.** g
7. mass of a truck **a.** km **b.** kg **c.** L

practice exercises

practice for example 1 (page 103)

Select the most reasonable unit. Choose mg, g, or kg.

1. The mass of a table tennis ball is about 2.5 _?_.
2. The mass of a spider is about 100 _?_.
3. The mass of an elephant is about 6300 _?_.
4. The mass of a bicycle is about 13.5 _?_.

practice for example 2 (page 103)

Complete.

5. 14 kg = _?_ g
6. 0.32 g = _?_ mg
7. 180 mg = _?_ g
8. 600 g = _?_ kg
9. 1.2 kg = _?_ g
10. 5983 g = _?_ kg

practice for example 3 (page 104)

Choose the letter of the appropriate unit for measuring the given object.

11. length of a jet **a.** L **b.** g **c.** m
12. capacity of a water pitcher **a.** L **b.** mm **c.** kg
13. mass of a whale **a.** kg **b.** km **c.** L
14. height of a person **a.** kg **b.** cm **c.** mL

mixed practice (pages 103–104)

Complete.

15. 3000 mg = ? g = ? kg

16. 150,000 mg = ? g = ? kg

17. 12.4 kg = ? g = ? mg

18. 2.61 kg = ? g = ? mg

19. 375 g = ? kg = ? mg

20. 49 g = ? kg = ? mg

Select the appropriate measure.

21. mass of a person:
70 m 70 kg 70 L

22. capacity of a soup can:
305 mL 305 cm 305 g

23. height of a table:
76 cm 76 L 76 kg

24. length of a hockey rink:
61 L 61 g 61 m

25. An egg contains about 6 g of fat. How many milligrams is this?

26. An orange contains about 54 mg of protein. How many grams is this?

review exercises

Estimate.

1. \$27.64 + \$12.97 **2.** 34,162 ÷ 72 **3.** 43 × 188 **4.** 71,140 − 43,605

5. 0.61 × 3.4 **6.** 56.3 × 2.09 **7.** 19.27 ÷ 0.4 **8.** 0.37 ÷ 6.15

calculator corner

You can use a calculator to solve problems involving metric units.

example How many 250-mL glasses can you fill from a 2-L bottle of water?

solution First change 2 L to milliliters, and then divide by 250 mL.

Enter 2 × 1000 ÷ 250 =. ← **× 1000 changes L to mL.**

The result is 8. You can fill 8 glasses.

Use a calculator to solve.

1. Julia cut a 12-m length of rope into 16 equal pieces. How many centimeters long is each piece?

2. Dennis bought nine 448-g packages of whole-grain cereal. How many kilograms of whole-grain cereal did he buy?

3. A case of fruit juice contains twenty-four 0.75-L bottles. How many 240-mL glasses can you fill using one case of juice?

PROBLEM SOLVING STRATEGY

6-4 IDENTIFYING PATTERNS

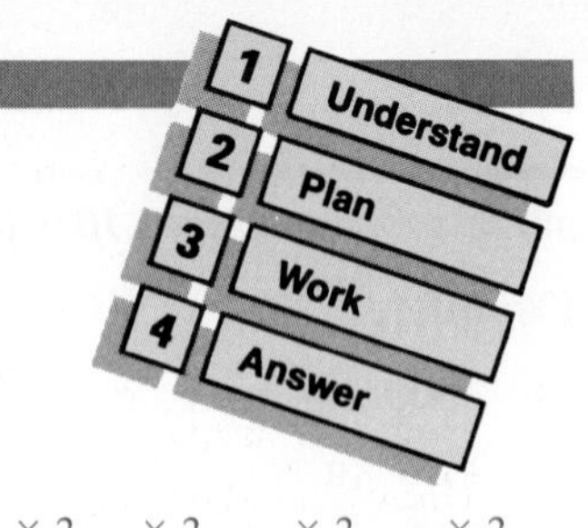

You can solve certain problems by identifying a number pattern. When searching for a pattern, look for one of the following.

The same operation is performed consistently. $1 \xrightarrow{\times 2} 2 \xrightarrow{\times 2} 4 \xrightarrow{\times 2} 8 \xrightarrow{\times 2} 16$

Different operations are performed alternately. $1 \xrightarrow{+6} 7 \xrightarrow{-2} 5 \xrightarrow{+6} 11 \xrightarrow{-2} 9$

One operation is used, but the number involved changes. $1 \xrightarrow{+6} 7 \xrightarrow{+12} 19 \xrightarrow{+18} 37 \xrightarrow{+24} 61$

example

Tina's Restaurant has a Tuesday special. Dinner for one person costs $13. Dinner for a party of two people costs $22. Dinner for a party of three people costs $31. If this pattern continues, how much does dinner for a party of seven people cost?

solution

Make a table to organize the given information. Look for a pattern.

number of people	1	2	3
cost of dinner	$13	$22	$31

+ $9 + $9

The cost of dinner increases by $9 for each additional person.

Now extend the table until 7 appears in the first row.

number of people	1	2	3	4	5	6	7
cost of dinner	$13	$22	$31	$40	$49	$58	$67

+ $9 + $9 + $9 + $9 + $9 + $9

Dinner for a party of seven people costs $67.

problems

1. The cost of shipping a 1-lb (pound) package is $3. The cost of shipping a 2-lb package is $5. The cost of shipping a 3-lb package is $7. If this pattern continues, what is the cost of shipping a 9-lb package?
2. The cost of a 1-min (minute) phone call from Bowtown to Farmville is 39¢. The cost of a 2-min call is 55¢. The cost of a 3-min call is 71¢. If this pattern continues, what is the cost of a 12-min call?

3. Sandy runs the 200-m dash for her track team. Two weeks ago her time was 41 s. Last week her time was 39 s. This week her time was 37 s. If this pattern continues, what will her time be next week?
4. There are 45 seats in the first row of an auditorium. There are 51 seats in the second row, 63 seats in the third row, and 81 seats in the fourth row. If this pattern continues, how many seats are in the tenth row?
5. The Bensons are saving money to remodel a room. They had $50 in March, $75 in April, $125 in May, and $200 in June. If this pattern continues, how much money will the Bensons have in October?
6. A bus leaves the station with 33 passengers. At the first stop 6 people get off. At the second stop 3 people get on. At the third stop 6 people get off. At the fourth stop 3 people get on. If this pattern continues, after how many stops will the bus have no passengers?
7. The numbers 1, 1, 2, 3, 5, 8, 13, 21, 34 form a pattern. Give the twelfth number of this pattern.
8. The numbers 3, 6, 9 form a pattern. Explain why the next number could be either 12 or 15.
9. Arrange the numbers 8, 18, 36, 6, and 16 to form a pattern. Describe the pattern. What are the next five numbers in your pattern?
10. Arrange the numbers 6, 4, 2, 10, 5, and 3 to form a pattern. Describe the pattern. What are the next five numbers in your pattern?

review exercises

1. Paul Jones earns $4.75 per hour as a cashier. Last week he earned $80.75. How many hours did Paul work last week?
2. Elizabeth Hoguto drives a bus and makes eight round trips per day. She makes ten stops per round trip, and the length of each round trip is 18.8 mi (miles). How many miles does Elizabeth drive the bus each day?
3. A towing company charges $25 to tow a car for the first five miles. The company charges $3 per mile for each mile after five miles. How much would the company charge to tow a car nine miles?

SKILL REVIEW

Select the most reasonable unit. Choose mm, cm, m, or km.

6-1

1. The length of a textbook page is about 23.5 __?__.
2. The distance from London to Paris is about 343 __?__.

Measure each line segment in: a. **millimeters** b. **centimeters**

3. |——————————————————|
4. |————————————————————————|

Complete.

5. 145 cm = __?__ m
6. 2530 mm = __?__ cm
7. 46.3 km = __?__ m

Select the more reasonable unit. Choose mL or L.

6-2

8. A carton of cream holds about 237 __?__.
9. A bottle of liquid dish detergent holds about 0.66 __?__.

Complete.

10. 31 L = __?__ mL
11. 0.257 L = __?__ mL
12. 3404 mL = __?__ L

Select the most reasonable unit. Choose mg, g, or kg.

6-3

13. The mass of a jumbo jet is about 375,000 __?__.
14. The mass of a baseball is about 141.7 __?__.

Complete.

15. 4.5 kg = __?__ g
16. 36 g = __?__ mg
17. 422 mg = __?__ g

Choose the letter of the appropriate unit for measuring the given object.

18. mass of a telephone a. km b. kg c. L
19. capacity of a teacup a. mm b. mg c. mL

Solve.

6-4

20. The Shermans are saving money to make a down payment on a new car. They had $165 in January, $215 in February, $315 in March, and $465 in April. If they continue this pattern, how much money will the Shermans have in July?
21. A supermarket display has 1 can in the top row, 3 cans in the second row, 5 cans in the third row, and 7 cans in the fourth row. If this pattern continues, how many cans are in the ninth row?

APPLICATION

6-5 CONSERVING ELECTRICITY

The amount of electricity an appliance uses is measured in **kilowatt-hours** (kW·h). One kilowatt-hour is used each time 1000 W (watts) of electricity are used for one hour. Electric companies charge customers for the number of kilowatt-hours used. To conserve electricity and save money, you should use appliances that operate efficiently.

example

A color television that is on for five hours each day uses about 42 kW·h of electricity each month. A black-and-white television that is on for the same amount of time uses about 31 kW·h each month. The electric company charges $.10 per kW·h. Find the amount of money you save each month by using a black-and-white television instead of a color television.

solution

monthly cost of using a color television: $42 \times \$.10 = \$4.20$

monthly cost of using a black-and-white television: $31 \times \$.10 = \$3.10$

Subtract: $\$4.20 - \$3.10 = \$1.10$

You save about $1.10 each month by using a black-and-white television.

exercises

1. One model of refrigerator uses about 113 kW·h each month. A second model uses about 88 kW·h each month. The electric company charges $.10 per kW·h. Find the amount of money you save each month by using the second model instead of the first.
2. One model of air conditioner uses about 385 kW·h each month. A more efficient model uses about 340 kW·h each month. The electric company charges $.11 per kW·h. Find the amount of money you save each month by using the more efficient model.
3. An electric heater has three heat settings. On the low setting, it uses about 270 kW·h each month. On the middle setting, it uses about 405 kW·h each month. On the high setting, it uses about 675 kW·h each month. The electric company charges $.08 per kW·h. Find the amount of money you save each month if you run the heater on the low setting instead of the high setting.

6-6 NUTRITION

Rosie is taking a nutrition course. Her project is to check that she is receiving adequate nutrition. Rosie makes a record of the foods she eats every day. She uses a chart that lists the amounts of nutrients each food contains. The table below shows Rosie's food intake for last Friday.

Food	Protein (g)	Vitamin C (mg)	Thiamine (mg)	Calcium (mg)	Iron (mg)
Breakfast					
2 eggs	12	0	0.08	56	2.0
2 slices wheat toast	6	0	0.18	48	1.6
200 mL orange juice	2	120	0.23	25	0.4
Lunch					
225 g lowfat yogurt	12	2	0.10	415	0.2
1 apple	0	6	0.04	10	0.4
250 mL whole milk	8	2	0.09	291	0.1
Dinner					
1 slice roast beef	17	0	0.06	8	2.2
100 g green beans	1	5	0.03	21	0.3
1 baked potato	4	31	0.15	14	1.1
250 mL whole milk	8	2	0.09	291	0.1
100 g popcorn	1	0	0.05	1	0.3

To check that she is receiving adequate nutrition, Rosie totals each column in the table. She then checks the totals against the Recommended Daily Dietary Allowance (RDA), shown below, for her age group.

	Protein	Vitamin C	Thiamine	Calcium	Iron
RDA	46 g	60 mg	1.1 mg	1200 mg	18 mg

example

a. How many grams of protein did Rosie have on Friday?

b. Did Rosie have enough protein on Friday?

solution

a. Add the amounts shown in the protein column.

$12 + 6 + 2 + 12 + 0 + 8 + 17 + 1 + 4 + 8 + 1 = 71$ ← **A calculator may be helpful.**

Rosie had 71 g of protein on Friday.

b. The RDA for protein is 46 g. Since Rosie had 71 g of protein, she had more than enough.

exercises

Solve. A calculator may be helpful.

1. a. How many milligrams of calcium did Rosie have on Friday?
   b. Did Rosie have enough calcium on Friday?
2. a. How many milligrams of thiamine did Rosie have on Friday?
   b. Did Rosie have enough thiamine on Friday?
3. How many grams of protein did Rosie have at lunch?
4. How many milligrams of iron did Rosie have at dinner?
5. How many milligrams of vitamin C above the RDA did Rosie have on Friday?
6. How many more milligrams of iron did Rosie need on Friday to meet the RDA?
7. If Rosie's only source of calcium was whole milk, estimate how many 250-mL glasses she would need to drink to meet her RDA.
8. If Rosie had not had any orange juice on Friday, would she still have met her RDA of vitamin C? Explain.
9. What food gave Rosie the greatest amount of iron?
10. What food gave Rosie the least amount of thiamine?
11. Find the RDA for various age groups, including your own.
12. Get a chart that lists the amounts of nutrients in food. Make a table of your food intake for a day. Did you meet the RDA for your age group?

Suppose you had $1,000,000 and gave away $1 every minute. About how many years would it take you to give away all your money?

APPLICATION

6-7 UTILITY BILLS

A telephone bill contains more information than just the amount of money you owe. It also shows the account number, a list of itemized calls and charges, billing dates, and federal taxes.

ACCOUNT NO. 617 555 0000 350
BILLING PERIOD: SEPTEMBER 3–OCTOBER 2, 1988

Monthly Charges **21.71**
Calling Services
Itemized Calls

NO.	DATE	TIME	PLACE	AREA-NUMBER	MIN	AMOUNT
1.	SEP 2	945PM	SCOVILLE	617 555-9876 ED	17	2.28
2.	SEP 7	242PM	CHARLETON	617 555-0987 DD	1	.31
3.	SEP 8	221PM	MERIDEN	617 555-2013 DD	2	.60
4.	SEP 12	117PM	REDMOND	617 555-1234 DD	3	.69
5.	SEP 12	710PM	ENDICOTT	617 555-7465 ED	6	.90
6.	SEP 16	521PM	SUDBURY	617 555-6574 ED	1	.12
7.	SEP 16	523PM	MARLBORO	617 555-1029 ED	1	.16
8.	SEP 23	829AM	OLDTON	617 555-3333 DD	1	.48
9.	SEP 30	549PM	GRANDON	617 555-5049 ED	3	.45

OPERATOR ASSISTED CALLS TO DIRECTORY ASSISTANCE 2
DIRECTLY DIALED CALLS TO DIRECTORY ASSISTANCE 6

DD *DAY DIAL* ED *EVENING DIAL*

TOTAL OF CALLING SERVICES 5.99
Total Tax—FEDERAL .... .69
TOTAL NEW CHARGES FOR COMMUNITY TELEPHONE 28.39

example

Use the telephone bill shown to find the following.

a. the total cost of calling services

b. the total number of calls to directory assistance

solution

a. The total cost of calling services is $5.99.

b. There were 2 operator assisted and 6 directly dialed calls.

$$2 + 6 = 8$$

There were 8 calls to directory assistance.

exercises

Use the telephone bill to find the following.

1. the cost of the call to Sudbury
2. the length of the call to Scoville
3. the total number of itemized calls
4. the date and time of the call to Redmond
5. the length of the longest call
6. the total cost of the call to Endicott
7. the date and time of the call to Oldton
8. the length of the call to Grandon
9. the monthly charges
10. the total federal tax
11. the total new charges
12. the billing period

13. Many local telephone companies offer optional special services, such as call waiting. Find out what special services your local company offers and the monthly charges for these services.
14. Telephone service between different local service areas is provided by long distance telephone companies. Find out what services these companies offer and the cost for each service in your area.

Consumers also receive utility bills for electricity, water, natural gas, and home heating oil. These bills usually show meter readings, billing dates, amounts used, unpaid balances, previous payments, and rates.

Rate	From	To	Reading	Constant	kW·h used	Description	Amount
R1	04/10/88	05/12/88	27112	1	772		$64.79

Your account number	Apts.	Bill includes cost of fuel Per kW·h	Total	Billing date	Total amount due
08 8417 272650	1	$.022470	$17.35	May 15, 1988	$64.79

TOTAL PAYMENTS APPLIED $46.36	RATE R1 SCHEDULE SERVICE CHARGE $6.15 PER MONTH $.053479 PER KW·H FOR ALL USE

Use the electric bill shown above to find the following.

15. the number of kilowatt-hours used
16. the cost per kilowatt-hour for all use
17. the cost per kilowatt-hour for fuel
18. the total cost of fuel
19. the service charge per month
20. the meter reading
21. the total amount due
22. the amount paid on the last bill
23. the account number
24. the billing date
25. the time period covered by the bill
26. the number of apartments
27. The cost of fuel may vary every month. Find out how electric companies determine this cost.
28. Some electric companies do not read the meter every month. They send estimated bills instead. Find out if your electric company does this. If so, how does the company estimate the number of kilowatt-hours used?

DATE	METER READINGS		HUNDRED CUBIC FEET USED	WATER		SEWER		INTEREST	TOTAL AMOUNT DUE
	PRESENT	PREVIOUS		SERVICE	UNPAID SERVICE BALANCE	RENTAL	UNPAID RENTAL BALANCE		
7/30/88	12683	12652	31	$21.70	$.00	$9.30	$.00	$.00	$31.00

MAKE PAYMENTS → AND MAIL TO — City of Morriston, Water Division, 73 Center Avenue

RATES: WATER $.70 SEWER $.30 PER 100 CUBIC FEET

MINIMUM RATES: WATER $1.75 SEWER $.75 PER MONTH

Use the water bill shown above to find the following.

29. the service charge for the water

30. the previous meter reading

31. the date of the bill

32. the sewer rate per hundred cubic feet

33. the sewer rental charge

34. the total amount of the bill

35. the present meter reading

36. the water rate per hundred cubic feet

37. the minimum rate for water

38. the minimum rate for sewer

39. the interest charge

40. the number of hundred cubic feet used

41. Water and sewer rates vary from community to community. Find the rates for your community and compare them to those of two surrounding communities.

42. Sometimes water bills are sent every other month or every third month. Find the billing period for your community.

calculator corner

You can use a calculator to check the total amount due on an electric bill.

example kW·h used: 854
cost per kW·h for fuel: $.024150
cost per kW·h for all use: $.067345
service charge: $5.95

solution Enter 0.024150 + 0.067345 = × 854 + 5.95 =.
(0.024150 + 0.067345: total cost per kW·h)

The result is 84.08673.
To the nearest cent, the total amount due is $84.09.

Use a calculator to find the total amount due.

1. kW·h used: 789
cost per kW·h for fuel: $.030451
cost per kW·h for all use: $.073294
service charge: $6.25

2. kW·h used: 1047
cost per kW·h for fuel: $.026432
cost per kW·h for all use: $.066474
service charge: $6.50

CHAPTER REVIEW

vocabulary vo·cab·u·lar·y

Choose the correct word to complete each sentence.

1. The basic unit of mass in the metric system is the *(meter, gram)*.
2. In the metric system, you can measure long distances in *(kilometers, kilograms)*.
3. In the metric system, you can measure liquid capacity in *(milliliters, milligrams)*.
4. In the metric system, the meter is a measure of *(capacity, length)*.

skills

Select the most reasonable measure.

5. length of a bicycle:
   2 cm   2 m   2 km
6. capacity of a mug:
   1.8 mL   18 mL   180 mL
7. mass of a pencil:
   15 mg   15 g   15 kg
8. length of an envelope:
   2.4 cm   24 cm   240 cm

Complete.

9. 164 mm = ? cm
10. 2.3 km = ? m
11. 22,413 mL = ? L
12. 0.097 L = ? mL
13. 0.25 kg = ? g
14. 2409 mg = ? g

Measure each line segment in: a. **millimeters** b. **centimeters**

15. ├────────────┤
16. ├──────────────────┤

Solve.

17. A commuter train ticket in Zone 1 costs $2.00. A ticket in Zone 2 costs $2.40. A ticket in Zone 3 costs $2.80. If this pattern continues, what is the cost of a ticket in Zone 7?
18. A regular freezer uses about 125 kW·h each month. A frost-free freezer uses about 188 kW·h each month. The electric company charges $.09 per kW·h. Find the amount of money you save each month by operating a regular freezer instead of a frost-free freezer.
19. Use the table on page 110. How many milligrams of calcium did Rosie have at dinner?
20. Use the telephone bill on page 112. How much did the call to Charleton cost?

CHAPTER TEST

Select the most reasonable unit. Choose mm, cm, m, or km.

1. The length of a flashlight is about 31.5 __?__. **6-1**

2. The length of a passenger train is about 180 __?__.

Measure each line segment in: a. millimeters b. centimeters

3. |————————————————|

4. |——————————|

Complete.

5. 8920 cm = __?__ m **6.** 3.04 m = __?__ mm **7.** 2.8 km = __?__ m

Select the more reasonable unit. Choose mL or L.

8. A bottle holds about 205 __?__. **9.** A sink holds about 12.8 __?__. **6-2**

Complete.

10. 7074 mL = __?__ L **11.** 0.61 L = __?__ mL **12.** 3.2 L = __?__ mL

Select the most reasonable unit. Choose mg, g, or kg.

13. The mass of a tractor is about 3000 __?__. **6-3**

14. The mass of a softball is about 177.2 __?__.

Complete.

15. 84 g = __?__ kg **16.** 2.87 kg = __?__ g **17.** 4753 mg = __?__ g

Choose the letter of the appropriate unit for measuring the given object.

18. mass of a camera **a.** m **b.** L **c.** kg

19. height of a building **a.** kg **b.** m **c.** mL

20. The Park-Rite Garage charges $5 to park for one hour, $7 to park for two hours, and $9 to park for three hours. If this pattern continues, how much does the garage charge to park for six hours? **6-4**

21. A water heater uses about 228 kW·h each month. Another water heater uses about 330 kW·h each month. The electric company charges $.10 per kW·h. Find the amount of money you save each month by using the first water heater instead of the second. **6-5**

22. Use the table on page 110. How many milligrams of thiamine did Rosie have at dinner? **6-6**

23. Use the water bill on page 114. What is the unpaid service balance for water? **6-7**

COMPUTER

From ENIAC to Microcomputers

The first electronic digital computer, called ENIAC, was built in the 1940's. It was a huge machine that weighed about thirty tons. ENIAC had over 18,000 vacuum tubes that enabled information to be stored and processed electronically. The tubes created problems, however, because they heated up quickly and burned out frequently. Every time a program was to be run, the operators had to connect the electrical wires manually. By today's standards, ENIAC was not a very efficient machine.

In the 1960's, the *integrated circuit* was developed. A circuit is a path through which electricity flows. An **integrated circuit** joins many tiny electronic components into a single circuit. Integrated circuits were then etched on a thin *chip* made of silicon, which is a common element found in beach sand. These **silicon chips** are about a quarter of an inch square. A **microprocessor** is a chip that contains all the integrated circuits necessary to process information and run programs that previously required large computers, such as ENIAC.

The development of the microprocessor revolutionized the computer industry. It paved the way for **microcomputers,** or **personal computers.** Microcomputers work faster and are more powerful than ENIAC, yet they are small enough to fit on a desk or even a person's lap.

exercises

1. Why is ENIAC considered to be a costly and inefficient machine?
2. What made it possible to greatly reduce the size of computers?
3. Find out what the letters ENIAC stand for.

COMPETENCY TEST

Choose the letter of the correct answer.

1. **Round 23.487 to the nearest hundredth.**

   A. 23.4
   B. 23.48
   C. 23.49
   D. 23.5

2. **Add.** 9684 + 276

   A. 9850
   B. 9960
   C. 10,960
   D. 12,444

3. **Subtract.**

$$\begin{array}{r} 15{,}714 \\ -\quad 879 \\ \hline \end{array}$$

   A. 14,835
   B. 15,165
   C. 15,945
   D. 16,593

4. **Choose the best estimate.**

   41,812 + 39,665

   A. about 8000
   B. about 70,000
   C. about 80,000
   D. about 90,000

5. **Multiply.**

$$\begin{array}{r} 212 \\ \times 52 \\ \hline \end{array}$$

   A. 1484
   B. 10,024
   C. 10,924
   D. 11,024

6. **Which number is greatest?**

   A. 0.4735
   B. 0.04735
   C. 0.4730
   D. 0.47035

7. **Add.** 0.75 + 2.9 + 8.6

   A. 4.41
   B. 12.25
   C. 19.0
   D. 45.1

8. **Subtract.** 44.56 − 3.665

   A. 7.91
   B. 40.905
   C. 40.895
   D. 0.791

9. **Multiply.** 0.32×81.4

   A. 25.948
   B. 26.048
   C. 259.48
   D. 260.48

10. **Jared bought 4 towels at $7.99 each and 2 washcloths at $2.99 each. How much did he spend altogether?**

    A. $10.98
    B. $21.96
    C. $37.94
    D. $43.92

11. **Divide.** $5794 \div 24$

A. 24 R10
B. 24
C. 241 R10
D. 241

12. **Choose the best estimate.**

$19\overline{)84,615}$

A. about 40,000
B. about 4000
C. about 400
D. about 40

13. **Complete the pattern.**

1, 4, 8, 11, 22, 25, ?, ?, ?

A. 28, 56, 59
B. 50, 53, 106
C. 75, 77, 231
D. 27, 81, 83

14. **Choose the best unit to measure the capacity of a bathtub.**

A. kg
B. g
C. L
D. m

15. **Complete.**

3.5 km = ? m

A. 0.035
B. 0.0035
C. 350
D. 3500

16. **Divide.** $0.15\overline{)36.45}$

A. 2.43
B. 2043
C. 24,300
D. 243

17. **Choose the best estimate.**

61×889

A. about 4800
B. about 48,000
C. about 5400
D. about 54,000

18. **Evaluate.** $24 \div 8 + 4 \times 3^2$

A. 63
B. 18
C. 39
D. 27

19. **Complete.**

1,780,000,000 = 1.78 ?

A. hundred thousand
B. million
C. hundred million
D. billion

20. **Felicity earned $29.40 for 3 h of work. How much did she earn per hour?**

A. $25.90
B. $10.29
C. $88.20
D. $9.80

CUMULATIVE REVIEW CHAPTERS 1–6

Find each answer.

1. \$372 + \$88 + \$43
2. \$84 × 112
3. 4 × 172 × 25
4. 4217 − 653
5. 6 × (5 − 3) ÷ 4
6. 5544 ÷ 36
7. 2.13 × 8.8
8. $0.34\overline{)1.87}$
9. 55 × 73
10. 7.5 + 9.9 + 2.5
11. $16\overline{)48.8}$
12. 12 − 8.34
13. 48 ÷ 6 + 2 × 3
14. 16.2 × 3
15. \$17.48 + \$9.88
16. 0.04 × 5.15
17. \$9.72 ÷ 9
18. $3^2 \times 2^3$
19. Round \$137.546 to the nearest cent.
20. Write the word form of 8,205,046.
21. Write 32 thousand, 185 and 15 thousandths in numeral form.
22. Round to the place of the underlined digit: 42,$\underline{6}$05,312
23. Find the quotient to the nearest hundredth: 1.42 ÷ 3.6

Estimate.

24. 822 + 991 + 376
25. \$82.50 − \$63.29
26. 812 × 498
27. 3578 ÷ 58
28. 0.547 × 5897
29. $71\overline{)47,957}$

Complete.

30. 452 cm = _?_ m
31. 84 L = _?_ mL
32. 4.57 kg = _?_ g

Write each set of numbers in order from least to greatest.

33. 0.0351; 0.351; 0.035
34. 1258; 1528; 1285; 1582

Select the most reasonable unit for measuring the given object.

35. mass of a toothpick
    mm mL mg
36. capacity of a teaspoon
    mL cm mm
37. height of a flagpole
    mm m km
38. The market value of a stock was \$74,120,000. Write this amount in the short word form that includes decimals.
39. A taxi company bought nine taxis for \$98,576. About how much did the company pay per taxi?
40. Louise bought three notebooks at \$2.29 each and six pencils at \$.24 each. How much did she spend altogether?
41. Bennie Schiavo left home at 7:43 A.M. He arrived at work at 8:01 A.M. Find the elapsed time.
42. Jose Diaz earns \$7.50/h. He works 38 h each week. Find his weekly gross pay.

In planning a garden, fractions can help you decide what part of the available space you want to use for each kind of flower or vegetable.

CHAPTER 7

NUMBER THEORY AND FRACTION CONCEPTS

7-1 DIVISIBILITY TESTS

A number is **divisible** by a second number if the remainder is zero when the first number is divided by the second.

$$\begin{array}{r} 12 \\ 7\overline{)84} \\ \underline{7} \\ 14 \\ \underline{14} \\ 0 \end{array} \leftarrow$$ **84 is *divisible* by 7.**

$$\begin{array}{r} 16 \\ 5\overline{)84} \\ \underline{5} \\ 34 \\ \underline{30} \\ 4 \end{array} \leftarrow$$ **84 is *not divisible* by 5.**

You can use the following *divisibility tests* to determine if one number is divisible by another number.

Divisibility Tests

Divisible by	Test
2	The last digit is 0, 2, 4, 6, or 8.
3	The sum of the digits is divisible by 3.
4	The number named by the last two digits is divisible by 4.
5	The last digit is 0 or 5.
8	The number named by the last three digits is divisible by 8.
9	The sum of the digits is divisible by 9.
10	The last digit is 0.

example 1

Is each number divisible by 5? by 10? by 2? Write *Yes* or *No*.

a. 976 **b.** 580 **c.** 155 **d.** 7327

solution

a. 976
Divisible:
by 5? *No*
by 10? *No*
by 2? *Yes*

b. 580
Divisible:
by 5? *Yes*
by 10? *Yes*
by 2? *Yes*

c. 155
Divisible:
by 5? *Yes*
by 10? *No*
by 2? *No*

d. 7327
Divisible:
by 5? *No*
by 10? *No*
by 2? *No*

your turn

Is each number divisible by 5? by 10? by 2? Write *Yes* or *No*.

1. 55 **2.** 343 **3.** 4620 **4.** 1,273,984

example 2

Is each number divisible by 4? by 8? Write *Yes* or *No*.

a. 71,640 **b.** 43,260

solution

a. 71,640 Divisible by 4? Yes, because 40 is divisible by 4.
71,640 Divisible by 8? Yes, because 640 is divisible by 8.

b. 43,260 Divisible by 4? Yes, because 60 is divisible by 4.
43,260 Divisible by 8? No, because 260 is not divisible by 8.

your turn

Is each number divisible by 4? by 8? Write *Yes* or *No*.

5. 3384 **6.** 4140 **7.** 21,905 **8.** 231,408

example 3

Is each number divisible by 3? by 9? Write *Yes* or *No*.

a. 160,431 **b.** 89,451

solution

a. $1 + 6 + 0 + 4 + 3 + 1 = 15$
Divisible by 3? Yes, because 15 is divisible by 3.
Divisible by 9? No, because 15 is not divisible by 9.

b. $8 + 9 + 4 + 5 + 1 = 27$
Divisible by 3? Yes, because 27 is divisible by 3.
Divisible by 9? Yes, because 27 is divisible by 9.

your turn

Is each number divisible by 3? by 9? Write *Yes* or *No*.

9. 2706 **10.** 42,739 **11.** 9,104,409 **12.** 563,857

practice exercises

practice for example 1 (page 122)

Is each number divisible by 5? by 10? by 2? Write *Yes* or *No*.

1. 26 **2.** 125 **3.** 320 **4.** 5311
5. 36,050 **6.** 621,392 **7.** 8,473,655 **8.** 79,651,400

practice for example 2 (page 123)

Is each number divisible by 4? by 8? Write *Yes* or *No*.

9. 2804 **10.** 4638 **11.** 4320 **12.** 9900
13. 17,000 **14.** 673,440 **15.** 867,155 **16.** 1,287,154

practice for example 3 (page 123)

Is each number divisible by 3? by 9? Write *Yes* or *No*.

17. 256 **18.** 594 **19.** 6126 **20.** 4413
21. 24,972 **22.** 30,421 **23.** 101,448 **24.** 9,206,013

mixed practice (pages 122–123)

Is the first number divisible by the second number? Write *Yes* or *No*.

25. 69; 3 **26.** 384; 2 **27.** 481; 4 **28.** 905; 5
29. 4956; 8 **30.** 3744; 8 **31.** 53,427; 9 **32.** 8,230,560; 10

A number is divisible by 6 if it is divisible by both 2 and 3. Is each number divisible by 6? Write *Yes* or *No*.

33. 78 **34.** 102 **35.** 1442 **36.** 2865 **37.** 1,436,972

Any number that is divisible by 2 is an *even number*. Any number that is not divisible by 2 is an *odd number*. Tell whether each number is *even* or *odd*.

38. 84 **39.** 117 **40.** 6481 **41.** 43,986 **42.** 5,631,040

43. Josh is buying three items. The price is the same for each item. Is it possible that the total cost of the three items is $2.19?

44. Mr. Bosner donated $46,500 to six different charities. Is it possible that he donated the same amount to each charity?

45. Give an example to show that a number that is divisible by both 2 and 4 is *not* necessarily divisible by 8.

46. Insert the same digit in each blank to make this number divisible by 3: ■8■. (*Hint:* There is more than one correct answer.)

review exercises

Find each answer.

1. 5^3 **2.** 4^0 **3.** 1^5 **4.** 9^2
5. $3^3 \times 4$ **6.** $2^4 \times 10^2$ **7.** $7^2 \times 6^1$ **8.** 8×4^2

7-2 FACTORS, MULTIPLES, AND PRIMES

When one whole number is divisible by a second whole number, the second number is a **factor** of the first.

45 is divisible by 9. 9 is a *factor* of 45.

A **prime number** is a whole number greater than 1 that has exactly two factors, 1 and the number itself. For instance, 5 is *prime* because its only factors are 1 and 5. When you write a whole number as the product of prime numbers, you are finding the **prime factorization** of the number.

example 1

a. Write the prime factorization of 20.

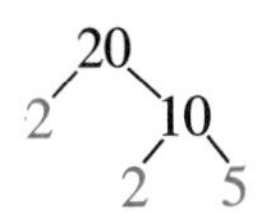

← This is called a factor tree.

$20 = 2 \times 2 \times 5$
$= 2^2 \times 5$

b. Write the prime factorization of 54.

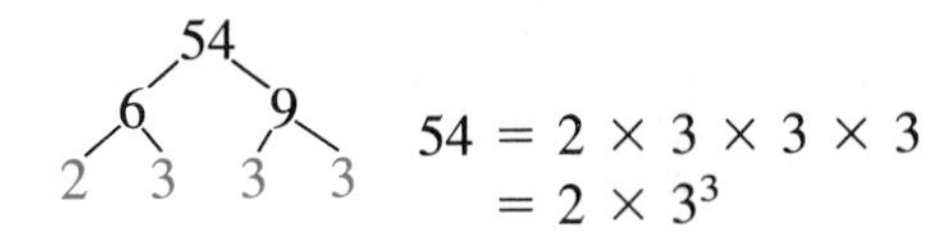

$54 = 2 \times 3 \times 3 \times 3$
$= 2 \times 3^3$

your turn

Write the prime factorization of each number.

1. 70 **2.** 66 **3.** 36 **4.** 24 **5.** 51 **6.** 105

The factors of 20 and 50 are listed at the right. Using these lists, you can see that the **common factors** of 20 and 50 are 1, 2, 5, and 10. The **greatest common factor (GCF)** is 10.

factors of 20: 1, 2, 4, 5, 10, 20
factors of 50: 1, 2, 5, 10, 25, 50

You can also find the GCF of two numbers by using their prime factorizations. The GCF is the product of the lesser power of each *common* prime factor.

example 2

a. Find the GCF of 20 and 50.

$20 = 2 \times 2 \times 5 = 2^2 \times 5$
$50 = 2 \times 5 \times 5 = 2 \times 5^2$
$\text{GCF} = 2 \times 5 = 10$

b. Find the GCF of 24 and 60.

$24 = 2 \times 2 \times 2 \times 3 = 2^3 \times 3$
$60 = 2 \times 2 \times 3 \times 5 = 2^2 \times 3 \times 5$
$\text{GCF} = 2^2 \times 3 = 12$

your turn

Find the GCF.

7. 10 and 15 **8.** 16 and 24 **9.** 11 and 16 **10.** 8, 12, and 36

When a number is multiplied by a nonzero whole number, the product is a **multiple** of the given number.

multiples of 9: 9, 18, 27, 36, 45, 54, 63, 72, . . . ← **Multiply the number by 1, 2, 3, 4, and so on.**
multiples of 12: 12, 24, 36, 48, 60, 72, 84, 96, . . .

The numbers 36 and 72 are called **common multiples** of 9 and 12. The **least common multiple (LCM)** of 9 and 12 is 36. To find the LCM of two numbers, you can list the multiples as shown above, or you can use their prime factorizations. The LCM is the product of the greater power of each prime factor that occurs in *either*.

example 3

a. Find the LCM of 8 and 12.

$8 = 2 \times 2 \times 2 = 2^3$
$12 = 2 \times 2 \times 3 = 2^2 \times 3$
$\text{LCM} = 2^3 \times 3 = 24$

b. Find the LCM of 15 and 18.

$15 = 3 \times 5$
$18 = 2 \times 3 \times 3 = 2 \times 3^2$
$\text{LCM} = 2 \times 3^2 \times 5 = 90$

your turn

Find the LCM.

11. 4 and 6 **12.** 9 and 12 **13.** 4 and 5 **14.** 6, 15, and 30

practice exercises

practice for example 1 (page 125)

Write the prime factorization of each number.

1. 35 **2.** 82 **3.** 63 **4.** 75
5. 250 **6.** 270 **7.** 102 **8.** 143

practice for example 2 (page 125)

Find the GCF.

9. 27 and 36 **10.** 54 and 81 **11.** 60 and 105 **12.** 56 and 84
13. 26 and 85 **14.** 35 and 36 **15.** 52, 78, and 80 **16.** 18, 30, and 45

practice for example 3 (page 126)

Find the LCM.

17. 5 and 10 **18.** 14 and 56 **19.** 9 and 15 **20.** 32 and 48
21. 16 and 18 **22.** 21 and 28 **23.** 8, 15, and 25 **24.** 6, 8, and 12

mixed practice (pages 125–126)

A *composite number* has more than two factors. Tell whether each number is *prime* or *composite*.

25. 13 **26.** 29 **27.** 51 **28.** 97 **29.** 102 **30.** 123

Write the prime factorization of each number.

31. 110 **32.** 85 **33.** 242 **34.** 300 **35.** 168 **36.** 148

Find the GCF and the LCM.

37. 3 and 27 **38.** 4 and 18 **39.** 15 and 25 **40.** 13 and 10
41. 28 and 70 **42.** 16 and 24 **43.** 85 and 34 **44.** 96 and 72
45. 3, 6, and 7 **46.** 5, 10, and 12 **47.** 36, 54, and 72 **48.** 30, 75, and 100

49. List all the multiples of 8 between 51 and 90.

50. List all the prime numbers between 40 and 54.

51. Are any even numbers prime? Explain.

52. Are all odd numbers prime? Explain.

53. Jim wants to buy some soup bowls and some dinner plates. The soup bowls come in boxes of eight, and the dinner plates come in boxes of six. What is the least number of soup bowls and dinner plates that Jim can buy to get an equal number of each?

54. Mr. Began can arrange his entire class into groups of 2, 5, or 6 students with no one left out. What is the least number of students that Mr. Began can have in order to do this?

review exercises

Complete.

1. 3 m = _?_ mm **2.** 14.7 cm = _?_ m **3.** 0.25 L = _?_ mL
4. 130 kg = _?_ g **5.** 63 g = _?_ mg **6.** 10.4 mL = _?_ L
7. 900 g = _?_ kg **8.** 13 m = _?_ km **9.** 22.4 cm = _?_ mm
10. 3.9 L = _?_ mL **11.** 7204 mg = _?_ g **12.** 0.6 km = _?_ m

puzzle corner

If the golf balls in a bucket are counted by twos, threes, fives, and sevens, there is exactly one left over each time. What is the least number of golf balls that could be in the bucket?

7-3 EQUIVALENT FRACTIONS

In the figures below, the same part of each circle is shaded. The circles show that $\frac{1}{2}$, $\frac{2}{4}$, $\frac{3}{6}$, and $\frac{4}{8}$ represent the same amount. Fractions that represent the same amount are called **equivalent fractions.**

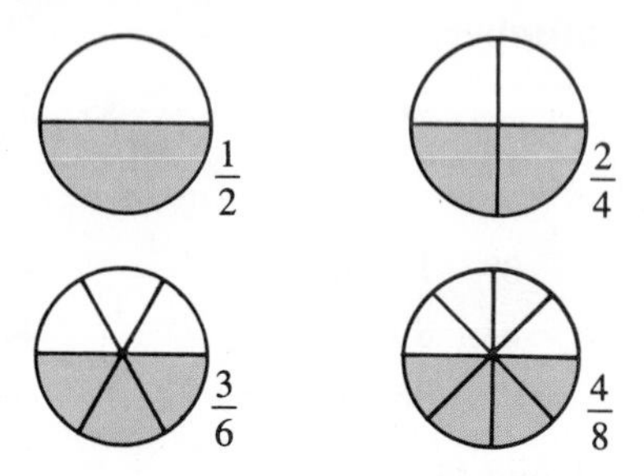

number of shaded parts in the circle → / total number of parts in the circle → $\frac{1}{2} = \frac{2}{4} = \frac{3}{6} = \frac{4}{8}$ ← **numerator** / ← **denominator**

To find a fraction that is equivalent to a given fraction, you can *multiply* or *divide* both the numerator and the denominator by the same nonzero number.

example 1

Replace each __?__ with the number that will make the fractions equivalent.

a. $\frac{2}{3} = \frac{?}{12}$ **b.** $\frac{28}{35} = \frac{4}{?}$

solution

a. $\frac{2}{3} = \frac{2 \times 4}{3 \times 4} = \frac{8}{12}$ ← **You see that 12 = 3 × 4, so multiply both the numerator and the denominator by 4.**

b. $\frac{28}{35} = \frac{28 \div 7}{35 \div 7} = \frac{4}{5}$ ← **You see that 4 = 28 ÷ 7, so divide both the numerator and the denominator by 7.**

your turn

Replace each __?__ with the number that will make the fractions equivalent.

1. $\frac{6}{7} = \frac{?}{21}$ **2.** $\frac{3}{8} = \frac{24}{?}$ **3.** $\frac{33}{55} = \frac{3}{?}$ **4.** $\frac{12}{38} = \frac{?}{19}$

A fraction is in **lowest terms** if the greatest common factor (GCF) of the numerator and the denominator is 1. You can write a fraction in lowest terms by dividing the numerator and denominator by their GCF.

example 2

Write each fraction in lowest terms.

a. $\frac{10}{15}$ b. $\frac{24}{42}$

solution

a. The GCF of 10 and 15 is 5. → $\frac{10}{15} = \frac{10 \div 5}{15 \div 5} = \frac{2}{3}$

b. $\frac{24}{42} = \frac{24 \div 6}{42 \div 6} = \frac{4}{7}$

your turn

Write each fraction in lowest terms.

5. $\frac{25}{50}$ **6.** $\frac{9}{21}$ **7.** $\frac{6}{22}$ **8.** $\frac{12}{20}$ **9.** $\frac{28}{35}$ **10.** $\frac{45}{80}$

practice exercises

practice for example 1 (page 128)

Replace each ? with the number that will make the fractions equivalent.

1. $\frac{3}{9} = \frac{?}{3}$ **2.** $\frac{16}{20} = \frac{4}{?}$ **3.** $\frac{?}{90} = \frac{15}{18}$ **4.** $\frac{14}{?} = \frac{7}{12}$

5. $\frac{?}{44} = \frac{1}{11}$ **6.** $\frac{6}{12} = \frac{?}{48}$ **7.** $\frac{?}{27} = \frac{2}{9}$ **8.** $\frac{18}{?} = \frac{2}{3}$

9. $\frac{18}{36} = \frac{2}{?}$ **10.** $\frac{52}{54} = \frac{?}{27}$ **11.** $\frac{35}{42} = \frac{5}{?}$ **12.** $\frac{39}{54} = \frac{?}{18}$

practice for example 2 (page 129)

Write each fraction in lowest terms.

13. $\frac{5}{10}$ **14.** $\frac{24}{36}$ **15.** $\frac{35}{49}$ **16.** $\frac{4}{52}$ **17.** $\frac{35}{42}$ **18.** $\frac{15}{30}$

19. $\frac{64}{72}$ **20.** $\frac{36}{38}$ **21.** $\frac{22}{121}$ **22.** $\frac{18}{45}$ **23.** $\frac{48}{60}$ **24.** $\frac{88}{100}$

mixed practice (pages 128–129)

Replace each ? with the number that will make the fractions equivalent.

25. $\frac{11}{21} = \frac{?}{63}$ **26.** $\frac{54}{72} = \frac{6}{?}$ **27.** $\frac{60}{100} = \frac{?}{25}$ **28.** $\frac{169}{?} = \frac{13}{12}$

29. $\frac{?}{120} = \frac{8}{15}$ **30.** $\frac{66}{72} = \frac{11}{?}$ **31.** $\frac{1000}{2000} = \frac{?}{20}$ **32.** $\frac{75}{?} = \frac{15}{16}$

Write each fraction in lowest terms.

33. $\frac{48}{64}$ 34. $\frac{21}{28}$ 35. $\frac{40}{90}$ 36. $\frac{36}{96}$ 37. $\frac{24}{30}$ 38. $\frac{18}{32}$

39. $\frac{15}{20}$ 40. $\frac{10}{40}$ 41. $\frac{26}{52}$ 42. $\frac{20}{55}$ 43. $\frac{34}{34}$ 44. $\frac{100}{100}$

45. Write four fractions that are equivalent to $\frac{1}{8}$.
46. Write four fractions that are equivalent to $\frac{36}{54}$.
47. Of the 120 members of the computer club, 110 have a personal computer at home. The part of the club having personal computers at home is $\frac{110}{120}$. Write this fraction in lowest terms.
48. Eight of the twenty-six members of the track-and-field team won first place at the track meet. The part of the team that won first place was $\frac{8}{26}$. Write this fraction in lowest terms.
49. Sue Ellen cut an apple into fourths. Bryan ate one half the apple. How many pieces of the apple did he eat?

review exercises

Find the LCM.

1. 4 and 6 2. 7 and 8 3. 3 and 27 4. 10 and 15
5. 14 and 21 6. 12 and 18 7. 4 and 9 8. 24 and 120

Compare. Replace each ? with >, <, or =.

9. 3.81 ? 38.1 10. 1711 ? 1171 11. 202.2 ? 202.2
12. 0.0032 ? 0.031 13. 0.99 ? 0.9 14. 1.005 ? 11.005
15. 4269 ? 420 16. 1345 ? 1345 17. 95,789 ? 9578

mental math

Find each answer mentally.

1. 2×99 2. $2\overline{)368}$ 3. $3 \times \$1.25$ 4. $4\overline{)644}$
5. $4 \times \$4.60$ 6. $50 \times 27 \times 2$ 7. $40 \times 12 \times 10$ 8. $6963 \div 3$
9. $245 + 148$ 10. $8.7 + 5.5$ 11. $\$5.20 - \$3.60$ 12. $\$9.25 - \$4.85$

7-4 COMPARING FRACTIONS

To compare fractions, use the following method.

1. If the fractions have a *common* denominator, just compare the numerators.
2. If the fractions have *different* denominators, rewrite the fractions as equivalent fractions with a common denominator. Then compare the numerators.

example 1

Compare. Replace each $\underline{?}$ with >, <, or =.

a. $\frac{6}{7} \underline{?} \frac{5}{7}$

$6 > 5$, so $\frac{6}{7} > \frac{5}{7}$.

b. $\frac{4}{9} \underline{?} \frac{11}{18}$ ← **The LCM of 9 and 18 is 18, so the *least common denominator* (LCD) is 18.**

↓ ↓

$\frac{8}{18} \underline{?} \frac{11}{18}$ → $\frac{8}{18} < \frac{11}{18}$, so $\frac{4}{9} < \frac{11}{18}$.

your turn

Compare. Replace each $\underline{?}$ with >, <, or =.

1. $\frac{5}{13} \underline{?} \frac{10}{13}$ **2.** $\frac{18}{23} \underline{?} \frac{36}{46}$ **3.** $\frac{6}{7} \underline{?} \frac{19}{21}$ **4.** $\frac{1}{4} \underline{?} \frac{1}{6}$

example 2

Compare. Replace each $\underline{?}$ with >, <, or =.

a. $2\frac{5}{9} \underline{?} \frac{22}{9}$ ← **Write the fraction as a mixed number.**

↓ ↓

$2\frac{5}{9} \underline{?} 2\frac{4}{9}$

$$\begin{array}{r} 2 \\ 9\overline{)22} \\ \underline{18} \\ 4 \end{array}$$

$2\frac{5}{9} > 2\frac{4}{9}$

So $2\frac{5}{9} > \frac{22}{9}$.

b. $\frac{14}{11} \underline{?} 1\frac{8}{11}$ ← **Write the mixed number as a fraction.**

↓ ↓

$\frac{14}{11} \underline{?} \frac{19}{11}$

$1\frac{8}{11} = \frac{1 \times 11 + 8}{11} = \frac{19}{11}$

$\frac{14}{11} < \frac{19}{11}$

So $\frac{14}{11} < 1\frac{8}{11}$.

your turn

Compare. Replace each $\underline{?}$ with >, <, or =.

5. $3\frac{1}{4} \underline{?} \frac{15}{4}$ **6.** $\frac{17}{5} \underline{?} 3\frac{4}{5}$ **7.** $1\frac{7}{8} \underline{?} \frac{7}{4}$ **8.** $\frac{9}{4} \underline{?} 2\frac{1}{6}$

example 3

Write the fractions in order from least to greatest: $\frac{5}{9}, \frac{3}{4}, \frac{2}{3}$

solution

$\frac{5}{9} = \frac{20}{36}$ ← **The LCM of 9, 4, and 3 is 36, so the LCD is 36.**

$\frac{3}{4} = \frac{27}{36}$

$\frac{2}{3} = \frac{24}{36}$

$\frac{20}{36} < \frac{24}{36} < \frac{27}{36}$

least to greatest: $\frac{5}{9}, \frac{2}{3}, \frac{3}{4}$

your turn

Write each set of numbers in order from least to greatest.

9. $\frac{3}{7}, \frac{1}{7}, \frac{5}{7}$
10. $\frac{5}{6}, \frac{3}{5}, \frac{1}{2}$
11. $\frac{1}{6}, \frac{2}{3}, \frac{5}{9}$
12. $1\frac{29}{39}, 1\frac{4}{13}, 1\frac{1}{3}$

practice exercises

practice for example 1 (page 131)

Compare. Replace each $\underline{?}$ with >, <, or =.

1. $\frac{15}{16} \underline{?} \frac{3}{16}$
2. $\frac{5}{14} \underline{?} \frac{9}{14}$
3. $\frac{26}{29} \underline{?} \frac{78}{87}$
4. $\frac{97}{103} \underline{?} \frac{99}{103}$
5. $\frac{2}{5} \underline{?} \frac{3}{10}$
6. $\frac{7}{9} \underline{?} \frac{2}{3}$
7. $\frac{2}{5} \underline{?} \frac{3}{4}$
8. $\frac{11}{12} \underline{?} \frac{7}{8}$

practice for example 2 (page 131)

Compare. Replace each $\underline{?}$ with >, <, or =.

9. $1\frac{9}{10} \underline{?} \frac{17}{10}$
10. $\frac{41}{22} \underline{?} 1\frac{21}{22}$
11. $2\frac{3}{4} \underline{?} \frac{15}{4}$
12. $4\frac{4}{9} \underline{?} \frac{39}{9}$
13. $\frac{18}{5} \underline{?} 3\frac{7}{10}$
14. $\frac{65}{12} \underline{?} 5\frac{3}{4}$
15. $4\frac{6}{7} \underline{?} \frac{19}{4}$
16. $7\frac{1}{3} \underline{?} \frac{36}{5}$

practice for example 3 (page 132)

Write each set of numbers in order from least to greatest.

17. $\frac{1}{4}, \frac{3}{4}, \frac{2}{5}$
18. $\frac{7}{8}, \frac{3}{4}, \frac{5}{8}$
19. $\frac{7}{10}, \frac{5}{6}, \frac{7}{15}$
20. $\frac{7}{20}, \frac{2}{5}, \frac{3}{8}$
21. $\frac{9}{11}, \frac{5}{22}, \frac{6}{11}$
22. $\frac{1}{20}, \frac{2}{25}, \frac{9}{10}$
23. $1\frac{1}{2}, 1\frac{3}{4}, 1\frac{2}{9}$
24. $2\frac{1}{2}, 2\frac{4}{7}, 1\frac{9}{35}$

mixed practice (pages 131–132)

Compare. Replace each $\underline{\ ?\ }$ with >, <, or =.

25. $\frac{4}{17} \underline{\ ?\ } \frac{15}{17}$
26. $\frac{23}{29} \underline{\ ?\ } \frac{21}{29}$
27. $\frac{3}{4} \underline{\ ?\ } \frac{1}{3}$
28. $\frac{5}{8} \underline{\ ?\ } \frac{7}{12}$
29. $\frac{25}{6} \underline{\ ?\ } 4\frac{5}{6}$
30. $\frac{34}{15} \underline{\ ?\ } 2\frac{2}{15}$
31. $4\frac{8}{52} \underline{\ ?\ } 4\frac{2}{13}$
32. $2\frac{9}{10} \underline{\ ?\ } 2\frac{11}{15}$

Write each set of numbers in order from least to greatest.

33. $\frac{5}{9}, \frac{2}{9}, \frac{7}{9}$
34. $\frac{4}{11}, \frac{1}{2}, \frac{7}{11}$
35. $\frac{2}{3}, \frac{1}{6}, \frac{1}{3}$
36. $\frac{2}{5}, \frac{5}{6}, \frac{4}{5}$
37. $\frac{5}{7}, \frac{3}{4}, \frac{1}{2}$
38. $\frac{5}{8}, \frac{6}{7}, \frac{1}{2}$
39. $2\frac{1}{3}, 2\frac{3}{5}, 1\frac{4}{5}$
40. $9\frac{8}{11}, 8\frac{6}{7}, 8\frac{2}{3}$

41. Every month Toni deposits one third of her salary in her checking account and one sixth of her salary in her savings account. In which account does she deposit more money?
42. Bill spends $\frac{1}{6}$ of his workday commuting to work and $\frac{1}{5}$ of his workday commuting from work. Does Bill spend more time commuting *to* or *from* work?

review exercises

Estimate.

1. 791.4 − 43.2
2. 0.03 × 96.4
3. 3371 ÷ 80
4. $5.12\overline{)28.8}$
5. 18.6 × 0.32
6. 724 + 214 + 366
7. $9.1\overline{)801.6}$
8. 9 × 327

calculator corner

You can use a calculator to compare fractions by finding their decimal equivalents.

example Compare $\frac{9}{13}$ and $\frac{12}{17}$.

solution Enter 9 ÷ 13 =. The result is 0.6923076.

Then enter 12 ÷ 17 =. The result is 0.7058823.

$0.6923076 < 0.7058823$, so $\frac{9}{13} < \frac{12}{17}$.

Use a calculator to compare. Replace each $\underline{\ ?\ }$ with >, <, or =.

1. $\frac{2}{7} \underline{\ ?\ } \frac{5}{16}$
2. $\frac{9}{19} \underline{\ ?\ } \frac{5}{11}$
3. $\frac{7}{16} \underline{\ ?\ } \frac{11}{23}$
4. $\frac{5}{38} \underline{\ ?\ } \frac{25}{190}$
5. $\frac{1}{21} \underline{\ ?\ } \frac{1}{24}$

7-5 ESTIMATING FRACTIONS

You can estimate a fraction by writing a simpler fraction that is close in value to the original fraction.

example 1

Write a simpler fraction as an estimate of each fraction: **a.** $\frac{6}{29}$ **b.** $\frac{39}{59}$

solution

a. $\frac{6}{29}$ ← **Because 6 × 5 = 30, the denominator is about 5 times the numerator.**

$\frac{6}{29} \rightarrow$ about $\frac{1}{5}$

b. $\frac{39}{59}$ is close to $\frac{40}{60}$

$\frac{39}{59} \rightarrow$ about $\frac{2}{3}$

your turn

Write a simpler fraction as an estimate of each fraction.

1. $\frac{9}{35}$ **2.** $\frac{7}{50}$ **3.** $\frac{15}{32}$ **4.** $\frac{30}{41}$ **5.** $\frac{98}{302}$ **6.** $\frac{79}{100}$

A fraction is close to:

0 when the numerator is very small compared to the denominator.

$\frac{1}{2}$ when the denominator is about twice the numerator.

1 when the numerator and denominator are nearly the same.

example 2

Tell whether each fraction is closest to 0, $\frac{1}{2}$, or 1: **a.** $\frac{2}{51}$ **b.** $\frac{8}{17}$ **c.** $\frac{31}{32}$

solution

a. 2 is very small compared to 51, so $\frac{2}{51}$ is closest to 0.

b. 17 is about twice as large as 8, so $\frac{8}{17}$ is closest to $\frac{1}{2}$.

c. 31 and 32 are nearly the same, so $\frac{31}{32}$ is closest to 1.

your turn

Tell whether each fraction is closest to 0, $\frac{1}{2}$, or 1.

7. $\frac{8}{9}$ **8.** $\frac{2}{21}$ **9.** $\frac{50}{52}$ **10.** $\frac{12}{25}$ **11.** $\frac{33}{64}$ **12.** $\frac{3}{100}$

practice exercises

practice for example 1 (page 134)

Write a simpler fraction as an estimate of each fraction.

1. $\frac{5}{14}$ **2.** $\frac{8}{33}$ **3.** $\frac{9}{17}$ **4.** $\frac{13}{27}$ **5.** $\frac{13}{36}$ **6.** $\frac{12}{55}$

7. $\frac{29}{40}$ **8.** $\frac{22}{34}$ **9.** $\frac{21}{100}$ **10.** $\frac{59}{80}$ **11.** $\frac{5}{39}$ **12.** $\frac{400}{601}$

practice for example 2 (page 134)

Tell whether each fraction is closest to 0, $\frac{1}{2}$, or 1.

13. $\frac{8}{15}$ **14.** $\frac{6}{7}$ **15.** $\frac{11}{23}$ **16.** $\frac{3}{19}$ **17.** $\frac{7}{15}$ **18.** $\frac{4}{75}$

19. $\frac{48}{50}$ **20.** $\frac{39}{81}$ **21.** $\frac{202}{203}$ **22.** $\frac{3}{104}$ **23.** $\frac{14}{15}$ **24.** $\frac{17}{35}$

mixed practice (page 134)

Choose the best estimate from: **a.** 0 **b.** $\frac{1}{4}$ **c.** $\frac{1}{2}$ **d.** $\frac{3}{4}$ **e.** 1

25. $\frac{15}{62}$ **26.** $\frac{117}{121}$ **27.** $\frac{1}{15}$ **28.** $\frac{297}{405}$ **29.** $\frac{9}{121}$ **30.** $\frac{25}{102}$

31. $\frac{24}{51}$ **32.** $\frac{53}{57}$ **33.** $\frac{52}{102}$ **34.** $\frac{9}{118}$ **35.** $\frac{60}{119}$ **36.** $\frac{27}{35}$

37. About what part of a year is 42 weeks?

38. This season Taylor Johnson played in 79 out of 162 baseball games. About what part of the baseball games did Taylor play?

39. The distance from Los Angeles to San Francisco is about 403 mi (miles). About what part of the trip have you completed after traveling 54 mi?

40. George Baylor's gross pay is $35,798 per year. George has received $12,381 of his gross pay so far this year. About what part of his gross pay has George received?

review exercises

Write the numeral form of each number.

1. 61 and 407 ten-thousandths

2. 280 million, 14 thousand, 9

3. one and fifty-two thousandths

4. four billion, seven hundred eleven

5. 75.6 thousand

6. 6.8 billion

SKILL REVIEW

Is the first number divisible by the second number? Write *Yes* or *No*. 7-1

1. 885; 5
2. 9732; 8
3. 100,640; 10
4. 66,548; 3
5. 3030; 9
6. 13,238; 2
7. 554,120; 4
8. 7,997,922; 3

Write the prime factorization of each number. 7-2

9. 30
10. 42
11. 210
12. 104
13. 130
14. 98

Find the GCF and the LCM.

15. 5 and 36
16. 21 and 39
17. 6, 10, and 12
18. 25, 75, and 100
19. A soap company puts a coupon for a free bar of soap in every 15th bar and a coupon for two free bars in every 21st bar. In which bar of soap will there be both coupons?

Replace each __?__ with the number that will make the fractions equivalent. 7-3

20. $\frac{?}{7} = \frac{16}{28}$
21. $\frac{10}{?} = \frac{100}{110}$
22. $\frac{?}{25} = \frac{2}{5}$
23. $\frac{36}{?} = \frac{3}{8}$

Write each fraction in lowest terms.

24. $\frac{4}{20}$
25. $\frac{49}{56}$
26. $\frac{24}{32}$
27. $\frac{15}{18}$
28. $\frac{99}{108}$
29. $\frac{75}{90}$
30. Of the 40 members of the math club, 24 are seniors. The part of the club that consists of seniors is $\frac{24}{40}$. Write this fraction in lowest terms.

Compare. Replace each __?__ with >, <, or =. 7-4

31. $\frac{5}{21}$ __?__ $\frac{8}{21}$
32. $\frac{6}{7}$ __?__ $\frac{43}{49}$
33. $\frac{15}{4}$ __?__ $3\frac{1}{4}$
34. $\frac{25}{6}$ __?__ $2\frac{7}{8}$

Write each set of numbers in order from least to greatest.

35. $\frac{5}{9}, \frac{2}{9}, \frac{2}{3}$
36. $\frac{11}{12}, \frac{5}{6}, \frac{1}{4}$
37. $\frac{2}{13}, \frac{2}{52}, \frac{3}{26}$
38. $2\frac{4}{7}, 3\frac{2}{7}, 2\frac{9}{14}$

Tell whether each fraction is closest to 0, $\frac{1}{2}$, or 1. 7-5

39. $\frac{108}{109}$
40. $\frac{7}{13}$
41. $\frac{67}{68}$
42. $\frac{3}{100}$
43. $\frac{76}{149}$
44. $\frac{11}{200}$
45. Of the 89 employees of Hart Construction, 57 are enrolled in the dental plan. About what fraction of the number of employees is enrolled in this plan?

APPLICATION

7-6 READING RECIPES

Jeannette is planning to bake pumpkin bread. To read the recipe she needs to be familiar with these abbreviations:

c (cup)   tsp (teaspoon)

Jeannette must check her ingredients to make sure she has enough available to make the recipe.

Pumpkin Bread

$3\frac{1}{3}$ c flour	1 c oil
2 tsp baking soda	4 eggs
$1\frac{1}{2}$ tsp salt	$\frac{2}{3}$ c water
$2\frac{1}{8}$ tsp nutmeg	2 c pumpkin
3 c sugar	$\frac{3}{4}$ c walnuts
$1\frac{3}{4}$ tsp cinnamon	

example

Jeannette has $3\frac{1}{2}$ c of flour. Does she have enough flour to make the pumpkin bread?

solution

Jeannette needs $3\frac{1}{3}$ c of flour to make the recipe. Compare $3\frac{1}{3}$ and $3\frac{1}{2}$. The whole numbers are the same, so compare $\frac{1}{3}$ and $\frac{1}{2}$.

$$\frac{1}{3} \;\underline{\ ?\ }\; \frac{1}{2} \quad \leftarrow \text{The LCM of 3 and 2 is 6, so the LCD is 6.}$$

$$\downarrow \qquad \downarrow$$

$$\frac{2}{6} \;\underline{\ ?\ }\; \frac{3}{6} \rightarrow \frac{2}{6} < \frac{3}{6}, \text{ so } 3\frac{1}{3} < 3\frac{1}{2}.$$

Jeannette has enough flour to make the recipe.

exercises

Use the recipe above.

1. Jeannette has $2\frac{3}{4}$ tsp of nutmeg. Does the recipe require *more* or *less* nutmeg?
2. Jeannette has an already-opened package of walnuts that contains $\frac{3}{8}$ c of walnuts and an unopened package that contains 2 c of walnuts. Will she need to open the 2-c package to make the recipe?
3. Jeannette thought she put $1\frac{3}{4}$ tsp cinnamon into the mixture. She actually put in $1\frac{5}{8}$ tsp cinnamon. Did she put in *too much* or *too little* cinnamon?
4. Jeannette's measuring cup is marked in fourths of a cup. She filled the cup to the $\frac{3}{4}$ mark. To get the right amount of water, should she add a little more water to the measuring cup or should she pour a little out? Explain.

APPLICATION

7-7 PHOTOGRAPHY

Photographers always try to achieve good lighting in their photographs. Good lighting depends on two controls. One control is the **shutter speed.** The shutter speed controls the amount of time that light is admitted to the film. On a camera, the speeds are usually marked with a series of numbers such as 1000, 500, 250, 125, 60, 30, 15, 8, 4, 2, and 1. Think of these numbers as denominators of fractions. For example, "30" means one thirtieth of a second.

When photographing something moving very fast, you use a speed of $\frac{1}{500}$ or $\frac{1}{1000}$. To photograph something standing still, you use a slower speed of $\frac{1}{125}$ or $\frac{1}{60}$.

←faster slower→

shutter speed	$\frac{1}{1000}$	$\frac{1}{500}$	$\frac{1}{250}$	$\frac{1}{125}$	$\frac{1}{60}$	$\frac{1}{30}$	$\frac{1}{15}$	$\frac{1}{8}$

The other control that affects good lighting is the **aperture.** This is the size of the lens opening. Aperture is measured in **f-stops,** or **f-numbers.** The f-numbers can also be thought of as the denominators of fractions. The number f4 ($\frac{1}{4}$) represents a larger aperture than the number f16 ($\frac{1}{16}$). The f-stops control how much of the photograph will be in sharp focus. You can blur a background by using a large aperture, like f1.8. If you want to put as much as possible in focus, use one of the smaller apertures, like f16.

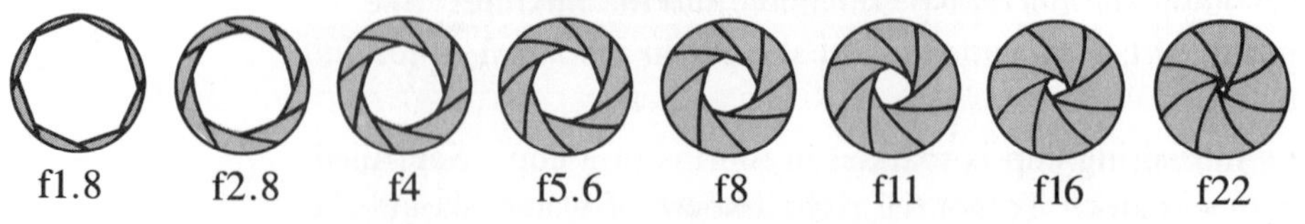

f1.8 f2.8 f4 f5.6 f8 f11 f16 f22

Some cameras have fixed controls for shutter speed and aperture. Other cameras automatically adjust the controls.

example

Robert wants to take a picture of a baseball player running toward first base. Should Robert use a shutter speed of $\frac{1}{1000}$ or a speed of $\frac{1}{30}$?

solution

To capture the feeling of motion, he will need a fast shutter speed. The shutter speed Robert should use is $\frac{1}{1000}$.

exercises

1. Kathleen Ross wants to get a picture of her daughter Lisa at the height of Lisa's jump. Should Kathleen Ross try a shutter speed of $\frac{1}{500}$ or a speed of $\frac{1}{15}$? Explain.
2. Kimberly has finally managed to have her entire family pose for a picture. Should she use a shutter speed of $\frac{1}{30}$ or a speed of $\frac{1}{1000}$? Explain.
3. Roland plans to take a picture of a young woman sitting on a park bench. Should he use a slow shutter speed to photograph her? Explain.
4. One of Jeannie's photographs of a football player in midair is blurred. She used a shutter speed of $\frac{1}{125}$. Next time, should Jeannie try a speed of $\frac{1}{500}$ or a speed of $\frac{1}{30}$? Explain.
5. Erica Baldachino would like to photograph a moving train. Should she use a shutter speed of $\frac{1}{1000}$ or a speed of $\frac{1}{60}$? Explain.
6. Jeffrey Fraser wants a picture of a crowded city street. He would like to put as much activity as possible in focus. Should Jeffrey set the aperture at f4 or f16? Explain.
7. Tung is adjusting his camera to photograph both his brand new sports car and his girlfriend, who is standing behind the car. Should he use a lens opening of f2.8 or f11? Explain.
8. Mark Gregory is planning to take a spectacular picture of the fireworks finale. Should Mark set the aperture at f1.8 or f8? Explain.

CAREER

AUTO MECHANIC

Sal Rivera is an auto mechanic. He often works with a ratchet-and-socket set. The sockets are made in sizes that differ by sixteenths of an inch.

example

Sal is installing a new carburetor. The $\frac{5}{8}$-in. (inch) socket is too large and the $\frac{1}{2}$-in. socket is too small. What size socket does Sal need?

solution

Since $\frac{5}{8} = \frac{10}{16}$ and $\frac{1}{2} = \frac{8}{16}$, Sal needs a $\frac{9}{16}$-in. socket.

exercises

1. Sal is putting a new water pump in a van. If the $\frac{7}{16}$-in. socket is too small and the $\frac{9}{16}$-in. socket is too large, what size socket does Sal need?

2. Sal bought a six-piece socket set that included sizes $\frac{15}{16}$ through $1\frac{1}{4}$ in. Is there a socket in this set that he can use for a $\frac{1}{4}$-in. bolt?

3. Sal wants to install shock absorbers on a car. The $\frac{5}{8}$-in. socket is too small and the $\frac{3}{4}$-in. socket is too large. What size socket should Sal try?

4. Sal has a special set of sockets that range from $\frac{1}{16}$ in. to $\frac{1}{2}$ in. Their sizes in inches are $\frac{3}{16}$, $\frac{1}{2}$, $\frac{1}{16}$, $\frac{3}{8}$, $\frac{1}{4}$, $\frac{5}{16}$, $\frac{7}{16}$, $\frac{1}{8}$. Arrange these sizes in order from smallest to largest.

CHAPTER REVIEW

vocabulary vo·cab·u·lar·y

Choose the correct word or phrase to complete each sentence.

1. For the numbers 24 and 36, 12 is the *(greatest common factor, least common multiple)*.
2. A whole number greater than 1 that has exactly two factors is *(prime, composite)*.

skills

Is the first number divisible by the second number? Write *Yes* or *No*.

3. 936; 9 **4.** 1040; 5 **5.** 742; 10 **6.** 14,248; 8

Find the GCF and the LCM.

7. 24 and 48 **8.** 9 and 21 **9.** 3 and 8 **10.** 10 and 15

11. 42 and 63 **12.** 100 and 125 **13.** 14, 28, and 98 **14.** 10, 12, and 30

Write each fraction in lowest terms.

15. $\frac{4}{8}$ **16.** $\frac{8}{12}$ **17.** $\frac{25}{55}$ **18.** $\frac{9}{66}$ **19.** $\frac{24}{72}$ **20.** $\frac{14}{84}$

Write each set of numbers in order from least to greatest.

21. $\frac{5}{7}, \frac{1}{3}, \frac{1}{5}$ **22.** $\frac{5}{6}, \frac{1}{2}, \frac{3}{5}$ **23.** $2\frac{1}{4}, 2\frac{7}{10}, 2\frac{2}{5}$

Write a simpler fraction as an estimate of each fraction.

24. $\frac{6}{11}$ **25.** $\frac{5}{26}$ **26.** $\frac{4}{17}$ **27.** $\frac{49}{76}$ **28.** $\frac{17}{33}$ **29.** $\frac{11}{34}$

30. Edward must place 68 chairs on a stage for a meeting. He wants to have 9 chairs in each row. Is it possible for him to place the same number of chairs in each row?

31. Arnaud wants to take a picture of some dolphins jumping out of the water. Should he use a shutter speed of $\frac{1}{20}$ or a speed of $\frac{1}{500}$?

32. Monica wants to take a picture of her cousin's large flower garden. Should she use a lens opening of f4 or f11?

33. Rita's cranberry cake recipe calls for $2\frac{1}{4}$ c of flour. She has already measured $2\frac{1}{3}$ c of flour. Does she need *more* or *less* flour?

CHAPTER TEST

Is the first number divisible by the second number? Write *Yes* or *No*.

7-1

1. 63; 3
2. 280; 5
3. 3264; 9
4. 81,162; 2
5. 4995; 10
6. 192; 4
7. 1400; 8
8. 608,000; 3
9. Carol has 132 record albums. She wants to arrange the albums in stacks so that each stack will have the same number of albums. Is it possible for Carol to have exactly five albums in each stack?

Write the prime factorization of each number.

7-2

10. 33
11. 60
12. 78
13. 150
14. 165
15. 330

Find the GCF and the LCM.

16. 9 and 15
17. 6 and 9
18. 8 and 9
19. 25 and 20
20. 18 and 12
21. 18, 24, and 28
22. 5, 9, and 15
23. 12, 18, and 21

Replace each __?__ with the number that will make the fractions equivalent.

7-3

24. $\frac{4}{5} = \frac{?}{15}$
25. $\frac{3}{4} = \frac{21}{?}$
26. $\frac{?}{7} = \frac{40}{56}$
27. $\frac{9}{81} = \frac{1}{?}$

Write each fraction in lowest terms.

28. $\frac{12}{24}$
29. $\frac{16}{30}$
30. $\frac{35}{49}$
31. $\frac{18}{27}$
32. $\frac{36}{128}$

Compare. Replace __?__ with >, <, or =.

7-4

33. $\frac{3}{7}$ __?__ $\frac{5}{7}$
34. $\frac{20}{21}$ __?__ $\frac{100}{105}$
35. $\frac{7}{10}$ __?__ $\frac{2}{5}$
36. $\frac{14}{15}$ __?__ $\frac{5}{6}$
37. $\frac{13}{7}$ __?__ $1\frac{19}{21}$
38. $1\frac{3}{4}$ __?__ $\frac{5}{2}$
39. $5\frac{2}{9}$ __?__ $\frac{52}{9}$
40. $\frac{33}{16}$ __?__ $2\frac{7}{32}$
41. To receive a passing grade, Randall needed to get at least $\frac{2}{3}$ of the answers right on the history exam. He got $\frac{3}{4}$ of the answers right. Did Randall receive a passing grade?

Tell whether each fraction is closest to 0, $\frac{1}{2}$, or 1.

7-5

42. $\frac{210}{213}$
43. $\frac{21}{39}$
44. $\frac{46}{47}$
45. $\frac{2}{301}$
46. $\frac{59}{109}$

7-6

47. Marvin's banana bread recipe calls for $\frac{1}{2}$ c chopped nuts. He has already measured $\frac{2}{3}$ c. Does he need *more* or *less* chopped nuts?

7-7

48. A photographer wants to take a picture of a figure skater in midair. Should he use a shutter speed of $\frac{1}{500}$ or $\frac{1}{60}$?

To deal with measurements efficiently, a person working in the building trades needs to be good at multiplying and dividing with fractions.

CHAPTER 8

MULTIPLYING AND DIVIDING FRACTIONS

8-1 MULTIPLYING WITH FRACTIONS AND MIXED NUMBERS

To multiply with fractions, use the following method.

1. Write any whole number as a fraction.
2. Multiply the numerators.
3. Multiply the denominators.
4. Write the product in lowest terms.

$$\frac{1}{3} \times \frac{2}{3} = \frac{1 \times 2}{3 \times 3} = \frac{2}{9}$$

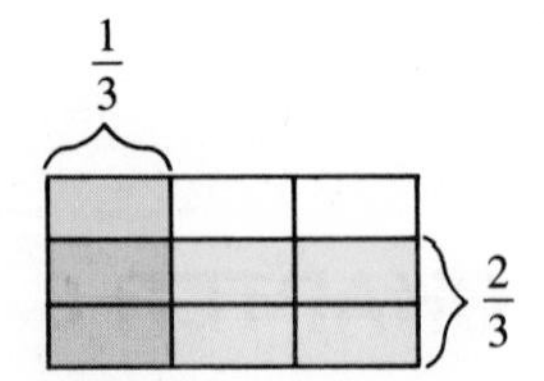

When a numerator and a denominator have a common factor, you can divide them by this factor to simplify the multiplication.

example 1

Write each product in lowest terms: **a.** $\frac{4}{9} \times \frac{6}{7}$ **b.** $4 \times \frac{3}{5}$

solution

a. $\frac{4}{9} \times \frac{6}{7} = \frac{4 \times \overset{2}{\cancel{6}}}{\underset{3}{\cancel{9}} \times 7} = \frac{8}{21}$ ← **Divide 6 and 9 by 3.**

b. $4 \times \frac{3}{5} = \frac{4}{1} \times \frac{3}{5} = \frac{4 \times 3}{1 \times 5} = \frac{12}{5} = 2\frac{2}{5}$ ← **Write 4 as $\frac{4}{1}$. Write the answer as a mixed number.**

your turn

Write each product in lowest terms.

1. $\frac{4}{7} \times \frac{5}{9}$ **2.** $\frac{5}{12} \times \frac{8}{15}$ **3.** $5 \times \frac{2}{3}$ **4.** $\frac{1}{8} \times 8$

To multiply with mixed numbers, first write each mixed number as a fraction.

example 2

Write each product in lowest terms: **a.** $8 \times 3\frac{3}{4}$ **b.** $2\frac{1}{10} \times 1\frac{1}{3}$

solution

a. $8 \times 3\frac{3}{4}$

$$\frac{8}{1} \times \frac{15}{4} = \frac{\overset{2}{\cancel{8}} \times 15}{1 \times \underset{1}{\cancel{4}}} = \frac{30}{1} = 30$$

b. $2\frac{1}{10} \times 1\frac{1}{3}$

$$\frac{21}{10} \times \frac{4}{3} = \frac{\overset{7}{\cancel{21}} \times \overset{2}{\cancel{4}}}{\underset{5}{\cancel{10}} \times \underset{1}{\cancel{3}}} = \frac{14}{5} = 2\frac{4}{5}$$

your turn

Write each product in lowest terms.

5. $\frac{4}{5} \times 3\frac{1}{3}$ **6.** $1\frac{3}{8} \times 4$ **7.** $6\frac{2}{5} \times 1\frac{3}{4}$ **8.** $10\frac{1}{2} \times \frac{2}{3}$

practice exercises

practice for example 1 (page 144)

Write each product in lowest terms.

1. $\frac{2}{7} \times \frac{4}{5}$ **2.** $\frac{7}{10} \times \frac{3}{4}$ **3.** $5 \times \frac{1}{6}$ **4.** $\frac{2}{7} \times 3$

5. $\frac{2}{5} \times \frac{10}{13}$ **6.** $\frac{5}{16} \times \frac{4}{9}$ **7.** $\frac{2}{9} \times 18$ **8.** $3 \times \frac{5}{6}$

9. $\frac{2}{7} \times \frac{7}{12}$ **10.** $\frac{10}{21} \times \frac{3}{4}$ **11.** $\frac{1}{20} \times 28$ **12.** $16 \times \frac{0}{8}$

practice for example 2 (pages 144–145)

Write each product in lowest terms.

13. $\frac{1}{2} \times 1\frac{3}{5}$ **14.** $2\frac{4}{5} \times \frac{1}{7}$ **15.** $12 \times 2\frac{1}{4}$ **16.** $2\frac{5}{6} \times 6$

17. $4\frac{1}{2} \times 1\frac{2}{3}$ **18.** $1\frac{2}{15} \times 3\frac{1}{3}$ **19.** $1\frac{7}{12} \times 6$ **20.** $\frac{2}{5} \times 12\frac{1}{2}$

21. $1\frac{1}{6} \times 1\frac{3}{7}$ **22.** $3\frac{1}{3} \times 4\frac{4}{5}$ **23.** $6\frac{3}{4} \times 1\frac{1}{3}$ **24.** $1\frac{5}{18} \times 1\frac{11}{16}$

mixed practice (pages 144–145)

Write each product in lowest terms.

25. $\frac{5}{7} \times \frac{1}{3}$ **26.** $\frac{4}{9} \times 24$ **27.** $1\frac{5}{9} \times 1\frac{4}{5}$ **28.** $18 \times 2\frac{1}{6}$

29. $10 \times 2\frac{1}{4}$ **30.** $\frac{5}{9} \times \frac{2}{3}$ **31.** $\frac{2}{33} \times \frac{11}{16}$ **32.** $\frac{5}{9} \times 8$

33. $1\frac{7}{12} \times 2\frac{2}{5}$ **34.** $1\frac{1}{5} \times 3\frac{1}{10}$ **35.** $7 \times \frac{1}{7}$ **36.** $6\frac{2}{3} \times \frac{3}{4}$

37. $\frac{13}{40} \times 8$ **38.** $1\frac{1}{8} \times \frac{8}{9}$ **39.** $3\frac{3}{4} \times 1\frac{3}{5}$ **40.** $\frac{5}{8} \times \frac{2}{5}$

41. $12 \times \frac{0}{4}$ **42.** $3\frac{1}{9} \times 1\frac{5}{7}$ **43.** $5\frac{3}{5} \times \frac{7}{8}$ **44.** $\frac{4}{15} \times \frac{0}{3}$

45. $\frac{3}{10} \times \frac{2}{7} \times \frac{5}{12}$ **46.** $\frac{3}{4} \times \frac{3}{5} \times \frac{28}{33}$ **47.** $6 \times 1\frac{17}{30} \times \frac{2}{3}$ **48.** $7\frac{7}{10} \times 2\frac{1}{2} \times 2\frac{8}{11}$

Write each product in lowest terms.

49. a. $\frac{1}{5} \times 15$
 b. $\frac{2}{5} \times 15$
 c. $\frac{3}{5} \times 15$
 d. $\frac{4}{5} \times 15$

50. a. $\frac{1}{7} \times 42$
 b. $\frac{2}{7} \times 42$
 c. $\frac{3}{7} \times 42$
 d. $\frac{4}{7} \times 42$

51. a. $\frac{1}{8} \times 32$
 b. $\frac{3}{8} \times 32$
 c. $\frac{5}{8} \times 32$
 d. $\frac{7}{8} \times 32$

52. a. $\frac{1}{12} \times 60$
 b. $\frac{5}{12} \times 60$
 c. $\frac{7}{12} \times 60$
 d. $\frac{11}{12} \times 60$

53. JoAnn works at a service station. She has to change the oil in six cars, each of which holds $4\frac{1}{2}$ qt (quarts) of oil. How many quarts of oil will JoAnn need?

54. Kelly's Supermarket has 24 employees, $\frac{3}{8}$ of whom are high-school students. Two thirds of these students work on Saturday. How many high-school students work on Saturday?

review exercises

Write a simpler fraction as an estimate of each fraction.

1. $\frac{4}{23}$ 2. $\frac{7}{13}$ 3. $\frac{5}{32}$ 4. $\frac{24}{99}$ 5. $\frac{6}{59}$ 6. $\frac{31}{42}$

Tell whether each fraction is closest to 0, $\frac{1}{2}$, or 1.

7. $\frac{5}{8}$ 8. $\frac{3}{19}$ 9. $\frac{4}{5}$ 10. $\frac{11}{13}$ 11. $\frac{11}{100}$ 12. $\frac{31}{60}$

mental math

To multiply a whole number by a mixed number mentally, use the distributive property.

$$4 \times 5\frac{1}{2} = 4 \times \left(5 + \frac{1}{2}\right) = (4 \times 5) + \left(4 \times \frac{1}{2}\right) = 20 + 2 = 22$$

$$3 \times 2\frac{1}{5} = 3 \times \left(2 + \frac{1}{5}\right) = (3 \times 2) + \left(3 \times \frac{1}{5}\right) = 6 + \frac{3}{5} = 6\frac{3}{5}$$

Multiply mentally.

1. $6 \times 3\frac{1}{6}$ 2. $8 \times 4\frac{1}{4}$ 3. $5 \times 2\frac{1}{7}$ 4. $3 \times 2\frac{1}{10}$

5. $2 \times 1\frac{1}{8}$ 6. $3 \times 9\frac{1}{12}$ 7. $9 \times 1\frac{2}{3}$ 8. $10 \times 7\frac{3}{5}$

8-2 ESTIMATING PRODUCTS OF FRACTIONS AND MIXED NUMBERS

To estimate a product of mixed numbers, round each factor *to the nearest whole number*. If the fractional part of a mixed number is greater than or equal to $\frac{1}{2}$, round up. Otherwise, round down.

example 1

Estimate by rounding: a. $3\frac{1}{2} \times 6\frac{5}{7}$ b. $5\frac{2}{9} \times 1\frac{3}{4}$

solution

a. $3\frac{1}{2} \times 6\frac{5}{7} \rightarrow \underbrace{4 \times 7}_{\text{about 28}}$

b. $5\frac{2}{9} \times 1\frac{3}{4} \rightarrow \underbrace{5 \times 2}_{\text{about 10}}$

your turn

Estimate by rounding.

1. $1\frac{2}{5} \times 8\frac{3}{7}$
2. $6\frac{5}{8} \times 5\frac{1}{3}$
3. $2\frac{3}{10} \times 5\frac{1}{2}$
4. $11\frac{2}{3} \times 3\frac{5}{9}$

If one factor is a fraction less than 1, you can often use *compatible numbers* to estimate the product.

example 2

Estimate using compatible numbers: a. $\frac{1}{4} \times 22\frac{5}{6}$ b. $\frac{21}{30} \times 294$

solution

a. $\frac{1}{4} \times 22\frac{5}{6}$

$\downarrow \quad \downarrow$

$\underbrace{\frac{1}{4} \times 24}_{\text{about 6}}$ ← **24 is a whole number close to $22\frac{5}{6}$ that is compatible with $\frac{1}{4}$.**

b. $\frac{21}{30} \times 294$

$\downarrow \quad \downarrow$

$\underbrace{\frac{2}{3} \times 300}_{\text{about 200}}$ ← **Choose compatible numbers that are close to the given factors.**

your turn

Estimate using compatible numbers.

5. $\frac{1}{5} \times 33\frac{2}{7}$
6. $\frac{3}{17} \times 302$
7. $41 \times \frac{20}{49}$
8. $55\frac{1}{3} \times \frac{5}{34}$

practice exercises

practice for example 1 (page 147)

Estimate by rounding.

1. $7\frac{1}{4} \times 2\frac{1}{2}$
2. $9\frac{1}{2} \times 2\frac{1}{3}$
3. $1\frac{5}{11} \times 4\frac{3}{8}$
4. $2\frac{2}{7} \times 5\frac{7}{9}$
5. $7\frac{2}{3} \times 8\frac{3}{5}$
6. $4\frac{1}{5} \times 12\frac{3}{10}$
7. $10\frac{7}{12} \times 9\frac{4}{9}$
8. $6\frac{3}{7} \times 3\frac{4}{5}$

practice for example 2 (page 147)

Estimate using compatible numbers.

9. $\frac{1}{5} \times 393$
10. $\frac{2}{3} \times 28$
11. $28 \times \frac{74}{99}$
12. $\frac{6}{23} \times 392$
13. $\frac{1}{6} \times 43\frac{2}{9}$
14. $13\frac{3}{5} \times \frac{1}{7}$
15. $\frac{8}{21} \times 5\frac{1}{3}$
16. $\frac{7}{16} \times 22\frac{1}{3}$

mixed practice (page 147)

Choose the letter of the best estimate for each product.

17. $3\frac{3}{5} \times 5\frac{2}{5}$ **a.** 15 **b.** 9 **c.** 20 **d.** 24
18. $9\frac{2}{7} \times 6\frac{2}{3}$ **a.** 54 **b.** 63 **c.** 16 **d.** 70
19. Louis roasts a turkey $\frac{1}{4}$ h for each pound of turkey. About how long will he roast a $15\frac{1}{2}$-lb (pound) turkey?
20. Members of the Westfield Walking Club walk at an average speed of $3\frac{1}{2}$ mi/h (miles per hour). About how far will they walk in $2\frac{1}{4}$ h?
21. In recent years about two thirds of the senior class have gone to the class outing. There are 381 seniors this year. If the trend continues, about how many seniors will go to the outing this year?
22. A recipe for 12 muffins requires $1\frac{1}{4}$ c of whole-wheat flour. You want to make 30 muffins. About how many cups of whole-wheat flour will you need?

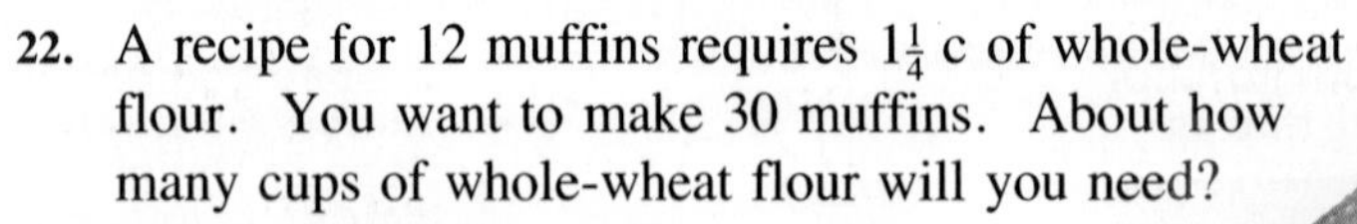

review exercises

Write each fraction in lowest terms.

1. $\frac{8}{48}$
2. $\frac{21}{35}$
3. $\frac{38}{100}$
4. $\frac{15}{250}$

8-3 DIVIDING WITH FRACTIONS AND MIXED NUMBERS

Reciprocals are two numbers whose product is 1. You find the reciprocal of a fraction by interchanging the numerator and the denominator. Every number except zero has exactly one reciprocal. Zero has no reciprocal.

example 1

Write the reciprocal of each number: **a.** $\frac{4}{5}$ **b.** 16 **c.** $2\frac{3}{7}$

solution

a. $\frac{4}{5} \rightarrow \frac{5}{4}$

$\frac{5}{4}$ is the reciprocal.

b. $16 = \frac{16}{1} \rightarrow \frac{1}{16}$

$\frac{1}{16}$ is the reciprocal.

c. $2\frac{3}{7} = \frac{17}{7} \rightarrow \frac{7}{17}$

$\frac{7}{17}$ is the reciprocal.

your turn

Write the reciprocal of each number, if possible.

1. $\frac{1}{8}$ **2.** $\frac{3}{10}$ **3.** 0 **4.** 9 **5.** $1\frac{5}{8}$ **6.** $5\frac{2}{3}$

To divide with fractions, use the following method.

1. Multiply the dividend by the reciprocal of the divisor.

2. Write the product in lowest terms.

example 2

Write each quotient in lowest terms: **a.** $\frac{3}{4} \div \frac{1}{3}$ **b.** $12 \div \frac{9}{10}$

solution

a. $\frac{3}{4} \div \frac{1}{3} = \frac{3}{4} \times \frac{3}{1}$ ← **The divisor is $\frac{1}{3}$. The reciprocal of $\frac{1}{3}$ is 3.**

$= \frac{9}{4} = 2\frac{1}{4}$

b. $12 \div \frac{9}{10} = \frac{12}{1} \times \frac{10}{9}$ ← **The reciprocal of $\frac{9}{10}$ is $\frac{10}{9}$.**

$= \frac{\overset{4}{\cancel{12}}}{1} \times \frac{10}{\underset{3}{\cancel{9}}} = \frac{40}{3} = 13\frac{1}{3}$

your turn

Write each quotient in lowest terms.

7. $\frac{2}{9} \div \frac{4}{5}$ **8.** $\frac{7}{12} \div \frac{7}{18}$ **9.** $\frac{6}{11} \div 8$ **10.** $5 \div \frac{1}{4}$

To divide with mixed numbers, first write each mixed number as a fraction. Then divide.

example 3

Write each quotient in lowest terms: **a.** $3\frac{3}{4} \div 2\frac{1}{2}$ **b.** $1\frac{5}{9} \div 7$

solution

a. $3\frac{3}{4} \div 2\frac{1}{2} = \frac{15}{4} \div \frac{5}{2}$

$= \frac{\overset{3}{\cancel{15}}}{\underset{2}{\cancel{4}}} \times \frac{\overset{1}{\cancel{2}}}{\underset{1}{\cancel{5}}} = \frac{3}{2} = 1\frac{1}{2}$

b. $1\frac{5}{9} \div 7 = \frac{14}{9} \div \frac{7}{1}$

$= \frac{\overset{2}{\cancel{14}}}{9} \times \frac{1}{\underset{1}{\cancel{7}}} = \frac{2}{9}$

your turn

Write each quotient in lowest terms.

11. $8 \div 2\frac{2}{5}$ **12.** $1\frac{7}{8} \div \frac{9}{16}$ **13.** $3\frac{2}{3} \div 1\frac{1}{4}$ **14.** $5\frac{4}{9} \div 3\frac{1}{2}$

practice exercises

practice for example 1 (page 149)

Write the reciprocal of each number, if possible.

1. $\frac{2}{9}$ **2.** $\frac{1}{6}$ **3.** $\frac{12}{5}$ **4.** $\frac{24}{7}$ **5.** $3\frac{1}{2}$

6. $9\frac{3}{10}$ **7.** 4 **8.** 8 **9.** 0 **10.** 1

practice for example 2 (page 149)

Write each quotient in lowest terms.

11. $12 \div \frac{6}{7}$ **12.** $10 \div \frac{3}{4}$ **13.** $\frac{7}{8} \div \frac{4}{5}$ **14.** $\frac{8}{9} \div 6$ **15.** $\frac{7}{8} \div 7$

16. $\frac{3}{13} \div \frac{1}{13}$ **17.** $\frac{5}{6} \div \frac{1}{2}$ **18.** $\frac{16}{25} \div \frac{4}{5}$ **19.** $\frac{16}{27} \div \frac{8}{15}$ **20.** $\frac{3}{5} \div \frac{3}{5}$

practice for example 3 (page 150)

Write each quotient in lowest terms.

21. $1\frac{7}{8} \div 8$ **22.** $\frac{5}{6} \div 3\frac{3}{4}$ **23.** $2\frac{3}{4} \div \frac{7}{8}$ **24.** $33 \div 5\frac{1}{2}$

25. $4\frac{1}{5} \div 1\frac{2}{5}$ **26.** $6\frac{3}{4} \div 1\frac{1}{2}$ **27.** $10\frac{5}{6} \div 1\frac{7}{18}$ **28.** $1\frac{1}{8} \div 2\frac{13}{16}$

mixed practice (pages 149–150)

Write each quotient in lowest terms.

29. $4 \div \frac{1}{5}$ **30.** $6 \div \frac{1}{7}$ **31.** $12 \div \frac{1}{2}$ **32.** $8 \div \frac{1}{3}$

33. In Exercises 29–32, you divided a whole number by a *unit fraction,* a fraction with numerator 1. Are the quotients fractions, whole numbers, or mixed numbers?

34. Write two exercises similar to Exercises 29–32. Find the quotients.

35. You need to use $1\frac{1}{2}$ oz (ounces) of a certain type of oatmeal to make one serving. How many servings are there in a 15-oz box?

36. Wanda works five days per week. Each day she works $7\frac{1}{2}$ h. She earns $300 per week. How much does Wanda earn per hour?

review exercises

Find each answer.

1. 65×302 **2.** $15.42 \div 6$ **3.** $\$106 - \$29.14$ **4.** $16\overline{)15,088}$

5. 0.07×10.4 **6.** $0.17\overline{)11.798}$ **7.** $3842 + 293$ **8.** 2.9×5.37

calculator corner

Some calculators allow you to perform operations with fractions in fraction form. If you do not have this type of calculator, however, you can still multiply fractions. Just remember that multiplication and division are done in the order in which they occur.

example $\frac{2}{3} \times \frac{9}{10}$

solution Enter 2 ÷ 3 × 9 ÷ 10 =.

The result is either 0.5999999 or 0.6.

If your calculator gives the result as 0.5999999, then enter 3 × 10 = M+ 2 × 9 ÷ MR =.

The result is 0.6.

Multiply using a calculator.

1. $\frac{7}{10} \times \frac{1}{2}$ **2.** $\frac{2}{5} \times \frac{7}{16}$ **3.** $\frac{2}{7} \times \frac{14}{25}$ **4.** $\frac{17}{18} \times \frac{9}{20}$

5. $\frac{3}{4} \times \frac{1}{8}$ **6.** $\frac{8}{15} \times \frac{3}{4}$ **7.** $\frac{1}{6} \times 4\frac{4}{5}$ **8.** $2\frac{1}{8} \times \frac{3}{5}$

8-4 ESTIMATING QUOTIENTS OF FRACTIONS AND MIXED NUMBERS

You can estimate a quotient of mixed numbers by first rounding the divisor to the nearest whole number and then using compatible numbers.

example 1

Estimate using compatible numbers: a. $9\frac{1}{7} \div 5\frac{1}{4}$ b. $35\frac{3}{8} \div 3\frac{1}{2}$

solution

a. $9\frac{1}{7} \div 5\frac{1}{4}$ ← **Round $5\frac{1}{4}$ to 5.**

$10 \div 5$ ← **10 is compatible with 5.**

about 2

b. $35\frac{3}{8} \div 3\frac{1}{2}$ ← **Round $3\frac{1}{2}$ to 4.**

$36 \div 4$ ← **36 is compatible with 4.**

about 9

your turn

Estimate using compatible numbers.

1. $26\frac{3}{10} \div 6\frac{3}{4}$
2. $6\frac{4}{5} \div 2\frac{11}{12}$
3. $35\frac{1}{9} \div 6\frac{1}{6}$
4. $7\frac{1}{3} \div 1\frac{1}{8}$

If the divisor is a fraction, write a simpler fraction as an estimate. Then round the dividend to the nearest whole number and divide.

example 2

Estimate: a. $4\frac{3}{10} \div \frac{5}{14}$ b. $10\frac{3}{4} \div \frac{4}{23}$

solution

a. $4\frac{3}{10} \div \frac{5}{14}$ ← **$\frac{1}{3}$ is an estimate of $\frac{5}{14}$. Round $4\frac{3}{10}$ to 4.**

$4 \div \frac{1}{3}$, or 4×3

about 12

b. $10\frac{3}{4} \div \frac{4}{23}$ ← **$\frac{1}{6}$ is an estimate of $\frac{4}{23}$. Round $10\frac{3}{4}$ to 11.**

$11 \div \frac{1}{6}$, or 11×6

about 66

your turn

Estimate.

5. $2\frac{1}{2} \div \frac{1}{5}$
6. $15\frac{1}{3} \div \frac{5}{19}$
7. $6\frac{2}{5} \div \frac{12}{25}$
8. $19\frac{5}{6} \div \frac{3}{26}$

practice exercises

practice for example 1 (page 152)

Estimate using compatible numbers.

1. $5\frac{2}{3} \div 1\frac{6}{7}$
2. $16\frac{1}{3} \div 5\frac{1}{4}$
3. $89\frac{3}{13} \div 10\frac{1}{3}$
4. $5\frac{9}{11} \div 4\frac{1}{2}$
5. $51\frac{2}{5} \div 6\frac{7}{10}$
6. $15\frac{1}{7} \div 2\frac{2}{7}$
7. $55\frac{3}{4} \div 11\frac{3}{8}$
8. $23\frac{1}{9} \div 3\frac{5}{8}$

practice for example 2 (page 152)

Estimate.

9. $2\frac{1}{2} \div \frac{1}{9}$
10. $3\frac{5}{6} \div \frac{10}{31}$
11. $1\frac{5}{12} \div \frac{7}{12}$
12. $11\frac{3}{10} \div \frac{7}{50}$
13. $4\frac{9}{10} \div \frac{9}{53}$
14. $9\frac{1}{4} \div \frac{19}{101}$
15. $8\frac{1}{3} \div \frac{4}{39}$
16. $11\frac{3}{5} \div \frac{7}{27}$

mixed practice (page 152)

Estimate.

17. $14\frac{2}{3} \div 3\frac{8}{11}$
18. $2\frac{5}{6} \div 1\frac{3}{7}$
19. $2\frac{4}{7} \div \frac{1}{8}$
20. $6\frac{1}{2} \div \frac{9}{28}$
21. $36\frac{1}{8} \div 5\frac{1}{4}$
22. $25\frac{2}{3} \div 11\frac{1}{2}$
23. $16\frac{3}{8} \div \frac{51}{101}$
24. $30\frac{1}{6} \div \frac{4}{19}$

Choose the division that will give you the best estimate.

25. $24\frac{1}{2} \div 4\frac{5}{7}$ **a.** $24 \div 4$ **b.** $25 \div 4$ **c.** $20 \div 5$ **d.** $25 \div 5$
26. $19\frac{2}{5} \div 3\frac{1}{4}$ **a.** $21 \div 3$ **b.** $18 \div 3$ **c.** $20 \div 3$ **d.** $20 \div 4$
27. An auto mechanic can inspect a car in about $\frac{1}{3}$ h. About how many cars can the mechanic inspect in $7\frac{1}{4}$ h?
28. A coil of wire is $9\frac{1}{2}$ ft (feet) long. Garth needs pieces of wire that are $1\frac{3}{4}$ ft long. About how many pieces can Garth cut from the coil?

review exercises

Is the first number divisible by the second number? Write *Yes* or *No*.

1. 819; 3
2. 3148; 8
3. 225; 10
4. 774; 2
5. 165; 5
6. 45,612; 9
7. 626; 4
8. 8530; 10

PROBLEM SOLVING STRATEGY

8-5 INTERPRETING THE ANSWER

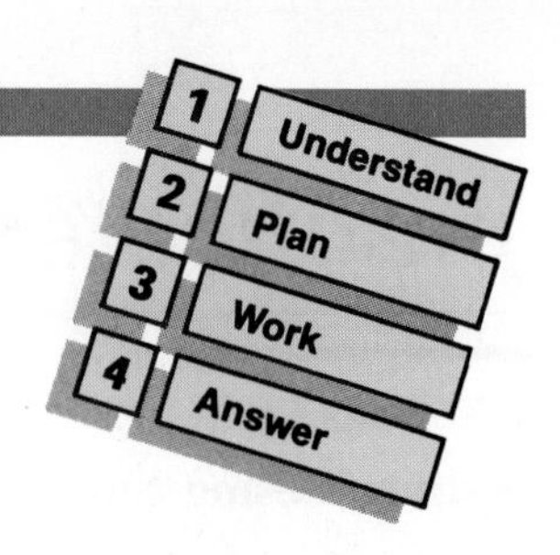

When you use division to solve a problem, you may need to interpret the results before you can give a reasonable answer. You need to decide whether the answer should be a whole number, a fraction, or a decimal.

example

a. The 21 students in gym class are running relay races. There are four students on a team. How many teams can there be?

b. Apples are sold in packages of four. Mr. Page wants to give one apple to each of the 21 students in class. How many packages of apples does he need to buy?

c. The length of a hike is 21 mi (miles). Heidi walks at an average speed of 4 mi/h (miles per hour). How many hours will she take to finish the hike?

d. A package of four blank cassette tapes costs $21. If the tapes were sold individually, how much would one tape cost?

solution

a.
$$\begin{array}{r} 5\text{ R}1 \\ 4\overline{)21} \\ \underline{20} \\ 1 \end{array}$$
← **You can make only a whole number of teams, so round down.**

There can be 5 teams.

b.
$$\begin{array}{r} 5\text{ R}1 \\ 4\overline{)21} \\ \underline{20} \\ 1 \end{array}$$
← **He needs to buy more than 5 packages, so round up.**

Mr. Page needs to buy 6 packages.

c.
$$\begin{array}{r} 5\tfrac{1}{4} \\ 4\overline{)21} \\ \underline{20} \\ 1 \end{array}$$
← **The answer is a number of hours, so write the answer as a mixed number.**

Heidi takes $5\frac{1}{4}$ h to finish.

d.
$$\begin{array}{r} \$5.25 \\ 4\overline{)\$21.00} \\ \underline{20} \\ 1\,0 \\ \underline{8} \\ 20 \\ \underline{20} \\ 0 \end{array}$$
← **The answer is an amount of money, so write the answer as a decimal.**

One tape would cost $5.25.

problems

1. A caterer must divide 39 lb (pounds) of raisins equally among 12 boxes. How many pounds should be put into each box?
2. If 4 pairs of socks are sold for $7, how much would one pair cost?
3. A recipe calls for 4 c of flaked coconut. A bag of coconut contains $2\frac{3}{4}$ c. How many bags would you need to make the recipe?

4. Leigh has \$15 to spend on books. How many books costing \$4 each can she buy?
5. The school band and its conductor are going by bus to a football game. There are 21 freshmen, 34 sophomores, 37 juniors, and 28 seniors in the band. Each bus holds 45 people. How many buses will they need?
6. Ann wants to bake 10 loaves of bread for a charity bake sale. One dozen eggs cost \$1.19, and the recipe calls for 3 eggs per loaf. How many dozen eggs must she buy?
7. A game company sold \$500,000 worth of *Game A* last year. The company gives a bonus worth $\frac{1}{100}$ of the sales to the entire team of people who worked on the game. If 16 people developed *Game A,* how much does each person receive as a bonus?
8. The toaster at a local breakfast restaurant takes $1\frac{1}{2}$ min to toast bread. The toaster can hold four slices of bread at the same time. How many slices of bread can be toasted in 20 min?

Write a reasonable word problem for each division.

9. $50\overline{)320}$ with quotient $6\frac{2}{5}$

   Answer: $6\frac{2}{5}$

10. $7\overline{)60}$ with quotient 8 R4

    Answer: 8

review exercises

1. Sue is saving money to buy a stereo system that costs \$670. She had \$40 in savings in January, \$70 in February, and \$130 in March. If this pattern continues, in which month will she have saved enough to buy the stereo system?
2. LeRoy works 35 h per week at an hourly rate of \$4.25. Weekly deductions of \$23.31 for income tax and \$11.17 for social security tax are taken from his pay. What was LeRoy's total take-home pay last week?

SKILL REVIEW

Write each product in lowest terms.

8-1

1. $\frac{1}{3} \times \frac{5}{7}$
2. $\frac{3}{5} \times \frac{1}{2}$
3. $7 \times \frac{1}{10}$
4. $\frac{5}{18} \times 9$
5. $\frac{2}{3} \times \frac{9}{10}$
6. $4 \times 3\frac{1}{2}$
7. $2\frac{1}{6} \times 6$
8. $\frac{3}{4} \times 2\frac{2}{3}$
9. $3\frac{1}{3} \times \frac{5}{8}$
10. $1\frac{1}{6} \times 4\frac{1}{2}$
11. $2\frac{1}{2} \times 1\frac{1}{15}$
12. $1\frac{13}{14} \times 2\frac{1}{3}$
13. A recipe calls for $\frac{3}{4}$ c of walnuts. If you triple the recipe, how many cups of walnuts will you need?

Estimate.

8-2

14. $4\frac{2}{3} \times 2\frac{1}{5}$
15. $9\frac{3}{8} \times 3\frac{5}{6}$
16. $11\frac{1}{6} \times 8\frac{1}{9}$
17. $6\frac{7}{8} \times 49\frac{4}{5}$
18. $\frac{1}{3} \times 11\frac{2}{5}$
19. $\frac{3}{4} \times 25\frac{1}{3}$
20. $\frac{61}{90} \times 31$
21. $149 \times \frac{10}{51}$

Write the reciprocal of each number, if possible.

8-3

22. $\frac{5}{9}$
23. $\frac{1}{4}$
24. 0
25. 7
26. $4\frac{2}{3}$

Write each quotient in lowest terms.

27. $8 \div \frac{5}{7}$
28. $\frac{4}{5} \div 6$
29. $\frac{7}{10} \div \frac{1}{20}$
30. $\frac{8}{9} \div \frac{16}{27}$
31. $1\frac{9}{11} \div 35$
32. $\frac{2}{9} \div 1\frac{3}{5}$
33. $2\frac{1}{5} \div 1\frac{7}{15}$
34. $2\frac{5}{8} \div 1\frac{1}{6}$
35. The length of the school track is $\frac{3}{4}$ mi (mile). During practice, Lisa ran 15 mi. How many times around the track did she run?

Estimate.

8-4

36. $21\frac{2}{3} \div 9\frac{7}{9}$
37. $40\frac{2}{5} \div 5\frac{5}{7}$
38. $8\frac{3}{8} \div \frac{7}{16}$
39. $6\frac{3}{4} \div \frac{5}{34}$
40. $39\frac{1}{5} \div 9\frac{3}{10}$
41. $35\frac{1}{2} \div 3\frac{3}{8}$
42. $2\frac{5}{8} \div \frac{11}{30}$
43. $5\frac{9}{20} \div \frac{4}{31}$

8-5

44. Ramon Cuervo works 40 h per week. His weekly pay is $290. How much does he earn per hour?
45. There are 230 students going to the class picnic. Each student needs one paper plate for dinner and one for dessert. Paper plates are sold in packages of fifty. How many packages must the class buy?

APPLICATION

8-6 STOPPING DISTANCE

Stopping distance is the distance you need to stop your car after you notice an obstacle in your path. The stopping distance depends on the speed of the car and the condition of the road. The following formula gives an estimate of the distance needed to stop on a dry concrete road.

$$\underset{\text{in feet}}{\text{stopping distance}} = \left(1\frac{1}{10} \times \underset{\text{in miles per hour}}{\text{speed}}\right) + \left(\frac{1}{20} \times \text{speed}^2\right)$$

You will need a greater distance to stop when the road conditions are poor or when the weather is bad.

example

Robert James is driving his car at 40 mi/h (miles per hour) when he sees a traffic control officer in the road ahead. Find the stopping distance of his car, assuming that Robert is driving on a dry concrete road.

solution

$$\begin{aligned}\text{stopping distance} &= \left(1\frac{1}{10} \times 40\right) + \left(\frac{1}{20} \times 40^2\right)\\ &= \left(\frac{11}{10} \times 40\right) + \left(\frac{1}{20} \times 1600\right) \quad \leftarrow \quad 40^2 = 40 \times 40 = 1600\\ &= \quad 44 \quad + \quad 80 \quad = 124\end{aligned}$$

The stopping distance of his car is about 124 ft (feet).

exercises

Solve. Assume that the car is traveling on a dry concrete road. A calculator may be helpful.

1. Find the stopping distance for a car traveling at 10 mi/h.
2. Find the stopping distance for a car traveling at 30 mi/h.
3. How many more feet does it take to stop at 30 mi/h than at 20 mi/h?
4. Some highways have a 65 mi/h speed limit. One car on the highway is traveling at 60 mi/h. Another is traveling at 50 mi/h. Find the difference between the stopping distances for these two cars.
5. About how many times greater is the stopping distance at 60 mi/h than at 20 mi/h?
6. If the speed of your car is doubled, will the stopping distance be doubled, *more than* doubled, or *less than* doubled?

APPLICATION

8-7 COST OF OPERATING AN AUTOMOBILE

Operating an automobile involves many costs. **Fixed costs** remain about the same from year to year. **Variable costs** change as the age of the car and the total number of miles driven increase. You can use these costs to determine the expense per mile of driving your automobile.

Fixed Costs		Variable Costs	
loan payment	$2160.67	gasoline and oil	$685.00
insurance	$579.00	maintenance and repairs	$235.89
license fee	$6.25	tires	$214.50
registration fee	$24.00	parking fees and tolls	$360.00
taxes and other fees	$175.00		
Total	$2944.92	Total	$1495.39

example

The table above is a record of Suzanne Dobner's automobile expenses for the past year. At the beginning of the year, her car's odometer showed 15,684 mi (miles). At the end it showed 27,803 mi.

a. How much did she spend last year to operate her automobile?

b. What was the cost per mile for the year?

solution

a. Add the totals of the fixed costs and the variable costs to find the cost for the year.

$$\begin{array}{r} \$2944.92 \\ +\$1495.39 \\ \hline \$4440.31 \end{array}$$

Suzanne spent $4440.31 last year to operate her automobile.

b. Subtract the odometer readings to find the number of miles driven during the year.

$$\begin{array}{r} 27{,}803 \\ -15{,}684 \\ \hline 12{,}119 \end{array}$$

← **She drove the car 12,119 mi.**

Divide the total cost by the number of miles driven to find the cost per mile.

$$\$4440.31 \div 12{,}119 \approx \$.366$$

To the nearest cent, the cost for the year was $.37 per mile.

exercises

Solve. A calculator may be helpful.

1. Janet Smoren registers her car once every two years at a cost of $50. How much does registration cost per year?
2. Troy Hoopman washes his car at a carwash once every two weeks. If the wash costs $3.50, how much does he spend for the year?
3. Each workday LeVon Hobbes spends $2 to park at a commuter lot. He also pays a $.35 toll driving to and from the lot. How much does he spend per month if he works 21 days each month?
4. Last year the total fixed costs for operating Helen Levy's car were $3913. The total variable costs were $2186. Helen drove the car 14,974 mi (miles). What was the cost per month? per mile?
5. Jim Watley pays $375 every six months for insurance, $184.52 each month for a car loan, $25 every four years for his driver's license, $20 each year for car registration, and $55 each year for taxes. How much does he spend each year for these fixed costs?
6. Use the example on page 158. At the end of this year, Suzanne determined the following costs for the year: gasoline, $720; repairs, $474; tires, $0; parking fees, $375. She drove the car 11,645 mi during the year. Did operating her car cost more per mile this year or last year? How much more?
7. Find out the cost of a driver's license in your state.
8. Find out the cost of car registration in your state. How often do you have to renew the registration?
9. Some states require that all cars be inspected for items such as emission control, brakes, and lights. Does your state require any type of car inspection? How much does the inspection cost, and how often is it required?
10. Why is gasoline considered a variable cost?

At the right is a division problem in which two denominators are missing. Find the denominators having the least value that make the statement true.

$$\frac{4}{\blacksquare} \div \frac{8}{\blacksquare} = \frac{5}{6}$$

8-8 AUTOMOBILE INSURANCE

Liability insurance covers injury and property damage resulting from an automobile accident. Insurance policies indicate coverage as follows.

25/50/15

- $25,000 maximum to any *one* person injured
- $50,000 maximum for *all* persons injured
- $15,000 maximum for any property damaged

A **premium** is the amount you pay for insurance coverage. Premiums are based on factors such as the driver's age and driving record and the area in which the driver lives. Sample premiums appear in the table below.

Liability Coverage	Yearly Premium
25/50/15	1.21 × basic rate
50/100/50	1.40 × basic rate
100/300/100	1.63 × basic rate

Basic Rates

teen female: $140
teen male: $275

Collision insurance covers damage done to your car in an accident. **Comprehensive** insurance covers damage resulting from fire, theft, and acts of nature. These types of insurance contain a **deductible.** When there is a repair bill, you pay the deductible amount and the insurance company pays the rest. The table below shows sample yearly premiums.

Driver	Collision premium $50 ded.	Collision premium $100 ded.	Comprehensive premium $0 ded.	Comprehensive premium $50 ded.
teen female	$170	$135	$67	$53
teen male	$290	$225	$100	$66

example

Alicia Stephens is 16 years old. She purchases 50/100/50 liability, $100-deductible collision, and $50-deductible comprehensive insurance. What is Alicia's total yearly premium?

solution

Use the tables above.
Liability: The basic rate for a teen female is $140: 1.40 × $140 = $196
Collision: The premium in the *$100 ded.* column is $135.
Comprehensive: The premium in the *$50 ded.* column is $53.
Find the sum of the premiums: $196 + $135 + $53 = $384
Alicia's total yearly premium is $384.

exercises

Solve. Use the tables on page 160. A calculator may be helpful.

1. Maida Wasmer is 16 years old. Because she has an 8-year-old car, she decides to purchase only liability insurance. What is her yearly premium for 100/300/100 liability insurance?
2. Doug Basehart is 17 years old. He purchases 50/100/50 liability, $50-deductible collision, and $50-deductible comprehensive insurance. What is his total yearly premium?
3. Walter D'Agin has not had an accident in five years and is entitled to a "safe-driver discount." His yearly premium is $483.35, but with the discount he pays only $418.35. How much is the discount?
4. Becky Curtin's yearly premium is $355. She has just completed a driver education course and is entitled to a discount of $\frac{1}{4}$ off her yearly premium. How much does she save? How much does she pay each year for insurance?
5. Sean Toole's insurance company offers a "good-student discount" to students having at least a B average in school. The discount is $\frac{1}{5}$ off the yearly premium. Sean qualifies and has a yearly premium of $540 before the discount. How much does he pay each year?
6. Emily Pang is 17 years old and has 25/50/15 liability, $50-deductible collision, and $0-deductible comprehensive insurance. She pays an equal part of the yearly premium each month instead of the entire amount at the beginning of the year. Emily must also pay a $3.00 fee each month. What is her monthly premium?
7. What does a teen female pay for collision insurance with a $50 deductible? with a $100 deductible? Find out why a $100-deductible premium costs less than a $50-deductible premium.
8. Find out what other types of coverage are available for liability insurance besides the three types listed in the table on page 160.
9. Some states have *no-fault* insurance. What is no-fault insurance? Does your state offer this type of insurance?

APPLICATION

8-9 DEPRECIATION

At the end of year	Depreciation rate
1	$\frac{1}{3}$
2	$\frac{2}{5}$
3	$\frac{1}{2}$
4	$\frac{11}{20}$
5	$\frac{2}{3}$
6	$\frac{3}{4}$

Depreciation is a decrease in the value of your car because of its age and condition. Although you do not actually spend money for depreciation, depreciation is considered one of the costs of operating an automobile.

The table at the right shows sample depreciation rates. The **trade-in value** of your car is the difference between the original price of the car and the depreciation.

example

A new car costs $10,780. What is its trade-in value at the end of four years?

solution

Use the table above.

Depreciation: $\frac{11}{20} \times \$10{,}780 = \$5929$

Trade-in value: $\$10{,}780 - \$5929 = \$4851$

The trade-in value of the car at the end of four years is $4851.

exercises

Solve. Use the table above. A calculator may be helpful.

1. Find the depreciation of a $9510 car at the end of five years.
2. Find the depreciation of a $12,600 car at the end of one year. What is its trade-in value at that time?
3. Find the trade-in value of a $11,720 car at the end of six years.
4. Sarah bought a car for $14,800. How much more is the trade-in value at the end of two years than at the end of three years?
5. Two factors that affect the trade-in value of a car are its age and condition. Find out what other factors affect the trade-in value.
6. Usually you can receive a higher price if you sell your car yourself instead of trading it in to a car dealer. List some advantages and disadvantages of these two ways of selling a car.
7. Choose a make and model of a car. Find out how much such a new car would cost. Then find out the trade-in value of a three-year-old car of the same make and model having the same options. What is the difference between the cost of the new car and the trade-in value of the older car?

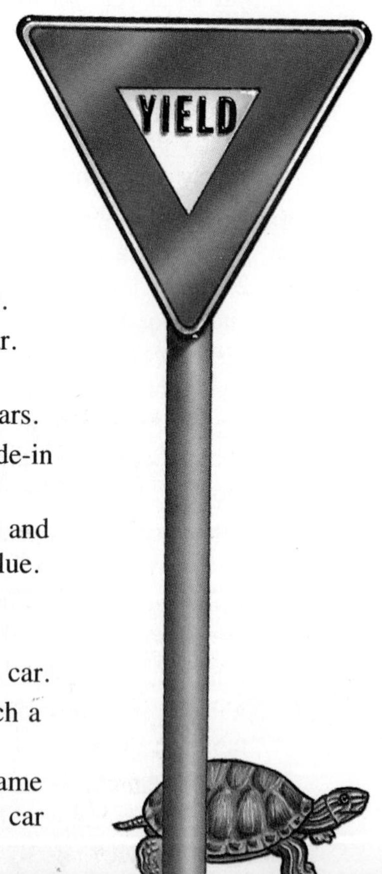

CHAPTER REVIEW

vocabulary vo·cab·u·lar·y

Choose the correct word or phrase to complete each sentence.

1. Two numbers whose product is 1 are *(mixed numbers, reciprocals).*
2. To estimate a product of mixed numbers, you can often use *(compatible numbers, mixed numbers).*

skills

Write each product or quotient in lowest terms.

3. $\frac{3}{11} \times \frac{1}{2}$
4. $\frac{9}{16} \times \frac{4}{5}$
5. $\frac{2}{7} \times 12$
6. $14 \times \frac{5}{7}$
7. $2\frac{2}{9} \times 3$
8. $3\frac{1}{3} \times 1\frac{3}{10}$
9. $\frac{6}{7} \times 2\frac{1}{3}$
10. $2\frac{5}{8} \times 1\frac{7}{9}$
11. $\frac{2}{5} \div \frac{4}{5}$
12. $\frac{3}{10} \div \frac{1}{4}$
13. $\frac{2}{3} \div 5$
14. $12 \div \frac{6}{7}$
15. $\frac{3}{4} \div 1\frac{4}{11}$
16. $10\frac{1}{2} \div 9$
17. $3\frac{1}{5} \div 1\frac{13}{15}$
18. $1\frac{5}{7} \div 2\frac{2}{9}$

Estimate.

19. $4\frac{1}{3} \times 5\frac{3}{4}$
20. $\frac{6}{29} \times 39$
21. $16\frac{3}{8} \times \frac{25}{49}$
22. $11\frac{5}{8} \times 3\frac{12}{25}$
23. $18\frac{9}{10} \div \frac{11}{31}$
24. $7\frac{9}{16} \div \frac{25}{101}$
25. $65\frac{1}{4} \div 8\frac{2}{5}$
26. $23\frac{2}{3} \div 10\frac{3}{4}$
27. Lili Jackson works at a bicycle shop assembling bicycles. Spokes come in boxes of 100, and each bicycle uses a total of 64 spokes. Lili has 359 spokes left. How many bicycles can Lili assemble?
28. Use the formula on page 157. Find the stopping distance for a car traveling at 10 mi/h (miles per hour). Assume that the car is traveling on a dry concrete road.
29. A driver's license costs $30 every four years. Automobile registration costs $40 every two years. What is the total cost per year for license and registration?
30. Use the tables on page 160. Pierre Hebert is 17 years old. He has 100/300/100 liability, $100-deductible collision, and $50-deductible comprehensive insurance. What is Pierre's yearly premium?
31. Use the table on page 162. Martha bought a new car for $6600. What is the trade-in value of the car at the end of four years?

CHAPTER TEST

Write each product in lowest terms.

1. $\frac{7}{8} \times \frac{4}{5}$ **2.** $\frac{3}{22} \times \frac{2}{3}$ **3.** $\frac{5}{6} \times 10$ **4.** $\frac{3}{10} \times 7$ **8-1**

5. $21 \times 1\frac{2}{7}$ **6.** $3\frac{1}{3} \times \frac{2}{5}$ **7.** $1\frac{4}{5} \times 4\frac{1}{6}$ **8.** $2\frac{2}{5} \times 1\frac{1}{4}$

Estimate.

9. $7\frac{9}{16} \times 2\frac{2}{11}$ **10.** $201 \times \frac{29}{41}$ **11.** $16\frac{7}{10} \times \frac{1}{3}$ **12.** $6\frac{3}{4} \times 2\frac{1}{2}$ **8-2**

Write each quotient in lowest terms.

13. $\frac{3}{5} \div \frac{1}{25}$ **14.** $\frac{3}{8} \div \frac{21}{32}$ **15.** $\frac{3}{11} \div 6$ **16.** $8 \div \frac{8}{9}$ **8-3**

17. $\frac{4}{9} \div 1\frac{7}{9}$ **18.** $4\frac{2}{3} \div 7$ **19.** $3\frac{3}{4} \div 2\frac{1}{2}$ **20.** $6\frac{2}{3} \div 1\frac{7}{8}$

Estimate.

21. $26\frac{3}{4} \div 9\frac{7}{16}$ **22.** $32\frac{5}{9} \div 1\frac{7}{8}$ **23.** $8\frac{1}{3} \div \frac{11}{50}$ **24.** $3\frac{9}{11} \div \frac{6}{53}$ **8-4**

25. A carpenter has a board that is 20 ft (feet) long. How many pieces that are 3 ft long can the carpenter cut from the board? **8-5**

26. Toni wants to jog 25 mi (miles) each week. She jogs around a lake on a road that is $4\frac{1}{2}$ mi long. How many times should she jog around the lake each week?

27. Use the formula on page 157. Find the stopping distance for a car traveling at 50 mi/h (miles per hour). Assume that the car is traveling on a dry concrete road. **8-6**

28. An oil change costs $16.95. Lori LaScala has an oil change done on her car 5 times per year. How much should she budget each month in order to pay for this work? **8-7**

29. Use the tables on page 160. Will is 16 years old. He purchases 25/50/15 liability, $100-deductible collision, and $0-deductible comprehensive insurance. What is his total yearly premium? **8-8**

30. Use the table on page 162. A new car sells for $13,890. What is its trade-in value at the end of five years? **8-9**

Yarn is used for many purposes, so it comes in a variety of colors and sizes. In planning a knitted article, you often add or subtract fractions.

CHAPTER 9

ADDING AND SUBTRACTING FRACTIONS

9-1 ADDING AND SUBTRACTING FRACTIONS

To add or subtract fractions with a *common* denominator, use the following method.

1. Add or subtract the numerators.
2. Write the sum or difference over the common denominator.
3. Write the answer in lowest terms.

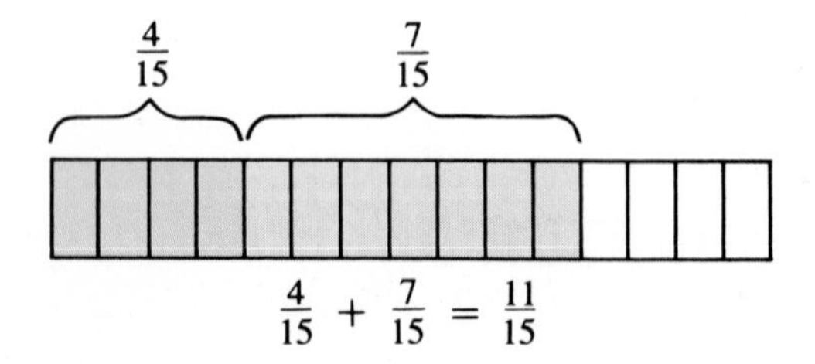

example 1

Write each sum or difference in lowest terms.

a. $\frac{1}{5} + \frac{3}{5}$ **b.** $\frac{7}{8} - \frac{3}{8}$ **c.** $\frac{4}{7} + \frac{6}{7}$

solution

a. $\frac{1}{5} + \frac{3}{5} = \frac{1+3}{5} = \frac{4}{5}$

b. $\frac{7}{8} - \frac{3}{8} = \frac{7-3}{8} = \frac{4}{8} = \frac{1}{2}$

c. $\frac{4}{7} + \frac{6}{7} = \frac{4+6}{7} = \frac{10}{7} = 1\frac{3}{7}$

your turn

Write each sum or difference in lowest terms.

1. $\frac{6}{17} + \frac{7}{17}$ **2.** $\frac{7}{9} - \frac{5}{9}$ **3.** $\frac{3}{10} + \frac{7}{10}$ **4.** $\frac{11}{12} - \frac{5}{12}$

To add or subtract fractions with *different* denominators, rewrite one fraction or both as equivalent fractions using the least common denominator (LCD). Then add or subtract as shown above.

example 2

Write each sum or difference in lowest terms: **a.** $\frac{5}{9} + \frac{2}{3}$ **b.** $\frac{5}{6} - \frac{3}{8}$

solution

a.
$$\begin{array}{rcr} \frac{5}{9} & = & \frac{5}{9} \\ +\frac{2}{3} & = & +\frac{6}{9} \\ \hline & & \frac{11}{9} = 1\frac{2}{9} \end{array}$$

The LCM of 9 and 3 is 9, so the LCD is 9.

b.
$$\begin{array}{rcr} \frac{5}{6} & = & \frac{20}{24} \\ -\frac{3}{8} & = & -\frac{9}{24} \\ \hline & & \frac{11}{24} \end{array}$$

The LCM of 6 and 8 is 24, so the LCD is 24.

your turn

Write each sum or difference in lowest terms.

5. $\frac{1}{4} + \frac{5}{12}$

6. $\frac{5}{8} - \frac{1}{2}$

7. $\frac{4}{7} + \frac{1}{2}$

8. $\frac{9}{14} - \frac{1}{4}$

practice exercises

practice for example 1 (page 166)

Write each sum or difference in lowest terms.

1. $\frac{1}{11} + \frac{3}{11}$
2. $\frac{2}{5} + \frac{2}{5}$
3. $\frac{2}{3} - \frac{1}{3}$
4. $\frac{6}{7} - \frac{4}{7}$
5. $\frac{13}{15} + \frac{7}{15}$
6. $\frac{9}{17} + \frac{8}{17}$
7. $\frac{3}{4} - \frac{3}{4}$
8. $\frac{8}{9} - \frac{2}{9}$
9. $\frac{19}{20} - \frac{7}{20}$
10. $\frac{5}{8} + \frac{7}{8}$

practice for example 2 (pages 166–167)

Write each sum or difference in lowest terms.

11. $\frac{3}{10} + \frac{2}{5}$
12. $\frac{2}{3} + \frac{5}{9}$
13. $\frac{2}{3} - \frac{7}{24}$
14. $\frac{11}{12} - \frac{1}{4}$
15. $\frac{1}{4} + \frac{17}{18}$
16. $\frac{4}{15} + \frac{8}{9}$
17. $\frac{6}{7} - \frac{3}{4}$
18. $\frac{5}{8} - \frac{1}{3}$
19. $\frac{7}{8} - \frac{5}{12}$
20. $\frac{2}{5} + \frac{3}{8}$

mixed practice (pages 166–167)

Write each sum or difference in lowest terms.

21. $\frac{2}{5} + \frac{1}{3}$
22. $\frac{4}{7} + \frac{5}{8}$
23. $\frac{3}{4} - \frac{3}{8}$
24. $\frac{16}{25} - \frac{3}{5}$
25. $\frac{5}{6} + \frac{7}{18}$
26. $\frac{1}{8} + \frac{3}{8}$
27. $\frac{7}{64} + \frac{9}{64}$
28. $\frac{5}{12} - \frac{1}{8}$
29. $\frac{5}{6} - \frac{3}{10}$
30. $\frac{17}{18} - \frac{7}{18}$
31. $\frac{11}{40} - \frac{3}{40}$
32. $\frac{8}{9} + \frac{1}{6}$
33. $\frac{5}{14} + \frac{3}{4}$
34. $\frac{7}{15} + \frac{11}{15}$
35. $\frac{1}{2} - \frac{2}{22}$
36. $\frac{4}{7} - \frac{2}{5}$
37. $\frac{8}{25} - \frac{0}{25}$
38. $\frac{2}{7} - \frac{2}{7}$
39. $\frac{4}{9} + \frac{0}{9} + \frac{5}{9}$
40. $\frac{1}{4} + \frac{2}{3} + \frac{5}{8}$

41. Tonia started a trip with the gas tank of her car $\frac{7}{8}$ full. At the end of the trip, the tank was $\frac{1}{4}$ full. How much of the tank did she use?

42. Lynn Dolby finished her mathematics homework in $\frac{2}{3}$ h. She finished her English homework in $\frac{3}{4}$ h and her science homework in $\frac{1}{2}$ h. How many hours did Lynn take to finish her homework?

43. During practice, Warren Kelson jogged $\frac{3}{4}$ mi (mile), walked $\frac{1}{2}$ mi, and then jogged $\frac{1}{8}$ mi. How many more miles did he jog than walk?

44. After baking yesterday, Mike had $\frac{3}{4}$ c of coconut left. Today he is making one recipe that requires $\frac{1}{3}$ c of coconut and another that requires $\frac{1}{4}$ c. How much coconut will he have after baking today?

45. Using fractions, write one addition exercise and one subtraction exercise, each having an answer of $\frac{2}{5}$.

46. Find two fractions whose sum is $\frac{3}{4}$ and whose difference is 0.

review exercises

Write a simpler fraction as an estimate of each fraction.

1. $\frac{4}{17}$　2. $\frac{15}{32}$　3. $\frac{19}{31}$　4. $\frac{3}{29}$　5. $\frac{5}{19}$　6. $\frac{32}{41}$

Tell whether each fraction is closest to 0, $\frac{1}{2}$, or 1.

7. $\frac{3}{8}$　8. $\frac{11}{12}$　9. $\frac{1}{6}$　10. $\frac{4}{51}$　11. $\frac{21}{44}$　12. $\frac{83}{100}$

calculator corner

Some calculators have a fraction key, [a b/c]. You can use this key to perform operations with fractions. Your answers will be in fraction form. For instance, the number $2\frac{1}{3}$ is displayed as 2 ⌋ 1 ⌋ 3.

example $\frac{3}{10} + \frac{1}{6}$

solution Enter 3 [a b/c] 10 + 1 [a b/c] 6 =.
The answer appears as 7 ⌋ 15. ← **This represents $\frac{7}{15}$.**

Write each answer as it would appear on a calculator having a fraction key.

1. $\frac{3}{5} + \frac{1}{4}$　2. $\frac{7}{8} + \frac{1}{6}$　3. $\frac{6}{7} - \frac{1}{2}$　4. $\frac{2}{3} - \frac{7}{24}$　5. $\frac{5}{6} \times \frac{8}{15}$

6. $5\frac{1}{2} \times 2\frac{2}{3}$　7. $\frac{7}{8} \div \frac{4}{5}$　8. $2\frac{2}{9} \div 1\frac{1}{3}$　9. $\frac{1}{3} + \frac{5}{11}$　10. $\frac{9}{10} - \frac{1}{15}$

9-2 ESTIMATING SUMS AND DIFFERENCES OF FRACTIONS AND MIXED NUMBERS

You can estimate a sum or difference of mixed numbers by rounding to the nearest whole number.

example 1

Estimate by rounding.

a. $15\frac{1}{8} + 9\frac{4}{7}$ ← If the fraction part of a mixed number is greater than or equal to $\frac{1}{2}$, round up.

$15 + 10$

about 25

b. $13\frac{5}{9} - 6\frac{2}{5}$

$14 - 6$

about 8

your turn

Estimate by rounding.

1. $12\frac{3}{4} + 6\frac{7}{10}$
2. $4\frac{1}{2} + 7\frac{1}{3}$
3. $19\frac{1}{4} - 8\frac{4}{9}$
4. $5\frac{5}{11} - 2\frac{2}{3}$

Sometimes you can get a better estimate by adding the whole numbers and then adjusting.

example 2

Estimate by adding the whole numbers and then adjusting.

a. $\frac{3}{7} + 4\frac{6}{11} + 6\frac{1}{9}$

$\frac{1}{2} + 4 + \frac{1}{2} + 6 + 0$ ← Add the whole numbers. Add the fractions, estimating if necessary.

about $10 + 1$, or 11

b. $1\frac{5}{18} + 3\frac{5}{6} + 8 + 2\frac{2}{3}$

$1 + \frac{1}{3} + 3 + 1 + 8 + 2 + \frac{2}{3}$

about $14 + 2$, or 16

your turn

Estimate by adding the whole numbers and then adjusting.

5. $3\frac{1}{2} + 5\frac{9}{10} + \frac{4}{7}$
6. $\frac{7}{20} + 6\frac{1}{3} + 2\frac{6}{19}$
7. $4\frac{5}{9} + 1\frac{1}{25} + 7\frac{6}{13}$

practice exercises

practice for example 1 (page 169)

Estimate by rounding.

1. $1\frac{1}{7} + 3\frac{3}{5}$ **2.** $6\frac{5}{9} + 2\frac{2}{3}$ **3.** $4\frac{4}{5} - 1\frac{3}{7}$ **4.** $3\frac{4}{9} - 2\frac{1}{3}$

5. $12\frac{4}{15} + 10\frac{3}{8}$ **6.** $4\frac{5}{6} + 5\frac{1}{5}$ **7.** $12\frac{3}{4} - 10\frac{1}{2}$ **8.** $27\frac{1}{4} - 3\frac{5}{8}$

practice for example 2 (page 169)

Estimate by adding the whole numbers and then adjusting.

9. $1 + \frac{5}{11} + 3\frac{4}{9}$ **10.** $4\frac{5}{18} + \frac{1}{3} + 2\frac{4}{11}$ **11.** $1\frac{10}{31} + 8\frac{1}{9} + 5\frac{9}{26} + 2\frac{1}{3}$

12. $\frac{11}{12} + 7\frac{7}{12} + 2\frac{1}{6} + 3\frac{5}{8}$ **13.** $1\frac{5}{6} + 2\frac{8}{9} + 6\frac{2}{11}$ **14.** $5\frac{2}{3} + 10 + 4\frac{8}{25}$

mixed practice (page 169)

Estimate.

15. $2\frac{2}{11} + 4\frac{1}{7}$ **16.** $6\frac{1}{3} - 2\frac{7}{8}$ **17.** $5\frac{7}{8} + 4\frac{3}{20} + 3$ **18.** $2\frac{9}{16} + 4\frac{3}{7} + 1\frac{2}{25}$

19. $10\frac{5}{6} - 6\frac{8}{15}$ **20.** $5\frac{1}{2} + 7\frac{4}{9}$ **21.** $6\frac{7}{15} + \frac{9}{10} + 8\frac{13}{24}$ **22.** $\frac{2}{3} + 3\frac{1}{9} + 11\frac{4}{13}$

Choose the letter of the best estimate.

23. $5\frac{1}{2} + 2\frac{4}{9}$ **a.** 6 **b.** 8 **c.** 7

24. $15\frac{4}{9} - 4\frac{5}{11}$ **a.** 12 **b.** 10 **c.** 11

25. $8\frac{3}{20} - 6\frac{7}{9}$ **a.** 3 **b.** 2 **c.** 1

26. $7\frac{7}{8} + 1\frac{5}{12} + 9\frac{9}{19}$ **a.** 18 **b.** 19 **c.** 20

27. $2\frac{1}{3} + 4\frac{1}{5}$ **a.** greater than 6 **b.** less than 6

28. $9\frac{7}{8} + 7\frac{3}{4}$ **a.** greater than 18 **b.** less than 18

29. $6 + 10\frac{7}{20} + 8\frac{1}{3} + \frac{12}{37}$ **a.** greater than 24 **b.** less than 24

30. $4\frac{19}{22} + 2 + \frac{30}{59} + 2\frac{6}{13}$ **a.** greater than 9 **b.** less than 9

31. Wendy skated $11\frac{2}{5}$ h last week. This week she skated $5\frac{7}{10}$ h. About how many more hours did Wendy skate last week than this week?

32. The distance from Joanne's house to school is $1\frac{9}{10}$ mi, from school to the mall is $2\frac{3}{4}$ mi, and from the mall to Joanne's house is $3\frac{1}{5}$ mi. Joanne went to the mall after school today. About how many miles was her trip home from school today?

33. To make a banner, the Key Club needs $1\frac{7}{8}$ yd (yards) of white fabric, $\frac{1}{2}$ yd of yellow, and $2\frac{3}{8}$ yd of blue. If the fabric costs about \$3 per yard, about how much will the fabric cost altogether?

34. Rafael bought $3\frac{3}{4}$ lb (pounds) of mixed nuts. He gave a friend a jar containing one fourth of these nuts. About how many pounds of nuts did he have for himself?

review exercises

Find the GCF.

1. 14 and 21
2. 16 and 64
3. 36 and 81

Find the LCM.

4. 3 and 15
5. 5 and 6
6. 8 and 10

Replace each ___?___ with the number that will make the fractions equivalent.

7. $\frac{3}{5} = \frac{?}{10}$
8. $\frac{4}{16} = \frac{?}{4}$
9. $\frac{2}{3} = \frac{14}{?}$
10. $\frac{12}{30} = \frac{2}{?}$
11. $\frac{?}{28} = \frac{3}{14}$
12. $\frac{?}{9} = \frac{8}{36}$
13. $\frac{5}{?} = \frac{15}{18}$
14. $\frac{30}{?} = \frac{3}{4}$

puzzle corner

A *unit fraction* is a fraction whose numerator is 1. An interesting fact is that you can write any given fraction as the sum of two or more *different* unit fractions. For instance, $\frac{3}{5} = \frac{1}{2} + \frac{1}{10}$, and $\frac{7}{8} = \frac{1}{2} + \frac{1}{4} + \frac{1}{8}$.

1. Write $\frac{3}{4}$ as the sum of *two* different unit fractions.
2. Write $\frac{19}{20}$ as the sum of *three* different unit fractions.

9-3 ADDING MIXED NUMBERS

To add mixed numbers, use the following method.

1. If necessary, rewrite one fraction or both as equivalent fractions using the LCD.
2. Add the fractions.
   Add the whole numbers.
3. Write the answer in lowest terms.

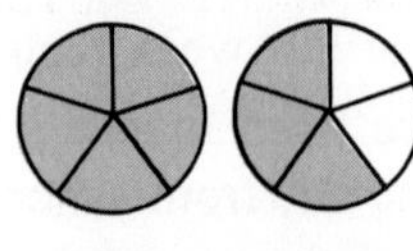

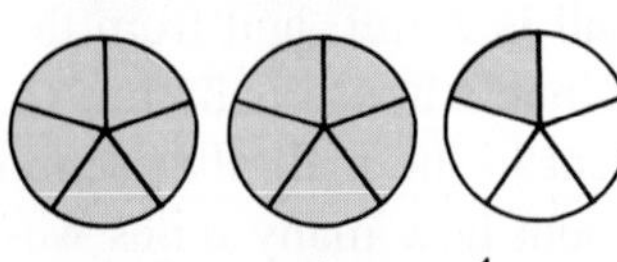

$1\frac{3}{5} + 2\frac{1}{5} = 3\frac{4}{5}$

example 1

Write each sum in lowest terms: **a.** $2\frac{1}{8} + 5\frac{5}{8}$ **b.** $12\frac{5}{6} + 9$ **c.** $3\frac{1}{6} + 4\frac{2}{9}$

solution

a.
$$\begin{array}{r} 2\frac{1}{8} \\ +5\frac{5}{8} \\ \hline 7\frac{6}{8} = 7\frac{3}{4} \end{array}$$

b.
$$\begin{array}{r} 12\frac{5}{6} \\ +\ 9 \\ \hline 21\frac{5}{6} \end{array}$$

c.
$$\begin{array}{rcr} 3\frac{1}{6} & = & 3\frac{3}{18} \\ +4\frac{2}{9} & = & +4\frac{4}{18} \\ \hline & & 7\frac{7}{18} \end{array}$$

← **The LCD is 18.**

your turn

Write each sum in lowest terms.

1. $4\frac{2}{5} + 1\frac{1}{5}$
2. $12 + 3\frac{7}{9}$
3. $6\frac{1}{2} + \frac{3}{10}$
4. $7\frac{1}{6} + 7\frac{2}{15}$

example 2

Write each sum in lowest terms: **a.** $2\frac{4}{5} + 4\frac{2}{3}$ **b.** $4\frac{7}{9} + 3\frac{2}{9}$

solution

a.
$$\begin{array}{rcr} 2\frac{4}{5} & = & 2\frac{12}{15} \\ +4\frac{2}{3} & = & +4\frac{10}{15} \\ \hline & & 6\frac{22}{15} = 6 + 1\frac{7}{15} = 7\frac{7}{15} \end{array}$$

← **The fraction part should be in lowest terms *and* less than 1.**

b.
$$\begin{array}{r} 4\frac{7}{9} \\ +3\frac{2}{9} \\ \hline 7\frac{9}{9} = 7 + 1 = 8 \end{array}$$

your turn

Write each sum in lowest terms.

5. $4\frac{5}{7} + 6\frac{1}{2}$
6. $2\frac{5}{6} + 3\frac{4}{9}$
7. $1\frac{3}{4} + 1\frac{5}{12}$
8. $8\frac{1}{5} + 5\frac{4}{5}$

practice exercises

practice for example 1 (page 172)

Write each sum in lowest terms.

1. $10 + 13\frac{4}{5}$

2. $2\frac{2}{7} + 6\frac{3}{7}$

3. $3\frac{5}{18} + 1\frac{11}{18}$

4. $8\frac{1}{6} + 4\frac{7}{12}$

5. $5\frac{1}{7} + 1\frac{6}{35}$

6. $\frac{2}{5} + 9\frac{3}{8}$

7. $8\frac{3}{4} + \frac{2}{9}$

8. $7\frac{7}{10} + 2\frac{1}{4}$

9. $12\frac{5}{12} + 6\frac{3}{8}$

practice for example 2 (page 172)

Write each sum in lowest terms.

10. $4\frac{2}{3} + 1\frac{1}{2}$

11. $6\frac{3}{5} + 7\frac{5}{9}$

12. $2\frac{7}{12} + 3\frac{5}{12}$

13. $5\frac{3}{7} + 1\frac{4}{7}$

14. $10\frac{1}{3} + 8\frac{17}{21}$

15. $2\frac{13}{16} + 9\frac{3}{4}$

16. $\frac{9}{10} + 4\frac{1}{3}$

17. $15\frac{3}{7} + \frac{23}{28}$

18. $9\frac{5}{8} + 3\frac{7}{10}$

mixed practice (page 172)

Write each sum in lowest terms.

19. $7\frac{3}{10} + 2\frac{3}{10}$

20. $5\frac{1}{2} + 7$

21. $9\frac{1}{6} + 1\frac{5}{6}$

22. $15\frac{7}{8} + 20\frac{4}{7}$

23. $7\frac{2}{9} + 10\frac{5}{12}$

24. $3\frac{1}{4} + 8\frac{6}{11}$

25. $\frac{1}{2} + 5\frac{3}{14}$

26. $6 + 7\frac{7}{8}$

27. $1\frac{7}{9} + 8\frac{2}{9}$

28. $6\frac{21}{32} + \frac{5}{8}$

29. $9\frac{4}{5} + 9\frac{16}{25}$

30. $2\frac{4}{13} + 4\frac{9}{13}$

31. $13\frac{5}{6} + 5\frac{3}{10}$

32. One bread recipe calls for $6\frac{3}{4}$ c of flour. Another calls for $7\frac{1}{2}$ c. How many cups of flour are needed to make both recipes?

33. During his first day of hiking, Kevin hiked $3\frac{1}{4}$ mi (miles) before stopping for lunch. After lunch he hiked another $5\frac{3}{8}$ mi. On the second day Kevin hiked $10\frac{1}{2}$ mi in all. How many miles did Kevin hike the first day?

Use the time card at the right.

34. Find the total number of hours Kim worked on Tuesday and Thursday.
35. Find the total number of hours Kim worked on Monday, Wednesday, and Friday.
36. Find the total number of hours Kim worked during the week.
37. If Kim earns \$4 per hour, what was her weekly pay before deductions?
38. If Kim works more than 40 h in one week, she is entitled to overtime pay. Does she qualify for overtime this week? Explain.

TIME CARD	Name: Kim Lucas		
Day	**In**	**Out**	**Hours**
M	8:02	3:59	8
T	10:58	5:15	$6\frac{1}{4}$
W	8:16	4:00	$7\frac{3}{4}$
TH	1:57	9:29	$7\frac{1}{2}$
F	8:30	6:14	$9\frac{3}{4}$

review exercises

Compare. Replace each __?__ with >, <, or =.

1. $\frac{10}{21}$ __?__ $\frac{13}{21}$
2. $\frac{6}{8}$ __?__ $\frac{3}{4}$
3. $\frac{1}{7}$ __?__ $\frac{1}{10}$
4. $\frac{8}{9}$ __?__ $\frac{7}{9}$
5. $1\frac{2}{3}$ __?__ $1\frac{5}{8}$
6. $\frac{17}{4}$ __?__ $5\frac{1}{4}$
7. $\frac{16}{7}$ __?__ $2\frac{2}{7}$
8. $3\frac{3}{5}$ __?__ $\frac{7}{2}$

mental math

The commutative and associative properties of addition can help you to add fractions and mixed numbers mentally.

$$\frac{4}{7} + \underbrace{\frac{7}{12} + \frac{5}{12}}_{1} = 1 + \frac{4}{7} = 1\frac{4}{7}$$

$$2\frac{1}{2} + 4\frac{1}{3} + 1\frac{1}{2} = 4 + 4\frac{1}{3} = 8\frac{1}{3}$$

(In the second example, $2\frac{1}{2}$ and $1\frac{1}{2}$ are joined by an arrow labeled 4.)

Add mentally.

1. $\frac{4}{5} + \frac{1}{5} + \frac{3}{8}$
2. $\frac{6}{11} + \frac{1}{4} + \frac{5}{11}$
3. $\frac{11}{12} + \frac{2}{9} + \frac{7}{9}$
4. $\frac{7}{15} + \frac{5}{9} + \frac{8}{15}$
5. $3\frac{1}{4} + 2\frac{3}{4} + 2\frac{7}{10}$
6. $1\frac{7}{18} + \frac{2}{3} + 4\frac{11}{18}$
7. $4\frac{5}{8} + 3\frac{13}{21} + \frac{8}{21}$
8. $1\frac{2}{3} + 2\frac{5}{14} + 3\frac{9}{14}$

9-4 SUBTRACTING MIXED NUMBERS

To subtract mixed numbers, use the following method.

1. If necessary, rewrite one fraction or both as equivalent fractions using the LCD.
2. If necessary, rename a whole number or mixed number.
3. Subtract the fractions. Subtract the whole numbers.
4. Write the answer in lowest terms.

example 1

Write each difference in lowest terms: **a.** $5\frac{7}{10} - 2\frac{3}{10}$ **b.** $7\frac{4}{5} - 3\frac{1}{3}$

solution

a.
$$\begin{array}{r} 5\frac{7}{10} \\ -2\frac{3}{10} \\ \hline 3\frac{4}{10} = 3\frac{2}{5} \end{array}$$

b.
$$\begin{array}{rcr} 7\frac{4}{5} & = & 7\frac{12}{15} \\ -3\frac{1}{3} & = & -3\frac{5}{15} \\ \hline & & 4\frac{7}{15} \end{array}$$

your turn

Write each difference in lowest terms.

1. $12\frac{4}{9} - 12\frac{1}{9}$
2. $4\frac{17}{18} - 2\frac{2}{3}$
3. $5\frac{5}{8} - 3\frac{1}{6}$
4. $65\frac{11}{14} - 59\frac{2}{7}$

example 2

Write each difference in lowest terms: **a.** $3\frac{1}{6} - 1\frac{2}{3}$ **b.** $14 - 5\frac{7}{8}$

solution

a.
$$\begin{array}{rcrcr} 3\frac{1}{6} & = & 3\frac{1}{6} & = & 2\frac{7}{6} \\ -1\frac{2}{3} & = & -1\frac{4}{6} & = & -1\frac{4}{6} \\ \hline & & & & 1\frac{3}{6} = 1\frac{1}{2} \end{array}$$

← $3\frac{1}{6} = 2 + 1\frac{1}{6}$
$= 2 + \frac{7}{6}$
$= 2\frac{7}{6}$

b.
$$\begin{array}{rcr} 14 & = & 13\frac{8}{8} \\ -5\frac{7}{8} & = & -5\frac{7}{8} \\ \hline & & 8\frac{1}{8} \end{array}$$

← $14 = 13 + 1$
$= 13 + \frac{8}{8}$
$= 13\frac{8}{8}$

your turn

Write each difference in lowest terms.

5. $2\frac{3}{10} - 1\frac{7}{10}$
6. $6\frac{3}{8} - 3\frac{1}{2}$
7. $11 - 4\frac{5}{16}$
8. $8\frac{1}{4} - 5\frac{3}{5}$

practice exercises

practice for example 1 (page 175)

Write each difference in lowest terms.

1. $10\frac{6}{7} - 3\frac{2}{7}$
2. $5\frac{9}{16} - 2\frac{3}{16}$
3. $4\frac{8}{15} - \frac{8}{15}$
4. $3\frac{4}{5} - 3\frac{4}{5}$
5. $9\frac{3}{5} - 4\frac{13}{30}$
6. $7\frac{8}{9} - 6\frac{2}{3}$
7. $13\frac{7}{8} - 7\frac{1}{3}$
8. $6\frac{2}{7} - 4\frac{1}{6}$
9. $29\frac{5}{6} - 21\frac{5}{9}$
10. $53\frac{3}{4} - 48\frac{7}{10}$

practice for example 2 (page 175)

Write each difference in lowest terms.

11. $5 - 2\frac{1}{2}$
12. $7 - 6\frac{2}{3}$
13. $9\frac{3}{8} - 1\frac{7}{8}$
14. $4\frac{1}{6} - 2\frac{5}{6}$
15. $6\frac{7}{15} - 4\frac{4}{5}$
16. $8\frac{3}{20} - 3\frac{1}{2}$
17. $8\frac{1}{6} - 4\frac{7}{10}$
18. $15\frac{3}{8} - \frac{11}{12}$
19. $29\frac{1}{4} - 17\frac{3}{7}$
20. $88\frac{2}{9} - 64\frac{3}{5}$

mixed practice (page 175)

Write each difference in lowest terms.

21. $6\frac{11}{12} - 1\frac{7}{12}$
22. $4\frac{1}{2} - 2\frac{1}{8}$
23. $8 - 5\frac{6}{7}$
24. $7 - 2\frac{3}{4}$
25. $3\frac{11}{36} - 1\frac{5}{6}$
26. $9\frac{3}{10} - 8\frac{9}{10}$
27. $22\frac{2}{3} - 18\frac{6}{11}$
28. $46\frac{9}{10} - 39\frac{1}{15}$
29. $33\frac{1}{6} - 18\frac{3}{4}$
30. $30\frac{2}{9} - 25\frac{7}{8}$
31. A floppy disk for one type of computer is $5\frac{1}{4}$ in. (inches) wide. A floppy disk for another type of computer is $3\frac{1}{2}$ in. wide. How many inches wider is the larger floppy disk?
32. In 1900 the winning height for the men's Olympic high jump was $74\frac{3}{4}$ in. Sixty years later, the winning height was 85 in. How much higher did the winner in 1960 jump than the winner in 1900?
33. Using mixed numbers, write two subtraction exercises, each having an answer of $2\frac{3}{10}$. Include renaming in one exercise.

34. Sometime during a 15-min period, a disk jockey must read a $1\frac{1}{2}$-min weather forecast. The disk jockey also wants to play songs that are $3\frac{1}{2}$ min, $4\frac{3}{4}$ min, and $3\frac{2}{3}$ min long. Can these songs be played if $1\frac{1}{4}$ min must be left for commercials? Explain.

35. Lester usually cooks a roast $\frac{1}{2}$ h for each pound that the roast weighs. He cooks potatoes for 50 min. Lester is cooking a 5-lb (pound) roast today. He wants the roast and potatoes to be done at the same time. How long after the roast is put in the oven should Lester start cooking the potatoes?

review exercises

Estimate.

1. $72 + 58 + 19$
2. $6720 - 3813$
3. $11\frac{2}{5} \div 3\frac{6}{7}$
4. $\frac{1}{4} \times 29\frac{1}{8}$
5. $10\frac{1}{10} - 3\frac{7}{9}$
6. $1\frac{5}{9} + \frac{3}{7} + 5\frac{11}{12}$
7. 843×57
8. $982 \div 54$
9. 0.73×68.1
10. $\$6.38 + \$4.65$
11. $7.2\overline{)36.6}$
12. $936.6 - 148.2$
13. $17\frac{2}{3} - 9\frac{2}{13}$
14. $7\frac{1}{3} \div \frac{7}{12}$
15. $3\frac{1}{9} + 11\frac{7}{8}$
16. $1\frac{5}{7} \times 13\frac{2}{5}$

mental math

To subtract $6\frac{1}{9} - 2\frac{5}{9}$ mentally, *add on* from $2\frac{5}{9}$ to $6\frac{1}{9}$.

$2\frac{5}{9} + \frac{4}{9} = 3$ $\quad 3 + 3\frac{1}{9} = 6\frac{1}{9}$ $\quad \frac{4}{9} + 3\frac{1}{9} = 3\frac{5}{9}$ $\quad$ So $6\frac{1}{9} - 2\frac{5}{9} = 3\frac{5}{9}$.

Subtract mentally.

1. $6\frac{1}{3} - 3\frac{2}{3}$
2. $12\frac{4}{15} - 4\frac{8}{15}$
3. $9\frac{3}{8} - 2\frac{5}{8}$
4. $10\frac{1}{4} - 8\frac{3}{4}$
5. $4 - 1\frac{2}{5}$
6. $10 - 5\frac{1}{4}$
7. $8 - 4\frac{3}{7}$
8. $7 - 5\frac{4}{9}$

SKILL REVIEW

Write each sum or difference in lowest terms.

9-1

1. $\frac{4}{15} + \frac{2}{15}$ 2. $\frac{7}{10} + \frac{7}{10}$ 3. $\frac{6}{7} - \frac{2}{7}$ 4. $\frac{11}{14} - \frac{5}{14}$ 5. $\frac{1}{2} + \frac{3}{4}$
6. $\frac{4}{5} + \frac{1}{3}$ 7. $\frac{3}{4} - \frac{5}{28}$ 8. $\frac{5}{8} - \frac{2}{9}$ 9. $\frac{3}{10} + \frac{13}{15}$ 10. $\frac{5}{6} - \frac{3}{4}$
11. At a music store, $\frac{1}{10}$ of the albums feature male vocalists. Another $\frac{1}{8}$ feature female vocalists. What fraction of the albums feature vocalists?

Estimate by rounding.

9-2

12. $4\frac{2}{5} + 2\frac{1}{3}$ 13. $9\frac{7}{12} + 5\frac{5}{11}$ 14. $8\frac{1}{4} - 6\frac{8}{9}$ 15. $13\frac{1}{2} - 4\frac{3}{4}$

Estimate by adding the whole numbers and then adjusting.

16. $1\frac{1}{15} + 5\frac{4}{9} + 2\frac{12}{25}$ 17. $4\frac{1}{3} + \frac{7}{20} + 2\frac{10}{33}$ 18. $3\frac{6}{11} + 4 + 3\frac{19}{20} + \frac{7}{13}$
19. Renaldo worked $6\frac{3}{4}$ h on Monday, $7\frac{1}{4}$ h on Tuesday, and $9\frac{1}{2}$ h on Wednesday. About how many more hours did he work on Wednesday than on Monday?

Write each sum in lowest terms.

9-3

20. $6 + 2\frac{7}{8}$ 21. $3\frac{5}{14} + 1\frac{3}{14}$ 22. $2\frac{1}{2} + 5\frac{1}{3}$ 23. $4\frac{2}{3} + \frac{4}{21}$
24. $\frac{5}{9} + 3\frac{3}{4}$ 25. $8\frac{11}{18} + 10\frac{7}{18}$ 26. $4\frac{4}{5} + 4\frac{8}{15}$ 27. $12\frac{5}{6} + 7\frac{3}{8}$
28. For an endurance race, a participant must swim $3\frac{2}{3}$ mi (miles), ride a bicycle $15\frac{1}{4}$ mi, and run $26\frac{1}{2}$ mi. What is the total length of the endurance race?

Write each difference in lowest terms.

9-4

29. $3\frac{5}{6} - 2\frac{1}{6}$ 30. $8\frac{13}{18} - 5\frac{5}{18}$ 31. $9\frac{2}{3} - 1\frac{2}{3}$ 32. $5\frac{1}{8} - 3\frac{7}{8}$
33. $6\frac{23}{24} - 4\frac{7}{12}$ 34. $10\frac{1}{7} - 6\frac{2}{3}$ 35. $5 - \frac{6}{7}$ 36. $13 - 3\frac{15}{22}$
37. $4\frac{5}{6} - 4\frac{2}{5}$ 38. $12\frac{1}{2} - 2\frac{11}{16}$ 39. $7\frac{7}{12} - 2\frac{4}{9}$ 40. $16\frac{3}{10} - 9\frac{3}{4}$
41. At the end of one baseball season, the Orioles were 9 games behind the division leader. The Brewers were $6\frac{1}{2}$ games behind the leader. How many games behind the Brewers were the Orioles?

APPLICATION

9-5 STOCK MARKET

When you buy a **share** of a company's **stock,** you become a part owner of that company. You can follow the progress of your stock daily in a newspaper stock report. In the report shown, prices per share are listed in eighths of a dollar.

Stock	Div	PE	Sales 100s	High	Low	Close	Net Chg
ABC	.76	21	9453	$25\frac{7}{8}$	$25\frac{1}{4}$	$25\frac{1}{2}$	$-\frac{1}{2}$
AdEn	.22	14	1544	$35\frac{3}{4}$	$34\frac{1}{8}$	$34\frac{1}{4}$	$+1\frac{7}{8}$
AltCo	.33	12	59	$57\frac{3}{8}$	$56\frac{1}{2}$	$57\frac{3}{8}$	$+\frac{1}{4}$
Avion	.75	12	37	25	$23\frac{1}{4}$	$23\frac{1}{2}$	$-\frac{3}{4}$

example

What is the difference between the highest price paid per share and the lowest price paid per share for AdEn stock?

solution

Read down the *Stock* column to find AdEn.

Read across to the *High* column: $35\frac{3}{4} = 35\frac{6}{8}$

Read across to the *Low* column: $-34\frac{1}{8} = -34\frac{1}{8}$

Subtract the prices: $1\frac{5}{8}$

The difference between the prices is $\$1\frac{5}{8}$.

exercises

Use the table above. Find the difference between the highest price paid per share and the lowest price paid per share for each stock.

1. ABC
2. AltCo
3. Avion
4. John Huchra bought 100 shares of ABC stock at its lowest price during the day. How much did he pay in all for the stock?
5. Kay Li bought 120 shares of AltCo stock at its highest price. How much would she have saved in all by buying at the lowest price?
6. The number in the *Close* (closing price) column is the price paid for the last sale today. The number in the *Net Chg* (net change) column is the difference between today's and yesterday's closing prices. What was the closing price of AltCo stock yesterday?
7. The column titled *Div* in the stock report indicates *stock dividends*. Find out what a stock dividend is.
8. Choose a stock that is listed in an available newspaper. Follow the stock's progress for three days. How much has the price changed?

9-6 TECHNICAL DRAWINGS

A **technical drawing** shows the size and shape of an object. A draftsperson often makes a technical drawing of an object that is to be manufactured. The manufacturer then refers to the technical drawing when producing the object.

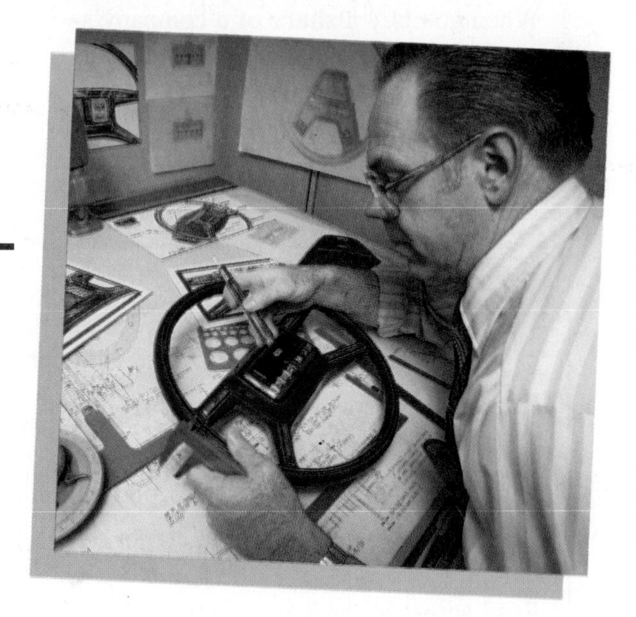

example

Find the missing length in the drawing of the hook below.

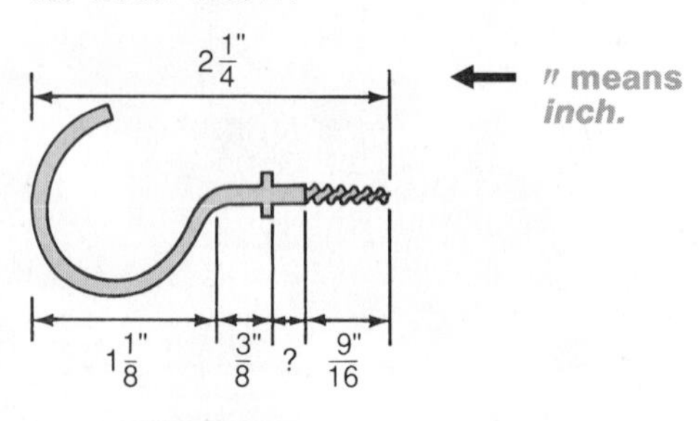

solution

First add the partial lengths that you know.

$$\begin{aligned} 1\tfrac{1}{8} &= 1\tfrac{2}{16} \\ \tfrac{3}{8} &= \tfrac{6}{16} \\ +\tfrac{9}{16} &= +\tfrac{9}{16} \\ \hline & \quad 1\tfrac{17}{16} = 2\tfrac{1}{16} \end{aligned}$$

Then subtract from the total length.

$$\begin{aligned} 2\tfrac{1}{4} &= 2\tfrac{4}{16} \\ -2\tfrac{1}{16} &= -2\tfrac{1}{16} \\ \hline & \quad \tfrac{3}{16} \end{aligned}$$

The missing length is $\frac{3}{16}$ in.

exercises

Find the missing length in each drawing.

1. 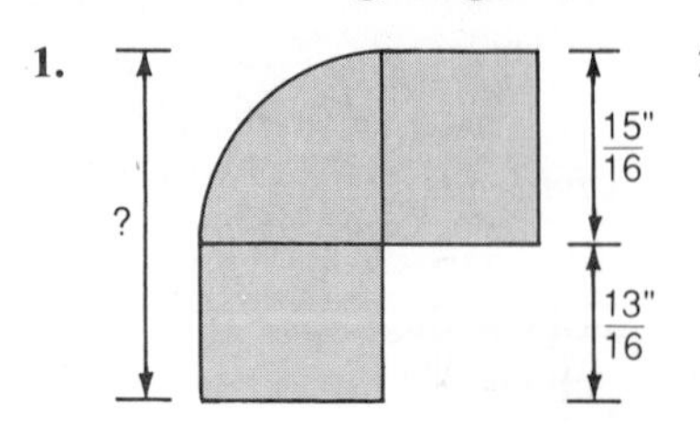

2. 1" — 11"/16 — ?

3. 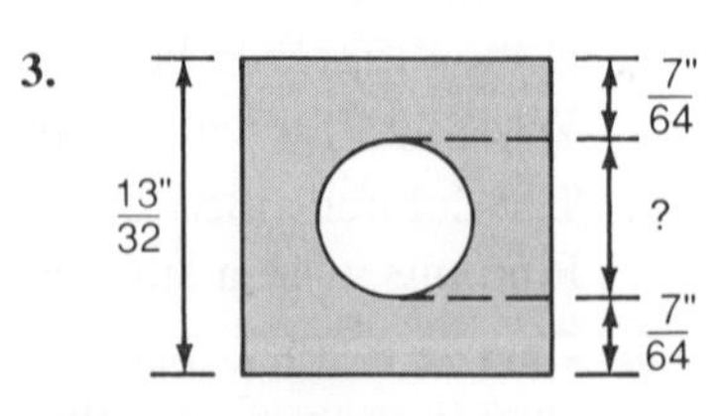

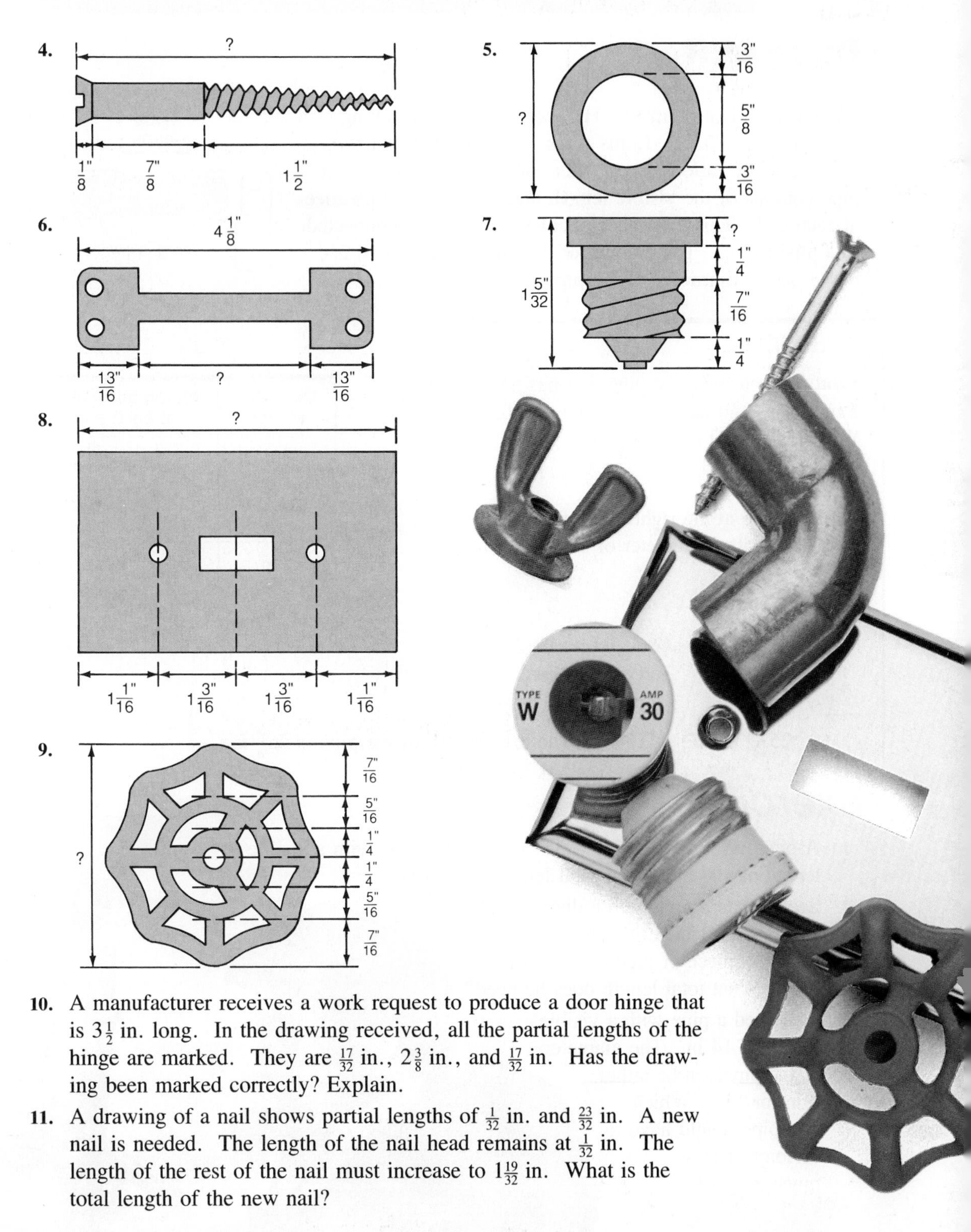

10. A manufacturer receives a work request to produce a door hinge that is $3\frac{1}{2}$ in. long. In the drawing received, all the partial lengths of the hinge are marked. They are $\frac{17}{32}$ in., $2\frac{3}{8}$ in., and $\frac{17}{32}$ in. Has the drawing been marked correctly? Explain.

11. A drawing of a nail shows partial lengths of $\frac{1}{32}$ in. and $\frac{23}{32}$ in. A new nail is needed. The length of the nail head remains at $\frac{1}{32}$ in. The length of the rest of the nail must increase to $1\frac{19}{32}$ in. What is the total length of the new nail?

CAREER

PLUMBER

Scott Tokarz is a plumber. His work involves installing pipes for water lines. In his work, Scott needs to know the lengths of various pipes. The total length of a straight pipe consists of the **visible length** and the **fitting allowance** at each end. This allowance enables a pipe to be connected to a pipe fitting. The amount of the allowance needed is based on the **diameter** of the pipe.

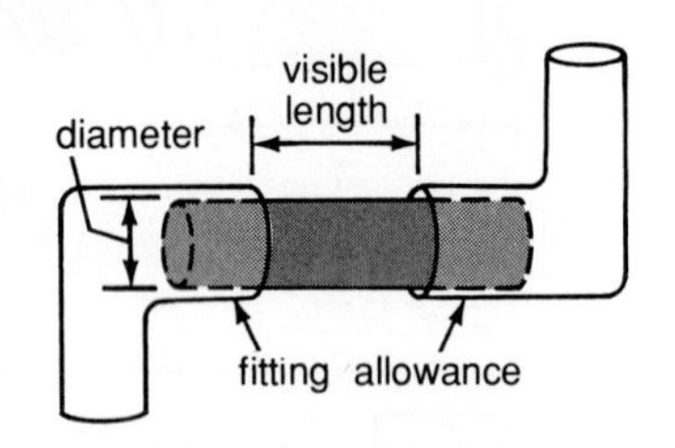

example

Scott needs to install a pipe that has a diameter of 1 in. (inch) and has a visible length of $25\frac{1}{2}$ in. What is the total length of the pipe?

solution

Use the table at the right. The fitting allowance for a pipe with a diameter of 1 in. is $\frac{5}{8}$ in.

Diameter of pipe	Fitting allowance at each end
$\frac{1}{2}$ in.	$\frac{1}{2}$ in.
$\frac{3}{4}$ in.	$\frac{1}{2}$ in.
1 in.	$\frac{5}{8}$ in.
$1\frac{1}{4}$ in.	$\frac{5}{8}$ in.

$$25\frac{1}{2} = 25\frac{4}{8}$$

$$\frac{5}{8} = \frac{5}{8}$$

$$+\frac{5}{8} = +\frac{5}{8}$$

Add $\frac{5}{8}$ in. for each end of the pipe.

$$25\frac{14}{8} = 26\frac{6}{8} = 26\frac{3}{4}$$

The total length of the pipe is $26\frac{3}{4}$ in.

exercises

1. A pipe has a diameter of $\frac{1}{2}$ in. You need a pipe with a visible length of $10\frac{5}{8}$ in. What total length do you need?
2. Scott needs a pipe with a diameter of $1\frac{1}{4}$ in. and a visible length of $17\frac{3}{4}$ in. He has a piece of pipe that is 24 in. long. What total length does he need?
3. You need a pipe with a visible length of 14 in. The diameter of the pipe can be either 1 in. or $\frac{3}{4}$ in. Which size pipe would have to be longer? How much longer?

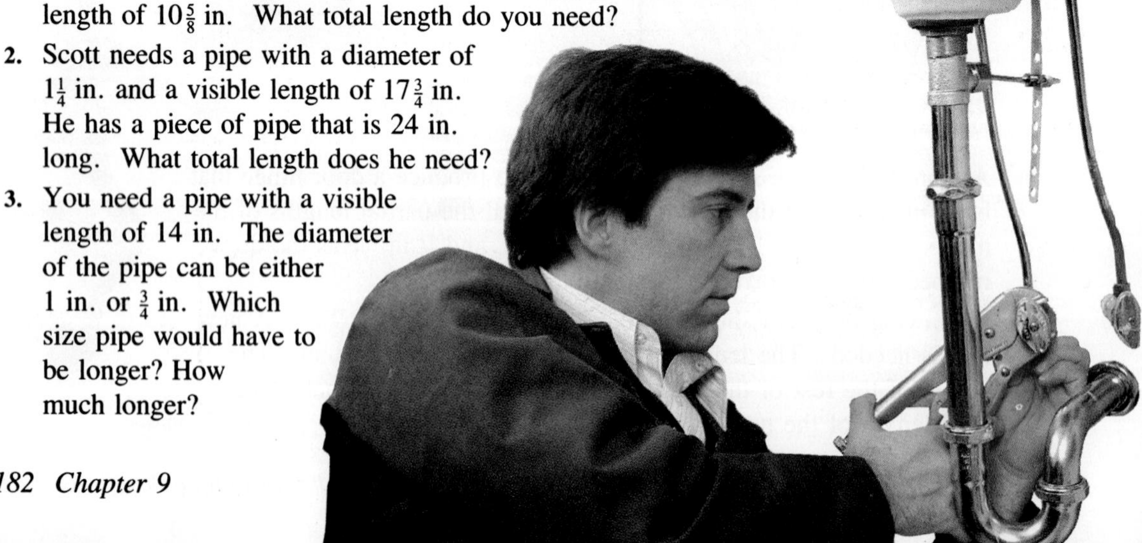

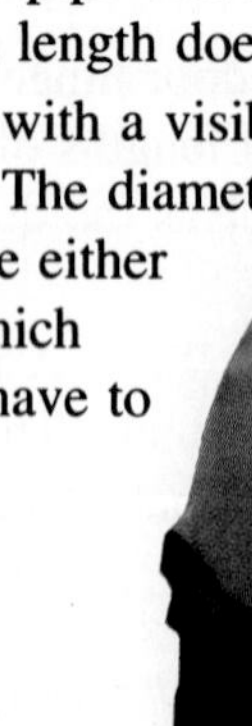

CHAPTER REVIEW

vocabulary vo·cab·u·lar·y

Choose the correct phrase to complete each sentence.

1. To add or subtract fractions with different denominators, rewrite the fractions as *(equivalent fractions, mixed fractions)*.
2. When you rewrite $3\frac{1}{8}$ as $2\frac{9}{8}$, you are *(writing $3\frac{1}{8}$ in lowest terms, renaming $3\frac{1}{8}$)*.

skills

Write each sum or difference in lowest terms.

3. $\frac{1}{6} + \frac{5}{6}$
4. $\frac{7}{8} - \frac{2}{5}$
5. $1\frac{7}{10} + 8\frac{9}{10}$
6. $5\frac{1}{12} + \frac{2}{3}$
7. $\frac{5}{7} + \frac{19}{42}$
8. $\frac{11}{12} - \frac{4}{9}$
9. $7\frac{23}{26} - 7\frac{5}{13}$
10. $8 - \frac{2}{9}$
11. $4\frac{3}{8} + 3\frac{1}{6}$
12. $15\frac{3}{4} + 6\frac{4}{7}$
13. $24\frac{2}{9} - \frac{1}{2}$
14. $22\frac{3}{5} - 8\frac{14}{15}$

Estimate.

15. $3\frac{8}{9} + 4\frac{5}{6}$
16. $6\frac{1}{9} - 2\frac{2}{3}$
17. $12\frac{4}{5} - 5\frac{3}{10}$
18. $2\frac{5}{11} + 3\frac{1}{6} + \frac{7}{12} + 4\frac{7}{8}$
19. After school, Carlos had baseball practice for $2\frac{1}{2}$ h. Later he played his trumpet for $\frac{3}{4}$ h and then studied for $1\frac{2}{3}$ h. Find the total number of hours he spent on all three activities.
20. Wanda noticed that the gas gauge on her car showed $\frac{3}{8}$ full. How much of the tank was empty?
21. Curtis needs $1\frac{1}{2}$ c of milk to make one recipe and $\frac{2}{3}$ c of milk to make another. He buys a quart of milk. A quart contains four cups. How many cups of milk does he have after making the recipes?
22. The highest price paid for MolTx stock during the day was $\$32\frac{1}{2}$ per share. The lowest price paid was $\$29\frac{7}{8}$ per share. What is the difference between these two prices?
23. A technical drawing of an electrical staple is shown at the right. Find the missing length.

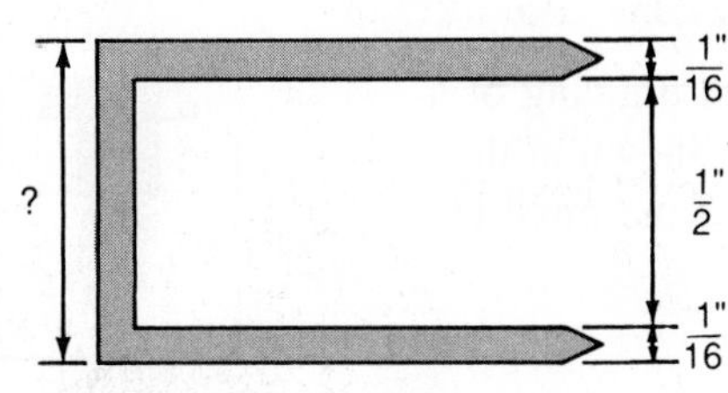

CHAPTER TEST

Write each sum or difference in lowest terms.

1. $\frac{5}{8} + \frac{3}{8}$ 2. $\frac{8}{9} - \frac{5}{9}$ 3. $\frac{7}{9} + \frac{1}{2}$ 4. $\frac{11}{15} - \frac{1}{5}$ 5. $\frac{1}{6} + \frac{9}{10}$ **9-1**

6. On Sunday, $\frac{1}{8}$ in. (inch) of rain fell on the city of Waterville. Another $\frac{1}{4}$ in. fell on Thursday and $\frac{1}{2}$ in. fell on Friday. Find the total number of inches of rain that fell during these three days.

Estimate by rounding.

7. $2\frac{9}{10} + 5\frac{1}{8}$ 8. $10\frac{3}{4} - 4\frac{4}{7}$ 9. $7\frac{5}{11} - 5\frac{4}{5}$ **9-2**

Estimate by adding the whole numbers and then adjusting.

10. $3\frac{7}{16} + \frac{1}{10} + 1\frac{5}{9}$ 11. $2\frac{5}{6} + 6\frac{4}{7} + \frac{10}{21}$ 12. $\frac{6}{19} + 2\frac{8}{23} + 3\frac{1}{7} + 1\frac{3}{10}$

13. On a trip from New Orleans to Birmingham, Charles drove for $1\frac{3}{4}$ h. Darryl drove for $4\frac{1}{5}$ h. About how many hours did they drive in all?

Write each sum or difference in lowest terms.

14. $5 + 7\frac{2}{3}$ 15. $8\frac{3}{16} + 3\frac{9}{16}$ 16. $4\frac{3}{4} + 4\frac{1}{4}$ **9-3**

17. $\frac{9}{20} + 13\frac{4}{5}$ 18. $2\frac{1}{8} + 1\frac{5}{12}$ 19. $21\frac{1}{4} + 9\frac{8}{9}$

20. $7\frac{5}{6} - 3\frac{1}{6}$ 21. $15 - 6\frac{9}{14}$ 22. $4\frac{6}{7} - 1\frac{2}{3}$ **9-4**

23. $5\frac{12}{25} - 2\frac{17}{25}$ 24. $10\frac{1}{6} - \frac{4}{9}$ 25. $6\frac{3}{8} - 5\frac{7}{16}$

26. A small can contains $1\frac{7}{8}$ oz (ounces) of pepper. A large can contains 4 oz of pepper. How much more pepper does the large can contain?

27. The highest price paid for VenSys stock during the day was $\$18\frac{3}{8}$ per share. The lowest price paid was $\$17\frac{3}{4}$ per share. What is the difference between these two prices? **9-5**

28. A technical drawing of a fastener is shown at the right. Find the missing length. **9-6**

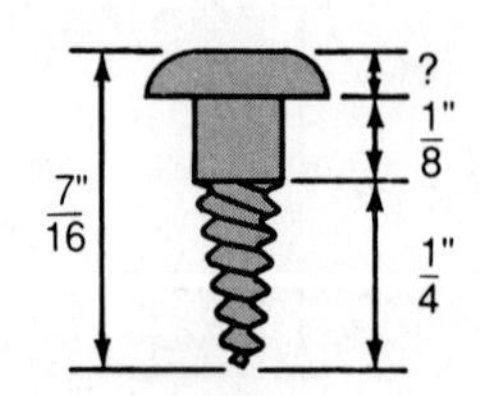

Making your own clothes can be fun. Both fractions and decimals are involved in working with a pattern and figuring costs.

CHAPTER 10

FRACTIONS AND DECIMALS

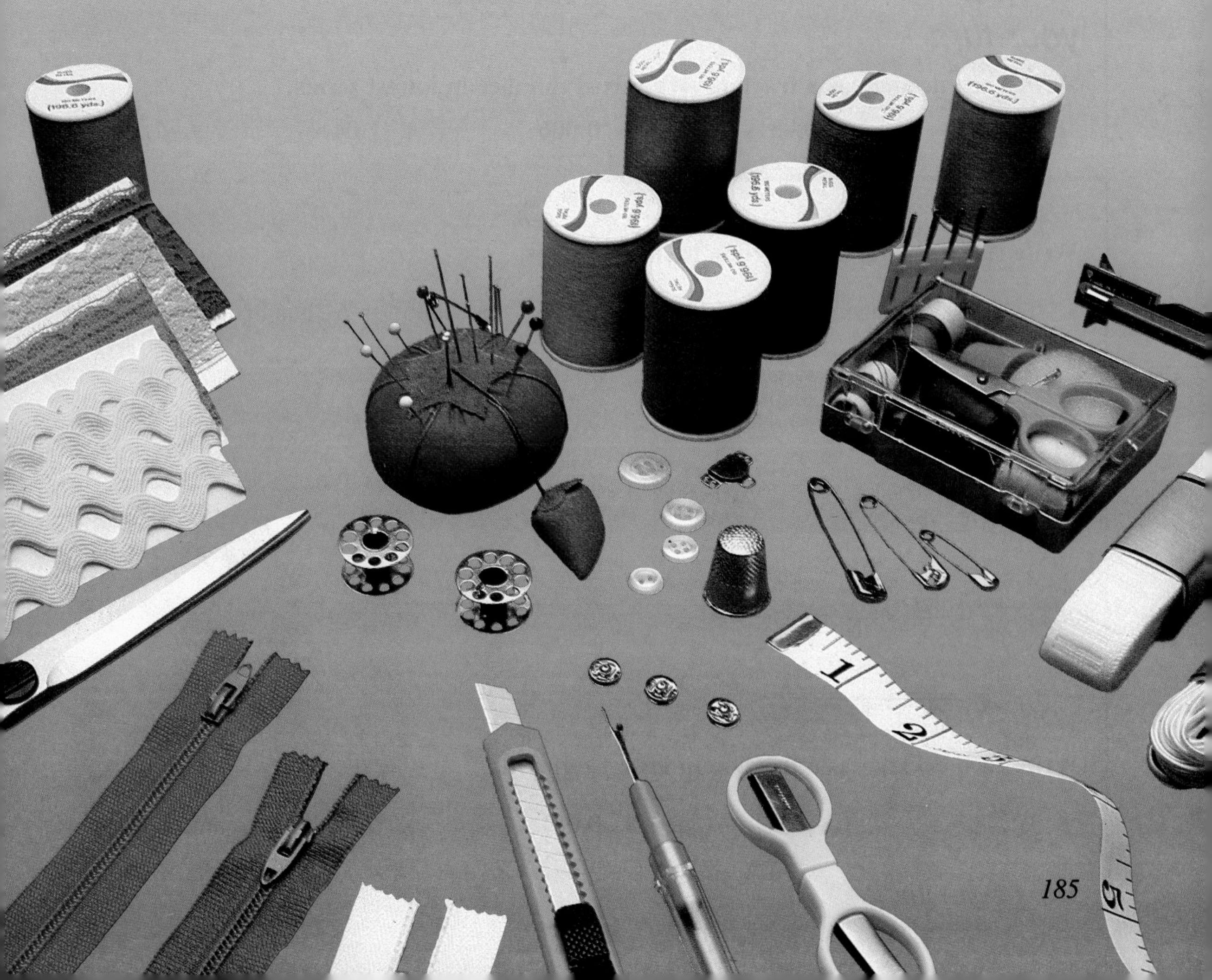

10-1 WRITING DECIMALS AS FRACTIONS

To write a decimal as a fraction, use the following method.

1. Use place value to write the decimal as a fraction whose denominator is a power of ten (10, 100, 1000, and so on).
2. Write the fraction in lowest terms.

example 1

Write each decimal as a fraction or a mixed number in lowest terms.

a. $0.625 = 625$ thousandths

$= \frac{625}{1000}$

$= \frac{5}{8}$ ← **Write the fraction in lowest terms.**

b. $1.6 = 1$ and 6 tenths

$= 1\frac{6}{10}$ ← **6 tenths** $= \frac{6}{10}$

$= 1\frac{3}{5}$

your turn

Write each decimal as a fraction or a mixed number in lowest terms.

1. 0.3
2. 0.45
3. 0.006
4. 1.04
5. 2.125

example 2

Write each decimal as a fraction in lowest terms.

a. $0.41\frac{2}{3} = 41\frac{2}{3}$ hundredths

$= \frac{41\frac{2}{3}}{100}$ ← **Write the decimal as a fraction.**

$= 41\frac{2}{3} \div 100$ ← **Write the fraction as a division.**

$= 41\frac{2}{3} \times \frac{1}{100}$

$= \frac{\overset{5}{\cancel{125}}}{3} \times \frac{1}{\underset{4}{\cancel{100}}} = \frac{5}{12}$

b. $0.06\frac{2}{3} = 6\frac{2}{3}$ hundredths

$= \frac{6\frac{2}{3}}{100}$

$= 6\frac{2}{3} \div 100$

$= 6\frac{2}{3} \times \frac{1}{100}$

$= \frac{\overset{1}{\cancel{20}}}{3} \times \frac{1}{\underset{5}{\cancel{100}}} = \frac{1}{15}$

your turn

Write each decimal as a fraction in lowest terms.

6. $0.33\frac{1}{3}$
7. $0.66\frac{2}{3}$
8. $0.08\frac{1}{2}$
9. $0.61\frac{1}{9}$
10. $0.26\frac{2}{3}$

practice exercises

practice for example 1 (page 186)

Write each decimal as a fraction or a mixed number in lowest terms.

1. 0.7 2. 0.5 3. 0.008 4. 0.375 5. 1.2
6. 4.8 7. 3.65 8. 9.08 9. 2.064 10. 5.15

practice for example 2 (page 186)

Write each decimal as a fraction in lowest terms.

11. $0.22\frac{2}{9}$ 12. $0.18\frac{2}{3}$ 13. $0.62\frac{1}{2}$ 14. $0.82\frac{1}{2}$ 15. $0.05\frac{1}{6}$
16. $0.04\frac{1}{3}$ 17. $0.91\frac{2}{3}$ 18. $0.38\frac{8}{9}$ 19. $0.13\frac{1}{3}$ 20. $0.72\frac{2}{9}$

mixed practice (page 186)

Write each decimal as a fraction or a mixed number in lowest terms.

21. 0.2 22. 0.6 23. 0.275 24. 1.032 25. 3.85
26. $0.05\frac{1}{2}$ 27. $0.83\frac{1}{3}$ 28. $0.37\frac{1}{2}$ 29. $0.04\frac{3}{4}$ 30. $0.26\frac{2}{5}$

Write each decimal as a fraction. Then find the answer in lowest terms.

31. $0.5 + \frac{2}{3}$ 32. $\frac{1}{2} - 0.1$ 33. $0.25 \times \frac{8}{9}$ 34. $0.2 \times \frac{4}{5}$ 35. $\frac{5}{6} \div 0.4$

Solve. Write each answer as a mixed number in lowest terms.

36. The weather bureau recorded 9.4 in. (inches) of snowfall in December and $8\frac{1}{4}$ in. of snowfall in January. What was the total snowfall recorded by the weather bureau for these two months?
37. Mrs. Moreira bought two pieces of material. One piece measured 1.75 yd (yards) and the other piece measured $1\frac{1}{8}$ yd. How much material did Mrs. Moreira buy in all?

review exercises

Write each answer in lowest terms.

1. $\frac{3}{8} \times \frac{9}{16}$ 2. $3\frac{1}{3} \times 2\frac{2}{5}$ 3. $9 \div 2\frac{1}{4}$ 4. $6\frac{1}{2} \div 2\frac{3}{4}$
5. $\frac{4}{5} + \frac{1}{6}$ 6. $\frac{9}{10} - \frac{3}{5}$ 7. $1\frac{1}{6} + 2\frac{2}{3}$ 8. $10\frac{1}{3} - 6\frac{1}{2}$

10-2 WRITING FRACTIONS AS DECIMALS

To write a fraction as a decimal, use the following method.

1. Write the fraction as a division.
2. Place a decimal point after the dividend.
3. Annex zeros as needed to divide.

$$\frac{3}{4} \rightarrow 4\overline{)3.} \rightarrow \begin{array}{r} 0.75 \\ 4\overline{)3.00} \\ \underline{2\,8} \\ 20 \\ \underline{20} \\ 0 \end{array}$$

example 1

Write each fraction or mixed number as a decimal.

a. $\frac{1}{4} \rightarrow \begin{array}{r} 0.25 \\ 4\overline{)1.00} \\ \underline{8} \\ 20 \\ \underline{20} \\ 0 \end{array}$ So $\frac{1}{4} = 0.25$.

b. $1\frac{1}{4} = 1 + \frac{1}{4}$
$= 1 + 0.25$
$= 1.25$

your turn

Write each fraction or mixed number as a decimal.

1. $\frac{2}{5}$
2. $\frac{3}{20}$
3. $\frac{3}{8}$
4. $1\frac{1}{2}$
5. $2\frac{3}{4}$

example 2

Write $\frac{5}{6}$ as a decimal.

$$\frac{5}{6} \rightarrow \begin{array}{r} 0.833\ldots \\ 6\overline{)5.000} \\ \underline{4\,8} \\ 20 \\ \underline{18} \\ 20 \\ \underline{18} \\ 2 \end{array}$$

$0.833\ldots = 0.8\overline{3}$ ← **The bar over the "3" tells you that 3 repeats. $0.8\overline{3}$ is called a *repeating decimal.***

2 ← **The remainder will always be 2.**

So $\frac{5}{6} = 0.8\overline{3}$.

your turn

Write each fraction or mixed number as a repeating decimal.

6. $\frac{1}{3}$
7. $1\frac{2}{9}$
8. $\frac{7}{15}$
9. $2\frac{1}{6}$
10. $\frac{3}{11}$

You can compare a decimal and a fraction by changing either the fraction to a decimal or the decimal to a fraction.

example 3

Compare. Replace each __?__ with >, <, or =.

a. $0.33 \ \underline{?}\ \frac{2}{5}$ ← **Write the fraction as a decimal.**

$\downarrow \quad \downarrow$

$0.33 < 0.4$, so $0.33 < \frac{2}{5}$.

b. $\frac{3}{5}\ \underline{?}\ 0.4$ ← **Write the decimal as a fraction.**

$\downarrow \quad \downarrow$

$\frac{3}{5} > \frac{2}{5}$, so $\frac{3}{5} > 0.4$.

your turn

Compare. Replace each __?__ with >, <, or =.

11. $0.45\ \underline{?}\ \frac{1}{3}$ **12.** $0.92\ \underline{?}\ \frac{9}{10}$ **13.** $\frac{5}{6}\ \underline{?}\ 0.75$ **14.** $1\frac{4}{5}\ \underline{?}\ 1.8$

practice exercises

practice for example 1 (page 188)

Write each fraction or mixed number as a decimal.

1. $\frac{1}{5}$ **2.** $\frac{1}{2}$ **3.** $\frac{5}{8}$ **4.** $\frac{4}{5}$ **5.** $\frac{1}{8}$

6. $\frac{3}{40}$ **7.** $1\frac{7}{8}$ **8.** $3\frac{13}{25}$ **9.** $9\frac{1}{40}$ **10.** $14\frac{7}{20}$

practice for example 2 (page 188)

Write each fraction or mixed number as a repeating decimal.

11. $\frac{2}{3}$ **12.** $\frac{5}{9}$ **13.** $\frac{11}{30}$ **14.** $\frac{17}{18}$ **15.** $\frac{8}{11}$

16. $\frac{2}{11}$ **17.** $3\frac{7}{9}$ **18.** $4\frac{1}{3}$ **19.** $1\frac{1}{6}$ **20.** $2\frac{11}{18}$

practice for example 3 (page 189)

Compare. Replace each __?__ with >, <, or =.

21. $0.55\ \underline{?}\ \frac{1}{2}$ **22.** $\frac{1}{8}\ \underline{?}\ 0.25$ **23.** $\frac{3}{40}\ \underline{?}\ 0.1$ **24.** $0.75\ \underline{?}\ \frac{9}{16}$

25. $\frac{5}{8}\ \underline{?}\ 0.625$ **26.** $\frac{11}{12}\ \underline{?}\ 0.5$ **27.** $2\frac{2}{3}\ \underline{?}\ 2.5$ **28.** $4\frac{3}{5}\ \underline{?}\ 4.6$

mixed practice (pages 188–189)

Write each fraction or mixed number as a decimal. If the decimal is a repeating decimal, use a bar to show the repeating part of the decimal.

29. $\frac{2}{9}$ **30.** $\frac{2}{5}$ **31.** $\frac{9}{200}$ **32.** $\frac{14}{25}$ **33.** $\frac{19}{20}$

34. $\frac{4}{15}$ **35.** $\frac{14}{15}$ **36.** $2\frac{5}{6}$ **37.** $1\frac{8}{11}$ **38.** $3\frac{7}{20}$

Compare. Replace each ? with >, <, or =.

39. $0.1 \underline{\ ?\ } \frac{1}{9}$ **40.** $\frac{2}{3} \underline{\ ?\ } 0.8$ **41.** $0.57 \underline{\ ?\ } \frac{3}{8}$ **42.** $3.6 \underline{\ ?\ } 3\frac{3}{5}$

43. $0.49 \underline{\ ?\ } \frac{37}{50}$ **44.** $4\frac{9}{10} \underline{\ ?\ } 4.86$ **45.** $5\frac{1}{4} \underline{\ ?\ } 5.25$ **46.** $9\frac{4}{5} \underline{\ ?\ } 9.7$

Choose the calculator display that best represents each fraction.

47. $\frac{1}{4}$ **a.** 0.25 **b.** 0.5 **c.** 0.025

48. $\frac{4}{9}$ **a.** 0.5555555 **b.** 0.4444444 **c.** 0.07777777

49. $\frac{7}{11}$ **a.** 0.6363636 **b.** 0.8181818 **c.** 0.2727272

50. $\frac{7}{8}$ **a.** 0.375 **b.** 0.625 **c.** 0.875

51. The weight of a baseball is about $5\frac{1}{8}$ oz (ounces). Write this mixed number as a decimal.

52. The length of an airplane is about $153\frac{1}{5}$ ft (feet). Write this mixed number as a decimal.

53. Sondra Giacobbe needs $2\frac{1}{2}$ lb (pounds) of hamburger for a recipe. She selects a package marked 2.27 lb. Is this more or less than she will need? Explain.

54. Valerie Richdale ordered $1\frac{3}{4}$ lb of sliced turkey. The package was labeled 1.79 lb. Is this more or less than she ordered? Explain.

review exercises

Find each answer.

1. $4.127 - 0.29$ **2.** 35×208 **3.** $82{,}106 - 12{,}477$ **4.** $728 + 96 + 3051$

5. $9338 \div 23$ **6.** $1.4\overline{)3.99}$ **7.** 0.08×1250 **8.** $4.9 + 14.6 + 1.8$

10-3 ESTIMATING WITH FRACTIONS AND DECIMALS

Estimating the answer to a problem before you solve the problem helps you decide whether your answer is reasonable. In problems that involve decimals, rounding is only one way of estimating an answer. Sometimes estimating is easier if you substitute a fraction for the decimal. You should choose a fraction that is close in value and that is easier for you to use.

You should memorize the fraction equivalents of commonly used decimals. The chart below lists some of these.

Equivalent Decimals and Fractions			
$0.2 = \frac{1}{5}$	$0.25 = \frac{1}{4}$	$0.125 = \frac{1}{8}$	$0.1\overline{6} = \frac{1}{6}$
$0.4 = \frac{2}{5}$	$0.5 = \frac{1}{2}$	$0.375 = \frac{3}{8}$	$0.\overline{3} = \frac{1}{3}$
$0.6 = \frac{3}{5}$	$0.75 = \frac{3}{4}$	$0.625 = \frac{5}{8}$	$0.\overline{6} = \frac{2}{3}$
$0.8 = \frac{4}{5}$		$0.875 = \frac{7}{8}$	$0.8\overline{3} = \frac{5}{6}$

You know that $\frac{2}{3} = 0.\overline{6}$, so $\frac{2}{3}$ is often a good estimate for decimals such as 0.68, 0.65, and 0.671.

example 1

Estimate each decimal as a simple fraction or mixed number.

a. 0.23 **b.** 0.498 **c.** 0.762 **d.** 2.6791

solution

a. about $\frac{1}{4}$ **b.** about $\frac{1}{2}$ **c.** about $\frac{3}{4}$ **d.** about $2\frac{2}{3}$

your turn

Estimate each decimal as a simple fraction or mixed number.

1. 0.53 **2.** 0.74 **3.** 0.19 **4.** 0.1659 **5.** 1.337

When you are estimating a product, you can sometimes use *compatible numbers*. As shown on the next page, first choose an appropriate fraction. Then choose a number that is compatible with the fraction and is close to the other factor.

example 2

Estimate using compatible numbers.

a. 0.19×249 b. 3548×0.762

solution

a. 0.19×249 ← **0.19 is about 0.2. You can use the fraction equivalent of 0.2.**

$\frac{1}{5} \times 250$ ← **250 is compatible with $\frac{1}{5}$ and is close to 249.**

about 50

b. 3548×0.762

$3600 \times \frac{3}{4}$

about 2700

your turn

Estimate using compatible numbers.

6. 0.339×605
7. 0.2453×1196
8. $\$25,165 \times 0.84$
9. 567×0.87

practice exercises

practice for example 1 (page 191)

Estimate each decimal as a simple fraction or mixed number.

1. 0.68
2. 0.512
3. 0.999
4. 0.989
5. 0.17
6. 0.6248
7. 5.749
8. 6.326
9. 10.1996
10. 25.262

practice for example 2 (page 192)

Estimate using compatible numbers.

11. $0.498 \times \$179$
12. 0.12×3056
13. 0.1643×4818
14. $\$35.99 \times 0.329$
15. 0.67×8953
16. $0.763 \times \$17,034$
17. 4122×0.585
18. 8821×0.627
19. $\$137 \times 0.839$
20. 345×0.41
21. $0.3765 \times \$65.19$
22. $0.7923 \times \$9836$

mixed practice (pages 191–192)

Match each decimal with a simple fraction estimate.

23. 0.51
24. 0.76
25. 0.341
26. 0.396
27. 0.772
28. 0.421
29. 0.534
30. 0.327

A. $\frac{1}{3}$ B. $\frac{1}{2}$ C. $\frac{3}{4}$ D. $\frac{2}{5}$

Estimate using compatible numbers.

31. 0.123×238
32. $\$314 \times 0.59$
33. $\$4487 \times 0.81$
34. $0.516 \times \$2052$
35. $\$1966 \times 0.261$
36. 0.668×2175
37. 0.878×15.79
38. 43.09×0.1651

39. Ellis wants to buy a bag of shrimp that weighs 0.495 lb (pound). The shrimp costs $7.89 per pound. About how much will the bag of shrimp cost?
40. Jackie bought a 0.74-lb block of cheese that cost $2.95 and a 14-oz (ounce) box of wheat crackers that cost $1.89. About how much did the two items cost?
41. A package of pears weighs 3.15 lb and costs $2.80. About how much do the pears cost per pound?
42. The length of a remnant of fabric is 0.65 yd (yard). The cost of the fabric is $5.99 per yard. About how much will the remnant cost?

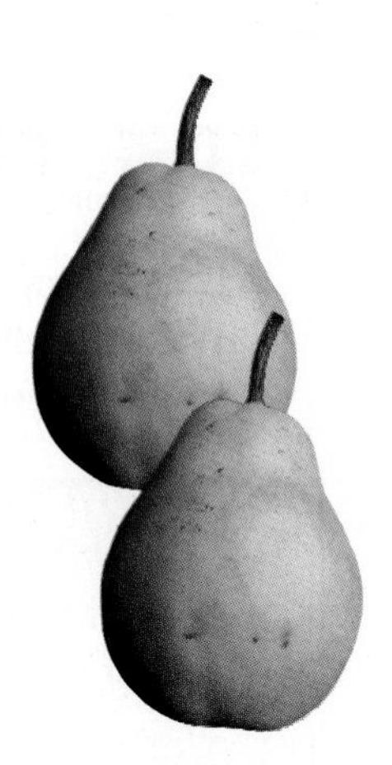

review exercises

Find each answer.

1. $6 \times 3 + 25 \div 5$
2. $(52 + 11) \div (7 \times 3)$
3. $30 - 2 \times 6 \div 4$
4. $2 \times (30 + 47)$
5. $4 \times 89 \times 25$
6. $5 \times (120 - 39)$

mental math

Sometimes rewriting a decimal as a fraction makes a computation easier to do. An exercise that looks difficult becomes one that you can do mentally.

0.25×360

↓ ↓

$\frac{1}{4} \times 360 = 90$

Multiply mentally.

1. 484×0.25
2. 88×0.375
3. 3.2×0.5
4. 575×0.2
5. $3.93 \times 0.\overline{3}$

calculator corner

You can use a calculator with a fraction key (a b/c) to compute with fractions and decimals. The answer will be in decimal form.

example $4\frac{3}{8} - 1.7$

solution Enter 4 (a b/c) 3 (a b/c) 8 − 1.7 =. The answer is 2.675.

Use a calculator to find each answer. Use a calculator with a fraction key if you have one.

1. $\frac{1}{8} + 1.23$
2. $13\frac{3}{4} - 7.1$
3. $0.75 + 3\frac{1}{8}$
4. $5.3 \times 2\frac{2}{5}$

PROBLEM SOLVING STRATEGY

10-4 TRIAL AND ERROR

1 Understand
2 Plan
3 Work
4 Answer

When solving a problem, you can learn from your mistakes. Different trials can lead you to find the correct answer. You should look to see what errors were made when solving the problem.

example

At the ski shop, the total cost of a lift ticket and ski rental is $42. If the lift ticket costs $12 more than the ski rental, how much does each cost?

solution

Make a guess for the cost of the ski rental. Add $12 to figure the cost of the lift ticket. Check for the correct total.

	ski rental	lift ticket	total cost	
first trial	$20	$20 + $12 = $32	$20 + $32 = $52	too high
second trial	$10	$10 + $12 = $22	$10 + $22 = $32	too low
third trial	$15	$15 + $12 = $27	$15 + $27 = $42	correct

The ski rental costs $15 and the lift ticket costs $27.

problems

1. Gino weighs 17 lb (pounds) more than his brother Rusty. If their total weight is 313 lb, how much does each brother weigh?
2. John bought some sweaters for $35 each and some pants for $50 per pair. He spent $240. How many of each did he buy?
3. Eva charges $7 for the rental of a bicycle and $18 for a moped. Her rentals totaled $174. How many rentals of each did she have?
4. Clark sold fans for $29.95 each and air conditioners for $425 each. If his sales totaled $2244.80, how many of each did he sell?
5. Tran needs to buy 24 paintbrushes. The cost of the paintbrushes is 5 for $12.95 or 7 for $17.50. Describe how he should make his purchase for the least possible cost.

Use the menu to answer the questions.

LUNCHEON SPECIALS	
Sandwiches:	
tuna salad	$3.10
ham and cheese	$3.85
Beverages:	
milk	$ .60
juice	$ .75
Desserts:	
yogurt	$ 1.00
fruit	$ .65

6. Chris bought a sandwich, a dessert, and two cartons of milk. He spent $5.70. What kind of sandwich did Chris buy? What kind of dessert?
7. Patrick ordered a sandwich, a beverage, and a dessert. If his lunch cost $5.60, what did he buy?
8. Jolene ordered five sandwiches. The total cost was $17.75. How many of each kind of sandwich did she order?
9. Can you make an order that costs $4.95 and includes one sandwich? If so, which items would be in the order?
10. Jackie Supino bought a ham and cheese sandwich. She also bought three other items. If the total cost was $6.60, which other items did Jackie buy?
11. Beverly Hitchcock bought a sandwich, a beverage, and a dessert. The total cost was $4.35. Which items did she order?

review exercises

If possible, solve. If it is not possible to solve, tell what additional information is needed.

1. Don Masters works at a car wash. He washes one car in 18 min. How many cars can Don wash in 4 h?
2. On Friday, Sydney bought three packages of AA batteries for $8.85. Each package contained four batteries. On Saturday, she returned to the same store to buy one additional package of batteries. About how much did one package cost?
3. Liza and Spencer played tennis on Monday. If their court time began at 4:00 P.M., what time did they finish?

An expression such as the one shown at the right is called a *continued fraction*. Find the value of this continued fraction.

$$1 + \frac{1}{1 + \frac{1}{1}}$$

SKILL REVIEW

Write each decimal as a fraction or a mixed number in lowest terms.

10-1

1. 0.9
2. 0.2
3. 0.72
4. 0.018
5. 0.004
6. 5.77
7. 3.375
8. 14.125
9. $0.83\frac{1}{3}$
10. $0.04\frac{1}{6}$
11. $0.37\frac{1}{2}$
12. $0.14\frac{2}{7}$
13. On a cross-country course, Anna skied 0.75 mi (mile) before she stopped to rest. Then she skied $1\frac{1}{2}$ mi. How many miles did Anna ski in all? Write the answer as a mixed number in lowest terms.

Write each fraction or mixed number as a decimal. If the decimal is a repeating decimal, use a bar to show the repeating part of the decimal.

10-2

14. $\frac{2}{25}$
15. $\frac{11}{50}$
16. $\frac{7}{25}$
17. $4\frac{1}{2}$
18. $2\frac{1}{8}$
19. $\frac{7}{18}$
20. $\frac{5}{11}$
21. $5\frac{2}{3}$
22. $3\frac{4}{33}$
23. $8\frac{5}{6}$

Compare. Replace each __?__ with >, <, or =.

24. 0.15 __?__ $\frac{1}{10}$
25. $3\frac{2}{5}$ __?__ 3.79
26. $\frac{1}{8}$ __?__ 0.125
27. 6.23 __?__ $6\frac{1}{4}$
28. The diameter of a tennis ball is $2\frac{5}{8}$ in. (inches), and the diameter of a racquetball is 2.25 in. Which ball has the larger diameter? Explain.

Estimate each decimal as a simple fraction or mixed number.

10-3

29. 0.324
30. 0.745
31. 0.811
32. 0.3748
33. 2.52
34. 23.26
35. 5.598
36. 44.397

Estimate using compatible numbers.

37. 0.671×302
38. $\$41.04 \times 0.498$
39. 158×0.6249
40. $\$19,897 \times 0.59$
41. $0.3329 \times \$11,942$
42. $\$1999.89 \times 0.247$
43. A container of shrimp salad weighs 0.595 lb (pound) and costs $5.19 per pound. About how much will the container of salad cost?

10-4

44. Elise bought some Jersey cows at $1100 each and some Hereford cows at $900 each. She spent a total of $7100. How many of each did Elise buy?
45. In gym class, Bret did a total of 47 push-ups and sit-ups. If Bret did 17 more sit-ups than push-ups, how many of each did he do?
46. At the resort, Paul rented chairs for $10 each and rafts for $7.50 each. The total cost was $35. How many of each did Paul rent?

APPLICATION

10-5 COMMUTING COSTS

Costs associated with traveling to and from work are called **commuting costs.** These costs can include highway tolls, gasoline, and parking fees. One way to reduce these expenses is to form a car pool. The commuting costs are then shared by the members of the car pool.

example

Elisa, John, Bernie, David, Eleanor, and Po Ling decided to form a car pool. To help determine the weekly cost of commuting, Elisa made the chart at the right.

Expense	Cost	
gasoline	$2.00	per day
turnpike toll	$ .65	each way
bridge toll	$ .30	per day
parking space	$7.50	per day

Determine the weekly cost of commuting for each member of the car pool.

solution

First determine the total commuting costs per day.

$$\begin{array}{r} \$2.00 \\ 1.30 \\ .30 \\ +7.50 \\ \hline \$11.10 \end{array}$$

← 2 × $.65 (for the 1.30 line)

Next determine commuting costs per week.

$$\begin{array}{r} \$11.10 \\ \times 5 \\ \hline \$55.50 \end{array}$$

← 5 work days

Then find out the cost per person.

$\frac{1}{6} \times \$55.50 = \$9.25$ ← There are six people, so the cost per person is $\frac{1}{6}$ of the total.

The weekly cost for each member of the car pool is $9.25.

exercises

1. Jeremiah, Manuel, Francisco, Pete, Colleen, and Marsha have formed a car pool. Their daily commuting expenses are $.40 each way for a bridge toll, $3.25 for gasoline, and $6.75 per day for parking. What is the weekly commuting cost for each person?
2. Shawn, Oliver, Ethel, and Julian all work in the same building. In order to reduce commuting costs, they form a car pool. Their expenses include $2.70 per day for gasoline, $.45 each way for the turnpike toll, and $6.00 per day for parking. Determine the weekly commuting cost for each member of the car pool.
3. When Kim drove to work alone, her commuting costs were $16 per week. She joined a car pool with four other people. The total weekly commuting cost for all the people in the car pool is $28.75. How much is Kim saving each week by belonging to the car pool?
4. The weekly commuting cost for each of the four members of a car pool is $7.50. One of the four members goes on vacation for a week. How much more will it cost each of the other three members to commute that week?

Another way of commuting is to use public transportation. For many people this is the most economical way to travel.

5. Using public transportation, Howard's commute costs him $2.70 per day. Howard works an average of 20 days per month. Is it less expensive to buy a monthly pass that costs $40 or to continue to pay each day? Explain.
6. Riley travels to work on a commuter train. The cost of a one-way train ticket is $1.75. Riley can buy a monthly pass that costs $56. Riley works 12 days per month. Would it be more or less expensive for him to buy a pass? Explain.
7. Traveling to work, Louise takes an express bus that costs $.75. To get home, she takes a train that costs $.60. Louise works six days per week. What is Louise's weekly commuting cost?
8. Rheta pays $1.50 per day to park her car at the train station. She then pays $.85 each way to ride the train to and from work. Rheta works 20 days per month. What is Rheta's daily commuting cost?
9. Find out what types of public transportation are available in your city.
   a. What is the cost of a bus ride? a train ticket?
   b. What is the cost of a monthly pass for a bus? for the train?
   c. Suppose you worked 20 days per month. Would it be more economical to travel by bus or by train? Would it be more economical to buy single tickets or a monthly pass?

APPLICATION

10-6 CLOTHING COSTS

Stephanie Broderick is planning to make a new skirt for the school dance. She needs the following items to make the skirt.

Item	Cost
pattern	\$3.25
$2\frac{1}{2}$ yd (yards) material	\$6.50 per yard
seam binding	\$.49
$\frac{3}{4}$ yd elastic	\$.80 per yard
thread	\$.79

example

a. Determine the cost of making the skirt.

b. In a department store a skirt of similar quality costs \$37.99. How much will Stephanie save by making the skirt?

solution

a. Add to find the cost of making the skirt.

$$\begin{array}{r} \$\ 3.25 \\ 16.25 \\ .49 \\ .60 \\ +\ .79 \\ \hline \$21.38 \end{array}$$

16.25 ← 2.5 × \$6.50

.60 ← 0.75 × \$.80

The cost of making the skirt is \$21.38.

b. Subtract the cost of making the skirt from the department store cost.

\$37.99 − \$21.38 = \$16.61

Stephanie will save \$16.61 by making the skirt.

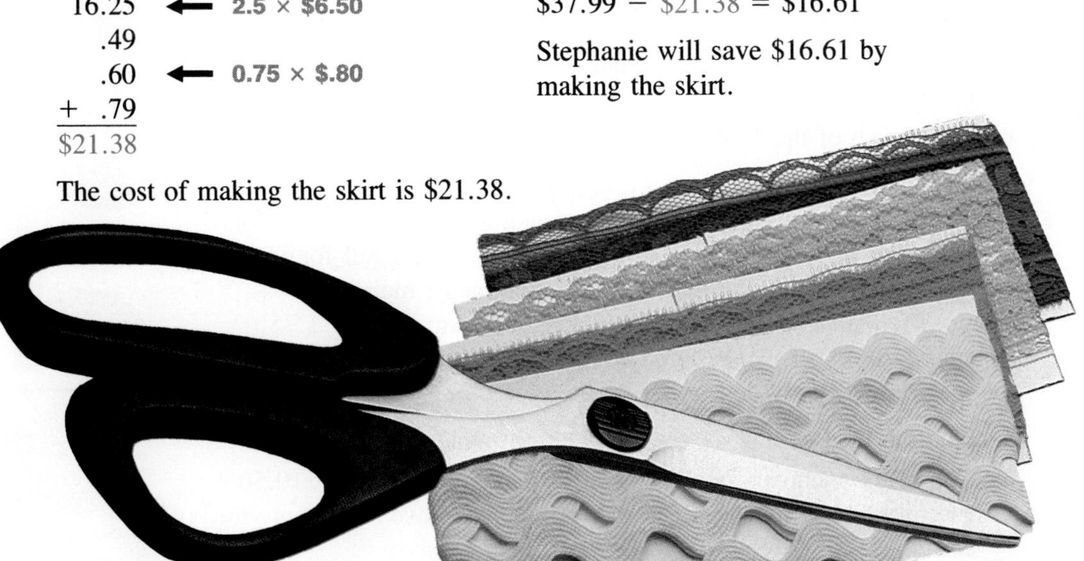

exercises

Find the total cost.

1.

Item	Cost
pattern	\$4.25
$6\frac{3}{4}$ yd material	\$3.24 per yard
0.5 yd elastic	\$.90 per yard
seam binding	\$.60
thread	\$1.10

2.

Item	Cost
pattern	\$3.75
$10\frac{1}{4}$ yd material	\$5.44 per yard
2 yd ribbon	\$1.19 per yard
seam binding	\$.95
thread	\$.89

3. The Lincoln High School baseball coach has decided to have all his players make their own baseball shirts this year. The pattern costs \$4.50, the material \$6 per yd, the buttons \$1.40, a package of seam binding \$.70, and thread \$1.18. David Sweeney, the pitcher, determines that he will need $1\frac{1}{2}$ yd of material.

   a. Find the total cost of making his baseball shirt.

   b. In a sports store, a baseball shirt of similar quality costs \$30.50. How much will David save by making his own shirt?

4. Maria Ruiz has saved \$800 for her bridal dress and her bridesmaid's dress, but this is not enough to buy both dresses. After looking at her budget, Maria decides to make her bridesmaid's dress and buy her bridal dress. Maria determined the following prices for the bridesmaid's dress. The pattern costs \$9, the material \$13.50 per yard, satin buttons \$4, a package of seam binding \$.50, shoulder pads \$5, and thread \$3.20.

   a. If Maria needs $6\frac{1}{2}$ yd of material, find the total cost of making the bridesmaid's dress.

   b. How much of the \$800 will be left for Maria to spend on her bridal dress?

5. The members of a high school ballet troupe need new costumes for an upcoming performance. In one store the costumes sell for \$55 each. Each member can either buy the costume or make the costume. The pattern costs \$3.50, a package of seam binding \$.75, ribbon \$2.20, elastic \$.50, thread \$.89, and snaps \$.79. The cost of the material will vary depending on the type used.

   a. Find the total cost of all the items except material.

   b. One of the members bought $3\frac{1}{2}$ yd of material that costs \$10.50 per yard. How much money did the member save by making the costume?

CHAPTER REVIEW

vocabulary vo·cab·u·lar·y

Choose the correct word or phrase to complete each sentence.

1. A number such as $0.\overline{1}$ is called a *(continued fraction, repeating decimal)*.
2. In estimation *(compatible numbers, mixed numbers)* are used because they multiply or divide easily.

skills

Write each decimal as a fraction or mixed number in lowest terms.

3. 0.4
4. 6.1
5. 0.375
6. 7.15
7. 0.28
8. 5.12
9. 0.225
10. 1.025
11. $0.19\frac{1}{5}$
12. $0.03\frac{1}{8}$
13. $0.48\frac{1}{3}$
14. $0.51\frac{2}{3}$

Write each fraction or mixed number as a decimal. If the decimal is a repeating decimal, use a bar to show the repeating part of the decimal.

15. $\frac{8}{25}$
16. $\frac{7}{50}$
17. $8\frac{3}{5}$
18. $1\frac{17}{40}$
19. $\frac{1}{33}$
20. $\frac{9}{11}$
21. $2\frac{14}{15}$
22. $\frac{4}{9}$

Estimate each decimal as a simple fraction or mixed number.

23. 0.511
24. 0.657
25. 0.2499
26. 3.63

Estimate using compatible numbers.

27. 0.4976×399
28. 0.876×2377
29. 451×0.198
30. 8054×0.24

31. Ray has a total of 45 cassette tapes and compact discs. He has four times as many tapes as discs. How many of each does he have?
32. Susan used to pay $42 per month in commuting costs when she drove her car to work. Now she takes the train to work for $1.10 per day. Susan works 20 days per month. How much money does Susan save each month by taking the train to work?
33. Mei Ling has decided to make her prom dress. She figured the following costs for the prom dress. The pattern costs $4.50, the material $7.96 per yd, lace $19.99, buttons $8.75, seam binding $.75, and thread $1.50. If Mei Ling needs to buy $4\frac{3}{4}$ yd of material, find the total cost of making the prom dress.

CHAPTER TEST

Write each decimal as a fraction or a mixed number in lowest terms.

10-1

1. 1.6 2. 3.444 3. 0.805 4. 0.41 5. 6.75
6. $0.12\frac{1}{2}$ 7. $0.18\frac{1}{3}$ 8. $0.02\frac{1}{4}$ 9. $0.08\frac{1}{6}$ 10. $0.66\frac{2}{3}$

Write each fraction or mixed number as a decimal. If the decimal is a repeating decimal, use a bar to show the repeating part of the decimal.

10-2

11. $\frac{1}{20}$ 12. $11\frac{1}{4}$ 13. $2\frac{19}{50}$ 14. $3\frac{17}{20}$ 15. $7\frac{7}{8}$
16. $\frac{5}{18}$ 17. $3\frac{5}{6}$ 18. $5\frac{8}{11}$ 19. $1\frac{4}{9}$ 20. $4\frac{1}{3}$

Compare. Replace each __?__ with >, <, or =.

21. $3\frac{3}{8}$ __?__ 3.59 22. $\frac{3}{4}$ __?__ 0.38 23. 2.8 __?__ $2\frac{4}{5}$ 24. 0.14 __?__ $\frac{7}{50}$

Estimate each decimal as a simple fraction or mixed number.

10-3

25. 0.53 26. 0.199 27. 3.34 28. 9.81
29. 0.657 30. 7.39 31. 10.767 32. 0.999

Estimate using compatible numbers.

33. 2099×0.67 34. $\$238 \times 0.2489$ 35. 0.4011×549 36. 879×0.1248

10-4

37. Sally bought some skirts for $26 each and some blouses for $23 each. She spent $173. How many of each did Sally buy?
38. Robert sold rakes for $5.55 and lawnmowers for $395. If his sales totaled $828.85, how many of each did he sell?

10-5

39. Roy belongs to a car pool. The weekly cost for each member is $7.75. When Roy commuted to work alone, his commuting costs were $18.50 per week. How much is Roy saving each week?
40. Five friends form a car pool. Their costs per day are $1.60 for gas, $1.45 for tolls, and $6 for parking. What is the weekly cost for each person?

10-6

41. Annie is planning on making a shirt. She determines the following prices for the shirt. The costs are $3.75 for the pattern, $5.48 per yd (yard) for the material, and $3.50 for a package of buttons. If she needs $1\frac{1}{2}$ yd of material, how much will it cost her altogether to make the shirt?

This robot arm can do many jobs, from welding to sewing. Robots are often used to perform tasks that require very precise measurements.

CHAPTER 11

U.S. CUSTOMARY MEASUREMENT

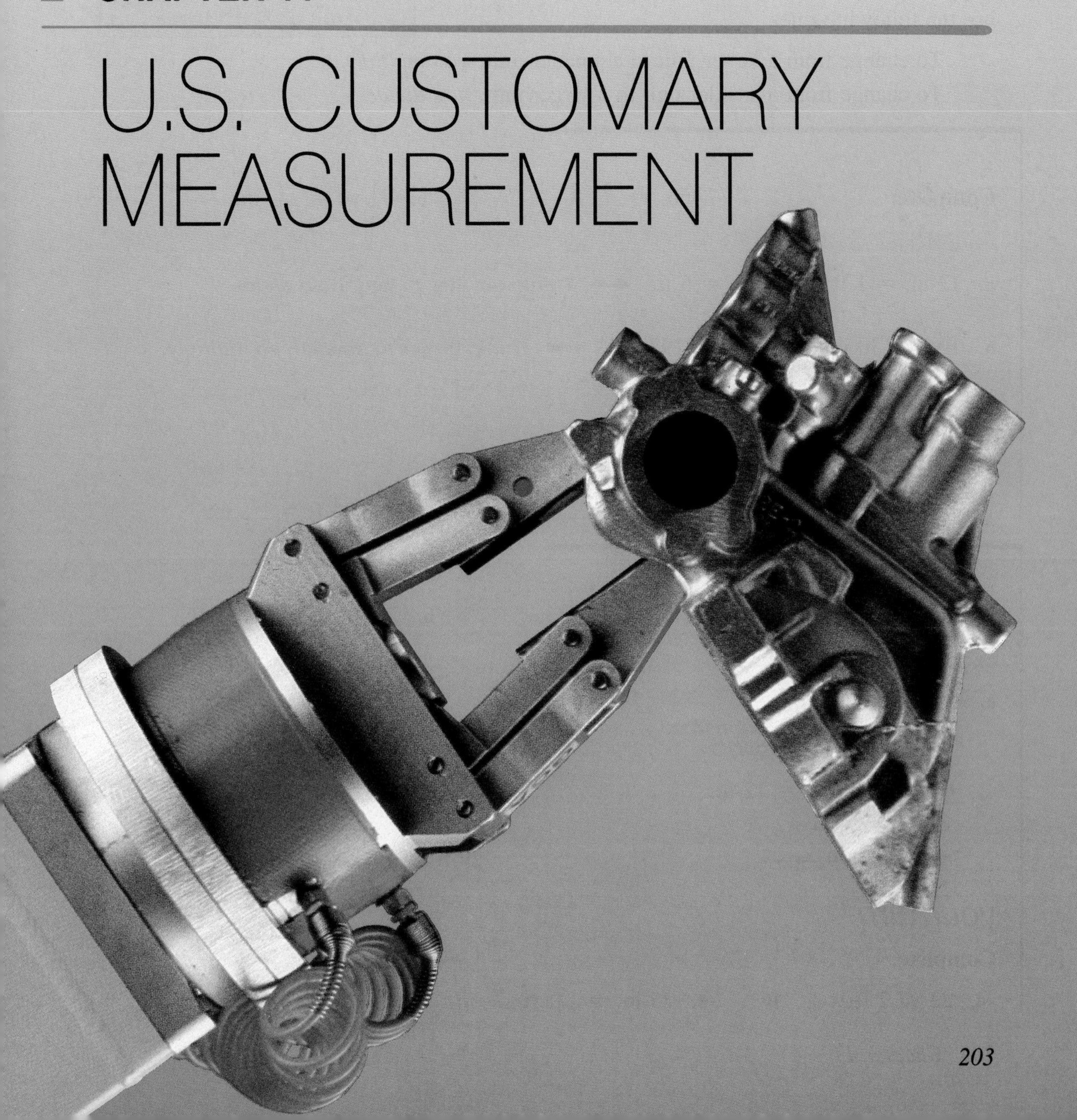

11-1 U.S. CUSTOMARY UNITS OF LENGTH

The commonly used units of length in the United States Customary system of measurement are the **inch, foot, yard,** and **mile.** The chart at the right shows how these units are related.

To change from one unit of length to another, use the following rules.

U.S. Customary Measures
12 inches (in.) = 1 foot (ft)
3 feet = 1 yard (yd)
36 inches = 1 yard
5280 feet = 1 mile (mi)
1760 yards = 1 mile

To change from a larger unit to a smaller unit, you *multiply*.

To change from a smaller unit to a larger unit, you *divide*.

example 1

Complete: **a.** 72 in. = _?_ ft **b.** $5\frac{1}{2}$ yd = _?_ in.

solution

a. 12 in. = 1 ft, so 72 in. = 6 ft (÷ 12) ← ***Think:* smaller to larger, so *divide.***

b. 1 yd = 36 in., so $5\frac{1}{2}$ yd = 198 in. (× 36) ← ***Think:* larger to smaller, so *multiply.***

your turn

Complete.

1. 39 ft = _?_ yd **2.** 3 mi = _?_ ft **3.** $3\frac{1}{4}$ ft = _?_ in.

example 2

Complete: **a.** 7 ft 8 in. = _?_ in. **b.** 29 ft = _?_ yd _?_ ft

solution

a.

$$\begin{array}{r} 12 \\ \times 7 \\ \hline 84 \\ +\ 8 \\ \hline 92 \end{array}$$

← ***Think:* larger to smaller, so *multiply.***

1 ft = 12 in. (× 12)

7 ft 8 in. = 92 in.

b.

$$\begin{array}{r} 9 \\ 3\overline{)29} \\ 27 \\ \hline 2 \end{array}$$

← ***Think:* smaller to larger, so *divide.***

3 ft = 1 yd (÷ 3)

29 ft = 9 yd 2 ft

your turn

Complete.

4. 22 yd 2 ft = _?_ ft **5.** 75 in. = _?_ ft _?_ in. **6.** 15,900 ft = _?_ mi _?_ ft

You can use a customary ruler to measure lengths in fractions of an inch. Many customary rulers separate the inch into sixteen equal parts.

example 3

Measure the line segment to the nearest $\frac{1}{16}$ inch.

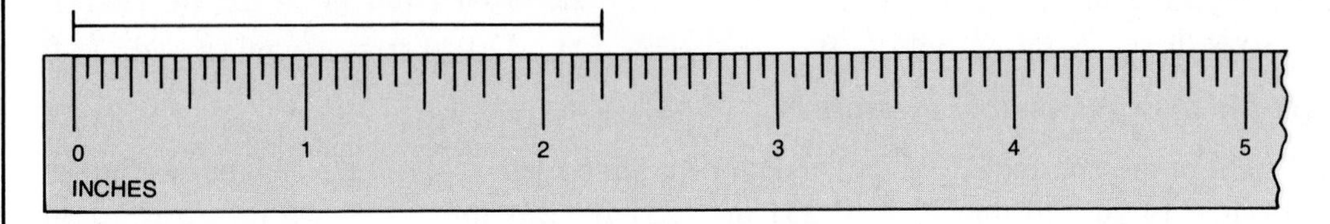

solution

Place the ruler so the zero mark lines up with one end of the line segment. Read the mark nearest the other end of the line segment. To the nearest $\frac{1}{16}$ inch, the length of the line segment is $2\frac{4}{16}$ in. $= 2\frac{1}{4}$ in.

your turn

Measure each line segment to the nearest $\frac{1}{16}$ inch.

7. ├──────────┤ **8.** ├──────────────┤

practice exercises

practice for example 1 (page 204)

Complete.

1. 9 yd = _?_ in. **2.** 16 yd = _?_ ft **3.** 84 in. = _?_ ft
4. 8800 yd = _?_ mi **5.** $7\frac{1}{3}$ ft = _?_ in. **6.** $2\frac{1}{2}$ mi = _?_ ft

practice for example 2 (page 204)

Complete.

7. 11 ft 7 in. = _?_ in. **8.** 18 yd 2 ft = _?_ ft **9.** 6 mi 87 yd = _?_ yd
10. 86 in. = _?_ ft _?_ in. **11.** 77 ft = _?_ yd _?_ ft **12.** 4000 yd = _?_ mi _?_ yd

practice for example 3 (page 205)

Measure each line segment to the nearest $\frac{1}{16}$ inch.

13. ├─────────────┤ **14.** ├───────┤

mixed practice (pages 204–205)

Complete.

15. 7040 yd = _?_ mi
16. $14\frac{2}{3}$ yd = _?_ ft
17. $9\frac{1}{4}$ ft = _?_ in.
18. 327 in. = _?_ ft _?_ in.
19. 249 in. = _?_ yd _?_ in.
20. 4932 yd = _?_ mi _?_ yd
21. 12 yd 2 ft 3 in. = _?_ in.
22. 9 yd 1 ft 7 in. = _?_ in.
23. 15,866 ft = _?_ mi _?_ yd _?_ ft
24. 42,944 ft = _?_ mi _?_ yd _?_ ft

Select the most reasonable measure.

25. height of an oak tree
    14 in. 14 yd 14 mi
26. length of a jumbo jet
    231 in. 231 ft 231 mi
27. height of a horse
    5 in. 5 ft 5 yd
28. width of a door
    1 ft 1 yd 1 mi
29. length of a bus
    45 in. 45 ft 45 yd
30. length of a pencil
    6 in. 6 ft 6 yd

Draw a line segment of the given length.

31. 3 in.
32. $5\frac{1}{4}$ in.
33. $4\frac{5}{8}$ in.
34. $3\frac{11}{16}$ in.
35. The Marianas Trench in the Pacific Ocean is about 36,200 ft deep. About how many miles is this?
36. Tomas is 5 ft 8 in. tall. Willis is 71 in. tall. Who is taller? How many inches taller?
37. Estimate the length of the cover of this book in inches.
38. Estimate the length in feet of the chalkboard in your classroom. Measure the length to check your estimate.

review exercises

Write each fraction or mixed number as a decimal. If the decimal is a repeating decimal, use a bar to show the repeating part of the decimal.

1. $\frac{5}{8}$
2. $2\frac{1}{3}$
3. $\frac{4}{5}$
4. $4\frac{3}{4}$
5. $\frac{2}{9}$
6. $3\frac{3}{10}$
7. $\frac{7}{9}$
8. $1\frac{2}{3}$

mental math

Find each answer mentally.

1. $8\frac{2}{9} - 5\frac{4}{9}$
2. $5\frac{1}{8} + 2\frac{3}{4} + 3\frac{7}{8}$
3. $\frac{5}{7} + \frac{3}{11} + \frac{8}{11}$
4. $7 - 5\frac{1}{4}$
5. $4\frac{1}{3} \times 9$
6. $4 \times 2\frac{1}{5}$
7. $6\frac{1}{2} + 3\frac{1}{3} + \frac{1}{2}$
8. $9 - 4\frac{3}{4}$

11-2 U.S. CUSTOMARY UNITS OF WEIGHT

U.S. Customary Measures
16 ounces (oz) = 1 pound (lb) 2000 pounds = 1 ton (t)

The commonly used units of weight in the United States Customary system of measurement are the **ounce, pound,** and **ton.** The chart at the right shows how these units are related.

To change from one unit of weight to another, use the following rules.

To change from a larger unit to a smaller unit, you *multiply*.
To change from a smaller unit to a larger unit, you *divide*.

example 1

Complete: **a.** 32 oz = ? lb **b.** $3\frac{1}{2}$ t = ? lb

solution

a. 16 oz = 1 lb, so 32 oz = 2 lb (÷ 16) ← ***Think:*** **smaller to larger, so *divide.***

b. 1 t = 2000 lb, so $3\frac{1}{2}$ t = 7000 lb (× 2000) ← ***Think:*** **larger to smaller, so *multiply.***

your turn

Complete.

1. 64 oz = ? lb **2.** 6000 lb = ? t **3.** $7\frac{1}{4}$ lb = ? oz

example 2

Complete: **a.** 5 lb 4 oz = ? oz **b.** 36 oz = ? lb ? oz

solution

a.

$$\begin{array}{r} 16 \\ \times 5 \\ \hline 80 \\ +\ 4 \\ \hline 84 \end{array}$$

← ***Think:*** **larger to smaller, so *multiply.***

1 lb = 16 oz (× 16)

5 lb 4 oz = 84 oz

b.

$$\begin{array}{r} 2 \\ 16\overline{)36} \\ \underline{32} \\ 4 \end{array}$$

← ***Think:*** **smaller to larger, so *divide.***

16 oz = 1 lb (÷ 16)

36 oz = 2 lb 4 oz

your turn

Complete.

4. 10 lb 9 oz = ? oz **5.** 5 t 350 lb = ? lb **6.** 454 oz = ? lb ? oz

practice exercises

practice for example 1 (page 207)

Complete.

1. 5 lb = ? oz
2. 11 t = ? lb
3. 176 oz = ? lb
4. 16,000 lb = ? t
5. $6\frac{1}{4}$ t = ? lb
6. $8\frac{1}{2}$ lb = ? oz

practice for example 2 (page 207)

Complete.

7. 9 lb 8 oz = ? oz
8. 5 lb 13 oz = ? oz
9. 4 t 1125 lb = ? lb
10. 8 t 1453 lb = ? lb
11. 270 oz = ? lb ? oz
12. 7950 lb = ? t ? lb

mixed practice (page 207)

Complete.

13. 14,000 lb = ? t
14. $4\frac{3}{4}$ lb = ? oz
15. $5\frac{2}{5}$ t = ? lb
16. 142 oz = ? lb ? oz
17. 8 lb 5 oz = ? oz
18. 4788 lb = ? t ? lb

Compare. Replace each ? with >, <, or =.

19. 60 oz ? 5 lb
20. 9 lb ? 144 oz
21. $2\frac{1}{2}$ t ? 5000 lb
22. 25,000 lb ? 12 t
23. 15,500 lb ? 7 t 500 lb
24. 316 oz ? 19 lb 14 oz
25. A bowling ball weighs about 256 oz. How many pounds is this?
26. A bus weighs about $13\frac{2}{5}$ t. How many pounds is this?

review exercises

Write each decimal as a fraction or mixed number in lowest terms.

1. 0.04
2. 3.5
3. 0.17
4. 16.8
5. 0.025
6. $0.24\frac{1}{2}$
7. $0.09\frac{1}{3}$
8. $0.55\frac{5}{9}$
9. $0.87\frac{1}{2}$
10. $0.16\frac{2}{3}$

puzzle corner

A grocer has five weights. Using these weights, the grocer can weigh quantities in pounds from 1 lb to 31 lb. How many pounds is each weight?

11-3 U.S. CUSTOMARY UNITS OF CAPACITY

The commonly used units of liquid capacity in the United States Customary system of measurement are the **fluid ounce, cup, pint, quart,** and **gallon.** The chart at the right shows how these units are related.

To change from one unit of capacity to another, use the following rules.

U.S. Customary Measures
8 fluid ounces (fl oz) = 1 cup (c)
2 cups = 1 pint (pt)
2 pints = 1 quart (qt)
4 quarts = 1 gallon (gal)
8 pints = 1 gallon

To change from a larger unit to a smaller unit, you *multiply*.

To change from a smaller unit to a larger unit, you *divide*.

example 1

Complete: **a.** 32 fl oz = _?_ c **b.** $4\frac{1}{2}$ qt = _?_ pt

solution

a. 8 fl oz = 1 c, so 32 fl oz = 4 c (÷ 8) ← ***Think:*** **smaller to larger, so *divide*.**

b. 1 qt = 2 pt, so $4\frac{1}{2}$ qt = 9 pt (× 2) ← ***Think:*** **larger to smaller, so *multiply*.**

your turn

Complete.

1. 18 c = _?_ pt
2. 26 c = _?_ fl oz
3. $7\frac{1}{4}$ gal = _?_ qt

example 2

Complete: **a.** 7 gal 3 qt = _?_ qt **b.** 17 pt = _?_ qt _?_ pt

solution

a.

$$\begin{array}{r} 7 \\ \times 4 \\ \hline 28 \\ +\ 3 \\ \hline 31 \end{array}$$

← ***Think:*** **larger to smaller, so *multiply*.**

1 gal = 4 qt (× 4)

7 gal 3 qt = 31 qt

b.

$$\begin{array}{r} 8 \\ 2\overline{)17} \\ 16 \\ \hline 1 \end{array}$$

← ***Think:*** **smaller to larger, so *divide*.**

2 pt = 1 qt (÷ 2)

17 pt = 8 qt 1 pt

your turn

Complete.

4. 16 qt 1 pt = _?_ pt
5. 34 qt = _?_ gal _?_ qt
6. 21 c = _?_ pt _?_ c

Length, capacity, and weight are different types of measurement. You must be able to determine the appropriate unit of measurement.

example 3

Choose the letter of the appropriate unit for measuring the given object.

the weight of a bird **a.** fl oz **b.** oz **c.** in.

solution

You must find the *weight* of an object. Weight can be measured in ounces. Choose **b.**

your turn

Choose the letter of the appropriate unit for measuring the given object.

7. the length of a car **a.** gal **b.** lb **c.** ft

8. the capacity of a bucket **a.** qt **b.** yd **c.** oz

practice exercises

practice for example 1 (page 209)

Complete.

1. 24 pt = __?__ gal **2.** 28 pt = __?__ qt **3.** 14 c = __?__ fl oz

4. 11 gal = __?__ pt **5.** $3\frac{1}{2}$ gal = __?__ qt **6.** $6\frac{1}{4}$ c = __?__ fl oz

practice for example 2 (page 209)

Complete.

7. 27 pt = __?__ qt __?__ pt **8.** 60 fl oz = __?__ c __?__ fl oz **9.** 50 qt = __?__ gal __?__ qt

10. 3 gal 1 pt = __?__ pt **11.** 18 pt 1 c = __?__ c **12.** 11 gal 3 qt = __?__ qt

practice for example 3 (page 210)

Choose the letter of the appropriate unit for measuring the given object.

13. the length of a bridge **a.** lb **b.** gal **c.** ft

14. the capacity of a water tank **a.** mi **b.** gal **c.** t

15. the length of a football field **a.** lb **b.** qt **c.** yd

16. the weight of a table **a.** lb **b.** ft **c.** pt

mixed practice (pages 209–210)

Complete.

17. 18 pt = ? c

18. $9\frac{1}{2}$ qt = ? pt

19. $2\frac{3}{4}$ c = ? fl oz

20. 5 gal 3 qt = ? qt

21. 25 pt = ? qt ? pt

22. 22 qt = ? gal ? qt

Select the most reasonable measure.

23. length of a house
50 qt 50 ft 50 lb

24. capacity of a sink
20 oz 20 qt 20 in.

25. weight of a cat
14 in. 14 pt 14 lb

26. capacity of a mug
12 oz 12 ft 12 fl oz

27. weight of a tennis ball
2 fl oz 2 in. 2 oz

28. length of a hockey rink
66 lb 66 yd 66 gal

29. Lou bought 6 pt of orange juice. How many quarts did he buy?

30. A standard barrel of oil holds 42 gal. How many quarts is this?

31. A case of fruit juice contains 24 cans. Each can holds 12 fl oz. How many quarts of juice are in a case?

32. How many 8-fl-oz glasses of water can be filled from a 5-gal jug?

review exercises

Find each answer.

1. 3^4 **2.** 8^2 **3.** 1^{12} **4.** 10^0 **5.** 2^6 **6.** 7^1 **7.** 4^5 **8.** 5^3

calculator corner

You can use a calculator, mental math, or paper-and-pencil to change from one United States Customary unit to another.

example **Complete:** **a.** 12 mi = ? yd **b.** 4 t = ? lb

solution

a. 1 mi = 1760 yd
Use a calculator.
Enter 12 × 1760 =.
The result is 21,120.
So 12 mi = 21,120 yd.

b. 1 t = 2000 lb
Use mental math.
Think: 4 × 2000
The result is 8000.
So 4 t = 8000 lb.

Tell whether you would complete each exercise using a *calculator, mental math,* or *paper-and-pencil.* Then complete.

1. 244 yd = ? in.

2. 6 t = ? lb

3. 12 gal = ? pt

4. 20 c = ? fl oz

5. 27 yd = ? ft

6. 485 lb = ? oz

11-4 COMPUTING WITH CUSTOMARY UNITS

When you compute with United States Customary units, you may need to rename a measurement.

example 1

Write each sum or product in simplest form.

a. 17 ft 9 in. + 24 ft 7 in.

b. 9 lb 7 oz × 5

solution

a.

$$\begin{array}{r} 17 \text{ ft } \ 9 \text{ in.} \\ +24 \text{ ft } \ 7 \text{ in.} \\ \hline 41 \text{ ft } 16 \text{ in.} \end{array} = 41 \text{ ft} + 1 \text{ ft } 4 \text{ in.} = 42 \text{ ft } 4 \text{ in.}$$

b.

$$\begin{array}{r} 9 \text{ lb } \ 7 \text{ oz} \\ \times \quad 5 \\ \hline 45 \text{ lb } 35 \text{ oz} \end{array} = 45 \text{ lb} + 2 \text{ lb } 3 \text{ oz} = 47 \text{ lb } 3 \text{ oz}$$

your turn

Write each sum or product in simplest form.

1. 29 ft 8 in. + 12 ft 2 in.
2. 5 t 1798 lb + 955 lb
3. 31 qt 1 pt + 22 qt 1 pt
4. 12 ft 8 in. × 4
5. 4 t 1200 lb × 3
6. 8 c 6 fl oz × 7

example 2

Write each difference or quotient in simplest form.

a. 6 lb 7 oz − 3 lb 11 oz

b. 10 qt 1 pt ÷ 3

solution

a.

$$\begin{array}{r} 6 \text{ lb } \ 7 \text{ oz} \\ -\ 3 \text{ lb } 11 \text{ oz} \\ \hline \end{array} = \begin{array}{r} 5 \text{ lb } 23 \text{ oz} \\ -\ 3 \text{ lb } 11 \text{ oz} \\ \hline 2 \text{ lb } 12 \text{ oz} \end{array}$$

← **6 lb 7 oz = 5 lb + 16 oz + 7 oz = 5 lb 23 oz**

b.

$$3\overline{)10 \text{ qt } 1 \text{ pt}} = 3\overline{)9 \text{ qt } 3 \text{ pt}} \quad \text{quotient: } 3 \text{ qt } 1 \text{ pt}$$

← **10 qt 1 pt = 9 qt + 2 pt + 1 pt = 9 qt 3 pt**

your turn

Write each difference or quotient in simplest form.

7. 24 gal 3 qt − 15 gal 1 qt
8. 86 ft 3 in. − 42 ft 10 in.
9. 37 lb − 24 lb 13 oz
10. 16 yd 2 ft ÷ 5
11. 25 gal 2 qt ÷ 2
12. 14 lb 8 oz ÷ 4

practice exercises

practice for example 1 (page 212)

Write each sum or product in simplest form.

1. 12 mi 237 yd + 28 mi 562 yd
2. 54 gal 3 qt + 14 gal 2 qt
3. 31 lb 12 oz + 66 lb 10 oz
4. 76 ft 8 in. + 28 ft 9 in.
5. 12 ft 5 in. × 3
6. 5 t 400 lb × 7
7. 13 c 6 fl oz × 6
8. 9 lb 14 oz × 4

practice for example 2 (page 212)

Write each difference or quotient in simplest form.

9. 9 c 6 fl oz − 7 c 5 fl oz
10. 8 yd 2 ft − 5 yd 1 ft
11. 26 lb 8 oz − 15 lb 13 oz
12. 34 ft − 18 ft 9 in.
13. 28 ft 9 in. ÷ 3
14. 50 yd 2 ft ÷ 8
15. 31 lb 2 oz ÷ 6
16. 23 c 5 fl oz ÷ 7

mixed practice (page 212)

Write each answer in simplest form.

17. 24 c 6 fl oz + 11 c 5 fl oz
18. 13 pt 1 c × 5
19. 35 ft 4 in. − 23 ft 9 in.
20. 20 t 200 lb ÷ 6
21. 7 mi 540 yd × 7
22. 11 t − 6 t 1649 lb
23. 31 gal 2 qt ÷ 6
24. 47 yd 2 ft + 51 yd 2 ft
25. 30 gal 2 qt 1 pt × 3
26. 50 yd 1 ft 3 in. ÷ 3
27. The women's world record for the javelin throw is 247 ft 4 in. The men's world record is 343 ft 10 in. How much farther is the men's world record?
28. Carlos Santiago is practicing for a marathon by running on a 440-yd track. In the morning he ran 8 mi 880 yd. Later he ran 6 mi 1320 yd. How far did Carlos run altogether?

review exercises

Complete.

1. 3 g = ? mg
2. 8.4 cm = ? mm
3. 29 mL = ? L
4. 4800 m = ? km
5. 90 L = ? mL
6. 112 mm = ? m
7. 0.16 kg = ? g
8. 5.2 m = ? cm

11-5 PRECISION AND SIGNIFICANT DIGITS

Every measurement is an approximation. The **precision** of a measurement is determined by the smallest unit on the measuring instrument. When a measurement is recorded, the **unit of precision** is indicated by the place of the last **significant digit.** Precision is important in scientific work.

Significant Digits	Not Significant Digits
• all nonzero digits • all zeros between significant digits • all zeros at the end of a decimal • underlined zeros	• zeros that immediately follow or precede the decimal point in a decimal less than 1 • zeros ending a whole number, unless underlined

example 1

Give the number of significant digits and the unit of precision.

a. 307.04 m ← **All digits are significant.**
5 significant digits; 0.01 m

b. 0.062 L ← **The zeros are not significant.**
2 significant digits; 0.001 L

c. $18,\underline{0}00$ mi ← **An underlined zero is significant.**
3 significant digits; 100 mi

d. 6.50 kg ← **A zero at the end of a decimal is significant.**
3 significant digits; 0.01 kg

your turn

Give the number of significant digits and the unit of precision.

1. 1050 mi
2. $11,0\underline{0}0$ lb
3. 0.08 km
4. 15.70 L

In a computation that involves measurements, the result cannot be more precise than the least precise measurement involved.

example 2

Find each answer. Round to the correct number of significant digits.

a. 1510 lb + 274 lb

b. 82.9 m − 0.32 m

solution

a.
1510 lb ← **precise to tens**
+ 274 lb ← **precise to ones**
1784 lb ≈ 1780 lb ← **precise to tens**

b.
82.9 m ← **precise to tenths**
− 0.32 m ← **precise to hundredths**
82.58 m ≈ 82.6 m ← **precise to tenths**

your turn

Find each answer. Round to the correct number of significant digits.

5. 8600 yd − 3450 yd
6. 7.84 kg + 15.9 kg
7. 235.7 L − 17.04 L

practice exercises

practice for example 1 (page 214)

Give the number of significant digits and the unit of precision.

1. 306 yd
2. 2040 lb
3. 17.01 cm
4. 19.0 mL
5. 1100 ft
6. $15\underline{0}0$ gal
7. $20\underline{0}0$ mi
8. 0.084 kg

practice for example 2 (pages 214–215)

Find each answer. Round to the correct number of significant digits.

9. 409 ft + 4810 ft
10. 5442 gal + 230 gal
11. 37.4 km + 108 km
12. 14,000 yd − 5700 yd
13. 7100 lb − 475 lb
14. 1.04 kg − 0.073 kg

mixed practice (pages 214–215)

Choose the less precise measurement.

15. 14,500 ft; 750 ft
16. 1.34 g; 0.008 g
17. 9.021 m; 50.24 m
18. $11{,}0\underline{0}0$ mi; 1800 mi
19. 43.0 mg; 43 mg
20. $27\underline{0}0$ gal; $270\underline{0}$ gal
21. 37.03 m; 37.3 m
22. 0.040 L; 0.04 L
23. 707 mi; 1270 mi
24. 1901 lb; $28\underline{0}0$ lb
25. 6.02 kg; 6.020 kg
26. 1400 gal; 280 gal
27. The Panama Canal is about 89,000 yd long. The Houston Canal is about 75,680 yd long. Graham said that the Panama Canal is 13,320 yd longer than the Houston Canal. Is Graham's statement scientifically correct? Explain.
28. The Sears Tower in Chicago is 1454 ft tall. The Chrysler Building in New York is 1127 ft tall. Renee said that the Sears Tower is 327 ft taller than the Chrysler Building. Is Renee's statement scientifically correct? Explain.

review exercises

Is the first number divisible by the second number? Write *Yes* or *No*.

1. 2865; 5
2. 924; 8
3. 717; 3
4. 1023; 10
5. 615; 2
6. 1992; 3
7. 10,720; 4
8. 883; 9

SKILL REVIEW

Complete. **11-1**

1. 96 in. = _?_ ft
2. 43 yd = _?_ ft
3. $6\frac{1}{2}$ yd = _?_ in.
4. 9 ft 7 in. = _?_ in.
5. 3 mi 94 ft = _?_ ft
6. 8595 yd = _?_ mi _?_ yd

Measure each line segment to the nearest $\frac{1}{16}$ inch.

7. |——————————————————|
8. |————————|

9. The Verrazano-Narrows bridge in New York is 4260 ft long. How many yards is this?

Complete. **11-2**

10. 208 oz = _?_ lb
11. 5 t = _?_ lb
12. $8\frac{1}{8}$ lb = _?_ oz
13. 15 lb 7 oz = _?_ oz
14. 6 t 794 lb = _?_ lb
15. 79 oz = _?_ lb _?_ oz
16. Each stone of the Great Pyramid in Egypt weighs about $2\frac{1}{2}$ t. About how many pounds does each stone weigh?

Complete. **11-3**

17. 136 fl oz = _?_ c
18. 14 gal = _?_ qt
19. $7\frac{1}{4}$ c = _?_ fl oz
20. 23 c = _?_ pt _?_ c
21. 33 pt = _?_ qt _?_ pt
22. 19 gal 1 qt = _?_ qt

Choose the letter of the appropriate unit for measuring the given object.

23. the weight of a baseball — **a.** oz **b.** c **c.** in.
24. the length of an umbrella — **a.** qt **b.** in. **c.** lb

Write each answer in simplest form. **11-4**

25. 4 t 675 lb + 9 t 1855 lb
26. 72 ft 4 in. − 48 ft 5 in.
27. 23 qt 1 pt × 3
28. 30 lb 10 oz ÷ 7
29. The Baltimore Harbor tunnel is 6336 ft long. The Hampton Roads tunnel is 1 mi 2112 ft long. How many feet longer is the Hampton Roads tunnel than the Baltimore Harbor tunnel?

Give the number of significant digits and the unit of precision. **11-5**

30. 780 ft
31. 11.08 L
32. 15,0$\underline{0}$0 yd
33. 0.060 km

Find each answer. Round to the correct number of significant digits.

34. 12,450 ft + 4700 ft
35. 540.57 kg − 317 kg

11-6 GASOLINE MILEAGE

The number of miles per gallon of gasoline that a car gets is called its **gasoline mileage.** To find gasoline mileage, you use the odometer readings when the gas tank is full.

example

When Mia last filled the gas tank of her car, the odometer showed 8263.5 mi. When she filled the tank this time, the odometer showed 8645.3 mi. Mia bought 15.8 gal of gasoline. Find the gasoline mileage to the nearest tenth of a mile per gallon.

solution

Find the number of miles traveled.

$8645.3 - 8263.5 = 381.8$

Divide the number of miles traveled by the number of gallons of gasoline used.

$381.8 \div 15.8 \approx 24.16$ ← **Divide to the hundredths' place.**

The car's gasoline mileage is about 24.2 mi/gal (miles per gallon).

exercises

Complete. If necessary, round to the nearest tenth of a mile per gallon. A calculator may be helpful.

	ODOMETER READING FIRST	ODOMETER READING SECOND	GALLONS OF GASOLINE USED	GASOLINE MILEAGE
1.	5041.8	5230.5	11.1	?
2.	12,345.6	12,829.2	15.6	?
3.	18,413.3	18,794.1	13.2	?
4.	44,444.4	44,654.2	10.8	?
5.	88,816.3	89,178.7	14.7	?
6.	69,835.2	70,158.9	12.5	?

7. The gasoline mileage of a car may be different at certain times. List some of the circumstances that may cause this difference.

8. Each year the federal government publishes the gasoline mileage for all new car models. Find out how the government determines these gasoline mileages.

APPLICATION

11-7 CATALOGUE SHOPPING

When you shop from a catalogue, you must fill out an order form. Often shipping charges are added to the cost of your order. These charges may be determined by the weight of your order and the distance you live from the catalogue merchandise distribution center. These shipping charges are usually listed in a chart like the one shown at the right.

SHIPPING WEIGHT	APPROXIMATE DISTANCE FROM CATALOGUE MERCHANDISE DISTRIBUTION CENTER					
	Local in CMDC city	Not over 150 miles	151 to 300 miles	301 to 450 miles	451 to 600 miles	Over 600 miles
1 oz to 8 oz	$ 1.94	$ 2.23	$ 2.23	$ 2.23	$ 2.23	$ 2.23
9 oz to 15 oz	2.36	2.69	2.69	2.69	2.69	2.69
1 lb to 2 lb	2.68	3.26	3.26	3.26	3.26	3.26
2 lb 1 oz to 3 lb	2.75	3.48	3.48	3.71	3.71	4.14
3 lb 1 oz to 5 lb	2.82	3.73	3.73	4.03	4.03	4.87
5 lb 1 oz to 10 lb ...	3.08	4.69	4.69	5.01	5.01	5.66
10 lb 1 oz to 15 lb ..	3.90	5.13	6.98	6.98	6.98	7.47
15 lb 1 oz to 25 lb ..	4.40	6.10	8.04	8.22	8.33	9.09
25 lb 1 oz to 45 lb ..	5.07	8.27	10.11	10.35	10.64	13.95
45 lb 1 oz to 55 lb ..	6.17	10.43	12.70	13.59	14.77	16.75
55 lb 1 oz to 70 lb ..	7.30	11.39	13.72	15.67	16.27	18.99

example

Amy decides to order the television, video cassette recorder, and table. Complete the order form. Use the chart of shipping charges above. Amy lives 189 mi from the catalogue merchandise distribution center.

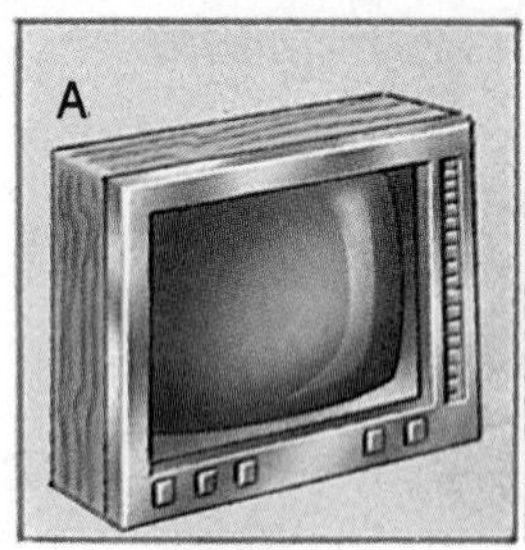

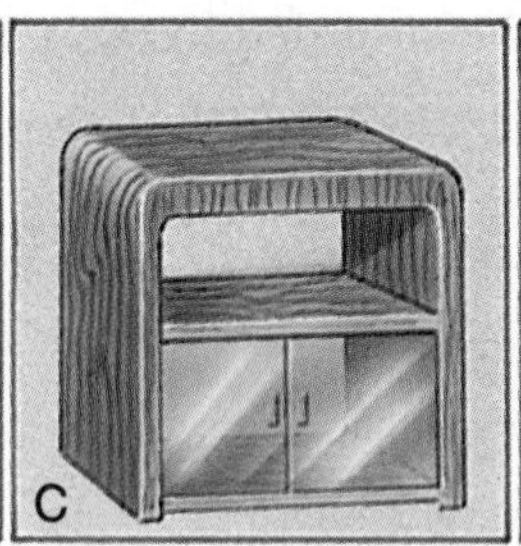

A. 13 Inch Color TV
#3-124 $179.00
shipping weight: 25 lb 8 oz

B. Video Cassette Recorder
#6-842 $249.00
shipping weight: 21 lb 12 oz

C. TV Table
#5-090 $29.95
shipping weight: 14 lb 6 oz

solution

Item no.	How many	Ship wgt.	Price each	Total price
3-124	1	25 lb 8 oz	$179.00	$179.00
6-842	1	21 lb 12 oz	$249.00	$249.00
5-090	1	14 lb 6 oz	$29.95	$29.95
			Item total	$457.95
	Total ship wgt.	61 lb 10 oz	Shipping	$ 13.72
			Total due	$471.67

← **List each item separately.**

← **Add the prices of the items.**

← **Use the chart to find the shipping charges.**

← **Add the item total and the shipping charges.**

exercises

Copy and complete each order form. Use the chart of shipping charges on page 218. Assume the customers live 179 mi from the catalogue merchandise distribution center.

1.

Item no.	How many	Ship wgt.	Price each	Total price
4307	1	10 lb 8 oz	$199.95	?
6342	1	8 lb 4 oz	$ 34.95	?
			Item total	?
	Total ship wgt.	?	Shipping	?
			Total due	?

2.

Item no.	How many	Ship wgt.	Price each	Total price
5-1101	1	24 lb 9 oz	$ 99.95	?
5-1164	2	12 lb 6 oz	$ 32.50	?
			Item total	?
	Total ship wgt.	36 lb 15 oz	Shipping	?
			Total due	?

Use the catalogue page at the right and the chart of shipping charges on page 218.

3. Rico orders the electronic flash, the wide-angle lens, and the zoom lens. He lives 97 mi from the catalogue merchandise distribution center. Find the total cost of his order.
4. Lisa orders the 35-mm camera, the tripod, and the camera case. She lives 506 mi from the catalogue merchandise distribution center. Find the total cost of her order.
5. Benton ordered a chair from the catalogue. The chair cost $399.95. The shipping weight was 24 lb 10 oz. Benton lives 217 mi from the catalogue merchandise distribution center. Find the total cost of his order.
6. Shipping charges may be determined by other methods. Find out what some of these methods are.

A. Tripod #2-198
shipping weight:
4 lb 8 oz $37.99

B. Camera Case
#2-050
shipping weight:
1 lb 14 oz
$19.99

C. 35 mm Camera
#2-008
shipping weight:
3 lb 5 oz
$299.00

D. Zoom Lens
#2-093
shipping weight:
1 lb 2 oz $47.95

E. Wide-Angle Lens
#2-094
shipping weight:
1 lb 4 oz $47.95

F. Electronic Flash
#2-103
shipping weight:
1 lb 3 oz $34.50

A B C D E F

CAREER

CATERER

Sam Tung manages a small catering company. To determine how much of each ingredient he needs to purchase for his recipes, Sam may need to change from one United States Customary unit to another.

example

A recipe requires 4 c of tomato juice. How many 6-fl-oz cans of tomato juice should Sam purchase?

solution

Change cups to fluid ounces.

1 c = 8 fl oz, so 4 c = 32 fl oz
(× 8) (× 8)

Divide to find the number of cans.

$32 \div 6 = 5\frac{1}{3}$

Since Sam needs *more than* 5 cans, he should purchase 6 cans.

exercises

1. A recipe requires 5 c of custard-style yogurt. How many 6-fl-oz containers should the chef purchase?
2. A recipe requires 3 pt of a baking sauce. The sauce is packaged in 8-fl-oz containers. How many containers of sauce should the chef purchase?
3. A chef is going to use a recipe that requires 6 lb of fresh tomatoes. How many 12-oz packages of fresh tomatoes should the chef purchase?
4. A chef is going to use a recipe that requires 4 lb of canned tomatoes. How many 28-oz cans of tomatoes should the chef purchase?
5. A chef needs $2\frac{1}{2}$ lb of mushrooms for a recipe. How many 12-oz packages of mushrooms should the chef purchase?
6. A chef needs $8\frac{3}{4}$ lb of ground beef for a recipe. How many 48-oz packages of ground beef should the chef purchase?

CHAPTER REVIEW

vocabulary vo·cab·u·lar·y

Choose the correct word to complete each sentence.

1. A commonly used customary unit of length is the *(ton, yard)*.
2. The ounce is a commonly used customary unit of *(weight, capacity)*.

skills

Complete.

3. 18 ft = __?__ yd
4. 34 yd 1 ft = __?__ ft
5. 5400 yd = __?__ mi __?__ yd
6. 8000 lb = __?__ t
7. 127 oz = __?__ lb __?__ oz
8. 3 t 225 lb = __?__ lb
9. 88 pt = __?__ gal
10. 9 gal 2 qt = __?__ qt
11. 23 pt = __?__ qt __?__ pt
12. $8\frac{1}{2}$ ft = __?__ in.
13. $10\frac{1}{4}$ lb = __?__ oz
14. $6\frac{1}{2}$ c = __?__ fl oz

Measure each line segment to the nearest $\frac{1}{16}$ inch.

15. |———————|
16. |——————————————|

Choose the letter of the appropriate unit for measuring the given object.

17. the height of a flagpole **a.** lb **b.** ft **c.** qt
18. the weight of a whale **a.** mi **b.** gal **c.** t

Write each answer in simplest form.

19. 14 ft 7 in. + 12 ft 9 in.
20. 12 c 2 fl oz − 7 c 6 fl oz
21. 17 t 975 lb × 6
22. 29 ft 3 in. ÷ 3

Give the number of significant digits and the unit of precision.

23. 12.7 m
24. 4.90 L
25. $420\underline{0}$ ft
26. 0.04 kg

27. When Levi filled the gas tank of his car last week, the odometer showed 8901.6 mi. When Levi filled the gas tank this week, the odometer showed 9252.3 mi. Levi put 14 gal of gas into the tank. Find the gasoline mileage to the nearest tenth of a mile per gallon.
28. Use the chart of shipping charges on page 218. Carmelita ordered a computer from a catalogue. The computer cost $1599. The shipping weight was 53 lb 6 oz. Carmelita lives 214 mi from the catalogue merchandise distribution center. Find the total cost of her order.

CHAPTER TEST

Complete.

1. 78 ft = _?_ yd 2. $3\frac{1}{2}$ yd = _?_ in. **11-1**

3. 12,500 ft = _?_ mi _?_ ft 4. 16 ft 7 in. = _?_ in.

Measure each line segment to the nearest $\frac{1}{16}$ inch.

5. ├──────────────┤ 6. ├────────┤

Complete.

7. 10,000 lb = _?_ t 8. $9\frac{1}{4}$ lb = _?_ oz **11-2**

9. 4 t 1265 lb = _?_ lb 10. 58 oz = _?_ lb _?_ oz

Complete.

11. 28 qt = _?_ gal 12. $11\frac{1}{4}$ c = _?_ fl oz **11-3**

13. 12 gal 3 qt = _?_ qt 14. 19 c = _?_ pt _?_ c

Choose the letter of the appropriate unit for measuring the given object.

15. the weight of a truck tire **a.** pt **b.** ft **c.** lb
16. the capacity of a juice bottle **a.** fl oz **b.** oz **c.** in.

Write each answer in simplest form.

17. 21 ft 4 in. + 17 ft 5 in. 18. 13 mi − 7 mi 2765 ft **11-4**
19. 17 c 7 fl oz × 6 20. 65 lb 10 oz ÷ 3

Give the number of significant digits and the unit of precision.

21. 132.7 g 22. 0.085 m 23. $4\underline{0}0$ yd 24. 1.08 L **11-5**

Find each answer. Round to the correct number of significant digits.

25. 12.746 km + 4.37 km 26. 18,500 gal − 7845 gal

27. The last time Akiro filled the gas tank of her car the odometer showed 32,396.8 mi. When she filled the gas tank this time, the odometer showed 32,673.1 mi. If Akiro put 9 gal into the tank, find the gasoline mileage to the nearest tenth of a mile per gallon. **11-6**

28. Use the chart of shipping charges on page 218. Allen purchased a stereo from a catalogue. The stereo cost $419.50. The shipping weight was 36 lb 7 oz. Allen lives 603 mi from the catalogue merchandise distribution center. Find the total cost of his order. **11-7**

How Computers Work

A computer system is made up of *software* and *hardware*. **Software** consists of sets of instructions, called programs, that the computer is to follow. **Hardware** is the physical equipment that runs the programs. There are four major types of hardware devices: input, process, memory, and output.

You use an **input** device to enter information into the computer. The most common input device for a personal computer is the keyboard. You may also be familiar with other input devices such as a mouse, a light pen, or a graphics tablet.

The **central processing unit (CPU)** is often called the "heart" or "brain" of the computer. The CPU performs arithmetic operations and controls the flow of information through the entire computer system. The CPU is usually contained on a single microprocessor chip inside the computer. One or more other chips contain the **memory,** which stores information until the CPU is ready to process it.

After information has been processed, the result is communicated to the user through an **output** device. For output that is to be read, the most commonly used output devices are the display screen and the printer. Output may also be in the form of sound.

exercises

1. Name the four major types of computer hardware devices.
2. Find out the difference between *read only memory (ROM)* and *random access memory (RAM)*.
3. Memory size is measured by the number of *bytes* that can be stored. A byte can be a letter, a number, or some other symbol. Find out the meaning of *kilobyte (K)* and *megabyte (M)*.

COMPETENCY TEST

Choose the letter of the correct answer.

1. **Which number is greatest?**

   A. $\frac{5}{8}$ B. $\frac{7}{8}$

   C. $\frac{3}{4}$ D. $\frac{7}{9}$

2. **Choose the best estimate.**

   $62\frac{7}{15} + 57\frac{15}{32}$

   A. about 100
   B. about 120
   C. about 140
   D. about 200

3. **Add.** $\frac{3}{4} + \frac{1}{3} + \frac{1}{2}$

   A. $1\frac{1}{3}$ B. $1\frac{7}{12}$

   C. $\frac{19}{36}$ D. $\frac{5}{9}$

4. **198 campers are separated into groups of equal size. How many campers are in each group?**

   A. 10 B. 8
   C. 6 D. 4

5. **On Wednesday STK stock closed at $\$57\frac{7}{8}$. On Thursday the stock closed at $\$54\frac{3}{4}$. Find the difference between these prices.**

   A. $\$2\frac{7}{8}$ B. $\$3\frac{1}{2}$

   C. $\$3\frac{1}{8}$ D. $\$3\frac{7}{8}$

6. **Multiply.** $4\frac{2}{3} \times 2\frac{1}{2}$

   A. $8\frac{1}{3}$ B. $1\frac{13}{15}$

   C. $7\frac{1}{3}$ D. $11\frac{2}{3}$

7. **Choose the best estimate.**

   $118 \div \frac{6}{17}$

   A. about 40
   B. about 60
   C. about 200
   D. about 300

8. **Choose the best estimate for 4.329.**

   A. about $4\frac{1}{4}$ B. about $4\frac{1}{3}$

   C. about $4\frac{1}{2}$ D. about $4\frac{2}{5}$

9. **Find the difference rounded to the correct number of significant digits.**
   7827 mi − 3200 mi

   A. 4627 mi B. 4630 mi
   C. 4600 mi D. 5000 mi

10. **Choose the only true statement.**

    A. $\frac{12}{35} > \frac{5}{7}$

    B. $\frac{3}{8} < \frac{5}{16}$

    C. $3\frac{5}{8} > 3\frac{14}{32}$

    D. $2\frac{15}{21} = 2\frac{3}{7}$

11. **Divide.** $6\frac{3}{5} \div 3\frac{3}{4}$

A. $1\frac{19}{25}$ B. $2\frac{4}{5}$
C. $1\frac{27}{65}$ D. $24\frac{3}{4}$

16. **Subtract.** $12\frac{5}{9} - 7\frac{2}{3}$

A. $4\frac{8}{9}$ B. $5\frac{2}{3}$
C. $5\frac{1}{9}$ D. $20\frac{2}{9}$

12. **Complete.** 512 in. = __?__ ft __?__ in.

A. 14 ft 8 in.
B. 170 ft 2 in.
C. 51 ft 2 in.
D. 42 ft 8 in.

17. **Add.** 22 lb 9 oz + 8 lb 11 oz

A. 31 lb 4 oz
B. 31 lb 8 oz
C. 32 lb
D. 32 lb 4 oz

13. **Express $\frac{7}{11}$ as a decimal.**

A. $0.\overline{636}$
B. $0.63\overline{6}$
C. $0.6\overline{3}$
D. $0.\overline{63}$

18. **Choose the best estimate.**

$\frac{5}{19} \times 387$

A. about 10
B. about 100
C. about 130
D. about 1600

14. **Which unit should be used to measure the capacity of a vase?**

A. oz B. lb
C. in. D. fl oz

19. **Which number is divisible by both 5 and 8?**

A. 7240 B. 5340
C. 8925 D. 5128

15. **Andrew washed three more cars than Jody at the Service Club car wash. Together they washed nineteen cars. How many cars did each person wash?**

A. Andrew: 10; Jody: 7
B. Andrew: 12; Jody: 9
C. Andrew: 11; Jody: 8
D. Andrew: 12; Jody: 7

20. **It costs Sarah Weatherton $28 per week to drive to work. A subway pass would cost her $52 per month. How much can she save per week by using the subway?**

A. about $60
B. about $47
C. about $15
D. about $7

CUMULATIVE REVIEW CHAPTERS 1-11

Find each answer. Write fractions in lowest terms.

1. $9 - 3\frac{5}{7}$	**2.** $3\frac{1}{2} \times 2\frac{3}{4}$	**3.** $\frac{24}{35} \times 0.7$
4. $0.73\overline{)61.685}$	**5.** $5814 - 389$	**6.** $\$31.48 + \$19.96$
7. $\frac{3}{5} \times \frac{7}{18}$	**8.** $5\frac{1}{2} + 2\frac{7}{8}$	**9.** $8 \div \frac{2}{3}$
10. $24 - 4^2 \div 2$	**11.** $5 \times 712 \times 20$	**12.** $32\overline{)544}$
13. $\frac{3}{8} \div \frac{4}{5}$	**14.** $\frac{1}{2} + \frac{2}{5} + \frac{1}{4}$	**15.** $2\frac{1}{2} \div 1\frac{3}{4}$

16. 5 gal 2 qt − 3 gal 3 qt

17. 74 ft 8 in. + 15 ft 9 in.

18. Write $\frac{15}{40}$ as a fraction in lowest terms and as a decimal.

19. Find the quotient to the nearest hundredth: $21.54 \div 8.1$

20. Find the elapsed time from 9:48 P.M. to 11:10 P.M.

21. Is 872 divisible by 4? by 3?

22. Round to the correct number of significant digits: 46.54 m + 32 m

23. Which number is greater, 7.3 or $7\frac{1}{3}$?

24. Find the GCF and the LCM of 15 and 100.

25. Complete: **a.** 6 ft 8 in. = _?_ in. **b.** 7.9 m = _?_ cm

Estimate.

26. $\frac{11}{32} \times 911$	**27.** $8\frac{4}{9} + 5\frac{1}{8} + 6\frac{7}{13}$	**28.** $48\overline{)9654}$
29. $\$71.83 - \$28.65$	**30.** $19 \times 41{,}851$	**31.** $803 \div 0.21$
32. $2154 + 7854$	**33.** $8\frac{1}{5} \div \frac{5}{19}$	**34.** $17\frac{1}{12} - 11\frac{7}{8}$

Write each set of numbers in order from least to greatest.

35. $\frac{5}{6}; \frac{3}{5}; \frac{2}{3}$

36. 1070; 701; 1007

37. 0.321; 0.3021; 0.0321

38. Abe earns $8 per hour. He worked $6\frac{1}{2}$ h on Monday and $8\frac{3}{4}$ h on Tuesday. How many more hours did he work on Tuesday?

39. The Zantows' fixed costs for operating their car last year were $2168.54. Their variable costs were $2418.80. They drove a total of 12,952 mi. Find their cost per mile to the nearest cent.

40. The Trainors bought sleeping bags at $52 each and backpacks at $24 each. They spent $204. How many of each did they buy?

The idea of ratio and proportion is used in almost every step of creating a new building—from plan to model to finished structure.

CHAPTER 12

RATIO AND PROPORTION

12-1 RATIO AND RATE

A **ratio** is a comparison of two numbers by division. You can write a ratio three ways.

5 to 14 $\qquad$ 5 : 14 $\qquad$ $\frac{5}{14}$

example 1

Write each ratio as a fraction in lowest terms.

a. 8 to 12 $\qquad$ **b.** 22 : 7 $\qquad$ **c.** $\frac{12}{6}$

solution

a. $\frac{8}{12} = \frac{2}{3}$ $\qquad$ **b.** $\frac{22}{7}$ ← **Do not rewrite as a mixed number.** $\qquad$ **c.** $\frac{12}{6} = \frac{2}{1}$ ← **Keep the 1 in the denominator.**

your turn

Write each ratio as a fraction in lowest terms.

1. 12 to 35 $\qquad$ **2.** 4 : 16 $\qquad$ **3.** 50 : 40 $\qquad$ **4.** $\frac{21}{3}$

Sometimes a ratio compares measurements that are given in different units. To write such a ratio as a fraction in lowest terms, you must first write the measurements in the same unit.

example 2

Write the ratio as a fraction in lowest terms: 20 s to 2 min

solution

Change 2 min to seconds.

1 min = 60 s, so 2 min = 120 s. (× 60)

$$\frac{20}{120} = \frac{20 \div 20}{120 \div 20} = \frac{1}{6}$$

your turn

Write each ratio as a fraction in lowest terms.

5. 30 s to 4 min $\qquad$ **6.** 16 days to 7 weeks

7. \$3 : 25¢ $\qquad$ **8.** 2 yd : 2 ft

A **rate** is a ratio of two different types of measurement. A **unit rate** is a rate for one unit of a given measurement. Some examples of unit rates are *miles per gallon* and *feet per second.*

example 3

Write the unit rate: **a.** 200 mi in 5 h **b.** $45 for 6 lb

solution

a. $\frac{\text{miles} \rightarrow 200}{\text{hours} \rightarrow 5} = \frac{200 \div 5}{5 \div 5} = \frac{40}{1}$

The unit rate is 40 mi in 1 h, or 40 mi/h (miles per hour).

b. $\frac{\text{dollars} \rightarrow 45}{\text{pounds} \rightarrow 6} = \frac{45 \div 6}{6 \div 6} = \frac{7.5}{1}$

The unit rate is $7.50 for 1 lb, or $7.50/lb.

your turn

Write the unit rate.

9. 110 mi in 2 h **10.** 50 ft in 5 s **11.** $3 for 6 apples **12.** 12 m in 2.5 s

practice exercises

practice for example 1 (page 228)

Write each ratio as a fraction in lowest terms.

1. 6 to 39 **2.** 12 to 48 **3.** $\frac{50}{30}$ **4.** $\frac{24}{60}$

5. 150 : 50 **6.** 25 : 6 **7.** 5 to 78 **8.** 28 to 21

practice for example 2 (page 228)

Write each ratio as a fraction in lowest terms.

9. 10¢ : $2 **10.** 17 days : 3 weeks **11.** 4 h to 3 days

12. 10 cm to 7 m **13.** 15 ft to 8 in. **14.** 3 yd to 6 ft

practice for example 3 (page 229)

Write the unit rate.

15. 350 mi in 7 h **16.** 12 pets in 6 houses **17.** $21 for 6 tickets

18. $10.50 in 3 h **19.** 20.4 m in 2 s **20.** 245.6 mi on 8 gal

mixed practice (pages 228–229)

Write each ratio as a fraction in lowest terms.

21. 6 to 9 **22.** 40 to 33 **23.** 52 : 26 **24.** $\frac{65}{5}$

25. $10 to 50¢ **26.** 3 s : 5 min **27.** $5 to 89¢ **28.** 1 day : 24 h

Write the unit rate.

29. 6 quizzes in 2 weeks
30. 8 trips in 4 weeks
31. 76 mi in 4 days
32. 36 eggs for 12 omelets
33. 78 ft in 6 s
34. 96 mi on 3 gal
35. \$12 for 5 tickets
36. \$28.49 in 7 h
37. \$60 in 8 days
38. \$358 in 4 weeks
39. 17.5 m in 3.5 s
40. \$3.60 for 0.9 L
41. There are two dozen oak trees and eleven spruce trees on Main Street. Write the ratio of oaks to spruces in three different ways.
42. Last year four students won Merit Awards. This year ten students won Merit Awards. Write as a fraction in lowest terms the ratio of the number of last year's winners to this year's winners.
43. Yvonne LaPierre earned \$120 in March and \$150 in April. How much did she earn in these two months?
44. A plane took off at 2:00 P.M. and landed at 4:00 P.M. It traveled 1200 mi. What was its average speed in miles per hour?

review exercises

Replace each ? with the number that will make the fractions equivalent.

1. $\frac{5}{8} = \frac{?}{24}$
2. $\frac{9}{12} = \frac{3}{?}$
3. $\frac{2}{?} = \frac{10}{25}$
4. $\frac{?}{21} = \frac{2}{3}$
5. $\frac{1}{7} = \frac{4}{?}$
6. $\frac{4}{?} = \frac{32}{40}$
7. $\frac{6}{36} = \frac{?}{6}$
8. $\frac{?}{54} = \frac{7}{9}$

calculator corner

You can use a calculator to find a unit rate.

example Find the unit rate: 150 mi in 2 h 45 min

solution There are 60 min in an hour, so 2 h 45 min $= 2\frac{45}{60}$ h.
Enter 45 ÷ 60 + 2 = [M+] . ← **Puts the number of hours into memory.**
Then enter 150 ÷ [MR] =.
The result is 54.545454.
The unit rate is about 55 mi/h.

Use a calculator to find the unit rate. If necessary, round to the nearest whole number.

1. 350 mi in 6 h 40 min
2. 395 mi in 6 h 15 min
3. 40 mi in 1 h 10 min
4. 221 mi in 4 h 25 min
5. 71 mi in 1 h 35 min
6. 171 mi in 3 h 25 min

12-2 PROPORTION

A **proportion** is a statement that two ratios are equal.

You write: $\frac{3}{5} = \frac{15}{25}$ or $3:5 = 15:25$

You read: *3 is to 5 as 15 is to 25.*

The numbers 3, 5, 15, and 25 are called the **terms** of the proportion. A proportion can be either true or false. If a statement is a *true* proportion, the **cross products** of the terms are equal.

$$\frac{3}{5} \times \frac{15}{25}$$

$$3 \times 25 \stackrel{?}{=} 5 \times 15$$ ← **cross products**

$$75 = 75$$

$\frac{3}{5} = \frac{15}{25}$ is a true proportion.

example 1

Tell whether each proportion is *true* or *false*.

a. $\frac{4}{10} = \frac{2}{5}$

b. $\frac{7}{15} = \frac{0.3}{0.7}$

solution

a. $\frac{4}{10} \times \frac{2}{5}$

$$4 \times 5 \stackrel{?}{=} 10 \times 2$$

$$20 = 20$$

True

b. $\frac{7}{15} \times \frac{0.3}{0.7}$

$$7 \times 0.7 \stackrel{?}{=} 15 \times 0.3$$

$$4.9 = 4.5$$

False

your turn

Tell whether each proportion is *true* or *false*.

1. $\frac{16}{20} = \frac{4}{5}$
2. $\frac{9}{5} = \frac{27}{15}$
3. $\frac{0.3}{0.9} = \frac{5}{16}$
4. $\frac{6}{1.2} = \frac{8}{1.6}$

Sometimes one term of a proportion is a *variable*. A **variable** is a symbol, such as the letter *n,* that represents an unknown number. A number represented by the variable is called a **value** of that variable. To **solve a proportion,** you must find the value of the variable that makes the proportion true. You can use cross products when you solve a proportion.

example 2

Solve each proportion.

a. $\frac{2}{22} = \frac{8}{n}$

b. $\frac{a}{9} = \frac{20}{12}$

solution

a. $\frac{2}{22} \times \frac{8}{n}$

$$2 \times n = 22 \times 8$$
$$2 \times n = 176$$
$$\frac{2 \times n}{2} = \frac{176}{2}$$

← Divide by 2 on each side of the equals sign.

$$n = 88$$

b. $\frac{a}{9} \times \frac{20}{12}$

$$a \times 12 = 9 \times 20$$
$$a \times 12 = 180$$
$$\frac{a \times 12}{12} = \frac{180}{12}$$

← Divide by 12 on each side of the equals sign.

$$a = 15$$

your turn

Solve each proportion.

5. $\frac{2}{3} = \frac{36}{n}$
6. $\frac{6}{y} = \frac{2}{9}$
7. $\frac{h}{21} = \frac{2}{14}$
8. $\frac{90}{6} = \frac{c}{0.4}$

practice exercises

practice for example 1 (page 231)

Tell whether each proportion is *true* or *false*.

1. $\frac{6}{7} = \frac{30}{35}$
2. $\frac{18}{4} = \frac{8}{3}$
3. $\frac{25}{6} = \frac{7}{15}$
4. $\frac{2}{7} = \frac{5}{3}$
5. $\frac{21}{2} = \frac{12}{2}$
6. $\frac{40}{90} = \frac{8}{18}$
7. $\frac{24}{40} = \frac{0.3}{0.5}$
8. $\frac{6}{7} = \frac{0.7}{0.8}$
9. $\frac{1.1}{2.2} = \frac{10}{100}$
10. $\frac{100}{80} = \frac{1.5}{1.2}$
11. $\frac{12}{0.6} = \frac{18}{0.9}$
12. $\frac{0.4}{48} = \frac{0.7}{78}$

practice for example 2 (page 232)

Solve each proportion.

13. $\frac{3}{2} = \frac{6}{x}$
14. $\frac{a}{6} = \frac{5}{3}$
15. $\frac{k}{8} = \frac{45}{2}$
16. $\frac{10}{3} = \frac{400}{d}$
17. $\frac{8}{9} = \frac{16}{b}$
18. $\frac{y}{12} = \frac{24}{32}$
19. $\frac{24}{28} = \frac{d}{21}$
20. $\frac{12}{h} = \frac{28}{35}$
21. $\frac{7}{5} = \frac{49}{c}$
22. $\frac{120}{160} = \frac{3}{g}$
23. $\frac{1.8}{2.7} = \frac{m}{9}$
24. $\frac{p}{8.4} = \frac{5}{1.4}$

mixed practice (pages 231–232)

Tell whether each proportion is *true* or *false*.

25. $9:18 = 3:6$ **26.** $2:24 = 3:34$ **27.** $54:26 = 2:1$ **28.** $39:13 = 3:1$

29. $\frac{18}{5} = \frac{5.4}{1.5}$ **30.** $\frac{0.4}{0.9} = \frac{12}{27}$ **31.** $\frac{0.5}{0.6} = \frac{25}{36}$ **32.** $\frac{120}{30} = \frac{1.2}{0.2}$

33. $\frac{4}{4.2} = \frac{2}{2.5}$ **34.** $\frac{4.4}{2} = \frac{8.7}{4}$ **35.** $\frac{5}{75} = \frac{100}{1500}$ **36.** $\frac{8}{5} = \frac{240}{150}$

Solve each proportion.

37. $8:n = 8:1$ **38.** $\frac{a}{20} = \frac{16}{4}$ **39.** $\frac{45}{y} = \frac{5}{7}$ **40.** $10:1 = 100:b$

41. $76:y = 2:3$ **42.** $\frac{4.9}{h} = \frac{0.7}{4}$ **43.** $\frac{0.6}{0.8} = \frac{72}{p}$ **44.** $9:14 = w:84$

45. Write *seven is to fourteen as three is to six* in two different ways.

46. Replace just one of the terms of the following proportion to make a true proportion: $\frac{6}{18} = \frac{4}{24}$

review exercises

Estimate.

1. $312 + 96 + 577$ **2.** $\frac{2}{5} \times 24\frac{1}{2}$ **3.** $0.8\overline{)6.32}$ **4.** $15\frac{7}{8} - 8\frac{1}{10}$

5. $2\frac{1}{3} \times 6\frac{4}{5}$ **6.** $3.1 + 1.3 + 6.8$ **7.** $12\frac{1}{6} - 4\frac{1}{2}$ **8.** $584 \div 17$

mental math

In some proportions, one term is a multiple of another. You can usually solve these proportions mentally.

$\frac{3}{17} = \frac{6}{n}$ $\frac{3}{17} = \frac{6}{n}$ *Think:* $3 \times 2 = 6$, so $17 \times 2 = n$. $34 = n$

$\frac{x}{35} = \frac{3}{15}$ $\frac{x}{35} = \frac{3}{15}$ *Think:* $15 \div 5 = 3$, so $35 \div 5 = x$. $7 = x$

Solve each proportion mentally.

1. $\frac{1}{4} = \frac{m}{36}$ **2.** $\frac{6}{1} = \frac{48}{z}$ **3.** $\frac{7}{8} = \frac{a}{32}$ **4.** $\frac{9}{4} = \frac{y}{12}$

5. $\frac{15}{30} = \frac{c}{8}$ **6.** $\frac{24}{3} = \frac{80}{a}$ **7.** $\frac{36}{3} = \frac{w}{4}$ **8.** $\frac{z}{12} = \frac{5}{30}$

PROBLEM SOLVING STRATEGY

12-3 USING PROPORTIONS

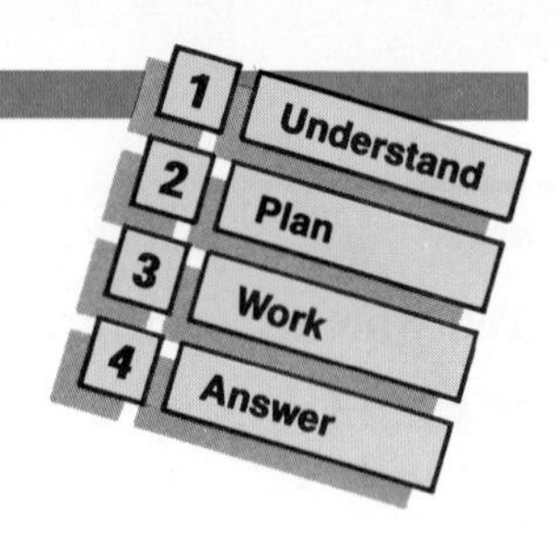

You can use proportions to solve many types of problems.

example

Last week Kirsten Sanderson drove 425 mi on 17 gal of gasoline. This week Kirsten expects to drive only 375 mi. At the same rate, how many gallons of gasoline should Kirsten expect to use this week?

solution

Step 1 Given: 425 mi on 17 gal last week
375 mi this week
Find: number of gallons this week

Step 2 Write a proportion.

$$\frac{\text{miles traveled}}{\text{gallons used}} \rightarrow \frac{425}{17} = \frac{375}{g}$$

← **Let g represent the number of gallons this week.**

Step 3 Solve the proportion.

$$425 \times g = 17 \times 375$$
$$425 \times g = 6375$$
$$\frac{425 \times g}{425} = \frac{6375}{425}$$

← **Divide by 425 on each side of the equals sign.**

$$g = 15$$

Step 4 *Check:*

$$\frac{425}{17} = \frac{375}{g}$$
$$\frac{425}{17} \stackrel{?}{=} \frac{375}{15}$$

← **Replace g with 15.**

$$425 \times 15 \stackrel{?}{=} 17 \times 375$$
$$6375 = 6375 \checkmark$$

Kirsten should expect to use 15 gal of gasoline.

problems

Solve using a proportion. A calculator may be helpful.

1. Last week Mark Mendez drove 800 mi and used 40 gal of gasoline. This week he plans to drive 1200 mi. At the same rate, how many gallons of gasoline should he expect to use?
2. One week Myra Oakes rode her bike 18 h and traveled 99 mi. At the same rate, how many miles would she travel if she rode for 20 h?

3. Zoe Pinter's heart beats 16 times in 15 s. At this rate, how many times does her heart beat in 60 s?
4. Thad Sepik will need 7 cartons of juice for every 4 people at the family reunion. Juice costs $.99 for 3 cartons. Thad expects 56 people at the reunion. How many cartons of juice should he buy?
5. The ratio of orange paint to white paint in a particular mixture is 2 : 1. Sharon Johnson has 3 gal of white paint. How many gallons of orange paint does she need for the mixture?
6. At the Shoppers Market three containers of yogurt cost $1.77. How much will eight containers of yogurt cost?
7. Jeanne Holden earns $16.50 for mowing three lawns. At this rate, how many lawns must she mow to earn $66?
8. Frank Cohn charges $3.50 for every 4 pages he types. How much does he charge to type 14 pages?
9. A recipe calls for 6 c of flour, 2 c of milk, and 6 eggs. Tom has only 3 c of flour. How much milk should he use?
10. Joan's family consumes 2 gal of milk in 3 days. At this rate, how much milk does the family consume in 30 days?
11. The label on a 1-gal can of Spreadright paint indicates that one gallon will cover 350 ft^2 (square feet). How many 1-gal cans of paint must you buy to cover 800 ft^2?
12. A 6-oz block of cheese costs $1.29. How much would a $\frac{1}{2}$-lb block of the same type of cheese cost?

review exercises

1. On Thursday and Friday, Ray typed a 28-page report for history class. He typed three times as many pages on Thursday as on Friday. How many pages did Ray type on Thursday?
2. Brenda Johnson rides her bicycle a total of 5 mi every 2 days. How many miles does she ride in 14 days?
3. On Friday evening, 408 students and adults attended a West High School basketball game. Only 139 adults attended the game. How many students attended?

Name all pairs of consecutive months that have 31 days in each month.

SKILL REVIEW

Write each ratio as a fraction in lowest terms. **12-1**

1. 6 to 7
2. $\frac{12}{18}$
3. 34 : 31
4. 15 : 10
5. $\frac{55}{11}$
6. $\frac{70}{35}$
7. 14 min to 3 h
8. 4 days to 4 weeks
9. 50¢ to $5
10. 9 in. to 3 ft
11. 6 m : 10 cm
12. $75 : 20¢

Write the unit rate.

13. 2250 mi in 6 weeks
14. 450 pages in 15 days
15. $26.40 in 4 h
16. $44.95 for 5 lb
17. 790 mi in 20 h
18. 273 mi on 13 gal
19. Susan Bently paid $3.87 for 3 spiral notebooks. How much did she pay per notebook?

Tell whether each proportion is *true* or *false*. **12-2**

20. $\frac{2}{40} = \frac{5}{100}$
21. $\frac{9}{18} = \frac{20}{40}$
22. $\frac{3}{1.2} = \frac{9}{3.6}$
23. $\frac{7}{44} = \frac{6}{40}$
24. $\frac{55}{75} = \frac{0.5}{0.7}$
25. $\frac{0.6}{0.5} = \frac{46}{34}$
26. $\frac{3}{9} = \frac{1.3}{3.9}$
27. $\frac{1.2}{5.4} = \frac{8}{36}$

Solve each proportion.

28. $\frac{3}{x} = \frac{1}{3}$
29. $\frac{2}{1} = \frac{12}{c}$
30. $\frac{y}{1} = \frac{14}{1.4}$
31. $\frac{76}{0.2} = \frac{d}{0.3}$
32. $\frac{3}{4} = \frac{27}{b}$
33. $\frac{40}{k} = \frac{10}{7}$
34. $\frac{y}{1} = \frac{140}{14}$
35. $\frac{54}{3} = \frac{d}{2}$

Solve using a proportion. **12-3**

36. Tom Perry's typing class is 50 min long. Tom types 46 words per minute. At that rate, how many words can he type in 10 min?
37. The cost of 4 posters is $14. How much do 6 posters cost?
38. Keiko Young uses 2 c of flour to make 36 rolls. How much flour does she need to make 126 rolls?
39. The city gardeners decided to plant trees in a ratio of 8 maple trees to 3 spruce trees. They planted 120 maple trees. How many spruce trees did they plant?
40. Scott Napier drove 245 mi on 7 gal of gasoline. At that rate, how much farther can he drive on two more gallons of gasoline?

APPLICATION

12-4 COMPARISON SHOPPING

When you shop, you may have the choice of buying an item in various sized packages or quantities. To determine which package or quantity is the best buy, you can use *unit pricing*. The **unit price** of an item is the cost of one of the units in which the product is measured. To find the unit price, you can write a proportion.

$$\frac{\text{unit price}}{\text{one unit}} = \frac{\text{total price}}{\text{number of units}}$$

example

At Excel Market, 3 lb of bananas cost \$1.29.
At Lyn Market, 4 lb of bananas cost \$1.69. If the quality is about the same, which is the better buy?

solution

Excel Market: 3 lb for \$1.29

$$\frac{p}{1} = \frac{\$1.29}{3}$$ ← **Let p represent the price of 1 lb of bananas.**

$$p \times 3 = 1 \times \$1.29$$

$$\frac{p \times 3}{3} = \frac{1 \times \$1.29}{3}$$

$$p = \$.43$$

unit price: \$.43/lb

Lyn Market: 4 lb for \$1.69

$$\frac{p}{1} = \frac{\$1.69}{4}$$

$$p \times 4 = 1 \times \$1.69$$

$$\frac{p \times 4}{4} = \frac{1 \times \$1.69}{4}$$

$$p = \$.4225 \approx \$.42$$

unit price: \$.42/lb

$\$.43 > \$.42$, so 4 lb of bananas for \$1.69 is the better buy.

exercises

Find the better buy. Assume the quality is about the same.

1. 2 qt of motor oil for \$1.62
   5 qt of motor oil for \$3.70
2. 3 lb of apples for \$1.53
   8 lb of apples for \$4.16
3. a 3-lb bag of grass seed for \$7.19
   a 10-lb bag of grass seed for \$24.99
4. a 2-lb bag of cement for \$4.90
   a 5-lb bag of cement for \$11.68
5. 3 cans of juice for \$1.05
   12 cans of juice for \$3.84
6. 3 combs for \$1.19
   10 combs for \$3.79
7. Kisha Smith finds that a 12-oz tube of shampoo sells for \$6.45, and a 4-oz tube of the same shampoo sells for \$2.30. How much does Kisha save if she buys the 12-oz tube instead of three 4-oz tubes?
8. Describe two situations in which it may *not* be preferable to buy an item that has the least unit price.

APPLICATION

12-5 MAPS

Most maps contain a **scale** that indicates the ratio of a distance on the map to actual distance. To estimate the actual distance between two places, first measure the distance between the points that represent the places on the map. Then use the scale to find the actual distance.

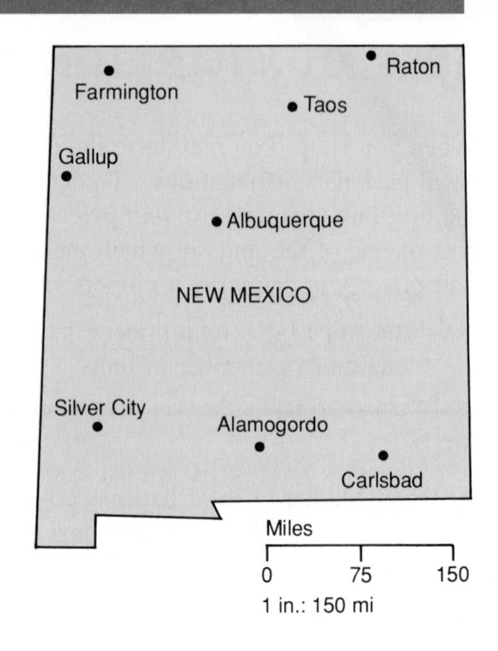

example

The scale of the map at the right is 1 in. : 150 mi. On the map, Albuquerque is about $\frac{3}{4}$ in. from Taos. Estimate the actual distance between Albuquerque and Taos.

solution

Write and solve a proportion.

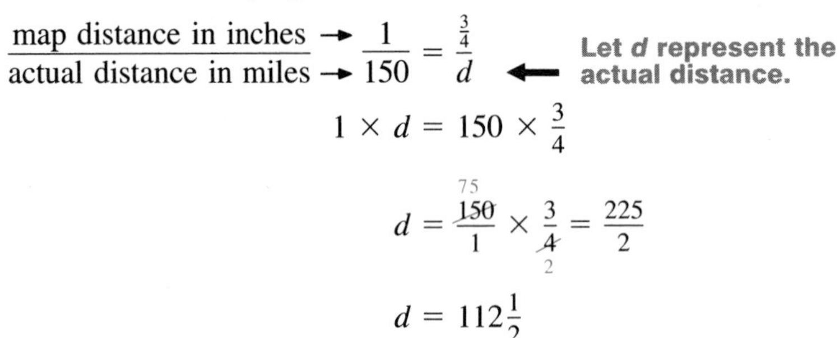

$$\frac{\text{map distance in inches} \rightarrow}{\text{actual distance in miles} \rightarrow} \frac{1}{150} = \frac{\frac{3}{4}}{d} \longleftarrow \text{Let } d \text{ represent the actual distance.}$$

$$1 \times d = 150 \times \frac{3}{4}$$

$$d = \frac{\overset{75}{\cancel{150}}}{1} \times \frac{3}{\underset{2}{\cancel{4}}} = \frac{225}{2}$$

$$d = 112\frac{1}{2}$$

$$d \approx 113 \longleftarrow \text{Round to the nearest mile.}$$

The actual distance between Albuquerque and Taos is about 113 mi.

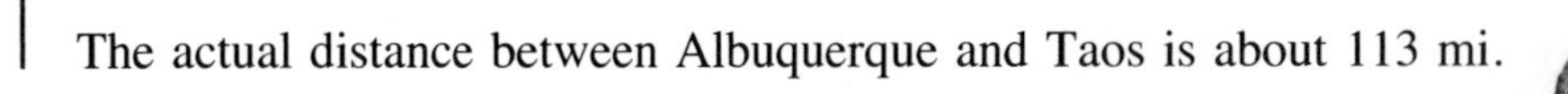

exercises

Use the map on page 238. Estimate the actual distance for each map distance.

1. Taos and Raton; $\frac{1}{2}$ in.
2. Albuquerque and Carlsbad; $1\frac{1}{2}$ in.
3. Albuquerque and Silver City; $1\frac{1}{4}$ in.
4. Farmington and Gallup; $\frac{5}{8}$ in.
5. The map distance between Taos and Gallup is $1\frac{1}{4}$ in. The map distance between Taos and Farmington is 1 in. About how much greater is the actual distance between Taos and Gallup than the actual distance between Taos and Farmington?
6. The map distance between Alamogordo and Silver City is $\frac{7}{8}$ in. The map distance between Alamogordo and Gallup is $1\frac{3}{4}$ in. About how much less is the actual distance between Alamogordo and Silver City than the actual distance between Alamogordo and Gallup?

7. The scale on a map is 1 in. : 8 mi. Estimate the actual distance between two cities that are 10 in. apart on this map.
8. The scale on a map is 1 cm : 25 km. Estimate the actual distance between two cities that are 6 cm apart on this map.
9. Find a map of the United States. Use the map scale and a proportion to estimate the distance between the state capital of your state and the nearest capital of another state.
10. What kind of book contains the following types of maps?
    a. a map of the world
    b. a map of the time zones in the United States
    c. a map of the streets of Rome, Italy

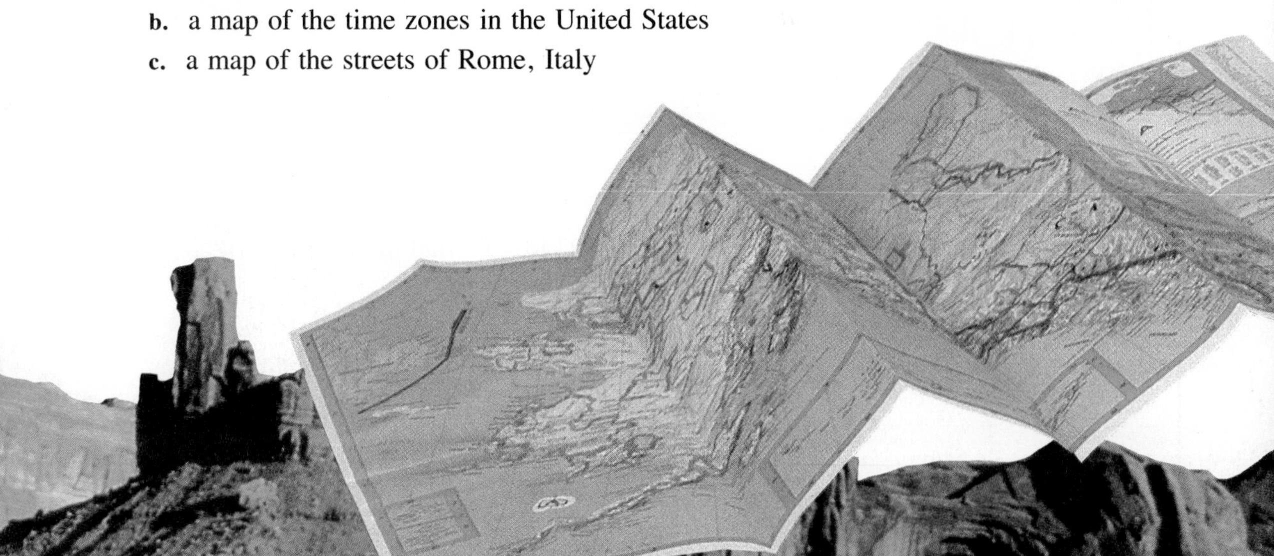

APPLICATION

12-6 SCALE DRAWINGS

A drawing that represents an actual object is called a **scale drawing**. The *scale* is the ratio of the size of the drawing to the actual size of the object. You can use the scale to find the actual measurements of the object.

example

The scale of the floor plan of this art museum is 1 in. : 16 ft. Find the actual length of the auditorium.

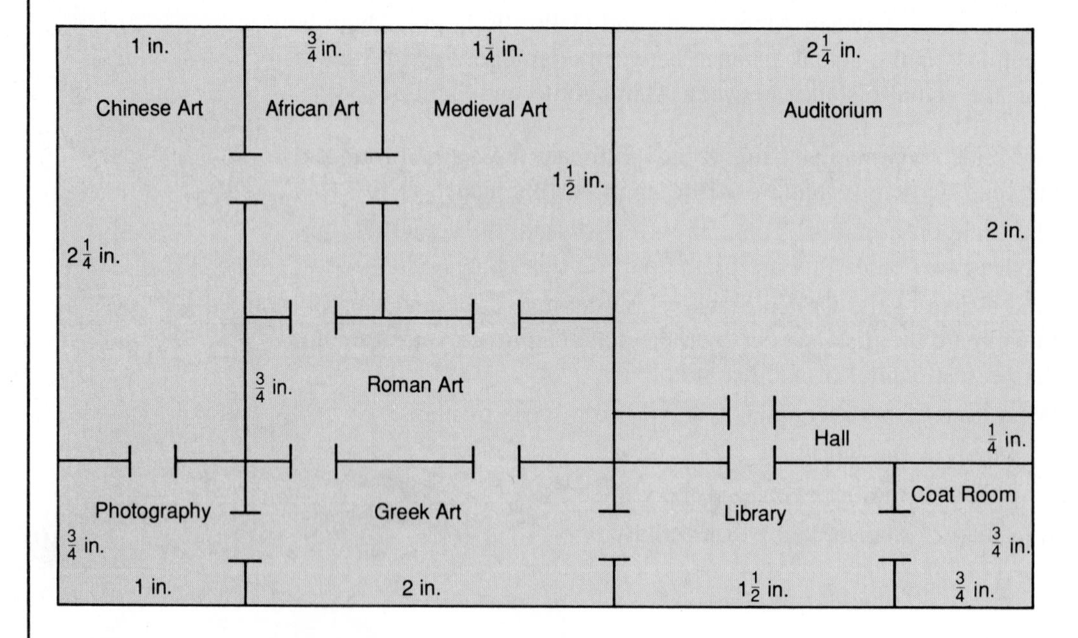

solution

The length of the auditorium in the floor plan is $2\frac{1}{4}$ in. To find the actual length, write and solve a proportion.

$$\frac{\text{floor plan length in inches}}{\text{actual length in feet}} \rightarrow \frac{1}{16} = \frac{2\frac{1}{4}}{f}$$

Let *f* represent the actual length in feet.

$$1 \times f = 16 \times 2\frac{1}{4}$$

$$f = \frac{\overset{4}{\cancel{16}}}{1} \times \frac{9}{\underset{1}{\cancel{4}}}$$

$$f = 36$$

The actual length of the auditorium is 36 ft.

exercises

Use the scale drawing on page 240. Find the actual width and length of each room.

1. Medieval Art room
2. Roman Art room
3. Photography room
4. African Art room
5. Chinese Art room
6. What is the difference between the actual length of the Greek Art room and the actual length of the Photography room?
7. What is the difference between the actual length of the Library and the actual length of the Coat Room?
8. The director of the Chinese Art Department wants to divide the Chinese Art room into two smaller sections. On the new floor plan, each section will be $1\frac{1}{8}$ in. long and 1 in. wide. How long and how wide will each section actually be?
9. The museum directors decide to build a room 32 ft long and 24 ft wide to store musical instruments and costumes. How long and how wide will the room be on the floor plan?
10. On the floor plan of a house, a living room is 3 in. wide and 5 in. long. The scale is 1 in. : 3 ft. Find the actual width and length of the living room.
11. On a floor plan, a kitchen is $1\frac{1}{4}$ in. wide and $1\frac{3}{4}$ in. long. The scale is 1 in. : 8 ft. Find the actual width and length of the kitchen.
12. On a floor plan, a master bathroom is $1\frac{1}{2}$ in. wide and 3 in. long. The scale is 1 in. : 5 ft. Find the actual width and length of the master bathroom.
13. Plan a "dream" room for your home, using the scale 1 in. : 2 ft. Make a scale drawing of the floor plan of the room. Include the scale, the furniture, and any special features.
14. Find out what blueprints are and how they are made.

CAREER

ARCHITECT

Lee Rogers is an architect. In his work he often makes *scale models* of projects. A **scale model** represents an actual object but is smaller.

example

Lee has designed a new stadium. He has built a scale model to present his plan to the developers. The actual building is 50 ft high. The scale is 1 in. : 4 ft. Find the height of the scale model.

solution

Write and solve a proportion.

$$\frac{\text{model height in inches} \rightarrow 1}{\text{actual height in feet} \rightarrow 4} = \frac{n}{50}$$ ← **Let *n* represent the height of the model.**

Solve the proportion: $1 \times 50 = 4 \times n$

$$\frac{50}{4} = \frac{4 \times n}{4}$$

$$12\frac{1}{2} = n$$

The height of the scale model is $12\frac{1}{2}$ in.

exercises

1. Sue Roth builds a scale model of a mall. The mall is 150 yd long. The scale is 1 in. : 5 yd. What is the length of the scale model?
2. City Transit hires Larry Bruce to build a shelter. The shelter will be 10 ft high. Larry builds a model with a scale of 1 in. : 2 ft. What is the height of the model?
3. Bell City is remodeling its City Hall. The proposed entrance is 30 ft wide and 18 ft high. The scale of the architect's model is 1 in. : 4 ft. Find the width and height of the entrance in the scale model.
4. Sam Farmer built a scale model of a new house. The house will be 20 ft high, 35 ft wide, and 40 ft long. The scale was 1 in. : 2 ft. Find the height, width, and length of the scale model.
5. Using the scale 1 in. : 1 ft, estimate the length, width, and height of a scale model of your classroom.

CHAPTER REVIEW

vocabulary vo·cab·u·lar·y

Choose the correct word to complete each sentence.

1. A ratio that compares two different types of measurement is called a *(rate, proportion)*.
2. A proportion is a statement that two *(ratios, terms)* are equal.

skills

Write each ratio as a fraction in lowest terms.

3. 9 : 17
4. 75¢ to $9
5. $\frac{34}{17}$
6. 40 min to 3 h

Write the unit rate.

7. $8.72 for 8 pens
8. 150 mi in 50 s
9. 7.4 km in 2 h
10. $510 for 40 h

Tell whether each proportion is *true* or *false*.

11. $\frac{70}{28} = \frac{5}{2}$
12. $\frac{24}{28} = \frac{6}{7}$
13. $\frac{0.2}{0.9} = \frac{4}{20}$
14. $\frac{3}{7} = \frac{2.5}{7.5}$

Solve each proportion.

15. $\frac{5}{n} = \frac{25}{10}$
16. $\frac{c}{55} = \frac{4}{5}$
17. $\frac{0.6}{0.4} = \frac{b}{26}$
18. $\frac{1.4}{21} = \frac{0.6}{n}$

Solve using a proportion.

19. Steve Ross buys 15 roses for $60. Find the cost of 12 roses.
20. Chef Koh uses 3 apples in a salad serving 18 people. How many apples should she use to make enough salad for 480 people?
21. Lynne Yasner reads 8 pages in 20 min. At that rate, how long does it take her to read 20 pages?

Solve.

22. Doug Farmer found a bottle of 100 vitamins selling for $2.25 and a bottle of 250 vitamins selling for $4.99. Find the better buy.
23. The map distance between two cities is 8 cm. The scale is 1 cm : 50 km. Estimate the actual distance between the cities.
24. The scale of the floor plan of a house is 1 in. : 4 ft. The living room on the plan is $4\frac{1}{2}$ in. long and $2\frac{3}{4}$ in. wide. Find the actual length and width of the living room.

CHAPTER TEST

Write each ratio as a fraction in lowest terms.

12-1

1. 17 to 45 2. 3 to 15 3. $\frac{34}{8}$ 4. $\frac{84}{4}$

5. 6 mm to 5 cm 6. 3 min to 2 s 7. 30 mL : 4 L 8. \$12 : 5¢

Write the unit rate.

9. 90 pages in 5 days 10. 168 mi in 6 h 11. 90.4 ft in 20 s

Tell whether each proportion is *true* or *false*.

12-2

12. $\frac{5}{9} = \frac{25}{40}$ 13. $\frac{4}{6} = \frac{34}{51}$ 14. $\frac{70}{0.5} = \frac{28}{0.2}$ 15. $\frac{3}{6} = \frac{3.2}{6.2}$

Solve each proportion.

16. $\frac{6}{1} = \frac{24}{y}$ 17. $\frac{a}{15} = \frac{4}{6}$ 18. $\frac{1.2}{x} = \frac{0.5}{10}$ 19. $\frac{0.3}{0.9} = \frac{n}{24}$

Solve using a proportion.

12-3

20. Steve Carson works 30 h and earns \$195. At this rate, how much does he earn if he works 35 h?
21. The Fine and Fancy Caterers charge \$9 to prepare 12 spicy meatballs. The Reiger family's order for spicy meatballs cost \$63. How many meatballs did the caterers prepare for the Reigers?

Find the better buy. Assume the quality is about the same.

12-4

22. 4 pens for \$7.16
    6 pens for \$11.34
23. 12 issues for \$30
    24 issues for \$57.60
24. 2 rolls of film for \$5.49
    3 rolls of film for \$8
25. 3 cans of soup for \$.79
    5 cans of soup for \$1.24

12-5

26. Use the map on page 238. The map distance between Farmington and Silver City is $1\frac{7}{8}$ in. Estimate the actual distance.
27. On a map, the distance between Tampa and Savannah is 10 in. The scale is 1 in. : 30 mi. Estimate the actual distance between Tampa and Savannah.

12-6

28. Jon Sylvia has the floor plan for a patio. The scale is 1 in. : 8 ft. The plan length of the patio is $1\frac{3}{4}$ in. Find the actual length of the patio.
29. Emily Parsons made a scale drawing of her garden. The scale is 1 in. : 2 ft. In the drawing, the carrots are in a section $2\frac{1}{2}$ in. long and $1\frac{1}{2}$ in. wide. Find the actual length and width of the carrot section.

About thirty percent of the energy produced in the United States comes from natural gas. These pipes convey materials in a natural gas plant.

CHAPTER 13

PERCENT

13-1 PERCENTS AND DECIMALS

A **percent** represents a ratio that compares a number to 100. Percent means *per hundred, hundredths,* or *out of every hundred.* The symbol % is read "percent."

$$1\% = 1 \text{ hundredth} = \frac{1}{100} = 0.01$$

To write a percent as a decimal, use the following method.

1. Move the decimal point two places to the left. $6\% = 0.06$ ← **Insert zeros as placeholders if necessary.**
2. Remove the percent symbol.

example 1

Write each percent as a decimal.

a. 45%   b. 7.9%   c. 130%   d. $33\frac{1}{3}\%$ e. $12\frac{1}{2}\%$

solution

a. 0.45   b. 0.079   c. 1.3   d. $0.33\frac{1}{3}$ e. $12\frac{1}{2}\% = 12.5\% = 0.125$

your turn

Write each percent as a decimal.

1. 85%   2. 4%   3. 2.7%   4. 270%   5. $27\frac{2}{3}\%$

To write a decimal as a percent, use the following method.

1. Move the decimal point two places to the right. $0.2 = 20\%$ ← **Insert zeros as placeholders if necessary.**
2. Write the percent symbol.

example 2

Write each decimal as a percent.

a. 0.96   b. 0.03   c. 0.705   d. 2.4   e. $0.16\frac{3}{8}$

solution

a. 96%   b. 3%   c. 70.5%   d. 240%   e. $16\frac{3}{8}\%$

your turn

Write each decimal as a percent.

6. 0.8   7. 0.002   8. 0.64   9. $0.05\frac{1}{2}$ 10. 3.25

practice exercises

practice for example 1 (page 246)

Write each percent as a decimal.

1. 32%
2. 86%
3. 6.75%
4. 1.09%
5. 157%
6. 100%
7. 0.7%
8. 0.2%
9. $37\frac{2}{5}\%$
10. $10\frac{3}{4}\%$
11. $66\frac{2}{3}\%$
12. $83\frac{1}{3}\%$

practice for example 2 (page 246)

Write each decimal as a percent.

13. 0.01
14. 0.06
15. 0.806
16. 0.298
17. 2.45
18. 1.81
19. 0.3
20. 0.7
21. 0.005
22. 0.009
23. $0.62\frac{1}{2}$
24. $0.08\frac{1}{3}$

mixed practice (page 246)

Write each percent as a decimal. Write each decimal as a percent.

25. 8%
26. 11.5%
27. 0.43
28. 0.662
29. 230%
30. 0.25%
31. 2.59
32. 1.1
33. 0.003
34. $0.09\frac{2}{3}$
35. $87\frac{1}{4}\%$
36. $16\frac{2}{3}\%$

Use the table at the right.

National Basketball Association			
Team	W	L	Standing
Hawks	57	25	.695
Pistons	52	30	.634
Bucks	50	32	.610
Pacers	41	41	.500
Bulls	40	42	.488
Cavaliers	31	51	.378

37. The Pistons won 0.634 of their games. Write this decimal as a percent.
38. What percent of their games did the Cavaliers win?
39. Which team won exactly 50% of its games?
40. Which team(s) won more than $\frac{2}{3}$ of its games?
41. Rita received a pay raise of $6\frac{1}{2}\%$. Write this percent as a decimal.
42. Five years ago the Johnsons bought a house. Its present value is 1.15 times the price they paid. Write this decimal as a percent.

review exercises

Write each fraction or mixed number as a decimal.

1. $\frac{9}{10}$
2. $\frac{1}{4}$
3. $2\frac{4}{5}$
4. $\frac{3}{8}$
5. $1\frac{9}{16}$
6. $\frac{17}{20}$

13-2 PERCENTS AND FRACTIONS

To write a percent as a fraction, use the following method.

1. Write the percent as a ratio of a number to 100.
2. Write the ratio in lowest terms.

$$40\% = \frac{40}{100} = \frac{40 \div 20}{100 \div 20} = \frac{2}{5}$$

example 1

Write each percent as a fraction or a mixed number in lowest terms.

a. 136% b. 3.9%

solution

a. $136\% = \frac{136}{100} = 1\frac{36}{100} = 1\frac{9}{25}$

b. $3.9\% = \frac{3.9}{100} = \frac{3.9 \times 10}{100 \times 10} = \frac{39}{1000}$ ← **The numerator should be a whole number.**

your turn

Write each percent as a fraction or a mixed number in lowest terms.

1. 9% 2. 15% 3. 250% 4. 88.4% 5. 0.04%

example 2

Write each percent as a fraction in lowest terms: a. $\frac{2}{3}\%$ b. $4\frac{1}{2}\%$

solution

a. $\frac{2}{3}\% = \frac{\frac{2}{3}}{100} = \frac{2}{3} \div 100 = \frac{2}{3} \times \frac{1}{100} = \frac{\overset{1}{\cancel{2}}}{3} \times \frac{1}{\underset{50}{\cancel{100}}} = \frac{1}{150}$

b. $4\frac{1}{2}\% = \frac{4\frac{1}{2}}{100} = 4\frac{1}{2} \div 100 = \frac{9}{2} \times \frac{1}{100} = \frac{9}{200}$

your turn

Write each percent as a fraction in lowest terms.

6. $\frac{7}{8}\%$ 7. $\frac{2}{5}\%$ 8. $33\frac{1}{3}\%$ 9. $10\frac{3}{4}\%$ 10. $29\frac{3}{5}\%$

You can write a fraction or mixed number as a percent by first finding an equivalent fraction with a denominator of 100. If that is not easy to do, first write the fraction or mixed number as a decimal. Then write the decimal as a percent.

example 3

Write each fraction or mixed number as a percent.

a. $\frac{3}{5}$ **b.** $1\frac{1}{4}$ **c.** $\frac{1}{16}$

solution

a. $\frac{3}{5} = \frac{3 \times 20}{5 \times 20} = \frac{60}{100} = 60\%$

b. $1\frac{1}{4} = \frac{5}{4} = \frac{5 \times 25}{4 \times 25} = \frac{125}{100} = 125\%$

c. $\frac{1}{16} \rightarrow 16\overline{)1.00}$ gives $0.06\frac{4}{16}$ (96, remainder 4) ← **Divide to the hundredths' place.**

So $\frac{1}{16} = 0.06\frac{4}{16} = 6\frac{4}{16}\% = 6\frac{1}{4}\%$.

your turn

Write each fraction or mixed number as a percent.

11. $\frac{21}{50}$ **12.** $\frac{2}{3}$ **13.** $\frac{5}{8}$ **14.** $2\frac{7}{10}$ **15.** $1\frac{4}{9}$

practice exercises

practice for example 1 (page 248)

Write each percent as a fraction or a mixed number in lowest terms.

1. 7% **2.** 4% **3.** 45% **4.** 19% **5.** 125% **6.** 248%

7. 9.3% **8.** 0.9% **9.** 67.2% **10.** 42.5% **11.** 0.06% **12.** 8.45%

practice for example 2 (page 248)

Write each percent as a fraction in lowest terms.

13. $\frac{1}{3}\%$ **14.** $\frac{5}{6}\%$ **15.** $\frac{4}{9}\%$ **16.** $\frac{6}{7}\%$ **17.** $7\frac{2}{3}\%$ **18.** $8\frac{3}{4}\%$

19. $23\frac{1}{2}\%$ **20.** $43\frac{1}{3}\%$ **21.** $12\frac{1}{2}\%$ **22.** $16\frac{2}{3}\%$ **23.** $66\frac{2}{3}\%$ **24.** $87\frac{1}{2}\%$

practice for example 3 (page 249)

Write each fraction or mixed number as a percent.

25. $\frac{1}{2}$ **26.** $\frac{9}{20}$ **27.** $\frac{47}{50}$ **28.** $\frac{18}{25}$ **29.** $1\frac{3}{10}$ **30.** $2\frac{2}{5}$

31. $\frac{2}{9}$ **32.** $\frac{3}{8}$ **33.** $\frac{5}{6}$ **34.** $\frac{13}{15}$ **35.** $2\frac{4}{7}$ **36.** $3\frac{27}{40}$

mixed practice (pages 248–249)

Write each percent as a fraction or a mixed number in lowest terms. Write each fraction or mixed number as a percent.

37. 35%　**38.** 280%　**39.** $\frac{3}{4}$　**40.** $\frac{9}{16}$　**41.** $\frac{4}{5}\%$　**42.** $2\frac{3}{4}\%$

43. $37\frac{1}{2}\%$　**44.** $91\frac{2}{3}\%$　**45.** 0.42%　**46.** 22.5%　**47.** $1\frac{12}{25}$　**48.** $2\frac{7}{8}$

49. Residents of the states that border the Gulf of Mexico total about $\frac{3}{20}$ of the people living in the United States. Write this fraction as a percent.

50. A car dealer finances the purchase of a new car at a $9\frac{3}{4}\%$ interest rate. Write this percent as a fraction.

51. In 1980, 24.8% of the population of the United States was between the ages of 5 and 19 years. What fraction of the population was not between those ages?

52. Each day 1385 airplanes arrive at TravelWay Airport. Usually 91% of the planes are *on time,* which means that they arrive within 15 min of the scheduled time. On Monday 1197 planes were on time. How many airplanes arrived late on Monday?

review exercises

Find each answer.

1. $\frac{11}{15} \times \frac{5}{7}$　**2.** $20 \times 6\frac{1}{5}$　**3.** $1\frac{2}{3} \times 3\frac{3}{10}$　**4.** $\frac{5}{8} \div 10$

5. $\frac{4}{5} \div \frac{4}{25}$　**6.** $5\frac{2}{3} \div 1\frac{1}{2}$　**7.** 1.07×4.2　**8.** 0.015×600

9. 0.72×133　**10.** $68.1 \div 0.3$　**11.** $15\overline{)160.5}$　**12.** $0.43\overline{)1.677}$

Erik, Manuel, and Willie are friends. One of them lives on Elm Street, one on Main Street, and one on West Street. Main Street is closest to school. Willie must pass Erik's house to get home from school. Willie and the person who lives on West Street walk to school together. Manuel lives closer to school than Erik. On which street does each live?

13-3 FINDING A PERCENT OF A NUMBER

You should be able to memorize certain common percents and their decimal and fraction *equivalents*. The chart below lists some of these.

Equivalent Percents, Decimals, and Fractions				
$20\% = 0.2 = \frac{1}{5}$	$25\% = 0.25 = \frac{1}{4}$	$12\frac{1}{2}\% = 0.125 = \frac{1}{8}$	$16\frac{2}{3}\% = 0.1\overline{6} = \frac{1}{6}$	$100\% = 1$
$40\% = 0.4 = \frac{2}{5}$	$50\% = 0.5 = \frac{1}{2}$	$37\frac{1}{2}\% = 0.375 = \frac{3}{8}$	$33\frac{1}{3}\% = 0.\overline{3} = \frac{1}{3}$	
$60\% = 0.6 = \frac{3}{5}$	$75\% = 0.75 = \frac{3}{4}$	$62\frac{1}{2}\% = 0.625 = \frac{5}{8}$	$66\frac{2}{3}\% = 0.\overline{6} = \frac{2}{3}$	
$80\% = 0.8 = \frac{4}{5}$		$87\frac{1}{2}\% = 0.875 = \frac{7}{8}$	$83\frac{1}{3}\% = 0.8\overline{3} = \frac{5}{6}$	

To find a percent of a number, use the following method.

1. Write a number sentence. Let n represent the unknown number.
2. Write the percent as a fraction or as a decimal.
3. Find the number that n represents.

example 1

Find each answer.

a. What number is 80% of 135?

b. $12\frac{1}{2}\%$ of 248 is what number?

solution

a. What number is 80% of 135?

$$n = 80\% \times 135$$

$$n = \frac{\overset{1}{\cancel{4}}}{\underset{1}{\cancel{5}}} \times \frac{\overset{27}{\cancel{135}}}{1}$$

$$n = 108$$

So 108 is 80% of 135.

b. $12\frac{1}{2}\%$ of 248 is what number?

$$12\frac{1}{2}\% \times 248 = n$$

$$\frac{1}{\underset{1}{\cancel{8}}} \times \frac{\overset{31}{\cancel{248}}}{1} = n$$

$$31 = n$$

So $12\frac{1}{2}\%$ of 248 is 31.

your turn

Find each answer.

1. What number is 40% of 75?
2. What number is $33\frac{1}{3}\%$ of 258?
3. $83\frac{1}{3}\%$ of 140 is what number?
4. 25% of 234 is what number?

example 2

Find each answer.

a. What number is 35% of 21.2?

b. 5.2% of 375 is what number?

solution

a. What number is 35% of 21.2?

$n = 35\% \times 21.2$

$n = 0.35 \times 21.2$

$n = 7.42$

So 7.42 is 35% of 21.2.

b. 5.2% of 375 is what number?

$5.2\% \times 375 = n$

$0.052 \times 375 = n$

$19.5 = n$

So 5.2% of 375 is 19.5.

your turn

Find each answer.

5. What number is 54% of 191?
6. What number is 10.25% of 280?
7. 120% of 37.5 is what number?
8. 0.9% of 146 is what number?

practice exercises

practice for example 1 (page 251)

Find each answer.

1. What number is 20% of 85?
2. 75% of 336 is what number?
3. $66\frac{2}{3}\%$ of 105 is what number?
4. What number is $16\frac{2}{3}\%$ of 468?
5. What number is 60% of 203?
6. $87\frac{1}{2}\%$ of 524 is what number?

practice for example 2 (page 252)

Find each answer.

7. 1% of 2700 is what number?
8. What number is 64% of 350?
9. What number is 11.5% of 36?
10. 0.8% of 116.25 is what number?
11. 135% of 87.5 is what number?
12. What number is 4.75% of 490?

mixed practice (pages 251–252)

Give the decimal and fraction equivalent of each percent.

13. 25%
14. 60%
15. 75%
16. 40%
17. 20%
18. 50%
19. $33\frac{1}{3}\%$
20. $66\frac{2}{3}\%$
21. $83\frac{1}{3}\%$
22. $12\frac{1}{2}\%$
23. $62\frac{1}{2}\%$
24. 100%

Find each answer.

25. What number is 50% of 68?
26. What number is 0.5% of 20.9?
27. 215% of 7.6 is what number?
28. $62\frac{1}{2}$% of 420 is what number?
29. What number is 16.4% of 132?
30. What number is 90% of 4370?
31. $37\frac{1}{2}$% of 88 is what number?
32. 100% of 47.15 is what number?
33. Forty people were asked if they would vote for a certain candidate. The results were recorded as 35% *Yes*, 57% *No*, and 18% *Undecided*. Can these results be accurate? Explain.
34. In one high school, $16\frac{2}{3}$% of the students are 14 years old. What fraction of the students are not exactly 14 years old?
35. In 1986 the New York Mets won $66\frac{2}{3}$% of the 162 games that they played. How many games did the Mets win?
36. The Walters pay $600 per month to rent their apartment. On June 1, the landlord is raising the rent 5%. How much will the Walters pay each month for rent after June 1?

review exercises

Estimate each decimal as a simple fraction or mixed number.

1. 0.24
2. 0.496
3. 1.67
4. 2.113

Estimate using compatible numbers.

5. $\frac{2}{3} \times 61$
6. $53\frac{1}{2} \times \frac{1}{6}$
7. $398 \times \frac{7}{34}$
8. $\frac{31}{40} \times 35\frac{2}{3}$
9. 0.26×319
10. 0.341×8.89
11. 4019×0.74
12. 0.506×79.10

calculator corner

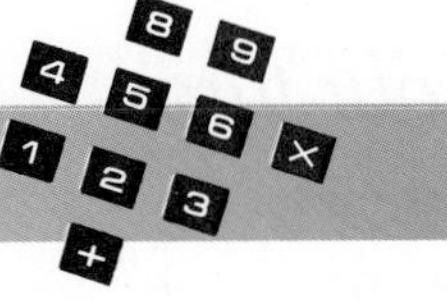

Most calculators have a percent key, [%]. You can use this key to find the percent of a number.

example Find 18% of 245.

solution Enter 245 × 18 [%]. ← **Some calculators require you to enter = after the [%] key.**
The result is 44.1.

Use a calculator to find each answer.

1. 9% of 218
2. 36% of 258
3. 77% of 1380
4. 123% of 45.99
5. 4.75% of 29
6. 38.5% of 620
7. $10\frac{1}{2}$% of 1500
8. $6\frac{1}{4}$% of 279.8

13-4 ESTIMATING A PERCENT OF A NUMBER

You can use *compatible numbers* to estimate a percent of a number.

example 1

Estimate.

a. 24% of 331 ← **24% is about 25%. You can use the fraction equivalent of 25%.**

$\frac{1}{4} \times 320$

about 80

b. 61% of 247

$\frac{3}{5} \times 250$

about 150

c. 67% of $382.79

$\frac{2}{3} \times \$390$

about $260

your turn

Estimate.

1. 81% of 19
2. 49% of 593
3. 74% of $241
4. 33.5% of $219.19

example 2

Estimate.

a. 29% of $5.89 ← **29% is about 30%. You can use the decimal equivalent of 30%.**

$0.3 \times \$6$

about $1.80

b. 8.75% of $423

$0.09 \times \$400$

about $36

your turn

Estimate.

5. 92% of 48
6. 7.7% of 5928
7. 9.8% of $212.85
8. $2\frac{1}{4}$% of $42,766

practice exercises

practice for example 1 (page 254)

Estimate.

1. 20% of 348
2. 25% of $277
3. 73% of 166
4. 42% of 458
5. 61% of 39
6. 65% of 17
7. 32% of 208
8. 12.8% of 156
9. 51.9% of $283
10. 16% of $603
11. 79% of $102.75
12. 23% of $88.99

practice for example 2 (page 254)

Estimate.

13. 89% of 823 **14.** 31% of $41 **15.** 3.8% of $58.95 **16.** 7.2% of $915.49

17. 5.15% of $784.97 **18.** 9.75% of $19.13 **19.** $3\frac{1}{3}$% of $52,035 **20.** $5\frac{3}{4}$% of 693

mixed practice (page 254)

Estimate.

21. 19% of $289 **22.** 67% of 272 **23.** 76% of 291 **24.** 58% of $24

25. 29% of 795 **26.** 1.25% of 209 **27.** 9.9% of $407 **28.** $7\frac{1}{2}$% of $61.95

29. There is a total of 443,802 books in the Waton Public Library. Of these books, 21% are nonfiction. About how many nonfiction books are in the library?

30. A baseball stadium has a seating capacity of 59,146. For one sold-out game, 65% of the tickets were bought before the season started, 24% were bought after the season started but before the day of the game, and 11% were bought on game day. About how many more tickets were bought before the season started than on game day?

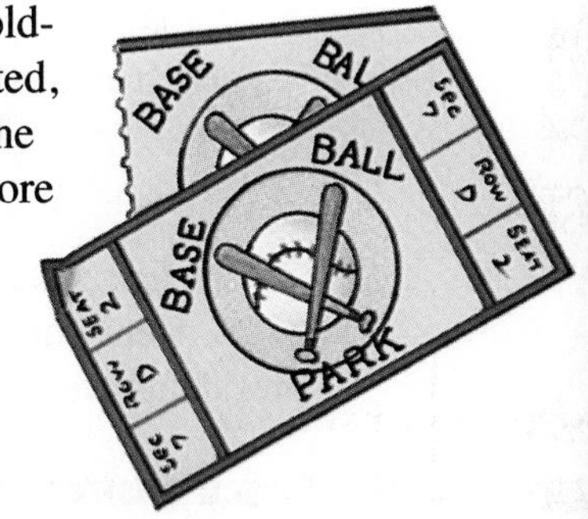

review exercises

Write each sum or difference in lowest terms.

1. $\frac{1}{10} + \frac{7}{10}$ **2.** $1\frac{5}{8} + 2\frac{3}{8}$ **3.** $3\frac{5}{6} - \frac{1}{3}$ **4.** $10 - 6\frac{1}{9}$

5. $4\frac{3}{4} + \frac{2}{5}$ **6.** $\frac{7}{8} - \frac{5}{6}$ **7.** $9 + 3\frac{4}{7}$ **8.** $5\frac{3}{10} - 2\frac{7}{15}$

mental math

You can use 10% of a number to find 20% or 5% of a number mentally.

10% of 320 = 32

20% = 2 × 10%, so 20% of 320 = 2 × 32 = 64.

5% = $\frac{1}{2}$ × 10%, so 5% of 320 = $\frac{1}{2}$ × 32 = 16.

Find each answer mentally.

1. 20% of $210 **2.** 5% of $62.80 **3.** 20% of 3.4 **4.** 5% of 106

5. 30% of $730 **6.** 40% of 2.2 **7.** 50% of $3.10 **8.** 60% of 610

SKILL REVIEW

Write each percent as a decimal.

13-1

1. 4%
2. 0.8%
3. 173%
4. $59\frac{2}{3}\%$
5. $14\frac{1}{5}\%$

Write each decimal as a percent.

6. 0.3
7. 0.462
8. 2.61
9. 0.005
10. $0.03\frac{1}{3}$
11. At the end of the baseball season, George Carlos had a batting average of 0.291. During the season George was at bat 592 times. Write the batting average as a percent.

Write each percent as a fraction or a mixed number in lowest terms.

13-2

12. 5%
13. 182%
14. 13.6%
15. $\frac{1}{3}\%$
16. $11\frac{3}{7}\%$

Write each fraction or mixed number as a percent.

17. $\frac{9}{10}$
18. $\frac{1}{6}$
19. $\frac{7}{16}$
20. $1\frac{4}{5}$
21. $2\frac{1}{8}$
22. Of the first forty people who were presidents of the United States, $17\frac{1}{2}\%$ were born in the twentieth century. What fraction of these presidents were born before the twentieth century?

Find each answer.

13-3

23. What number is 50% of 81.8?
24. What number is 40% of 415?
25. $16\frac{2}{3}\%$ of 333 is what number?
26. $66\frac{2}{3}\%$ of 27 is what number?
27. What number is $87\frac{1}{2}\%$ of 128?
28. What number is 200% of 8.5?
29. 0.4% of 628 is what number?
30. 31.5% of 54.3 is what number?
31. What number is 93% of 80?
32. 6% of 239.95 is what number?
33. At a local business, 40% of the employees are women. There are 3 department heads, 8 supervisors, and 34 other workers. How many women does the business employ?

Estimate.

13-4

34. 26% of 39
35. 81% of 102
36. 92% of $53.40
37. 59.4% of 43.92
38. 9.8% of 59,913
39. 33.25% of $247
40. 4.25% of $79.99
41. $6\frac{7}{8}\%$ of $312
42. $2\frac{1}{5}\%$ of $387
43. A sweater sells for $38.99 and a shirt for $21.49. Both items are on sale for 65% of the original price. About how much is the total sale price for the sweater and shirt?

APPLICATION

13-5 ROYALTIES

Authors, composers, artists, playwrights, and inventors may receive *royalties*. A **royalty** is a payment based on a percent of the *net receipts* resulting from the sale or performance of a work. The percent is called the **royalty rate.**

example

Dana Dane is an author. She receives royalties of $2\frac{1}{2}\%$ from the publisher of her book. The net receipts for a six-month period were \$543,803. How much did Dana receive in royalties for this period?

solution

Find $2\frac{1}{2}\%$ of \$543,803.

$0.025 \times \$543{,}803 = \$13{,}595.075$ ← **Write the percent as a decimal.**

To the nearest cent, Dana received \$13,595.08 in royalties.

exercises

Complete. If necessary, round to the nearest cent. A calculator may be helpful.

	NET RECEIPTS	ROYALTY RATE	ROYALTIES
1. author	\$228,543	3%	?
2. artist	\$59,675	9%	?
3. playwright	\$460,931	$10\frac{1}{2}\%$	?
4. inventor	\$31,517	11.5%	?
5. composer	\$79,750	2.2%	?
6. artist	\$166,390	$7\frac{3}{4}\%$	?

7. Tom Boswain is an inventor. He receives royalties of 7% from the manufacturer of his invention. The net receipts for the first six months are \$82,136. Tom had already received a \$3500 *advance,* which is to be subtracted from his royalties for the first six-month period. What will be the amount of Tom's check for this period?

8. Susan Daniels is a playwright and receives royalties for both the published version and the theatrical performance of her play. She receives 1.6% from the book publisher and 4% from the theater owner. The net receipts were \$19,285 for the book and \$327,914 for the performances. How much did Susan receive in royalties?

APPLICATION

13-6 SALES TAX

Many states collect a **sales tax** on certain items or services that you buy. The sales tax helps pay the cost of managing the state government. The **sales tax rate** is given as a percent. The *total cost* of an item is the sum of the price of the item and the sales tax.

5% Sales Tax Table

Amount of Sale	Tax	Amount of Sale	Tax
0.00– 0.09	0.00	5.10– 5.29	0.26
0.10– 0.29	0.01	5.30– 5.49	0.27
0.30– 0.49	0.02	5.50– 5.69	0.28
0.50– 0.69	0.03	5.70– 5.89	0.29
0.70– 0.89	0.04	5.90– 6.09	0.30
0.90– 1.09	0.05	6.10– 6.29	0.31
1.10– 1.29	0.06	6.30– 6.49	0.32
1.30– 1.49	0.07	6.50– 6.69	0.33
1.50– 1.69	0.08	6.70– 6.89	0.34
1.70– 1.89	0.09	6.90– 7.09	0.35
1.90– 2.09	0.10	7.10– 7.29	0.36
2.10– 2.29	0.11	7.30– 7.49	0.37
2.30– 2.49	0.12	7.50– 7.69	0.38
2.50– 2.69	0.13	7.70– 7.89	0.39
2.70– 2.89	0.14	7.90– 8.09	0.40
2.90– 3.09	0.15	8.10– 8.29	0.41
3.10– 3.29	0.16	8.30– 8.49	0.42
3.30– 3.49	0.17	8.50– 8.69	0.43
3.50– 3.69	0.18	8.70– 8.89	0.44
3.70– 3.89	0.19	8.90– 9.09	0.45
3.90– 4.09	0.20	9.10– 9.29	0.46
4.10– 4.29	0.21	9.30– 9.49	0.47
4.30– 4.49	0.22	9.50– 9.69	0.48
4.50– 4.69	0.23	9.70– 9.89	0.49
4.70– 4.89	0.24	9.90–10.09	0.50
4.90– 5.09	0.25	10.10–10.29	0.51

example

The price of a record album is $7.87.

a. Find the total cost when the tax rate is 5%.

b. Find the total cost when the tax rate is $5\frac{3}{4}\%$.

solution

a. Use the 5% Sales Tax Table shown at the right. Find $7.87 in the *Amount of Sale* column. The number 7.87 is in the interval 7.70–7.89. Read across to the *Tax* column to find the sales tax: $.39

Add to find the total cost.

$7.87 + $.39 = $8.26

The total cost is $8.26.

b. Multiply to find the amount of sales tax. A calculator may be helpful.

$5\frac{3}{4}\%$ of $7.87 → 0.0575 × $7.87 ← **$5\frac{3}{4}\% = 5.75\% = 0.0575$**

= $.452525

≈ $.45 ← **Round to the nearest cent.**

Add to find the total cost.

$7.87 + $.45 = $8.32

The total cost is $8.32.

exercises

Use the table above. Find the 5% sales tax on each purchase.

1. $2.39 **2.** $5.99 **3.** $.18 **4.** $9.49

Find the sales tax on each purchase with the given sales tax rate. If necessary, round to the nearest cent.

5. $6.50; 4% **6.** $12.99; 3% **7.** $128.40; 7.5% **8.** $22.56; $4\frac{3}{4}\%$

Complete. If necessary, round to the nearest cent.

	ITEM	PRICE	SALES TAX RATE	SALES TAX	TOTAL COST
9.	notebook	\$1.99	6%	?	?
10.	shirt	\$16.20	4%	?	?
11.	watch	\$59.50	3.5%	?	?
12.	stereo	\$600.00	$4\frac{3}{4}$%	?	?

Solve. If necessary, round to the nearest cent.

13. At a baseball stadium the price of a season ticket to all 81 home games is \$720. The sales tax rate is 5.5%. What is the total cost of a season ticket?

14. At Circle Bike Shop the price of a ten-speed bicycle is \$219. The sales tax rate is $6\frac{1}{2}$%. What is the total cost of the bicycle?

15. Samantha Sanchez bought a new car priced at \$11,854. The sales tax rate is $5\frac{1}{4}$%. What was the total cost of the car?

16. Lynn Holland bought three blank video cassette tapes at \$4.79 each. The sales tax rate is 5%. She paid with a \$20 bill. How much change should Lynn receive?

17. Tony Tiezzi bought two pairs of high-top sneakers at \$38.99 per pair. The sales tax rate is 4.75%. How much did Tony spend for the sneakers?

18. The price of a haircut at HairCuts is \$14. The price at Salon Beauty is \$22. The sales tax rate is 6.5%. How much more is the total cost of a haircut at Salon Beauty than at HairCuts?

19. Find the sales tax rate in your state.

20. Find the states in which there is no sales tax.

21. In some cities there is a city sales tax in addition to the state sales tax. Find out if there is a city sales tax in your city.

APPLICATION

13-7 STATE AND LOCAL INCOME TAXES

Many states require that employers withhold money from employees' paychecks to pay state income taxes. The state government often uses the money collected to operate state colleges and maintain state roads. Many state agencies, such as the motor vehicles department and state police, also are financed by a state income tax. As with the federal tax, employees must fill out a state income tax return each year.

example

Debbie Cohen has an annual taxable income of \$25,730. The state income tax rate is shown in the table at the right. How much state income tax does she pay?

State Income Tax	
Annual Taxable Income	**Tax Rate**
first \$5000	3%
next \$5000	4%
over \$10,000	5%

solution

Find the tax for the first two rows in the table. A calculator may be helpful.

first \$5000 of income: 3% of \$5000 → 0.03 × \$5000 = \$150
next \$5000 of income: 4% of \$5000 → 0.04 × \$5000 = \$200

Subtract to find the part of Debbie's income that is over \$10,000.

\$25,730 − \$10,000 = \$15,730

income over \$10,000: 5% of \$15,730 → 0.05 × \$15,730 = \$786.50

Add: \$150 + \$200 + \$786.50 = \$1136.50

Debbie pays \$1136.50 for state income tax.

exercises

Find the state income tax. If necessary, round to the nearest cent.

1. *Taxable income:* \$18,734
   State tax rate:
   first \$3000: 2%
   next \$5000: 5%
   over \$8000: 6%

2. *Taxable income:* \$52,809
   State tax rate:
   first \$20,000: 2%
   next \$30,000: 2.5%
   over \$50,000: 3.5%

Use the table at the right. Find the state income tax. If necessary, round to the nearest cent.

If Taxable Income is:	State Tax Rate is:
\$0– \$2000	0%
\$2001–\$10,000	6%
\$10,001–\$20,000	8%
over \$20,000	9.5%

3. \$23,570
4. \$14,880
5. \$10,001
6. \$8942

Solve. If necessary, round to the nearest cent.

7. Gina Trozzi has an annual taxable income of \$16,278. The state tax rate is 7%. How much state income tax does she pay?
8. Roberto Ortego has an annual taxable income of \$33,955. The state tax rate is 4.6%. How much state income tax does he pay?
9. Alan Pickman had a total of \$1011.99 withheld from his paycheck during the year to pay state income tax. The state tax rate is $3\frac{2}{5}$%. Alan's taxable income for the year was \$21,492. Is there a refund or a balance due? Find the exact amount.
10. Find out if there is an income tax in your state. If there is, what is the state tax rate?

Some cities or towns also require that employers withhold money to pay local income taxes. Local income tax rates are usually a percent of the annual taxable income or a percent of state income tax owed.

11. Lori Kane lives in Springfield. The local tax rate in Springfield is 2.25% of the annual taxable income. Lori has an annual taxable income of \$22,809. How much local income tax does she pay?
12. Ray Chin lives in Simsbury and has an annual taxable income of \$14,880. The local tax rate in Simsbury is 35% of state income tax. Ray pays \$594 for state income tax. How much local income tax does he pay?
13. Walter Jackson lives in Midville and has an annual taxable income of \$25,850. The local tax rate in Midville is 50% of state income tax. The state tax rate is 7% of the annual taxable income. How much local income tax does he pay?
14. Find out if there is a local income tax where you live. If there is, what is the local tax rate?

CAREER

TAX ASSESSOR

Renee Miller is a tax assessor. She evaluates property and determines its **market value.** The market value is used to determine the **property tax.** Property owners pay taxes to the city, county, or town. The government uses the money collected to pay for local expenses, such as roads, parks, and schools, and to provide police and fire protection.

Property taxes are based on the *assessed value* of the property and on the **tax rate.** The **assessed value** is a percent of the market value. This percent is called the **assessment rate.**

example

Find the property tax on a house having a market value of \$175,000. The assessment rate is 45%, and the tax rate is \$2.62 per \$100.

solution

Multiply the market value by the assessment rate to find the assessed value.

$$45\% \text{ of } \$175{,}000 \rightarrow 0.45 \times \$175{,}000 = \$78{,}750$$

The tax rate is \$2.62 per \$100 of the assessed value.
Find the number of \$100's in \$78,750.

$$\$78{,}750 \div \$100 = 787.5$$

Multiply by \$2.62: $787.5 \times \$2.62 = \$2063.25$

The property tax is \$2063.25.

exercises

Complete. The tax rate is \$3.27 per \$100. If necessary, round to the nearest cent. A calculator may be helpful.

	MARKET VALUE	ASSESSMENT RATE	ASSESSED VALUE	PROPERTY TAX
1.	\$48,000	30%	?	?
2.	\$158,000	100%	?	?
3.	\$130,000	4.5%	?	?

4. The Changs own a house with a market value of \$94,000. The city assesses houses at a rate of 61.5%. The property tax rate is \$4.78 per \$100. How much property tax do they pay?
5. Sometimes the property tax rate is given in mills per \$1. Find out the meaning of *mill* in this context.

CHAPTER REVIEW

vocabulary vo·cab·u·lar·y

Choose the correct word or phrase to complete each sentence.

1. A (*royalty, percent*) represents a ratio that compares a number to 100.
2. A tax on certain items that you buy is (*a sales tax, an income tax*).

skills

Write each percent as a decimal. Write each decimal as a percent.

3. 1%
4. 9.75%
5. $46\frac{1}{4}\%$
6. 0.67
7. 1.8
8. $0.02\frac{1}{2}$

Write each percent as a fraction or a mixed number in lowest terms. Write each fraction or mixed number as a percent.

9. 140%
10. 6.25%
11. $7\frac{1}{5}\%$
12. $\frac{23}{50}$
13. $\frac{2}{7}$
14. $2\frac{8}{15}$

Find each answer.

15. What number is 75% of 502?
16. What number is $12\frac{1}{2}\%$ of 312?
17. $83\frac{1}{3}\%$ of 54 is what number?
18. 115% of 62.96 is what number?
19. What number is 23.5% of 150?
20. 4.25% of 3.8 is what number?

Estimate.

21. 39% of \$36.50
22. 51.2% of 1582
23. 5.75% of 309
24. $9\frac{1}{4}\%$ of \$71.29

Solve. If necessary, round to the nearest cent.

25. In the election for junior class president, JoEllen Riffle received 55% of the votes. There were 1280 votes in all. How many votes did JoEllen receive?
26. A team of three authors wrote a book. The team receives royalties of 9% from the publisher. The net receipts for a six-month period were \$407,665. If the royalties are divided equally, how much did each author receive for this period?
27. At Jurgen's Music Store the price of a trumpet is \$215. The sales tax rate is $7\frac{1}{2}\%$. What is the total cost of the trumpet?
28. Juan Delacruz lives in Canton. The local tax rate in Canton is 2.2% of the annual taxable income. Juan has an annual taxable income of \$31,898. How much local income tax does he pay?

CHAPTER TEST

Write each percent as a decimal. **13-1**

1. 60%
2. 3.5%
3. 218%
4. $44\frac{4}{9}\%$
5. $7\frac{1}{2}\%$

Write each decimal as a percent.

6. 0.77
7. 0.085
8. 1.2
9. 0.004
10. $0.30\frac{1}{2}$

Write each percent as a fraction or a mixed number in lowest terms. **13-2**

11. 4%
12. 275%
13. 0.5%
14. $\frac{3}{8}\%$
15. $2\frac{2}{5}\%$

Write each fraction or mixed number as a percent.

16. $\frac{3}{20}$
17. $\frac{1}{3}$
18. $\frac{5}{9}$
19. $1\frac{3}{25}$
20. $2\frac{7}{12}$

Find each answer. **13-3**

21. What number is 60% of 235?
22. What number is $33\frac{1}{3}\%$ of 156?
23. $62\frac{1}{2}\%$ of 48 is what number?
24. 72% of 41.5 is what number?
25. What number is 5.25% of 124?
26. 0.6% of 6.3 is what number?

Estimate. **13-4**

27. 67% of \$236.69
28. 19.7% of 253
29. $7\frac{3}{4}\%$ of \$591
30. Members of the high school band and orchestra wanted to sell 475 tickets for the spring concert. By the day of the concert, they had sold 76% of the tickets. About how many tickets were left to sell?

Solve. If necessary, round to the nearest cent.

31. Ed Sternberg is an artist. A poster company reproduces his painting as a poster. Ed receives royalties of 12% from the company. The net receipts for a four-month period were \$36,472. How much did Ed receive in royalties for this period? **13-5**

32. The price of a pair of shoes made from artificial leather is \$19.99. The price of a pair of shoes made from genuine leather is \$42. The sales tax rate is 6%. How much greater is the total cost of the genuine leather shoes than the artificial leather ones? **13-6**

33. Rachel O'Connor has an annual taxable income of \$26,547. The state income tax rate is $2\frac{1}{2}\%$ of the annual taxable income. How much state income tax does she pay? **13-7**

The many colors in this textbook are produced by combining different percents of a few basic colors. This scanner is used in the process.

CHAPTER 14

MORE PERCENT

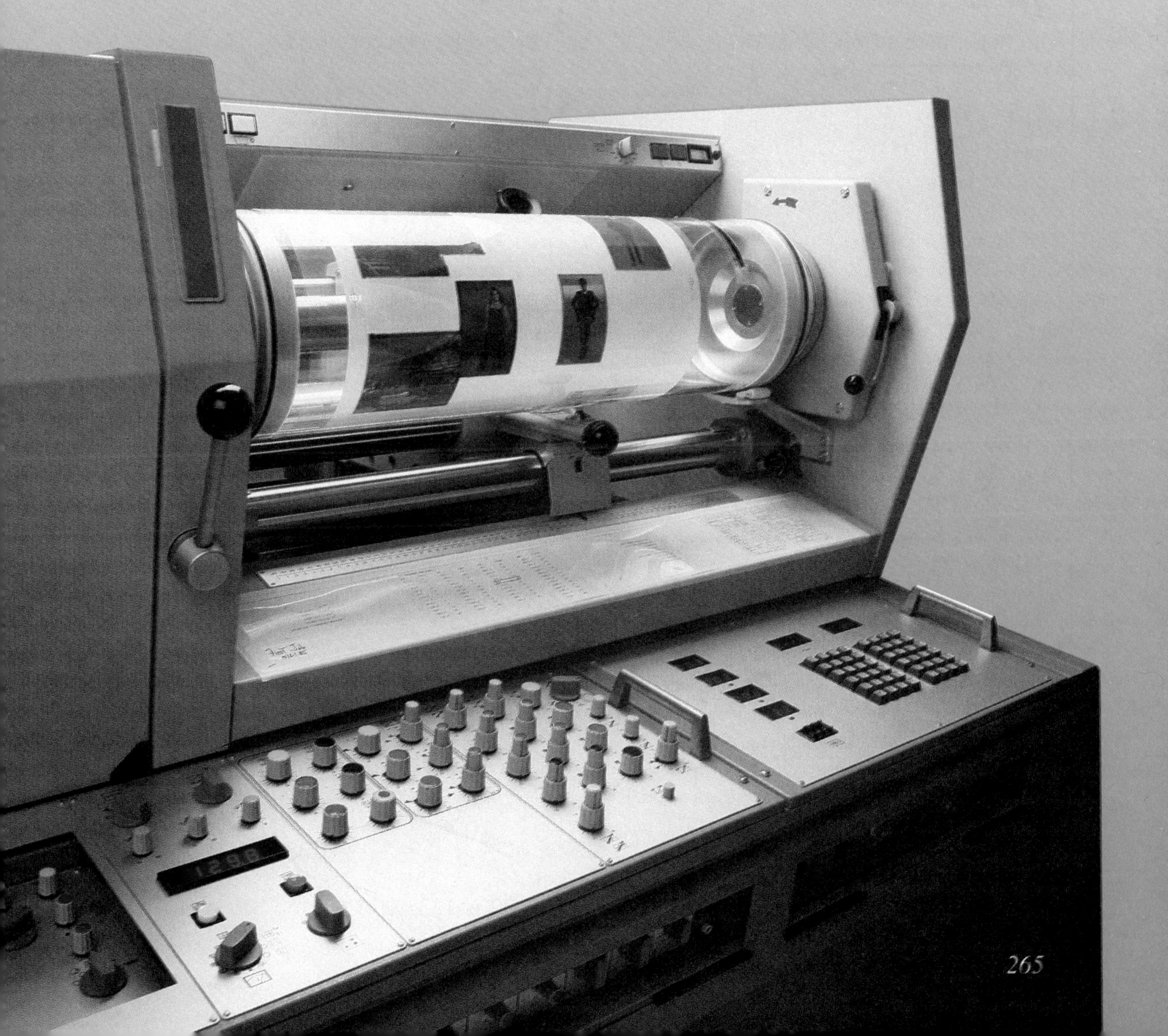

14-1 FINDING THE PERCENT ONE NUMBER IS OF ANOTHER

To find the percent one number is of another, use the following method.

1. Write a number sentence. Let n represent the unknown percent in fraction or decimal form.
2. Find the number that n represents.
3. Write n as a percent.

example 1

Find each answer.

a. What percent of 75 is 18?

$$\underbrace{\text{What percent}}_{n} \text{ of } 75 \text{ is } 18?$$

$$n \times 75 = 18$$

$$\frac{n \times 75}{75} = \frac{18}{75}$$ ← Divide by 75 on each side of the equals sign.

$$n = \frac{18}{75} = \frac{6}{25} = \frac{6 \times 4}{25 \times 4} = \frac{24}{100} = 24\%$$

So 18 is 24% of 75.

b. 11 is what percent of 5?

$$11 \text{ is } \underbrace{\text{what percent}}_{n} \text{ of } 5?$$

$$11 = n \times 5$$

$$\frac{11}{5} = \frac{n \times 5}{5}$$ ← Divide by 5 on each side of the equals sign.

$$n = \frac{11}{5} = \frac{11 \times 20}{5 \times 20} = \frac{220}{100} = 220\%$$

So 11 is 220% of 5.

your turn

Find each answer.

1. What percent of 52 is 39?
2. What percent of \$140 is \$84?
3. 35 is what percent of 125?
4. 20 is what percent of 16?

example 2

Find each answer.

a. What percent of 32 is 4?

$$\underbrace{\text{What percent}}_{n} \text{ of } 32 \text{ is } 4?$$

$$n \times 32 = 4$$

$$\frac{n \times 32}{32} = \frac{4}{32}$$ ← Divide by 32 on each side of the equals sign.

$$n = \frac{4}{32} = \frac{1}{8} = 12\tfrac{1}{2}\%$$

So 4 is $12\frac{1}{2}\%$ of 32.

b. 80 is what percent of 60?

$$80 \text{ is } \underbrace{\text{what percent}}_{n} \text{ of } 60?$$

$$80 = n \times 60$$

$$\frac{80}{60} = \frac{n \times 60}{60}$$ ← Divide by 60 on each side of the equals sign.

$$n = \frac{80}{60} = \frac{4}{3} = 1\tfrac{1}{3} = 1.33\tfrac{1}{3} = 133\tfrac{1}{3}\%$$

So 80 is $133\frac{1}{3}\%$ of 60.

your turn

Find each answer.

5. What percent of 16 is 14?
6. What percent of 84 is 70?
7. $13 is what percent of $8?
8. 55 is what percent of 40?

example 3

Find each answer.

a. What percent of 24 is 22?

b. 62.07 is what percent of 30?

solution

a. What percent of 24 is 22?

$$n \times 24 = 22$$

$$\frac{n \times 24}{24} = \frac{22}{24}$$ ← Divide by 24 on each side of the equals sign.

$$n = \frac{22}{24} = \frac{11}{12} \rightarrow 12\overline{)11.00}$$

Quotient: $0.91\frac{8}{12}$

```
    0.91 8/12
12)11.00
   10 8
      20
      12
       8
```

$$n = 0.91\frac{8}{12} = 0.91\frac{2}{3} = 91\frac{2}{3}\%$$

So 22 is $91\frac{2}{3}\%$ of 24.

b. 62.07 is what percent of 30?

$$62.07 = n \times 30$$

$$\frac{62.07}{30} = \frac{n \times 30}{30}$$ ← Divide by 30 on each side of the equals sign.

$$n = \frac{62.07}{30} \rightarrow 30\overline{)62.07}$$

Quotient: $2.06\frac{27}{30}$

```
   2.06 27/30
30)62.07
   60
    2 07
    1 80
      27
```

$$n = 2.06\frac{27}{30} = 2.06\frac{9}{10} = 206\frac{9}{10}\%$$

So 62.07 is $206\frac{9}{10}\%$ of 30.

your turn

Find each answer.

9. What percent of 120 is 9?
10. What percent of 400 is 109?
11. 13 is what percent of 15?
12. 66.43 is what percent of 21?

practice exercises

practice for example 1 (*page 266*)

Find each answer.

1. What percent of 20 is 19?
2. What percent of $10 is $7?
3. 16 is what percent of 80?
4. 6 is what percent of 40?
5. What percent of 80 is 32?
6. What percent of $48 is $60?

practice for example 2 (pages 266–267)

Find each answer.

7. What percent of 112 is 14?
8. What percent of 400 is 150?
9. \$24 is what percent of \$144?
10. \$35 is what percent of \$56?
11. What percent of 40 is 75?
12. What percent of 48 is 80?

practice for example 3 (page 267)

Find each answer.

13. What percent of \$16 is \$5?
14. What percent of \$300 is \$88?
15. 37 is what percent of 40?
16. 102 is what percent of 320?
17. What percent of 45 is 48?
18. What percent of 64 is 73.68?

mixed practice (pages 266–267)

Find each answer.

19. What percent of \$75 is \$24?
20. What percent of \$24 is \$21?
21. \$105 is what percent of \$840?
22. 300 is what percent of 300?
23. What percent of 15 is 6?
24. What percent of 60 is 62.82?
25. \$50.96 is what percent of \$28?
26. 210 is what percent of 180?
27. The Focus Company has 80 computers. Fifty of these are connected directly to a printer. What percent of the computers are connected directly to a printer?
28. Tech Sound requires a down payment of \$50 on a drum set that sells for \$200. What percent of the selling price is the down payment?
29. The Planner Company has 400 employees. Of these, 30 employees are accountants. Write as a fraction in lowest terms the ratio of the number of employees who are accountants to the total number of employees.

review exercises

Write each decimal or fraction as a percent.

1. 0.82
2. 0.09
3. 0.5
4. 0.76
5. 2.33
6. $\frac{3}{4}$
7. $\frac{2}{5}$
8. $\frac{1}{8}$
9. $\frac{2}{3}$
10. $\frac{5}{6}$

14-2 PERCENT OF INCREASE OR DECREASE

When the original quantity or price of an item increases, the amount of increase can be written as a percent. To find the **percent of increase,** use the following method.

1. Find the amount of increase.
2. Find what percent the amount of increase is of the original amount.

example 1

Find each percent of increase.

a. original score: 40
new score: 48

b. original length: 39 m
new length: 91 m

solution

a. amount of increase: $48 - 40 = 8$

What percent of 40 is 8?

$$n \times 40 = 8$$

$$\frac{n \times 40}{40} = \frac{8}{40}$$

$$n = \frac{8}{40} = \frac{1}{5} = 20\%$$

So the percent of increase is 20%.

b. amount of increase: $91 - 39 = 52$

What percent of 39 is 52?

$$n \times 39 = 52$$

$$\frac{n \times 39}{39} = \frac{52}{39}$$

$$n = \frac{52}{39} = \frac{4}{3} = 1\frac{1}{3} = 1.33\frac{1}{3} = 133\frac{1}{3}\%$$

So the percent of increase is $133\frac{1}{3}\%$.

your turn

Find each percent of increase.

1. original length: 460 m
new length: 598 m

2. original bill: \$75
new bill: \$165

3. original cost: \$60
new cost: \$160

4. original number of students: 180
new number of students: 330

When the original quantity or price of an item decreases, the amount of decrease can be written as a percent. To find the **percent of decrease,** use the following method.

1. Find the amount of decrease.
2. Find what percent the amount of decrease is of the original amount.

example 2

Find each percent of decrease.

a. original weight: 200 lb
new weight: 170 lb

b. original number of volunteers: 288
new number of volunteers: 252

solution

a. amount of decrease: $200 - 170 = 30$

What percent of 200 is 30?

$$n \times 200 = 30$$

$$\frac{n \times 200}{200} = \frac{30}{200}$$

$$n = \frac{30}{200} = \frac{15}{100} = 15\%$$

So the percent of decrease is 15%.

b. amount of decrease: $288 - 252 = 36$

What percent of 288 is 36?

$$n \times 288 = 36$$

$$\frac{n \times 288}{288} = \frac{36}{288}$$

$$n = \frac{36}{288} = \frac{1}{8} = 12\frac{1}{2}\%$$

So the percent of decrease is $12\frac{1}{2}\%$.

your turn

Find each percent of decrease.

5. original length: 65 yd
new length: 13 yd

6. original enrollment: 176
new enrollment: 88

7. original number of boxes: 192
new number of boxes: 32

8. original price: $240
new price: $150

practice exercises

practice for example 1 *(page 269)*

Find each percent of increase.

1. original donation: $25
new donation: $40

2. original salary: $165
new salary: $198

3. original distance: 32 km
new distance: 76 km

4. original number of books: 63
new number of books: 168

practice for example 2 *(page 270)*

Find each percent of decrease.

5. original price: $80
new price: $60

6. original cost: $525
new cost: $315

7. original length: 18 m
new length: 3 m

8. original number of pamphlets: 84
new number of pamphlets: 56

mixed practice (pages 269–270)

Tell whether there is an *increase* or *decrease*. Then find the percent of increase or decrease.

9. original length: 75 m
   new length: 15 m
10. original salary: $220
    new salary: $253
11. original weight: 72 lb
    new weight: 99 lb
12. original fare: $132
    new fare: $110
13. original distance: 100 yd
    new distance: 225 yd
14. original number of passengers: 144
    new number of passengers: 54
15. original bill: $25
    new bill: $69
16. original salary: $300
    new salary: $273
17. Last week Craig ran 36 mi. This week he ran 30 mi. Find the percent of decrease.
18. A tiger usually weighs 100 lb at age one. By age two, its weight is about 300 lb. Find the percent of increase.
19. The enrollment in a high school was 875 students. Five years later, the enrollment was 805 students. By how many students did the enrollment decrease?
20. Heather earned $20,600 in one year. The next year she earned $21,012. By what amount did her earnings increase?

review exercises

Give the decimal and fraction equivalents of each percent.

1. 25%
2. 40%
3. $33\frac{1}{3}\%$
4. $87\frac{1}{2}\%$
5. 60%
6. $66\frac{2}{3}\%$
7. $12\frac{1}{2}\%$
8. 75%
9. $83\frac{1}{3}\%$
10. 50%

There were 25 passengers riding a bus. After the first stop, the number of passengers increased by 100%. After the second stop, the number of passengers decreased by 100%. Find the number of passengers riding the bus after each of these two stops.

14-3 FINDING A NUMBER WHEN A PERCENT OF IT IS KNOWN

To find a number when a percent of it is known, use the following method.

1. Write a number sentence. Let n represent the unknown number.
2. Write the percent as a decimal or as a fraction.
3. Find the number that n represents.

example 1

Find each answer.

a. 6% of what number is 54?

b. 156 is 120% of what number?

solution

a. 6% of what number is 54?

$6\% \times n = 54$

$0.06 \times n = 54$

$\frac{0.06 \times n}{0.06} = \frac{54}{0.06}$ ← **Divide by 0.06 on each side of the equals sign.**

$n = \frac{54}{0.06} \rightarrow 0.06\overline{)54.00}$ = 900.

$n = 900$

So 54 is 6% of 900.

b. 156 is 120% of what number?

$156 = 120\% \times n$

$156 = 1.2 \times n$

$\frac{156}{1.2} = \frac{n \times 1.2}{1.2}$ ← **Divide by 1.2 on each side of the equals sign.**

$n = \frac{156}{1.2} \rightarrow 1.2\overline{)156.0}$ = 130.

$n = 130$

So 156 is 120% of 130.

your turn

Find each answer.

1. 62% of what number is 31?
2. 17% of what number is 1.02?
3. 105 is 140% of what number?
4. 46 is 200% of what number?
5. 57% of what number is 513?
6. 2.8% of what number is 84?

example 2

Find each answer.

a. 25% of what number is 76?

b. 49 is $87\frac{1}{2}\%$ of what number?

solution

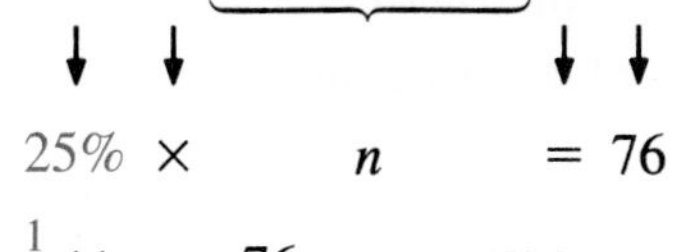

a. 25% of what number is 76?

$$25\% \times n = 76$$

$$\frac{1}{4} \times n = 76 \quad \leftarrow \text{Write the percent as a fraction.}$$

$$\frac{\frac{1}{4} \times n}{\frac{1}{4}} = \frac{76}{\frac{1}{4}}$$

$$n = 76 \div \frac{1}{4}$$

$$n = 76 \times 4$$

$$n = 304$$

So 76 is 25% of 304.

b. 49 is $87\frac{1}{2}\%$ of what number?

$$49 = 87\frac{1}{2}\% \times n$$

$$49 = \frac{7}{8} \times n \quad \leftarrow \text{Write the percent as a fraction.}$$

$$\frac{49}{\frac{7}{8}} = \frac{\frac{7}{8} \times n}{\frac{7}{8}}$$

$$49 \div \frac{7}{8} = n$$

$$49 \times \frac{8}{7} = n$$

$$\frac{\overset{7}{\cancel{49}}}{1} \times \frac{8}{\underset{1}{\cancel{7}}} = n \rightarrow n = 56$$

So 49 is $87\frac{1}{2}\%$ of 56.

your turn

Find each answer.

7. 20% of what number is 15?

8. $62\frac{1}{2}\%$ of what number is 60?

9. 35 is $87\frac{1}{2}\%$ of what number?

10. 8 is $33\frac{1}{3}\%$ of what number?

11. 168 is 40% of what number?

12. 90 is 75% of what number?

practice exercises

practice for example 1 (page 272)

Find each answer.

1. 4% of what number is 1?

2. 3% of what number is 4.5?

3. 931 is 76% of what number?

4. 299 is 65% of what number?

5. 744 is 18.6% of what number?

6. 805 is 32.2% of what number?

7. 125% of what number is 105?

8. 150% of what number is 336?

9. 694 is 200% of what number?

10. 500% of what number is 1565?

practice for example 2 *(page 273)*

Find each answer.

11. 80% of what number is 180?
12. 50% of what number is 481?
13. 60% of what number is 618?
14. $33\frac{1}{3}$% of what number is 99?
15. 57 is $12\frac{1}{2}$% of what number?
16. 5 is $16\frac{2}{3}$% of what number?
17. 235 is 25% of what number?
18. 1421 is $87\frac{1}{2}$% of what number?
19. $37\frac{1}{2}$% of what number is 216?
20. 85 is $83\frac{1}{3}$% of what number?

mixed practice *(pages 272–273)*

Find each answer.

21. 22 is 88% of what number?
22. 16.2 is 54% of what number?
23. $66\frac{2}{3}$% of what number is 34?
24. 590 is $62\frac{1}{2}$% of what number?
25. 7.5% of what number is 84?
26. 175% of what number is 210?
27. On a math test, Jean answered 86% of the questions correctly. If she answered 43 questions correctly, how many questions were on the test?
28. A car dealership has 387 cars that are last year's models. This represents 45% of its cars. How many cars does the car dealership have?
29. In 1974 the population of bald eagles was recorded as 8000. In 1987 the population was recorded as 14,000. What was the percent of increase?
30. The Sidecar originally charged $25 for a swordfish dinner. On July 15, the price of the dinner will be raised 8%. What will be the new price of the dinner?

review exercises

Write each ratio as a fraction in lowest terms.

1. 6 to 52
2. 120 to 100
3. 32 : 8
4. 25 : 75
5. 8 h : 45 min

Solve each proportion.

6. $\frac{3}{8} = \frac{27}{x}$
7. $\frac{a}{4} = \frac{6}{3}$
8. $\frac{12}{m} = \frac{20}{5}$
9. $\frac{3}{2} = \frac{b}{14}$
10. $\frac{105}{c} = \frac{35}{100}$

14-4 PERCENT AND PROPORTIONS

You can use a proportion to answer a percent question.

example 1

What number is 34% of 150?

solution

Rewrite the question as a proportion:
An unknown number is to 150 as 34 is to 100. ← You are looking for a percent of a number.

$$\frac{\textit{unknown}\text{ part}}{\text{whole}} \rightarrow \frac{n}{150} = \frac{34}{100} \leftarrow \frac{\text{part}}{\text{whole}}$$

Solve the proportion.

$$n \times 100 = 150 \times 34$$
$$n \times 100 = 5100$$
$$n = 51$$

So 51 is 34% of 150.

your turn

Find each answer using a proportion.

1. What number is 25% of 264?
2. What number is 35% of 42?
3. 8.1% of 150 is what number?
4. 140% of 75 is what number?

example 2

15 is what percent of 75?

solution

Rewrite the question as a proportion:
15 is to 75 as an unknown number is to 100. ← You are looking for what percent one number is of another.

$$\frac{\text{part}}{\text{whole}} \rightarrow \frac{15}{75} = \frac{n}{100} \leftarrow \frac{\textit{unknown}\text{ part}}{\text{whole}}$$

Solve the proportion.

$$15 \times 100 = 75 \times n$$
$$\frac{1500}{75} = \frac{75 \times n}{75}$$
$$20 = n$$

So 15 is 20% of 75.

your turn

Find each answer using a proportion.

5. What percent of 50 is 4?
6. What percent of 80 is 18.4?
7. 24 is what percent of 72?
8. 100 is what percent of 80?

example 3

144 is 75% of what number?

solution

Rewrite the question as a proportion:
144 is to an unknown number as 75 is to 100. ← **You are looking for a number when a percent of it is known.**

$$\frac{\text{part}}{\textit{unknown}\ \text{whole}} \begin{matrix}\rightarrow\\ \rightarrow\end{matrix} \frac{144}{n} = \frac{75}{100} \begin{matrix}\leftarrow\\ \leftarrow\end{matrix} \frac{\text{part}}{\text{whole}}$$

Solve the proportion.

$$144 \times 100 = n \times 75$$

$$\frac{14{,}400}{75} = \frac{n \times 75}{75}$$

$$192 = n$$

So 144 is 75% of 192.

your turn

Find each answer using a proportion.

9. 520 is 80% of what number?

10. 100.8 is 120% of what number?

11. 80% of what number is 98.4?

12. 5.5% of what number is 22?

practice exercises

practice for example 1 (page 275)

Find each answer using a proportion.

1. What number is 100% of 49?

2. What number is 20% of 945?

3. 75% of 30 is what number?

4. 230% of 8 is what number?

5. What number is 6.2% of 550?

6. What number is 300% of 54?

7. 8% of 192 is what number?

8. 2% of 99 is what number?

9. What number is 1.1% of 5000?

10. What number is 2.1% of 1900?

practice for example 2 (page 275)

Find each answer using a proportion.

11. 51 is what percent of 100?

12. 36 is what percent of 12?

13. What percent of 40 is 39.2?

14. What percent of 488 is 61?

15. 0.5 is what percent of 25?

16. 3.8 is what percent of 20?

17. What percent of 20 is 1?

18. What percent of 120 is 20.4?

19. 56 is what percent of 64?

20. 18 is what percent of 12?

practice for example 3 (page 276)

Find each answer using a proportion.

21. 70% of what number is 441?

22. 99% of what number is 99?

23. 2.45 is 35% of what number?

24. 1.2 is 6% of what number?

25. 12.5% of what number is 50?

26. 52 is 6.5% of what number?

27. 42% of what number is 21?

28. 120% of what number is 102?

29. 68% of what number is 37.4?

30. 40% of what number is 332?

mixed practice (pages 275–276)

Find each answer using a proportion.

31. What number is 5.8% of 6150?

32. What number is 400% of 179?

33. 60% of 114 is what number?

34. 7% of 174 is what number?

35. 21.6 is what percent of 45?

36. What percent of 90 is 47.7?

37. 161 is what percent of 70?

38. 15 is 30% of what number?

39. 78% of what number is 46.8?

40. 29% of what number is 237.8?

41. Videorama rented 72% of its stock of movies on Friday. If 324 movies were rented on Friday, how many movies did Videorama have in stock?

42. A market research company sent a survey to 10,000 consumers. Only 6250 surveys were returned. What percent of the surveys sent were returned?

review exercises

Round to the place of the underlined digit.

1. $3\underline{3}2.5$ **2.** $\$40.\underline{5}6$ **3.** $0.\underline{1}48$ **4.** $\underline{9}630$ **5.** $6.8\underline{4}1$ **6.** $\$5\underline{7}.83$

mental math

You can find the percent one number is of another mentally.

What percent of 40 is 20?

Think: $\frac{20}{40} \rightarrow \frac{1}{2} \rightarrow 50\%$

15 is what percent of 60?

Think: $\frac{15}{60} \rightarrow \frac{1}{4} \rightarrow 25\%$

Find each percent mentally.

1. What percent of 15 is 3?

2. What percent of 100 is 20?

3. 15 is what percent of 20?

4. 24 is what percent of 36?

PROBLEM SOLVING STRATEGY

14-5 WORKING BACKWARD

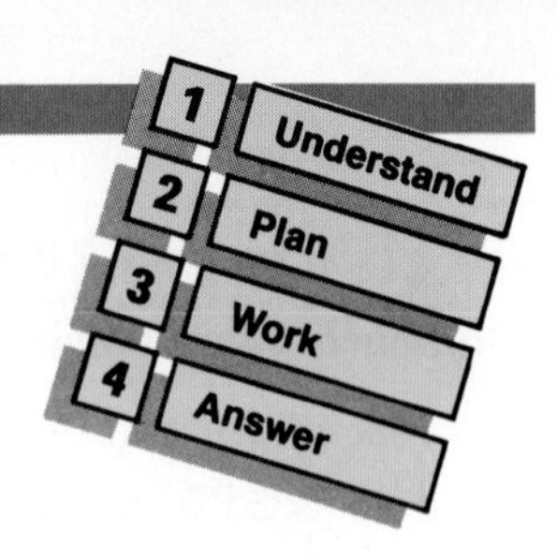

In some problems, you are given an end result and are asked to find a fact needed to achieve that result. One way to solve a problem of this type is by *working backward.*

example

Jerry Larkin is starting to save money for a trip to New York. He earns \$3.80/h working at a hardware store after school. The trip costs \$508, and he wants to take along \$100 as spending money. How many hours must he work to earn the total amount of money he needs?

solution

Step 1 Given: earns \$3.80 per hour
wants \$100 in spending money
needs \$508 to pay for the trip

Find: number of hours he must work

Step 2 *Work backward* from the cost of the trip.

Step 3 Start with the cost of the trip: \$508

Add the amount of spending money: +\$100

\$608 ← **This is the total amount of money Jerry needs.**

Divide by the amount per hour: $\$608 \div \$3.80 = 160$

Step 4 *Check:* $(160 \times \$3.80) - \$100 = \$508$ ✓

Jerry must work 160 h to earn the total amount of money he needs.

problems

1. Lynn Elbert wants to buy a sweater that costs \$25 and a skirt that costs \$35. She earns \$3.75/h. How many hours must she work to earn enough money to buy the sweater and the skirt?
2. The Cardozas want to buy a VCR. They are able to save \$35 per month for that purpose. The VCR costs \$300, and the sales tax rate is 5%. How many months will it take them to save enough money?
3. Maria Rodriguez earns \$6.50/h and works a 40-h week. She earns time-and-a-half for overtime. Last week her gross pay was \$318.50. For how many hours of work was she paid?
4. Jack Redbird paid \$15 for a taxi ride from the airport to his home, including a \$2 tip. According to the sign posted in the taxi, the fare is based on a charge of \$2.20 for the first mile and 15¢ for each additional one-sixth mile. How many miles was Jack charged for?

5. Lillian Novakowski has a job interview at 9:30 A.M., and she wants to be there 10 min early. She needs 20 min for travel time, and she wants to allow an hour and a half to dress and have breakfast. At what time should Lillian get up?
6. On a shopping trip, Debbie LeClair first spent half her money at Phone Values, then spent $19 at The Shoe Inn. Next she spent $4.50 for lunch. After lunch, she spent half her remaining money at Sweaterama, then spent her final $15 at Gifts Galore. How much money did Debbie have at the beginning of her shopping trip?
7. At the end of last week, LaVal Thomson had a balance of $1493.57 in his checking account. During the week he wrote a check for $89.99, deposited his paycheck of $498.34, wrote another check for $492.61, and made an electronic withdrawal of $75. What was the balance in LaVal's account at the beginning of the week?
8. At the end of this week, the closing price of XYZ stock was $\$46\frac{1}{2}$. During the week, the net change in the price of the stock was $+\frac{1}{4}$ on Monday, $-1\frac{1}{2}$ on Tuesday, $+\frac{3}{4}$ on Wednesday, $+1\frac{3}{4}$ on Thursday, and $-\frac{1}{4}$ on Friday. What was the closing price of the stock at the end of *last* week?

review exercises

1. Helen Rubel had $2 in her savings account at the end of January, $4 at the end of February, $8 at the end of March, and $16 at the end of April. If this pattern continues, how much money will be in her account at the end of December?
2. The Springfield Inn courtesy van can carry seven people. How many trips must the van make to transport 23 people?
3. Suppose that you start with a number, double it, add twelve, divide by five, then subtract eighteen. The result is 72. What was the original number?

SKILL REVIEW

Find each answer.

14-1

1. What percent of 25 is 9?
2. What percent of 10 is 17?
3. 41 is what percent of 50?
4. 9 is what percent of 60?
5. What percent of 88 is 55?
6. What percent of 90 is 30?
7. 500 is what percent of 600?
8. 18 is what percent of 11?
9. What percent of 13 is 40?
10. What percent of 36 is 31?

Tell whether there is an *increase* or a *decrease*. Then find the percent of *increase* or *decrease*.

14-2

11. original quantity: 100
    new quantity: 116
12. original distance: 75 km
    new distance: 66 km
13. original hours: 10
    new hours: 22
14. original score: 300
    new score: 150
15. original fare: \$50
    new fare: \$55
16. original depth: 5 ft
    new depth: 2 ft
17. original number: 150
    new number: 90
18. original points: 144
    new points: 24
19. original price: \$12
    new price: \$20
20. original weight: 24 lb
    new weight: 80 lb

Find each answer.

14-3

21. 500% of what number is 30?
22. 8% of what number is 16?
23. 117 is 13% of what number?
24. 78 is 25% of what number?
25. 50% of what number is 426?
26. 75% of what number is 213?
27. 210 is 21% of what number?
28. 54 is 7.2% of what number?
29. $12\frac{1}{2}\%$ of what number is 40?
30. $37\frac{1}{2}\%$ of what number is 150?

Find each answer using a proportion.

14-4

31. What number is 3.1% of 2000?
32. What number is 47% of 400?
33. 20% of 13 is what number?
34. 10 is what percent of 80?
35. 45 is what percent of 720?
36. What percent of 800 is 656?
37. What percent of 430 is 731?
38. 2.7 is 45% of what number?
39. 7.6 is 1% of what number?
40. 19% of what number is 17.1?

14-5

41. Daniel Ziegler is taking a flight that departs at 11:15 A.M. He wants to be at the airport 45 min in advance, and he needs to allow 50 min travel time from his house to the airport. At what time should Daniel leave his house?

APPLICATION

14-6 RENTING AN APARTMENT

Consumer agencies recommend that you budget about 25% of your gross income for **housing costs.** Housing costs include rent, various utilities, and optional renter's insurance.

example

Lorraine and Mark Leonard are looking for an apartment to rent. Their combined monthly gross income is \$3700. About how much money should they budget each month for housing costs?

solution

Find 25% of \$3700: $\frac{1}{4} \times \$3700 = \$925 \approx \$900$ ← **Round to the nearest hundred dollars.**

The Leonards should budget about \$900 each month for housing costs.

exercises

For each monthly gross income, find the amount that should be budgeted each month for housing costs.

1. \$1200
2. \$1500
3. \$2500
4. \$3900
5. Lynn and Ed are looking for a larger apartment. Each month, Lynn earns \$3000 and Ed earns \$2500. About how much money should they budget each month for housing costs?
6. Roger spends \$500 per month for rent. His monthly gross income is \$2500. What percent of his income does he spend on rent?
7. Brad was renting an apartment for \$400 per month. Recently, he moved in with Mike and Jeff. The three roommates now rent a house for \$750 per month and share each rent payment equally. How much does Brad now pay for rent each month? Find the percent of decrease for his monthly rent.
8. Find newspaper ads listing apartments for rent in your area. Choose three and decide the amount of income that is needed to afford them.

APPLICATION

14-7 HEALTH INSURANCE

If you become seriously ill, the cost of needed medical care can be high. **Health insurance** will help reduce these costs. A health insurance policy usually covers much of the cost of hospital services, doctors' care, and medicine.

Most health insurance policies have a **deductible** clause. This clause specifies an amount of money that you must pay toward your yearly medical expenses before your benefits begin. The insurance company usually pays a percent of the amount remaining after the deductible is subtracted from your covered expenses, and you pay the rest.

example

Mary Ireton has a health insurance policy that requires her to pay a \$250 deductible. The insurance company pays 80% of the remaining expenses. Last year Mary's covered medical expenses were \$3048.

a. How much did the insurance company pay?

b. How much did Mary pay?

solution

a. Subtract the deductible from the covered expenses: $\$3048 - \$250 = \$2798$

Find 80% of the difference: 80% of \$2798

$$0.8 \times \$2798 = \$2238.40$$

← **A calculator may be helpful.**

The insurance company paid \$2238.40.

b. Subtract the amount paid by the insurance company from the covered expenses: $\$3048 - \$2238.40 = \$809.60$

Mary paid \$809.60.

exercises

Complete. A calculator may be helpful.

	COVERED MEDICAL EXPENSES	AMOUNT OF DEDUCTIBLE	PERCENT PAID BY INSURANCE	AMOUNT PAID BY INSURANCE	AMOUNT PAID BY PATIENT
1.	\$550	\$50	90%	?	?
2.	\$2040	\$100	80%	?	?
3.	\$1946	\$250	70%	?	?
4.	\$5493	\$1000	80%	?	?

Solve. A calculator may be helpful.

5. Al Kier's health insurance policy requires him to pay a $220 deductible. The insurance company pays 80% of the remaining expenses. Last year his covered medical expenses were $850. How much did the insurance company pay for Al's medical expenses last year?
6. Last year Rasia Alexavich had covered medical expenses of $7050. She is required to pay a deductible of $100, and the insurance company pays 70% of the remaining expenses. How much did Rasia pay for these medical expenses last year?
7. Lee Hwang just began a new job and has chosen a health policy offered by his new employer. Mr. Hwang has a deductible of $100. His insurance company will pay 90% of the remaining expenses. If his covered medical expenses last year were $500, how much did Lee pay?
8. Ingrid and Sven Iverssen have a health insurance policy that requires a joint deductible of $200. Their insurance company will pay 80% of the remaining expenses. Last year Ingrid's covered medical expenses were $550, and Sven's were $700. What was the amount of their combined covered expenses? How much did Ingrid and Sven pay for these medical expenses last year?
9. Last year Jeannie Broderick's covered medical expenses were $1383 for doctors' bills and $1400 for hospital services. Jeannie's health insurance policy requires a $250 deductible. The insurance company will pay 80% of the remaining expenses. How much did Jeannie pay for these medical expenses last year?
10. Chuck Smith, a Hollywood stuntman, had covered medical expenses of $5050 last year. Chuck's health insurance policy requires a $1000 deductible. For the first $2500 of the remaining expenses, the insurance company will pay 70%. For the rest of the remaining expenses, the company will pay 100%. How much did the insurance company pay for Chuck's medical expenses last year?
11. What is meant by a *group plan* for health insurance? What are the advantages of a group plan over an *individual plan*?
12. What is an *HMO*? What are its advantages and disadvantages when compared to other health insurance policies?

APPLICATION

14-8 COMMISSION

Many salespeople are paid an amount of money based on their total sales. This amount, usually a percent of their total sales, is called a **commission.** The percent is called the **commission rate.** Some people are paid a commission in addition to their regular salary, or **base pay.** A person who is paid only a commission is paid on **straight commission.**

example

Edward Zito sells shoes. His weekly base pay is \$350. He also receives a 4.5% commission on his total sales. One week Edward Zito had sales that totaled \$5348. Find his gross pay for the week.

solution

Use the formula:

commission = total sales × commission rate
= \$5348 × 0.045 ← **Write the rate as a decimal.**
= \$240.66

The commission is \$240.66.

gross pay = base pay + commission
= \$350 + \$240.66
= \$590.66

Edward Zito's gross pay for the week is \$590.66.

exercises

Complete. If necessary, round to the nearest cent. A calculator may be helpful.

	BASE PAY	TOTAL SALES	COMMISSION RATE	COMMISSION EARNED	GROSS PAY
1.	\$250	\$412	4%	?	?
2.	\$450	\$1600	6%	?	?
3.	\$500	\$15,000	10%	?	?
4.	\$285	\$623	13.6%	?	?
5.	\$490	\$315	3.5%	?	?
6.	\$375	\$80,000	$2\frac{3}{4}$%	?	?
7.	none	\$26,437	$15\frac{1}{2}$%	?	?
8.	none	\$124,000	5.25%	?	?

Solve. If necessary, round to the nearest cent. A calculator may be helpful.

9. Thomas Lightfoot earns a base pay of \$250 per week selling boots. His commission rate on all sales is 4%. This week his total sales were \$416. Find his commission and his gross pay for the week.
10. Lisa is an artist's agent. She earns a straight commission of $15\frac{1}{4}$%. She sold a painting for \$9575. Find her commission on this sale.
11. Juanita Juarez sells real estate on a straight commission basis. Her commission rate is 3.5%. Juanita sold a house for \$135,000. Find her commission on this sale.
12. Lawrence Baker earns a weekly base pay of \$423. His commission rate on all sales is 2.5%. One week he had sales that totaled \$5512. Find his gross pay for the week.
13. Paul Duvale is paid a 20% commission on all homeowner's insurance policies that he sells. Paul received a commission of \$160. Find his total sales.
14. Last week, Jill's sales totaled \$2400, and her commission on these sales was \$300. What was her commission rate?
15. Claudette Williams sells automobile insurance. Her commission rate on all sales is 6.8%. If her sales for last month were \$34,672, what was her commission?
16. Caroline O'Malley sells cosmetics. She earns either \$560 per month or $8\frac{1}{2}$% of her total sales, whichever is greater. Find her gross pay for a month in which her total sales were \$6542.
17. Jeffrey Hauser is paid a monthly salary plus a commission. His commission rate depends upon his total sales. He receives a commission of 4% on the first \$5000 of his sales and 5% on all sales over \$5000. Last month Jeffrey's sales totaled \$24,000. If his base pay is \$875, what was his gross pay?
18. Suppose you had a sales job. Would you prefer a straight commission or base pay plus commission? What factors might affect your preference?

APPLICATION

14-9 DISCOUNTS AND MARKUPS

When you buy an item, the price tag may show the words **list price.** This is the price at which the manufacturer recommends that the item be sold. When an item is on sale, the list price is decreased. The amount by which the list price is decreased is called the **discount.** The **rate of discount** is the percent the list price is decreased.

The amount that a store pays for an item is the **wholesale price.** To make a profit stores must sell an item at a *markup*. The **markup** is the amount of money added to the wholesale price. The **rate of markup** is the percent the wholesale price is increased.

example

a. Elizabeth Gallant bought a novel that had a list price of \$18.99. The list price was discounted 20%. Find the discount and the selling price of the novel.

b. Admar Auto Sales paid \$12,500 for a 123X Special. Then a 6% markup was added to the price of the car. Find the markup and the selling price of the car.

solution

a. Use the formula: discount = list price × rate of discount

$$= \$18.99 \times 0.2$$ ← **Write the rate as a decimal.**

$$= \$3.798$$

$$\approx \$3.80$$ ← **Round to the nearest cent.**

The discount is \$3.80.

Subtract the discount from the list price to find the selling price.

$$\begin{array}{r} \$18.99 \\ -\ \ 3.80 \\ \hline \$15.19 \end{array}$$

The selling price is \$15.19.

b. Use the formula: markup = wholesale price × rate of markup

$$= \$12{,}500 \times 0.06$$ ← **Write the rate as a decimal.**

$$= \$750$$

The markup is \$750.

Add the markup to the wholesale price to find the selling price.

$$\begin{array}{r} \$12{,}500 \\ +\ \ \ \ 750 \\ \hline \$13{,}250 \end{array}$$

The selling price is \$13,250.

exercises

Complete. If necessary, round to the nearest cent. A calculator may be helpful.

	ITEM	LIST PRICE	RATE OF DISCOUNT	DISCOUNT	SELLING PRICE
1.	exercise bicycle	$130	10%	?	?
2.	table tennis set	$40	20%	?	?
3.	microwave oven	$250	25%	?	?
4.	picture frame	$12.95	20%	?	?
5.	sweater	$45	$33\frac{1}{3}$%	?	?
6.	running shoes	$35.99	15%	?	?

Complete. If necessary, round to the nearest cent. A calculator may be helpful.

	ITEM	WHOLESALE PRICE	RATE OF MARKUP	MARKUP	SELLING PRICE
7.	figure skates	$50	35%	?	?
8.	spotlight	$70	50%	?	?
9.	compact disc system	$250	60%	?	?
10.	hockey gloves	$99.99	40%	?	?
11.	alarm clock	$75	100%	?	?
12.	silk dress	$140	110%	?	?

13. Alby Mastrocola bought a record album that has a list price of $12.65. The list price was discounted 30%. Find the discount and the selling price of the record album. Round to the nearest cent.

14. Jacqueline bought a pair of figure skates that has a list price of $79.50. The list price was discounted 15%. Find the discount and the selling price of the figure skates. Round to the nearest cent.

15. A flower shop sells roses for $48.50 per dozen. As a holiday special, the price is discounted 45%. Find the discount and the selling price of a dozen roses. Round to the nearest cent.

Solve. A calculator may be helpful.

16. Computer Inc. paid $11,200 for a new TechnixII computer and printer. Then Computer Inc. added a 9% markup to the cost of the computer system. Find the markup and the selling price.
17. Happy Health natural food store bought bulk soybeans for $3 per pound. The store added a 5% markup to the cost of the bulk soybeans. Find the markup and the selling price for one pound of soybeans.
18. Pamela Stevens bought a leather handbag for $24. The list price was $32. Find the rate of discount.
19. Giuseppe Iuliano bought a pair of leather cowboy boots for $114. The wholesale price was $60. Find the rate of markup.
20. The list price of a designer blouse at a mall boutique is $60. The boutique is now selling the blouse at 30% off the list price. A fashion discount outlet bought a large quantity of these same blouses from the manufacturer for $17.20 per blouse. The discount outlet will sell the blouse at a markup of 60%. Which store sells the blouse at the lower price? How much lower?
21. Employees of certain stores receive a discount on the items that they buy. Go to two stores and find out the rate of discount offered to the employees of the stores.

calculator corner

You can use a calculator to compute a selling price.

example Find the selling price: list price: $64.50
discount: 30%

solution Enter 64.50 × 30 [%] [M+]. ← **Puts discount into memory.**
Enter 64.50 − [MR] =. ← **Subtracts discount from list price.**
The result is 45.15. The selling price is $45.15.

Use a calculator to find each selling price. If necessary, round to the nearest cent.

1. list price: $22.80
   discount: 40%
2. list price: $193
   discount: 25%
3. list price: $50.50
   discount: 20%
4. list price: $89.42
   discount: 50%
5. list price: $154
   discount: 15%
6. list price: $99.99
   discount: 33%

CHAPTER REVIEW

vocabulary vo·cab·u·lar·y

Choose the correct word to complete each sentence.

1. The amount that a store pays for an item is the (*list, wholesale*) price.
2. Many salespeople are paid a (*commission, deductible*) based on their total sales.

skills

Find each answer.

3. What percent of 50 is 29?
4. 13 is what percent of 20?
5. 70% of what number is 70?
6. 6 is what percent of 45?
7. 600 is 120% of what number?
8. What percent of 96 is 12?
9. 8.5% of what number is 17?
10. 28 is $33\frac{1}{3}$% of what number?

Tell whether there is an *increase* or a *decrease*. Then find each percent of increase or decrease.

11. original quantity: 80
new quantity: 104
12. original length: 50 km
new length: 16 km
13. original score: 90
new score: 60

Find each answer using a proportion.

14. What number is 3.5% of 200?
15. What number is 70% of 270?
16. 20% of what number is 77?
17. 48 is what percent of 60?

18. At the end of last week, Cindy Hooper had a balance of \$153.92 in her checking account. During the week, she wrote a check for \$25.49 and made a deposit of \$33. What was the balance in her account at the beginning of the week?
19. Stella spends \$780 per month for rent. Her monthly gross income is \$3000. What percent of her income does she spend on rent?
20. Larry earns a commission of 12% on his total sales. Last week he received a commission of \$138. Find his total sales for the week.
21. Betty Vecchione has a health insurance policy that requires her to pay a \$100 deductible. The insurance company pays 70% of the remaining expenses. Last year Betty's covered medical expenses were \$874. How much did the insurance company pay?
22. Amy bought a purse that has a list price of \$48. The list price was discounted 30%. Find the discount and the selling price.

CHAPTER TEST

Find each answer.

14-1

1. What percent of 500 is 210?
2. 48 is what percent of 96?
3. 9 is what percent of 8?
4. What percent of 21 is 44?

Tell whether there is an *increase* or *decrease*. Then find each percent of increase or decrease.

14-2

5. original quantity: 10
   new quantity: 1
6. original number: 28
   new number: 49
7. original price: \$320
   new price: \$240
8. original weight: 48 lb
   new weight: 66 lb

Find each answer.

14-3

9. 2% of what number is 3?
10. 21 is 300% of what number?
11. 32 is 80% of what number?
12. $12\frac{1}{2}$% of what number is 99?

Find each answer using a proportion.

14-4

13. What percent of 100 is 92?
14. 36 is 60% of what number?
15. 5.5% of 1200 is what number?
16. What number is 62% of 300?

14-5

17. During the next twelve months, the Leongs want to save enough money to buy a new sofa. The sofa costs \$750, and the sales tax rate is 4%. How much money must they save each month in order to buy the sofa?

14-6

18. Dane's monthly gross income is \$1900. About how much money should he budget each month for housing costs if he wants to spend 25% of his monthly gross income on housing?

14-7

19. Rick Clark's health insurance policy requires him to pay a \$250 deductible. The insurance company pays 80% of the remaining expenses. Last year his covered medical expenses were \$1350. How much did Rick pay for these medical expenses last year?

14-8

20. Joan earns a straight commission of 8% on total sales. Last month, her total sales were \$44,100. Find Joan's gross pay for last month.

14-9

21. A store manager buys jackets at a wholesale price of \$11.50 each and then adds a 60% markup. Find the markup and the selling price for one jacket.

Savings and investments, taxes, spending plans—almost all our personal finances require an understanding of percents.

CHAPTER 15

USING PERCENT

15-1 SIMPLE INTEREST

Interest is money paid for the use of money. When you borrow money, you *pay* interest. When you deposit or invest money, the money *earns* interest. **Principal** is money on which interest is paid. **Simple interest** is interest paid only on the principal. You can use the following formula to find the simple interest paid or earned on the principal.

$$\text{Interest} = \text{Principal} \times \text{rate} \times \text{time}$$
$$I = P \times r \times t$$

In the formula, r is the rate of interest per year and t is the time in years.

example 1

Find the interest earned on \$725 invested at a rate of 4.5% per year for 6 months.

solution

Use the formula:

$I = P \times r \times t$

$I = \$725 \times r \times t$

$I = \$725 \times 0.045 \times t$ ← **Write the rate as a decimal.**

$I = \$725 \times 0.045 \times 0.5$ ← **6 months = 0.5 year**

$I = \$16.3125 \approx \$16.31$ ← **Round to the nearest cent.**

The interest earned is \$16.31.

your turn

Find the interest. If necessary, round to the nearest cent.

1. \$1200 at 6.5% per year for 1 year
2. \$80 at $5\frac{1}{2}\%$ per year for 1 year
3. \$950 at 5% per year for 3 years
4. \$375 at 7% per year for 6 months

When you repay borrowed money, the **amount due** is the principal *plus* the interest.

example 2

Find the amount due: \$800 at 11.5% per year for 1 year

solution

Interest $= P \times r \times t$
$= \$800 \times 0.115 \times 1$
$= \$92$

Amount due = Principal + Interest
$= \$800 + \$92$
$= \$892$

The amount due is \$892.

your turn

Find the amount due. If necessary, round to the nearest cent.

5. $2000 at 12% per year for 1 year
6. $1200 at 10% per year for 6 months
7. $5000 at $9\frac{1}{2}$% per year for 2 years
8. $375 at 15.5% per year for 3 years

practice exercises

practice for example 1 (page 292)

Find the interest. If necessary, round to the nearest cent.

1. $2000 at 5% per year for 1 year
2. $4500 at 4% per year for 1 year
3. $275 at 6.1% per year for 1 year
4. $385 at 7.3% per year for 1 year
5. $1000 at $12\frac{1}{2}$% per year for 1 year
6. $500 at $10\frac{1}{2}$% per year for 1 year
7. $2500 at 5% per year for 3 years
8. $3600 at 5% per year for 2 years
9. $2000 at 7% per year for 6 months
10. $3175 at 5% per year for 6 months
11. $6000 at 5.4% per year for 6 months
12. $4550 at 7.3% per year for 6 months

practice for example 2 (pages 292–293)

Find the amount due. If necessary, round to the nearest cent.

13. $600 at 13% per year for 1 year
14. $800 at 10% per year for 1 year
15. $5000 at 12% per year for 6 months
16. $7000 at 16% per year for 6 months
17. $1000 at 8.5% per year for 1 year
18. $1200 at 9.5% per year for 2 years
19. $8000 at 8.4% per year for 2 years
20. $9000 at 9.3% per year for 2 years
21. $2000 at $8\frac{1}{2}$% per year for 3 years
22. $1000 at $9\frac{1}{2}$% per year for 4 years
23. $725 at 7% per year for 6 months
24. $675 at 5% per year for 6 months

mixed practice (pages 292–293)

Complete. If necessary, round to the nearest cent.

	PRINCIPAL	RATE PER YEAR	TIME	INTEREST	AMOUNT DUE
25.	$700	5%	1 year	?	?
26.	$400	6%	1 year	?	?
27.	$1255	8.3%	3 years	?	?
28.	$1555	9.1%	5 years	?	?
29.	$4500	$12\frac{1}{2}$%	6 months	?	?
30.	$2750	$14\frac{1}{2}$%	6 months	?	?

In each exercise, for which loan would you pay more interest in 1 year?

31. \$225 at $5\frac{1}{2}$% per year
 \$230 at 5% per year

32. \$6000 at 7.1% per year
 \$5900 at 7.4% per year

33. \$7500 at 6.6% per year
 \$7000 at 6.9% per year

34. \$4400 at $7\frac{1}{2}$% per year
 \$5400 at 7% per year

35. The Penfields lent their son Dan \$6000 at an interest rate of 7% per year for 6 months. How much interest must Dan pay?

36. Hal Bolter borrowed \$5000 for 1 year at an interest rate of 5.5% per year. How much interest did he pay?

37. Susan Forman borrowed \$3500 and agreed to repay the loan in 1 year at 16.5% interest per year. What is the total amount due?

38. Jon Wells wants to buy a television that costs \$690. He plans to make a \$90 down payment. He will borrow \$600 at an interest rate of $15\frac{1}{2}$% per year for 2 years. What is the total amount due on his loan?

review exercises

Find each answer.

1. \$5678 − \$3401
2. 56 + 21 + 49
3. 12 × \$417
4. 6231 ÷ 201
5. 0.25 × \$6.80
6. \$741 − \$69.83
7. 4.99 + 281
8. $0.4\overline{)1.96}$
9. $\frac{9}{10} \div \frac{4}{5}$
10. $\frac{2}{7} + \frac{3}{8}$
11. $3\frac{5}{6} \times 7\frac{1}{5}$
12. $14 - 6\frac{2}{5}$

calculator corner

You can use a calculator to compute simple interest.

example Find the interest: \$1450 at $9\frac{1}{8}$% per year for 6 months

solution Enter 1 ÷ 8 + 9 = [M+]. ← **Puts $9\frac{1}{8}$ into memory.**
Then enter 1450 × [MR] [%] × 0.5 =.

principal (1450) rate ([MR] [%]) time in years (0.5)

The result is 66.15625. The interest is \$66.16.

Find the interest.

1. \$2000 at $9\frac{1}{8}$% per year for 6 months
2. \$3000 at $9\frac{5}{8}$% per year for 6 months
3. \$2500 at $10\frac{1}{4}$% per year for 4 years
4. \$4500 at $11\frac{3}{4}$% per year for 3 years

15-2 COMPOUND INTEREST

Compound interest is interest paid on the principal *and* on any interest previously earned. A bank adds interest to savings at the end of each **interest period.** This sum becomes the new principal. Interest can be *compounded* annually, semiannually, quarterly, monthly, or daily.

example 1

The principal in a savings account is $800. The account earns 6.5% annual interest compounded semiannually (two times per year). Find the balance at the end of the second interest period.

solution

First interest period

Find the principal: .... $800.00 ← **original principal**

Find the interest using the interest formula.
A calculator may be helpful.

$P = \$800 \quad r = 6.5\% = 0.065 \quad t = 0.5$ year

$I = P \times r \times t$

$I = \$800 \times 0.065 \times 0.5$ .... + $26.00

Add the interest to the principal: .... $826.00 ← **new principal**

Second interest period

Find the principal: $826.00

Find the interest using the interest formula.

$P = \$826 \quad r = 6.5\% = 0.065 \quad t = 0.5$ year

$I = P \times r \times t$

$I = \$826 \times 0.065 \times 0.5$ .... + $26.85 ← **Round to the nearest cent.**

Add the interest to the principal: $852.85

The balance at the end of the second interest period is $852.85.

your turn

Find the balance at the end of the second interest period.

1. $500 at 5% annual interest compounded annually
2. $1000 at 7.5% annual interest compounded semiannually

Most financial institutions now use computers to find compound interest. You can also use a table to find the interest. A compound interest table similar to the one on page 296 shows how much you will receive for each $1 invested.

COMPOUND INTEREST TABLE

Number of Periods	INTEREST RATE PER PERIOD						
	1%	1.5%	2%	3%	4%	5%	6%
1	1.0100	1.0150	1.0200	1.0300	1.0400	1.0500	1.0600
2	1.0201	1.0302	1.0404	1.0609	1.0816	1.1025	1.1236
3	1.0303	1.0457	1.0612	1.0927	1.1249	1.1576	1.1910
4	1.0406	1.0614	1.0824	1.1255	1.1699	1.2155	1.2625
5	1.0510	1.0773	1.1041	1.1593	1.2167	1.2763	1.3382
6	1.0615	1.0934	1.1262	1.1941	1.2653	1.3401	1.4185
7	1.0721	1.1098	1.1487	1.2299	1.3159	1.4071	1.5036
8	1.0829	1.1265	1.1717	1.2668	1.3686	1.4775	1.5938

example 2

The principal in a savings account is \$350. The account earns 6% annual interest compounded quarterly (four times per year). Use the compound interest table to find the balance at the end of two years.

solution

Find the total number of interest periods.

4 periods per year × 2 years = 8 periods

Find the interest rate per period.

6% per year ÷ 4 periods per year = 1.5% per period

On the table, find the number across from *8* and under *1.5%*: 1.1265

Multiply the number from the table by the principal.

\$350 × 1.1265 = \$394.275 ≈ \$394.28 ← **Round to the nearest cent.**

The balance at the end of two years is \$394.28.

your turn

Use the compound interest table above to find the balance.

3. \$500 at 5% annual interest compounded annually for 3 years
4. \$30 at 6% annual interest compounded semiannually for 2 years

practice exercises

practice for example 1 (page 295)

Find the balance at the end of the second interest period.

1. \$6000 at 5% annual interest compounded annually
2. \$9000 at 6% annual interest compounded semiannually
3. \$2000 at 4.5% annual interest compounded annually
4. \$1000 at 6.5% annual interest compounded semiannually

practice for example 2 (page 296)

Use the compound interest table on page 296 to find the balance.

5. $400 at 6% annual interest compounded quarterly for 1 year
6. $1000 at 4% annual interest compounded quarterly for 1 year
7. $750 at 6% annual interest compounded semiannually for 3 years
8. $825 at 4% annual interest compounded semiannually for 2 years

mixed practice (pages 295–296)

Find the balance. If possible, use the table on page 296.

9. $1000 at 7.5% annual interest compounded annually for 2 years
10. $200 at 4% annual interest compounded semiannually for 2 years
11. $2450 at 4% annual interest compounded quarterly for 2 years
12. $3800 at 6% annual interest compounded quarterly for 2 years
13. $500 at 5.5% annual interest compounded semiannually for 1 year
14. $400 at 6.5% annual interest compounded semiannually for 1 year
15. Juan has $1200 in an account that earns 8% annual interest compounded quarterly. He makes no deposits or withdrawals for two years. Find the balance at the end of two years.
16. Joni has $400 in an account that earns 4.5% annual interest compounded semiannually. She makes no deposits or withdrawals for one year. Find the balance at the end of one year.
17. Which rate pays the greater interest on $1000 after one year: 16.2% compounded semiannually or 16% compounded quarterly?
18. Which rate pays the greater interest on $250 after one year: 6% simple interest or 5% annual interest compounded semiannually?

review exercises

Write the prime factorization of each number.

1. 56
2. 87
3. 125
4. 210
5. 144
6. 153

The New York Mets baseball team played its first season in 1962. The team celebrated its twenty-fifth season in 1986. However, 1986 − 1962 = 24. How could 1986 be the twenty-fifth season?

SKILL REVIEW

Find the interest. If necessary, round to the nearest cent.

15-1

1. $500 at 7% per year for 1 year
2. $800 at 9% per year for 1 year
3. $3500 at 12.5% per year for 5 years
4. $5500 at 13.5% per year for 3 years
5. $2055 at $9\frac{1}{2}$% per year for 1 year
6. $1005 at $8\frac{1}{2}$% per year for 1 year
7. $7500 at 14% per year for 6 months
8. $9500 at 15% per year for 6 months

Find the amount due. If necessary, round to the nearest cent.

9. $600 at 16% per year for 1 year
10. $200 at 15% per year for 1 year
11. $770 at $15\frac{1}{2}$% per year for 3 years
12. $3600 at $10\frac{1}{2}$% per year for 3 years
13. $2043 at 7% per year for 6 months
14. $1155 at 8% per year for 6 months
15. Meredith Jefferson borrowed $500 at an interest rate of $6\frac{1}{2}$% per year. She agreed to repay the loan in 1 year. How much interest did she pay?
16. Al Fry borrowed $200 for 6 months at an interest rate of $12\frac{1}{2}$% per year. What is the total amount due?

Find the balance. If possible, use the table on page 296. If necessary, round to the nearest cent.

15-2

17. $700 at 6% compounded semiannually for 1 year
18. $1000 at 7% compounded semiannually for 1 year
19. $8000 at 8.5% compounded annually for 2 years
20. $3000 at 9.5% compounded annually for 2 years
21. $750 at 6% compounded semiannually for 1 year
22. $3000 at 4% compounded semiannually for 1 year
23. $1200 at 6% compounded semiannually for 3 years
24. $1400 at 8% compounded semiannually for 4 years
25. $2000 at 4% compounded quarterly for 2 years
26. $2500 at 6% compounded quarterly for 2 years
27. Jason Tibbs has $2500 in an account that earns 4.5% annual interest compounded semiannually. He makes no deposits or withdrawals for one year. Find the balance at the end of one year.
28. General Savings Bank pays 6% interest compounded quarterly on its savings accounts. Mai Young has $300 in a savings account. Find the balance at the end of two years if she makes no other deposits or withdrawals.
29. Ethan Masterman has $4000 in a savings account that earns 6% annual interest compounded semiannually. Find the balance at the end of three years if he makes no other deposits or withdrawals.

APPLICATION

15-3 SAVINGS ACCOUNTS

Savings accounts earn compound interest at a rate determined by the bank. Savings accounts that require a *minimum balance* usually earn interest at a higher rate than accounts that require no minimum balance.

Many banks offer the use of an automated teller machine to customers who have a savings or checking account. Customers can make most banking transactions 24 hours a day at an automated teller machine. Sometimes there is a charge for this service.

example

Stu Olds has a balance of \$752 in his savings account. He withdraws \$20 and \$80. He then makes no other deposits or withdrawals for two months. His savings account earns 5.5% annual interest compounded monthly. Find his balance at the end of the two months.

solution

Find the balance after withdrawals.

Total withdrawals: \$20 + \$80 = \$100

Balance after withdrawals: \$752 − \$100 = \$652

Find the interest earned at the end of one month.

$P = \$652$

$r = 5.5\% = 0.055$

$t = \frac{30}{365} \approx 0.082$ ← **Banks often use a 30-day interest period for each month.**

$I = P \times r \times t$

$I = \$652 \times 0.055 \times 0.082$ ← **A calculator may be helpful.**

$I = \$2.94052 \approx \$2.94$

The interest at the end of one month is \$2.94

Add the interest to the balance to find the new balance.

\$652 + \$2.94 = \$654.94

Find the interest on the new balance.

$I = P \times r \times t$

$I = \$654.94 \times 0.055 \times 0.082$

$I = \$2.9537794 \approx \$2.95$

Add the interest to the new balance to find the balance at the end of two months.

\$654.94 + \$2.95 = \$657.89

Stu's balance at the end of two months is \$657.89.

exercises

For Exercises 1–10, use a 30-day interest period for each month.

Complete. If necessary, round each answer to the nearest cent. A calculator may be helpful.

	PRESENT BALANCE	INTEREST RATE PER YEAR	INTEREST PERIOD	FIRST MONTH INTEREST	NEW BALANCE	SECOND MONTH INTEREST	NEW BALANCE
1.	$300	5%	monthly	?	?	?	?
2.	$400	6%	monthly	?	?	?	?
3.	$5800	4%	monthly	?	?	?	?
4.	$7500	5%	monthly	?	?	?	?
5.	$1100	6.5%	monthly	?	?	?	?
6.	$1500	7.5%	monthly	?	?	?	?

7. Sandra Fisher has $1000 in a savings account that earns 6% annual interest compounded monthly. At the end of one month she withdraws $90. She makes no other deposits or withdrawals. How much money does she have in her account at the end of the next month?

8. Joe Hall deposits $500 in a savings account that earns 5.5% annual interest compounded monthly. How much more interest does he earn the first month than if he deposits his money in a savings account that earns 5% annual interest compounded monthly?

Some banks offer a checking account called a NOW account. With a NOW account, the checking privilege is free of charge as long as you maintain a minimum balance in either your savings account or your checking account. Unlike other checking accounts, a NOW account earns interest.

9. Lynn Mott has $1000 in a NOW checking account that earns 5.5% annual interest compounded monthly. She writes checks for $150 and $50. She then makes no deposits and writes no other checks for one month. How much interest does she earn in that month?

10. Jeff Fine has $600 in a savings account that earns 6% annual interest compounded monthly and $800 in a NOW account that earns 5.25% annual interest compounded monthly. He makes no deposits or withdrawals in either account. How much interest altogether does he earn in one month?

11. Find out how a local bank insures your savings.

12. Find out if a local bank has special minimum balance requirements for customers younger than 18 years or older than 65 years.

APPLICATION

15-4 INVESTING MONEY

There are many ways to invest money. Some of the more popular investments are savings accounts, certificates of deposit, stocks, and United States savings bonds.

A **certificate of deposit (CD)** is a written record of a special kind of savings account that earns at a higher interest rate than a regular account. You must leave your money on deposit for a specified period of time. If you withdraw your money before the specified period of time, you lose part or all of your interest. You may purchase certificates of deposit in varying minimum amounts, usually \$100 to \$2500.

example

Corey David has \$200 to invest. He considers a savings account that earns 5.25% annual interest. He also considers a CD that earns 7.45% annual interest. How much more money will he earn in one year by investing in the CD?

solution

Find the interest earned at 5.25% annual interest.

$\$200 \times 0.0525 \times 1 = \$10.50$ ← **A calculator may be helpful.**

Find the interest earned at 7.45% annual interest.

$\$200 \times 0.0745 \times 1 = \$14.90$

Subtract to find the difference.

$\$14.90 - \$10.50 = \$4.40$

Corey will earn \$4.40 more in one year by investing in the CD.

exercises

Solve. A calculator may be helpful.

1. Jim Duclos has \$900 to invest. He considers a savings account that earns 5.5% annual interest and a CD that earns 8% annual interest. How much more money will he earn in one year by investing in the CD?
2. Victoria Tobin wants to invest \$3500. She considers a savings account that earns 5.67% annual interest and a CD that earns 7.3% annual interest. How much more money will she earn in one year by investing in the CD?
3. Mary Rose MacCarthy has \$1000 to invest. She invests \$250 in a savings account that earns 7.1% annual interest. She also invests \$750 in a CD that earns 7.4% annual interest. How much total interest does she earn from these investments in the first year?

When you invest in a **savings bond,** you are loaning your money to the federal government for a period of time. You can purchase a *Series EE savings bond* with a face value ranging from $50 to $1000. The cost of a Series EE bond is 50% of its face value.

You can cash in, or *redeem,* a savings bond any time after six months. The **redemption value** of the bond is the cost of the bond plus the amount of interest earned in the time you owned it.

4. Nancy Stone redeemed a Series EE savings bond that had a face value of $1000. She had bought the bond as an investment to help pay for her daughter's college tuition. How much did the bond cost?
5. John Sullivan bought a savings bond that had a face value of $500. Six years later, its redemption value was $392.40. How much interest did the bond earn?

When planning to invest money, you have to consider your needs. One consideration is easy access to your money. For instance, you may earn a higher rate of interest with a CD than with a savings account. Yet, if you need to withdraw money from a CD before its specified period of time, the penalty of lost interest may not make it a wiser investment.

Security is another consideration. Savings accounts and CDs are insured. Stocks, however, are not insured. Stocks may earn more money than other forms of investment if the value of the stock increases at a higher rate. You have no guarantee, though, that the stock will increase in value.

6. Theresa Cardillo invests \$2000 in a CD that earns 8% annual interest compounded semiannually. She also invests \$2000 in a savings account that earns 6% annual interest compounded semiannually. How much more money does Theresa earn from the CD at the end of the 6 months?
7. John Barr decided to invest \$1000 in a CD that earns 7.52% annual interest. He chose not to invest in a savings account that earns 5% annual interest. How much more interest did his money earn from the CD than it would have at 5% interest in the first year?
8. Sue Carlman decided to invest \$2500 in a CD that earns 7.1% annual interest. She chose not to invest in a savings account that earns 5.3% annual interest. How much more interest did her money earn from the CD than it would have in the savings account in the first year?
9. Sam Baker invested \$2000 in a stock that earned \$200 one year. How much more did his investment earn that year than if he had invested the \$2000 in a CD that earns 7.2% annual interest?
10. Evan Nabors invested \$500 in a stock that lost \$35 one year. He could have invested the money in a savings account that earns 5% annual interest. How much more money would he have from his investment if he had deposited the \$500 in the savings account?
11. Bill Fox wants to buy Series EE savings bonds with a total face value of \$500. Explain why buying five bonds each with a face value of \$100 is a wiser way to invest than one bond for \$500.
12. Sherry Waltham has \$50 to invest. Will she be able to invest in a CD? Explain.
13. Find out the current rates of interest earned by CDs at a local bank.

mental math

You can mentally estimate a 15% tip by estimating 10% and 5%.

For a bill of \$38.25, round \$38.25 to \$38.00

10% of \$38.00 = \$3.80 ← **Move the decimal point one place to the left.**

$\frac{1}{2}$ of \$3.80 = \$1.90 ← **5% of \$38 is half of 10% of \$38.**

\$5.70

The 15% tip is about \$5.70.

Mentally estimate a 15% tip for the given amount.

1. \$24
2. \$18
3. \$12.20
4. \$14.10
5. \$25.99
6. \$31.89
7. \$16.31
8. \$22.14
9. \$24.85
10. \$28.75

APPLICATION

15-5 CREDIT CARDS

Using a **credit card** allows you to pay for goods and services without using cash. You agree to repay the amount charged either in total each month or in part over a period of time. If you do not repay the total amount due, called the **balance,** companies usually charge a *finance charge*. The **finance charge** is based on a percent of the unpaid balance.

example

Michael Cohen has a Complete Credit Card. He pays a monthly interest rate of 1.7% on the unpaid balance. Last month he had an unpaid balance of $150.10 and charged new purchases totaling $75. Find the finance charge and the new balance.

solution

Multiply the interest rate by the unpaid balance to find the finance charge.

$\$150.10 \times 0.017 = \$2.5517 \approx \$2.55$ ← **A calculator may be helpful.**

The finance charge is $2.55.

Add the finance charge and the amount of new purchases to the unpaid balance to find the new balance.

$\$150.10 + \$2.55 + \$75 = \$227.65$

The new balance is $227.65.

exercises

Complete. If necessary, round to the nearest cent. A calculator may be helpful.

	UNPAID BALANCE	MONTHLY INTEREST RATE	FINANCE CHARGE	NEW PURCHASES	NEW BALANCE
1.	$31.10	1.5%	?	$53.99	?
2.	$78.80	1.5%	?	$29.50	?
3.	$40.40	1.2%	?	none	?
4.	$25.66	1.3%	?	$71.26	?

5. Mimi Jespen has a Union Credit Card. She pays a monthly interest rate of 1.6% on the unpaid balance. Last month she had an unpaid balance of $259.84 and charged new purchases totaling $48.71. Find the finance charge and the new balance.
6. Ron Sung has a Great Charge Card. He pays a monthly interest rate of 1.8% on the unpaid balance. Last month he had an unpaid balance of $50.80 and made no new purchases. Find the finance charge and the new balance.

Interest rates vary with the amount owed, the state in which you live, and the institution issuing the card. You can find the monthly interest rate by dividing the annual rate by 12.

Use the annual interest rates and annual fees listed below for Exercises 7–12.

	ANNUAL INTEREST RATE	ANNUAL FEE
Fine Credit Card	12.6%	$15
Royal Credit Card	13.2%	$10
Global Credit Card	16.8%	none

7. Jim Sayles has a Global Credit Card. In December his unpaid balance was $56.50. Find his finance charge for December.
8. Rebecca Landers has a Fine Credit Card. In June her unpaid balance was $78.10. Find her finance charge for June.
9. Abby Letvin has a Royal Credit Card. In February her unpaid balance was $80.22. Find the finance charge and the new balance for February.
10. Steve Zachary has a Global Credit Card. In September his unpaid balance was $76.10. He charged new purchases totaling $30.40. Find the finance charge and the new balance for September.
11. People who always pay their monthly balance in full are called *convenience users* of credit cards. Convenience users benefit from a card that has no annual fee for usage. Which of the credit cards listed above would be a good choice for a convenience user?
12. People who always make partial payments are called *revolvers*. They carry a large unpaid balance and benefit from low interest rates. Which of the credit cards listed above would be a good choice for a revolver?
13. Find out how a financial institution decides who is eligible for credit.
14. Find out how you can establish a credit history.
15. If a financial institution denies you credit, how can you find out the reasons for credit denial?
16. Find out how you can replace a lost credit card.
17. Find out what to do if you find an error on your credit card bill.

APPLICATION

15-6 AUTOMOBILE LOANS

To buy an automobile, some people borrow money from a bank, credit union, or car dealer. People usually make a *down payment* toward the cost of the automobile at the time of purchase. A **down payment** is the initial amount of money paid. People take out an **automobile loan** to pay the rest of the cost. You repay an automobile loan in equal monthly payments over a specified period of time. Monthly payments include interest on the loan.

example

Richard Lee wants to buy a car that sells for \$7733. He makes a down payment of 20% and makes 24 monthly payments of \$264.53. Find the total amount paid for the car.

solution

To find the amount of Richard's down payment, multiply the percent of the down payment by the selling price.

$\$7733 \times 0.20 = \$1546.60$ ⟵ **A calculator may be helpful.**

Find the total of his 24 monthly payments.

$24 \times \$264.53 = \$6348.72$

Add the down payment to the total of the monthly payments.

$\$1546.60 + \$6348.72 = \$7895.32$

The total amount paid for the car is \$7895.32.

exercises

Solve. A calculator may be helpful.

1. Sarah Ying wants to buy a car that sells for \$9000. She makes a 30% down payment and 30 monthly payments of \$247.84. Find the total amount paid for the car.
2. Tom Fenwick wants to buy a car that sells for \$7200. He makes a 15% down payment and 12 monthly payments of \$553.81. Find the total amount paid for the car.
3. Amanda Hartman wants to buy a truck that sells for \$10,039. She makes a 20% down payment and 24 monthly payments of \$370.61. Find the total amount paid for the truck.
4. Ben DeWolfe wants to buy a truck that sells for \$7850. The manufacturer gives a \$500 rebate. Ben makes a 40% down payment on the price of the truck after the rebate. He makes 24 payments of \$209.66. Find the total amount paid for the truck.

5. Mike Shore wants to buy a car that sells for $10,800. The manufacturer gives a $500 rebate. Mike makes a 20% down payment on the price of the car after the rebate. He makes 36 monthly payments of $271.72. Find the total amount paid for the car.
6. Rae Hill wants to buy a car that sells for $5900. She trades in her old car and receives a $250 price reduction. She makes a 10% down payment on the new price. She makes 24 payments of $246.64. Find the total amount paid for the car.

Monthly payments include a finance charge on the amount of money borrowed. To find the total finance charge, subtract the amount of money borrowed from the total amount paid in monthly payments.

7. Roberto Bantisero arranged an automobile loan for $8000. He repaid the loan with 24 payments of $378.47 each. Find the finance charge.
8. Elizabeth Parkerman arranged an automobile loan for $10,000. She repaid the loan with 36 payments of $327.39 each. Find the finance charge.
9. Christine Petersen arranged an automobile loan for $5000. She repaid the loan with 24 monthly payments of $231.88 each. Find the finance charge.
10. Hamid Ghazarian arranged an automobile loan for $3000. He repaid the loan with 36 monthly payments of $96.80 each. Find the finance charge.
11. Find out if there is usually a higher interest rate charged for an automobile loan for a new car or for a used car.
12. Find out from a local bank the longest time period for repayment of an automobile loan. Is this time period for a new or used automobile?

APPLICATION

15-7 HOME MORTGAGE LOANS

Most people cannot pay the full purchase price of a home at one time. Usually people make a *down payment* and finance the rest by borrowing. Down payments are usually between 10% and 40% of the cost of a home. A loan to finance the purchase of a home is called a **home mortgage loan.** You repay a mortgage loan with monthly payments. Monthly payments include interest paid on the loan. Banks arrange mortgage loans for various lengths of time, usually 15 to 30 years.

example

The Taylors want to buy a house that sells for $95,000. They make a 20% down payment and arrange a mortgage loan. Find the amount of the down payment and the amount to be financed with a mortgage loan.

solution

To find the amount of the down payment, multiply the percent of the down payment by the selling price.

$95,000 × 0.20 = $19,000 ← **A calculator may be helpful.**

To find the amount of the mortgage loan, subtract the amount of the down payment from the selling price of the house.

$95,000 − $19,000 = $76,000

The amount of the down payment is $19,000. The amount to be financed with a mortgage loan is $76,000.

exercises

Complete. A calculator may be helpful.

	SELLING PRICE	PERCENT OF DOWN PAYMENT	AMOUNT OF DOWN PAYMENT	AMOUNT OF MORTGAGE LOAN
1.	$100,000	20%	?	?
2.	$78,000	15%	?	?
3.	$92,000	25%	?	?
4.	$150,000	20%	?	?
5.	$200,000	30%	?	?
6.	$70,000	25%	?	?
7.	$40,000	20%	?	?
8.	$50,000	15%	?	?
9.	$80,000	35%	?	?
10.	$155,000	40%	?	?

11. The Marshalls want to buy a house for \$115,000. They make a 30% down payment. Find the amount of the down payment and the amount of the mortgage loan.
12. The Jacksons plan to buy a house for \$135,000. They make a 20% down payment. Find the amount of the down payment and the amount of the mortgage loan.
13. Find out if there is a penalty for repaying a mortgage loan before the specified time.
14. Find out how lending institutions decide on whether a customer is eligible for a home mortgage loan.

In some states, banks allow customers to use the amount already paid on a mortgage loan to arrange another type of loan, a *home equity loan*. This is a comparatively low-cost way to cover large expenses, such as home improvement projects or college expenses. To find the maximum amount of money you can borrow, subtract the amount still owed on your mortgage loan from 80% of the value of your home.

Complete. A calculator may be helpful.

	VALUE OF HOME	80% OF VALUE OF HOME	AMOUNT STILL OWED ON MORTGAGE LOAN	MAXIMUM AMOUNT OF HOME EQUITY LOAN
15.	\$150,000	?	\$90,000	?
16.	\$250,000	?	\$100,000	?
17.	\$95,000	?	\$75,000	?
18.	\$80,000	?	\$10,000	?

19. Elizabeth Weingarten owns a house valued at \$100,000. She still owes \$8000 on her mortgage loan. Find the maximum amount of money that she can borrow on a home equity loan.
20. Pat Monahan owns a house valued at \$160,000. He still owes \$84,000 on his mortgage loan. Find the maximum amount of money that he can borrow on a home equity loan.

CAREER

DISCOUNT STOCKBROKER

Amy Gilman is a **discount stockbroker** for Reliable Discount Brokerage. She receives a salary for taking orders to buy and sell stocks and to make investments. She does not give investment advice to customers. Her company charges customers a commission on each *transaction,* but the charge is generally less than at a **full-service brokerage** that offers investment advice. The commission charged on a transaction is usually based on a set schedule like the one below.

TRANSACTION AMOUNT	COMMISSION
\$0–\$5000	\$32 + 0.6% of transaction amount
\$5001–\$15,000	\$32 + 0.5% of transaction amount
\$15,001–\$45,000	\$32 + 0.4% of transaction amount
over \$45,000	\$32 + 0.3% of transaction amount

example

Jo Orr asks Amy to buy \$7700 worth of a stock. Find the commission charged on the transaction.

solution

\$7700 is between \$5001 and \$15,000.
First find 0.5% of the transaction amount.

$0.005 \times \$7700 = \$38.50$

Commission $= \$32 + \$38.50 = \$70.50$

The commission charged is \$70.50.

exercises

1. The Bradley Investment Club asks Amy to buy \$30,000 worth of a stock. Find the commission charged on the transaction.
2. Hunter Phelps asks Amy to buy \$46,000 worth of a stock. Find the commission charged on the transaction.
3. Jim Talfer asks Amy to sell \$4005 worth of a stock. Find the commission charged on the transaction.
4. Dave Miller asks Amy to buy \$6000 worth of a stock and sell \$500 worth of another stock. Find the commission charged on the total transaction.
5. Amy receives an \$82 commission. What was the amount of the transaction?

CHAPTER REVIEW

vocabulary vo·cab·u·lar·y

Choose the correct word or phrase to complete each sentence.

1. Interest paid on the principal and on interest already earned is (*simple interest, compound interest*).
2. An initial sum of money paid toward the cost of a car or home is the (*down payment, mortgage*).

skills

Find the interest and amount due. If necessary, round to the nearest cent.

3. \$500 at 5% per year for 1 year
4. \$725 at 7% per year for 1 year
5. \$2555 at 6.5% per year for 3 years
6. \$800 at $15\frac{1}{2}$% per year for 3 years
7. \$3300 at 12% per year for 6 months
8. \$2400 at 0.7% per year for 6 months

Find the balance. If possible, use the table on page 296. If necessary, round to the nearest cent.

9. \$1000 at 8% annual interest compounded annually for 2 years
10. \$2500 at 6.5% annual interest compounded semiannually for 1 year
11. \$500 at 4% annual interest compounded semiannually for 2 years
12. \$450 at 6% annual interest compounded quarterly for 2 years
13. Lee Sims has \$3200 in a savings account that earns 6% annual interest compounded monthly. He makes no other deposits or withdrawals. Find the balance at the end of a 30-day month.
14. Angela Perella has \$4000 to invest. She considers a savings account that earns 5.5% annual interest and a CD that earns 7.25% annual interest. How much more money will she earn in one year by investing in the CD?
15. Karen Duffy has a Grand Credit Card. She pays a monthly interest rate of 1.6% on an unpaid balance. Last month her unpaid balance was \$80. She made no new purchases. Find the new balance.
16. Carl O'Day arranged an automobile loan for \$7500. He repaid the loan with 24 payments of \$347.81 each. Find the finance charge.
17. Sue Williams owns a home valued at \$100,000. She still owes \$60,000 on her mortgage loan. Find the maximum amount of money that she can borrow on a home equity loan.

CHAPTER TEST

Find the interest. If necessary, round to the nearest cent.

15-1

1. $350 at 4% per year for 1 year
2. $600 at 9% per year for 2 years
3. $3000 at $6\frac{1}{2}$% per year for 3 years
4. $105 at 9.5% per year for 6 months

Find the amount due. If necessary, round to the nearest cent.

5. $450 at 3% per year for 1 year
6. $4000 at 6% per year for 2 years
7. $875 at $4\frac{1}{2}$% per year for 3 years
8. $2850 at 7% per year for 6 months

Find the balance. If possible, use the table on page 296. If necessary, round to the nearest cent.

15-2

9. $800 at 8.5% annual interest compounded annually for 2 years
10. $7000 at 9.5% annual interest compounded annually for 2 years
11. $1000 at 6% annual interest compounded semiannually for 2 years
12. $1500 at 6% annual interest compounded quarterly for 2 years
13. $610 at 4% annual interest compounded semiannually for 4 years
14. $3400 at 4% annual interest compounded quarterly for 2 years

15-3

15. Nathaniel Sommers has $2000 in a savings account that earns 5.5% annual interest compounded monthly. He makes no deposits or withdrawals. Find his balance at the end of a 30-day month.

15-4

16. Dana Millis has decided to invest $3000 in a CD that earns 7.45% annual interest. She chose not to invest in a savings account that earns 5% annual interest. How much more interest did her money earn from the CD than it would have in the savings account in the first year?

15-5

17. Yolanda LaRoses has a Stately Credit Card. She pays a monthly interest rate of 1.5% on the unpaid balance. Last month her unpaid balance was $636.49, and she made no new purchases. Find the new balance.

15-6

18. Stuart Markham wants to buy a car that sells for $8500. He makes a 20% down payment and makes 36 monthly payments of $225.85. Find the total amount paid for the car.

15-7

19. The Howells want to buy a house that sells for $200,000. They make a 25% down payment. Find the amount of the down payment and the amount of the mortgage loan.

COMPUTER

Programming in BASIC

In order to perform a task, a computer must receive a set of instructions called a **program.** Programs must be written in a language that the computer can understand. Some of the more common programming languages are BASIC, COBOL, Pascal, and FORTRAN.

A program written in the BASIC language consists of numbered **statements.** Each statement is an instruction to the computer. The computer follows these statements in numerical order. In BASIC, the operations of arithmetic are represented by the following symbols.

+ addition − subtraction * multiplication / division

BASIC follows the same rules for order of operations that you learned in Chapter 4. The following is an example of a BASIC program.

```
10  PRINT "5 + 2/2 * 7 - 4 = ";
20  PRINT 5 + 2/2 * 7 - 4
30  END
RUN
5 + 2/2 * 7 - 4 = 8
```

← **The semicolon makes the computer put more than one result on a line.**

Lines 10 and 20 illustrate two different uses of **PRINT.** In line 10, PRINT instructs the computer to display exactly what is typed inside the quotation marks. Because there are no quotation marks in line 20, PRINT tells the computer to perform the calculation and display the result.

The **END** statement in line 30 is always the last line of a program. When you type **RUN** and press the *return* key, the computer will carry out the calculation.

exercises

Write the correct output.

1.
```
10  PRINT 9 * 2 - 6/3;
20  PRINT " FT"
30  END
```

2.
```
10  PRINT "THE SOLUTION IS ";
20  PRINT 3 * 4 - 3 + 4/8 * 2
30  END
```

3. Find out what is meant by the term *high-level language*.

4. Find out the meaning of the term *syntax error*.

COMPETENCY TEST

Choose the letter of the correct answer.

1. **Write the ratio as a fraction in lowest terms.** 15 s:3 min

   A. $\frac{5}{1}$ B. $\frac{1}{12}$
   C. $\frac{1}{4}$ D. $\frac{300}{1}$

2. **Find the percent of increase.**
   original length: 450 mi
   new length: 750 mi

   A. 40% B. 60%
   C. $166\frac{2}{3}\%$ D. $66\frac{2}{3}\%$

3. **55% of what number is 1320?**

   A. 24 B. 726
   C. 2400 D. 2640

4. **Amy Forand earns $3.50/h. She has already saved $56 toward a bike that costs $140. How many hours must she work to earn the rest of the money?**

   A. 16 h B. 24 h
   C. 40 h D. 56 h

5. **Louis Ravel's annual taxable income is $32,500. His state income tax rate is 3% of the first $5000 and 4.5% of the amount over $5000. Find his total state income tax.**

   A. $1462.50 B. $975
   C. $1050 D. $1387.50

6. **Choose the best estimate.**
   21.4% of 2484

   A. about 300 B. about 400
   C. about 500 D. about 700

7. **Which number is not equal to $25\frac{1}{2}\%$?**

   A. $\frac{255}{1000}$ B. $0.2\overline{5}$
   C. $\frac{51}{200}$ D. 0.255

8. **Solve the proportion.** $\frac{3}{8} = \frac{x}{24}$

   A. $x = 19$ B. $x = 9$
   C. $x = 64$ D. $x = 1$

9. **An 18-in. length of ribbon costs 42¢. How much would a $2\frac{1}{2}$-ft length of this ribbon cost?**

   A. 70¢ B. $2.10
   C. 28¢ D. 56¢

10. **Jacqueline and Andrew Stern bought a car that sells for $8500. They made a 25% down payment and 36 monthly payments of $207. Find the total amount paid for the car.**

    A. $9577 B. $10,625
    C. $9152 D. $13,827

11. **The map distance between two cities is $2\frac{1}{4}$ in. The scale is 1 in.:240 mi. Choose the best estimate of the actual distance between the cities.**

A. about 107 mi
B. about 160 mi
C. about 360 mi
D. about 540 mi

12. **What number is 75% of 180?**

A. 135 B. 240
C. 120 D. 126

13. **A computer system that has a list price of $2100 is discounted 35%. Find the selling price.**

A. $735 B. $2835
C. $1365 D. $1400

14. **Which account earns the most interest after one year?**

A. $5000 at 8% per year
B. $8000 at 5.2% per year
C. $6500 at 5.7% per year
D. $7000 at 6.5% per year

15. **Irma Ruiz swims 135 m in 90 s. Choose the proportion to determine how far she can swim in 3.5 min.**

A. $\frac{90}{135} = \frac{3.5}{x}$ B. $\frac{90}{135} = \frac{x}{3.5}$

C. $\frac{90}{135} = \frac{210}{x}$ D. $\frac{90}{135} = \frac{x}{210}$

16. **The Velezes made a 20% down payment on a house that sells for $99,000. Find the amount of the mortgage loan.**

A. $79,200
B. $19,800
C. $74,250
D. $18,800

17. **What percent of 720 is 90?**

A. 12.5% B. 80%
C. 64.8% D. 8%

18. **Find the balance at the end of one year: $900 at 8% annual interest compounded semiannually.**

A. $1044 B. $973.44
C. $972 D. $1049.76

19. **Write 8.5% as a fraction in lowest terms.**

A. $\frac{8.5}{100}$ B. $\frac{17}{20}$

C. $\frac{17}{200}$ D. $\frac{85}{100}$

20. **The monthly interest rate on a credit card's unpaid balance is 1.2%. Find the new balance.**

unpaid balance: $187.50
new purchases: $48.50

A. $236.58 B. $236
C. $238.83 D. $238.25

CUMULATIVE REVIEW CHAPTERS 1–15

Find each answer. Write fractions in lowest terms.

1. $4.2\overline{)8.694}$
2. $2715 + 88$
3. $25 \times 71 \times 4$
4. $\frac{2}{3} \times \frac{5}{8}$
5. $6\frac{3}{5} - 3\frac{1}{2}$
6. $15 \div \frac{5}{8}$
7. 78×105
8. $181.4 - 6.75$
9. $48 \div 12 + 2^2 \times 2$
10. $\frac{3}{8} + \frac{1}{2}$
11. $3150 \div 45$
12. $\$42.85 \times 8$
13. 3 yd 1 ft + 4 yd 2 ft
14. 14 lb 8 oz − 10 lb 11 oz
15. Write 375% as a decimal.
16. Write $\frac{17}{25}$ as a percent.
17. What percent of 188 is 47?
18. 462 is 55% of what number?
19. Find the GCF and the LCM of 14 and 49.
20. Write the ratio as a fraction in lowest terms: 12 s to 8 min
21. Complete: **a.** 19 kg = __?__ g **b.** 3400 mL = __?__ L
22. Solve: **a.** $\frac{n}{25} = \frac{18}{45}$ **b.** $\frac{4}{16} = \frac{x}{52}$

Estimate.

23. 19.9% of 148
24. 18×7981
25. $27\overline{)6177}$
26. $5289 + 3134$
27. 74% of 398
28. $\$52.78 + \$88.17$
29. $2\frac{1}{9} + 7\frac{8}{15} + 3\frac{6}{13}$
30. $\frac{15}{47} \times 897$
31. $19\frac{7}{12} - 11\frac{5}{8}$

Find the percent of increase or decrease.

32. original salary: \$350
    new salary: \$434
33. original number of books: 425
    new number of books: 255
34. Samuel kept \$2500 in the bank for one year. The interest rate was 8% per year. How much interest did the principal earn?
35. The principal in an account is \$1800. The account earns 6% annual interest, compounded semiannually. There were no other deposits or withdrawals. Find the balance at the end of one year.
36. Sally Rush wants to buy a car that sells for \$9800. She makes a 20% down payment and 36 monthly payments of \$250. Find the total amount paid for the car.
37. Farah Najafi's health insurance requires her to pay a \$100 deductible. The insurance company pays 80% of the remaining expenses. If her covered medical expenses were \$1250, how much did Farah pay?

Graphs are useful in biological research—for example, in studying how animal populations in a particular region change over the years.

CHAPTER 16

GRAPHING DATA

16-1 BAR GRAPHS

A **graph** is a picture that displays numerical facts. The numerical facts are called **data.** You can use a **bar graph** to compare data about different things at a given time. A bar graph has a **horizontal axis** (—) and a **vertical axis** (|). One of these *axes* is labeled with a numerical **scale.**

example 1

ESTIMATION

Use the bar graph at the right.
Estimate the height of each building.

a. Sears Tower **b.** General Motors Building

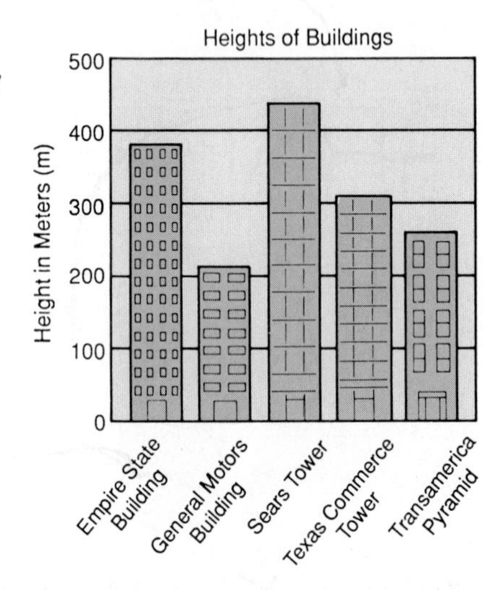

solution

a. Each space on the scale represents 100 m.
Locate the bar labeled *Sears Tower.*
The bar ends midway between 400 and 500.
The height of the building is about 450 m.

b. The bar labeled *General Motors Building* ends between 200 and 300, but closer to 200.
The height of the building is about 200 m.

your turn

Use the bar graph above. Estimate the height of each building.

1. Empire State Building **2.** Texas Commerce Tower **3.** Transamerica Pyramid

example 2

Use the *double bar graph* at the right.
How many girls chose jazz?

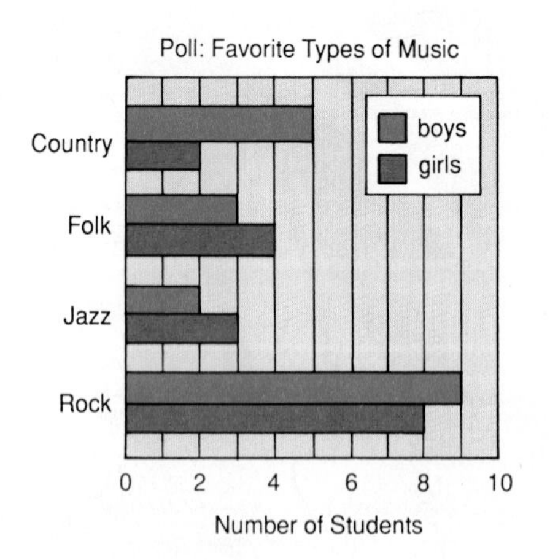

solution

Each space on the scale represents 1 student.
Locate the bars labeled *Jazz.* The *girls* bar is red.
Three girls chose jazz.

your turn

Use the double bar graph at the right.

4. How many boys chose rock music?

5. How many more boys than girls chose country music?

6. Which types of music were the favorite of fewer than seven boys?

example 3

Draw a bar graph to display the given data.

Areas of the Great Lakes

Name of Lake	Erie	Huron	Michigan	Ontario	Superior
Area in Thousands of Square Kilometers (km^2)	26	60	58	19	82

solution

- Draw the axes.
- The greatest number is 82 thousand, so draw and label a scale from 0 to 90. Let each space represent 10 thousand km^2.
- Draw and label one bar to represent the area of each lake. Estimate the length of the bar when the number is not a multiple of 10 thousand.
- Label the scale and title the graph.

Areas of the Great Lakes
Erie
Huron
Michigan
Ontario
Superior
0 20 40 60 80
Area in Thousands of Square Kilometers (km^2)

your turn

7. Draw a bar graph to display the given data.

Maximum Speeds of Animals

Animal	Cheetah	Elk	Giraffe	Pig	Zebra
Speed (mi/h)	70	45	32	11	40

practice exercises

practice for example 1 (page 318)

Use the bar graph at the right.
Estimate the average weight of each animal.

1. Sheep
2. Seal
3. Polar Bear
4. Horse
5. Which animal is the heaviest?
6. Which animal is the lightest?
7. On average, which animals are heavier than a seal?
8. On average, which animals weigh less than 800 lb?

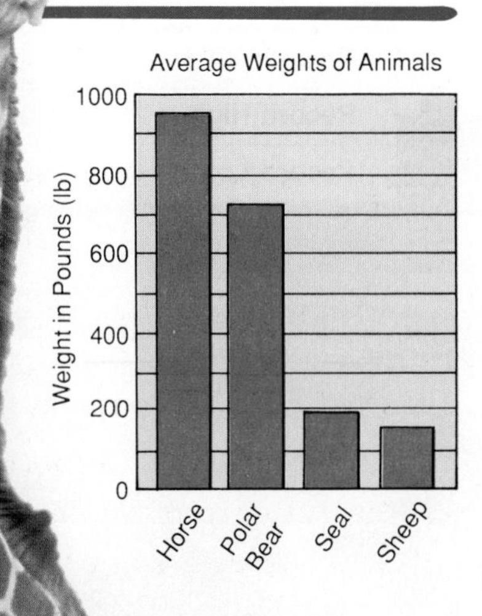

practice for example 2 (page 318)

Use the double bar graph at the right.

9. How many series did the Cubs win?
10. How many series did the Cardinals lose?
11. Which team won the most series?
12. Which team won as many series as it lost?
13. Which team won exactly six fewer series than the Cardinals won?
14. Which team lost twice as many series as the Red Sox lost?

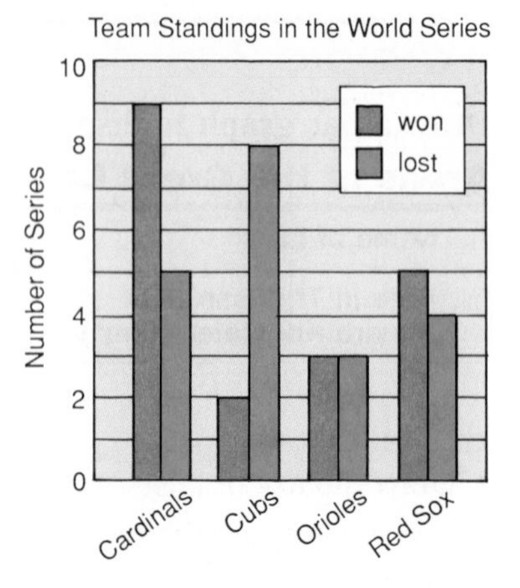

practice for example 3 (page 319)

Draw a bar graph to display the given data.

15. **National Basketball Association Playoff Champions**

Team	Bullets	Celtics	Knicks	Lakers	76ers	Warriors
Number of Championships	2	16	2	10	2	3

16. **Waterfalls of the World**

Name of Waterfall	Angel	Bridalveil	Kalambo	Sutherland	Vetisfoss
Height in Meters	1000	189	219	580	366

mixed practice (pages 318–319)

17. Draw a double bar graph to display the given data.

Record Temperatures in Degrees Fahrenheit (°F)

Name of City	Los Angeles	Miami	New York	Phoenix	Seattle
Record High	96	96	96	113	91
Record Low	41	44	8	36	20

Use the double bar graph that you drew for Exercise 17.

18. In which city is there the greatest difference between record high and low temperatures?

19. Which two cities have about the same record high temperature *and* about the same record low temperature?

20. Use an almanac or other reference to find the length in miles of each of these rivers: *Amazon, Danube, Mississippi, Nile, Volga.* Draw a bar graph to display these data.

21. Poll the students in your mathematics class about their favorite spectator sports. Have them choose from *baseball, basketball, football, hockey,* and *tennis*. Draw a bar graph to display the data.

review exercises

Write each set of numbers in order from least to greatest.

1. 73,521; 7359; 71,532

2. 8801; 8810; 8018

3. $985; $859; $958

4. 0.7; 0.007; 0.07

5. 0.03; 0.0033; 0.033

6. 79.2; 70.29; 7.92

7. $\frac{7}{11}; \frac{2}{11}; \frac{5}{11}$

8. $\frac{2}{3}; \frac{2}{5}; \frac{3}{5}$

9. $\frac{3}{4}; \frac{7}{10}; \frac{5}{8}$

10. $1\frac{1}{3}; 1\frac{1}{2}; 1\frac{4}{9}$

mental math

Find each answer mentally.

1. 20% of 60

2. 5% of $42

3. 30% of $12.20

4. 40% of 710

Find each percent mentally.

5. What percent of 44 is 11?

6. 8 is what percent of 12?

Mentally estimate a 15% tip for the given amount.

7. $22

8. $12.80

9. $44.19

10. $26.99

16-2 LINE GRAPHS

You can use a **line graph** to show changes in data over a period of time. A line graph shows both an *amount* and a *direction* of change. The direction of change may be an **increase** or a **decrease.**

example 1

Use the line graph at the right. What was the temperature at 6:00 A.M.?

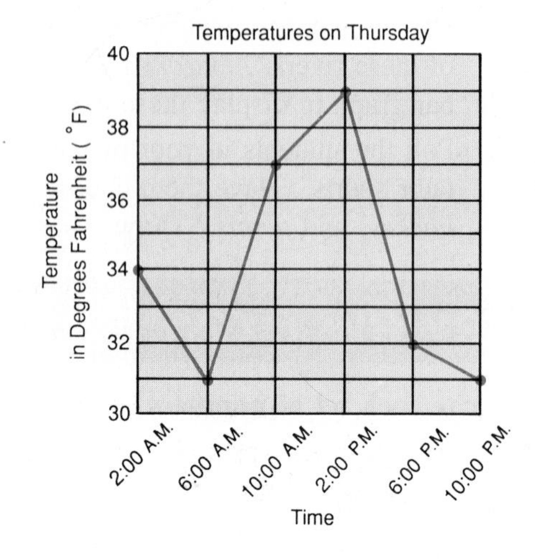

solution

Locate *6:00 A.M.* on the horizontal axis. Move up to the red point, then left to the vertical axis. Each space on the scale represents one degree Fahrenheit (1°F). The temperature was 31°F.

your turn

1. What was the temperature at 2:00 P.M.?
2. Between which two given times was there the greatest decrease?

example 2

Use the *double line graph* at the right. Estimate the urban population in 1980.

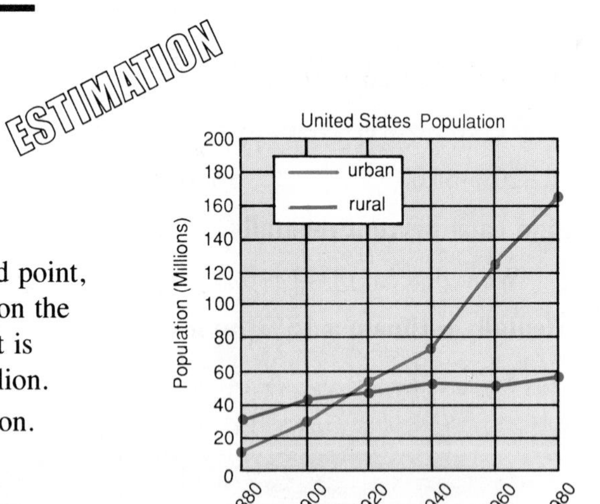

solution

Locate 1980 on the horizontal axis.

The *urban* line is red. Move up to the red point, then left to the vertical axis. Each space on the scale represents 20 million. The red point is midway between 160 million and 180 million. The urban population was about 170 million.

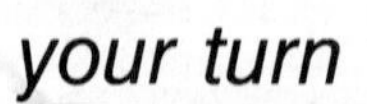

your turn

3. Estimate the rural population in 1900.
4. Between which two given years was there the greatest increase in urban population?
5. In which given year were the urban and rural populations about the same?
6. Estimate the total of the urban and rural populations in 1980.

example 3

Draw a line graph to display the given data.

Cable Television Subscribers

Year	1965	1970	1975	1980	1985	1990 (projected)
Subscribers (Millions)	2	5	10	18	35	46

solution

- Draw the axes.
- The greatest number is 46 million, so draw and label a scale from 0 to 50 on the vertical axis. Let each space represent 5 million subscribers.
- Draw vertical lines from the horizontal axis and label them with the given years.
- Place a point on the graph for each year and the corresponding number of subscribers.
- Connect the points from left to right.
- Label the axes and title the graph.

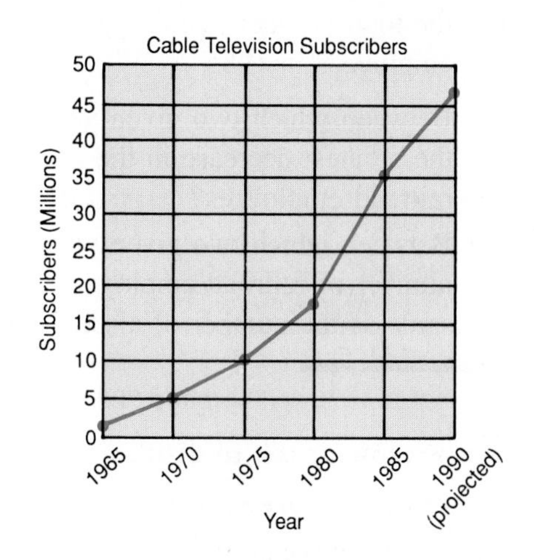

your turn

7. Draw a line graph to display the given data.

Telephones in the United States

Year	1930	1940	1950	1960	1970	1980
Number of Telephones (Millions)	20	22	43	74	105	157

practice exercises

practice for example 1 (page 322)

Use the line graph at the right. Determine the number of customers at the given time.

1. 12:00 noon
2. 8:00 P.M.
3. 4:00 P.M.
4. 2:00 P.M.
5. 10:00 A.M.
6. 6:00 P.M.
7. At which given time were there the fewest customers?
8. At which two given times do you think the manager of Video Village should have the most salespeople at work?

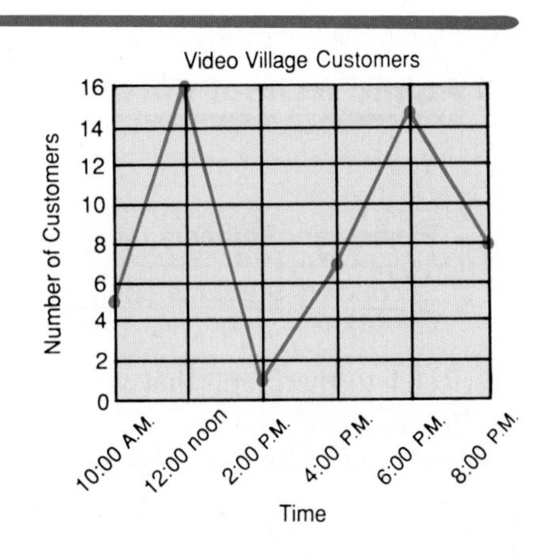

practice for example 2 (page 322)

Use the double line graph at the right to estimate.

9. the number of railroad employees in 1975
10. the number of airline employees in 1980
11. the total number of railroad and airline employees in 1960
12. the total number of railroad and airline employees in 1985
13. Between which two given years was there the greatest decrease in the number of railroad employees?
14. Between which two given years was the number of railroad employees about the same as the number of airline employees?

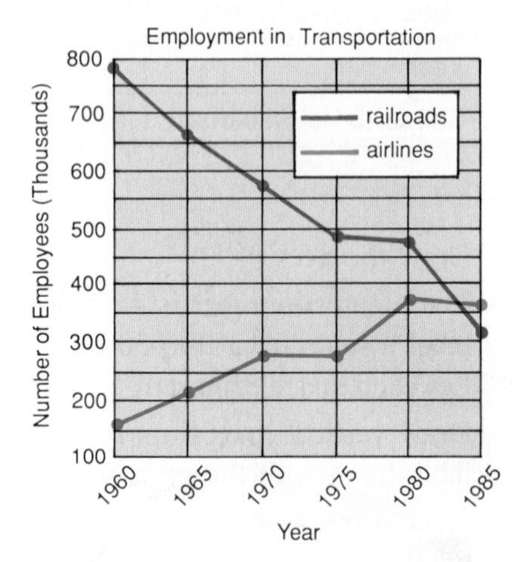

practice for example 3 (page 323)

Draw a line graph to display the given data.

15. **Daily Newspaper Circulation in the United States**

Year	1920	1930	1940	1950	1960	1970	1980
Circulation (Millions)	28	40	41	54	59	62	62

16. **Growth Record**

Age in Years	11	12	13	14	15	16	17	18
Height in Centimeters	141	146	152	158	165	170	174	174

mixed practice (pages 322–323)

17. Draw a double line graph to display the given data.

Public School Enrollment in the United States

Year	1965	1970	1975	1980	1985	1990 (projected)
Elementary Students (Millions)	27	28	26	24	24	27
Secondary Students (Millions)	16	18	19	17	15	14

18. Refer to the graph that you drew for Exercise 17. Was there an *increase* or a *decrease* in the elementary school enrollment between 1985 and 1990? Do you think there will be an *increase* or a *decrease* in the secondary school enrollment between 1990 and 1995?

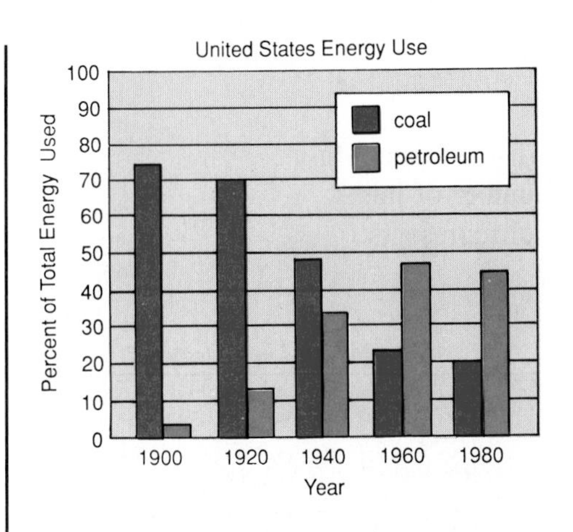

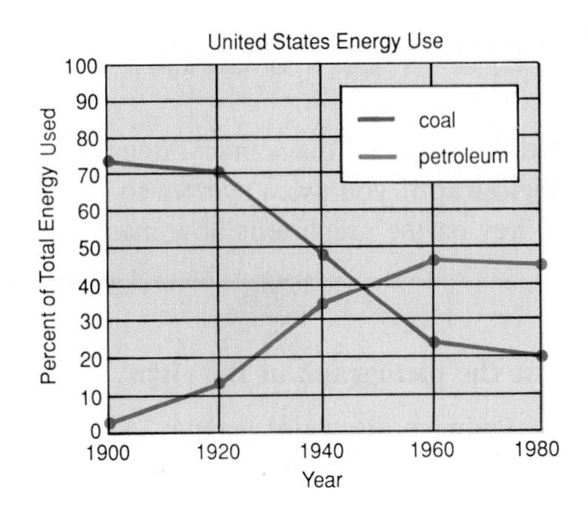

For Exercises 19–22, use the graphs above.

19. Between which two given years was the percent of energy used that was petroleum about equal to the percent that was coal?
20. Between which two given years was there a decrease in the percent of energy used that was petroleum *and* in the percent that was coal?
21. Estimate the percent of the energy used in 1920 that was coal.
22. Do you think that the *bar graph* or the *line graph* is more appropriate for displaying these data about energy use? Explain.

Tell whether it would be more appropriate to draw a *bar graph* or a *line graph* to display the given data. Then draw the graph.

23. **Buildings in Toronto**

Building	CN Tower	Palace Pier	Scotia Plaza	Simpson Tower
Height in Meters	555	138	270	144

24. **Average Temperatures for Toronto**

Month	January	April	July	October
Temperature in Degrees Fahrenheit	24	46	72	52

review exercises

Write the short word form of each number.

1. 8602
2. 720,005
3. 0.13
4. 107.6
5. 2.0071
6. 60.0507

16-3 PICTOGRAPHS

Newspapers and magazines often use *pictographs* to display data. In a **pictograph,** you use a symbol to represent a given number of items. A **key** on the graph tells how many items each symbol represents.

example 1

ESTIMATION

Use the pictograph at the right.

a. Estimate the major league baseball attendance in 1970.

b. Estimate the attendance in 1985.

c. About how many more people attended in 1980 than in 1975?

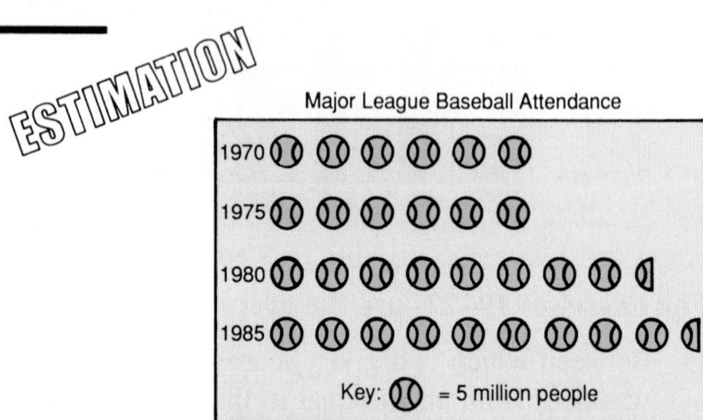

solution

a. Each symbol represents 5 million people. *1970* has 6 symbols. Multiply 6×5 million. In 1970 the attendance was about 30 million people.

b. *1985* has $9\frac{1}{2}$, or 9.5, symbols. Multiply 9.5×5 million. In 1985 the attendance was about 47.5 million people.

c. There are $2\frac{1}{2}$, or 2.5, more symbols for *1980* than for *1975*. Multiply 2.5×5 million. About 12.5 million more people attended in 1985 than in 1975.

your turn

Use the pictograph at the right.

1. Estimate the number of books that were published in 1960.
2. Estimate the number of books that were published in 1980.
3. About how many more books were published in 1980 than in 1940?
4. In which given year were about twice as many books published as in 1920?

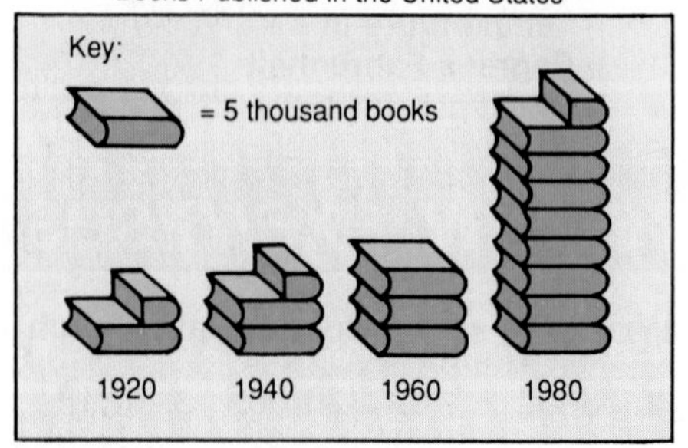

example 2

Draw a pictograph to display the given data.

Record Roundup Sales

Type of Music	Classical	Country	Folk	Jazz	Rock
Number of Albums Sold	25	45	20	65	80

solution

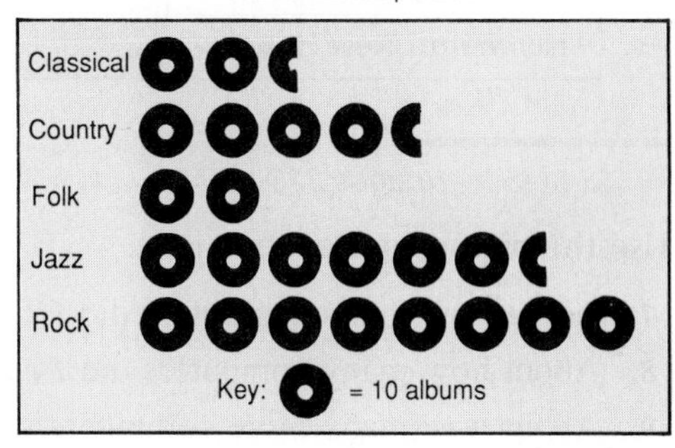

- Choose a symbol, ●, that resembles an album.
- Each number is a multiple of 5, so let ● represent 10 albums. Then half the symbol, ◖, represents 5 albums.
- Determine the correct number of symbols for each type of music.
- Draw a key on the graph.
- Title the graph.

your turn

5. Draw a pictograph to display the given data.

School Play Ticket Sales

Class	Freshmen	Sophomores	Juniors	Seniors
Number of Tickets Sold	150	275	325	450

practice exercises

practice for example 1 (page 326)

Use the pictograph at the right.

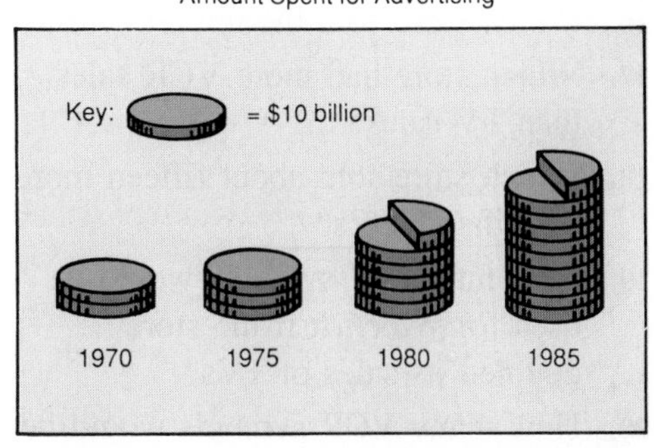

1. Estimate the amount spent for advertising in 1975.
2. Estimate the amount spent for advertising in 1985.
3. About how much more was spent for advertising in 1985 than in 1980?
4. In which given year was the amount spent for advertising about three times the amount spent in 1970?

practice for example 2 (page 327)

Draw a pictograph to display the given data.

5. **T-Shirt Sales**

Class	Freshmen	Sophomores	Juniors	Seniors
Number of T-Shirts Sold	70	55	90	45

6. **New Houses in the Metropolitan Area**

City or Town	Eastboro	Midtown	Northville	Southside	Weston
Number of New Houses	6	15	18	48	27

mixed practice (pages 326–327)

Use the pictograph at the right.

7. About how many computers did *HCN* sell?
8. About how many computers did *Pear* sell?
9. About how many more computers than *Stara* did *Pear* sell?
10. Which company sold about twice as many computers as *Stara*?
11. How many symbols would be used for *Stara* if the company tripled its sales?
12. How many symbols would be used for a company that sold 9500 computers?

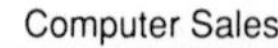
Computer Sales

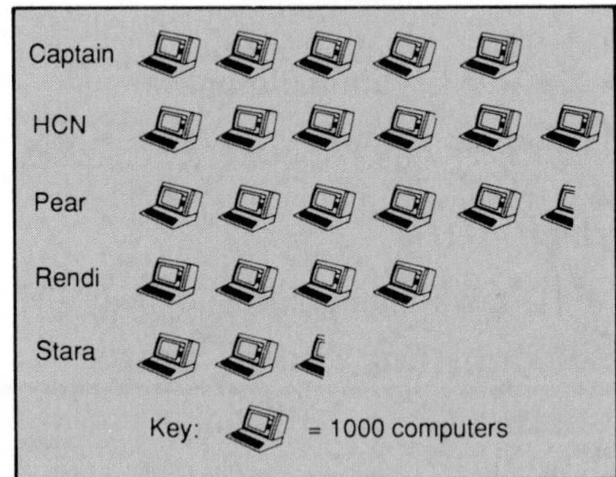

Use the *double pictograph* at the right.

13. About how many VCRs did *Watch-It* sell?
14. About how many TVs did *TV Time* sell?
15. Which store sold the most TVs?
16. Which store sold the fewest VCRs?
17. Which store had more VCR sales than TV sales?
18. Which store sold about fifteen more TVs than VCRs?
19. How many TV symbols would be used for *Watch-It* if the store doubled its sales of TVs?
20. How many VCR symbols would be used for a store that sold 115 VCRs?

TV and VCR Sales

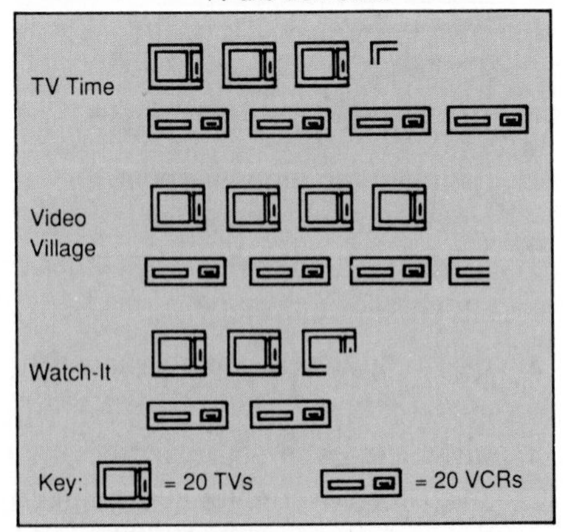

21. Draw a double pictograph to display the given data.

Radio and Television Stations

City	Radio Stations	Television Stations
Cincinnati	24	6
Houston	45	9
Kansas City	27	6
Los Angeles	72	18

22. Find out how many radio stations and how many television stations are in the capital of your state. Extend the pictograph that you drew for Exercise 21 to include these data.

23. Tell whether it would be more appropriate to draw a *bar graph* or a *pictograph* to display the given data. Then draw the graph.

Large Islands of the World

Island	Borneo	Greenland	Honshu	Sumatra
Area in Thousands of Square Miles (mi^2)	290	840	90	180

review exercises

Find each answer.

1. What number is 65% of 40?
2. 8 is what percent of 32?
3. 10 is 5% of what number?
4. What percent of 96 is 30?
5. 12.5% of 84 is what number?
6. 40% of what number is 150?
7. What percent of 46 is 69?
8. 18 is $66\frac{2}{3}\%$ of what number?

The figure at the right shows the top view of a set of wooden blocks. The blocks are packed in a box, and the box is completely filled. Each block is eight inches long, four inches wide, and two inches high. The box is eight inches deep. How many blocks are in the box?

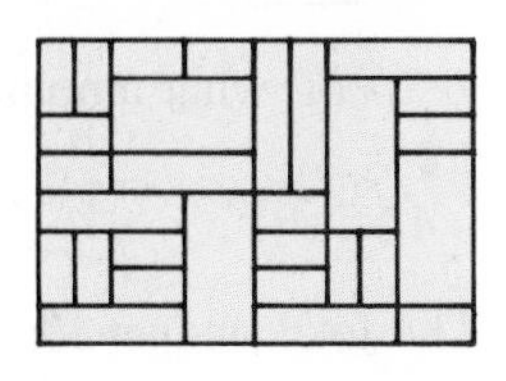

16-4 CIRCLE GRAPHS

You can use a **circle graph** to display data that are expressed as percents of a whole. The entire circle represents the whole, which is *100%*.

example 1

Use the circle graph at the right. If the population of the United States is about 240 million, estimate the number of people who live in the West.

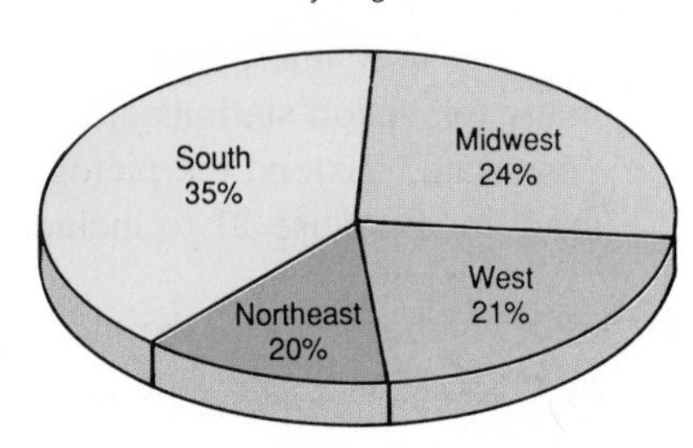

solution

About 21% of the population live in the West.

What number is 21% of 240 million?

$$n = 0.21 \times 240 \text{ million}$$
$$n = 50.4 \text{ million}$$

About 50.4 million people live in the West.

your turn

If the population of the United States is about 240 million, estimate the number of people who live in each region.

1. Northeast
2. South
3. Midwest
4. Estimate the fraction of the population who live in the Midwest.

example 2

The circle graph at the right shows how each dollar of an athletic department's budget is spent in a given year. If $3000 was budgeted for baseball one year, what was the total budget for that year?

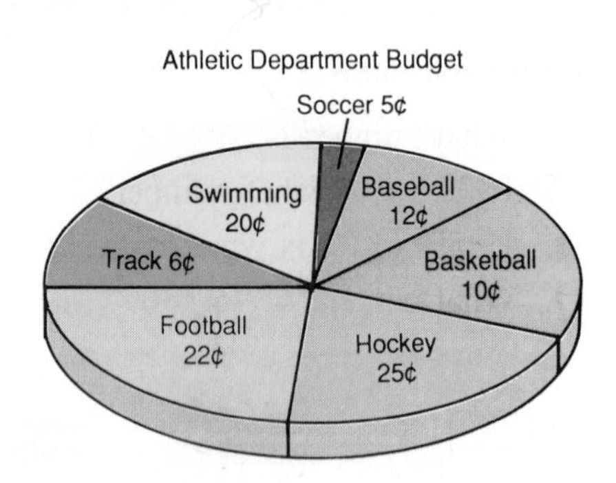

solution

12¢ of every dollar is budgeted for baseball.

cents budgeted for baseball → $\frac{12}{100} = 12\%$ ← total cents in a dollar

12% of what amount is $3000?

$0.12 \times n = \$3000$

$$\frac{0.12 \times n}{0.12} = \frac{\$3000}{0.12}$$
$$n = \$25{,}000$$

The total budget for the year was $25,000.

your turn

Use the circle graph at the bottom of page 330. Assume that the given amount was budgeted for each category. Find the total budget.

5. swimming: \$3000
6. soccer: \$1200
7. track: \$1350
8. If \$5500 was budgeted for football one year, how much was budgeted for basketball in that year?
9. If \$2400 was budgeted for baseball one year, how much was budgeted for the rest of the sports combined in that year?

practice exercises

practice for example 1 (page 330)

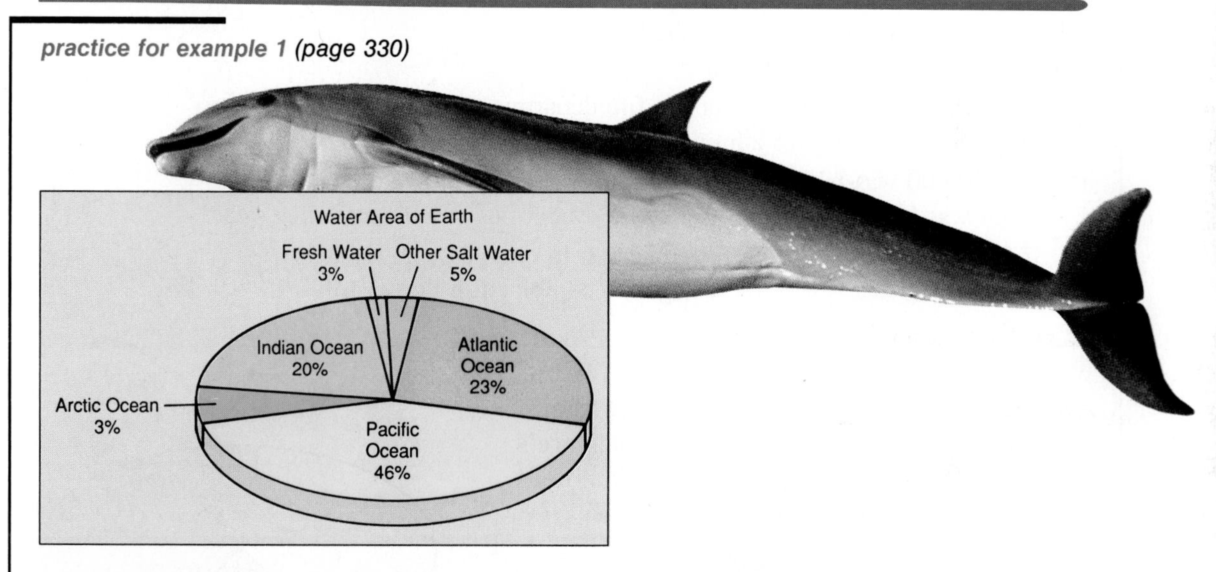

Use the circle graph above. If the total water area of Earth is about 140 million mi^2 (square miles), estimate the area covered by each ocean.

1. Indian Ocean
2. Arctic Ocean
3. Atlantic Ocean
4. Pacific Ocean
5. Which ocean covers about one fifth of the water area of Earth?
6. Which two oceans together cover about one half of the water area of Earth?
7. Estimate the fraction of the water area of Earth that is covered by the Atlantic Ocean.
8. Estimate the fraction of the water area of Earth that is covered by fresh water and other salt water combined.
9. Which two oceans together cover about the same area that is covered by the Pacific?
10. Which ocean covers about twice the area covered by the Atlantic?

practice for example 2 (pages 330–331)

The circle graph at the right shows how each dollar of a city's budget is spent in a given year. Assume that the given amount was budgeted for each category. Find the total budget.

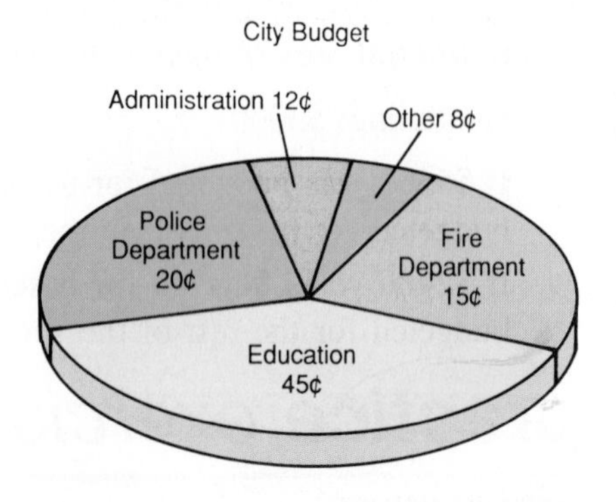

11. Police Department: $180,000
12. Fire Department: $150,000
13. Administration: $144,000
14. Education: $337,500
15. If $120,000 was budgeted for administration one year, how much was budgeted for the police department in that year?
16. If $112,500 was budgeted for the fire department one year, how much was budgeted for administration in that year?
17. If $100,000 was budgeted for the police department one year, how much was budgeted for the rest of the categories combined in that year?
18. If $360,000 was budgeted for education in one year, how much was budgeted for the rest of the categories combined in that year?

mixed practice (pages 330–331)

The circle graph at the right shows how the manager of a music store maintains the store's inventory of albums. If there are 5500 albums in all, find the number of albums of each type.

Music Store Inventory
Folk 15%
Rock 35%
Country 25%
Jazz 15%
Classical 10%

19. Classical
20. Country
21. Folk
22. Rock

Use the circle graph at the right. Assume that the inventory contains the given number of albums for each category. Find the total number of albums in the inventory.

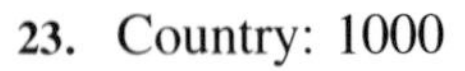

23. Country: 1000
24. Classical: 800

25. Rock: 2275
26. Jazz: 1140
27. If there are 800 country albums in the inventory, how many classical albums are there?
28. If there are 2100 rock albums in the inventory, how many jazz albums are there?
29. Which one type of album makes up one fourth of the inventory?
30. Estimate the fraction of the inventory that are rock albums.

31. Copy and complete the circle graph using the data given in the table.

Commuting Methods

Method	Percent
Mass Transit	40%
Automobile	27%
Car Pool	16%
Walking	10%
Other	7%

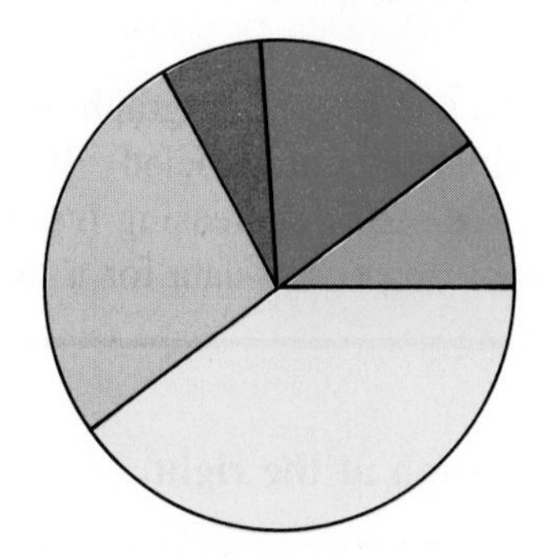

Another way to display data that are expressed as percents of a whole is to use a *divided-bar graph*. For Exercises 32–35, use the divided-bar graph shown below.

32. If the total number of registered motor vehicles in one year is about 180 million, about how many registered automobiles are there?
33. If the number of registered motorcycles in one year is about 5.25 million, estimate the total number of registered motor vehicles.
34. If the number of registered automobiles in one year is about 142.5 million, about how many registered trucks and buses are there?
35. Estimate the ratio of registered automobiles to registered trucks and buses.

36. Choose one school day and keep track of the number of hours that you spend on these activities: *Homework, Meals, School, Sleep, Other*. Draw a divided-bar graph to display these data.

review exercises

Complete.

1. 8 lb = _?_ oz
2. 16 qt = _?_ gal
3. 600 mL = _?_ L
4. 60 in. = _?_ ft
5. 11 pt = _?_ c
6. 14.7 mm = _?_ cm
7. 38 km = _?_ m
8. 5 c = _?_ fl oz
9. 6000 lb = _?_ t

PROBLEM SOLVING STRATEGY

16-5 USING GRAPHS

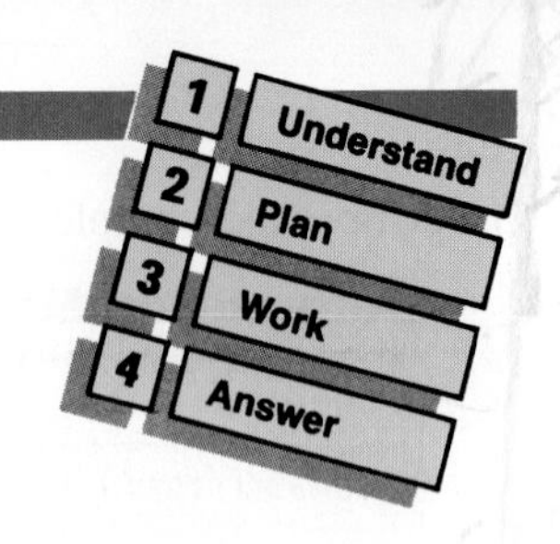

Sometimes you can use a line graph to *estimate* data between the times that are labeled. If the graph shows a steadily increasing or decreasing **trend,** you also may be able to use it to *predict* data for a future time.

example

Use the line graph at the right.

a. Estimate the number of women who were in the labor force in 1975.

b. Does the graph show an *increasing* or a *decreasing* trend?

c. Assume that the trend will continue. Predict the number of women who will be in the labor force in the year 2000.

solution

a. Locate 1975 on the horizontal axis. It is halfway between *1970* and *1980.* Move up to the red line, then left to the vertical axis.

Each space on the scale represents 10 million. The number for *1975* is between 30 million and 40 million, but closer to 40 million.

The number of women who were in the labor force in 1975 is about 40 million.

b. The graph shows an increasing trend.

c. Imagine a single straight line as close as possible to the given points.

Locate *2000* on the horizontal axis. Move up to the imaginary line, then move left to the vertical axis. The number for *2000* would be about midway between 60 million and 70 million.

If the trend continues, the number of women who will be in the labor force in the year 2000 is about 65 million.

problems

Use the line graph at the right. Estimate the number of people who were 65 years old and over in each year.

1. 1955 **2.** 1965 **3.** 1975

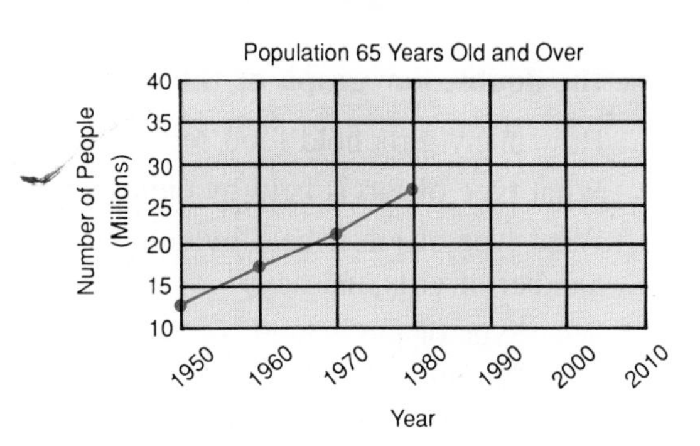

Use the line graph at the right. Assume that the trend will continue. Predict the number of people who will be 65 years old and over in each year.

4. 1990 **5.** 2000 **6.** 2010

Use the line graph at the right. Estimate the average temperature for Cleveland in each month.

7. February **8.** June

9. August **10.** December

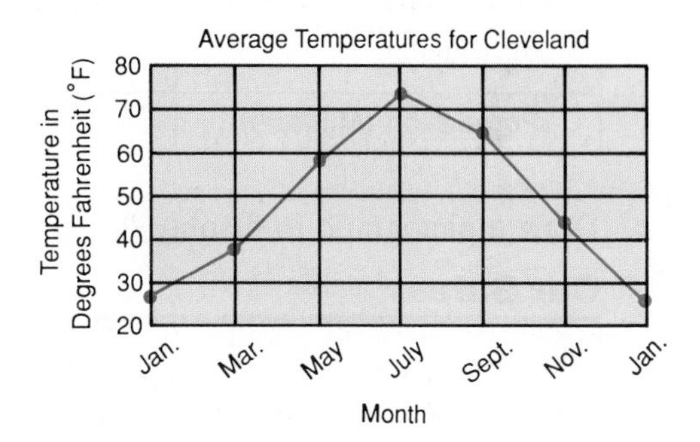

11. In which two months is the average temperature for Cleveland about 50°F?

12. Can you use this graph to predict what the temperature will be in Cleveland tomorrow? Explain.

Use the line graph at the right. Estimate the average height for boys at each age.

13. 5 **14.** 9 **15.** 11 **16.** 13

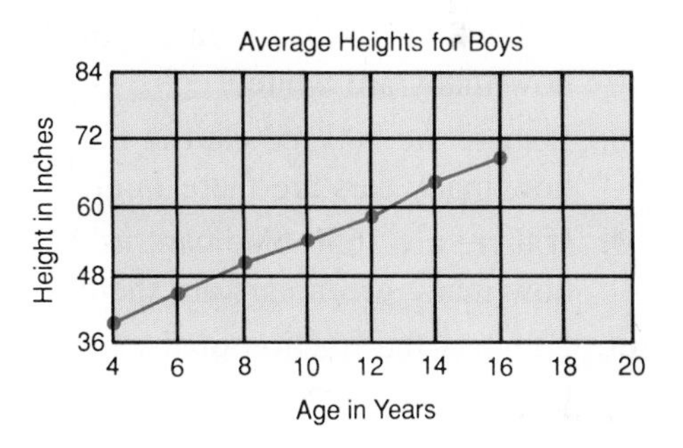

17. At which age is the average height for boys about 4 ft?

18. Can you use this graph to predict the average height for boys at age 20? Explain.

review exercises

1. Tom earns $25 for each tennis class he teaches. Each class contains five students. How many classes must Tom teach to earn $1000?

2. Jan jogs 8.3 mi around Clear Lake each day. How many miles in all does she jog around Clear Lake in a seven-day week?

SKILL REVIEW

Use the double bar graph at the right.

1. How many girls hold food service jobs?
2. What type of job is held by eight boys?
3. What type of job is held by an equal number of girls and boys?
4. What type of job is held by three more boys than girls?

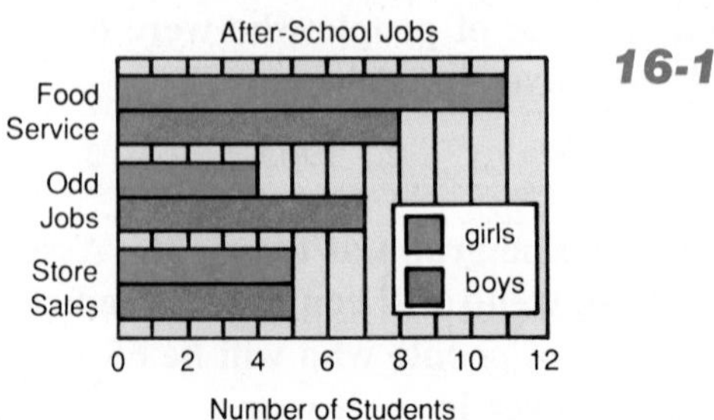

16-1

5. Draw a line graph to display the given data.

16-2

Population of Africa

Year	1950	1960	1970	1980
Population (Millions)	220	280	350	470

6. Draw a pictograph to display the given data.

16-3

Car Sales

Dealer	Auto World	Car Fair	Motor City	Wheel Deals
Number of Cars Sold	300	350	425	275

The circle graph at the right shows how a new car dealership maintains its inventory.

7. If there are 5000 cars in the inventory, how many are white?
8. If there are 900 red cars in the inventory, how many cars are there in all?
9. If there are 1650 blue cars in the inventory, how many green cars are there?
10. Estimate the fraction of the inventory that are gray cars.

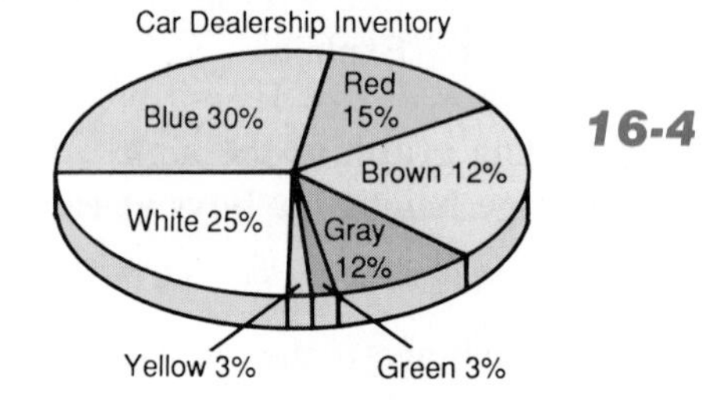

16-4

Use the line graph at the right. Estimate the median income for each year.

11. 1976 12. 1980 13. 1984

Assume that the trend will continue. Predict the median income for each year.

14. 1990 15. 1994 16. 1998

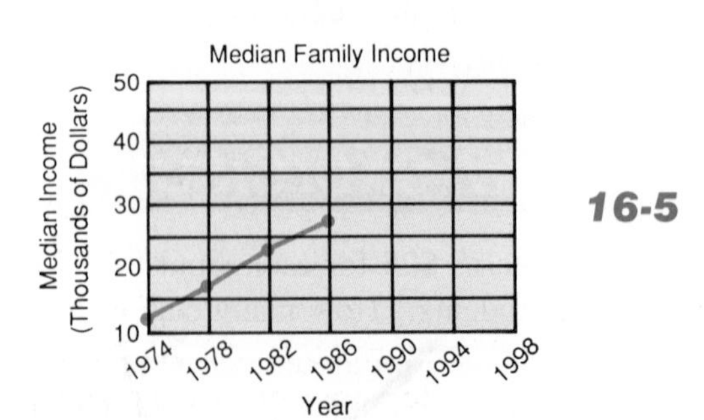

16-5

APPLICATION

16-6 MISLEADING GRAPHS

The person who draws a graph makes many choices when deciding upon the scale of the graph. Sometimes the way that these choices are made can mislead the reader by creating a false impression.

example

Both graphs below show the number of cans sold for three different brands of dog food.

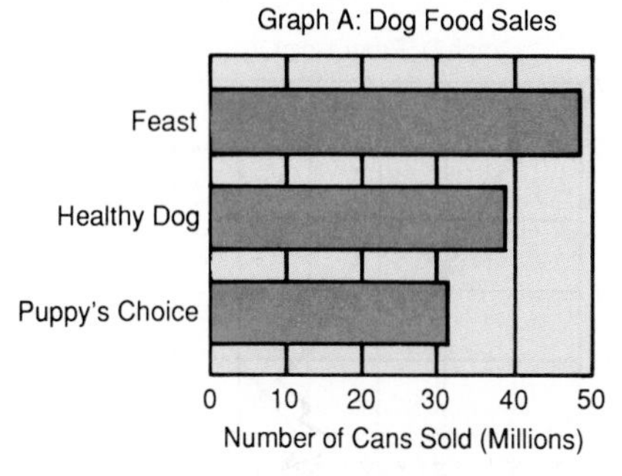

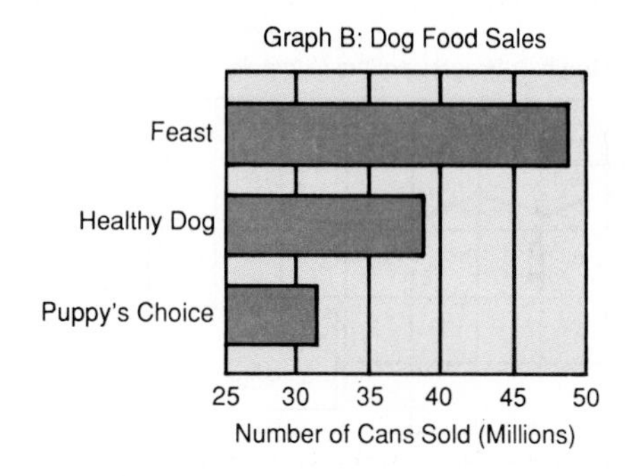

a. Use *Graph A* to estimate the number of cans sold for each brand.

b. Use *Graph B* to estimate the number of cans sold for each brand.

c. What is the difference between *Graph A* and *Graph B*?

d. If you were the manufacturer of *Feast* dog food, would you prefer *Graph A* or *Graph B*? Explain.

solution

a. *Feast:* about 50 million
Healthy Dog: about 40 million
Puppy's Choice: about 30 million

b. *Feast:* about 50 million
Healthy Dog: about 40 million
Puppy's Choice: about 30 million

c. On *Graph A*, the numerical scale begins at 0. On *Graph B*, it begins at 25. *Graph B* is misleading.

d. *Graph B*. It gives the impression that the number of cans of *Feast* sold was twice the number of cans of *Healthy Dog* and three times the number of cans of *Puppy's Choice*.

exercises

Exercises 1–4 refer to *Graph C* and *Graph D*, shown below.

1. Use *Graph C* to estimate the number of supporters for Carter and for Hooper in January and in September.
2. Use *Graph D* to estimate the number of supporters for Carter and for Hooper in January and in September.
3. What is the difference between *Graph C* and *Graph D*?
4. If you were Carter, would you prefer *Graph C* or *Graph D*? If you were Hooper, which graph would you prefer? Explain.

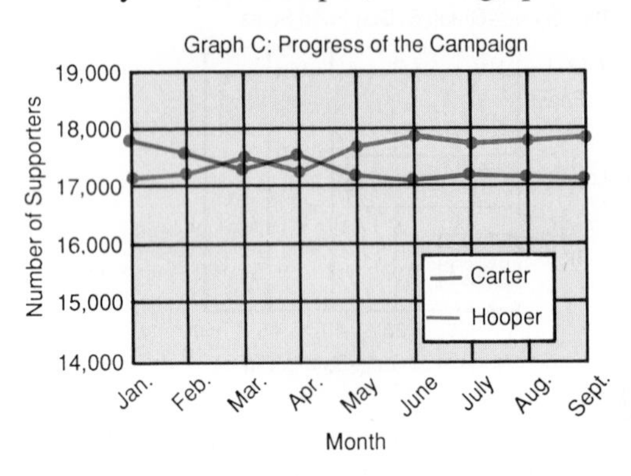

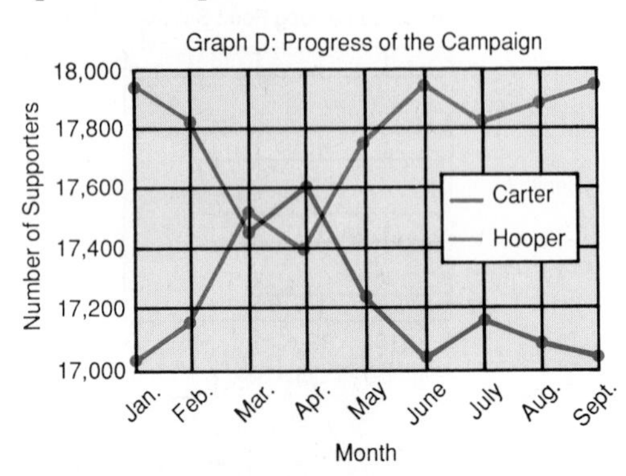

The following graphs have been drawn and titled to be misleading. In each case, draw a more appropriate graph. Give your graph a new title.

5.

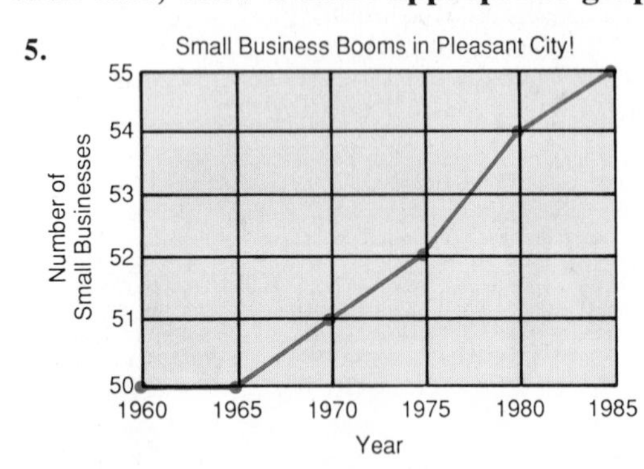

6.

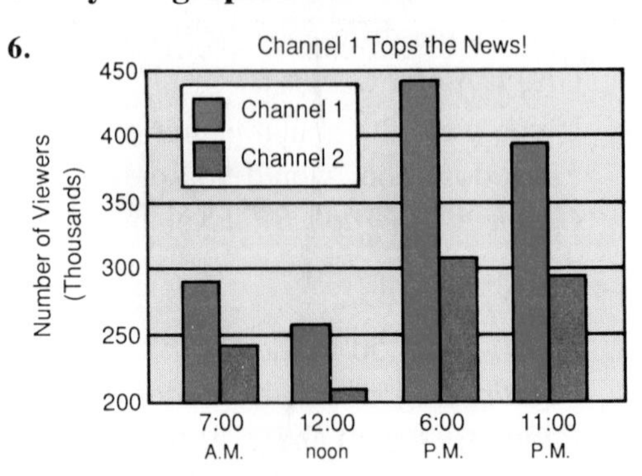

7. Draw a bar graph to display the given data. Compare your graph with that of a classmate. Are the graphs different? Can either graph be considered misleading? Explain.

Salaries at Local Companies

Company	ABC, Inc.	Best Company	Smith and Jones
Average Starting Salary	$19,000	$15,000	$16,500

APPLICATION

16-7 MAKING A BUDGET

A **budget** is a plan for managing your expenses. **Annual expenses,** such as automobile insurance, occur once each year. **Fixed expenses,** such as rent, are the same every month. **Variable expenses,** such as utility bills and entertainment, change from month to month.

example

Estimate the amount that should be budgeted monthly for each item.

a. Lamar Johnson's property taxes for the year are $1850.

b. Mary Coulter spent $72, $88, and $67 for home maintenance in the last three months.

solution

a. Divide the annual amount by the number of months in a year.

$\$1850 \div 12 \approx \$154.17$

Round $154.17 *up* to $160.

Lamar should budget about $160 monthly for property taxes.

b. Divide the total of the given amounts by the number of months.

$\$72 + \$88 + \$67 = \$227$

$\$227 \div 3 \approx \$75.67$

Mary should budget about $80 monthly for home maintenance.

exercises

Estimate the amount that should be budgeted monthly for each item. A calculator may be helpful.

1. Sarah Goldberg's auto insurance premium for the year is $585.
2. Brad Mason's life insurance premium for the year is $460.
3. Ed Maka spent $39, $46, $55, and $37 for clothes in the past four months.
4. The Ortagas saved $136, $109, $113, $92, and $96 in the past five months.
5. Cindy Franks spent $14.95, $43.24, $31.98, $11.98, and $29.42 for gifts in the past five months.
6. The Harts spent $432.88, $377.29, $454.35, $362.12, $458.68, and $427.77 for food in the past six months.

When you plan a budget, it is sometimes helpful to use a circle graph.

Golden Family Budget

Rent 25%
Other 10%
Commuting 10%
Clothing 10%
Food 20%
Insurance 4%
Savings 6%
Utilities 15%

The Golden family's monthly net income is $2400. Use the circle graph at the right to find the amount of money that they budget monthly for each item.

7. Rent
8. Food
9. Utilities
10. Insurance
11. Clothing
12. Commuting
13. Savings
14. Other

Ellen Li has budgeted her $2000 monthly net income as shown in Exercises 15–22. Estimate the *percent* of her income that she has budgeted for each item.

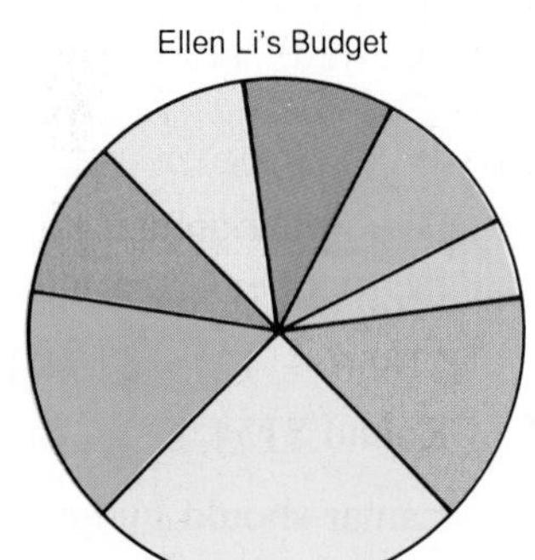
Ellen Li's Budget

15. Rent: $515
16. Food: $220
17. Utilities: $275
18. Insurance: $110
19. Clothing: $190
20. Commuting: $215
21. Savings: $200
22. Other: $275
23. Copy the circle graph for Ellen Li's budget, shown at the right. Use your answers to Exercises 15–22 to label the graph.
24. Record your expenses for two weeks. Include items such as food and entertainment. Use your record to plan a personal budget.

calculator corner

It is sometimes helpful to use a calculator in working with budgets that involve greater amounts of money.

Department Budget	
Salaries	$903,000
Benefits	$180,600
Travel	$241,500
Supplies	$136,500
Mailing	$211,400
Overhead	$567,000
TOTAL	$2,240,000

example Use the table at the right. Find the percent of the budget that is planned for salaries.

solution Enter 903000 ÷ 2240000 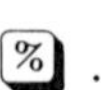 .

The result is 40.3125.

About 40.3% of the budget is planned for salaries.

Use a calculator to find the percent of the budget that is planned for each item. Round your answer to the nearest tenth of a percent.

1. Benefits
2. Travel
3. Supplies
4. Mailing
5. Overhead

CHAPTER REVIEW

vocabulary vo·cab·u·lar·y

Choose the correct word or phrase to complete each sentence.

1. A plan for managing your expenses is called a (*budget, circle graph*).
2. On a pictograph, a (*key, scale*) tells how many items each symbol represents.

skills

Use the double bar graph at the right.

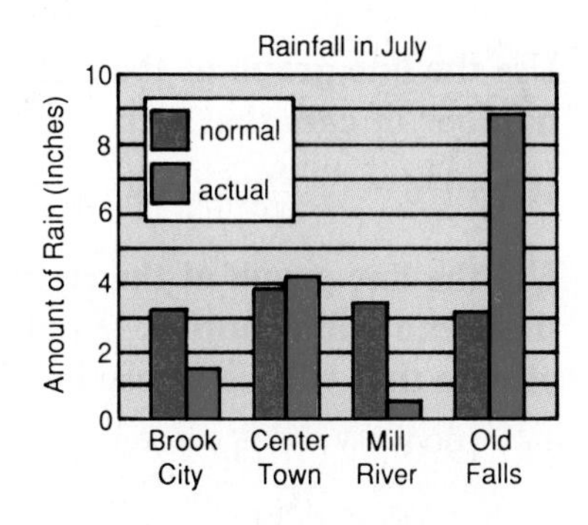

3. Estimate the normal amount of rainfall for Old Falls in July.
4. Estimate the actual amount of rainfall for Mill River in July.
5. In which cities was the actual amount of rainfall for July greater than the normal amount?
6. In which city was the normal amount of rainfall for July about twice the actual amount?

Use the double line graph at the right.

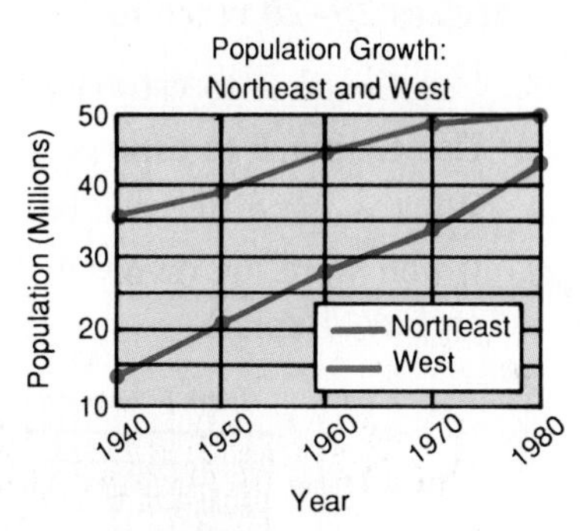

7. Estimate the population of the West in 1950.
8. Estimate the population of the Northeast in 1960.
9. Between which two given years was there the least increase in the population of the Northeast?
10. Estimate the difference between the population of the West and the population of the Northeast in 1940.

Use the pictograph at the right.

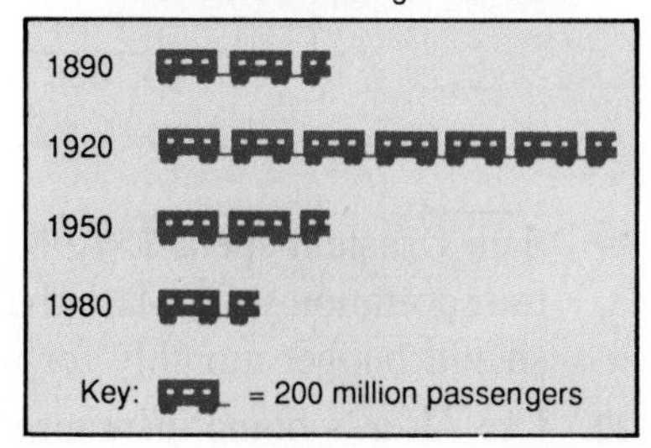

11. Estimate the number of railroad passengers that there were in 1890.
12. Estimate the number of railroad passengers that there were in 1980.
13. About how many more railroad passengers were there in 1920 than in 1950?
14. How many symbols would be used for a year in which there were about 900 million railroad passengers?

Use the circle graph at the right.

15. If the total amount spent for advertising was about $120 billion, estimate the amount that was spent for advertising on television.
16. If about $10.5 billion was spent for advertising on radio, estimate the total amount that was spent for advertising.
17. Which two types of advertising together make up about one eighth of the advertising dollar?
18. Estimate the fraction of the advertising dollar that is spent on advertising in newspapers.

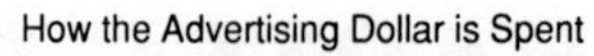

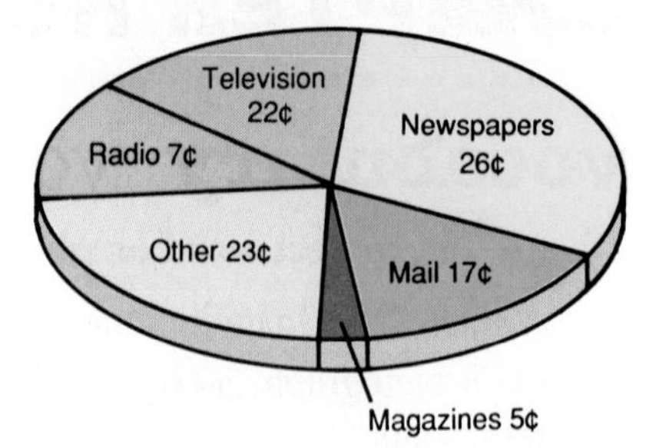

Use the line graph at the right. Estimate the number of cars that were in use in each year.

19. 1976 20. 1980 21. 1984

Use the line graph at the right. Assume that the trend will continue. Predict the number of cars that will be in use in each year.

22. 1990 23. 1994 24. 1998

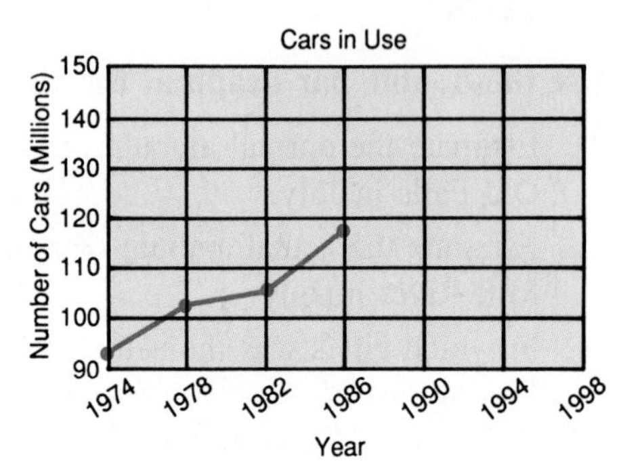

Exercises 25–28 refer to *Graph A* and *Graph B*, shown below.

25. Use *Graph A* to estimate the average cost of dinner at each restaurant.
26. Use *Graph B* to estimate the average cost of dinner at each restaurant.
27. What is the difference between *Graph A* and *Graph B*?
28. If you were the owner of Sandy's Kitchen, which graph would you prefer? Explain.

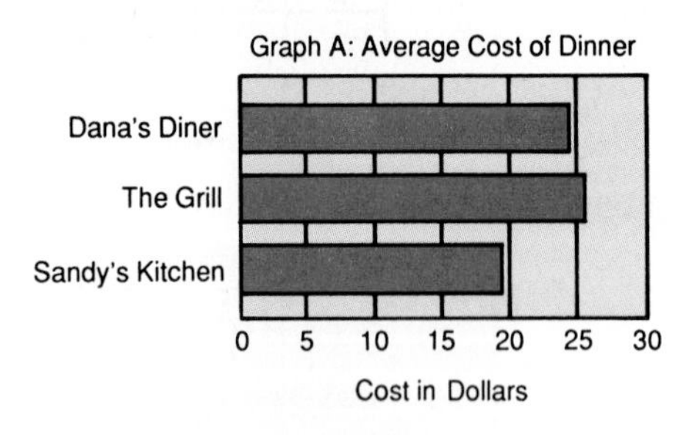

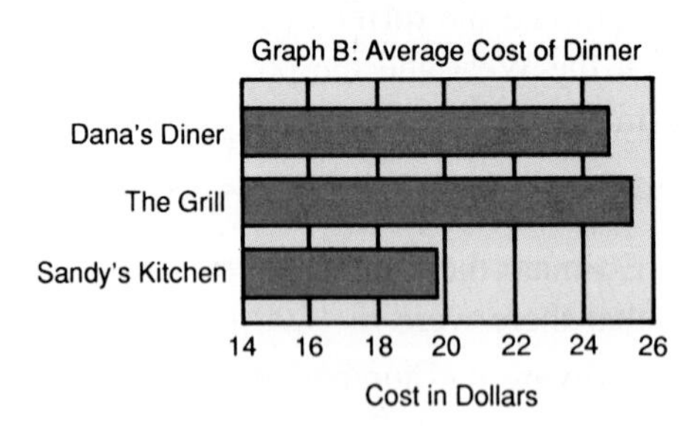

29. Sam Ornstein spent $54.96, $69.81, $65.37, $76.18, and $61.90 for transportation in the last five months. Estimate the amount that Sam should budget monthly for transportation.
30. Lee Allen's home insurance premium for the year is $912. Estimate the amount that Lee should budget monthly for home insurance.

CHAPTER TEST

Use the bar graph at the right. Estimate the length of each coastline.

16-1

1. Arctic
2. Atlantic
3. Gulf
4. Pacific
5. About how many miles longer than the Gulf coastline is the Pacific coastline?
6. Which coastline is about twice as long as the Arctic coastline?

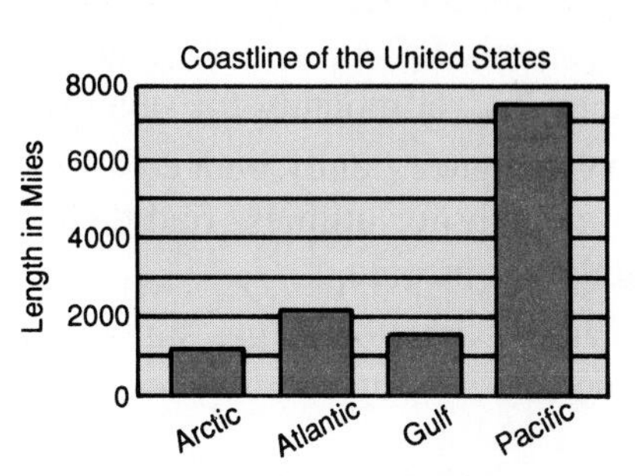

Draw a bar graph to display the given data.

7. **Small Town Populations**

Town	Estes	Midville	Newtown	Oldville
Population (Thousands)	24	49	12	36

Use the line graph at the right. Estimate the number of pieces of mail that were handled per day in each year.

16-2

8. 1970
9. 1975
10. 1980
11. 1985
12. In which two given years was there about the same amount of mail handled per day?
13. Between which two given years was there the greatest increase in the amount of mail handled per day?

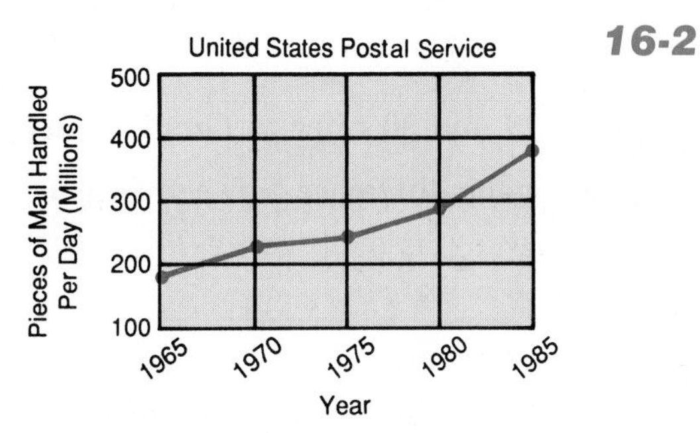

Use the pictograph at the right. Estimate the number of visits to the national park system in each year.

16-3

14. 1970
15. 1975
16. 1980
17. 1985
18. About how many more visits to the national park system were there in 1975 than in 1970?
19. In which given year were there about twice as many visits to the national park system as there were in 1970?

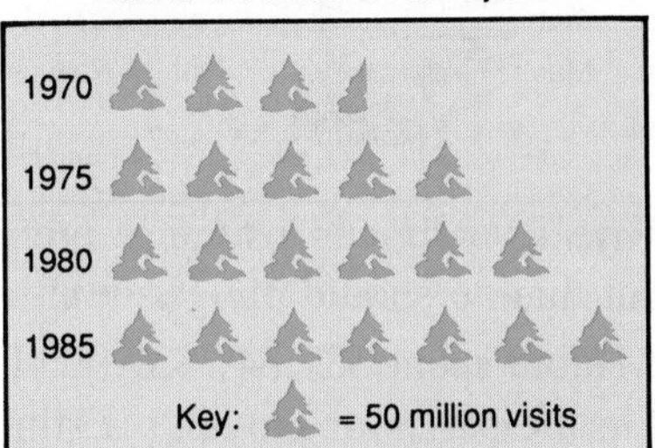

The circle graph at the right shows how the Rodriguez family budget their monthly net income.

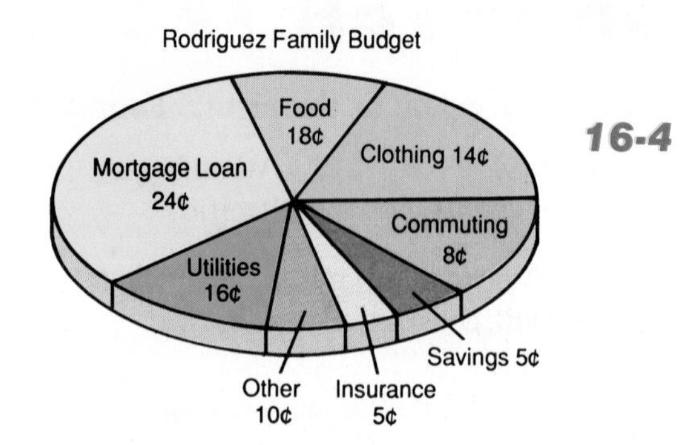

16-4

20. If the Rodriguez family's monthly net income is $2200, find the amount that they budget monthly for savings.
21. If the Rodriguez family budget $320 monthly for utilities, find their monthly net income.
22. Estimate the fraction of the family's income that is budgeted for food.

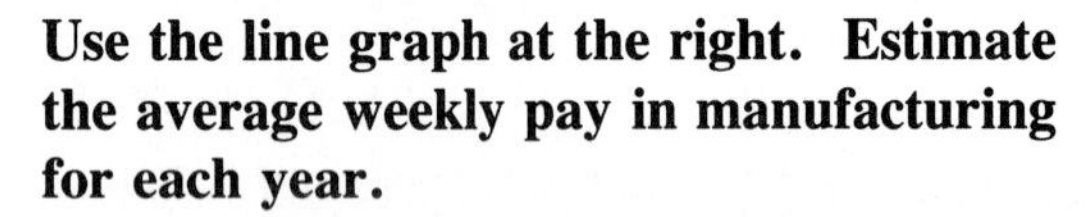

Use the line graph at the right. Estimate the average weekly pay in manufacturing for each year.

16-5

23. 1976 24. 1980 25. 1984

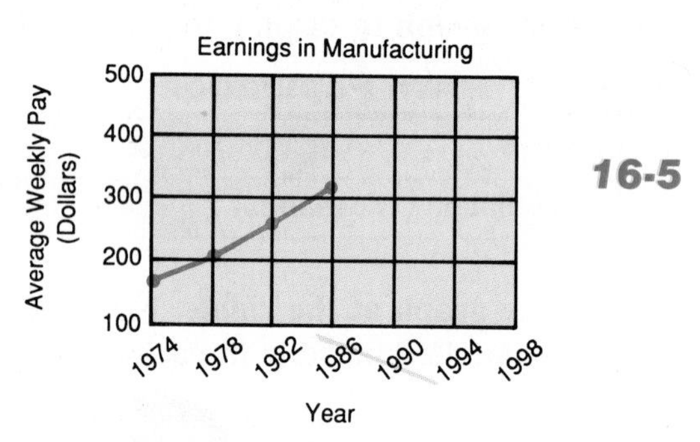

Use the line graph at the right. Assume that the trend will continue. Predict the average weekly pay for each year.

26. 1990 27. 1994 28. 1998

Exercises 29 and 30 refer to *Graph A* and *Graph B*, shown below.

16-6

29. What is the difference between *Graph A* and *Graph B*?
30. If you were the publisher of *TV Talk,* would you prefer *Graph A* or *Graph B*? Explain.

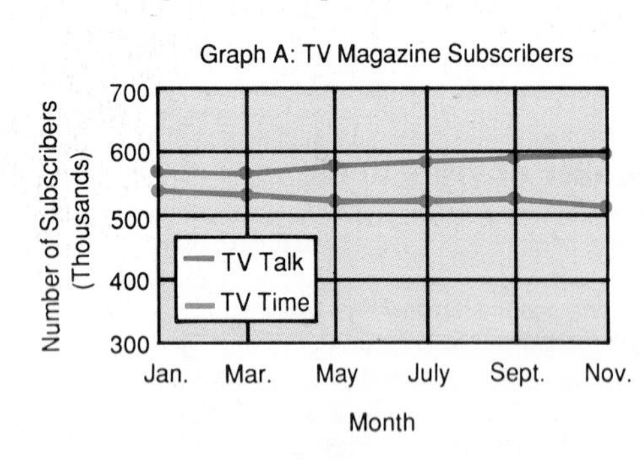

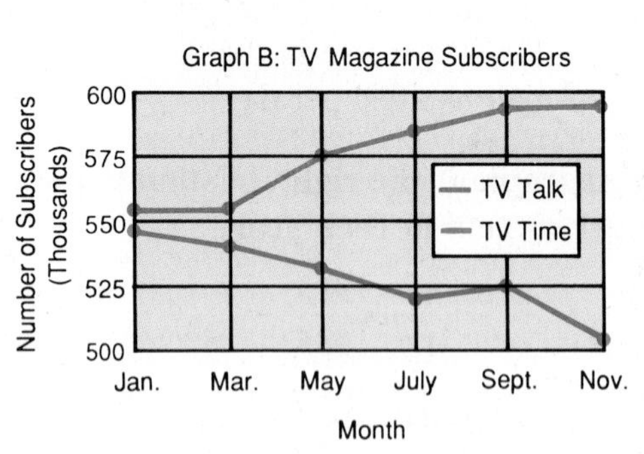

16-7

31. Bill Frye's yearly car insurance premium is $592. Estimate the amount that he should budget monthly for car insurance.
32. Ellen Habib spent $37.54, $29.90, $43.08, and $34.50 for entertainment in the past four months. Estimate the amount that she should budget monthly for entertainment.

Doctors use many types of instruments to help them collect data, but diagnosis and treatment depend on the correct interpretation of the data.

CHAPTER 17

INTERPRETING DATA

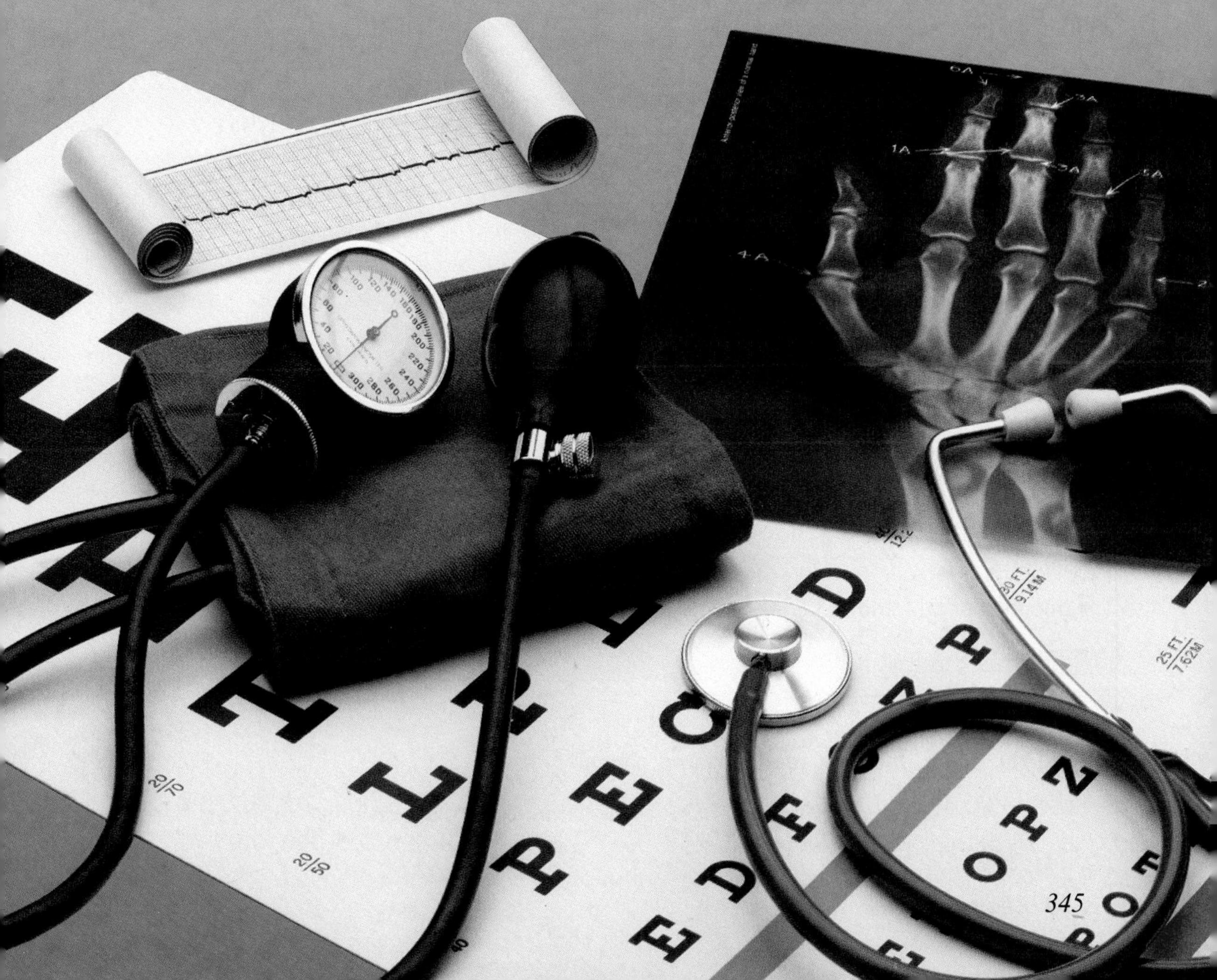

17-1 MEAN, MEDIAN, MODE, AND RANGE

Statistics is the branch of mathematics that deals with organizing and analyzing data. Four *statistical measures* used to describe a set of data are the *mean, median, mode,* and *range*.

Data: 84 68 68 72

The **mean** is the sum of the items in a set of data divided by the number of items. The mean is also called the *average*.

- mean $= \frac{84 + 68 + 68 + 72}{4}$ $= 73$

The **median** is the middle item when the data are arranged in numerical order. If the number of items is even, the median is the mean of the two middle items.

- 68 68 72 84

  median $= \frac{68 + 72}{2} = 70$

The **mode** is the item that appears most often. There can be more than one mode. If each item appears only once, there is no mode.

- 68 68 72 84

  mode = 68

The **range** is the difference between the greatest and least values of the data.

- range $= 84 - 68$ $= 16$

example 1

Find the mean, median, mode(s), and range.

4.5, 6.3, 5.5, 5.4, 6.5, 5.4, 6.3

solution

- $4.5 + 6.3 + 5.5 + 5.4 + 6.5 + 5.4 + 6.3 = 39.9$ ← **Find the sum. A calculator may be helpful.**

  mean $= \frac{39.9}{7} = 5.7$ ← **Divide by the number of items.**

- 4.5, 5.4, 5.4, 5.5, 6.3, 6.3, 6.5 ← **List the items in order.**

  median = 5.5

- mode(s) = 5.4 and 6.3 ← **Both 5.4 and 6.3 appear twice.**
- range $= 6.5 - 4.5 = 2.0$

your turn

Find the mean, median, mode(s), and range.

1. 94, 78, 89, 61, 78, 78, 61
2. 3.6, 4.1, 4.5, 8.0, 7.5, 3.7, 5.0
3. 57, 68, 65, 75, 57, 74
4. 4.7, 9.0, 8.6, 7.5, 8.6, 9.0

example 2

Find the mean, median, mode(s), and range.

\$10.70, \$8.30, \$9.25, \$10.55, \$9.65, \$8.80

solution

- $\$10.70 + \$8.30 + \$9.25 + \$10.55 + \$9.65 + \$8.80 = \$57.25$

  $\text{mean} = \frac{\$57.25}{6} \approx \$9.54$ ← **Round to the nearest cent.**

- \$8.30, \$8.80, \$9.25, \$9.65, \$10.55, \$10.70 ← **List the items in order.**

  $\text{median} = \frac{\$9.25 + \$9.65}{2} = \frac{\$18.90}{2} = \$9.45$

- mode = *none* ← **No item appears more than once.**
- $\text{range} = \$10.70 - \$8.30 = \$2.40$

your turn

Find the mean, median, mode(s), and range. Round the mean to the nearest cent or to the nearest tenth.

5. \$102, \$91, \$96, \$84, \$91

6. \$8.83, \$8.25, \$9.61, \$10.15, \$7.92

7. 36, 38, 44, 41, 49, 42

8. 15.1, 19.2, 17.3, 17.8, 19.2, 9.9

practice exercises

practice for example 1 (page 346)

Find the mean, median, mode(s), and range.

1. 66, 70, 68, 70, 62, 68, 72

2. 45, 65, 44, 45, 27, 45, 65

3. 98, 59, 42, 51, 41, 75

4. 36, 99, 47, 82, 43, 47

5. 13.4, 11.1, 15.5, 12.4, 12.6

6. 9.2, 16.4, 10.7, 13.8, 12.4

7. 8.4, 2.4, 3.6, 8.4, 3.6, 1.2

8. 10.0, 7.9, 3.4, 6.1, 3.4, 10.0

practice for example 2 (page 347)

Find the mean, median, mode(s), and range. Round the mean to the nearest cent or to the nearest tenth.

9. \$122, \$107, \$99, \$107, \$86, \$99

10. \$450, \$785, \$295, \$575, \$480, \$195

11. \$7.29, \$3.37, \$5.98, \$7.29, \$5.03

12. \$.98, \$.43, \$.82, \$.98, \$.43

13. 63, 68, 87, 84, 63, 74

14. 86, 32, 97, 57, 15, 32

15. 1.9, 1.3, 3.8, 1.8, 1.1, 3.7, 1.6

16. 3.6, 2.4, 1.7, 3.6, 2.4, 2.9, 3.6

mixed practice (pages 346–347)

Find the mean, median, mode(s), and range. If necessary, round the mean to the nearest cent or to the nearest tenth.

17. 5.8, 1.3, 5.9, 4.7, 7.7, 4.7

18. 50, 45, 50, 35, 45, 50

19. $150, $192, $173, $189, $192

20. $25, $58, $61, $59, $28, $34

21. 9.6, 9.4, 9.6, 9.4, 9.6, 9.4

22. 74, 74, 74, 74, 74

23. $.53, $.53, $.53, $.35, $.35, $.35

24. $2.29, $2.92, $9.22, $9.22

Salaries of the owner and employees of a video store are listed below.

owner	$72,000
manager	$34,000
assistant manager	$32,000
head clerk	$22,000
clerk	$16,000
clerk	$10,000
clerk	$10,000

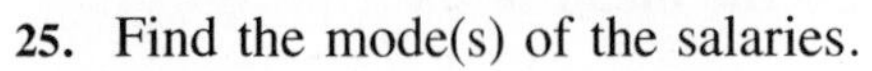

25. Find the mode(s) of the salaries.

26. What is the median salary?

27. What is the mean salary?

28. Which of the mode, median, and mean best represents the salaries? Explain.

In a shipment of 200 pumpkins, the mean weight of a pumpkin is $7\frac{1}{2}$ lb, the median weight is 7 lb, and the mode of the weights is 8 lb.

29. What would you expect four pumpkins to weigh?

30. A customer wants twenty pumpkins of about the same size. What would be the best weight to look for?

31. Suppose you selected the fifteen heaviest pumpkins in the shipment. What would you expect to be true of the weight of each?

32. Jan received scores of 92, 84, and 91 on her science tests. What score must she receive on the next test to have a mean score of 90?

review exercises

Write the numeral form of each number.

1. 3.5 million

2. 83.6 thousand

3. 100.1 billion

4. 62.25 million

5. 10 million, 320

6. 42 and 16 thousandths

7. thirty-eight ten-thousandths

8. six hundred thousand, two

17-2 FREQUENCY TABLES

Data can be organized in a **frequency table** like the one at the right. The **frequency** of an item is the number of times the item appears in the data. In a frequency table, a *tally mark* is made each time an item appears in the data. For every fifth tally of the same item, a mark is made diagonally across four (~~||||~~) to make counting easier.

example 1

Make a frequency table for the given data.

Quiz Scores

8 9 8 8 6 7 10
7 8 7 6 9 9 8

solution

- Set up a table organized into columns.
- Find the different numbers in the data and order them from least to greatest. List them in the first column.
- Make appropriate tally marks in the second column.
- Count the tally marks for each number and record the frequency in the third column.

Quiz Scores

Score	Tally	Frequency
6	||	2
7	|||	3
8	~~||||~~	5
9	|||	3
10	|	1

your turn

Make a frequency table for the given data.

1. **Number of Games Won**

9 11 9 10 8 10 12
9 12 11 9 10 10 12
10 9 10 11 9 9 10
11 8 12 12 11 12 12

2. **Typing Rates (Words/Min)**

65 70 65 55 65 65 70
55 65 50 70 70 55 55
55 55 70 50 70 65 70
45 65 70 60 65 70 55

Sometimes the range of a set of data is great or few numbers in the data repeat. You may then find it useful to group the data in equal *intervals* when you make a frequency table.

example 2

Make a frequency table for the given data.

Prices of Airline Tickets ($)

229	249	209	319	458	355	229
425	319	229	389	319	257	265

solution

There are 14 different numbers in the data.

- Determine the least price, $209, and the greatest price, $458.
- Decide upon equal intervals to include all the data. List the intervals in the first column.
- Complete the tallies and count the frequencies.

Prices of Airline Tickets ($)

Price	Tally	Frequency
200–299	卌 II	7
300–399	卌	5
400–499	II	2

your turn

Make a frequency table for the given data.

3. **Scores on a Spanish Test (%)**

87	64	73	94	77	79	83	82
68	77	91	80	75	83	79	74

(Use intervals such as 91–100.)

4. **Ages of Baseball Players**

27	34	37	41	29	22	27
31	23	40	34	28	25	31

(Use intervals such as 20–24.)

practice exercises

practice for example 1 (page 349)

Make a frequency table for the given data.

1. **Ages of Band Members**

15	18	17	15	15	16	17	17
18	15	16	16	17	16	18	16
16	16	15	17	16	15	16	15
17	16	17	18	18	17	18	17

2. **Student Heights (in.)**

62	64	61	65	60	63	65	61
65	62	65	60	63	64	61	62
60	64	65	63	62	61	63	61
64	64	63	62	65	61	62	61

practice for example 2 (page 350)

Make a frequency table for the given data.

3. **Lengths of Bridges (m)**

488	549	608	525	407	712
853	701	655	704	533	564

(Use intervals such as 501–600.)

4. **Number of Points Scored**

64	82	76	86	68	90	64	79
93	72	78	72	78	94	76	88

(Use intervals such as 60–64.)

mixed practice (pages 349–350)

Make a frequency table for the given data.

5. **Prices of Records ($)**

7.99	10.99	8.99	7.99	9.99
9.99	7.99	8.99	7.99	8.99
7.99	8.99	10.99	10.99	10.99

6. **Yards Gained Per Game**

48	47	44	52
43	73	34	25
36	55	62	85

7. **Seeds Per Packet**

25	27	28	25	25
26	27	26	25	28
25	25	26	26	26

8. **Winning Percentages**

35.4	54.9	52.2	72.0	34.1	45.1
68.5	62.0	37.8	53.0	48.8	77.3
57.0	36.6	65.8	52.2	56.0	38.9

9. Use an almanac or other reference to find the population of each of the fifty United States. Work together with a classmate to make a frequency table for these data.

review exercises

Tell whether it would be more appropriate to draw a *bar graph* or a *line graph* to display the given data. Then draw the graph.

1. **NCAA Football Championship Titles**

College	Alabama	Miami	Notre Dame	Oklahoma	Texas
Number of Titles	5	1	7	6	2

2. **Population of the United States**

Year	1820	1860	1900	1940	1980
Number of People Per Square Mile	6	10	26	44	63

mental math

The numbers 89, 85, 92, 94, and 88 are each close to 90. The data *cluster* around 90, so you can mentally estimate that the mean is about 90.

Mentally estimate the mean.

1. 73, 75, 66, 68, 71
2. 117, 115, 121, 126, 124
3. $23.88, $24.29, $23.12, $25.23
4. 20.4, 20.3, 21.2, 21.5, 22.1
5. $16\frac{2}{7}$, $16\frac{3}{7}$, $15\frac{5}{7}$, $15\frac{4}{7}$, $16\frac{4}{7}$
6. $3\frac{7}{9}$, $4\frac{2}{9}$, $4\frac{4}{9}$, $4\frac{1}{3}$, $3\frac{5}{9}$

17-3 HISTOGRAMS AND FREQUENCY POLYGONS

The data in frequency tables are often displayed in a graph. When a *bar graph* is used to show frequencies, it is called a **histogram.** A histogram is different from other bar graphs in that no space is left between the bars. For instance, the data in the frequency table below is displayed in the histogram to its right.

Quiz Scores

Score	Tally	Frequency
7	III	3
8	~~IIII~~ I	6
9	III	3
10	II	2

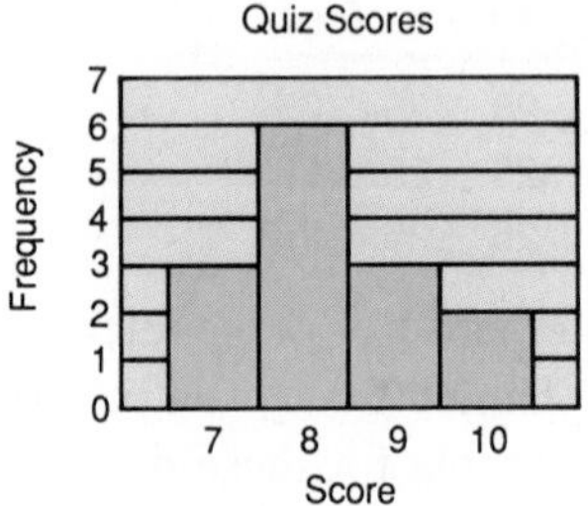

example 1

Use the histogram at the right.

a. How many scores are in the interval 91–100?

b. At what interval are there exactly six scores?

c. What is the increase in frequency from the interval 61–70 to the interval 71–80?

d. How many scores are less than 91?

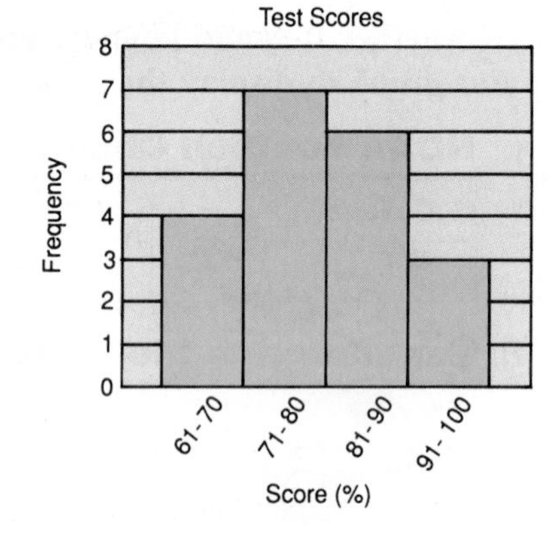

solution

a. There are three scores in the interval 91–100.

b. Exactly six scores occur at the interval 81–90.

c. The frequency increases from 4 to 7. This is an increase of 3.

d. The number of scores less than 91 is $4 + 7 + 6 = 17$.

your turn

Use the histogram above.

1. How many scores are in the interval 71–80?
2. At what interval are there exactly four scores?
3. What is the decrease in frequency from the interval 81–90 to the interval 91–100?
4. How many scores are greater than 80?

When a *line graph* is used to show frequencies, it is called a **frequency polygon.** A frequency polygon is different from other line graphs in that it is connected to the horizontal axis at both ends.

example 2

Use the frequency polygon below.

a. How many students are there in all?
b. What is the median height?
c. What is the mean height?
d. Find the mode(s) of the heights.

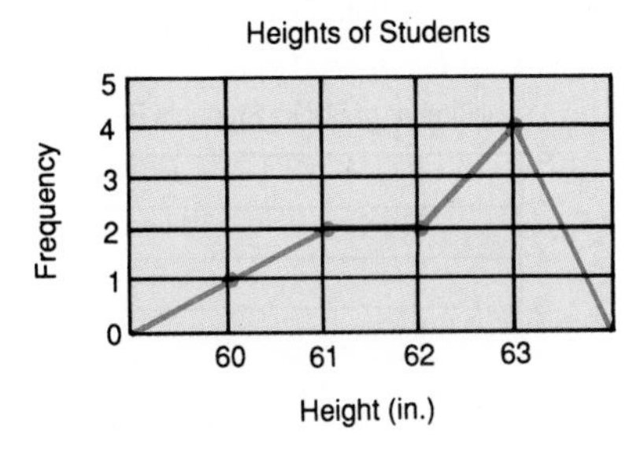

solution

a. Find the sum of the frequencies: $1 + 2 + 2 + 4 = 9$
There are 9 students in all.

b. There are 9 heights represented. The median is the fifth height.
The median height is 62 in.

c. Multiply each height by its frequency and add the products.
$1 \times 60 + 2 \times 61 + 2 \times 62 + 4 \times 63 = 558$ ← **A calculator may be helpful.**
Divide this sum by the sum of the frequencies.
$558 \div 9 = 62$
The mean height is 62 in.

d. The mode is the height that appears most frequently.
The mode of the heights is 63 in.

your turn

Use the frequency polygon at the right.

5. How many students are there in all?
6. What is the median age?
7. What is the mean age?
8. Find the mode(s) of the ages.

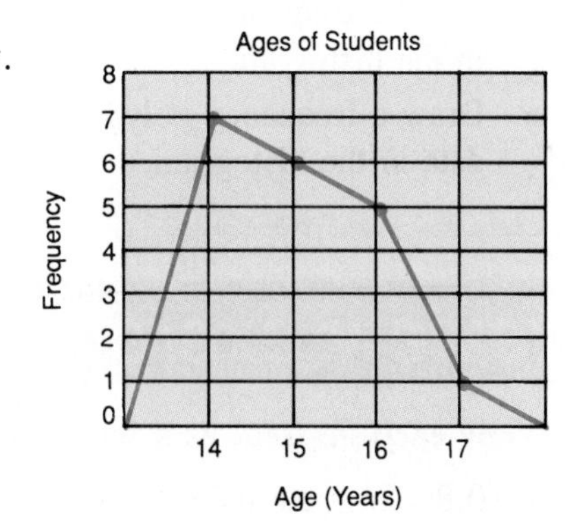

practice exercises

practice for example 1 (page 352)

Use the histogram at the right.

1. How many students own 11–20 tapes?
2. What is the decrease in frequency from the interval 11–20 to the interval 21–30?
3. How many students own more than ten tapes?
4. At which intervals do the same number of students own tapes?

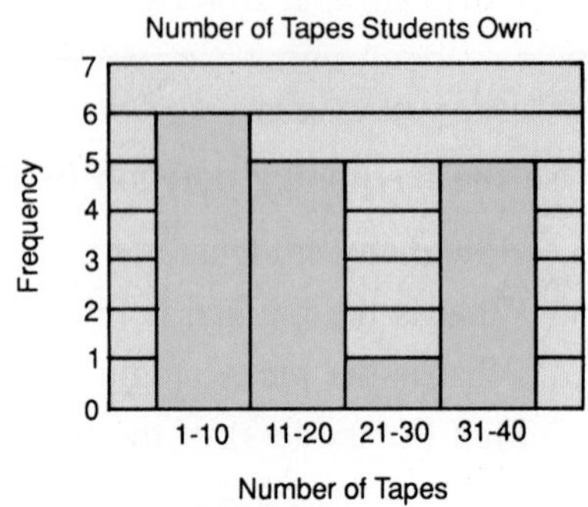

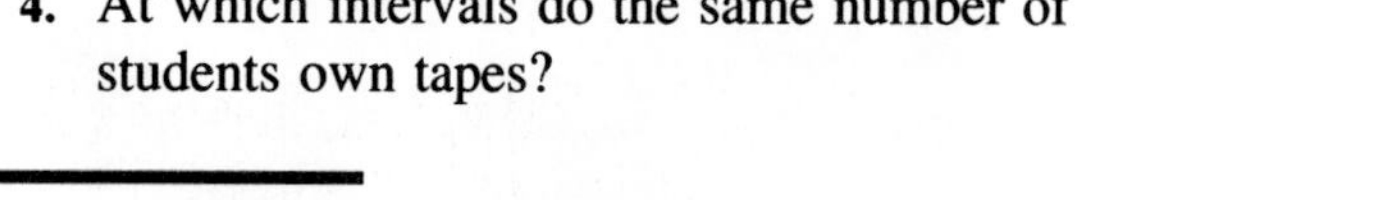

practice for example 2 (page 353)

Use the frequency polygon at the right.

5. How many students are there in all?
6. What is the median number of books read?
7. What is the mean number of books read?
8. Find the mode(s) of the number of books read.

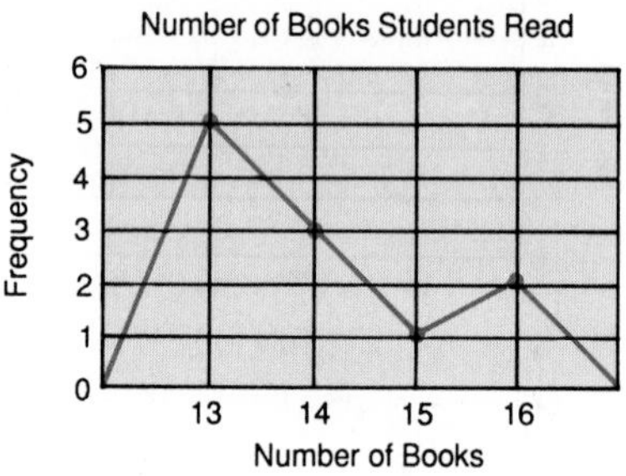

mixed practice (pages 352–353)

Use the histogram at the right.

9. How many teams are there in all?
10. How many teams scored exactly five runs?
11. What is the median number of runs?
12. What is the mean number of runs?
13. Find the mode(s) of the number of runs.
14. What is the range of the number of runs?
15. Make a frequency table for the data in the histogram.
16. Draw a frequency polygon for the data in the histogram.

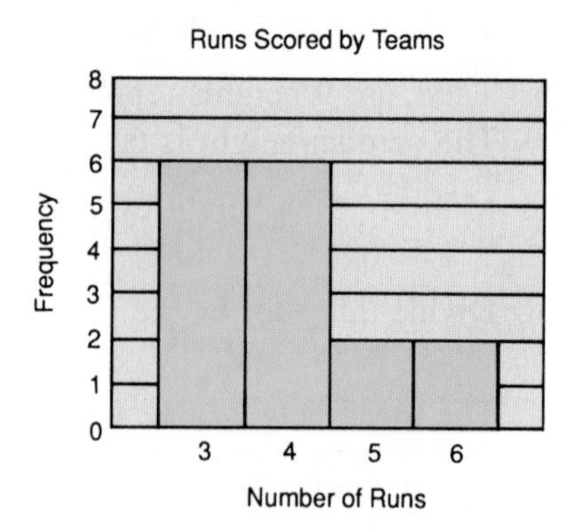

review exercises

Write each decimal as a fraction or mixed number in lowest terms.

1. 0.8
2. 0.03
3. 16.5
4. 0.011
5. 2.125
6. 7.625

17-4 SCATTERGRAMS

A **scattergram** is a graph that shows the relationship between *two* sets of data. Like a line graph, a scattergram has a numerical scale along each axis. A scattergram is different from a line graph because there can be more than one point for a given number on either scale. You do not connect the points in a scattergram.

example 1

Use the scattergram at the right.

a. What is the range of the heights?

b. What is the weight of the student who is 60 in. tall?

c. How many students weigh 169 lb?

d. Find the mode(s) of the weights.

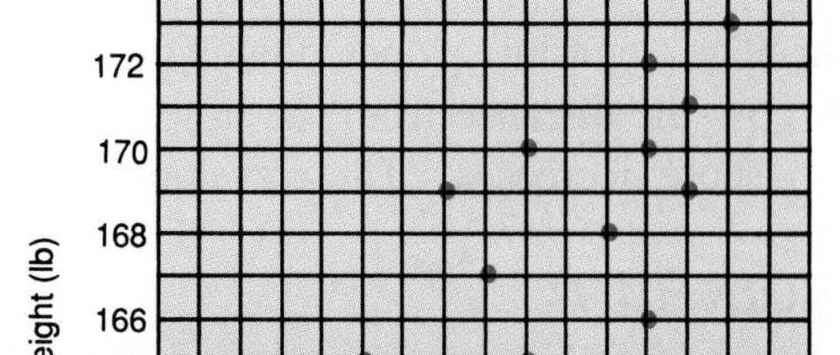
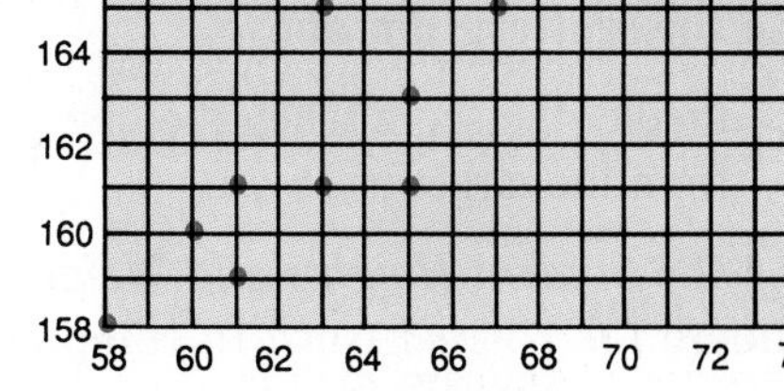

solution

a. The heights vary from 58 in. to 74 in.
The range is $74 - 58 = 16$ in.

b. Locate 60 on the horizontal axis.
Move up to the red point, then left to the vertical axis.
The student weighs 160 lb.

c. Locate 169 on the vertical axis. Move right and count the points.
Two students weigh 169 lb.

d. Three students weigh 161 lb. The mode is 161 lb.

your turn

Use the scattergram above.

1. What is the range of the weights?
2. What is the height of the student who weighs 173 lb?
3. How many students are 65 in. tall?
4. Find the mode(s) of the heights.

On some scattergrams the data lie close to a single line. This line is called the **trend line.** If the trend line slopes *upward* to the right, there is a **positive correlation** between the sets of data. If the trend line slopes *downward* to the right, there is a **negative correlation.** You can use the trend line to predict further information.

example 2

Use the scattergram at the right.

a. Predict the score when the distance from the target is 18 yd.

b. Is there a *positive* or *negative* correlation?

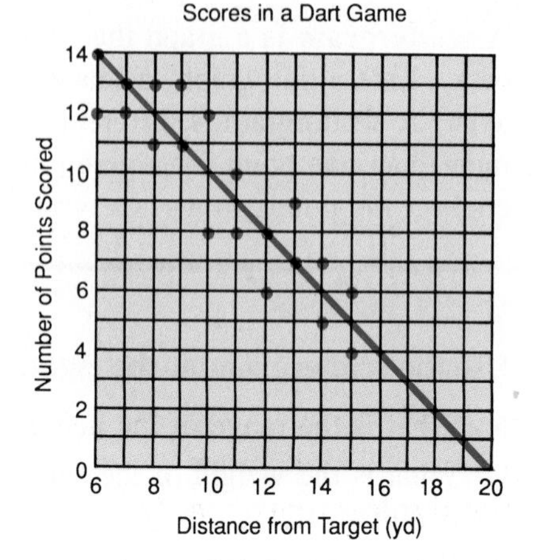

solution

a. Locate 18 on the horizontal axis.
Move up to the blue trend line, then left to the vertical axis.
The score should be about 2 points.

b. The trend line slopes downward to the right. As the distance from the target increases, the score decreases. The correlation is *negative*.

your turn

Use the scattergram above.

5. Predict the score when the distance from the target is 16 yd.

6. Predict the score when the distance from the target is 19 yd.

Use the *Student Heights and Weights* scattergram on page 355. Choose the correct word to complete each sentence.

7. As the heights increase, the weights (*increase, decrease*).

8. The correlation between the heights and weights is (*positive, negative*).

practice exercises

practice for example 1 (page 355)

Use the scattergram at the right.

1. What is the range of the number of gallons of gasoline used?

2. How many miles were traveled by the automobile that used nine gallons of gasoline?

3. How many automobiles traveled exactly 100 mi?

4. Find the mode(s) of the miles traveled.

5. Find the mode(s) of the gallons of gasoline used.

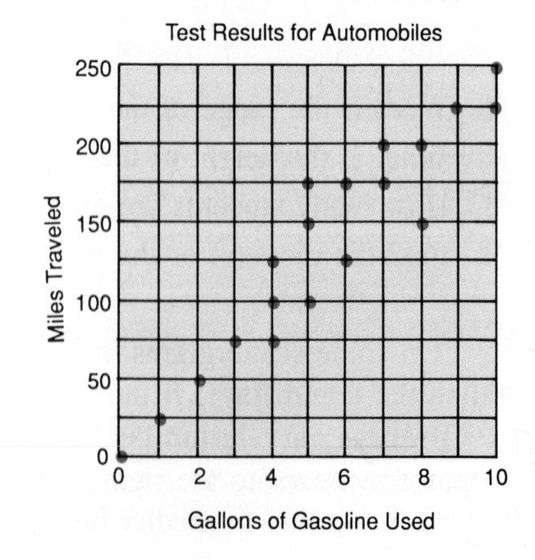

practice for example 2 (page 356)

Use the scattergram at the right.

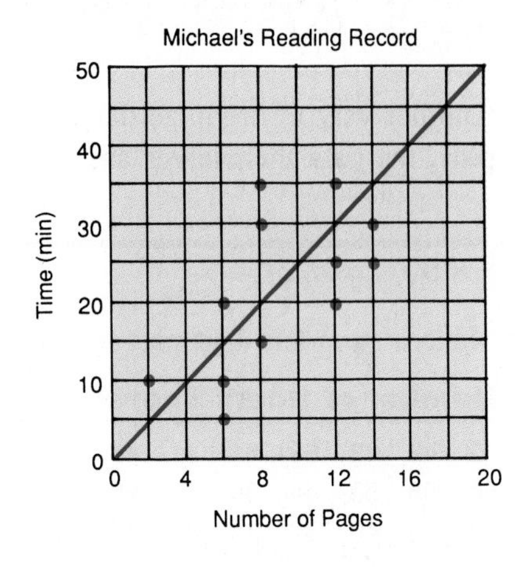

6. Predict the time spent reading 16 pages.
7. Predict the time spent reading 18 pages.

Choose the correct word to complete each sentence.

8. As the number of pages increases, the amount of time (*increases, decreases*).
9. The correlation between the number of pages and amount of time is (*positive, negative*).

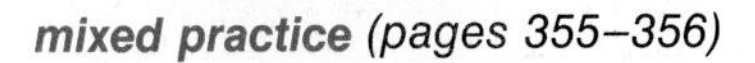

mixed practice (pages 355–356)

Use the scattergram at the right.

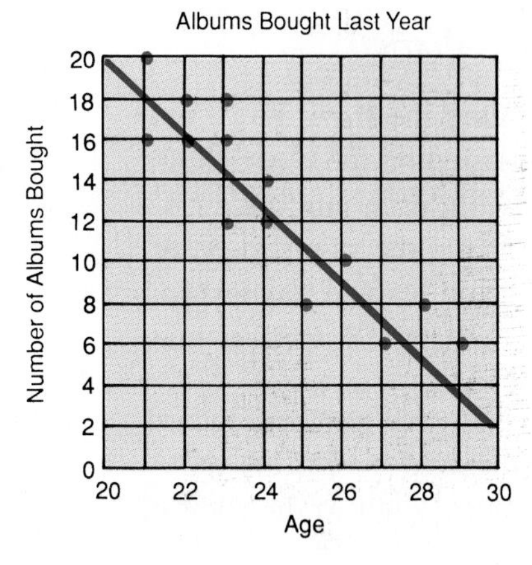

10. Find the mode(s) of the number of albums bought.
11. How old is the person who bought exactly fourteen albums?
12. How many albums did the 29-year-old buy?
13. How many people bought exactly six albums?
14. How many people were polled in all?
15. Use the trend line to predict how many albums a 30-year-old probably bought.
16. Is there a positive correlation or a negative correlation between the sets of data? Explain.

17. Draw a scattergram to display the *Ages and Heights* data given below. Is it possible to draw a trend line for this data?

Ages and Heights

Age (years)	50	10	20	30	20	40	50	10	30	30	30	40	40	10	50
Height (in.)	64	60	68	58	64	66	70	56	64	66	70	70	62	58	60

review exercises

Tell whether there is an *increase* or *decrease*. Then find the percent of increase or decrease.

1. original amount: 14
   new amount: 21
2. original price: $48
   new price: $30
3. original earnings: $95
   new earnings: $190
4. original length: 300 m
   new length: 200 m
5. original weight: 70 kg
   new weight: 56 kg
6. original salary: $500
   new salary: $625

17-5 STEM-AND-LEAF PLOTS

Another way to organize data is a *stem-and-leaf plot.* In a **stem-and-leaf plot,** the data themselves are used to make a frequency display.

example 1

Make a stem-and-leaf plot for the given data.

Number of Goals Scored

52	48	48	68	61	41	47	43
46	58	53	54	36	31	59	38
34	45	46	36	37	36	44	44

solution

Find the *stems* by dropping the ones' digit from each number. Write the stems in order from least to greatest to the left of a vertical line.

3
4
5
6

For each number, record the *leaf* by writing the ones' digit to the right of the stem.

stems	leaves
3	6, 1, 8, 4, 6, 7, 6
4	8, 8, 1, 7, 3, 6, 5, 6, 4, 4
5	2, 8, 3, 4, 9
6	8, 1

Rearrange the leaves in order from least to greatest. Title the stem-and-leaf plot.

Number of Goals Scored

3	1, 4, 6, 6, 6, 7, 8
4	1, 3, 4, 4, 5, 6, 6, 7, 8, 8
5	2, 3, 4, 8, 9
6	1, 8

your turn

Make a stem-and-leaf plot for the given data.

1. **Scores on a Science Test**

68	73	84	72	90	81	64	70	85	82
61	70	73	82	90	68	84	77	93	77

2. **Number of Touchdowns**

113	105	90	93	91	126	116	90	100	120

example 2

Use the stem-and-leaf plot at the right.

a. What is the greatest batting average?

b. How many batters had a batting average of 0.302?

c. What is the range of the batting averages?

d. How many batters had a batting average of 0.338?

Leading Batting Averages

Stem	Leaves
0.29	0, 1, 6, 8
0.30	0, 1, 2, 2, 5, 6, 9
0.31	0, 0, 2, 6
0.32	4, 6, 8
0.33	
0.34	
0.35	2, 7

solution

a. The greatest stem is 0.35. The greatest leaf after 0.35 is 7. The greatest batting average is 0.357.

b. There are two 2's after 0.30. Two batters had a batting average of 0.302.

c. The greatest batting average is 0.357. The least is 0.290. $0.357 - 0.290 = 0.067$. The range of the averages is 0.067.

d. There are no leaves to the right of 0.33. No batter had a batting average of 0.338.

your turn

Use the stem-and-leaf plot above.

3. How many batters had a batting average of 0.310?
4. How many batters had a batting average less than 0.320?
5. How many batters had a batting average between 0.340 and 0.350?
6. How many batters had a batting average of 0.345?

practice exercises

practice for example 1 (page 358)

Make a stem-and-leaf plot for the given data.

1. **Number of Home Runs Per Season**

49	37	37	36	40	37	31	48	48
38	36	47	40	48	49	36	47	54
50	61	51	59	58	60	54	47	40

2. **Indianapolis 500 Winning Speeds (mi/h)**

139	143	159	161	161	149	149	159	159	163
158	156	157	153	151	144	151	147	143	140
139	139	136	134	135	128	128	131	129	129

practice for example 2 (page 359)

Use the stem-and-leaf plot at the right.

Team Free-Throw Averages

0.75	4, 4, 6
0.76	0, 0, 0, 2, 2, 2, 6, 8, 9
0.77	2, 5, 6, 8
0.78	0, 3, 5
0.79	0, 6

3. How many teams averaged 0.762?
4. How many teams averaged 0.787?
5. What is the range of the free-throw averages?
6. How many teams averaged less than 0.770?

mixed practice (pages 358–359)

7. Make a stem-and-leaf plot for the given data.

Heights of World's Tallest Dams (m)

220	226	237	242	261	220
242	253	272	237	226	216
250	242	265	221	235	325
220	265	300	285	219	233

Exercises 8–13 refer to the stem-and-leaf plot that you made for Exercise 7.

8. How many dams have a height of 242 m?
9. How many dams have a height of 293 m?
10. What is the height of the tallest dam?
11. How many dams are more than 260 m tall?
12. How many dams are between 240 m and 280 m tall?
13. How many dams are listed in the stem-and-leaf plot?
14. Draw a histogram to display the *Team Free-Throw Averages* data above. Use intervals such as 0.750–0.759.
15. Refer to the histogram that you made for Exercise 14. What do you notice about the number of leaves on the stem-and-leaf plot and the length of the corresponding bar on the histogram?

review exercises

1. Rani borrows $6000 for two years at 12% simple interest per year. How much interest does she pay? What is the total amount due?
2. Allen has $1500 in a savings account that earns 7.5% annual interest compounded semiannually. He makes no deposits or withdrawals. Find the balance at the end of one year.

17-6 BOX-AND-WHISKER PLOTS

The median of a set of data divides the data into a lower half and an upper half. Each half has its own median. The median of the lower half is called the **first quartile.** The median of the upper half is called the **third quartile.** You can use these quartiles to display data in a **box-and-whisker plot.**

example 1

Make a box-and-whisker plot to display the given data.

Nutrition Index Numbers for Breakfast Cereals

65	32	54	42	66	48	56	55	33
49	68	38	54	27	62	50	36	

← **A higher index number means a higher nutritional value.**

solution

Arrange the data from least to greatest. Find the five numbers indicated.

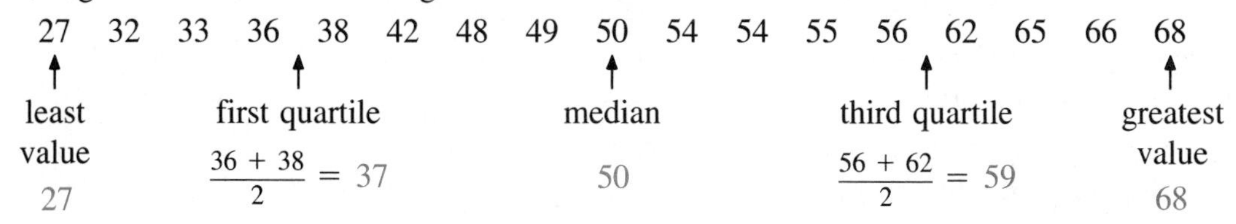

Show the five numbers as dots below a number line.

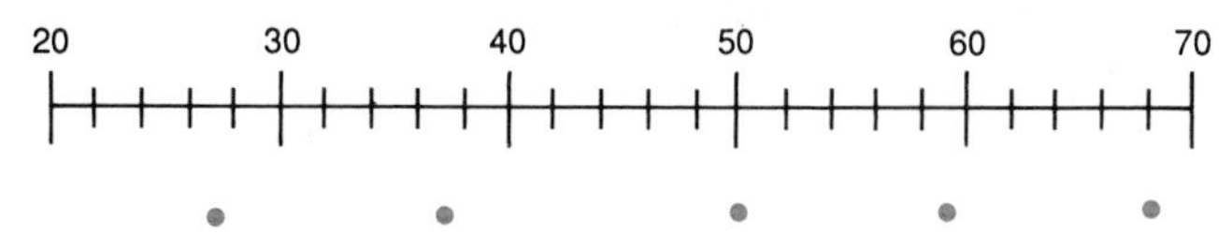

Draw a box with edges at the quartiles. Draw a line through the box at the median.

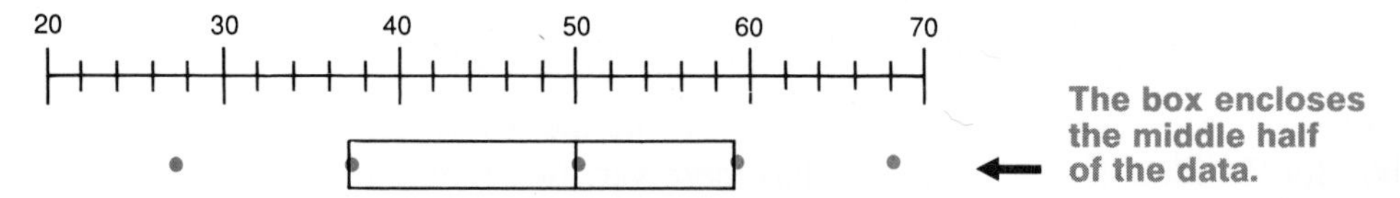

← **The box encloses the middle half of the data.**

Draw "whiskers" from the edges of the box to the points for the least and greatest values. Title the box-and-whisker plot.

Nutrition Index Numbers for Breakfast Cereals

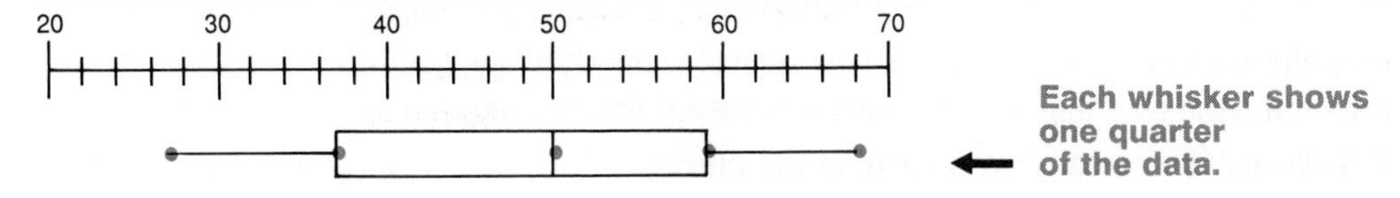

← **Each whisker shows one quarter of the data.**

your turn

Make a box-and-whisker plot to display the given data.

1. **Cost of Cordless Phones (Dollars)**

130	141	150	163	144	154
156	148	170	143	134	132
161	145	164	146	166	

2. **Energy Efficiency Ratios of Room Air Conditioners**

9.0	9.3	10.0	9.0	9.7
9.6	9.6	8.5	9.0	9.0
8.7	8.5	8.3	8.0	9.4

Box-and-whisker plots can also display more than one set of data.

example 2

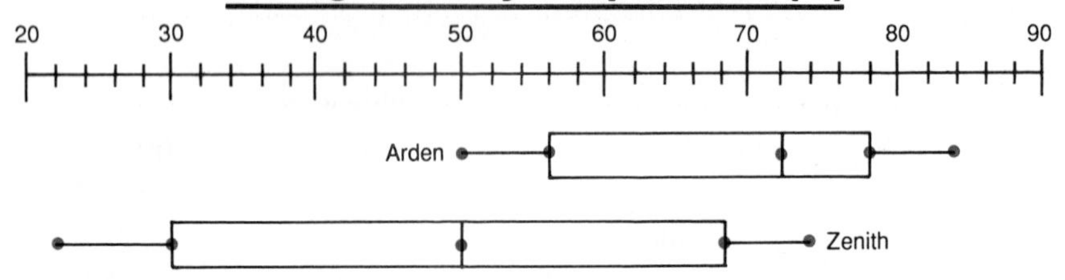

Use the box-and-whisker plot above.

a. Find the median of the data, the first quartile, the third quartile, the least temperature, and the greatest temperature for Arden.

b. Which city has more variation in its temperature?

solution

a. Examine the five dots in the *Arden* plot.
The middle dot is at 72; 72 is the median of the data.
The edges of the box are at 56 and 78; 56 is the first quartile and 78 is the third quartile.
The least temperature is 50. The greatest temperature is 84.

b. Since the "whiskers" are similar in length, compare the boxes. The box for Zenith is longer, so Zenith has more variation in its temperature.

your turn

Use the box-and-whisker plot above.

3. Find the median of the data, the first quartile, the third quartile, the least temperature, and the greatest temperature for Zenith.
4. Which city is generally warmer than the other?

practice exercises

practice for example 1 (pages 361–362)

Make a box-and-whisker plot to display the given data.

1. **Cost of Computer Printers ($)**

400	580	760	560	620	540
640	440	740	530	720	520
460	770	420	480	780	

2. **Gasoline Mileage (mi/gal)**

25.0	20.8	24.9	27.3	22.0
20.1	22.5	24.8	20.5	20.4
19.5	23.5	24.8	27.3	29.5

practice for example 2 (page 362)

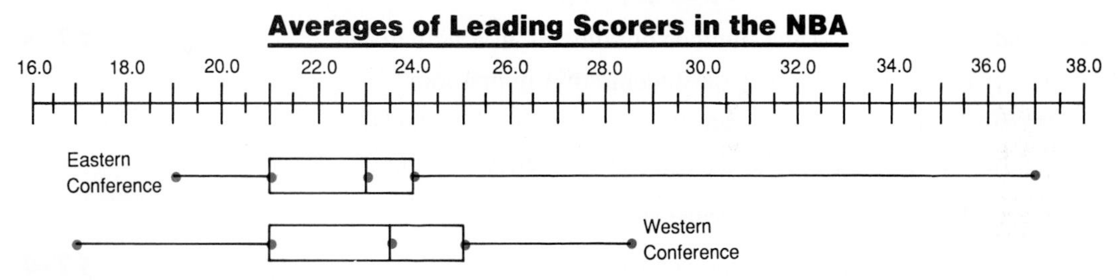

Use the box-and-whisker plot above.

3. Find the median of the data, the first quartile, and the third quartile for each conference.
4. Which conference had less variation in its averages?

mixed practice (pages 361–362)

5. In a recent survey, people were asked to rank several cities from best (1) to worst (215). The data for two cities are given. Make a box-and-whisker plot to display both sets of data.

Riverton

35	70	40	86	200	20	33
27	74	45	35	100	20	

Central Falls

85	148	140	124	188	95	80
82	131	190	170	150	149	

6. Refer to the box-and-whisker plot you made for Exercise 5. Which city do you think ranks better overall? Explain.

review exercises

Write each fraction in lowest terms.

1. $\frac{12}{20}$ 2. $\frac{32}{48}$ 3. $\frac{16}{24}$ 4. $\frac{21}{51}$ 5. $\frac{14}{35}$ 6. $\frac{15}{24}$

SKILL REVIEW

Find the mean, median, mode(s), and range. If necessary, round the mean to the nearest cent or to the nearest tenth.

17-1

1. 54, 38, 35, 38, 60
2. 80, 76, 43, 76, 80, 76
3. 1.4, 8.1, 3.6, 6.9, 8.1, 1.3
4. $4.60, $8.50, $7.37, $5.27

17-2

5. Make a frequency table for the given data.

Number of Cousins

6	8	8	6	7	9	6	8	7	8	7
6	7	6	6	9	7	8	6	6	8	9

Exercises 6–9 refer to the frequency polygon below.

17-3

6. Find the mode(s) of the number of students in a homeroom.
7. What is the mean number of students in a homeroom?
8. In how many homerooms are there less than 27 students?
9. Draw a histogram for the data in the frequency polygon.

Exercises 10–12 refer to the scattergram below.

17-4

10. Find the mode(s) of the time spent fishing.
11. When Joe caught just two fish, how many minutes did he spend fishing?
12. Use the trend line to predict how many fish Joe would catch if he spent 75 min fishing.

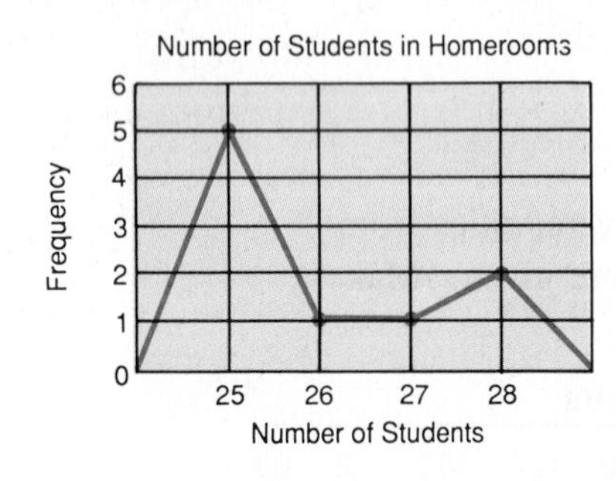

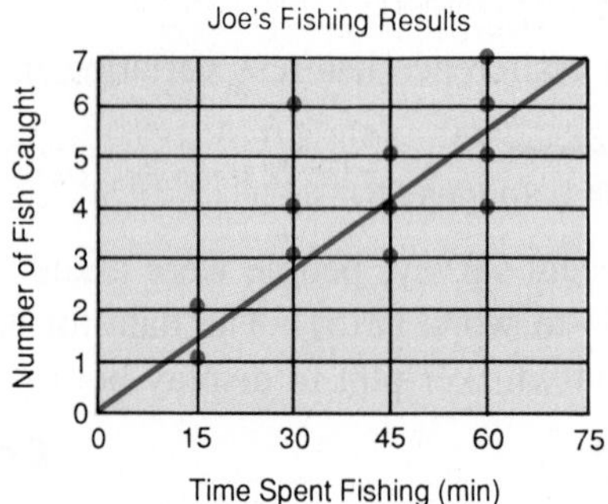

17-5

13. Make a stem-and-leaf plot to display the given data.

Record Weights of Fresh-Water Fish (lb)

22	26	12	22	21	18	58	58	32	54	50	38
51	11	27	31	22	12	12	11	14	36	32	41

17-6

14. Make a box-and-whisker plot to display the given data.

Cost of Shoes ($)

12	20	37	49	18	10	45	
36	15	40	26	50	25	35	30

APPLICATION

17-7 SURVEYS

A **survey** is used to obtain data about an entire group, or **population.** Survey results usually include as much data as possible. The United States Census is an example of a survey.

example

Sarah conducted a survey of the students in her homeroom. She asked how much money the students had earned each week from summer jobs. The results are shown in the table at the right. Find the median earnings.

Weekly Summer Earnings ($)

Earnings	Tally	Frequency
50	II	2
100	~~IIII~~ III	8
150	~~IIII~~	5
200	~~IIII~~	5

solution

Find the sum of the frequencies.

$2 + 8 + 5 + 5 = 20$

There are 20 items, so the median is the mean of the tenth and eleventh items.

tenth item = $100

eleventh item = $150

$$\frac{\$100 + \$150}{2} = \frac{\$250}{2} = \$125$$

The median earnings are $125.

exercises

Use the data in Sarah's survey.

1. Find the mean earnings.
2. Find the mode(s) of the earnings.
3. Find the range of the earnings.
4. Which of the mean, median, or mode best represents the data?

Conduct a survey of the students in your homeroom. Ask how many pets each student has.

5. Find the mean, the median, the mode(s), and the range of your data.
6. Combine your results with the results of students who are from other homerooms. Compare the mean, median, mode(s), and range of this larger survey with those of your survey. What differences, if any, occur in this data?

APPLICATION

17-8 MISLEADING STATISTICS

You can use the mean, median, and mode to interpret data. Because the mean, median, and mode help locate the "center" of a set of data, they are often called **measures of central tendency.** Sometimes one or more of these measures is used in a misleading manner.

example

Valley Insurance Agency has eight employees. Their salaries are \$175,000, \$125,000, \$50,000, \$20,000, \$16,000, \$16,000, \$16,000, and \$16,000. To attract new employees, the agency uses the mean to advertise that the average salary is \$54,250.

a. Is the mean a fair description of the pay scale?

b. Would the mode give a fair description of the pay scale?

c. Would the median give a fair description of the pay scale?

solution

a. No. Only one salary is near the mean.
Five of the eight salaries are much less than the mean.

b. Yes. The mode is \$16,000.
Five salaries are equal to or near the mode.

c. Yes. The median is \$18,000.
Five salaries are near the median.

exercises

Last month, seven houses were sold in Brook Acres. The selling prices were \$150,000, \$120,000, \$110,000, \$100,000, \$95,000, \$70,000, and \$70,000. A real estate agent used the mode and told a buyer that the most common selling price was \$70,000.

1. Is the mode a fair description of the selling prices? Explain.
2. Would the mean give a fair description of the selling prices? Explain.
3. Would the median give a fair description of the selling prices? Explain.
4. Another house just sold for \$125,000. Will the mean price increase or decrease?

Last month, the commissions for the salespeople at Auto World were \$2200, \$4800, \$2000, \$4800, \$2500, \$2400, \$4800, and \$2100. The dealership wants to hire more salespeople. The dealership uses the mean to advertise that the average monthly commission was \$3200.

5. Is the mean a fair description of the salespeople's monthly commissions? Explain.
6. Would the mode give a fair description of the salespeople's monthly commissions? Explain.
7. Would the median give a fair description of the salespeople's monthly commissions? Explain.

A real estate agency offers its services both to owners of rental property and to potential renters. In the past month, the agency has rented six apartments for rents of \$320, \$515, \$550, \$650, \$855, and \$950.

8. To attract new owners of rental property, is the agency likely to quote the mean rent or the median rent? Explain.
9. To attract new renters, is the agency likely to quote the mean rent or the median rent? Explain.

calculator corner

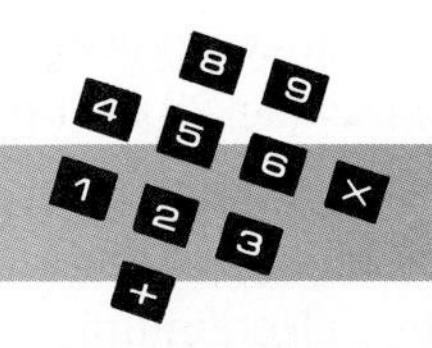

You can use the parentheses keys on a calculator to find the mean directly from a frequency table.

Pens Sold

Price	Tally	Frequency
\$.75	~~IIII~~ IIII	9
\$.85	~~IIII~~ I	6
\$.95	II	2

example Use the table at the right. Find the mean price of the pens that were sold.

solution Enter (9 × .75 + 6 × .85 + 2 × .95). ← **Finds the sum of the prices.**

Enter ÷ (9 + 6 + 2) =. ← **Divides the sum of the prices by the sum of the frequencies.**

The result is 0.808823529. The mean price of pens sold is about \$.81.

Use a calculator to find the mean.

1. **Points Scored Per Game**

Points	Tally	Frequency
16	~~IIII~~ IIII	9
17	~~IIII~~	5
18	III	3
19	III	3

2. **Houses Sold**

Price	Tally	Frequency
\$50,000	~~IIII~~ III	8
\$100,000	~~IIII~~ I	6
\$110,000	IIII	4
\$150,000	III	3

APPLICATION

17-9 LIFE INSURANCE

You purchase a **life insurance** policy to provide financial protection for your family in the event of your death. The **face value** of the policy is the amount of insurance you purchase. When you purchase a policy, you name a **beneficiary** who will receive the face value if you die. **Premiums** for life insurance are determined by using statistics about life expectancy at different ages.

A common type of life insurance is called **whole life,** or **straight life,** insurance. This type of policy remains in effect until the insured person dies or cancels it. Whole life policies have a **cash value** that the insured person receives if the policy is canceled. If you miss a payment, you can use part of the cash value to pay the premium. The **loan value,** usually the same as the cash value, is the amount the insurance company will loan you if you request a loan.

example

A 30-year-old nonsmoker purchases a $20,000 policy.

a. Use the table of annual premiums to find the annual premium.

b. Use the table of cash values to find the cash value of the policy after 10 years.

Annual Premiums per $1000 of Coverage

Age	25	30	35	40
Standard	$12.25	$14.10	$15.50	$17.40
Nonsmoker	$11.10	$12.80	$14.25	$16.85

Cash Value per $1000 of Coverage

Years after Issue	5	10	15	20	25	30
Cash Value ($)	39	121	196	287	352	473

solution

a. The premium per $1000 is $12.80. Since the policy is for $20,000, multiply by 20.

$20 \times \$12.80 = \$256.00$

The annual premium is $256.

b. After 10 years, the cash value per $1000 is $121. Since the policy is for $20,000, multiply by 20.

$20 \times \$121 = \$2420$

After 10 years, the cash value is $2420.

exercises

Complete. Use the tables on page 368.

	AGE	TYPE OF POLICY	FACE VALUE	ANNUAL PREMIUM	YEARS AFTER ISSUE	CASH VALUE
1.	30	nonsmoker	$40,000	?	15	?
2.	40	nonsmoker	$30,000	?	5	?
3.	25	standard	$10,000	?	20	?
4.	40	standard	$20,000	?	10	?
5.	25	nonsmoker	$45,000	?	30	?
6.	35	nonsmoker	$55,000	?	20	?

7. Sally Roh is 40 years old. She purchases a $35,000 standard policy. What is her annual premium?
8. Louise Russo is a 35-year-old nonsmoker. Sam Russo is a 30-year-old nonsmoker. Each has a $25,000 policy. Find the total annual premiums for both policies.
9. Warren Huntworth is 40 years old. He purchases a $40,000 standard policy. How much less would Warren have to pay over a 20-year period if he did not smoke?
10. Jason Hartman has had a $30,000 policy for 25 years. What is the cash value of the policy?
11. Ruth Stern has had a $45,000 policy for 5 years. What is the cash value of the policy?
12. Armand Jones is 35 years old and does not smoke. He purchases a $50,000 policy. What will be the cash value if Armand cancels his policy after 30 years?
13. Alan Mordell has had a $30,000 policy for 15 years. He decides to borrow $2000 from the insurance company. The company deducts the amount of the loan from the cash value. What is the cash value of his policy after the loan?
14. Janet Samar is a 45-year-old nonsmoker. She has had a $20,000 policy for 10 years. This year she asked the insurance company to deduct her annual premium from the cash value of her policy. What is the new cash value of the policy after the annual premium has been deducted?
15. Willie Hernandez is 50 years old. He has had a $40,000 standard policy for 20 years. This year he asked the insurance company to deduct his annual premium from the cash value of his policy. What is the new cash value of the policy after the annual premium has been deducted?

Term life insurance is bought for a specified number of years, or **term.** The beneficiary receives the face value of the policy only if the insured person dies during the term of the policy. Since term life insurance has no cash value or loan value, the premiums are much lower than those for whole life. Term life insurance can be renewed.

Annual Premiums per $1000 of Coverage

Age	25	30	35	40
Standard	$1.25	$1.25	$1.53	$1.91
Nonsmoker	$.90	$.90	$1.08	$1.36

Complete. Use the table above.

	AGE	TYPE OF POLICY	FACE VALUE	ANNUAL PREMIUM
16.	30	nonsmoker	$40,000	?
17.	25	nonsmoker	$30,000	?
18.	40	standard	$25,000	?
19.	35	standard	$50,000	?

20. Linda Ortiz is a 30-year-old nonsmoker. She purchases a $20,000 policy for a term of five years. Find the total amount of her premiums for the five years.

21. Ed Samms is 35 years old. He purchases a $35,000 standard policy for a term of ten years. Find the total amount of his premiums for the ten years.

22. Ann Sherman buys a $40,000 whole life policy. Beth Repast buys a $40,000 term life policy. Both are 25-year-old nonsmokers. Find the difference between their annual premiums.

23. Peter Westfall buys a $20,000 whole life policy. Gina Iorio buys a $20,000 term life policy. Both are 30-year-old nonsmokers. Find the difference between their annual premiums.

24. List the advantages of buying whole life insurance rather than term life insurance. List the advantages of buying term life insurance.

25. Find out the meaning of *endowment life insurance*.

A certain clock loses ten minutes every hour. This clock is set to the correct time at 10:30 A.M. What will the correct time be when the clock first shows 11:30 A.M. on that same day?

CHAPTER REVIEW

vocabulary vo·cab·u·lar·y

Choose the correct word to complete each sentence.

1. The (*mean, range*) of a set of data is also called the average.
2. One type of bar graph is called a (*histogram, frequency polygon*).

skills

3. Find the mean, median, mode(s), and range: 91, 66, 48, 37, 92, 56

4. Make a frequency table for the given data.

Rushing Yards Per Game

83	64	74	68	54	73	58	76	74	87	82	77

5. Use the histogram below. In how many games were fewer than 11 points scored?
6. Use the scattergram below. Predict the number of swimsuits sold when the temperature is 80°F.

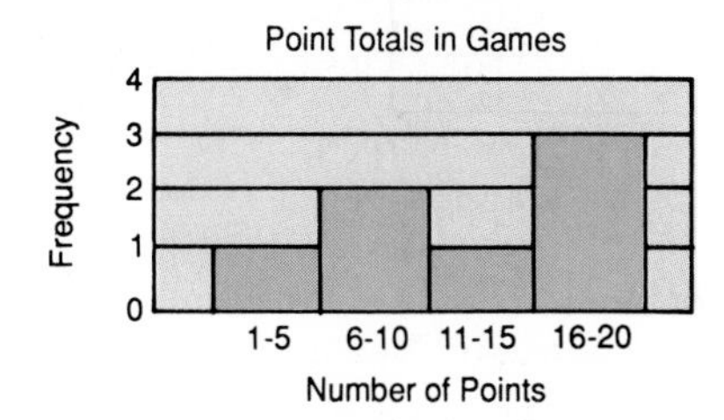

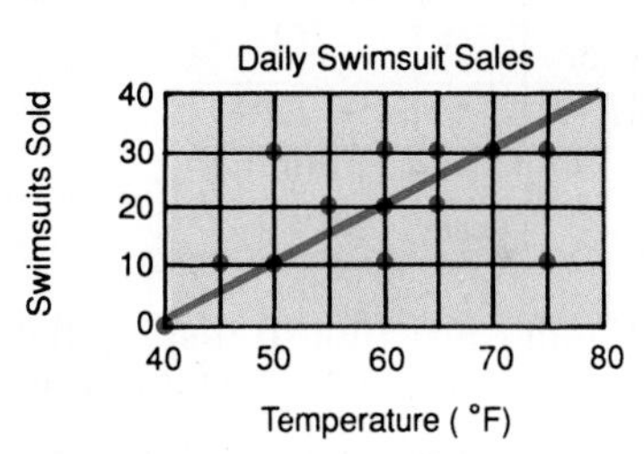

7. Make a stem-and-leaf plot to display the *Plane Fares* data below.
8. Make a box-and-whisker plot to display the *Pages Typed* data below.

Plane Fares (Dollars)

200	164	205	207	186	180
198	207	188	170	165	162

Pages Typed

22	12	16	17	28	15	10	30
18	13	26	19	20	22	27	

9. Jan took a survey of her friends. Two friends work 10 h per week and eight friends work 5 h per week. What is the mean number of hours worked?
10. The hourly wages at a company are \$8, \$8, \$29, \$11, \$10, \$9, \$30, and \$11. Is the median a fair description of the wages? Explain.
11. Use the table of premiums on page 370. A 25-year-old buys a \$20,000 term life standard policy. What is the annual premium?

CHAPTER TEST

Find the mean, median, mode(s), and range. If necessary, round the mean to the nearest cent or to the nearest tenth.

1. 47, 31, 24, 19, 24
2. \$5.23, \$1.29, \$7.58, \$4.07

17-1

Make a frequency table for the given data.

3. **Rebounds Per Game**

10	8	12	9	9	12	11
11	9	10	9	9	11	11

4. **Number of Tickets Sold**

11	25	30	23	21	35
22	20	24	32	13	15

(Use intervals such as 21–30.)

17-2

5. Use the frequency polygon below. What is the mean quiz score? **17-3**

6. Use the scattergram below. How many goals have been scored from a distance of 50 yd? **17-4**

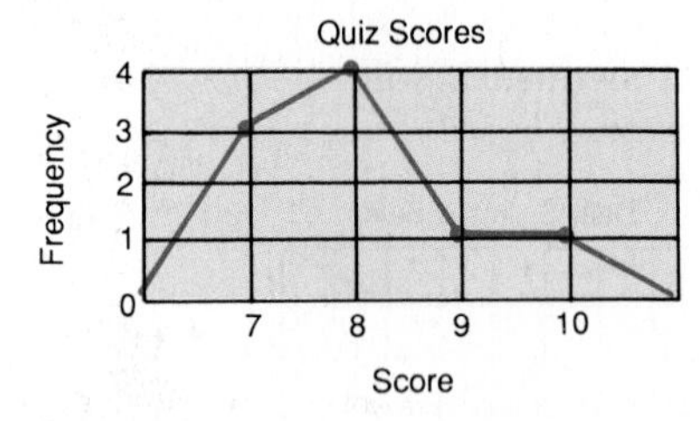

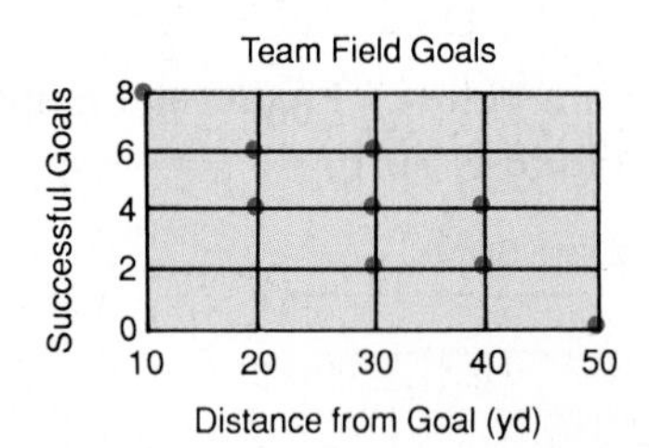

7. Make a stem-and-leaf plot to display the *Speeds of Animals* data below. **17-5**

8. Make a box-and-whisker plot to display the *Scores on a Quiz* data below. **17-6**

Speeds of Animals (mi/h)

11	30	70	40	25	35	32	41
61	20	18	53	45	40	30	32

Scores on a Quiz

19	21	15	10	19	22	12	20
12	14	25	19	15	18	19	

9. A survey of 20 food shoppers showed that 5 shoppers spent \$50/week, 6 shoppers spent \$60/week, and 9 shoppers spent \$70/week. What is the mean amount spent per week? **17-7**

10. The prices of six rings are \$80, \$80, \$995, \$950, \$795, and \$895. Is the mode a fair description of the prices? Explain. **17-8**

11. Use the tables on page 368. A 40-year-old nonsmoker purchases a \$30,000 whole life policy. What is the annual premium? **17-9**

Conservationists are working to protect many species of tropical birds and animals in order to increase their chances of survival.

CHAPTER 18

PROBABILITY

18-1 PROBABILITY AND ODDS

A *number cube* has six sides numbered 1 through 6. When you roll a number cube, there are six possible results, or **outcomes.** Each outcome is **equally likely** to occur. The **event** of rolling an even number has three **favorable outcomes:** 2, 4, and 6. You say that the **probability,** or chance, that this event will occur is $\frac{3}{6}$, or $\frac{1}{2}$.

When all outcomes are equally likely, you can use this formula to find the probability of an event E, written $P(E)$.

$$P(E) = \frac{\text{number of favorable outcomes}}{\text{number of possible outcomes}}$$

An event that will definitely occur is called a **certain** event. An event that will never occur is called an **impossible** event.

example 1

A number cube is rolled. Find each probability.

a. $P(3)$ **b.** $P(1 \text{ or } 5)$ **c.** $P(\text{not } 0)$ **d.** $P(\text{number} > 6)$

solution

a. $P(3) = \frac{1}{6}$ ← **The one favorable outcome is 3.** ← **The six possible outcomes are 1, 2, 3, 4, 5, and 6.**

b. $P(1 \text{ or } 5) = \frac{2}{6} = \frac{1}{3}$ ← **Write the fraction in lowest terms.**

c. $P(\text{not } 0) = \frac{6}{6} = 1$ ← **The probability of a certain event is 1.**

d. $P(\text{number} > 6) = \frac{0}{6} = 0$ ← **The probability of an impossible event is 0.**

your turn

A number cube is rolled. Find each probability.

1. $P(4)$ **2.** $P(\text{not } 2)$ **3.** $P(\text{number} < 3)$ **4.** $P(\text{odd or even number})$

When you roll a number cube, the event of rolling a number greater than 2 has four favorable outcomes: 3, 4, 5, and 6. So there are two **unfavorable outcomes:** 1 and 2. You say that the **odds** in favor of rolling a number greater than 2 are $\frac{4}{2}$, or $\frac{2}{1}$. This ratio is usually written 2 to 1. You can use this formula to find the odds in favor of an event E.

$$\text{odds in favor of event } E = \frac{\text{number of favorable outcomes}}{\text{number of unfavorable outcomes}}$$

example 2

A number cube is rolled. Find the odds in favor of each event.

a. 3 or 4 **b.** not 2

solution

a. odds $= \frac{2}{4} = \frac{1}{2}$ ← **The two favorable outcomes are 3 and 4.** ← **The four unfavorable outcomes are 1, 2, 5, and 6.**

The odds in favor of rolling a 3 or a 4 are 1 to 2.

b. odds $= \frac{5}{1}$ ← **The five favorable outcomes are 1, 3, 4, 5, and 6.** ← **The one unfavorable outcome is 2.**

The odds in favor of not rolling a 2 are 5 to 1.
You can also say that the odds *against* rolling a 2 are 5 to 1.

your turn

A number cube is rolled. Find the odds in favor of each event.

5. 5 or 6 **6.** number < 5 **7.** not 4 **8.** not even number

practice exercises

practice for example 1 (page 374)

Use the spinner at the right. Assume that each outcome is equally likely to occur. Find each probability.

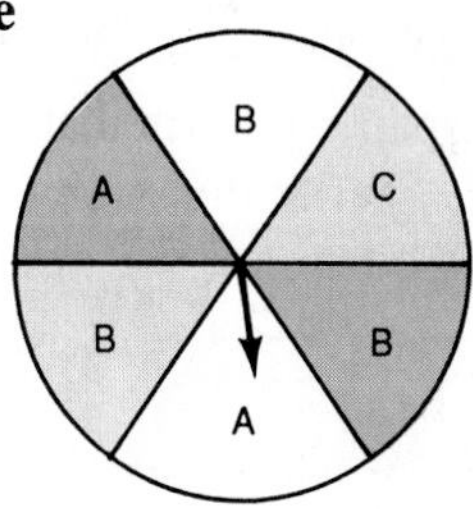

1. P(blue) **2.** P(C)
3. P(not B) **4.** P(not white)
5. P(yellow) **6.** P(not D)
7. P(A or B) **8.** P(red or C)
9. P(C or not blue) **10.** P(A or not A)

practice for example 2 (page 375)

Use the spinner at the right.
Find the odds in favor of each event.

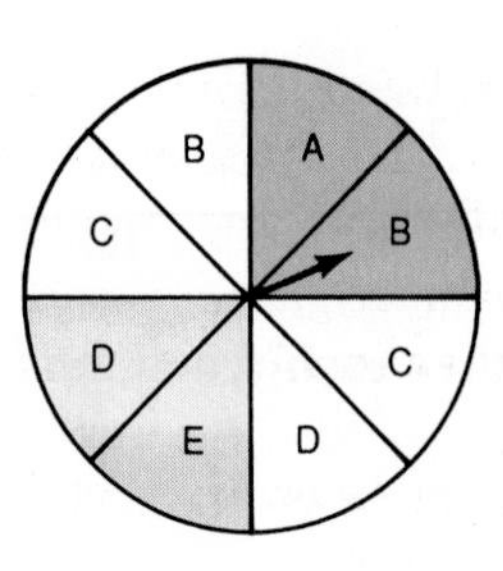

11. A **12.** vowel
13. not red **14.** not E
15. D or E **16.** red or white
17. B or blue **18.** white or A
19. vowel or C **20.** C or not white

mixed practice ***(pages 374–375)***

A drawer contains 14 black, 10 blue, 7 red, and 5 green pens. One pen is selected *at random,* which means that it is selected purely by chance. Find each probability.

21. P(black) **22.** P(not blue)

23. P(red or green) **24.** P(brown)

Use the pens described above. Find the odds in favor of each event.

25. red **26.** blue or green

27. not blue **28.** not black

A bag contains 3 green, 5 red, and 2 yellow marbles. One marble is selected at random. Find each probability and write it as a percent.

29. P(green) **30.** P(yellow) **31.** P(red or green) **32.** P(not red)

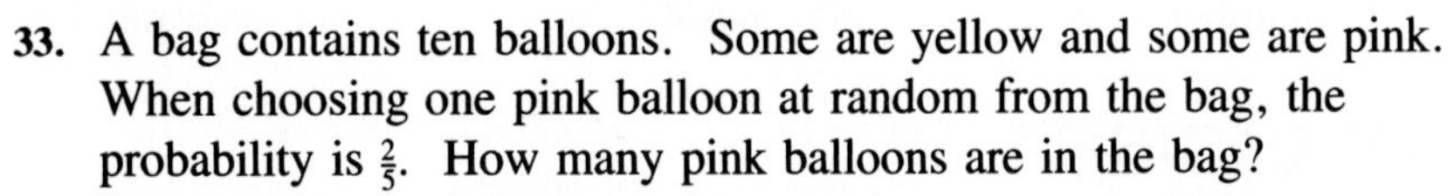

33. A bag contains ten balloons. Some are yellow and some are pink. When choosing one pink balloon at random from the bag, the probability is $\frac{2}{5}$. How many pink balloons are in the bag?

34. The odds in favor of winning a prize are 1 to 100. Find the odds *against* winning the prize.

35. The odds in favor of winning a race are 1 to 3. Find the probability of winning the race. Write this probability as a percent.

36. A newspaper reports that the probability that it will snow tomorrow is 80%. Write this probability as a fraction in lowest terms. Then find the odds in favor of snow tomorrow.

review exercises

Write each answer in lowest terms.

1. $\frac{1}{5} \times \frac{4}{5}$ **2.** $\frac{5}{8} \div \frac{1}{4}$ **3.** $\frac{4}{7} + \frac{4}{21}$ **4.** $\frac{3}{4} - \frac{7}{10}$

5. $1\frac{2}{3} + 5\frac{1}{3}$ **6.** $6\frac{5}{12} \div 11$ **7.** $14 - 7\frac{7}{9}$ **8.** $2\frac{1}{2} \times 3\frac{2}{15}$

9. $1\frac{1}{4} - \frac{5}{6}$ **10.** $\frac{8}{9} \times 1\frac{3}{4}$ **11.** $7\frac{3}{5} + \frac{1}{2}$ **12.** $3 \div \frac{1}{6}$

Tell whether each fraction is closest to 0, $\frac{1}{2}$, or 1.

13. $\frac{9}{10}$ **14.** $\frac{1}{5}$ **15.** $\frac{8}{15}$ **16.** $\frac{13}{28}$ **17.** $\frac{55}{61}$ **18.** $\frac{2}{99}$

18-2 TREE DIAGRAMS

A list of all possible outcomes is called a **sample space.** You can use a **tree diagram** to show a sample space.

example 1

A coin is tossed three times. Use a tree diagram to find the number of possible outcomes in the sample space.

solution

When you toss a coin, there are two possible outcomes: heads (H) and tails (T). A tree diagram for tossing a coin three times is shown at the right. The sample space is shown in red. There are 8 possible outcomes.

toss 1	toss 2	toss 3	outcomes
H	H	H	HHH
		T	HHT
	T	H	HTH
		T	HTT
T	H	H	THH
		T	THT
	T	H	TTH
		T	TTT

your turn

Use a tree diagram to find the number of possible outcomes in the sample space.

1. spinning the spinner twice
2. tossing a coin and spinning the spinner

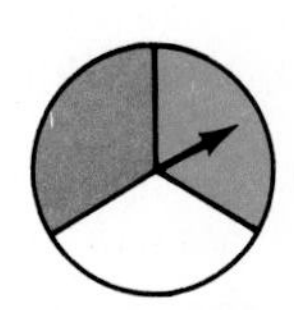

Often a tree diagram can help you to find the probability of an event.

example 2

When you toss a coin, heads and tails are equally likely outcomes. Use the tree diagram above to find each probability.

a. $P(\text{HTT})$ **b.** $P(\text{exactly one tail})$

solution

a. $P(\text{HTT}) = \frac{\text{number of favorable outcomes}}{\text{number of possible outcomes}} = \frac{1}{8}$

b. $P(\text{exactly one tail}) = \frac{3}{8}$ ← **The three favorable outcomes are HHT, HTH, and THH.**

your turn

Use the tree diagram above to find each probability.

3. $P(\text{TTH})$
4. $P(\text{exactly two tails})$
5. $P(\text{no tails})$
6. $P(\text{at least two heads})$

practice exercises

practice for example 1 (page 377)

Use a tree diagram to find the number of possible outcomes in the sample space.

1. rolling a number cube and tossing a coin
2. selecting a digit from 1 through 3 and a letter from A through C
3. buying a white or a black phone in a cordless, desk, or wall style
4. ordering orange, apple, or grape juice in a small, regular, large, or jumbo size
5. choosing soup and salad;
   soup: potato, tomato, beef, chicken
   salad: vegetable, fruit
6. selecting music;
   type: rock, country, jazz
   source: disk, record, cassette

practice for example 2 (page 377)

A coin is tossed three times. Use the tree diagram on page 377 to find each probability.

7. *P*(HTH)
8. *P*(exactly one head)
9. *P*(three tails)
10. *P*(no heads)
11. *P*(at least one head)
12. *P*(at least two tails)

mixed practice (page 377)

The tree diagram at the right shows the skirt/top outfits that Peggy can choose. Assume that each outfit is equally likely.

skirt	top	outfits
denim	blouse	denim skirt, blouse
	sweater	denim skirt, sweater
	T-shirt	denim skirt, T-shirt
khaki	blouse	khaki skirt, blouse
	T-shirt	khaki skirt, T-shirt

13. How many different skirt/top outfits can Peggy choose?
14. Which top does Peggy not wear with the khaki skirt?
15. Which skirt can Peggy use to make the most skirt/top outfits?
16. Peggy chooses one skirt/top outfit at random. Find the probability that she chooses the blouse.
17. Peggy chooses one skirt/top outfit at random. Find the probability that she does not choose the khaki skirt.
18. Peggy buys a cotton skirt that she wears with any of the three tops. How many different skirt/top outfits can she choose now?
19. Bob has four shirts: white, cream, blue, and pink. He has two ties: red and yellow. Bob does not wear the yellow tie with the cream shirt. He does not wear the red tie with the pink shirt. Find how many different shirt/tie sets Bob can choose.

20. A number cube is rolled three times. Explain why it would be unreasonable to use a tree diagram to find the number of possible outcomes.

21. A jeweler sells necklaces with either a gold or a silver chain and a pearl, a ruby, or a diamond pendant. How many types of necklaces does the jeweler sell?

22. The Lunchstop sells ham sandwiches made-to-order. For bread, you choose wheat or rye. You also choose lettuce, tomato, both, or neither. How many different kinds of ham sandwiches are possible at the Lunchstop?

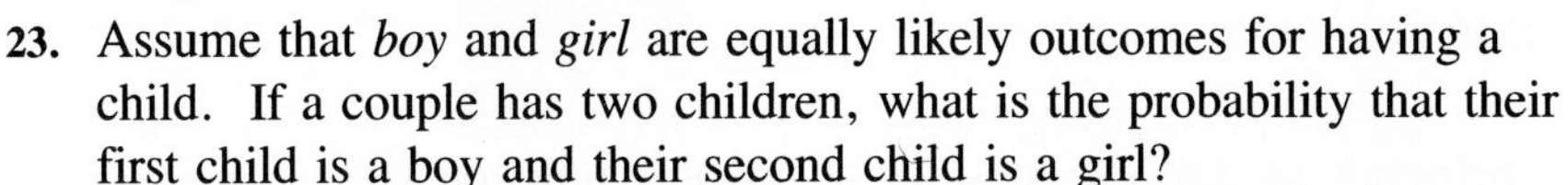

23. Assume that *boy* and *girl* are equally likely outcomes for having a child. If a couple has two children, what is the probability that their first child is a boy and their second child is a girl?

24. Lee is required to take a language course and a science course. The languages she can choose are Italian, French, and Spanish, and the sciences are biology and chemistry. What is the probability that she decides to take the Italian course and the biology course?

review exercises

Find each answer.

1. $(3 + 9) \times 11$
2. $60 \div (14 - 8)$
3. $8 \times (54 \div 6)$
4. $5 \times 4 - 2 \times 8$
5. $19 - 3 \times 4 \div 6$
6. $30 \div 6 + 8 - 7$

mental math

Solve each proportion mentally.

1. $\frac{1}{2} = \frac{45}{x}$
2. $\frac{4}{11} = \frac{20}{n}$
3. $\frac{y}{56} = \frac{3}{8}$
4. $\frac{t}{42} = \frac{6}{7}$

Mentally estimate the mean.

5. 21, 17, 19, 20, 22, 18, 21
6. 98.6, 99.1, 103.2, 101.3, 99.5

18-3 THE COUNTING PRINCIPLE

You can use the **counting principle** to find the number of possible outcomes without listing the outcomes in a tree diagram. You can also use the counting principle to find a probability. The counting principle is especially useful when a tree diagram would be unreasonably large.

example 1

Use the counting principle to find the number of possible outcomes when choosing a letter from A through Z and a digit from 0 through 9.

solution

Count the possible choices at each step, then multiply.

number of choices for a letter		number of choices for a digit	
26	$\times$	10	= 260 possible outcomes

your turn

Use the counting principle to find the number of possible outcomes.

1. choosing a letter from A through N and a digit from 1 through 5
2. selecting one month of the year and one year of the century
3. tossing seven nickels
4. rolling a number cube four times

example 2

Use the counting principle to find P(A or B, odd digit) when choosing a letter from A through Z and a digit from 0 through 9 at random.

solution

Count the *favorable outcomes* at each step, then multiply.

number of favorable outcomes for a letter		number of favorable outcomes for a digit	
2	$\times$	5	= 10 favorable outcomes

$$P(\text{A or B, odd digit}) = \frac{\text{number of favorable outcomes}}{\text{number of possible outcomes}} = \frac{10}{260} = \frac{1}{26}$$

your turn

Use the counting principle to find each probability when choosing a letter from A through Z and a digit from 0 through 9 at random.

5. P(M, 7)
6. P(L, even digit)
7. P(D or K, not 4)
8. P(not X, digit > 6)

practice exercises

practice for example 1 (page 380)

Use the counting principle to find the number of possible outcomes.

1. choosing a letter from L through T and a digit from 0 through 5
2. selecting a seat from 5 sections with 30 seats each
3. tossing ten dimes
4. rolling a number cube five times

practice for example 2 (page 380)

A number cube is rolled three times. Use the counting principle to find each probability.

5. $P(1, 2, 3)$
6. P(5, even number, 1 or 2)
7. P(no 4's)
8. P(three even numbers)

mixed practice (page 380)

Use spinners A, B, and C at the right. Use the counting principle to find the number of possible outcomes.

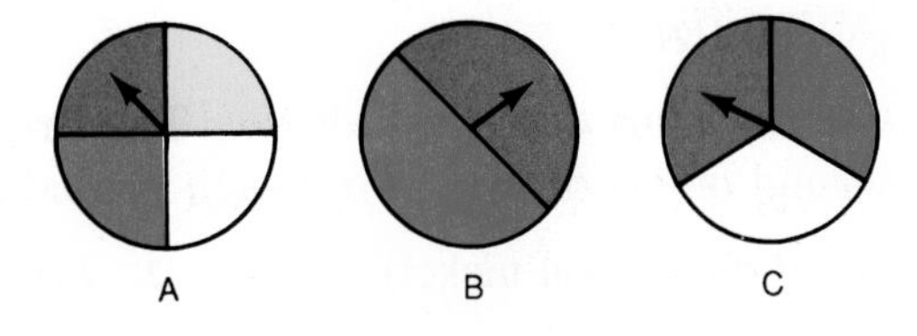

9. spinning A once and C once
10. spinning A three times
11. spinning B once and C twice
12. spinning C once and B four times

Spinners A, B, and C are each spun once. Use the counting principle to find each probability.

13. P(three blue)
14. P(no red)
15. P(no white)
16. P(three yellow)
17. How many different 3-letter sets of initials are possible when using the letters from A through Z and allowing letters to repeat?
18. A coin is tossed five times. Find P(all heads).
19. How many 7-digit phone numbers are possible if the first digit cannot be 0 or 1? Find P(765-4321).
20. What information is given by a tree diagram that is not given by the counting principle? Why might you not need to use a tree diagram?

review exercises

Is the first number divisible by the second number? Write *Yes* or *No*.

1. 916; 3
2. 110; 2
3. 3830; 5
4. 642; 4
5. 7155; 9
6. 4865; 10

18-4 INDEPENDENT AND DEPENDENT EVENTS

At times you need to find the probability that two events occur either at the same time or consecutively. If the occurrence of one event *does not* affect the occurrence of the other event, the two events are **independent.**

example 1

A purse contains 4 quarters, 1 dime, and 5 nickels. Two coins are drawn at random. The first coin is *replaced* before the second one is drawn. Find the probability of drawing a dime and then a quarter.

solution

$P(\text{dime, then quarter}) = P(\text{dime}) \times P(\text{quarter})$ ← **Multiply the probabilities of the events.**

$= \frac{1}{10} \times \frac{4}{10} = \frac{4}{100} = \frac{1}{25}$

your turn

Use the coins described above. The first coin is replaced before the second one is drawn. Find each probability.

1. *P*(dime, then nickel)
2. *P*(nickel, then quarter)
3. *P*(two quarters)

Two events are **dependent** if the occurrence of the first event *does* affect the occurrence of the second event.

example 2

Use the coins described above. Two coins are drawn at random. The first coin is *not* replaced before the second one is drawn. Find the probability of drawing a dime and then a quarter.

solution

$P(\text{dime, then quarter}) = P(\text{dime}) \times P(\text{quarter after dime})$

$= \frac{1}{10} \times \frac{4}{9} = \frac{4}{90} = \frac{2}{45}$ ← **After one coin is drawn, only 9 coins remain in the purse.**

your turn

Use the coins described above. The first coin is *not* replaced before the second one is drawn. Find each probability.

4. *P*(quarter, then nickel)
5. *P*(two nickels)
6. *P*(two dimes)

practice exercises

practice for example 1 (page 382)

Tom rolls the number cube and spins the spinner. Find each probability.

1. P(2, then red)
2. P(not 3, then blue)
3. P(number < 5, then white)
4. P(odd number, then not blue)
5. P(number > 6, then not red)
6. P(4 or 6, then red or blue)

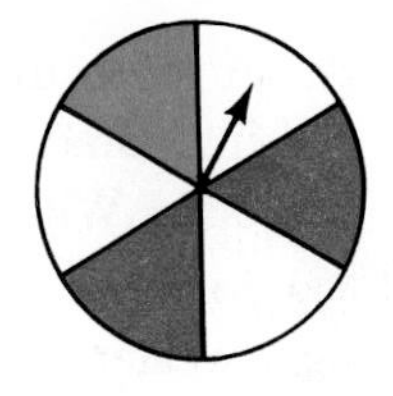

practice for example 2 (page 382)

A bag contains 2 yellow, 3 black, and 3 blue marbles. Two marbles are drawn at random. The first marble is *not* replaced before the second one is drawn. Find each probability.

7. P(black, then blue)
8. P(blue, then yellow)
9. P(two yellow)
10. P(two black)
11. P(not blue, then blue)
12. P(yellow, then not black)

mixed practice (page 382)

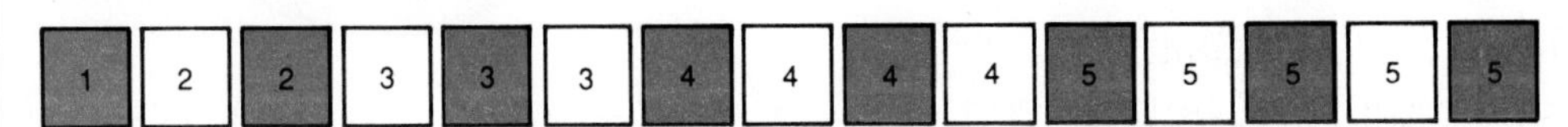

Two cards are drawn at random from those above. The first card is replaced before the second one is drawn. Find each probability.

13. P(2, then 3)
14. P(red, then white)
15. P(two 5's)
16. P(two red)
17. P(white, then not 4)
18. P(not 3, then 1)

Two cards are drawn at random from those above. The first card is *not* replaced before the second one is drawn. Find each probability.

19. P(1, then white)
20. P(4, then 2)
21. P(two red)
22. P(two 1's)
23. P(not 3, then 3)
24. P(white, then red)
25. Three cards are drawn at random from those above. Each card is replaced before the next one is drawn. Find P(5, then red, then 2).

Tell whether the two events are *independent* or *dependent*.

26. nickel: toss tails, then toss tails again
27. grapes in a bowl: choose and eat a red grape, then choose and eat a green grape
28. socks in a drawer: pick a black sock and put it back, then pick a brown sock
29. A relish tray contains 14 green and 11 black olives. Without looking, Brad takes an olive and eats it. Then he takes and eats another one. What is the probability that he eats two black olives?
30. Joan has 3 sweaters, and Lindsey has 4 sweaters. One of Joan's sweaters matches one of Lindsey's. Each plans to wear a sweater to school tomorrow. If they choose their sweaters at random, what is the probability that they wear matching sweaters?
31. One card is drawn at random from those shown on page 383. Find the probability and then the odds in favor of drawing a red 5.
32. Todd has a bucket of tennis balls. Some are yellow and some are orange. Give an example of two *independent* events and two *dependent* events relating to the bucket of tennis balls.

review exercises

Estimate.

1. 2813 − 2530	2. 18 × \$53	3. \$719 + \$677	4. 1481 ÷ 3
5. \$13.89 + \$5.75	6. 2.83 ÷ 0.42	7. 6.9 × \$4.05	8. 42.67 − 11.5
9. $\frac{1}{5} \times 24\frac{6}{11}$	10. $10\frac{7}{10} + 3\frac{6}{7}$	11. $15\frac{2}{9} - 9\frac{7}{8}$	12. $8\frac{1}{4} \div 1\frac{7}{8}$
13. 52% of \$198	14. 12.3% of 38	15. 76% of 79	16. 5.06% of \$617

Of 30 students in a math class, 16 play in band, 13 sing in choir, and 4 are in both band and choir. How many of the 30 students are in neither band nor choir? Hint: Copy and complete the *Venn diagram* at the right.

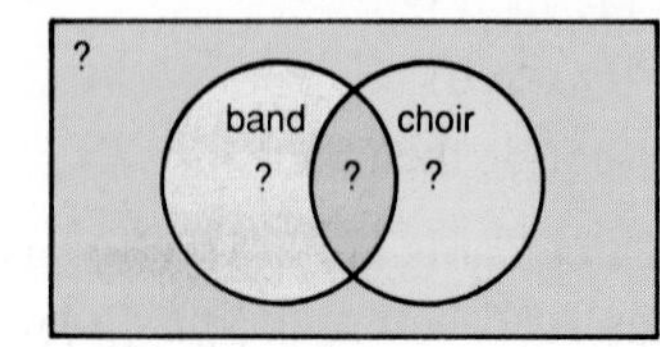

18-5 PERMUTATIONS AND COMBINATIONS

A **permutation** is an *arrangement* of objects in a particular order. You can use the counting principle to find the number of permutations of any group of objects.

example 1

Find the number of permutations of 5 runners in 5 lanes.

solution

Count the possible choices at each step, then multiply. As each lane is filled, the number of runners available for the next lane decreases by 1.

lane 1		lane 2		lane 3		lane 4		lane 5	
5	×	4	×	3	×	2	×	1	= 120

There are 120 permutations of 5 runners in 5 lanes.

your turn

Find the number of permutations.

1. 4 runners on a 4-person relay team
2. 8 cars in 8 parking spaces

In Example 1, you can write the multiplication $5 \times 4 \times 3 \times 2 \times 1$ as 5!, which you read as "five **factorial.**"

Sometimes permutations involve only *part* of a group of objects.

example 2

Find the number of permutations of 2 out of 5 runners in 2 lanes.

solution

number of choices for lane 1		number of choices for lane 2	
5	×	4	= 20

There are 20 permutations of 2 out of 5 runners in 2 lanes.

your turn

Find the number of permutations.

3. 4 out of 12 runners for a 4-person relay team
4. one representative and one alternate from 22 students

A **combination** is a group formed by part or all of the objects in a given group without regard to order.

example 3

Find the number of combinations of 2 out of 5 runners.

solution

Use what you have learned about permutations.

$$\frac{\text{number of permutations of 2 out of 5 runners}}{\text{number of permutations of 2 out of 2 runners}} = \frac{5 \times 4}{2 \times 1} = \frac{20}{2} = 10$$

There are 10 combinations of 2 out of 5 runners.

your turn

Find the number of combinations.

5. 4 out of 6 courses offered

6. 3 swimmers from a team of 8

practice exercises

practice for example 1 (page 385)

Find the number of permutations.

1. 6 workers for 6 shifts

2. 2 people for 2 places in line

3. the letters in the word ACT

4. the digits from 1 through 4

practice for example 2 (page 385)

Find the number of permutations.

5. 5 out of 8 people in 5 chairs

6. 6 out of 9 gifts in 6 boxes

7. 4 out of 7 students in a row of 4

8. 3 places in a contest of 40 candidates

practice for example 3 (page 386)

Find the number of combinations.

9. 3 out of 4 books on a shelf

10. 2 out of 18 pencils in a drawer

11. 4 colors from a dozen choices

12. 5 committee members from 10 students

mixed practice (pages 385–386)

Find the number represented by each factorial.

13. 4!

14. 2!

15. 7!

16. 6!

Find the number of permutations and the number of combinations.

17. 3 out of 5 people
18. 2 out of 6 books
19. 4 out of 5 shirts
20. 7 out of 7 dogs
21. There are 15 girls on a high school soccer team. In how many different ways can one girl be chosen as captain and a different girl be chosen as goalkeeper?
22. In how many different ways can 6 people sit in a 6-passenger car if they all can drive? If only one person can drive?
23. How many different combinations of 5 players can a coach select from a 12-member basketball team? In how many different ways can an announcer introduce these 5 players?

review exercises

Find the mean, median, mode(s), and range.

1. 13, 14, 9, 11, 10, 15
2. \$140, \$89, \$108, \$142, \$96
3. 50, 59, 51, 46, 53, 59
4. 3.9, 2.7, 2.1, 4.0, 2.7, 3.7, 4.0

calculator corner

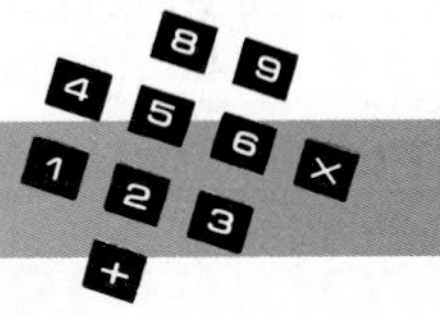

You can use a calculator to solve problems involving factorials. Some calculators even have a factorial key, x! .

example Find the number of permutations of 9 out of 9 books.

solution If your calculator has a factorial key, enter 9 x! .
The result is 362880.
If your calculator does not have a factorial key, enter
9 × 8 × 7 × 6 × 5 × 4 × 3 × 2 × 1 =.
The result is 362880. There are 362,880 permutations.

Use a calculator to find the number of permutations.

1. 8 people in 8 chairs
2. 10 bouquets in 10 vases
3. 11 books on a shelf
4. the digits from 2 through 7

PROBLEM SOLVING STRATEGY

18-6 MAKING AN ORGANIZED LIST

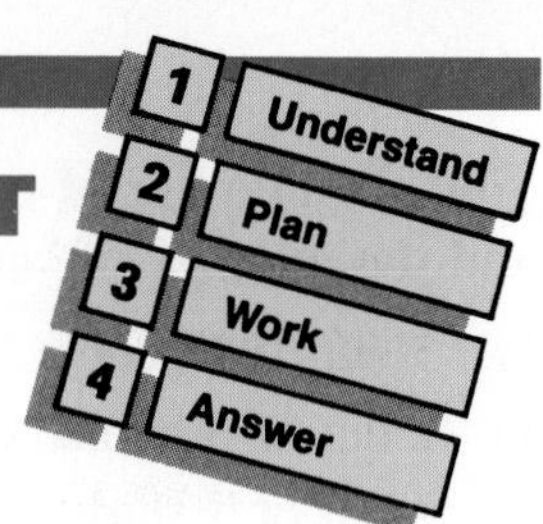

Sometimes a problem will require you to find all possibilities in a given situation. Making an **organized list** of the possibilities can help you to solve the problem.

example

Jeff Nelson scored 6 points in the championship basketball game. In how many different ways, without regard to order, could Jeff have scored the 6 points?

Basketball Scoring

Type of Shot	Number of Points
field goal	3
field goal	2
free throw	1

solution

Step 1 Given: 6 points scored
1, 2, or 3 possible points per shot
Find: the number of different ways to score 6 points

Step 2 Make an organized list according to the number of each type of shot made.

Step 3

number of 3-point field goals	2	1	1	0	0	0	0	← Multiply by 3.
number of 2-point field goals	0	1	0	3	2	1	0	← Multiply by 2.
number of 1-point free throws	0	1	3	0	2	4	6	← Multiply by 1.
Total number of *points*	6	6	6	6	6	6	6	

Step 4 Count the possibilities: 7

Jeff could have scored the 6 points in 7 different ways without regard to order.

problems

1. Cathleen Lento scored 8 points in a basketball game. In how many different ways, without regard to order, could she have scored the 8 points?
2. Mark White scored three times in a basketball game. How many different point totals could he have scored in the game?
3. There are ten 15¢ postage stamps and ten 25¢ postage stamps in a drawer. Ten stamps are taken from the drawer, and their total value is found. How many different totals are possible?
4. Joe has only quarters, dimes, and nickels. In how many different ways, without regard to order, can he make 40¢ in change?
5. In how many different ways, without regard to order, can you pay $100 with bills of $100, $50, $20, and $10?

6. Meredith has a coin bank filled with many pennies, nickels, and dimes. Meredith takes three coins from her bank and finds the total value of the three coins. How many different totals are possible?

7. Three darts are thrown at the target shown at the right. Each dart scores 2, 3, or 5 points. The three scores are added for a point total. How many different point totals are possible?

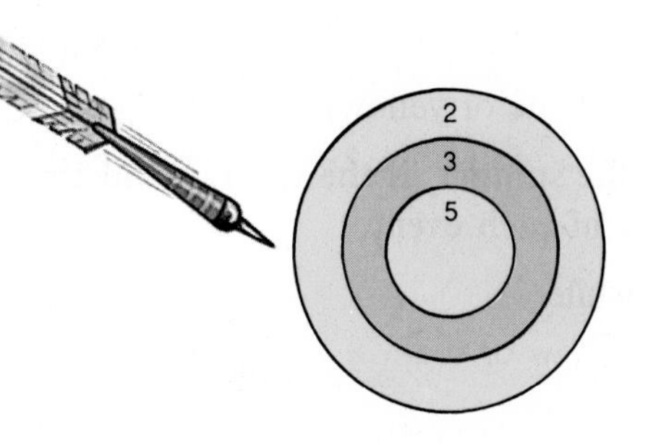

8. Each dart thrown at the target shown at the right scores 2, 3, or 5 points. The scores from the darts thrown are totaled. In how many different ways, without regard to order, can 15 points be scored?

9. In a high school football game against the Broncs, the Bears won by a score of 10 to 7. In how many different ways, without regard to order, could the Bears have scored their 10 points? Use the chart shown at the right.

10. The Capital High School football team scored twice in a game. List all the team's possible point totals for the game. Use the chart shown at the right.

Football Scoring

Type of Play	Number of Points
touchdown	6
conversion after touchdown	2
point after touchdown	1
field goal	3
safety	2

review exercises

1. Steven and Shelley Gillman went out for dinner last night. Steven's dinner cost $1.50 more than Shelley's. The total cost of the two dinners and the $2.25 tip was $16.25. How much did each dinner cost?

2. Rachel sold two-thirds of her fund-raising hats for $6.00 each. She has ten hats left to sell. How much money has Rachel collected from the hats that she has sold?

3. Michelle Maltais wants to buy a television set. She is able to save $60 per month for that purpose. The television set costs $400 and the sales tax rate is 5%. How many months will it take her to save enough money?

4. Ralph Monchick rolls two number cubes. He finds the sum of the numbers that are on the top sides of the cubes. How many different sums are possible?

SKILL REVIEW

Use the spinner at the right. Find each probability.

18-1

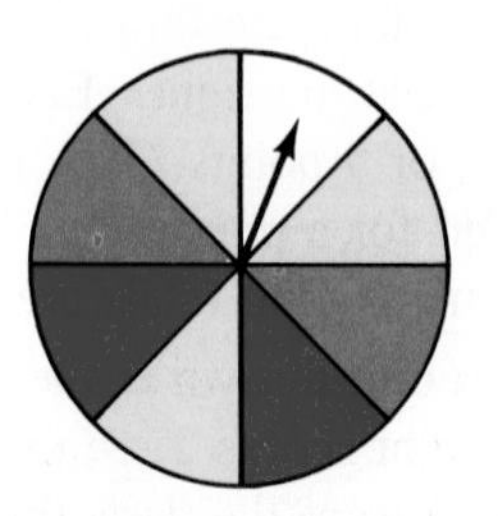

1. P(yellow)
2. P(not blue)
3. P(white or yellow)
4. P(red or not red)

Use the spinner at the right. Find the odds in favor of each event.

5. white
6. not red
7. yellow or red
8. red or blue

18-2

9. Use a tree diagram to find the number of possible outcomes when selecting a letter from X through Z and a digit from 0 through 3.
10. A coin is tossed three times. Use a tree diagram to find the probability that at least one toss will come up tails.

18-3

11. Use the counting principle to find the number of possible outcomes when selecting an outfit from 5 shirts and 6 pairs of pants.
12. Use the counting principle to find P(even digit, M or N) when choosing one digit from 0 through 9 and one letter from A through Z at random.

A number cube is rolled and a marble is drawn at random from those at the right. Find each probability.

18-4

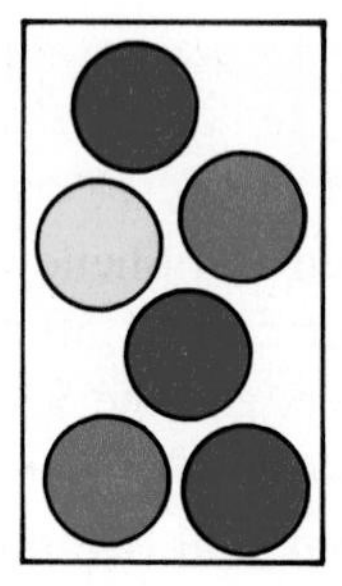

13. P(3, then red)
14. P(even number, then blue)
15. P(4 or 5, then yellow)
16. P(not 6, then not blue)

Two marbles are drawn at random from those at the right. The first marble is not replaced before the second one is drawn. Find each probability.

17. P(red, then blue)
18. P(not red, then red)
19. P(two blue)
20. P(two yellow)

18-5

21. Find the number of permutations of 4 drivers for 4 routes.
22. Find the number of permutations of 3 jobs for 6 people.
23. Find the number of combinations of 5 out of 7 employees.
24. Find the number represented by 8!.

18-6

25. Ramon Ruiz scored 7 points in a basketball game. In how many different ways, without regard to order, could Ramon have scored the 7 points? Use the chart titled *Basketball Scoring* on page 388.

APPLICATION

18-7 EXPERIMENTAL PROBABILITY

When you cannot assume that all outcomes are equally likely, you can do an experiment. You can then find the **experimental probability** of an event.

example

A paper cup is tossed 30 times. The results are shown in the frequency table at the right. Find the probability that the paper cup lands *up*.

Outcome	Tally	Frequency
up	ǀǀǀ	3
down	𝍸 ǀ	6
side	𝍸 𝍸 𝍸 𝍸 ǀ	21
	Total:	30

solution

$P(\text{up}) = \frac{3}{30} = \frac{1}{10}$. The probability that the paper cup lands *up* is $\frac{1}{10}$.

exercises

Use the frequency table for the experiment described above.

1. Find $P(\text{down})$.
2. Find $P(\text{side})$.
3. Find $P(\text{not up})$.
4. Find $P(\text{up or down})$.
5. Which outcome is seven times as likely as *up*?
6. Which outcome is half as likely as *down*?

7. Andy tossed a bottle cap 25 times. It landed up 14 times and down 11 times. Find the probability that the bottle cap lands *up*.
8. Carol tossed a thumbtack 20 times. It landed point up 15 times. Find the probability that the thumbtack does *not* land point up.

Toss a paper cup 30 times and record your results in a frequency table. Find each probability based on your experiment.

9. $P(\text{up})$
10. $P(\text{down})$
11. $P(\text{side})$
12. $P(\text{not up})$
13. How do your results differ from the results of the experiment in the example? Explain.
14. Give two reasons why your results might differ from the results of the experiment in the example.

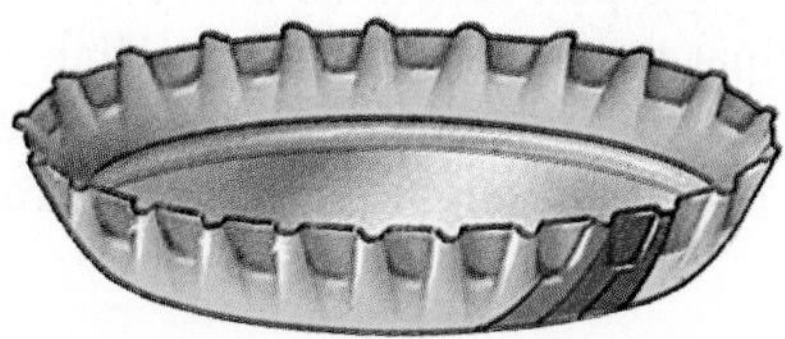

APPLICATION

18-8 SAMPLING AND MAKING PREDICTIONS

A **sample** is part of a group. If you select a sample purely by chance, you have a *random sample*. When you question or test a random sample, the process is called **sampling.** Sampling part of a group allows you to make **predictions** about the group as a whole. Sampling is used in taking opinion polls, projecting trends, and testing for quality control.

example

The Harris High School *Herald* polled a random sample of 100 students who attend the school. The students were asked to choose their favorite spectator sports. The results are shown in the frequency table at the right. How many of the 1300 students who attend the school would you expect to choose football?

Sport	Tally	Frequency
baseball	𝍸 𝍸 𝍸 I	16
basketball	𝍸 𝍸 𝍸 IIII	19
football	𝍸 𝍸 𝍸 𝍸	20
hockey	𝍸 𝍸 IIII	14
soccer	𝍸 𝍸 II	12
swimming	𝍸 III	8
tennis	𝍸 𝍸 I	11
		Total: 100

solution

$P(\text{football}) \times \text{number of students} = \text{expected number}$

$$\frac{20}{100} \times 1300 = \frac{1}{5} \times \frac{1300}{1} = 260$$

You would expect about 260 of the 1300 students to choose football.

exercises

Use the frequency table above. How many of the 1300 students would you expect to choose each of the following?

1. baseball
2. tennis
3. hockey or soccer
4. basketball or football
5. Using the results of the poll described above, the *Herald* concluded that hockey is the favorite spectator sport of 1.4% of the students at Harris High. What is wrong with this conclusion?
6. Using the results of the poll described above, the *Herald* predicted that more than 100 Harris High students would attend the upcoming state swim meet. Explain why this prediction might prove to be wrong.

A local newspaper polled a random sample of 250 registered voters one day before the election for mayor. Of those polled, 40 supported Leary, 90 supported Montoya, 100 supported Shigeta, and 20 were undecided. There are about 18,000 registered voters in the city.

7. Use the results of the poll. From how many registered voters in the city might each of the following expect support?

   a. Shigeta b. Leary c. Montoya

8. Use the results of the poll. How many registered voters in the city might the candidates expect to be undecided? Explain how the undecided voters can affect the outcome of the election.

9. A quality control inspector randomly tested 50 valves from a supply of 4500 valves. The inspector found 4 to be defective. Predict the number of defective valves in the supply.

10. A random sample of Red Lodge residents was asked to try a new chili. Of the sample, 72 residents said they would buy the chili and 28 said they would not. How many of the 10,000 Red Lodge residents would be expected to buy the new chili?

11. To predict the favorite rock band of the students in his school, Ted polled ten of his friends. Was this a random sample of the students in his school? Explain.

12. Marijo is test-marketing a new sparkling water. She gave a glassful to each of 30 swimmers at the beach. Each swimmer liked the water. She predicted that the water would sell very well. Why might her prediction prove to be wrong?

Tell whether it is appropriate to use sampling to make predictions about each of the following. If it is not appropriate, describe a more appropriate method.

13. the winner of an election
14. the winner of a marathon
15. the sales of a new album
16. the gasoline mileage of a new car
17. the weather for tomorrow
18. the quality of food at a restaurant

19. Ask a random sample of the students in your school the way in which they travel to and from school. Let them choose from *walk, drive a car, ride in a car, ride in a bus,* and *ride a bicycle*. Using your results, find out how many students in your school would be expected to walk to and from school.

20. Ask a random sample of the students in your school which radio station they prefer. Report your results, the size of your sample, and how you made sure that the sample was random.

CAREER

MARKET RESEARCHER

Ken Jefferson is a market researcher. In his work, Ken collects data about consumers' preferences, attitudes, and interests. Ken uses the data to predict which types of products people might buy. Companies use Ken's data and predictions to decide how to invest their time and money.

example

Ken chose 40 West High School students at random and asked which style of athletic shoe each preferred. The results are shown in the frequency table at the right. How many of the 800 West High students could he expect to prefer running shoes?

Style	Tally	Frequency
basketball	卌 卌 卌 I	16
running	卌 卌	10
tennis	卌 卌	10
walking	IIII	4
		Total: 40

solution

P(running) × number of students = expected number

$$\frac{10}{40} \times 800 = \frac{1}{4} \times 800 = 200$$

Ken could expect about 200 of the 800 students to prefer running shoes.

exercises

Use the frequency table above.

1. How many of the 800 students would Ken expect to prefer walking shoes?
2. How many of the 800 students would Ken expect to prefer basketball shoes?
3. Which style is most likely to sell to West High School students?
4. A shoe company receives the results of Ken's poll. Why might the company decide to spend *more* money than it does now on advertising its basketball shoes?
5. Suppose that a shopping mall developer hires Ken to conduct a poll. One question Ken might ask each shopper is, "Do you prefer to shop in a department store or in a discount store?" Write four other questions Ken might ask the shoppers.

CHAPTER REVIEW

vocabulary vo·cab·u·lar·y

Choose the correct word to complete each sentence.

1. If the occurrence of one event does not affect the occurrence of another event, the two events are (*dependent, independent*).
2. A (*permutation, combination*) is an arrangement of objects in a particular order.

skills

Use the spinner at the right. Find each probability.

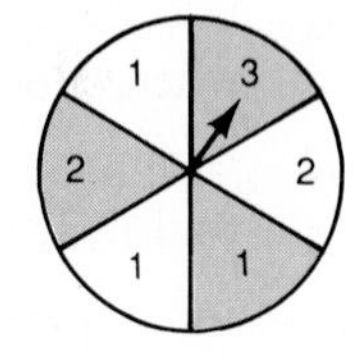

3. P(even number) 4. P(not blue) 5. P(3 or white)

Use the spinner at the right. Find the odds in favor of each event.

6. blue 7. 1 or 2 8. not odd number 9. 2 or 3

A purse contains 5 quarters, 3 dimes, and 4 nickels. Two coins are drawn at random. The first coin is replaced before the second one is drawn. Find each probability.

10. P(dime, then nickel) 11. P(quarter, then not dime) 12. P(two nickels)

Use the coins described above. Two coins are drawn at random. The first coin is not replaced before the second one is drawn. Find each probability.

13. P(nickel, then dime) 14. P(quarter, then nickel) 15. P(two dimes)

Find the number of permutations and the number of combinations.

16. 4 out of 8 people 17. 2 out of 10 books 18. 4 out of 4 cars

19. A coin is tossed three times. Use a tree diagram to find P(THH).
20. A number cube is rolled twice. Use the counting principle to find the probability that both numbers will be even.
21. Tina has only quarters, dimes, and nickels. In how many different ways, without regard to order, can she make 50¢ in change?
22. Dan tossed a bottle cap 30 times. It landed up 12 times and down 18 times. Find the probability that the bottle cap lands down.
23. In a production run of 5000 computer chips, 100 chips were tested. Only 1 of the 100 chips was defective. Predict the number of defective computer chips in the production run.

CHAPTER TEST

A number cube is rolled. Find each probability.

1. $P(5)$ **2.** $P(2 \text{ or } 3)$ **3.** $P(\text{not } 6)$ **4.** $P(\text{number} < 1)$ ***18-1***

A number cube is rolled. Find the odds in favor of each event.

5. 1 or 4 **6.** not 3 **7.** odd number **8.** number > 2

9. Use a tree diagram to find the number of possible outcomes when selecting a digit from 6 through 8 and a letter from P through R. ***18-2***

10. A coin is tossed three times. Use a tree diagram to find the probability that exactly two tosses will come up heads.

11. Use the counting principle to find the number of possible outcomes when choosing two cars from 4 models and 6 colors. ***18-3***

12. A number cube is rolled three times. Use the counting principle to find $P(2, \text{odd number}, \text{number} > 4)$.

D	R	A	W		A		C	A	R	D

Two cards are drawn at random from those above. The first card is replaced before the second one is drawn. Find each probability.

13. $P(\text{C, then R})$ **14.** $P(\text{two D's})$ **15.** $P(\text{A, then not A})$ ***18-4***

Two cards are drawn at random from those above. The first card is not replaced before the second one is drawn. Find each probability.

16. $P(\text{A, then W})$ **17.** $P(\text{two R's})$ **18.** $P(\text{not D, then D})$

19. Find the number of permutations of 3 out of 7 people. ***18-5***

20. Find the number of combinations of 3 out of 7 people.

21. Find the number represented by 6!.

22. In how many different ways, without regard to order, can you pay $80 with bills of $50, $20, and $10? ***18-6***

23. Chris tossed a magnet 50 times. It landed up 36 times and down 14 times. Find the probability that the magnet lands up. ***18-7***

24. Rona asked a random sample of 15 shoppers in Elm Supermarket which orange juice they preferred. Twelve preferred brand *A* and three preferred brand *B*. How many of the estimated 50 shoppers in the supermarket would Rona expect to prefer brand *B*? ***18-8***

Databases

A good way to organize information is to use a *database*. A **database** is simply a collection of data. A good database is well-organized, up-to-date, complete, and easy to use. A telephone directory is a good example of a database. It contains a well-organized listing of names, addresses, and telephone numbers.

When you use a computer to organize a database, you use a type of software that is called a **database management system (DBMS)**. A DBMS manages the database electronically. Using electronic files, you can store data more efficiently and in much less space.

The basic DBMS structure consists of *fields, records* and *files*. The individual pieces of information that you include in your database are called **fields.** A **record** is a collection of fields. A **file** is a collection of records. A personal database of friends' birthdates is shown below.

File: Birthdates of Friends		
Record	**Field 1**	**Field 2**
1	Aaron Cohen	August 17, 1975
2	Carol Davis	January 23, 1974
3	Paula Gravini	May 2, 1975
4	Robert Nashe	June 13, 1973
5	Jeffrey Miller	February 28, 1974

Once you have created a database, you should be able to spend less time updating, retrieving, and organizing information. A DBMS makes even a large and complicated database easier to use.

exercises

1. What is a database? List three characteristics of a good database.
2. What is a database management system?
3. Describe three sets of data for which you might want to create a personal database.

COMPETENCY TEST

Choose the letter of the correct answer.

1. 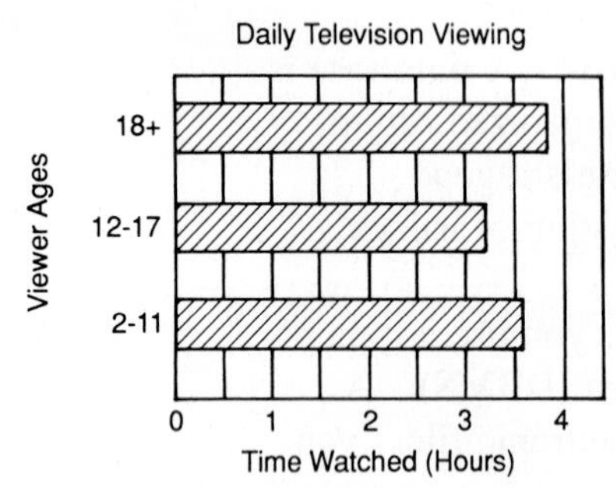

About how many hours of television does an average 15-year-old watch in one week?

A. 3 B. 4 C. 21 D. 28

2. **A marble is chosen at random from a bag containing 2 red, 3 blue, and 4 white marbles. Find *P*(blue or white).**

A. $\frac{7}{2}$ B. $\frac{7}{9}$ C. $\frac{1}{3}$ D. $\frac{4}{9}$

3. **Refer to Exercise 2. Find the odds *against* choosing a white marble.**

A. 4 to 9 B. 5 to 9
C. 4 to 5 D. 5 to 4

4. **Refer to Exercise 2. Two marbles are drawn. The first is not replaced before the second is drawn. Find *P*(white, then blue).**

A. $\frac{4}{9} \times \frac{3}{9}$ B. $\frac{4}{9} + \frac{3}{9}$
C. $\frac{4}{9} \times \frac{3}{8}$ D. $\frac{4}{9} + \frac{3}{8}$

5. **Test Scores**

6	7, 8
7	2, 2, 6, 8
8	0, 3, 4, 4, 5, 7, 8, 8
9	2, 2, 4, 5

The tests are graded as follows.
A: 95–100 C: 75–84
B: 85–94 D: 65–74

How many students received a B or better?

A. 4 B. 12
C. 7 D. 8

6. **Refer to Exercise 5. A histogram for the data has intervals 60–69, 70–79, 80–89, and 90–99. Which interval has the tallest bar?**

A. 90–99 B. 80–89
C. 70–79 D. 60–69

7. **Patty's temperature is recorded every 24 h. Which type of graph would best display the changes?**

A. line graph B. bar graph
C. pictograph D. circle graph

8. **A box-and-whisker plot is to be made for a set of data. Which number would *not* be used?**

A. first quartile
B. third quartile
C. mean
D. median

STATISTICS AND PROBABILITY

9.

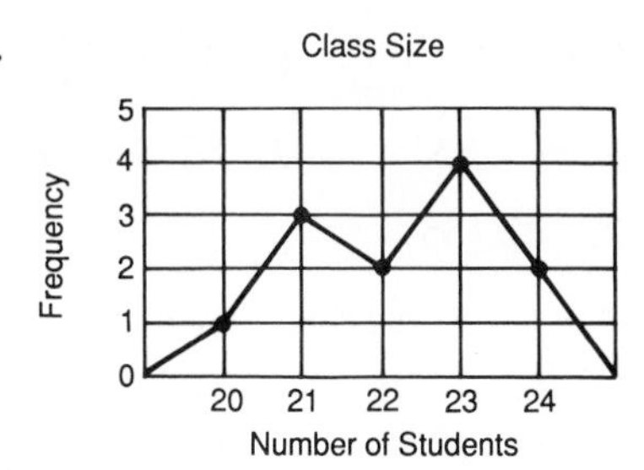

Find the mean class size.

A. 22 B. 22.25
C. 22.5 D. 23

10. Refer to Exercise 9. Find the median class size.

A. 22.5 B. 23
C. 22 D. 21

11. A company spends $300 million per year on advertising. How much is spent on TV advertising?

A. $52 million
B. $120 million
C. $156 million
D. $24 million

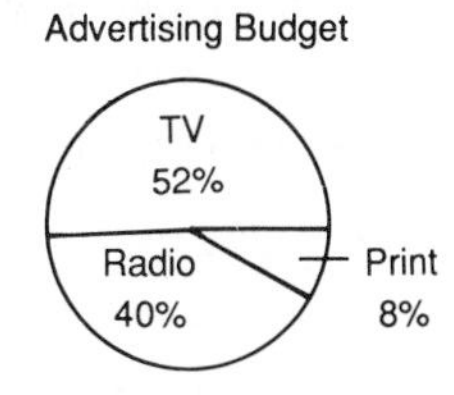

12. Refer to the graph for Exercise 11. Next year's percentages will be the same. The radio budget will be $128 million. Find the total advertising budget for next year.

A. $320 million B. $51.2 million
C. $120 million D. $428 million

13. Choose the most reasonable key for a pictograph displaying the data.

Ticket Sales

Day	M	T	W	TH	F
Number	120	80	90	50	140

A. ADMIT = 1 ticket
B. ADMIT = 5 tickets
C. ADMIT = 20 tickets
D. ADMIT = 100 tickets

14. In how many ways can 2 council members be chosen from a homeroom of 24 students?

A. 552 B. 12
C. 48 D. 276

15. A scattergram is best used to display:

A. changes in data over time
B. data expressed as percents of a whole
C. relationships between two sets of data
D. data from a frequency table

16. In how many ways can first, second, and third prizes be awarded in a contest of 30 people?

A. $\frac{30!}{3!}$ B. $3 \times 2 \times 1$
C. 30! D. $30 \times 29 \times 28$

CUMULATIVE REVIEW CHAPTERS 1–18

Find each answer. Write fractions in lowest terms.

1. \$29.25 − \$8.16
2. $2 \times 27 \times 5$
3. 0.52×6.8
4. $45 - 28 + 9$
5. $2\frac{3}{8} \times 4\frac{1}{2}$
6. $4\frac{1}{8} - 3\frac{2}{3}$
7. $\frac{3}{8} \div 1\frac{1}{2}$
8. $3\frac{2}{5} + 5\frac{3}{4}$
9. $24\overline{)1512}$
10. $18.2 \div 0.65$
11. $13.7 - 4.85$
12. $8795 - 918$
13. Find the mean, median, and mode: 43, 49, 84, 19, 22, 22, 55
14. Find the elapsed time from 11:15 P.M. to 10:00 A.M.
15. Write the ratio as a fraction in lowest terms: 6 in. to 8 ft
16. Round 43.79959 to the nearest hundredth.
17. Find the percent of increase: original cost: \$320; new cost: \$448
18. What percent of 85 is 17?
19. 34% of what number is 136?

Estimate.

20. $\frac{10}{29} \times 879$
21. 67% of 122
22. $38\overline{)1558}$
23. $22\frac{3}{4} - 8\frac{1}{8}$

A marble is drawn at random from those at the right.

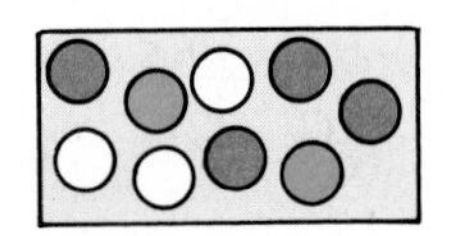

24. Find the odds in favor of drawing a blue or red marble.
25. The marble is *not* replaced. A second marble is drawn at random. Find *P*(white, then blue).
26. Trenton borrowed \$1500 for one year. The simple interest rate was 8.5% per year. How much interest did he pay?
27. The hourly wages at a company are \$5, \$7, \$26, \$8, \$27, \$5, \$6, and \$28. Is the mean a fair description of the wages? Explain.

Use the circle graph at the right.

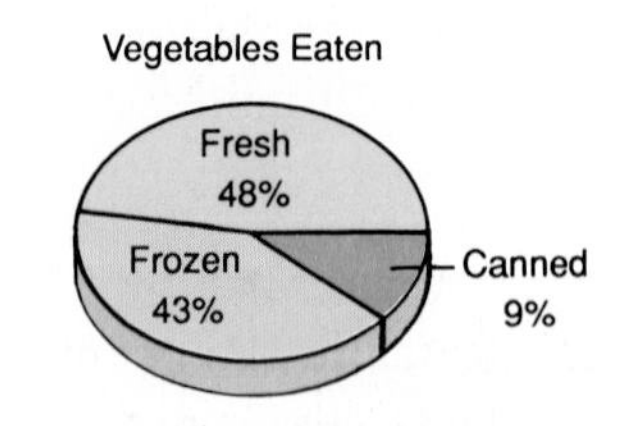

28. If the average American eats about 200 lb of vegetables per year, estimate the amount of these vegetables that are frozen.
29. Estimate the fraction of the vegetables consumed that are fresh vegetables.
30. Draw a line graph to display the given data.

United States Immigration

Year	1965	1970	1975	1980	1985
Number of Immigrants (Thousands)	297	373	386	531	570

A polar bear is comfortable in temperatures well below the freezing point. Negative numbers are used to express very low temperatures.

CHAPTER 19

RATIONAL NUMBERS

19-1 INTEGERS AND RATIONAL NUMBERS

In this book you have mainly used numbers that are *greater than* zero, or **positive numbers.** As you begin to study algebra, you will also use numbers that are *less than* zero, or **negative numbers.** The number zero itself is neither positive nor negative.

The numbers . . . , $^-3$, $^-2$, $^-1$, 0, 1, 2, 3, . . . are called **integers.** You use integers to represent quantities like distances above or below sea level, temperatures above or below zero, and gains or losses.

example 1

Write an integer that represents each phrase.

a. a loss of 7 yd

b. 40 ft above sea level

solution

a. $^-7$ ← **The negative *sign* indicates a number less than zero.**

b. 40 ← **You can also write 40 as $^+40$.**

your turn

Write an integer that represents each phrase.

1. a gain of 2 lb
2. 6°F below zero
3. a profit of $47

Integers are part of a set of numbers called *rational numbers.* A **rational number** can be written as a quotient of two integers, or $\frac{a}{b}$ where b does not equal 0. Each of these numbers is rational.

$$\frac{4}{5} \qquad 7 = \frac{7}{1} \qquad {}^-4 = \frac{^-4}{1} \qquad 6.3 = \frac{63}{10} \qquad 0.\overline{7} = \frac{7}{9} \qquad 0 = \frac{0}{5}$$

For any rational number, there are many ways you can show that the number is the quotient of two integers.

$$\frac{2}{3} = \frac{4}{6} = \frac{6}{9} = \frac{8}{12} = \ldots \qquad {}^-5 = \frac{^-5}{1} = \frac{^-10}{2} = \frac{^-15}{3} = \frac{^-20}{4} = \ldots$$

example 2

Rewrite each rational number as a quotient of two integers.

a. 9

b. $^-3\frac{1}{8}$

c. $0.1\overline{6}$

solution

a. $9 = \frac{9}{1}$

b. $^-3\frac{1}{8} = \frac{^-25}{8}$

c. $0.1\overline{6} = \frac{1}{6}$

your turn

Rewrite each rational number as a quotient of two integers.

4. $^{-}8$ **5.** $2\frac{1}{4}$ **6.** $0.8\overline{3}$ **7.** 3.99 **8.** 0

practice exercises

practice for example 1 (page 402)

Write an integer that represents each phrase.

1. a loss of $10
2. 150 ft below sea level
3. 75°F above zero
4. a gain of 15 yd
5. 4°F below zero
6. a gain of $2500

practice for example 2 (pages 402–403)

Rewrite each rational number as a quotient of two integers.

7. 15 **8.** 200 **9.** $^{-}44$ **10.** $6\frac{1}{8}$ **11.** $9\frac{2}{9}$
12. $^{-}0.7$ **13.** $^{-}1.16$ **14.** $0.\overline{5}$ **15.** $^{-}0.\overline{6}$ **16.** 0

mixed practice (pages 402–403)

Write a rational number that represents each phrase.

17. 23 ft above sea level
18. a loss of $120
19. 75°C below zero
20. 32 m below ground
21. 89°F above zero
22. 101°F above zero
23. a profit of $175.45
24. 42.5 m below sea level
25. a gain of $8\frac{1}{2}$ yd
26. a loss of $50.25
27. a gain of $\frac{1}{2}$ lb
28. a loss of $5\frac{1}{2}$ yd

Show two different ways to rewrite each rational number as a quotient of two integers.

29. 5 **30.** $^{-}6$ **31.** 0.24 **32.** $^{-}3\frac{1}{2}$ **33.** $0.\overline{2}$ **34.** 0

review exercises

Find the mean, median, mode(s), and range.

1. 28, 33, 32, 23, 25, 33
2. 8.5, 8.0, 5.8, 6.4, 6.6, 8.0, 6.4
3. Make a frequency table for the data at the right.

Bus Fares (Cents)

25	50	60	75	25
75	60	50	50	25
25	75	60	75	50
50	50	50	90	60
25	50	75	50	25

19-2 OPPOSITES AND ABSOLUTE VALUES

Often integers and rational numbers are represented as points on a **number line.** On a horizontal number line, positive numbers are usually to the right of zero and negative numbers are to the left of zero.

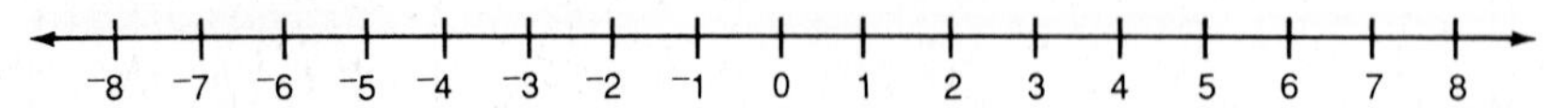

example 1

Use the number line at the right. Name the point that represents each number.

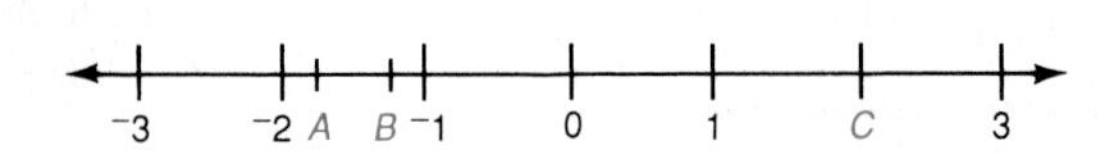

a. 2 **b.** $^{-}1\frac{3}{4}$ **c.** $^{-}1.25$

solution

a. 2 is between 1 and 3, so point *C* represents 2.

b. $^{-}1\frac{3}{4}$ is between $^{-}1$ and $^{-}2$ and is closer to $^{-}2$, so point *A* represents $^{-}1\frac{3}{4}$.

c. $^{-}1.25$ is between $^{-}1$ and $^{-}2$ and is closer to $^{-}1$, so point *B* represents $^{-}1.25$.

your turn

Use the number line at the right. Name the point that represents each number.

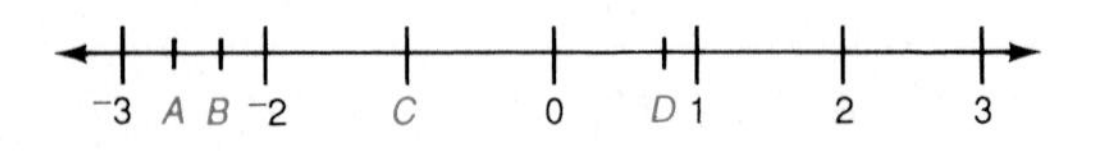

1. $^{-}1$ **2.** $^{-}2.\overline{3}$

3. $\frac{3}{4}$ **4.** $^{-}2\frac{2}{3}$

On a number line, numbers that are the same distance from zero but are on different sides of zero are called **opposites.** The opposite of zero is zero.

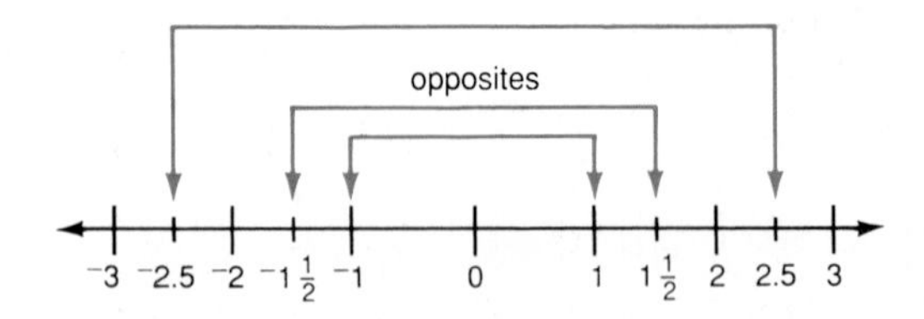

example 2

Write the opposite of each number: **a.** $^{-}1\frac{1}{2}$ **b.** 2.7 **c.** 0

solution

a. The opposite is $1\frac{1}{2}$. **b.** The opposite is $^{-}2.7$. **c.** The opposite is 0.

your turn

Write the opposite of each number.

5. 6 6. $1\frac{1}{2}$ 7. $^-3.2$ 8. 0

The **absolute value** of a number is the distance of the number from zero on a number line. The symbol | | indicates absolute value.

example 3

Find each absolute value.

a. $|^-2|$ b. $|5.7|$ c. $|0|$

solution

a. The distance from $^-2$ to 0 is 2 units. $|^-2| = 2$
b. The distance from 5.7 to 0 is 5.7 units. $|5.7| = 5.7$
c. The distance from 0 to 0 is 0 units. $|0| = 0$

your turn

Find each absolute value.

9. $|8|$ 10. $|^-9|$ 11. $|1.4|$ 12. $\left|^-6\frac{3}{5}\right|$

practice exercises

practice for example 1 (page 404)

Use the number line at the right. Name the point that represents each number.

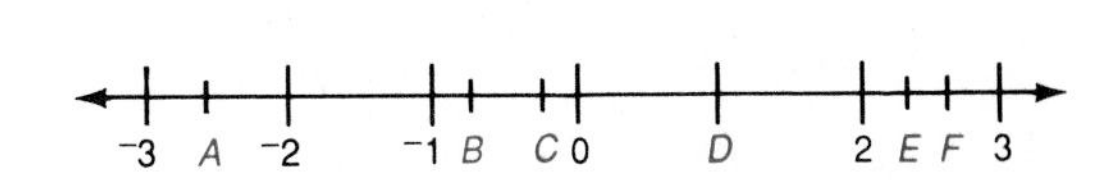

1. $\frac{^-1}{4}$ 2. 1 3. $\frac{^-3}{4}$

4. $2.\overline{6}$ 5. $2\frac{1}{3}$ 6. $^-2.\overline{6}$

practice for example 2 (pages 404–405)

Write the opposite of each number.

7. 15 8. 104 9. $^-100$ 10. $\frac{4}{5}$ 11. $\frac{^-3}{4}$

12. $3\frac{5}{6}$ 13. $^-7\frac{7}{8}$ 14. $^-31.5$ 15. 0.67 16. 0

practice for example 3 (page 405)

Find each absolute value.

17. $|5|$ **18.** $|23|$ **19.** $|{}^-41|$ **20.** $|{}^-77|$ **21.** $|0|$

22. $\left|{}^-\frac{6}{7}\right|$ **23.** $\left|\frac{3}{4}\right|$ **24.** $|3.2|$ **25.** $|{}^-4.99|$ **26.** $\left|{}^-15\frac{1}{8}\right|$

mixed practice (pages 404–405)

Use the number line below. Name the point that represents each number.

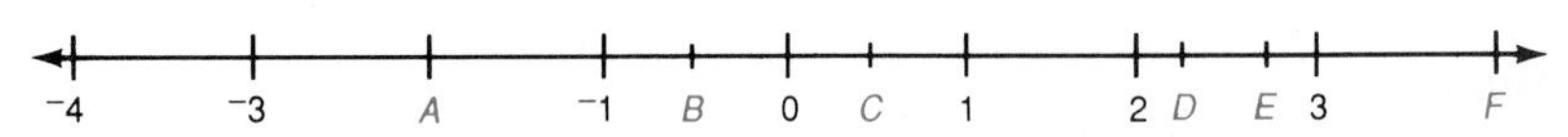

27. 4 **28.** $^-2$ **29.** $2\frac{3}{4}$ **30.** $\frac{1}{2}$ **31.** $^-0.5$ **32.** 2.25

Write the opposite and the absolute value of each number.

33. 95 **34.** $^-14$ **35.** $\frac{6}{17}$ **36.** $^-22\frac{3}{8}$ **37.** $^-5.87$ **38.** 0

Match each phrase to the appropriate number.

39. the opposite of $^-17$ **A.** 17

40. the opposite of 17 **B.** $^-17$

41. the absolute value of 17

42. the absolute value of $^-17$

review exercises

Compare. Replace each __?__ with >, <, or =.

1. 1.04 __?__ 10.4 **2.** 0.02 __?__ 0.003

3. $\frac{3}{4}$ __?__ $\frac{2}{3}$ **4.** $\frac{1}{9}$ __?__ $\frac{1}{10}$

5. 3.08 __?__ $3\frac{4}{5}$ **6.** $1\frac{3}{8}$ __?__ 1.375

7. Linda practices her violin for a total of 1 h in 2 days. At that rate, how many hours does she practice in 30 days?

8. Robert's coin bank contains 10 nickels, 12 dimes, and 7 quarters. He takes three coins from the bank and finds their total value. How many different totals are possible?

19-3 COMPARING RATIONAL NUMBERS

When you represent rational numbers on a horizontal number line, the greater number is to the right of the lesser number.

example 1

Compare. Replace each $\underline{?}$ with > or <.

a. $^{-}3 \underline{\ ?\ } {}^{-}1$ **b.** $1.5 \underline{\ ?\ } {}^{-}2.5$ **c.** $^{-}1 \underline{\ ?\ } {}^{-}3\frac{1}{2}$

solution

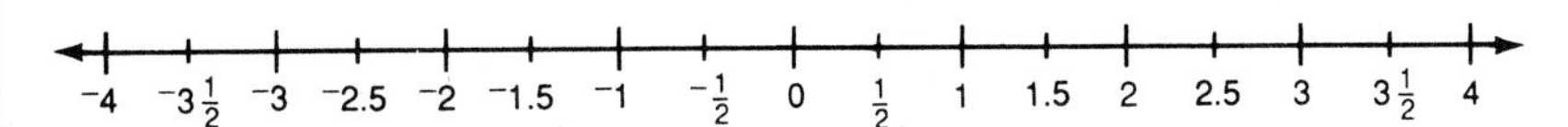

a. $^{-}3$ is to the *left* of $^{-}1$ on the number line, so $^{-}3$ is *less than* $^{-}1$, or $^{-}3 < {}^{-}1$.

b. 1.5 is to the *right* of $^{-}2.5$, so 1.5 is *greater than* $^{-}2.5$, or $1.5 > {}^{-}2.5$.

c. $^{-}1$ is to the *right* of $^{-}3\frac{1}{2}$, so $^{-}1$ is *greater than* $^{-}3\frac{1}{2}$, or $^{-}1 > {}^{-}3\frac{1}{2}$.

your turn

Compare. Replace each $\underline{?}$ with > or <.

1. $^{-}4 \underline{\ ?\ } {}^{-}2$ **2.** $0 \underline{\ ?\ } {}^{-}8$ **3.** $^{-}4.5 \underline{\ ?\ } 3.5$ **4.** $^{-}2 \underline{\ ?\ } {}^{-}3\frac{1}{6}$

example 2

Write each set of numbers in order from least to greatest.

a. $^{-}3$; $^{-}1$; 4; $^{-}4$ **b.** 1.6; $^{-}2.5$; $^{-}2.6$; 1.9 **c.** $\frac{3}{4}$; 0; $^{-}1\frac{1}{2}$; $^{-}2\frac{3}{4}$

solution

For each set of numbers, write the order in which they would appear *from left to right* on a horizontal number line.

a. $^{-}4$; $^{-}3$; $^{-}1$; 4 **b.** $^{-}2.6$; $^{-}2.5$; 1.6; 1.9 **c.** $^{-}2\frac{3}{4}$; $^{-}1\frac{1}{2}$; 0; $\frac{3}{4}$

your turn

Write each set of numbers in order from least to greatest.

5. $^{-}5$; 0; 9; $^{-}9$ **6.** 1.7; $^{-}3.3$; $^{-}3.4$; 1.8 **7.** $1\frac{1}{2}$; $^{-}1\frac{1}{2}$; $^{-}5\frac{1}{2}$; $2\frac{3}{4}$

practice exercises

practice for example 1 (page 407)

Compare. Replace each ? with > or <.

1. $^{-}3$? $^{-}2$
2. $^{-}5$? 0
3. 4 ? $^{-}3$
4. 8 ? $^{-}10$
5. $^{-}6.1$? 1.1
6. $^{-}4.2$? 4.4
7. $\frac{1}{3}$? $^{-}1\frac{1}{3}$
8. $^{-}3$? $^{-}2\frac{1}{2}$

practice for example 2 (page 407)

Write each set of numbers in order from least to greatest.

9. $^{-}1$; $^{-}2$; 3; $^{-}3$
10. $^{-}5$; 5; $^{-}6$; $^{-}8$
11. 1.6; $^{-}2.1$; 8.5; $^{-}4.5$
12. 2.7; $^{-}9.7$; 3.1; $^{-}6.3$
13. $\frac{3}{8}$; $\frac{^{-}7}{8}$; $^{-}1\frac{7}{8}$; $3\frac{5}{8}$
14. $\frac{^{-}3}{10}$; $\frac{9}{10}$; $\frac{1}{10}$; $^{-}5\frac{3}{10}$

mixed practice (page 407)

Tell whether each statement is *true* or *false*.

15. $^{-}9 < {^{-}12}$
16. $5 > 0$
17. $^{-}29 > {^{-}30}$
18. $^{-}11.6 < {^{-}12}$
19. $^{-}8 < \frac{^{-}1}{2} < \frac{1}{2}$
20. $4\frac{1}{5} < {^{-}6\frac{2}{5}}$
21. $0 > {^{-}2.1} > {^{-}2.6}$
22. $^{-}3.5 < 3.5$
23. Name two mixed numbers that are between $^{-}20$ and $^{-}21$.
24. Name two decimal numbers that are between $^{-}8.4$ and $^{-}9.4$.
25. The greatest recorded depth of a lake in Africa is 571 ft below sea level. In Australia, the greatest recorded depth of a lake is 38 ft below sea level. On which continent is the lake with the greater depth below sea level?

review exercises

Find each answer.

1. $19 + 206 + 73$
2. $1304 - 685$
3. $4.5 + 10.93$
4. $61.2 - 34.19$
5. $\frac{1}{10} + \frac{4}{5}$
6. $\frac{7}{8} - \frac{3}{8}$
7. $2\frac{1}{2} + 3\frac{3}{4}$
8. $1\frac{2}{3} - \frac{1}{6}$

19-4 ADDING RATIONAL NUMBERS

You can use arrows on a number line to represent the addition of two rational numbers.

example 1

Use a number line to find each sum: **a.** $^-3 + 4$ **b.** $1.5 + {}^-2.5$

solution

a. Start at $^-3$. Move 4 units to the right.

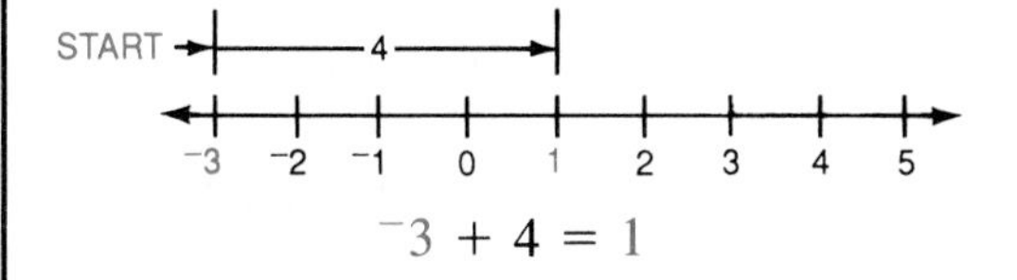

$^-3 + 4 = 1$

b. Start at 1.5. Move 2.5 units to the left.

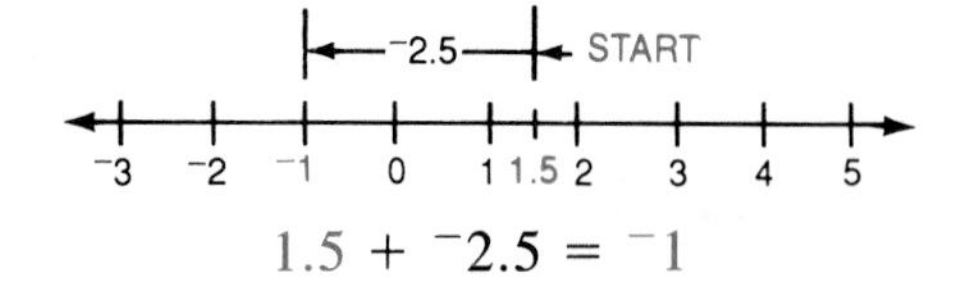

$1.5 + {}^-2.5 = {}^-1$

your turn

Use a number line to find each sum.

1. $4 + 5$ **2.** $5 + {}^-3$ **3.** $^-4.5 + 2$ **4.** $^-1\frac{1}{2} + {}^-3\frac{1}{2}$

You can also add rational numbers without drawing a number line. To add two numbers that have the *same* sign, use the following method.

1. Find the absolute value of each number.
2. Add the absolute values.
3. Give the sum the same sign as the numbers.

example 2

Find each sum: **a.** $8 + 4$ **b.** $^-4.3 + {}^-2.1$

solution

a. $|8| + |4| = 8 + 4 = 12$
8 and 4 are both *positive*.
$8 + 4 = 12$

b. $|{}^-4.3| + |{}^-2.1| = 4.3 + 2.1 = 6.4$
$^-4.3$ and $^-2.1$ are both *negative*.
$^-4.3 + {}^-2.1 = {}^-6.4$

your turn

Find each sum.

5. $2 + 7$ **6.** $^-30 + {}^-20$ **7.** $\frac{^-1}{3} + {}^-2\frac{1}{3}$ **8.** $2.2 + 10.9$

To add two numbers that have *different* signs, use the following method.

1. Find the absolute value of each number.
2. Subtract the lesser absolute value from the greater absolute value.
3. Give the sum the sign of the number with the greater absolute value.

example 3

Find each sum: **a.** $5 + {}^{-}12$ **b.** $7\frac{3}{5} + {}^{-}\frac{1}{5}$

solution

a. $|5| = 5$ and $|{}^{-}12| = 12$

$12 - 5 = 7$

$5 + {}^{-}12 = {}^{-}7$ ← **The sum is negative.**

b. $\left|7\frac{3}{5}\right| = 7\frac{3}{5}$ and $\left|{}^{-}\frac{1}{5}\right| = \frac{1}{5}$

$7\frac{3}{5} - \frac{1}{5} = 7\frac{2}{5}$

$7\frac{3}{5} + {}^{-}\frac{1}{5} = 7\frac{2}{5}$ ← **The sum is positive.**

your turn

Find each sum.

9. $3 + {}^{-}9$ **10.** ${}^{-}24 + 24$ **11.** $5.8 + {}^{-}1.7$ **12.** ${}^{-}4\frac{2}{3} + 1\frac{1}{3}$

practice exercises

practice for example 1 (page 409)

Use a number line to find each sum.

1. $8 + 1$ **2.** ${}^{-}4 + {}^{-}2$ **3.** ${}^{-}1 + 2$ **4.** $2 + {}^{-}3$

5. $1.5 + {}^{-}3.5$ **6.** ${}^{-}2.5 + 4$ **7.** ${}^{-}1\frac{1}{2} + 2\frac{1}{2}$ **8.** $3\frac{1}{2} + {}^{-}5$

practice for example 2 (page 409)

Find each sum.

9. $6 + 49$ **10.** $25 + 14$ **11.** ${}^{-}14 + {}^{-}21$ **12.** ${}^{-}18 + {}^{-}7$

13. ${}^{-}2.1 + {}^{-}4.3$ **14.** $1.9 + 5$ **15.** $1\frac{5}{6} + 4$ **16.** ${}^{-}\frac{1}{5} + {}^{-}3\frac{1}{5}$

practice for example 3 (page 410)

Find each sum.

17. ${}^{-}18 + 11$ **18.** ${}^{-}44 + 44$ **19.** $15 + {}^{-}7$ **20.** $28 + {}^{-}33$

21. ${}^{-}3.6 + 1$ **22.** $2.6 + {}^{-}6.2$ **23.** $4\frac{1}{3} + {}^{-}6\frac{2}{3}$ **24.** ${}^{-}\frac{1}{4} + 2$

mixed practice (pages 409–410)

Find each sum.

25. $40 + 20$ **26.** $^{-}3 + ^{-}51$ **27.** $4 + ^{-}58$ **28.** $^{-}7 + 57$

29. $10 + ^{-}10$ **30.** $^{-}2 + ^{-}2$ **31.** $2\frac{3}{7} + 6\frac{3}{7}$ **32.** $4\frac{1}{9} + ^{-}5\frac{5}{9}$

33. $^{-}6.2 + 7.4$ **34.** $^{-}5.5 + ^{-}2.9$ **35.** $3 + ^{-}6 + ^{-}7$ **36.** $^{-}7 + ^{-}2 + 20$

37. At 10:00 A.M. the temperature was 40°F above zero. At noon it was 10°F warmer. What was the temperature at noon?

38. Marcia and Sue dove to a depth of 30 ft underwater. After 15 s, they rose 14 ft. At what depth were they then?

39. Frank has 12 dozen stamps in his stamp collection. Jim has 140 stamps in his collection. Whose stamp collection contains more stamps? How many more?

review exercises

Write each percent as a decimal. Write each decimal as a percent.

1. 16% **2.** 70% **3.** 0.04 **4.** 0.31 **5.** 1.4% **6.** 212%

7. 1.5 **8.** 0.001 **9.** 0.2% **10.** $38\frac{1}{2}\%$ **11.** $0.11\frac{1}{3}$ **12.** 0.609

mental math

To add rational numbers mentally, look at all the addends before beginning. Look for opposites. Group positive and negative numbers.

$$4 + 7 + ^{-}5 + ^{-}15 + 2 + ^{-}4$$

opposites		positive numbers		negative numbers
$4 + ^{-}4$	+	$7 + 2$	+	$^{-}5 + ^{-}15$
0	+	9	+	$^{-}20$
			$^{-}11$	

Add mentally.

1. $^{-}5 + 7 + ^{-}2 + 5 + ^{-}8$ **2.** $6 + 9 + ^{-}6 + ^{-}3 + 8 + ^{-}9$

3. $6\frac{1}{3} + 1\frac{1}{4} + ^{-}6\frac{1}{3}$ **4.** $^{-}2\frac{1}{2} + ^{-}3\frac{1}{3} + 5\frac{1}{2} + 3\frac{1}{3}$

5. $1.2 + 6.7 + 9.1 + ^{-}1.2$ **6.** $4.9 + ^{-}2.2 + ^{-}4.9 + ^{-}3.8$

19-5 SUBTRACTING RATIONAL NUMBERS

For any subtraction, you can write a related addition.

example 1

Write a related addition.

a. $49 - 51$ **b.** $77 - {}^{-}10$

solution

a. $49 - 51$
$= 49 + {}^{-}51$ ← **Add the opposite of the number being subtracted.**

b. $77 - {}^{-}10$
$= 77 + 10$

your turn

Write a related addition.

1. $36 - 63$ **2.** $14 - {}^{-}20$ **3.** ${}^{-}29.1 - 17.5$ **4.** ${}^{-}6\frac{1}{3} - {}^{-}3\frac{1}{3}$

To subtract rational numbers, use the following method.

1. Write a related addition using the opposite of the number being subtracted.
2. Add, using the rules for addition of rational numbers.

example 2

Find each difference.

a. $8 - 12$ **b.** ${}^{-}23 - 14$ **c.** $\frac{1}{5} - {}^{-}2\frac{1}{5}$

solution

a. $8 - 12$
$= 8 + {}^{-}12$ ← **related addition**
$= {}^{-}4$

b. ${}^{-}23 - 14$
$= {}^{-}23 + {}^{-}14$
$= {}^{-}37$

c. $\frac{1}{5} - {}^{-}2\frac{1}{5}$
$= \frac{1}{5} + 2\frac{1}{5}$
$= 2\frac{2}{5}$

your turn

Find each difference.

5. $14 - 37$ **6.** ${}^{-}66 - 20$ **7.** $\frac{1}{9} - {}^{-}7\frac{4}{9}$ **8.** ${}^{-}7.5 - {}^{-}3.3$

practice exercises

practice for example 1 (page 412)

Write a related addition.

1. $21 - 57$
2. $210 - {}^-45$
3. ${}^-4 - 98$
4. ${}^-4 - {}^-12$
5. $49.5 - 32.7$
6. $16.8 - {}^-3.9$
7. ${}^-1\frac{1}{7} - 6\frac{2}{7}$
8. ${}^-3\frac{4}{15} - {}^-10\frac{4}{15}$

practice for example 2 (page 412)

Find each difference.

9. $6 - 9$
10. $44 - 18$
11. $52 - {}^-14$
12. $75 - {}^-19$
13. ${}^-2 - 78$
14. ${}^-6 - 33$
15. ${}^-18 - {}^-15$
16. ${}^-10 - {}^-26$
17. $34 - 43$
18. $5 - {}^-50$
19. ${}^-3 - 24$
20. ${}^-13 - {}^-17$
21. ${}^-2.2 - 1.1$
22. $4.6 - {}^-12.5$
23. $\frac{{}^-3}{11} - \frac{{}^-2}{11}$
24. $\frac{{}^-1}{4} - 1$

mixed practice (page 412)

Copy and complete the chart.

	Subtraction	Related Addition	Answer
25.	$6 - 15$	?	?
26.	${}^-8 - 20$	?	?
27.	${}^-5 - {}^-27$	?	?
28.	$7.3 - {}^-5.6$	?	?
29.	${}^-6\frac{1}{25} - 16\frac{1}{25}$	?	?
30.	?	${}^-1 + {}^-30$	?
31.	?	$9 + 12$	?
32.	?	${}^-5 + 100$	?

33. The temperature at noon in Rogers Pass, Montana, was ${}^-3$°F. By 6:00 P.M. the temperature had fallen 15°F. Find the temperature at 6:00 P.M.
34. Sara Devon had a balance of $63.15 in her checking account. She wrote a check for $14. Find her new balance.
35. On one play, a football player gained 3 yd and then lost 10 yd. Write one integer that represents the number of yards the player gained or lost on the play.

36. Mr. Simms found that the profit from his florist shop on June 3 was \$389.50. On June 2 the profit was \$350.89. How much greater was his profit on June 3?

37. Ann Boorstal wrote the first 110 pages of her novel. She then wrote 54 more pages. How many pages has Ann written so far?

38. Mr. Doherty coached 20 baseball games. His team, the Doherty Double Plays, won 9 of these games. What percent of the games played did the Double Plays win?

review exercises

Find each answer.

1. $3596 \div 58$
2. 4.6×1.09
3. 0.5×3.3
4. $3.9\overline{)0.819}$
5. $\frac{4}{5} \div 20$
6. 23×417
7. $\frac{3}{4} \times \frac{8}{9}$
8. 607×42
9. $0.64 \div 0.8$
10. $1\frac{1}{8} \times 3\frac{2}{3}$
11. $18\overline{)630}$
12. $12 \div 1\frac{5}{6}$

Give the number of significant digits and the unit of precision.

13. 103.4 m
14. 6.90 cm
15. 0.07 kg
16. 2100 mi

puzzle corner

A *magic square* is an arrangement of numbers whose rows, columns, and diagonals all have the same sum. For example, the magic square at the right uses the numbers 0, 1, 2, 3, 4, 5, 6, 7, and 8. The sum of any row, column, or major diagonal is 12.

1	8	3
6	4	2
5	0	7

Make a magic square using the numbers $^-5$, $^-4$, $^-3$, $^-2$, $^-1$, 0, 1, 2, and 3. The sum of any row, column, or major diagonal should be $^-3$.

19-6 MULTIPLYING AND DIVIDING RATIONAL NUMBERS

When you multiply rational numbers, parentheses are often used as a multiplication symbol.

$2 \times {}^{-}6$ can be written as $2({}^{-}6)$ or $(2)({}^{-}6)$.

Now consider the patterns shown below.

$2(6) = 12$
$1(6) = 6$ ← **Both factors are positive. The product is positive.**
$0(6) = 0$
$({}^{-}1)(6) = {}^{-}6$
$({}^{-}2)(6) = {}^{-}12$ ← **One factor is negative. One factor is positive. The product is negative.**

$2({}^{-}6) = {}^{-}12$
$1({}^{-}6) = {}^{-}6$ ← **One factor is positive. One factor is negative. The product is negative.**
$0({}^{-}6) = 0$
$({}^{-}1)({}^{-}6) = 6$
$({}^{-}2)({}^{-}6) = 12$ ← **Both factors are negative. The product is positive.**

The patterns above suggest the following rules for multiplying rational numbers.

The product of two positive or two negative numbers is positive.
The product of a positive number and a negative number is negative.

example 1

Find each product.

a. $3(7) = 21$ **b.** $({}^{-}2)({}^{-}4) = 8$ **c.** $5({}^{-}9) = {}^{-}45$ **d.** $({}^{-}6)(6) = {}^{-}36$

your turn

Find each product.

1. $({}^{-}7)(2)$ **2.** $2({}^{-}1.3)$ **3.** $({}^{-}8)\left({}^{-}\frac{1}{2}\right)$ **4.** $0({}^{-}4)$

example 2

Find each product.

a. $({}^{-}5)({}^{-}4)({}^{-}6)$
$20({}^{-}6)$ ← **Multiply two numbers at a time.**
${}^{-}120$

b. $({}^{-}3)(5.1)({}^{-}2)$
$({}^{-}15.3)({}^{-}2)$
30.6

your turn

Find each product.

5. $6({}^{-}3)(2)$ **6.** $({}^{-}8)({}^{-}2)(5)$ **7.** $\left(\frac{1}{4}\right)\left(\frac{5}{8}\right)\left({}^{-}\frac{2}{5}\right)$ **8.** $(0.1)(7)(5.5)$

You can divide rational numbers by using the relationship between division and multiplication.

$12 \div 6 = 2$ because $2(6) = 12$ $\quad$ $^-12 \div 6 = {}^-2$ because $(^-2)(6) = {}^-12$

$12 \div {}^-6 = {}^-2$ because $(^-2)(^-6) = 12$ $\quad$ $^-12 \div {}^-6 = 2$ because $2(^-6) = {}^-12$

The examples above suggest the following rules for dividing rational numbers.

The quotient of two positive or two negative numbers is positive.
The quotient of a positive number and a negative number is negative.

example 3

Find each quotient.

a. $14 \div 7 = 2$ **b.** $^-63 \div {}^-9 = 7$ **c.** $24 \div {}^-8 = {}^-3$ **d.** $^-12 \div 3 = {}^-4$

your turn

Find each quotient.

9. $^-48 \div {}^-6$ **10.** $^-2.5 \div 5$ **11.** $\frac{^-1}{3} \div \frac{^-3}{4}$ **12.** $0 \div {}^-5$

practice exercises

practice for example 1 (page 415)

Find each product.

1. $9(2)$ **2.** $(^-4)(7)$ **3.** $(^-5)(6)$ **4.** $(^-3)(0)$

5. $23(^-8)$ **6.** $(12)(17)$ **7.** $(^-2.6)(3.5)$ **8.** $\left(\frac{^-3}{4}\right)\left(\frac{^-5}{6}\right)$

practice for example 2 (page 415)

Find each product.

9. $6(^-5)(^-9)$ **10.** $4(2)(^-4)$ **11.** $(^-7)(^-8)(^-2)$ **12.** $(^-5)(5)(^-5)$

13. $\frac{22}{7}(4)(14)$ **14.** $(^-3)(^-4.5)(2.2)$ **15.** $\left(\frac{^-1}{2}\right)\left(\frac{4}{9}\right)\left(\frac{3}{4}\right)$ **16.** $7(^-3.1)(0)$

practice for example 3 (page 416)

Find each quotient.

17. $36 \div 4$ **18.** $^-42 \div 7$ **19.** $^-72 \div 9$ **20.** $0 \div {}^-6.5$

21. $56 \div {}^-2$ **22.** $80 \div 5$ **23.** $\frac{^-3}{8} \div \frac{5}{6}$ **24.** $^-19.2 \div {}^-8$

mixed practice (pages 415–416)

Find each answer.

25. $5(^-5)$
26. $^-27 \div {}^-3$
27. $^-96 \div 8$
28. $17(4)$
29. $(^-2.5)(^-8.2)$
30. $^-5.6 \div 7$
31. $\frac{2}{9} \div \frac{4}{5}$
32. $\frac{3}{10}\left(\frac{^-2}{3}\right)$
33. $0 \div {}^-2.5$
34. $8(^-5)(^-7)$
35. $(^-1.4)(4)(2)$
36. $\frac{5}{8}(0)(^-8)$
37. $(^-4)^2$
38. $(^-10)^3$
39. $(^-2)^3$
40. $(^-2)^4$
41. $(^-1)^4$
42. $(^-1)^7$
43. Scott read a thermometer four times. Each time he recorded a decrease of 5°F. What was the total change in the temperature?
44. Elizabeth read a thermometer four times. Each time she recorded an increase of 3°F. What was the total change in the temperature?
45. John noticed that his watch lost $3\frac{1}{2}$ min per day. After three days, how many minutes had his watch lost?
46. Mary sold six bracelets for $9.95 each and eight pairs of earrings for $2.75 per pair. What was the total amount of these sales?

review exercises

Write each percent as a fraction or a mixed number in lowest terms. Write each fraction or mixed number as a percent.

1. 5%
2. 160%
3. 3.4%
4. $\frac{3}{4}$
5. $\frac{1}{3}$
6. $2\frac{7}{8}$

calculator corner

You can use the [+/−] key to solve problems that involve rational numbers. When you enter a number and then press the [+/−] key, the calculator makes the number negative.

example $^-17 + {}^-34 - 28$

solution Enter 17 [+/−] + 34 [+/−] − 28 =.
The result is −79.

Use a calculator to find each answer.

1. $^-27 + {}^-66 - 103$
2. $^-51 - {}^-86 + {}^-34$
3. $^-22 \times 15 \div {}^-11$
4. $91 \div {}^-13 \times 45$
5. $^-7.3 + {}^-16.8 + 5.9$
6. $^-12.8 \times 1.5 \div {}^-9.6$

SKILL REVIEW

Write an integer that represents each phrase.

19-1

1. 30°F below zero 2. 20 ft below sea level 3. a gain of $20

Rewrite each rational number as a quotient of two integers.

4. 20 5. $^{-}4\frac{1}{4}$ 6. 35 7. 0.47 8. $0.\overline{1}$ 9. 0

Write the opposite and the absolute value of each number.

19-2

10. 8 11. $^{-}39$ 12. $^{-}2.6$ 13. 14.7 14. $^{-}0.9$

15. $\frac{^{-}1}{5}$ 16. $\frac{2}{9}$ 17. $15\frac{1}{4}$ 18. $^{-}37\frac{1}{5}$ 19. 0

Compare. Replace each __?__ with > or <.

19-3

20. $^{-}1$ __?__ $^{-}5$ 21. 2.7 __?__ 6.7 22. $^{-}5.8$ __?__ 0 23. 3 __?__ $\frac{^{-}1}{3}$

24. In Newtown the temperature is $^{-}5$°F. In Oldtown the temperature is $^{-}15$°F. In which town is the temperature higher?

Write each set of numbers in order from least to greatest.

25. $^{-}6$; $^{-}5$; 4; $^{-}3$ 26. 17; $^{-}17$; 16; $^{-}16$ 27. 4; $^{-}3.2$; $^{-}0.2$; 1.4

28. $^{-}4.9$; 9.4; $^{-}4.6$; 9.6 29. $\frac{^{-}4}{5}$; 0; $^{-}1\frac{1}{5}$; $\frac{4}{5}$ 30. $\frac{1}{12}$; $\frac{^{-}5}{12}$; $^{-}2\frac{1}{12}$; $\frac{7}{12}$

Find each sum.

19-4

31. 6 + 17 32. $^{-}3 + ^{-}27$ 33. $^{-}15 + 44$ 34. $18 + ^{-}23$

35. $5.2 + ^{-}6.5$ 36. $^{-}3.4 + 30.4$ 37. $^{-}2\frac{2}{7} + ^{-}11\frac{4}{7}$ 38. $^{-}4\frac{3}{5} + 4\frac{3}{5}$

39. One week the price paid for Cole Company stock decreased $\$5\frac{7}{8}$ per share. The next week the price decreased $2 per share. Find the total change in the price of the stock at the end of the two weeks.

Find each answer.

19-5

40. 3 − 7 41. $^{-}22 - ^{-}37$ 42. $^{-}58 - 30$ 43. $42 - ^{-}18$

44. $^{-}0.3 - ^{-}0.3$ 45. $^{-}6.5 - 2.1$ 46. $\frac{1}{7} - \frac{^{-}5}{7}$ 47. $\frac{2}{9} - \frac{7}{9}$

19-6

48. $(^{-}7)(7)$ 49. $(1.4)(8.5)$ 50. $^{-}8.5 \div 5$ 51. $^{-}54 \div ^{-}6$

52. $\left(\frac{^{-}2}{3}\right)\left(\frac{^{-}1}{2}\right)$ 53. $0 \div ^{-}9$ 54. $7(^{-}3)(^{-}5)$ 55. $\left(\frac{^{-}1}{3}\right)\left(\frac{9}{10}\right)\left(\frac{5}{7}\right)$

19-7 TEMPERATURE

Temperature can be measured in degrees Fahrenheit (°F) or degrees Celsius (°C). The two thermometers at the right show some common temperatures on both scales.

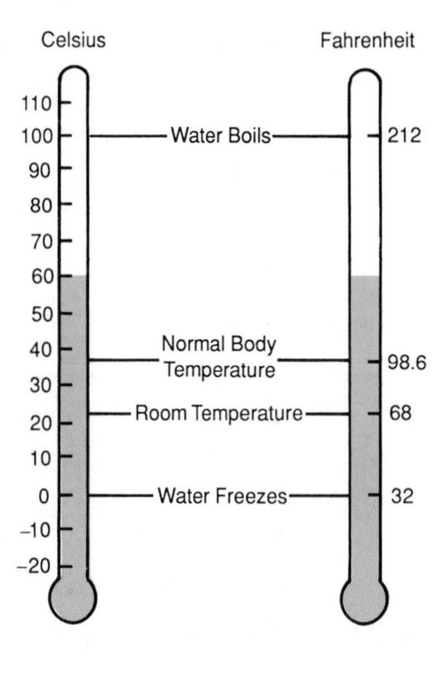

example

Select the more reasonable unit. Choose °F or °C.

a. Normal body temperature is about 37 __?__.

b. The temperature of a glass of cold milk is about 35 __?__.

solution

a. °C **b.** °F

exercises

Select the more reasonable unit. Choose °F or °C.

1. The temperature of an ice cube is about $^{-}1$ __?__.
2. The air temperature on a warm day is about 30 __?__.
3. The air temperature on a hot summer day is about 85 __?__.
4. A pleasant water temperature for a swimming pool is about 20 __?__.
5. The temperature of a person with a slight fever is about 100 __?__.
6. A comfortable air temperature for a room is about 22 __?__.
7. The temperature of a bowl of hot broth is about 175 __?__.
8. The temperature of a can of frozen orange juice is about 29 __?__.
9. Find out if your local newspaper reports the weather in the Fahrenheit scale, the Celsius scale, or both.

APPLICATION

19-8 SCIENTIFIC NOTATION

Scientists and engineers often write numbers in a simpler form called **scientific notation.** A number written in this form has two factors.

at least 1, but less than 10 → 2.3×10^6 ← **a power of 10**

2.3×10^6 is the number 2,300,000 written in scientific notation.

23×10^6 is *not* in scientific notation, because 23 is greater than 10.

2.3×6^6 is *not* in scientific notation, because 6^6 is not a power of 10.

example

a. Write 9.4×10^5 in numeral form.

b. Write 706.8 in scientific notation.

solution

a. $9.4 \times 10^5 = 940{,}000$ ← **Move the decimal point to the right as many places as the power of 10.**

5 places

b. $706.8 = 7.068 \times 10^2$

exercises

Is each number written in scientific notation? Write *Yes* or *No*.

1. 7.7×10^7 **2.** 3.044×10^9 **3.** 2.6×2.6^{10} **4.** 15×10^6

Write each number in scientific notation.

5. 1275 **6.** 439.8 **7.** 10,000,000 **8.** 275,000,000

Write each number in numeral form.

9. 6×10^2 **10.** 3.2×10^8 **11.** 8.7×10^{11} **12.** 2.012×10^5

Write each underlined number in scientific notation.

13. The temperature at the core of the sun is about 20,000,000°C.

14. The first American in space, Alan Shepard, Jr., flew to a peak altitude of 187 km.

15. The oldest fossil of a flowering plant that has been discovered is about 65,000,000 years old.

16. In 1987, the population of China was more than 1,062,000,000.

17. The height of the World Trade Center in New York City is 419 m.

18. The longest railroad tunnel in the world is in Japan. The tunnel is <u>53.5</u> km long.
19. In the national election of 1984, about <u>92,650,000</u> people voted.
20. The world's largest dam is in Arizona and has a volume of about <u>209,500,000</u> m^3 (cubic meters).
21. The record speed of an airplane over a measured straightaway course is <u>2193.16</u> mi/h.

Write each underlined number in numeral form.

22. One American company with a large number of stockholders has <u>2.782×10^6</u> stockholders.
23. The public library in Jackson, Mississippi, contains about <u>5×10^5</u> volumes.
24. A recent census showed that the population of Oregon was about <u>2.633×10^6</u>.
25. A teaspoon of salt contains about <u>2×10^3</u> mg of sodium.
26. Each year more than <u>5.3×10^7</u> passengers travel through O'Hare Airport in Chicago.

You can also use scientific notation as a simpler way of writing numbers that are between 0 and 1.

$1.6 \times 10^{-5} = 0.000016$ (5 places) ← **The negative exponent indicates that the decimal point should be moved *to the left*.**

$0.0082 = 8.2 \times 10^{-3}$

Is each number written in scientific notation? Write *Yes* or *No*.

27. 6.3×10^{-3} **28.** 4.4×4^{-4} **29.** 28×10^{-6} **30.** 8×10^{-5}

Write each number in scientific notation.

31. 0.3 **32.** 0.000005 **33.** 0.027 **34.** 0.000042

Write each number in numeral form.

35. 5×10^{-2} **36.** 7×10^{-4} **37.** 9.1×10^{-5} **38.** 2.44×10^{-3}

Write each underlined number in scientific notation.

39. The width of the smallest plant organism is about <u>0.03</u> mm.

40. The width of the smallest influenza virus is about <u>0.0001</u> mm.

41. A microsecond is equal to <u>0.000001</u> s.

42. The smallest insect has a mass of about <u>0.005</u> mg.

Write each underlined number in numeral form.

43. An average hummingbird has a mass of about $\underline{2 \times 10^{-3}}$ kg.

44. The mass of a house spider is about $\underline{1 \times 10^{-4}}$ kg.

45. A garden snail moves at a speed of about $\underline{3.13 \times 10^{-2}}$ mi/h.

46. The mass of the largest butterfly is about $\underline{2.464 \times 10^{-1}}$ g.

calculator corner

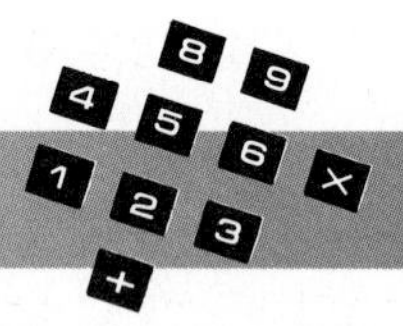

Some calculators have an [EXP] or [EE] key. You can use this key to perform operations with numbers that are written in scientific notation. Pressing the [EXP] or [EE] key enters the next number as a power of ten.

example $(4 \times 10^{-11}) \times (2 \times 10^{-7})$

solution Enter 4 [EXP] 11 [+/-] × 2 [EXP] 7 [+/-] =. ← **$^{-}11$ and $^{-}7$ are entered as powers of 10.**

The result is 8 − 18, or 8×10^{-18}.

Find each answer using a calculator.

1. $(1 \times 10^{8}) \times (5 \times 10^{7})$ **2.** $(1 \times 10^{-4}) \times (9 \times 10^{-15})$

3. $(2 \times 10^{12}) \times (3 \times 10^{4})$ **4.** $(3 \times 10^{-5}) \times (3 \times 10^{-16})$

Study your answers to Exercises 1–4. What is the pattern? Use your pattern to find these answers *without* a calculator.

5. $(1 \times 10^{2}) \times (7 \times 10^{4})$ **6.** $(1 \times 10^{-3}) \times (8 \times 10^{-14})$

7. $(1 \times 10^{10}) \times (6 \times 10^{-2})$ **8.** $(1 \times 10^{3}) \times (5 \times 10^{-10})$

CHAPTER REVIEW

vocabulary vo·cab·u·lar·y

Choose the correct word or phrase to complete each sentence.

1. On a number line, numbers that are the same distance from zero but on different sides of zero are called (*negative numbers, opposites*).
2. On a number line, the distance of a number from zero is the (*absolute value, scientific notation*) of the number.

skills

Write an integer that represents each phrase.

3. a loss of \$50　　**4.** 6°C below zero　　**5.** 45 m above sea level

Rewrite each rational number as a quotient of two integers.

6. $^{-}110$　　**7.** 49　　**8.** 0.31　　**9.** $0.\overline{8}$　　**10.** $^{-}7\frac{1}{2}$　　**11.** 0

Write the opposite of each number.

12. 59　　**13.** $^{-}78$　　**14.** 17.4　　**15.** $^{-}0.5$　　**16.** $\frac{^{-}1}{10}$　　**17.** $14\frac{1}{3}$

Find each absolute value.

18. $|^{-}36|$　　**19.** $|12|$　　**20.** $|^{-}7.6|$　　**21.** $|44.5|$　　**22.** $\left|3\frac{1}{12}\right|$　　**23.** $\left|\frac{^{-}13}{15}\right|$

Write each set of numbers in order from least to greatest.

24. $^{-}4$; $^{-}2$; 2; 0　　**25.** 1.8; $^{-}1.9$; 2.3; $^{-}3.1$　　**26.** $\frac{^{-}1}{4}$; $\frac{^{-}3}{4}$; $^{-}1\frac{3}{4}$; $2\frac{1}{4}$

Find each answer.

27. $^{-}11 + {}^{-}3$　　**28.** $7 - 15$　　**29.** $(^{-}6)(^{-}9)$　　**30.** $0 \div {}^{-}3$

31. $6.2 + {}^{-}6.2$　　**32.** $^{-}8.8 - {}^{-}14.2$　　**33.** $(^{-}7.5)(1.6)$　　**34.** $10.4 \div 8$

35. $\frac{^{-}2}{9} + 5\frac{7}{9}$　　**36.** $^{-}6\frac{3}{5} - 1\frac{1}{5}$　　**37.** $\left(\frac{^{-}3}{4}\right)(0)$　　**38.** $\frac{1}{2} \div \frac{^{-}5}{8}$

Select the more reasonable unit. Choose °F or °C.

39. a glass of cold juice: 40 __?__　　**40.** a cup of hot tea: 110 __?__

41. Write 80,000 in scientific notation.　　**42.** Write 1.7×10^{-1} in numeral form.

CHAPTER TEST

Write an integer that represents each phrase.

1. a loss of 18 yd **2.** 17 ft above sea level **3.** 5°F below zero ***19-1***

Rewrite each rational number as a quotient of two integers.

4. 87 **5.** $^{-}4\frac{1}{9}$ **6.** $^{-}50$ **7.** 0.43 **8.** $0.\overline{2}$ **9.** 0

Write the opposite and the absolute value of each number.

10. $^{-}4$ **11.** 8 **12.** $17\frac{11}{12}$ **13.** $^{-}4.5$ **14.** 0 ***19-2***

Compare. Replace each __?__ with > or <.

15. $^{-}15$ __?__ $^{-}10$ **16.** $^{-}7.9$ __?__ 2.1 **17.** $\frac{^{-}1}{5}$ __?__ $^{-}6$ **18.** 0 __?__ $\frac{^{-}2}{3}$ ***19-3***

Write each set of numbers in order from least to greatest.

19. $^{-}2$; 3; $^{-}4$; $^{-}5$ **20.** 0.2; 1.5; 0; $^{-}2.9$ **21.** $\frac{3}{10}$; $\frac{^{-}7}{10}$; $\frac{^{-}9}{10}$; $2\frac{1}{10}$

Find each sum.

22. 23 + 41 **23.** $^{-}5 + {}^{-}19$ **24.** $^{-}40 + 3$ **25.** $30 + {}^{-}18$ ***19-4***

26. $5.3 + {}^{-}5.3$ **27.** $^{-}4.7 + 3.2$ **28.** $^{-}3\frac{1}{3} + {}^{-}5\frac{1}{3}$ **29.** $3 + 5\frac{1}{6}$

Find each difference.

30. 6 − 24 **31.** $^{-}15 - 5$ **32.** $^{-}23 - {}^{-}27$ **33.** $19 - {}^{-}46$ ***19-5***

34. $3.1 - {}^{-}4.5$ **35.** 5.2 − 6.9 **36.** $^{-}4\frac{1}{5} - 2\frac{2}{5}$ **37.** $\frac{^{-}7}{8} - \frac{^{-}7}{8}$

Find each answer.

38. $(^{-}26)(^{-}7)$ **39.** 57 ÷ 3 **40.** $^{-}3.2 \div 8$ **41.** $(^{-}2)(0)$ ***19-6***

42. $\left(\frac{2}{5}\right)\left(\frac{^{-}3}{8}\right)$ **43.** $\frac{^{-}3}{4} \div \frac{^{-}6}{7}$ **44.** $(^{-}6)(^{-}3)(^{-}7)$ **45.** $(^{-}1.5)(4)(^{-}0.9)$

Select the more reasonable unit. Choose °F or °C.

46. Water boils at 100 __?__. **47.** Water freezes at 32 __?__. ***19-7***

48. A polar bear has a mass of 322 kg. Write this number in scientific notation. ***19-8***

49. A pin has a mass of 1.07×10^{-4} kg. Write this number in numeral form.

Chemists and other scientists, as well as mathematicians, make constant use of equations and inequalities in their work.

CHAPTER 20

EQUATIONS AND INEQUALITIES

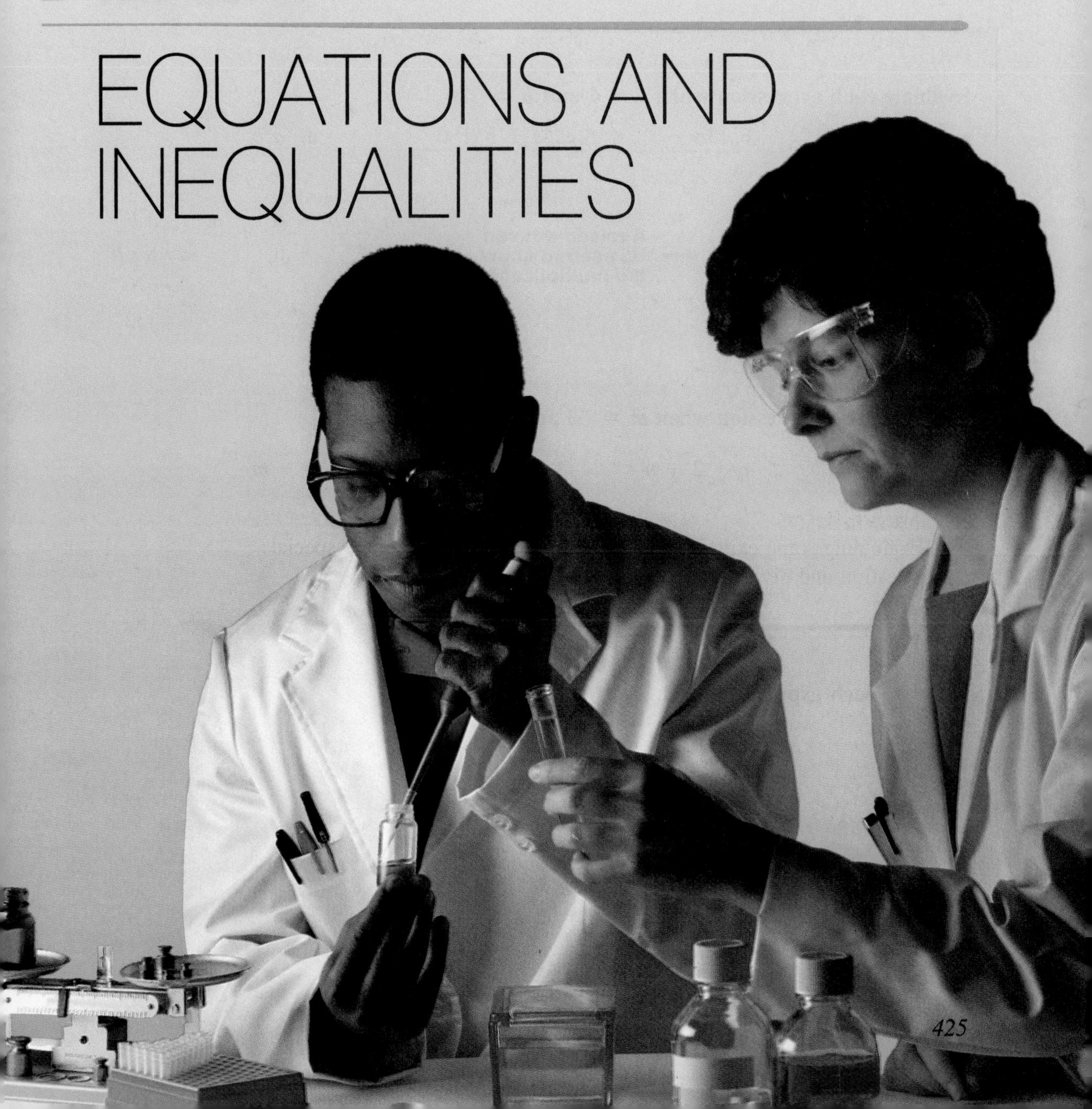

20-1 VARIABLE EXPRESSIONS

In Chapter 12, you learned that a **variable** is a symbol used to represent a number. An expression that contains a variable is called a **variable expression.** Sometimes you are given a **value** to replace a variable in a variable expression. Then you can **evaluate** the expression.

When you write a product that contains a variable, you usually omit the multiplication symbol.

example 1

Evaluate each expression when $a = 4$ and $b = {}^-2$.

a. $b - 15$ **b.** $3a$ **c.** $a + b + 6$ **d.** a^2b

solution

a. $b - 15$
$^-2 - 15$
$^-17$

b. $3a$
$3 \cdot a$ ← **A raised dot can be used to show the multiplication.**
$3 \cdot 4$
12

c. $a + b + 6$
$4 + {}^-2 + 6$
8

d. a^2b
$a \cdot a \cdot b$
$4 \cdot 4 \cdot {}^-2$
$^-32$

your turn

Evaluate each expression when $m = {}^-3$ and $n = 6$.

1. m^2 **2.** $n + 2 + n$ **3.** $\frac{n}{m}$ **4.** $n - m$

Evaluating some expressions may be easier if you know the special multiplication and division properties of 0 and 1.

example 2

Evaluate each expression when $x = 8$, $y = 0$, and $z = 1$.

a. ^-2z **b.** $5xy$ **c.** $\frac{0}{x}$ **d.** $\frac{16}{y}$

solution

a. $^-2z \rightarrow {}^-2(1) = {}^-2$ ← **The product of a number and 1 is the number itself.**

b. $5xy \rightarrow 5(8)(0) = 0$ ← **If any one of the factors is 0, the product is 0.**

c. $\frac{0}{x} \rightarrow \frac{0}{8} = 0$ ← **If you divide 0 by a nonzero number, the quotient is 0.**

d. $\frac{16}{y} \rightarrow \frac{16}{0}$; *undefined* ← **You cannot divide by 0.**

your turn

Evaluate each expression when $r = {}^{-}2$, $s = 1$, and $t = 0$.

5. $3rs$ 6. $^{-}4t$ 7. $\frac{0}{s}$ 8. $\frac{10}{t}$

example 3

Evaluate each expression when $a = 2$, $b = 4$, and $c = {}^{-}2$.

a. $3b - 7$ b. $2a + 5c$ c. $3 + \frac{b}{a}$

solution

a. $3b - 7$

$3 \cdot 4 - 7$

$12 - 7$

5

b. $2a + 5c$

$(2)(2) + (5)(^{-}2)$

$4 + {}^{-}10$

$^{-}6$

c. $3 + \frac{b}{a}$

$3 + \frac{4}{2}$

$3 + 2$

5

your turn

Evaluate each expression when $x = 6$, $y = {}^{-}3$, and $z = 2$.

9. $2x + 1$ 10. $3x + 10y$ 11. $4 + \frac{6}{z}$ 12. $z^2 - 14$

practice exercises

practice for example 1 (page 426)

Evaluate each expression when $p = 7$, $q = {}^{-}4$, and $r = 2$.

1. $p - 6$ 2. $r - 10$ 3. $^{-}25r$ 4. $^{-}6q$
5. $q + r + 8$ 6. $p + {}^{-}3 + r$ 7. p^2 8. qr^2
9. $\frac{q}{r}$ 10. $\frac{7}{p}$ 11. $q + q + {}^{-}2$ 12. $3 - q$

practice for example 2 (pages 426–427)

Evaluate each expression when $a = 0$, $b = 1$, and $c = {}^{-}3$.

13. $12b$ 14. $^{-}100b$ 15. $2bc$ 16. $^{-}4bc$
17. $^{-}14a$ 18. $13a$ 19. $5ab$ 20. $^{-}8ab$
21. $\frac{^{-}7}{a}$ 22. $\frac{15}{a}$ 23. $\frac{0}{b}$ 24. $\frac{0}{c}$

practice for example 3 (page 427)

Evaluate each expression when $w = 5$, $x = {}^-2$, and $y = 4$.

25. $2w - 4$
26. $6y - 30$
27. $49 + 10w$
28. $20 + 5x$
29. $7w + 5y$
30. $6x + 2y$
31. $2w - 3y$
32. $4y - 10x$
33. $3 + \frac{y}{x}$
34. $\frac{8}{y} + x$
35. $y^2 - 12$
36. $6 + w^2$

mixed practice (pages 426–427)

Evaluate each expression when $a = 0$, $b = {}^-10$, $c = 1$, and $d = 2$.

37. $d + 7$
38. $b - 10$
39. $4b$
40. $d^2 - 3$
41. ${}^-5c$
42. $\frac{5}{a}$
43. $4d - 20$
44. $\frac{80}{b}$
45. b^2d
46. $d + d + 6$
47. $\frac{0}{c}$
48. ac^2
49. $b + 5d$
50. $2 + \frac{b}{c}$
51. $4ab$
52. ${}^-2bc$

review exercises

Find each answer.

1. $16 + 18$
2. $7 - {}^-37$
3. $(12)({}^-8)$
4. ${}^-84 \div {}^-7$
5. $\left(\frac{{}^-1}{3}\right)\left(\frac{{}^-7}{10}\right)$
6. $\frac{{}^-1}{4} \div \frac{4}{15}$
7. ${}^-12.5 - {}^-10.5$
8. ${}^-5.2 + 6.3$
9. What number is 30% of 90?
10. 120% of 200 is what number?
11. 8 is what percent of 48?
12. What percent of 50 is 100?
13. Erik plans to address 25 envelopes. He has already addressed 19 envelopes. How many more envelopes does Erik need to address?
14. On Wednesday Rachel worked 3 h and earned \$9. On Friday she worked 4 h and earned \$12. What was the total amount that Rachel earned on Wednesday and Friday?

At the right you see two different ways that four 4's were used to create an expression equal to 1.

Show how four 4's can be used to create an expression equal to 2, an expression equal to 3, and an expression equal to 6. You can use parentheses and any of the four operations +, −, ×, and ÷.

$$\frac{4}{4} + 4 - 4 = 1$$

$$\frac{4 \times 4}{4} \div 4 = 1$$

20-2 SOLVING EQUATIONS: ADDITION AND SUBTRACTION

An **equation** is a statement that two numbers or quantities are equal. A **solution** of an equation involving one variable is a value of the variable that makes the equation true.

To **solve** an equation, you must get the variable alone on one *side* of the equals sign. You do this by using *inverse operations* that "undo" the operations in the equation.

To solve an equation that involves addition or subtraction, use the following method.

If a number has been *added* to a variable, you *subtract that number* from both sides.

If a number has been *subtracted* from a variable, you *add that number* to both sides.

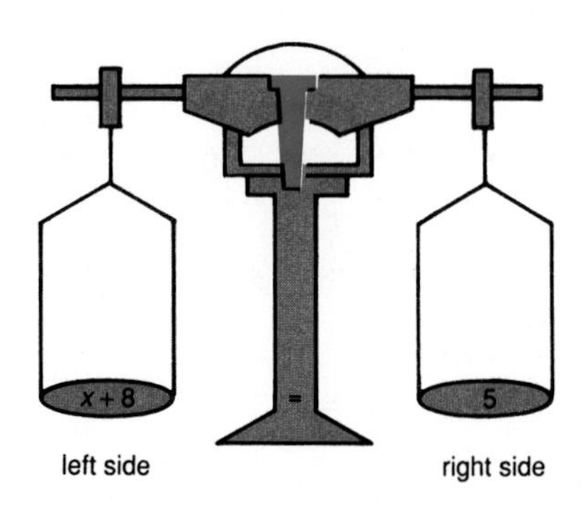

example 1

Solve each equation. Check your answers.

a. $x + 8 = 5$

b. $\frac{3}{7} = y + \frac{2}{7}$

solution

a.

$$x + 8 = 5$$

$$x + 8 - 8 = 5 - 8 \quad \leftarrow \text{Subtract 8 from both sides.}$$

$$x = {}^{-}3$$

The solution is $^{-}3$.

Check: $x + 8 = 5$

$$^{-}3 + 8 \stackrel{?}{=} 5 \quad \leftarrow \text{Replace } x \text{ with } {}^{-}3.$$

$$5 = 5 \checkmark$$

b.

$$\frac{3}{7} = y + \frac{2}{7}$$

$$\frac{3}{7} - \frac{2}{7} = y + \frac{2}{7} - \frac{2}{7}$$

$$\frac{1}{7} = y$$

The solution is $\frac{1}{7}$.

Check: $\frac{3}{7} = y + \frac{2}{7}$

$$\frac{3}{7} \stackrel{?}{=} \frac{1}{7} + \frac{2}{7}$$

$$\frac{3}{7} = \frac{3}{7} \checkmark$$

your turn

Solve each equation. Check your answers.

1. $d + 40 = 100$ **2.** $^{-}3 = 2 + m$ **3.** $y + \frac{3}{11} = \frac{5}{11}$ **4.** $20.6 = x + 17$

example 2

Solve each equation. Check your answers.

a. $r - 8 = {}^{-}36$

b. $19 = s - 0.5$

solution

a.

$$r - 8 = {}^{-}36$$
$$r - 8 + 8 = {}^{-}36 + 8 \quad \leftarrow \textbf{Add 8 to both sides.}$$
$$r = {}^{-}28$$

The solution is ${}^{-}28$.

Check:

$$r - 8 = {}^{-}36$$
$${}^{-}28 - 8 \stackrel{?}{=} {}^{-}36 \quad \leftarrow \textbf{Replace } r \textbf{ with } {}^{-}28.$$
$${}^{-}36 = {}^{-}36 \ \checkmark$$

b.

$$19 = s - 0.5$$
$$19 + 0.5 = s - 0.5 + 0.5$$
$$19.5 = s$$

The solution is 19.5.

Check:

$$19 = s - 0.5$$
$$19 \stackrel{?}{=} 19.5 - 0.5$$
$$19 = 19 \ \checkmark$$

your turn

Solve each equation. Check your answers.

5. $w - 15 = {}^{-}18$ **6.** $4 = p - 20$ **7.** $x - 0.3 = 24$ **8.** $1\frac{4}{9} = m - \frac{1}{9}$

practice exercises

practice for example 1 (page 429)

Solve each equation. Check your answers.

1. $y + 12 = 25$
2. $30 = z + 50$
3. $8 = 4 + a$
4. $20 + b = {}^{-}15$
5. $t + 10 = {}^{-}3$
6. $17 = q + 9$
7. $31 = 17 + s$
8. $45 + c = 25$
9. $d + 41 = {}^{-}2$
10. $5 = z + 5$
11. $h + 12 = 12.2$
12. $m + \frac{2}{9} = \frac{7}{9}$

practice for example 2 (page 430)

Solve each equation. Check your answers.

13. $10 = r - 4$
14. $m - 5 = 11$
15. $r - 4 = {}^{-}9$
16. ${}^{-}15 = z - 5$
17. $40 = a - 40$
18. $x - 6 = {}^{-}13$
19. $p - 14 = {}^{-}3$
20. $x - 7 = {}^{-}5$
21. ${}^{-}18 = f - 4$
22. ${}^{-}12 = b - 2$
23. $a - \frac{1}{7} = 5\frac{5}{7}$
24. $3 = r - 1.3$

mixed practice (pages 429–430)

Solve each equation. Check your answers.

25. $x + 7 = 20$

26. $17 = 18 + c$

27. $30 = t - 2$

28. $r - 7 = {}^{-}7$

29. ${}^{-}2 = 9 + m$

30. ${}^{-}6 = y - 11$

31. ${}^{-}4 = w - 3$

32. $z + 8 = 0$

33. ${}^{-}4 = a + 7$

34. $15 = 3 + b$

35. $p - 12 = 4$

36. $n - 7 = {}^{-}20$

37. ${}^{-}1 = d - 4$

38. $c + 8 = 7$

39. $r - 0.2 = 15$

40. ${}^{-}2.8 = q + 3$

41. $m + \frac{1}{9} = \frac{{}^{-}7}{9}$

42. $t - \frac{1}{7} = 3\frac{3}{7}$

Describe how you would solve and check each equation.

43. $3 + a = {}^{-}15$

44. $b - 14 = {}^{-}15$

review exercises

Write the reciprocal of each number.

1. $\frac{3}{4}$

2. $\frac{14}{15}$

3. 16

4. 0

5. $\frac{1}{8}$

6. $7\frac{2}{5}$

7. In a debate, Julio and Rosa are allowed a total of 15 min to speak. If Julio's speech takes $6\frac{1}{2}$ min, how many minutes are left for Rosa's speech?

8. Tim paid \$6 for an air filter for his car, \$10 for a fan belt, and \$4 for an oil filter. He began to repair his car at 3:30 P.M. and finished at 4:45 P.M. How much time did Tim spend repairing his car?

9. The Rowes want to buy a camera. They are able to save \$15 per month for that purpose. The camera costs \$200, and the sales tax rate is 5%. In how many months will the Rowes have saved enough money to buy the camera?

10. Sam bought some sweaters for \$39 each and some shirts for \$15 each. He spent \$201. How many of each did he buy?

20-3 SOLVING EQUATIONS: MULTIPLICATION AND DIVISION

To solve equations involving multiplication and division, use the following method.

If a variable has been *multiplied* by a number, you *divide* both sides by that number.

If a variable has been *divided* by a number, you *multiply* both sides by that number.

example 1

Solve each equation: **a.** $6a = {}^{-}12$ **b.** $5y = 2.5$

solution

a. $6a = {}^{-}12$

$\frac{6a}{6} = \frac{{}^{-}12}{6}$ ← **Divide both sides by 6.**

$a = {}^{-}2$ ← **The solution is ⁻2.**

Check: $6a = {}^{-}12$

$6({}^{-}2) \stackrel{?}{=} {}^{-}12$ ← **Replace a with ⁻2.**

${}^{-}12 = {}^{-}12$ ✓

b. $5y = 2.5$

$\frac{5y}{5} = \frac{2.5}{5}$

$y = 0.5$

Check: $5y = 2.5$

$5(0.5) \stackrel{?}{=} 2.5$

$2.5 = 2.5$ ✓

your turn

Solve each equation. Check your answers.

1. $140 = 7f$
2. $4y = {}^{-}16$
3. ${}^{-}6g = 7.8$
4. ${}^{-}1.5 = {}^{-}3b$

example 2

Solve each equation: **a.** $\frac{z}{{}^{-}7} = 2$ **b.** $\frac{a}{0.3} = 11$

solution

a. $\frac{z}{{}^{-}7} = 2$

$({}^{-}7)\left(\frac{z}{{}^{-}7}\right) = ({}^{-}7)(2)$ ← **Multiply both sides by ⁻7.**

$z = {}^{-}14$ ← **The solution is ⁻14.**

Check: $\frac{z}{{}^{-}7} = 2$

$\frac{{}^{-}14}{{}^{-}7} \stackrel{?}{=} 2$ ← **Replace z with ⁻14.**

$2 = 2$ ✓

b. $\frac{a}{0.3} = 11$

$(0.3)\left(\frac{a}{0.3}\right) = (0.3)(11)$

$a = 3.3$

Check: $\frac{a}{0.3} = 11$

$\frac{3.3}{0.3} \stackrel{?}{=} 11$

$11 = 11$ ✓

your turn

Solve each equation. Check your answers.

5. $\frac{d}{^-8} = 5$ **6.** $^-9 = \frac{r}{^-7}$ **7.** $^-6 = \frac{y}{2.1}$ **8.** $\frac{a}{^-15} = {}^-0.3$

In some equations, the variable is multiplied by a fraction. To solve these equations, you *multiply both sides by the reciprocal* of the fraction.

example 3

Solve each equation: **a.** $\frac{5}{6}y = 25$ **b.** $^-14 = \frac{^-2}{3}z$

solution

a.

$$\frac{5}{6}y = 25$$

$$\left(\frac{6}{5}\right)\left(\frac{5}{6}y\right) = \left(\frac{6}{5}\right)(25)$$ ← **Multiply both sides by $\frac{6}{5}$.**

$$y = 30$$ ← **The solution is 30.**

Check: $\frac{5}{6}y = 25$

$\frac{5}{6}(30) \stackrel{?}{=} 25$ ← **Replace y with 30.**

$25 = 25$ ✓

b.

$$^-14 = \frac{^-2}{3}z$$

$$\left(\frac{^-3}{2}\right)(^-14) = \left(\frac{^-3}{2}\right)\left(\frac{^-2}{3}z\right)$$

$$21 = z$$

Check: $^-14 = \frac{^-2}{3}z$

$^-14 \stackrel{?}{=} \frac{^-2}{3}(21)$

$^-14 = {}^-14$ ✓

your turn

Solve each equation. Check your answers.

9. $\frac{3}{4}a = 15$ **10.** $36 = \frac{^-3}{5}b$ **11.** $^-28 = \frac{4}{7}p$ **12.** $\frac{1}{3}m = {}^-7$

practice exercises

practice for example 1 (page 432)

Solve each equation. Check your answers.

1. $9z = {}^-72$ **2.** $^-8b = {}^-24$ **3.** $72 = {}^-6g$ **4.** $55 = 5k$

5. $^-2y = {}^-7.2$ **6.** $9p = {}^-4.5$ **7.** $0.42 = 14d$ **8.** $1.68 = {}^-21m$

practice for example 2 (pages 432–433)

Solve each equation. Check your answers.

9. $\frac{a}{^-4} = 50$ **10.** $\frac{t}{10} = 6$ **11.** $^-9 = \frac{s}{5}$ **12.** $^-7 = \frac{c}{^-11}$

13. $4 = \frac{m}{2.2}$ **14.** $3 = \frac{p}{^-4.3}$ **15.** $\frac{d}{^-0.4} = {}^-7$ **16.** $\frac{h}{2} = {}^-1.6$

practice for example 3 (page 433)

Solve each equation. Check your answers.

17. $\frac{5}{8}d = 10$ **18.** $\frac{4}{7}b = 16$ **19.** $12 = \frac{^-3}{5}g$ **20.** $20 = \frac{^-5}{6}p$

21. $^-5 = \frac{5}{12}w$ **22.** $^-14 = \frac{1}{8}c$ **23.** $\frac{^-1}{4}k = {}^-27$ **24.** $\frac{^-3}{25}a = {}^-9$

mixed practice (pages 432–433)

Solve each equation. Check your answers.

25. $5p = 40$ **26.** $^-24 = 4c$ **27.** $7 = \frac{y}{^-6}$ **28.** $\frac{m}{^-3} = {}^-13$

29. $\frac{^-4}{5}d = 16$ **30.** $\frac{1}{3}w = 18$ **31.** $^-21 = {}^-7a$ **32.** $32 = {}^-8n$

33. $^-9 = \frac{^-3}{10}r$ **34.** $^-8 = \frac{2}{5}m$ **35.** $\frac{z}{^-1.4} = 2$ **36.** $5 = \frac{c}{2.5}$

37. $9f = 8.1$ **38.** $7n = {}^-10.5$ **39.** $^-20 = \frac{m}{0.5}$ **40.** $\frac{^-4}{13}h = {}^-8$

review exercises

1. Steve and Jill need to pour 500 mL of orange juice and 750 mL of peach juice into one container. Will a 1-L bottle be large enough to contain both juices? Explain.
2. An inspector examined 900 lamps one day. The inspector found that 5% of the lamps did not pass inspection that day. How many lamps did pass inspection?

mental math

You can use mental math to solve some equations.

$5 + n = {}^-7$ *Think:* $5 + {}^-12 = {}^-7$ or $^-7 - 5 = {}^-12$ So $n = {}^-12$.

$\frac{z}{3} = {}^-5$ *Think:* $\frac{^-15}{3} = {}^-5$ or $(3)({}^-5) = {}^-15$ So $z = {}^-15$.

Use mental math to solve.

1. $a + 9 = {}^-2$ **2.** $^-3 + b = {}^-5$ **3.** $c - 4 = 12$ **4.** $d - 16 = {}^-3$

5. $\frac{z}{^-2} = {}^-5$ **6.** $\frac{y}{12} = {}^-3$ **7.** $2p = {}^-18$ **8.** $^-10q = {}^-30$

20-4 SOLVING TWO-STEP EQUATIONS

An equation such as $^-2n + 4 = 20$ involves two operations. To solve an equation like this, use the following method.

1. Add the same number to or subtract the same number from both sides.
2. Multiply or divide both sides by the same nonzero number.

example 1

Solve: $^-2n + 4 = 20$

$$^-2n + 4 = 20$$
$$^-2n + 4 - 4 = 20 - 4 \quad \leftarrow \text{Subtract 4 from both sides.}$$
$$^-2n = 16$$
$$\frac{^-2n}{^-2} = \frac{16}{^-2} \quad \leftarrow \text{Divide both sides by } ^-2.$$
$$n = {}^-8$$

The solution is $^-8$.

Check:
$$^-2n + 4 = 20$$
$$^-2(^-8) + 4 \stackrel{?}{=} 20$$
$$16 + 4 \stackrel{?}{=} 20$$
$$20 = 20 \checkmark$$

your turn

Solve each equation. Check your answers.

1. $^-3t + 6 = 15$
2. $\frac{n}{2} + 20 = 12$
3. $5.5 = 5w - 1$

example 2

Solve: $\frac{3}{4}h + 5 = 2$

$$\frac{3}{4}h + 5 = 2$$
$$\frac{3}{4}h + 5 - 5 = 2 - 5 \quad \leftarrow \text{Subtract 5 from both sides.}$$
$$\frac{3}{4}h = {}^-3$$
$$\left(\frac{4}{3}\right)\left(\frac{3}{4}h\right) = \left(\frac{4}{3}\right)(^-3) \quad \leftarrow \text{Multiply both sides by } \frac{4}{3}.$$
$$h = {}^-4$$

The solution is $^-4$.

Check:
$$\frac{3}{4}h + 5 = 2$$
$$\frac{3}{4}(^-4) + 5 \stackrel{?}{=} 2$$
$$^-3 + 5 \stackrel{?}{=} 2$$
$$2 = 2 \checkmark$$

your turn

Solve each equation. Check your answers.

4. $\frac{4}{5}x + 9 = 29$
5. $\frac{1}{3}y - 4 = {}^-10$
6. $^-8 = \frac{^-2}{7}a - 16$

practice exercises

practice for example 1 (page 435)

Solve each equation. Check your answers.

1. ${}^{-}4y + 3 = {}^{-}21$ **2.** ${}^{-}6t + 2 = 44$ **3.** ${}^{-}41 = 12g - 5$

4. $12 = \frac{a}{2} - 6$ **5.** $\frac{m}{3} + 2 = {}^{-}4$ **6.** $15.4 = 4r - 3$

practice for example 2 (page 435)

Solve each equation. Check your answers.

7. ${}^{-}35 = \frac{7}{8}c + 7$ **8.** $41 = \frac{3}{4}s + 8$ **9.** ${}^{-}21 = \frac{{}^{-}4}{5}z - 9$

10. $\frac{1}{2}r - 1 = {}^{-}8$ **11.** $\frac{{}^{-}2}{9}a - 13 = {}^{-}1$ **12.** $\frac{{}^{-}2}{5}y + 4 = 44$

mixed practice (page 435)

Solve each equation. Check your answers.

13. ${}^{-}5z + 20 = 25$ **14.** ${}^{-}2y - 10 = 30$ **15.** $9.9 = 3k + 6$

16. $\frac{z}{{}^{-}6} - 1 = 19$ **17.** $\frac{1}{4}p - 8 = 23$ **18.** ${}^{-}10 = \frac{{}^{-}4}{7}q + 2$

19. ${}^{-}16 = \frac{3}{4}d - 7$ **20.** $\frac{2}{3}b - 13 = {}^{-}15$ **21.** $\frac{r}{3} + 15 = {}^{-}25$

22. $0.8 = 6a - 1$ **23.** ${}^{-}19 = \frac{{}^{-}5}{8}m - 9$ **24.** ${}^{-}2 = \frac{{}^{-}3}{5}p + 16$

Choose the letters of all the equations that have the same solution as the given equation.

25. $2x + 6 = 12$ **A.** $2x = 2$ **B.** $2x = 6$ **C.** $x = 4$ **D.** $x = 3$

26. ${}^{-}6 = \frac{a}{4} - 3$ **A.** ${}^{-}3 = \frac{a}{4}$ **B.** ${}^{-}9 = \frac{a}{4}$ **C.** ${}^{-}36 = a$ **D.** ${}^{-}12 = a$

review exercises

1. Joe McClellan charges $2.50 for every 3 pages he types. How much does he charge to type 18 pages?
2. An 11-oz block of cheese costs $1.65. How much would a 2-lb block of the same type of cheese cost?
3. The ratio of carnations to roses in a bouquet is 3 to 1. There are 3 roses in the bouquet. How many carnations are in the bouquet?

20-5 SOLVING INEQUALITIES

An **inequality** is a sentence that has an *inequality symbol* between two numbers or quantities. Commonly used inequality symbols are $<$, $>$, $\leq$, and $\geq$.

$<$ means *is less than.*
$>$ means *is greater than.*
$\leq$ means *is less than or equal to.*
$\geq$ means *is greater than or equal to.*

The **solution of an inequality** is the set of all the values of the variable that make the inequality true. To *solve* an inequality, use inverse operations that "undo" the operations in the inequality.

example 1

Solve: $x + 5 > 4$

solution

$$x + 5 > 4$$
$$x + 5 - 5 > 4 - 5$$
$$x > {}^{-}1$$

Check: $x + 5 > 4$
$$2 + 5 \overset{?}{>} 4$$
$$7 > 4 \ \checkmark$$

← **Replace *x* with *any number* greater than ⁻1.**

your turn

Solve each inequality. Check your answers.

1. $y + 8 < 10$ **2.** $s - 3 \leq {}^{-}9$ **3.** $a - 5 \geq 15$

example 2

Solve: $\frac{b}{3} < 2$

solution

$$\frac{b}{3} < 2$$
$$(3)\left(\frac{b}{3}\right) < (3)(2)$$
$$b < 6$$

Check: $\frac{b}{3} < 2$
$$\frac{3}{3} \overset{?}{<} 2$$
$$1 < 2 \ \checkmark$$

← **Replace *b* with *any number* less than 6.**

your turn

Solve each inequality. Check your answers.

4. $\frac{x}{4} > {}^{-}4$ **5.** $5a \geq {}^{-}10$ **6.** $4y \leq 12$

A number line is often used to show the **graph of an inequality.** To make the graph, you use a dot and an arrow to indicate all the points included in the solution of the inequality.

example 3

Graph each inequality.

a. $x + 2 > 5$

b. 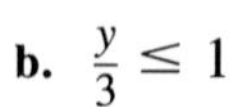 $\frac{y}{3} \leq 1$

solution

a. Solve the inequality.

$$x + 2 > 5$$

$$x + 2 - 2 > 5 - 2$$

$$x > 3$$

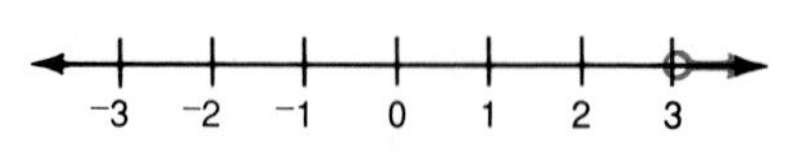

An *open dot* shows that 3 is *not* a solution.

b. Solve the inequality.

$$\frac{y}{3} \leq 1$$

$$(3)\left(\frac{y}{3}\right) \leq (3)(1)$$

$$y \leq 3$$

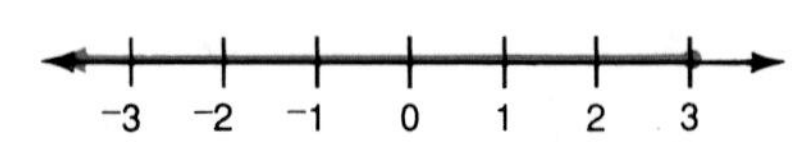

A *closed dot* shows that 3 *is* a solution.

your turn

Graph each inequality.

7. $a + 3 > {}^{-}1$ **8.** $m - 3 \leq 2$ **9.** $\frac{z}{2} \geq 2$ **10.** $2b < {}^{-}8$

practice exercises

practice for example 1 (page 437)

Solve each inequality. Check your answers.

1. $z + 10 > {}^{-}5$ **2.** $t - 4 > 10$ **3.** $a - 3 < 21$ **4.** $d + 17 < 7$

5. $w + 6 \geq {}^{-}13$ **6.** $b - 6 \geq {}^{-}1$ **7.** $c - 7 \leq {}^{-}2$ **8.** $y + 4 \leq 2$

practice for example 2 (page 437)

Solve each inequality. Check your answers.

9. $\frac{z}{4} > 50$ **10.** $6t > {}^{-}54$ **11.** $3c \leq {}^{-}9$ **12.** $\frac{w}{7} \leq 6$

13. $8b \geq 24$ **14.** $\frac{z}{5} \geq {}^{-}8$ **15.** $\frac{s}{2} < {}^{-}7$ **16.** $9r < 81$

practice for example 3 (page 438)

Graph each inequality.

17. $y + 12 > 14$ **18.** $b - 2 \geq 1$ **19.** $z - 3 \geq {}^-4$ **20.** $w + 3 \leq {}^-2$

21. $\frac{r}{6} < 1$ **22.** $\frac{a}{2} > {}^-2$ **23.** $8c \leq {}^-32$ **24.** $6m < 18$

mixed practice (pages 437–438)

Solve and graph each inequality.

25. $a + 7 > 8$ **26.** $6p < 12$ **27.** $q - 8 \geq {}^-9$ **28.** $w + 3 \leq {}^-1$

29. $\frac{r}{2} < {}^-3$ **30.** $z - 3 > {}^-6$ **31.** $9b \leq 27$ **32.** $\frac{m}{4} \geq 1$

Choose all the numbers that make the given inequality true.

33. $x + 3 > {}^-4$ **A.** $^-10$ **B.** $^-7$ **C.** $^-3$ **D.** 1 **E.** 0

34. $5r \leq 15$ **A.** 3 **B.** 15 **C.** $^-1$ **D.** 2 **E.** 5

Match each inequality with its graph.

35. $y + 10 \geq 12$

36. $2y < 4$

37. $\frac{y}{2} \leq 1$

38. $y - 5 > {}^-3$

A. −3 −2 −1 0 1 2 3

B. −3 −2 −1 0 1 2 3

C.

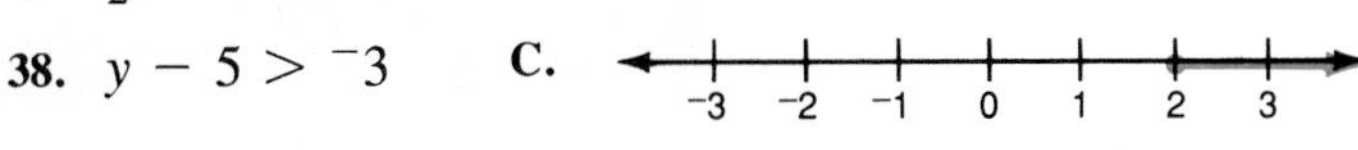

D.

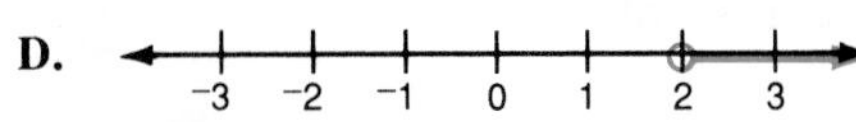

Write an inequality that represents each situation.

39. Joe used less than 18 gal of gasoline on his trip. Let g represent the number of gallons he used.

40. Mary rode her bicycle for more than 30 min. Let t represent the number of minutes she rode her bicycle.

review exercises

Complete.

1. 4000 lb = __?__ t **2.** 2.9 kg = __?__ g

3. 74 in. = __?__ ft __?__ in. **4.** 1174 m = __?__ km

5. 2 mi = __?__ ft **6.** 6 L = __?__ mL

PROBLEM SOLVING STRATEGY

20-6 USING EQUATIONS

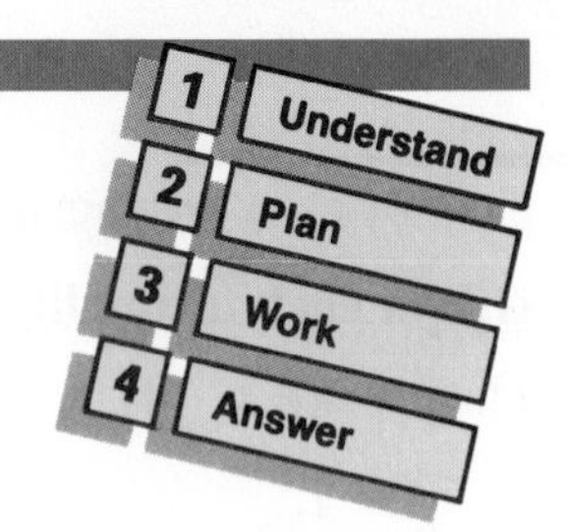

Sometimes an equation can be useful when solving a word problem. First choose a variable to represent the unknown number. Next use this variable to write an equation that represents the facts of the problem. Then solve the equation.

example

Faye Cohen plans to plant 52 bulbs. She has one bag of twelve daffodil bulbs. She plans to buy tulip bulbs in bags that contain four bulbs each. How many bags of tulip bulbs should she buy?

solution

Step 1 Given: total of 52 bulbs
12 daffodil bulbs
4 tulips bulbs per bag
Find: number of bags of tulip bulbs

Step 2 Write an equation to show that the sum of daffodil bulbs and tulip bulbs is 52. Let the variable *b* represent the unknown number of bags of tulip bulbs.

4	times	number of bags of tulips	plus	12	is	52
↓	↓	↓	↓	↓	↓	↓
4	$\times$	b	$+$	12	$=$	52

Step 3

$$4b + 12 = 52$$

$$4b + 12 - 12 = 52 - 12$$ ← **Subtract 12 from both sides.**

$$4b = 40$$

$$\frac{4b}{4} = \frac{40}{4}$$ ← **Divide both sides by 4.**

$$b = 10$$

Step 4 *Check:*

$$4b + 12 = 52$$

$$4(10) + 12 \stackrel{?}{=} 52$$ ← **Replace *b* with 10.**

$$52 = 52 \ \checkmark$$

Faye needs to buy 10 bags of tulip bulbs.

problems

Each of Exercises 1–4 can be solved using one of the given equations. Choose the correct equation. Then solve the problem.

$y + 3 = 12$ $\quad y - 3 = 12$ $\quad 3y = 12$ $\quad \frac{y}{3} = 12$

1. Jonathan lent three tapes to his friends. He still has twelve tapes left. How many tapes did Jonathan have before he lent the tapes to his friends?
2. Sarah deposited $3 in her savings account. She now has $12. How much money did she have before the deposit?
3. Frank, Tony, and Rob shared the cost of a present equally. Each paid $12. What was the cost of the present?
4. Rosa bought a package of three video cassettes for $12. How much did each cassette cost?

Solve using an equation.

5. Charles bought six spark plugs for $9. How much did each spark plug cost?
6. Melanie buys a book for $10. She has $4.50 left. How much money did she have before buying the book?
7. Jacklyn subscribes to a magazine that costs $24.60 per year. She makes an initial payment of $15. She then must make four equal payments. What is the amount of each payment?
8. Darren needs 124 plastic forks. At the store he finds five boxes that contain eight forks each. He then finds boxes that contain twelve forks each. How many of these boxes does he need to buy?

Write two word problems that can be solved using each equation.

9. $b - 5 = 7$
10. $4w = 16$

review exercises

Use the line graph at the right. Estimate the number of airports in operation in each year.

1. 1945
2. 1965
3. 1955
4. 1975

Assume the trend will continue. Predict the number of airports in operation in each year.

5. 1990
6. 2000

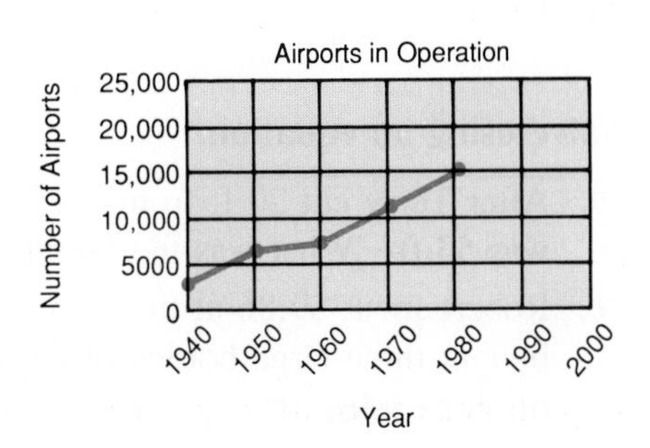

SKILL REVIEW

Evaluate each expression when $p = {}^-3$, $q = 0$, $r = 12$, and $s = 1$. **20-1**

1. $r - p$
2. $3p$
3. $r + 13 + r$
4. rp^2
5. ^-4s
6. $6qr$
7. $r^2 - 100$
8. $4s + 2p$
9. $\frac{q}{7}$
10. $\frac{9}{q}$
11. $^-8 + \frac{p}{s}$
12. $6 - \frac{12}{r}$

Solve each equation. Check your answers.

20-2

13. $a + 2 = {}^-48$
14. $15 = r - 8$
15. $2 = c + 6$
16. $c - 3 = {}^-19$
17. $h + 2 = 12.4$
18. $7\frac{1}{3} = t - \frac{1}{3}$

20-3

19. $5y = {}^-50$
20. $10.8 = 6m$
21. $4 = \frac{t}{^-5}$
22. $\frac{b}{0.4} = 9$
23. $\frac{^-1}{4}z = 12$
24. $^-10 = \frac{^-5}{12}r$

20-4

25. $^-6b + 4 = 40$
26. $8y - 14 = {}^-22$
27. $3.2 = 7y - 1$
28. $4.2 = 2a + 4$
29. $28 = \frac{2}{9}p + 10$
30. $12 = \frac{^-3}{7}z + 27$
31. $\frac{c}{^-3} + 29 = 19$
32. $20 = \frac{m}{4} - 30$
33. $\frac{1}{3}w - 8 = {}^-41$
34. $\frac{4}{5}h - 9 = {}^-13$
35. $96 = \frac{^-5}{6}s + 61$
36. $^-2 = \frac{^-9}{14}r + 16$

Solve each inequality. **20-5**

37. $z + 3 > 4$
38. $x - 2 \geq {}^-5$
39. $4a \leq 16$
40. $\frac{c}{3} < 1$
41. $\frac{t}{6} > {}^-4$
42. $y - 23 \leq {}^-19$

Graph each inequality.

43. $a + 5 < 2$
44. $r - 2 \leq 4$
45. $b - 4 \leq {}^-5$
46. $6w > 18$
47. $\frac{x}{2} > 2$
48. $\frac{y}{2} > {}^-1$

Solve using an equation. **20-6**

49. After Terry cut 20 ft from a wire, the length of the remaining wire was 53 ft. What was the length of the original wire?
50. Robert spent \$9.86 at the store. He spent \$2.99 on oranges and bought three large bottles of apple juice. How much did each bottle of apple juice cost?

APPLICATION

20-7 USING FORMULAS

A **formula** is an equation that describes a relationship between two or more quantities. Often each quantity is represented by a variable. The variable used is often the first letter of the word it represents.

For instance, the following formula shows the relationship between a temperature reading on the Celsius scale and the equivalent reading on the Fahrenheit scale.

$$\text{Fahrenheit temperature} = \frac{9}{5} \times \text{Celsius temperature} + 32$$

$$F = \frac{9}{5}C + 32$$

example

The temperature is 15°C. Find the equivalent Fahrenheit temperature.

solution

Write the formula given above. Replace C with 15.

$$F = \frac{9}{5}C + 32$$

$$F = \frac{9}{5}(15) + 32$$

$$F = 27 + 32 = 59$$

The equivalent Fahrenheit temperature is 59°F.

exercises

Change each Celsius temperature to the equivalent Fahrenheit temperature.

1. 90°C 2. 40°C 3. $^-$20°C 4. 0°C

Write the given rule as a formula.

5. A baseball player's batting average (A) equals the number of hits (h) divided by the number of times at bat (b).
6. The total cost (C) equals the number of items (n) times the price per item (p).
7. A player's field goal percentage (p) equals the number of field goals made (m) divided by the number of field goals attempted (a).
8. A normal blood pressure reading (P) equals a person's age (a) divided by 2, plus 110.

The chart below shows formulas both in words and symbols.

Formula in Words	Symbols
distance = rate × time	$D = rt$
distance (in feet) from a thunderstorm = 1100 × time (in seconds) between lightning and thunder	$D = 1100t$
ground speed = air speed − head wind speed	$g = a - h$

Solve. Use the formulas in the chart above.

9. Jeremiah counts 3 s between a flash of lightning and the clap of thunder. Find the distance Jeremiah is from the thunderstorm.
10. A plane flying at 650 mi/h takes about 4 h to fly from Los Angeles to Honolulu. Find the distance between the two cities.
11. An airplane is flying into a head wind of 20 mi/h. The plane's air speed is 600 mi/h. What is the ground speed?
12. An airplane with an air speed of 650 mi/h is flying into a head wind of 30 mi/h. What is the ground speed?
13. The first solar-powered airplane flew at a rate of 30 mi/h for $5\frac{1}{2}$ h. How many miles did the airplane fly?
14. Kent drives at 50 mi/h for 3 h. He leaves home at 2:15 P.M. How many miles does he drive?
15. An airplane is flying into a head wind of 70 km/h. The plane's air speed is 1126 km/h. What is the ground speed?
16. Katrina counts 1 min 15 s between a flash of lightning and the clap of thunder. Find the distance Katrina is from the thunderstorm.

calculator corner

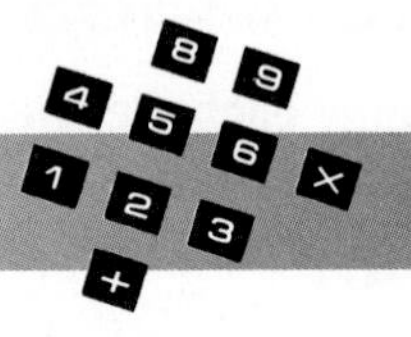

You can use a calculator when you work with formulas.

example The temperature is 68°F. Find the equivalent Celsius temperature. Use the formula $C = \frac{5}{9}(F - 32)$.

solution Enter 68 − 32 = M+. ← **Enters the quantity within parentheses into memory.**
Then enter 5 ÷ 9 = × MR =.
The result is 19.999999 or 20.
Rounded to the nearest degree, the Celsius temperature is 20°C.

Change each Fahrenheit temperature to the equivalent Celsius temperature. Use the formula $C = \frac{5}{9}(F - 32)$.

1. 50°F
2. 212°F
3. 32°F
4. 98.6°F

APPLICATION

20-8 TRANSFORMING FORMULAS

Sometimes you need to *transform* a formula before you can use it. When you **transform** a formula, you use inverse operations to get a particular variable alone on one side of the equals sign.

example

Arnie Thayer has a job interview in a town 175 mi from his home. His average driving speed is 50 mi/h. How much time should it take him to drive this distance?

solution

Write the formula for distance. Use inverse operations to get the variable t alone on one side of the equals sign.

$$D = rt$$

$$\frac{D}{r} = \frac{rt}{r} \quad \leftarrow \text{Divide both sides by } r.$$

$$\frac{D}{r} = t$$

$$\frac{175}{50} = t \quad \leftarrow \text{Replace } D \text{ with 175 and } r \text{ with 50.}$$

$$3.5 = t \quad \leftarrow \text{Divide. A calculator may be helpful.}$$

It should take him $3\frac{1}{2}$ h to drive this distance.

exercises

Complete. Use the formula $D = rt$.

	DISTANCE (D)	RATE (r)	TIME (t)
1.	3400 km	850 km/h	? h
2.	817 km	76 km/h	? h
3.	120 mi	? mi/h	3 h
4.	264 ft	? ft/s	48 s

5. Jim and his friend can ride their bicycles at an average speed of 6 mi/h. How much time should it take them to travel 18 mi?

6. Janet ran a 100-m race in 16 s. At what average speed did she run the race?

Transform each formula to get the variable in color alone on one side of the equals sign.

7. $P = 4s$

8. $C = \frac{22}{7}d$

9. $C = np$

10. $A = lw$

11. $A = \frac{t}{n}$

12. $E = \frac{I}{R}$

13. $A = \frac{1}{2}bh$

14. $V = \frac{1}{3}Bh$

15. $D = 2s + r$

16. $P = 2l + 2w$

The amount of work (*W*) needed to move an object is defined as the product of the amount of force (*F*) exerted on the object and the distance (*d*) that the object is moved. Work is measured in foot-pounds (ft-lb). Use the formula $W = Fd$ to complete.

	WORK (W)	FORCE (F)	DISTANCE (d)
17.	90,000 ft-lb	2000 lb	? ft
18.	200 ft-lb	? lb	25 ft

19. When Jack lifted a 75-lb crate, he did 375 ft-lb of work. How high did he lift the crate?

20. A crane does 11,200 ft-lb of work raising a piano 32 ft off the ground. Find the weight of the piano.

The amount of simple interest (*I*) paid or earned on a loan is determined by finding the product of the principal (*P*), the rate of interest (*r*), and the time in years (*t*). Use the formula $I = Prt$ to complete.

	INTEREST (I)	PRINCIPAL (P)	RATE (r)	TIME (t)
21.	\$100	?	5%	1 year
22.	\$60	\$500	?	2 years
23.	\$200	\$1000	4%	? years

24. Jennifer borrowed \$3000 from her parents. After one year she repaid the loan with \$120 interest. What was the rate of interest?

25. Mr. Darms paid \$1950 in interest on a \$10,000 loan that earned 6.5% simple interest. For how many years was the loan?

26. To pay for a stereo system, Lee borrowed money from a friend at a rate of 6% simple interest per year. After two years, he repaid the loan with \$300 interest. What was the amount of the loan?

27. Mary borrowed \$500 from a friend. At the end of one year, she paid the friend \$515. What was the rate of interest on the loan?

CHAPTER REVIEW

vocabulary vo·cab·u·lar·y

Choose the correct word to complete each sentence.

1. A symbol used to represent a number is a (*fraction, variable*).
2. A (*solution, formula*) is an equation that describes the relationship between two or more quantities.

skills

Evaluate each expression when $a = 4$, $b = 1$, $c = 0$, and $d = {}^{-}8$.

3. $b + 10 + d$
4. $3ac$
5. a^2d
6. $2a + 3b$

Solve each equation. Check your answers.

7. $1 = y + 8$
8. ${}^{-}30 = w - 5$
9. $12 = \frac{{}^{-}2}{3}a$
10. $b + 3 = 14.1$
11. ${}^{-}9d = {}^{-}54$
12. $\frac{k}{0.4} = 7$
13. ${}^{-}8x - 4 = 4.8$
14. $3 = \frac{{}^{-}1}{7}m + 14$
15. $z - \frac{1}{5} = 3\frac{2}{5}$

Solve each inequality.

16. $z + 6 < 4$
17. $6p > 18$
18. $\frac{c}{4} \leq {}^{-}8$

Graph each inequality.

19. $t - 2 > {}^{-}5$
20. $y + 2 < 1$
21. $\frac{d}{2} \geq 2$

Solve using an equation.

22. Meredith and Joan equally share a monthly rental payment of $650. How much rent does each pay?
23. Donald bought a used car for $3300. He made a down payment of $500. He repaid a car loan for the remaining amount in ten equal payments. What was the amount of each payment?
24. Use a formula from the chart on page 444. An airplane is flying into a head wind of 25 mi/h. The plane's air speed is 625 mi/h. Find the plane's ground speed.
25. Sue drives 180 mi on the highway. Her average speed is 45 mi/h. Use the formula $D = rt$. Find how many hours Sue drives.

CHAPTER TEST

Evaluate each expression when $w = 1$, $x = {}^{-}6$, $y = 0$, and $z = 2$.

20-1

1. $10w$
2. x^2z
3. $2w - 11$
4. $3xy$
5. $7 + \frac{x}{z}$
6. $14 + x + z$
7. $\frac{30}{y}$
8. $\frac{x}{^{-}3}$

Solve each equation. Check your answers.

20-2

9. $t + 10 = 3$
10. $s - 4 = 5$
11. $21 = a + 20$
12. $^{-}16 = b - 6$
13. $\frac{8}{9} = \frac{4}{9} + m$
14. $q - 0.4 = 11$

20-3

15. $8t = {}^{-}48$
16. $\frac{c}{^{-}0.7} = {}^{-}3$
17. $\frac{2}{3}y = 6$
18. $16.5 = 3z$
19. $^{-}5 = \frac{a}{6}$
20. $16 = \frac{^{-}4}{5}m$

20-4

21. $^{-}4a + 31 = 43$
22. $^{-}4 = \frac{^{-}1}{8}w + 7$
23. $6 = \frac{s}{^{-}5} - 4$
24. $\frac{4}{7}x + 32 = 64$
25. $\frac{4}{15}z - 3 = {}^{-}43$
26. $16 = 3d - 17$

Solve each inequality.

20-5

27. $m + 5 < 3$
28. $2q \geq 20$
29. $\frac{r}{3} > {}^{-}6$

Graph each inequality.

30. $h + 10 > 11$
31. $4g \leq 16$
32. $\frac{y}{2} \geq 1$

Solve using an equation.

20-6

33. Jared bought a package of eight pencils for $.64. How much did each pencil cost?
34. Sarah lent Jan $2.50. Sarah still has $7.27. How much money did Sarah have before she lent Jan the money?

20-7

35. Use a formula from the chart on page 444. Jo counts 55 s between a flash of lightning and the clap of thunder. Find her distance from the storm.

20-8

36. Jeremy borrowed $200 from his brother. After one year he repaid the loan and $6 interest. Use the formula $I = Prt$ to find the rate of interest.

This electronic instrument, which is called an oscilloscope, can create a graph that represents the motion of electrons.

CHAPTER 21

GRAPHING IN THE COORDINATE PLANE

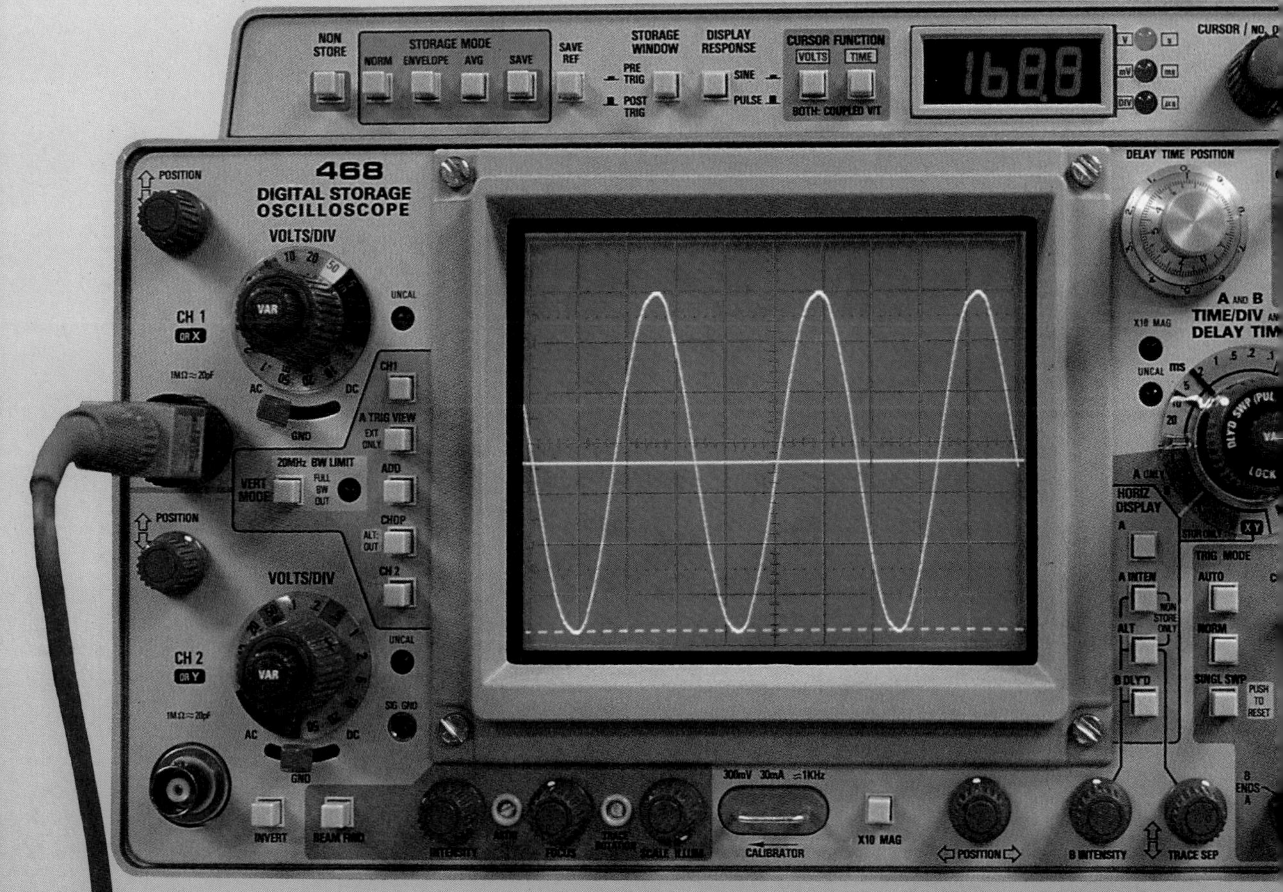

21-1 THE COORDINATE PLANE

The grid at the right is called a **coordinate plane.** The *horizontal* number line is the ***x*-axis.** The *vertical* number line is the ***y*-axis.** These *axes* meet at point *O*, which is called the **origin.**

You can assign an **ordered pair** of numbers to any point on a coordinate plane. The *first* number of an ordered pair is the ***x*-coordinate.** The *second* number is the ***y*-coordinate.** The origin has coordinates (0, 0).

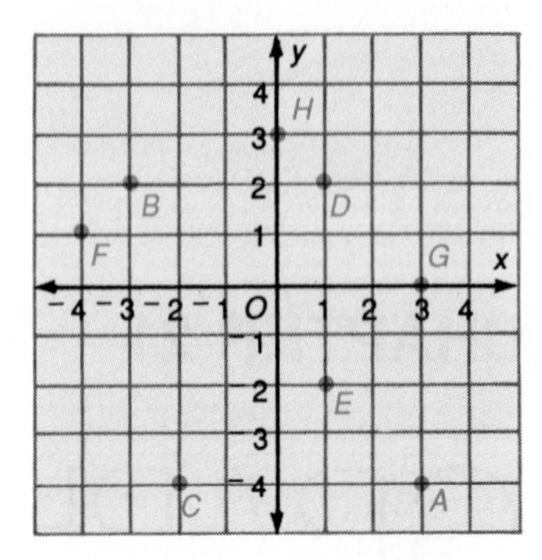

example 1

Use the coordinate plane above. Give the coordinates of each point.

a. *A* **b.** *B*

solution

a. Start at the origin.
Point *A* is 3 units right (positive) and 4 units down (negative).
The coordinates are $(3, {}^-4)$.

b. Start at the origin.
Point *B* is 3 units left (negative) and 2 units up (positive).
The coordinates are $({}^-3, 2)$.

your turn

Use the coordinate plane above. Give the coordinates of each point.

1. *C* **2.** *D* **3.** *E* **4.** *F* **5.** *G* **6.** *H*

The **graph of an ordered pair** is the point that is assigned to it.

example 2

Graph the point $M({}^-4, 3)$ on a coordinate plane.

solution

Start at the origin. Move 4 units to the left and then 3 units up.

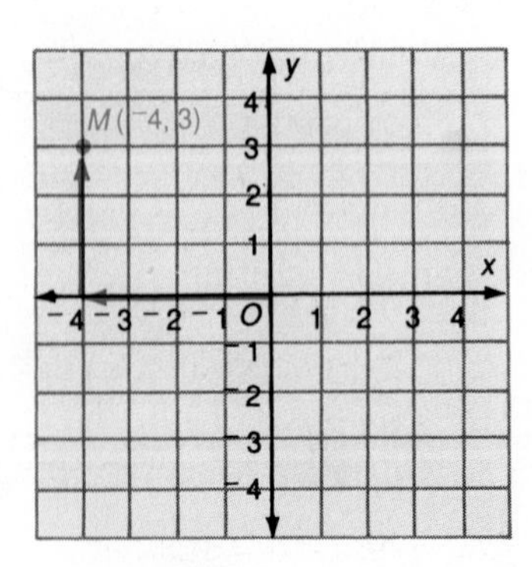

your turn

Graph the given points on a coordinate plane.

7. $N(2, 3)$ **8.** $P({}^-4, {}^-1)$ **9.** $Q(1, {}^-3)$
10. $R({}^-4, 2)$ **11.** $S({}^-2, 0)$ **12.** $T(0, {}^-1)$

practice exercises

practice for example 1 (page 450)

Use the coordinate plane at the right. Give the coordinates of each point.

1. *A*	**2.** *B*	**3.** *C*	**4.** *D*
5. *E*	**6.** *F*	**7.** *G*	**8.** *H*
9. *J*	**10.** *K*	**11.** *L*	**12.** *O*

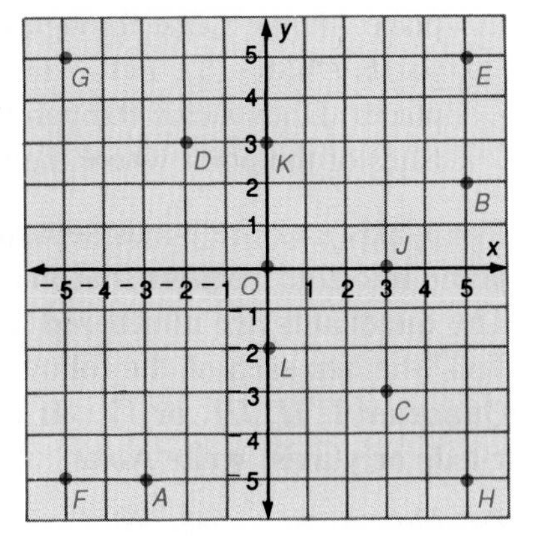

practice for example 2 (page 450)

Graph the given points on a coordinate plane.

13. $A(4, 1)$	**14.** $B(1, 4)$
15. $C(^-3, ^-3)$	**16.** $D(^-5, 4)$
17. $E(2, ^-4)$	**18.** $F(^-4, ^-2)$
19. $G(^-5, 2)$	**20.** $H(^-2, ^-4)$
21. $J(2, 0)$	**22.** $K(^-3, 0)$
23. $L(0, 5)$	**24.** $M(0, ^-2)$

mixed practice (page 450)

Use the coordinate plane at the right. Write the letter of the point with the given coordinates.

25. $(0, 4)$	**26.** $(4, 0)$	**27.** $(^-4, 0)$
28. $(0, ^-4)$	**29.** $(3, ^-4)$	**30.** $(^-3, ^-4)$
31. $(3, 4)$	**32.** $(4, ^-3)$	**33.** $(^-4, 3)$
34. $(4, 3)$	**35.** $(^-4, ^-3)$	**36.** $(0, 0)$

Name five different ordered pairs that satisfy the given condition. Then graph the ordered pairs. Use a different coordinate plane for each exercise.

37. The y-coordinate is equal to the x-coordinate.

38. The y-coordinate is equal to the opposite of the x-coordinate.

39. The y-coordinate is one greater than the x-coordinate.

40. The x-coordinate is twice the y-coordinate.

41. The y-coordinate is 2.

42. The x-coordinate is $^-3$.

43. The y-coordinate is 0.

44. The x-coordinate is 0.

45. Graph the points $R(^-4, 0)$, $S(2, 3)$, and $T(4, 4)$ on a coordinate plane. Draw a line through the three points. Name the coordinates of two other points that appear to lie on this line.

46. Graph the points $A(^-5, 1)$, $B(^-3, ^-1)$, and $C(1, ^-5)$ on a coordinate plane. Draw a line through these three points. Then graph the points $D(^-5, ^-3)$, $E(1, ^-1)$, and $F(4, 0)$ on the same coordinate plane. Draw a line through these three points. Name the coordinates of the point where the two lines meet.

The x-axis and the y-axis separate a coordinate plane into four regions that are called *quadrants*. The quadrants are numbered as shown at the right. Tell whether each of the following points lies in Quadrant *I, II, III,* or *IV*. If the point lies on the x-axis or y-axis, write *None*.

47. $M(6, 8)$	48. $N(^-3, ^-11)$	49. $O(0, 0)$
50. $P(0, ^-10)$	51. $Q(^-7, 8)$	52. $R(9, ^-2)$
53. $S(8, 0)$	54. $T(0, ^-9)$	55. $U(10, ^-3)$
56. $V(^-7, ^-9)$	57. $W(10, 10)$	58. $Z(^-9, 8)$

Solve. Making a drawing may be helpful.

59. Joe leaves his house and walks 5 blocks due north, 4 blocks due east, 9 blocks due south, 9 blocks due west, and 4 blocks due north. At that point, where is Joe in relation to his house?

60. Alicia leaves school and bicycles 7 blocks due west and then 4 blocks due south to the market. When she leaves the market, she bicycles 7 blocks due east and then 6 blocks due south to her house. Where is Alicia's house in relation to school?

review exercises

Solve each equation. Check your answers.

1. $m + 9 = ^-2$
2. $12 = r - 0.4$
3. $^-4y = 56$
4. $^-1.05 = ^-7d$
5. $5w + 11 = 1$
6. $2t - 12 = ^-6$
7. $\frac{n}{5.2} = ^-3$
8. $10 = \frac{2}{5}k + 20$

21-2 EQUATIONS WITH TWO VARIABLES

The equation $2x + 6 = 12$ contains one variable. This equation has one solution, which is 3. Some equations, such as $y = 2x + 5$, contain two variables. A solution of this type of equation is a value of x and a value of y that *together* make the equation true. You write a solution of an equation with two variables as an ordered pair of numbers.

example 1

Is each ordered pair a solution of $y = 2x + 5$?

a. (4, 13)

b. ($^-3$, 11)

solution

a. Replace x with 4 and y with 13.

$y = 2x + 5$

$13 \stackrel{?}{=} 2(4) + 5$

$13 = 13$

Yes, (4, 13) is a solution of $y = 2x + 5$.

b. Replace x with $^-3$ and y with 11.

$y = 2x + 5$

$11 \stackrel{?}{=} 2(^-3) + 5$

$11 \neq {}^-1$

$\neq$ means *is not equal to.*

No, ($^-3$, 11) is not a solution of $y = 2x + 5$.

your turn

Is each ordered pair a solution of $y = 3x + 1$? Write *Yes* or *No*.

1. (3, 10) **2.** (4, 11) **3.** ($^-2$, $^-5$) **4.** ($^-4$, $^-13$) **5.** ($\frac{1}{3}$, 2)

example 2

Complete each ordered pair so that it is a solution of $y = 4x - 1$.

a. (2, ?)

b. ($^-3$, ?)

solution

a. Replace x with 2. Solve for y.

$y = 4x - 1$

$y = 4(2) - 1$

$y = 7$

The solution is (2, 7).

b. Replace x with $^-3$. Solve for y.

$y = 4x - 1$

$y = 4(^-3) - 1$

$y = {}^-13$

The solution is ($^-3$, $^-13$).

your turn

Complete each ordered pair so that it is a solution of $y = 2x - 3$.

6. (5, ?) **7.** (1, ?) **8.** ($^-4$, ?) **9.** (0, ?) **10.** ($\frac{1}{2}$, ?)

An equation with two variables has *infinitely many* solutions. Often some of the solutions are organized into a table.

example 3

Make a table of ordered pairs that are solutions of $y = 5x - 3$. Use $^-2$, $^-1$, 0, 1, and 2 as values for x.

solution

x	y = 5x − 3	y	solution
$^-2$	$y = 5(^-2) - 3$	$^-13$	$(^-2, ^-13)$
$^-1$	$y = 5(^-1) - 3$	$^-8$	$(^-1, ^-8)$
0	$y = 5(0) - 3$	$^-3$	$(0, ^-3)$
1	$y = 5(1) - 3$	2	$(1, 2)$
2	$y = 5(2) - 3$	7	$(2, 7)$

your turn

For each equation, make a table of ordered pairs that are solutions. Use $^-2$, $^-1$, 0, 1, and 2 as values for x.

11. $y = x + 5$ **12.** $y = ^-2x$ **13.** $y = 5x - 2$ **14.** $y = ^-3x + 1$

practice exercises

practice for example 1 (page 453)

Is each ordered pair a solution of $y = 6x - 2$? Write *Yes* or *No*.

1. (2, 10) **2.** (3, 18) **3.** $(^-1, 4)$ **4.** $(^-3, ^-20)$
5. $(0, ^-2)$ **6.** (0, 0) **7.** $(\frac{1}{2}, ^-1)$ **8.** $(\frac{1}{3}, 0)$

practice for example 2 (page 453)

Complete each ordered pair so that it is a solution of $y = 4x + 3$.

9. (2, ?) **10.** (1, ?) **11.** $(^-1, ?)$ **12.** $(^-3, ?)$
13. $(^-4, ?)$ **14.** (0, ?) **15.** $(\frac{1}{2}, ?)$ **16.** $(\frac{1}{4}, ?)$

practice for example 3 (page 454)

For each equation, make a table of ordered pairs that are solutions. Use $^-2$, $^-1$, 0, 1, and 2 as values for x.

17. $y = x + 3$ **18.** $y = x - 6$ **19.** $y = 4x$ **20.** $y = ^-3x$
21. $y = 3x - 1$ **22.** $y = 4x + 2$ **23.** $y = ^-2x + 2$ **24.** $y = ^-5x - 4$

mixed practice (pages 453–454)

Choose the letters of all the ordered pairs that are solutions of the given equation.

25. $y = x - 5$ a. $(0, 5)$ b. $(1, {}^-4)$ c. $({}^-2, {}^-7)$ d. $({}^-5, 0)$

26. $y = {}^-3x$ a. $(0, {}^-3)$ b. $({}^-1, 3)$ c. $({}^-2, {}^-6)$ d. $(\frac{1}{3}, {}^-9)$

27. $y = {}^-2x + 5$ a. $(0, 5)$ b. $(1, 7)$ c. $(2, 1)$ d. $({}^-2, 9)$

28. $y = 4x - 8$ a. $(0, {}^-8)$ b. $(1, 4)$ c. $({}^-2, 2)$ d. $(\frac{1}{2}, {}^-10)$

Copy and complete each table.

29.

$y = {}^-2x$		
x	y	(x, y)
$^-10$	20	$({}^-10, 20)$
$^-5$	?	$({}^-5, ?)$
0	?	$(?, ?)$
5	?	$(?, ?)$
?	$^-20$	$(?, {}^-20)$

30.

$y = x - 4$		
x	y	(x, y)
$^-4$	?	$({}^-4, ?)$
$^-2$	$^-6$	$(?, ?)$
?	?	$(0, {}^-4)$
2	?	$(?, ?)$
?	0	$(?, 0)$

31.

$y = 2x + 7$		
x	y	(x, y)
$^-6$	$^-5$	$(?, ?)$
$^-3$	?	$({}^-3, ?)$
?	7	$(?, 7)$
3	?	$(?, 13)$
6	?	$(?, ?)$

For each equation, make a table of five ordered pairs that are solutions.

32. $x + y = 9$ **33.** $x - y = 2$ **34.** $x + y = {}^-10$ **35.** $x - y = {}^-1$

review exercises

Write the opposite and the absolute value of each number.

1. 24 **2.** $^-73$ **3.** 0 **4.** 1.8 **5.** $^-0.75$ **6.** $\frac{1}{4}$

7. Gabriella earns $24,700 annually. What is her weekly salary?

8. To make one banner, Alan needs a piece of burlap that is 45 in. wide and $1\frac{3}{4}$ yd long. How many banners can he make from a burlap remnant that is 45 in. wide and $10\frac{1}{2}$ yd long?

mental math

Add mentally.

1. $6 + {}^-8 + 9 + {}^-2 + {}^-5 + 11$ **2.** ${}^-3.5 + {}^-4.7 + 3.5 + {}^-11.2$

Solve mentally.

3. ${}^-2 + t = {}^-7$ **4.** $m - 12 = {}^-9$ **5.** ${}^-8z = 48$ **6.** $\frac{a}{3} = {}^-12$

21-3 GRAPHING EQUATIONS WITH TWO VARIABLES

The **graph of an equation** with two variables is the set of all points whose coordinates are solutions of the equation. To graph an equation with two variables, use the following method.

1. Make a table of ordered pairs that are solutions.
2. Graph the ordered pairs as points on a coordinate plane.
3. Connect the points with a straight line.

example 1

Graph $y = x - 3$ on a coordinate plane.

solution

x	y = x − 3	y	solution
$^{-}2$	$y = {}^{-}2 - 3$	$^{-}5$	$(^{-}2, {}^{-}5)$
0	$y = 0 - 3$	$^{-}3$	$(0, {}^{-}3)$
2	$y = 2 - 3$	$^{-}1$	$(2, {}^{-}1)$

← **Choose reasonable values for x.**

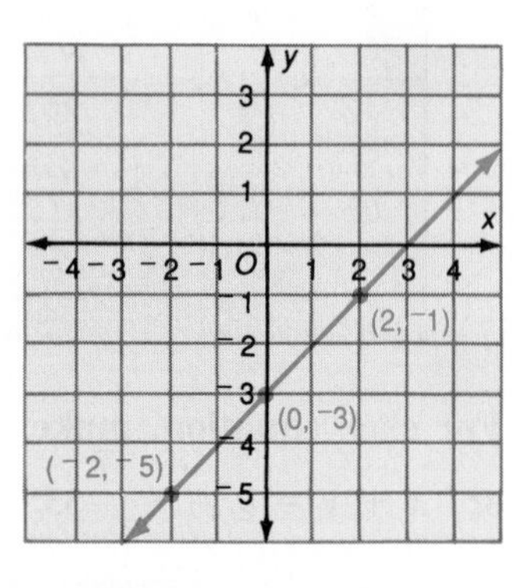

your turn

Graph each equation on a coordinate plane. Use $^{-}2$, 0, 2 as values for x.

1. $y = x + 3$ **2.** $y = x - 5$ **3.** $y = 2x - 1$ **4.** $y = 2x$

example 2

Graph $y = -x - 3$ on a coordinate plane.
Use $^{-}2$, 0, 2 as values for x.

solution

x	y = −x − 3	y	solution
$^{-}2$	$y = {}^{-}1(^{-}2) - 3$	$^{-}1$	$(^{-}2, {}^{-}1)$
0	$y = {}^{-}1(0) - 3$	$^{-}3$	$(0, {}^{-}3)$
2	$y = {}^{-}1(2) - 3$	$^{-}5$	$(2, {}^{-}5)$

← **$-x$ is equal to $^{-}1(x)$.**

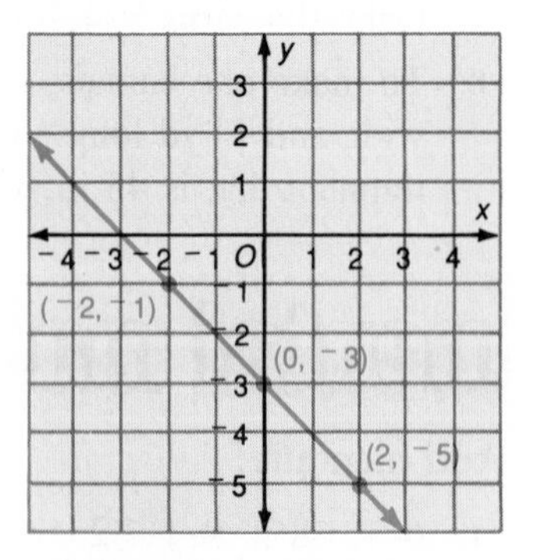

your turn

Graph each equation on a coordinate plane. Use $^{-}2$, 0, 2 as values for x.

5. $y = -x + 3$ **6.** $y = -x - 4$ **7.** $y = {}^{-}2x + 3$ **8.** $y = {}^{-}2x$

practice exercises

practice for example 1 (page 456)

Graph each equation on a coordinate plane. Use $^{-}2, 0, 2$ as values for x.

1. $y = x + 2$
2. $y = x + 5$
3. $y = x - 1$
4. $y = x - 6$
5. $y = 2x + 3$
6. $y = 3x - 2$
7. $y = 3x$
8. $y = x$

practice for example 2 (page 456)

Graph each equation on a coordinate plane. Use $^{-}2, 0, 2$ as values for x.

9. $y = -x + 6$
10. $y = -x + 1$
11. $y = -x - 5$
12. $y = -x - 2$
13. $y = {}^{-}2x + 5$
14. $y = {}^{-}3x - 1$
15. $y = {}^{-}3x$
16. $y = -x$

mixed practice (page 456)

Choose the letter of the correct equation for each graph. (Not all the equations will be used.)

17.

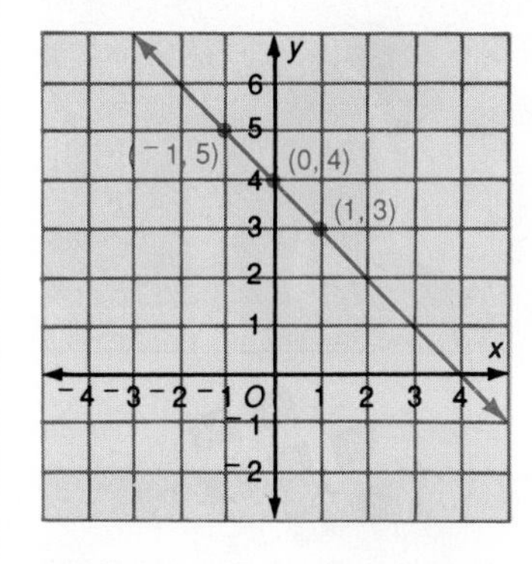

18.

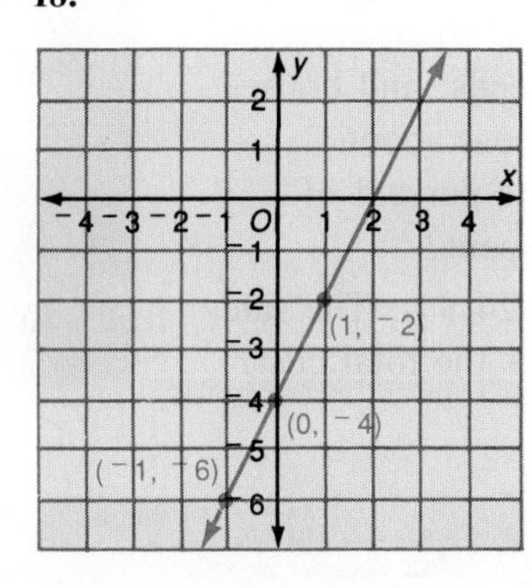

A. $y = x + 4$
B. $y = x - 4$
C. $y = -x + 4$
D. $y = -x - 4$
E. $y = 2x + 4$
F. $y = 2x - 4$
G. $y = {}^{-}2x + 4$
H. $y = {}^{-}2x - 4$

19.

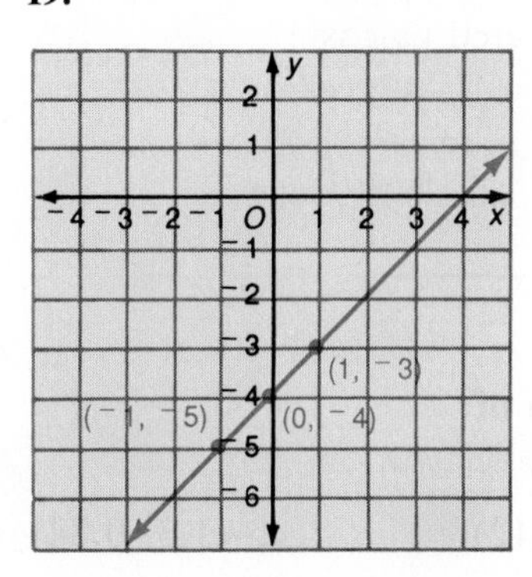

20.

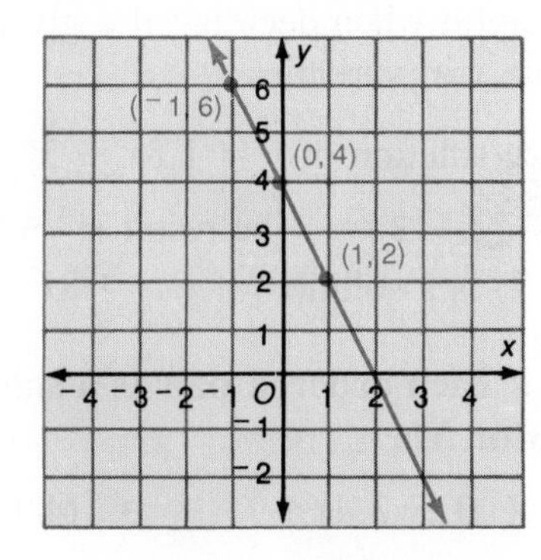

Graph each equation on a coordinate plane. Use the given values for x.

21. $y = x - 8$; 1, 3, 5

22. $y = -x + 9$; 2, 4, 6

23. $y = {}^{-}5x$; $^{-}1$, 0, 1

24. $y = 4x + 7$; $^{-}2$, $^{-}1$, 0

Graph each equation on a coordinate plane. Choose three reasonable values for x.

25. $y = x - 7$

26. $y = x + 10$

27. $y = -x + 10$

28. $y = -x - 8$

29. $y = 4x$

30. $y = {}^{-}5x$

31. $y = {}^{-}5x + 6$

32. $y = 6x + 9$

Graph each pair of equations on the same coordinate plane.

33. $y = x + 4$
$y = x - 1$

34. $y = 2x + 2$
$y = 2x - 3$

35. $y = x - 1$
$y = 2x - 1$

36. $y = -x + 2$
$y = 3x + 2$

review exercises

Write the unit rate.

1. 144 mi in 3 h
2. $260 for 4 days
3. 24 m in 5 s
4. 97.8 mi on 3 gal
5. In 1970, Americans spent about $2 billion for flowers, seeds, and house plants. In 1985, they spent about $6 billion. What was the percent of increase in the amount spent?
6. The price of a radio is $24.99. The sales tax rate is $5\frac{1}{4}$%. What is the total cost?

calculator corner

Sometimes a calculator is helpful when deciding if a given ordered pair is a solution of an equation with two variables.

example Is ($^{-}1.5$, $^{-}5.7$) a solution of $y = 1.8x - 3$?

solution Enter 1.8 × 1.5 [+/−] − 3 =. The result is $^{-}5.7$.
Yes, ($^{-}1.5$, $^{-}5.7$) is a solution of $y = 1.8x - 3$.

Use a calculator to decide if each ordered pair is a solution of $y = {}^{-}3.2x + 5$. Write *Yes* or *No*.

1. (0.6, 3.08)
2. ($^{-}0.5$, 3.4)
3. (4.1, 18.12)
4. (1.3, 0.84)

21-4 SYSTEMS OF EQUATIONS

Two equations that have the same variables can form a **system of equations.** An ordered pair that is a solution of *both* equations is called a **solution of the system.**

example 1

Is (2, 1) a solution of each system?

a. $y = 2x - 3$
$y = x - 1$

b. $y = -x + 3$
$y = 4x - 5$

solution

In each equation, replace x with 2 and y with 1.

a.

$y = 2x - 3$	$y = x - 1$
$1 \stackrel{?}{=} 2(2) - 3$	$1 \stackrel{?}{=} 2 - 1$
$1 = 1$	$1 = 1$

Yes, (2, 1) is a solution.

b.

$y = -x + 3$	$y = 4x - 5$
$1 \stackrel{?}{=} {}^{-}1(2) + 3$	$1 \stackrel{?}{=} 4(2) - 5$
$1 = 1$	$1 \neq 3$

No, (2, 1) is not a solution.

your turn

Is the given ordered pair a solution of the system? Write *Yes* or *No*.

1. (2, 1)
$y = 3x - 5$
$y = 6x - 11$

2. (3, 5)
$y = x + 2$
$y = 3x - 1$

3. $({}^{-}1, 4)$
$y = 4x + 8$
$y = x + 3$

4. (0, 6)
$y = 9x + 6$
$y = {}^{-}9x + 6$

You graph a system of equations by graphing all the equations in the system on the same coordinate plane.

If the graphs of the equations meet at a point, the ordered pair for this point is the solution of the system.

If the graphs of the equations will never meet, then the system has no solution.

Solution: (1, 2)

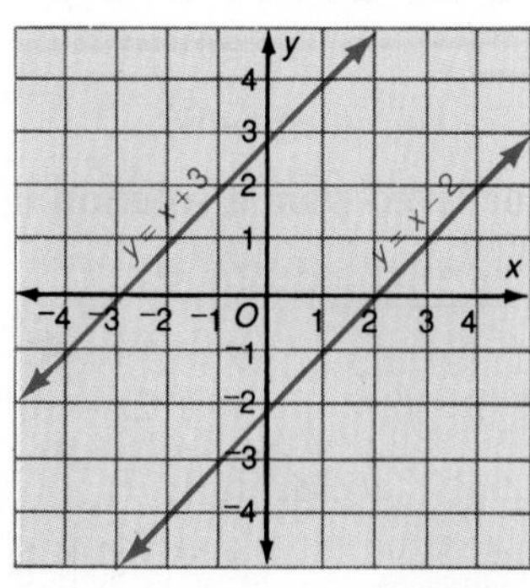

No solution

To solve a system by graphing, use the following method.

1. Make a table of ordered pairs for each equation.
2. Use the ordered pairs to graph both equations on the same coordinate plane.
3. Identify the point where the graphs of the equations meet.
4. If the graphs will not meet, the system has *no solution*.

example 2

Solve the system: $y = x + 1$
$y = 2x - 1$

solution

x	y = x + 1	y	solution
$^-1$	$y = {^-1} + 1$	0	$(^-1, 0)$
0	$y = 0 + 1$	1	$(0, 1)$
1	$y = 1 + 1$	2	$(1, 2)$

x	y = 2x − 1	y	solution
$^-1$	$y = 2(^-1) - 1$	$^-3$	$(^-1, {^-3})$
0	$y = 2(0) - 1$	$^-1$	$(0, {^-1})$
1	$y = 2(1) - 1$	1	$(1, 1)$

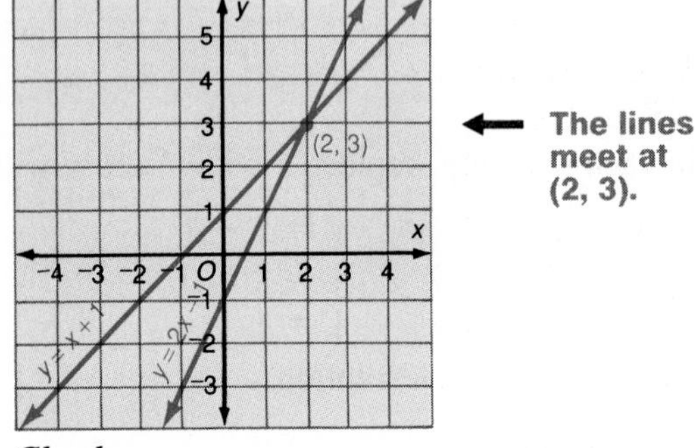

Check:

$y = x + 1$ $\quad$ $y = 2x - 1$

$3 \stackrel{?}{=} 2 + 1$ $\quad$ $3 \stackrel{?}{=} 2(2) - 1$

$3 = 3$ $\quad$ $3 = 3$

The solution of the system is (2, 3).

your turn

Solve each system by graphing.

5. $y = x - 1$
   $y = 3x - 5$
6. $y = 2x + 1$
   $y = 2x - 3$
7. $y = 3x + 1$
   $y = -x - 7$
8. $y = {^-4}x - 2$
   $y = 2x - 2$

practice exercises

practice for example 1 (page 459)

Is the given ordered pair a solution of the system? Write *Yes* or *No*.

1. (1, 5)
   $y = 6x - 1$
   $y = 2x + 3$
2. (6, 2)
   $y = x - 4$
   $y = 2x - 6$
3. (0, 5)
   $y = 7x + 5$
   $y = {^-3}x + 5$
4. (2, 7)
   $y = 4x - 1$
   $y = 3x + 1$
5. $(^-2, 5)$
   $y = {^-4}x - 3$
   $y = -x + 3$
6. $(^-1, 7)$
   $y = 2x + 5$
   $y = {^-3}x + 4$
7. $(^-3, 1)$
   $y = {^-2}x + 5$
   $y = 2x + 7$
8. $(^-1, {^-6})$
   $y = 5x - 1$
   $y = 7x + 1$

COMPUTER

Ethics in Computing

Computers have rapidly become an important part of our everyday lives. This fact has created the need for a set of standards, or rules, that govern the proper use of computers. Such standards are often referred to as **computer ethics.**

Many issues of computer ethics center around *individual privacy.* The term **individual privacy** refers to your right to have personal information kept secure from use or control by other people. Many questions of privacy arise because of the existence of vast computer **data banks,** which are electronic ''storehouses'' containing millions of pieces of information about individuals. For instance, do you have a right to see information about you that is stored in one of these data banks? Which others have the right to see this information, and when?

One attempt to address such questions is the *Privacy Act of 1974.* This act of the United States Congress outlines a set of rules for the creation and management of government data banks. The act allows you access to your personal files in these data banks, and it prohibits the sharing of your files with others without your permission.

Entering a computer data bank without permission in order to change information or to find out information about others is called **computer trespassing.** To discourage trespassing, many computer systems require a user to enter a *password* before gaining access to the system. Other computer systems try to prevent trespassing by using *cryptography,* in which the information itself is entered in a coded form that is only understandable to a user who knows the code.

exercises

1. Name two ways in which computer systems may attempt to discourage computer trespassing.
2. Computer trespassing is only one of several ethical problems that have arisen from the growing use of computers in our society. Find out what is meant by *software piracy* and *computer viruses.*

COMPETENCY TEST

Choose the letter of the correct answer.

1. **Which phrase could *not* be represented by the integer $^{-}12$?**

   A. 12° C below zero
   B. a loss of 12 lb
   C. The opposite of $|^{-}12|$
   D. the distance between $^{-}4$ and 8 on a number line

2. **Find the product.** $^{-}5(^{-}2)(^{-}3)$

   A. $^{-}10$ B. $^{-}30$
   C. 30 D. 10

3. **Which point has coordinates $(^{-}2, 1)$?**

   A. M
   B. A
   C. T
   D. H

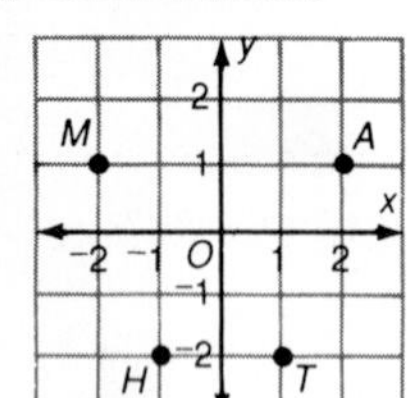

4. **Find the quotient.** $24 \div {}^{-}6$

   A. $^{-}4$
   B. $\frac{^{-}1}{4}$
   C. 18
   D. $\frac{1}{4}$

5. **Lisa took out a loan for \$150. After one year she repaid the loan and \$7.50 interest. Use the formula $I = Prt$ to find the interest rate.**

   A. 20% B. 5%
   C. 0.2% D. 0.05%

6. **Which describes all the points on the x-axis?**

   A. The x-coordinate is 0.
   B. The y-coordinate is 0.
   C. The x-coordinate is equal to the y-coordinate.
   D. The y-coordinate is positive.

7. **Solve.** $y - 2 \geq {}^{-}3$

   A. $y \geq {}^{-}5$ B. $y \geq {}^{-}1$
   C. $y \geq 1$ D. $y \geq 5$

8. **Identify the equation whose graph is shown.**

   A. $y = x - 1$
   B. $y = -x - 1$
   C. $y = -x + 1$
   D. $y = x + 1$

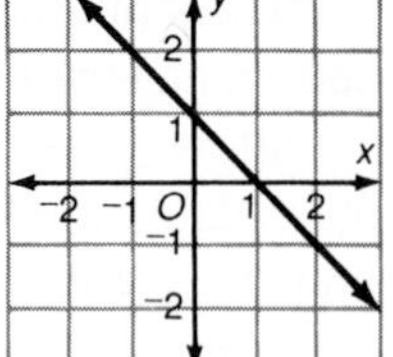

9. **Solve.** $2y - 5 = {}^{-}10$

   A. $y = {}^{-}7\frac{1}{2}$
   B. $y = {}^{-}15$
   C. $y = {}^{-}25$
   D. $y = {}^{-}2\frac{1}{2}$

10. **Which ordered pair is the solution of the given system?**

$$y = 2x - 1$$
$$y = 3x + 2$$

   A. (2, 3) B. $(^{-}3, {}^{-}7)$
   C. $(^{-}1, 1)$ D. (3, 5)

11. Rhea's age, r, is one more than twice her brother's age b. Which function rule describes the relationship?

A. $r = 1 - 2b$
B. $r = 2b - 1$
C. $r = 2b + 1$
D. $b = 2r + 1$

12. Which ordered pair is *not* a solution of $y = {}^{-}2x + 4$?

A. $({}^{-}1, 2)$ B. $(0, 4)$
C. $(4, {}^{-}4)$ D. $(2, 0)$

13. Which number is greatest?

A. ${}^{-}2.8$ B. ${}^{-}2$
C. ${}^{-}2\frac{3}{4}$ D. ${}^{-}3$

14. Solve. $\frac{1}{3}x - 4 = 29$

A. $x = 91$
B. $x = 75$
C. $x = 11$
D. $x = 99$

15. Which has an average temperature of (1.08×10^5)°F?

A. the human body
B. a preheated baking oven
C. the surface of Earth
D. the surface of the sun

16. Pat spent \$6.73 at the grocery store. He spent \$2.78 for chicken and bought five cans of soup. Choose an equation to find how much each can of soup cost.

A. $c + 2.78 = 6.73$
B. $5(c + 2.78) = 6.73$
C. $2.78c + 5 = 6.73$
D. $5c + 2.78 = 6.73$

17. Find the sum. ${}^{-}3\frac{2}{3} + 4\frac{1}{3}$

A. $\frac{2}{3}$ B. $1\frac{1}{3}$
C. 2 D. ${}^{-}8$

18. Find the difference. $7.5 - 8.4$

A. ${}^{-}0.9$ B. 1.1
C. 0.9 D. ${}^{-}1.1$

19. Identify the inequality whose graph is shown.

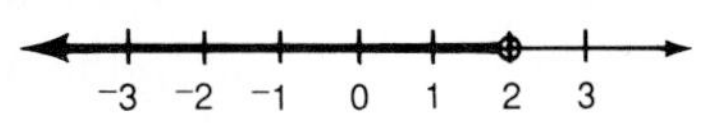

A. $z \leq 2$ B. $z < 2$
C. $z \geq 2$ D. $z > 2$

20. Evaluate the expression $\frac{a}{2} - 5b$ when $a = 0$ and $b = {}^{-}6$.

A. 28 B. ${}^{-}30$
C. 30 D. ${}^{-}3$

CUMULATIVE REVIEW CHAPTERS 1–21

Find each answer. Write fractions in lowest terms.

1. $^{-}4.5 - {}^{-}5.2$
2. $|{}^{-}8|$
3. $^{-}3({}^{-}18)$
4. $\$11.56 \div 17$
5. $^{-}2.5 + {}^{-}7.3$
6. $^{-}105 \div 7$
7. $32\overline{)14{,}016}$
8. $2^3 \times 3^2 \times 8^0$
9. $\left(\frac{3}{5}\right)\left({}^{-}9\frac{2}{3}\right)$
10. $5\frac{1}{4} - 3\frac{2}{3}$
11. $4\frac{1}{2} \div \frac{3}{7}$
12. $\frac{12}{25}(450)$
13. Let y = your age and x = your grade on a quiz. Is y a function of x?
14. Evaluate $4a + 5b$ when $a = 3$ and $b = {}^{-}4$.
15. Write the word form of 2,000,028,415.
16. Write in scientific notation. **a.** 8,260,000 **b.** 0.0045
17. Complete. **a.** 583 mm = _?_ cm **b.** 12 h 3 min = _?_ min
18. What percent of 204 is 136?
19. 520 is 65% of what number?

Estimate.

20. 49% of 177
21. 22.2×59.2
22. $1287 + 887 + 1043$
23. $68\overline{)55{,}231}$
24. $\$9152.34 - \$3219.66$
25. $2\frac{7}{15} + 5\frac{1}{20} + 1\frac{5}{9}$

Solve.

26. $\frac{t}{50} = \frac{9}{30}$
27. $\frac{1}{7}y = {}^{-}3$
28. $z - 4 \geq {}^{-}1$
29. $3x + 5 = 23$
30. A census determines the populations of the five largest states. Is a bar graph or a line graph more appropriate for displaying these data?
31. Each of the letters G, A, R, D, E, N is written on a card. One card is drawn at random.
    a. Find P(vowel). **b.** Find the odds against drawing a vowel.
32. The Science Club trip to the museum will cost \$250. If 29 students go on the trip, about how much will each student have to pay?
33. Make a table of ordered pairs that are solutions of $y = {}^{-}3x + 1$. Use $^{-}2$, $^{-}1$, 0, 1, and 2 as values for x.
34. Lou bought a new car for \$8400. He made a down payment of \$840 and paid the remaining amount in 36 equal payments. Write and solve an equation to find the amount of each payment.
35. Carlos has \$2500 in an account that earns 8% annual interest compounded semiannually. He makes no other deposits or withdrawals for one year. Find the balance at the end of one year.
36. Solve by graphing: $y = -x + 5$
    $y = 2x - 1$

In planning the best use of land, a farmer has to take into account the areas of the fields to be used for different crops.

CHAPTER 22

BASIC GEOMETRY, PERIMETER, AND AREA

22-1 POINTS, LINES, PLANES, AND ANGLES

A **point** is an exact location in space. • A point A

A **plane** is a set of points that forms a flat surface that extends without end. plane W

A **line** is a set of points that extends without end in opposite directions. line XY, written $\overleftrightarrow{XY}$

A **line segment** is a part of a line. It consists of two **endpoints** and all points between them. line segment MN, written $\overline{MN}$

A **ray** is part of a line. It has one endpoint and extends without end in one direction. When you name a ray, the endpoint is named first. ray BC, written $\overrightarrow{BC}$

An **angle** is formed by two rays that have the same endpoint. The endpoint is called the **vertex** of the angle. The rays are called the **sides**. When you name an angle, the vertex is named by the middle letter.

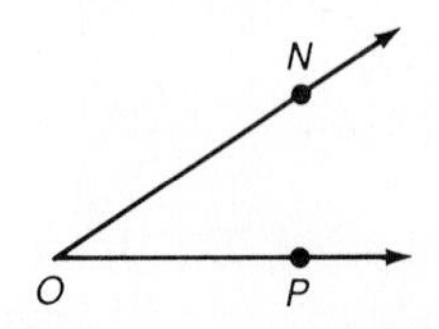

angle NOP, written $\angle NOP$

example 1

Use the figure at the right. Name as many of the following as are shown.

a. points **b.** planes
c. lines **d.** angles
e. line segments

solution

a. $H, R, S, B, C, P, V, T, M, N$

b. Z

c. $\overleftrightarrow{BC}, \overleftrightarrow{RS}$

d. $\angle PVT$

e. $\overline{BC}, \overline{RS}, \overline{MN}, \overline{VP}, \overline{VT}$

your turn

Use the figure at the right. Name as many of the following as are shown.

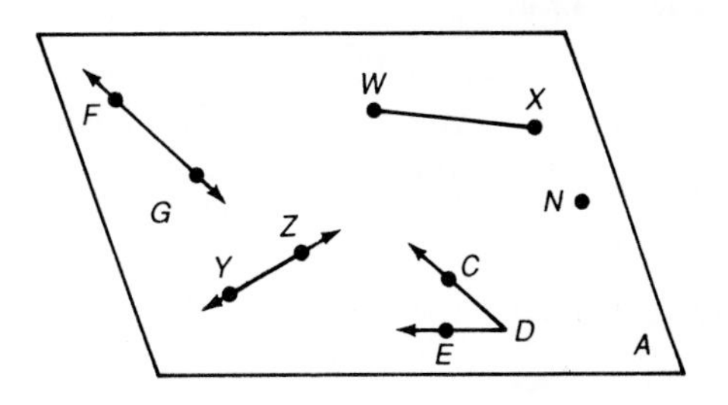

1. points
2. planes
3. lines
4. angles
5. line segments

Lines that meet in one point are called **intersecting lines**. Two lines that intersect form pairs of *adjacent angles* and *vertical angles*. The **adjacent angles** are those angles that share a common side. The angles that are not adjacent are called **vertical angles.**

example 2

Use the figure at the right. Identify each pair of angles as *adjacent* or *vertical*.

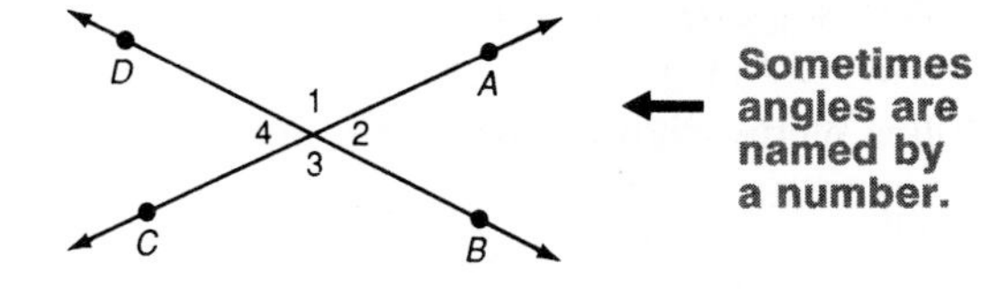

Sometimes angles are named by a number.

a. $\angle 1$ and $\angle 2$

b. $\angle 2$ and $\angle 4$

solution

a. $\angle 1$ and $\angle 2$ share a common side. They are *adjacent*.

b. $\angle 2$ and $\angle 4$ do *not* share a common side. They are *vertical*.

your turn

Use the figure above. Identify each pair of angles as *adjacent* or *vertical*.

6. $\angle 2$ and $\angle 3$
7. $\angle 1$ and $\angle 3$
8. $\angle 3$ and $\angle 4$
9. $\angle 1$ and $\angle 4$

practice exercises

practice for example 1 (pages 478–479)

Use the figure at the right. Name as many of the following as are shown.

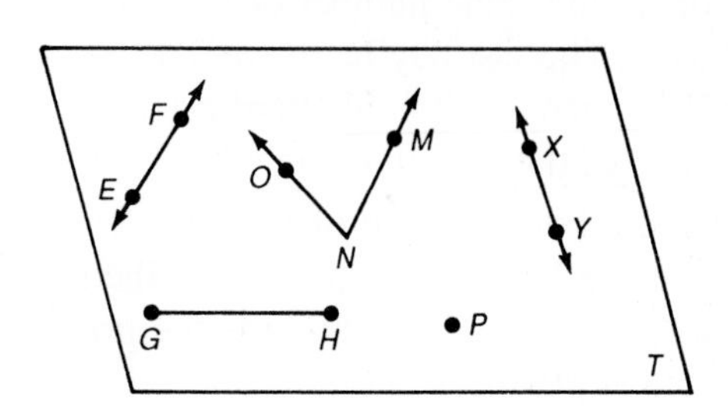

1. points
2. planes
3. lines
4. angles
5. line segments

practice for example 2 (page 479)

Use the figure at the right. Identify each pair of angles as *adjacent* or *vertical*.

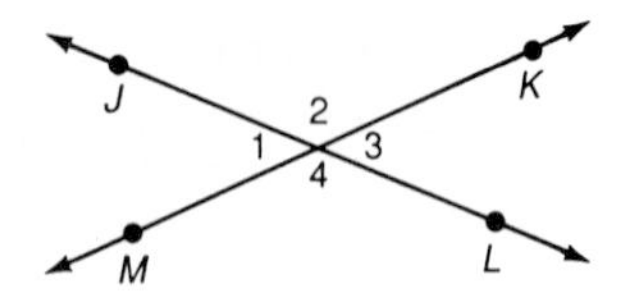

6. $\angle 2$ and $\angle 4$
7. $\angle 1$ and $\angle 4$
8. $\angle 2$ and $\angle 3$
9. $\angle 1$ and $\angle 3$

mixed practice (pages 478–479)

Use the figure at the right to name the following.

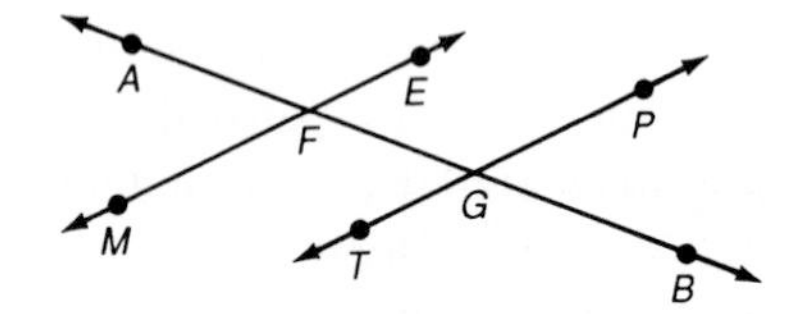

10. three points
11. three lines
12. three line segments
13. four rays
14. two pairs of vertical angles

Make a drawing to represent each of the following.

15. plane *Z*
16. $\overrightarrow{MN}$
17. $\overline{XY}$
18. $\angle XYZ$
19. intersecting lines *JK* and *MN*
20. vertical angles *ANC* and *BND*

Use the figure at the right to name the following.

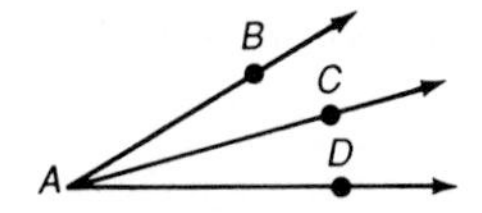

21. four points
22. three rays
23. three line segments
24. three angles

review exercises

Solve by using an equation.

1. It took Tia and Lori 8 min to get to school this morning. They waited $2\frac{1}{2}$ min for a *Walk* signal before crossing a street. How many minutes did they walk?
2. Ward drove 1845 mi in four days on his trip. For the first three days he drove the same number of miles each day. The last day he drove 480 mi. How many miles did Ward drive on each of the first three days?
3. Jacqui bought a can of juice for \$1.37, and she bought five peaches. She spent a total of \$3.12. What was the price of one peach?

22-2 DRAWING AND MEASURING ANGLES

An angle is measured in units called **degrees** (°). You can use a **protractor** to measure or draw an angle.

example 1

Use a protractor to measure $\angle ABC$.

solution

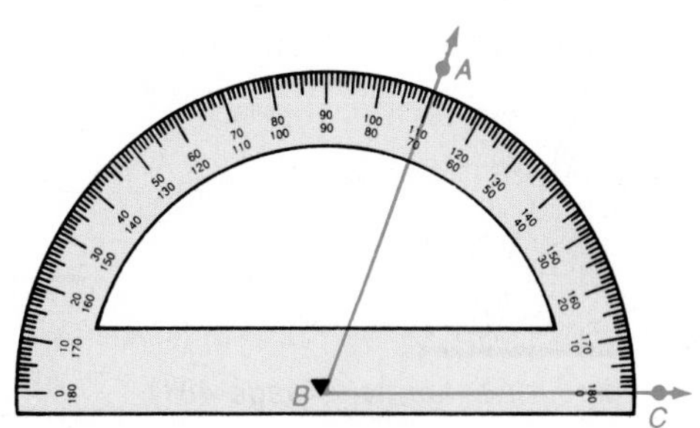

Put the center of the protractor on the vertex (B).
Place the 0° mark on one side ($\overrightarrow{BC}$).
Read the number where the other side ($\overrightarrow{BA}$) crosses the scale.
The measure of $\angle ABC$ is 70°.
You write: $m\ \angle ABC = 70°$

your turn

Use a protractor to measure each angle.

1.

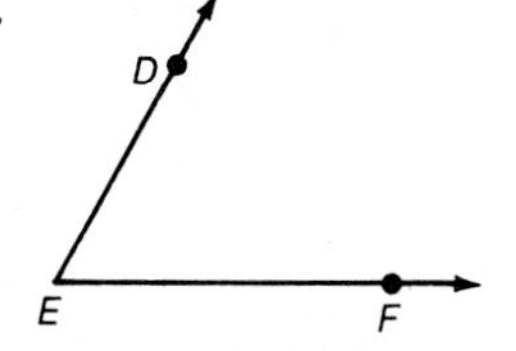

2.

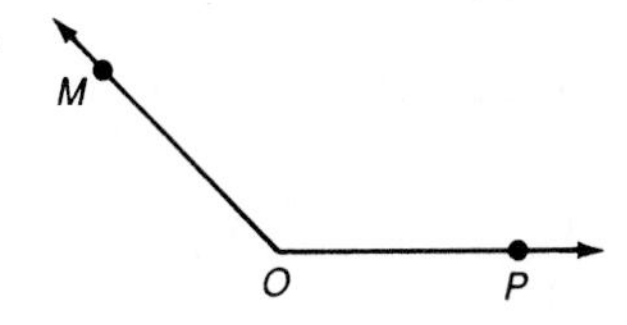

3.

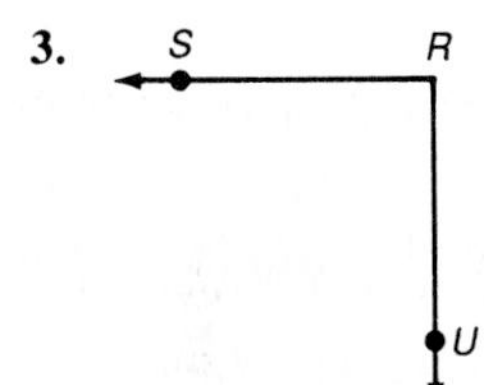

example 2

Use a protractor to draw $\angle WBZ$ with measure 150°.

solution

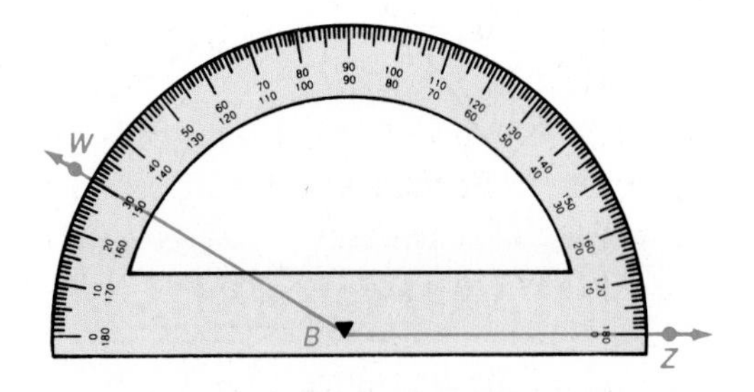

Draw a ray ($\overrightarrow{BZ}$) to represent one side.
Put the center of the protractor on the endpoint (B) with the 0° mark on the ray.
Mark the number of degrees (150°) and remove the protractor.
Draw a ray ($\overrightarrow{BW}$) through the mark.

your turn

Use a protractor to draw an angle of the given measure.

4. 20° **5.** 140° **6.** 90°

practice exercises

practice for example 1 (page 481)

Use a protractor to measure each angle.

1. 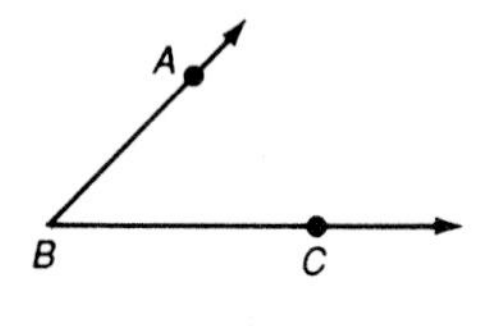

2. 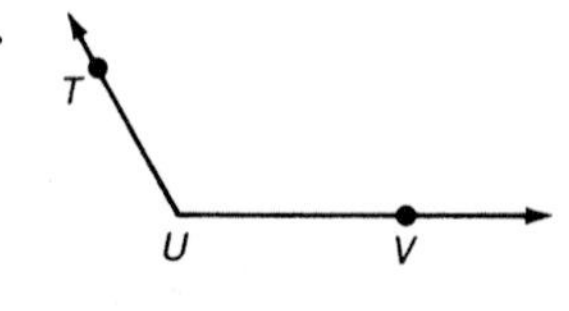

3. 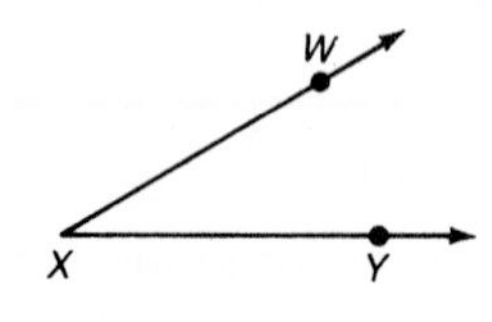

4. 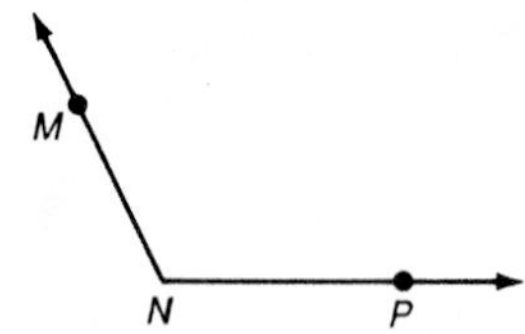

5. 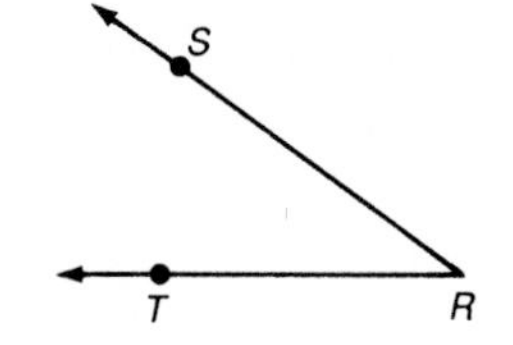

6. 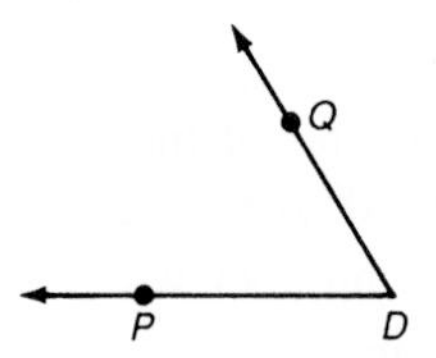

practice for example 2 (page 481)

Use a protractor to draw an angle of the given measure.

7. 60° **8.** 110° **9.** 75° **10.** 160° **11.** 35° **12.** 50°

mixed practice (page 481)

Use a protractor to measure the given angle in the figure at right.

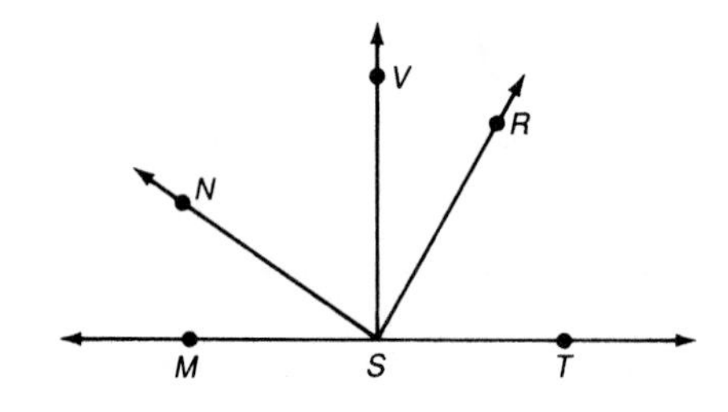

13. $\angle RST$ **14.** $\angle NSM$
15. $\angle VSM$ **16.** $\angle MSR$
17. $\angle NST$ **18.** $\angle MST$

Estimate the measure of each angle.

19. 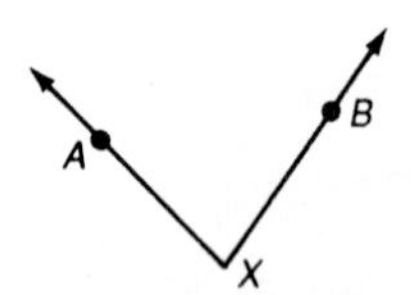

20. 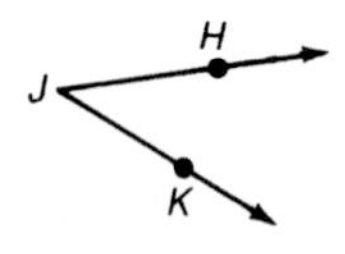

21. 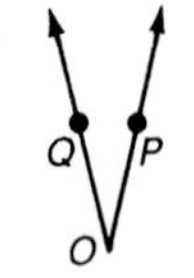

22. 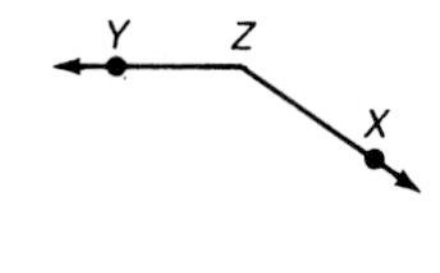

review exercises

Find the interest and amount due.

1. \$500 at 6% per year for 1 year
2. \$950 at 12% per year for 1 year
3. \$12,000 at $10\frac{1}{2}$% per year for 6 months
4. \$770 at 14.2% per year for 6 months
5. \$230 at $7\frac{1}{2}$% per year for 2 years
6. \$25,000 at 9.4% per year for 2 years

22-3 CLASSIFYING ANGLES AND LINES

Angles may be classified according to their measures.

A **straight angle** has a measure of 180°.

A **right angle** has a measure of 90°.

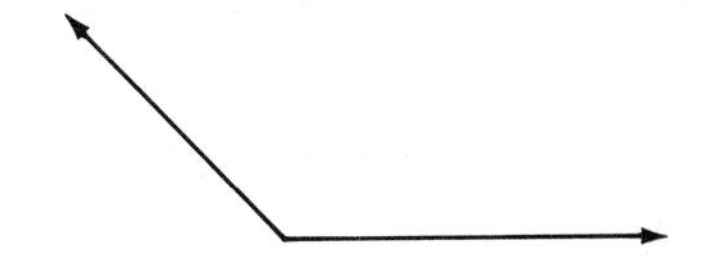

An **acute angle** has a measure between 0° and 90°.

An **obtuse angle** has a measure between 90° and 180°.

example 1

Classify an angle with the given measure as *straight, right, acute,* or *obtuse.*

a. 129° **b.** 34° **c.** 90° **d.** 17°

solution

a. obtuse **b.** acute **c.** right **d.** acute

your turn

Classify an angle with the given measure as *straight, right, acute,* or *obtuse.*

1. 14° **2.** 180° **3.** 146° **4.** 90°

Sometimes you can classify a *pair* of angles by the sum of their measures.

When the sum of the measures of two angles is 90°, the angles are **complementary.**

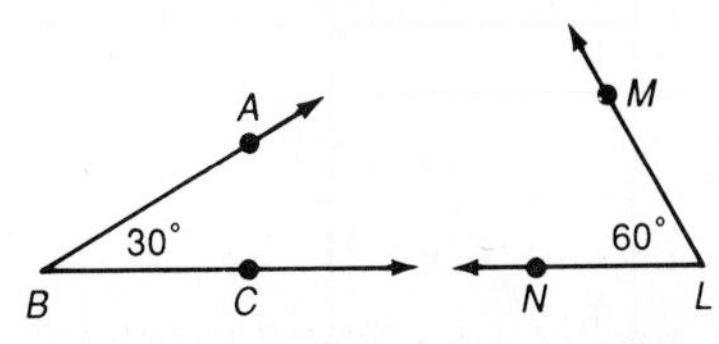

$m\angle ABC + m\angle MLN = 30° + 60° = 90°$
$\angle ABC$ and $\angle MLN$ are complementary.

When the sum of the measures of two angles is 180°, the angles are **supplementary.**

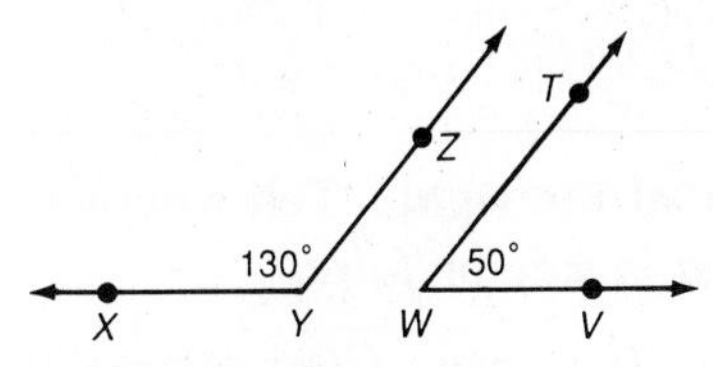

$m\angle XYZ + m\angle TWV = 130° + 50° = 180°$
$\angle XYZ$ and $\angle TWV$ are supplementary.

example 2

Tell whether two angles with the given measures are *complementary*, *supplementary*, or *neither*.

a. 13°, 77° **b.** 125°, 55° **c.** 24°, 46°

solution

a. 13° + 77° = 90°; *complementary*

b. 125° + 55° = 180°; *supplementary*

c. 24° + 46° = 70°; *neither*

your turn

Tell whether two angles with the given measures are *complementary*, *supplementary*, or *neither*.

5. 33°, 57° **6.** 156°, 34° **7.** 105°, 75° **8.** 45°, 45°

When two lines intersect to form a right angle, the lines are **perpendicular.**

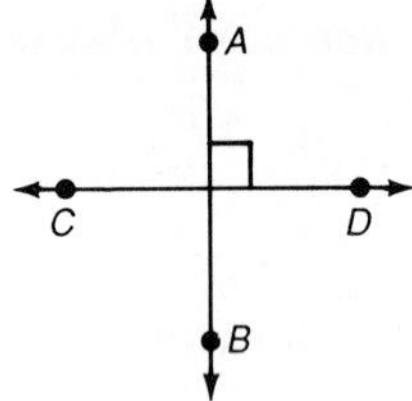

$\overleftrightarrow{AB}$ is perpendicular to $\overleftrightarrow{CD}$.

You write: $\overleftrightarrow{AB} \perp \overleftrightarrow{CD}$

When two lines in a plane do not intersect, the lines are **parallel.**

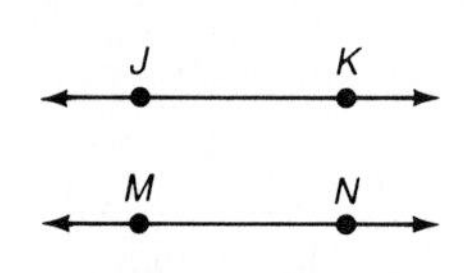

$\overleftrightarrow{JK}$ is parallel to $\overleftrightarrow{MN}$.

You write: $\overleftrightarrow{JK} \parallel \overleftrightarrow{MN}$

example 3

Use the figure at the right. Tell whether each statement is *true* or *false*.

a. $\overleftrightarrow{AB} \parallel \overleftrightarrow{CD}$ **b.** $\overleftrightarrow{EF} \perp \overleftrightarrow{GH}$

solution

a. True **b.** False

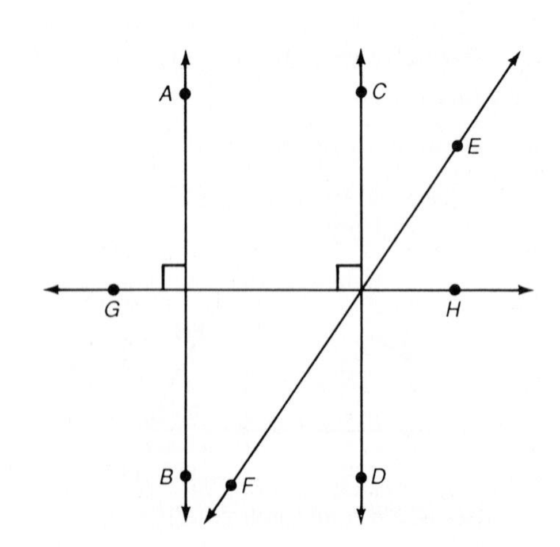

your turn

Use the figure at the right. Tell whether each statement is *true* or *false*.

9. $\overleftrightarrow{AB} \parallel \overleftrightarrow{EF}$ **10.** $\overleftrightarrow{CD} \parallel \overleftrightarrow{AB}$

11. $\overleftrightarrow{CD} \perp \overleftrightarrow{GH}$ **12.** $\overleftrightarrow{GH} \perp \overleftrightarrow{EF}$

practice exercises

practice for example 1 (page 483)

Classify an angle with the given measure as *straight, right, acute,* or *obtuse*.

1. 33° **2.** 55° **3.** 106° **4.** 144° **5.** 87°

6. 93° **7.** 90° **8.** 150° **9.** 13° **10.** 180°

practice for example 2 (page 484)

Tell whether two angles with the given measures are *complementary, supplementary,* or *neither*.

11. 82°, 98° **12.** 20°, 70° **13.** 42°, 48° **14.** 100°, 28°

15. 111°, 99° **16.** 141°, 29° **17.** 173°, 7° **18.** 55°, 35°

practice for example 3 (page 484)

Use the figure at the right. Tell whether each statement is *true* or *false*.

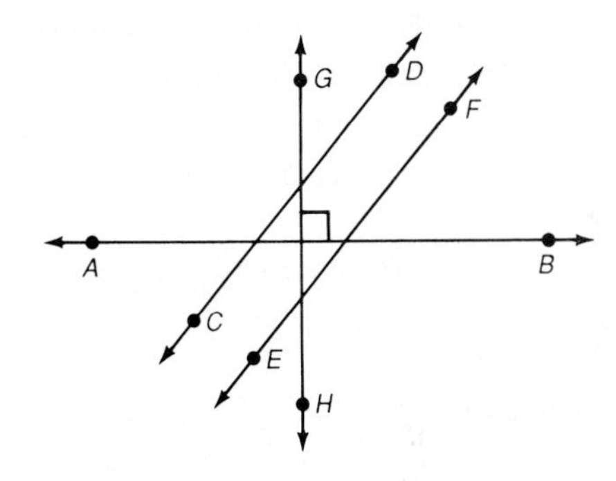

19. $\overleftrightarrow{AB} \parallel \overleftrightarrow{GH}$ **20.** $\overleftrightarrow{CD} \parallel \overleftrightarrow{EF}$

21. $\overleftrightarrow{GH} \perp \overleftrightarrow{AB}$ **22.** $\overleftrightarrow{EF} \perp \overleftrightarrow{AB}$

23. $\overleftrightarrow{CD} \parallel \overleftrightarrow{GH}$ **24.** $\overleftrightarrow{EF} \parallel \overleftrightarrow{GH}$

25. $\overleftrightarrow{CD} \perp \overleftrightarrow{AB}$ **26.** $\overleftrightarrow{AB} \perp \overleftrightarrow{GH}$

mixed practice (pages 483-484)

Use the figure at the right. Choose the correct word to complete each statement.

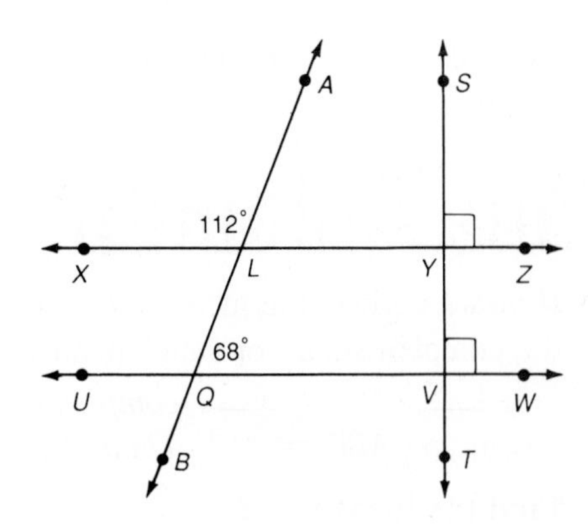

27. $\overleftrightarrow{XZ}$ is (*parallel, perpendicular*) to $\overleftrightarrow{UW}$.

28. $\overleftrightarrow{UW}$ is (*parallel, perpendicular*) to $\overleftrightarrow{ST}$.

29. $\angle YVU$ is a(n) (*acute, right*) angle.

30. $\angle UVW$ is a(n) (*obtuse, straight*) angle.

31. $\angle ALX$ is an (*obtuse, acute*) angle.

32. $\angle BQU$ is an (*acute, obtuse*) angle.

33. $\angle SYZ$ and $\angle YVW$ are a pair of (*complementary, supplementary*) angles.

34. $\angle YVW$ and $\angle UVT$ are a pair of (*complementary, vertical*) angles.

35. $\angle UQB$ and $\angle BQV$ are a pair of (*adjacent, vertical*) angles.

Classify each angle as *acute* or *obtuse*. Then use a protractor to measure the angle.

36. 

37. 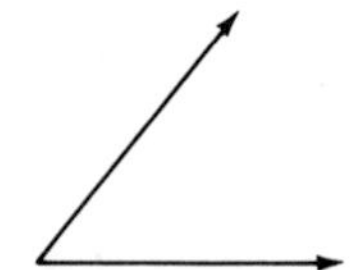

38. 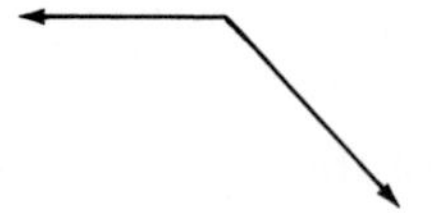

39. 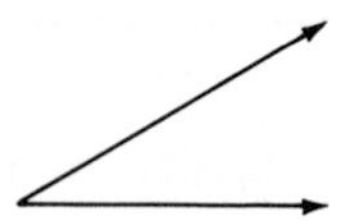

Make a drawing to represent each of the following. Label each drawing.

40. a pair of parallel lines

41. a pair of perpendicular lines

42. an obtuse angle

43. an acute angle

44. a pair of complementary angles

45. a pair of supplementary angles

46. An angle measures 78°. Find the measure of an angle that is complementary to it.

47. An angle measures 124°. Find the measure of an angle that is supplementary to it.

48. Two complementary angles are equal in measure. What is the measure of each angle?

49. Two supplementary angles are equal in measure. What is the measure of each angle?

review exercises

A drawer contains 10 red, 4 green, 1 purple, and 5 blue pencils. One pencil is selected at random. Find each probability.

1. *P*(blue) **2.** *P*(green or red) **3.** *P*(not purple) **4.** *P*(orange)

Use the pencils described above. Find the odds in favor of each event.

5. red **6.** purple **7.** not blue **8.** not green

mental math

If you are given the measure of an angle, you can *subtract* to find the measure of the complementary or supplementary angle. Often you can do this mentally.

	complementary angle	*supplementary angle*
Given: $m\angle ABC = 30°$	*Think:* 90° − 30° = 60°	*Think:* 180° − 30° = 150°

Find the measure of the complementary angle.

1. 60° **2.** 45° **3.** 84°

Find the measure of the supplementary angle.

4. 90° **5.** 140° **6.** 100°

22-4 POLYGONS

A **polygon** is a closed plane figure formed by joining three or more line segments *at their endpoints*. Exactly two line segments meet at each endpoint. The line segments are called **sides** of the polygon. The points where the sides intersect are called **vertices** (plural of vertex). A polygon is identified by its number of sides.

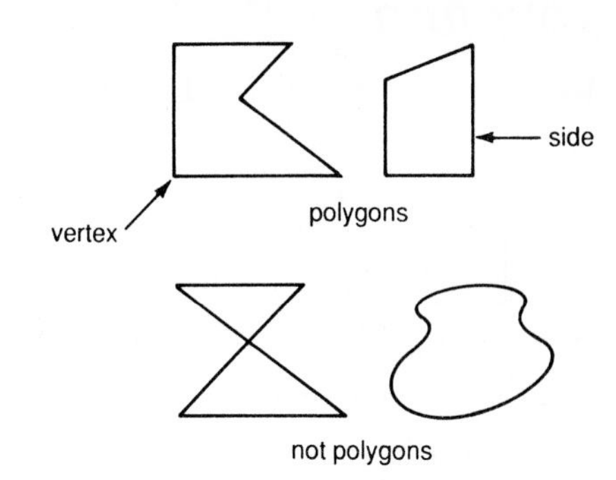

Name of Polygon	Prefix	Meaning of Prefix	Number of Sides
triangle	tri	three	3
quadrilateral	quadri	four	4
pentagon	penta	five	5
hexagon	hexa	six	6
heptagon	hepta	seven	7
octagon	octa	eight	8
nonagon	nona	nine	9
decagon	deca	ten	10

A **diagonal** of a polygon is a line segment that joins two vertices and *is not a side* of the polygon. In the pentagon shown at the right, $\overline{AD}$ and $\overline{BE}$ are diagonals.

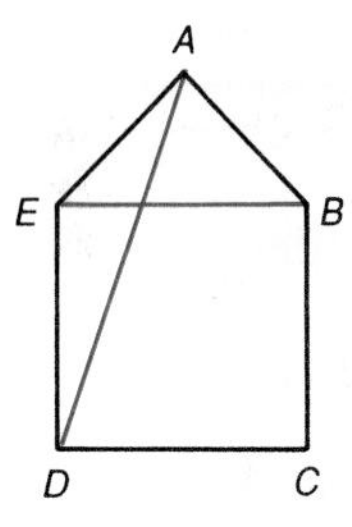

Two line segments are **congruent** if they have the same length. Two angles are congruent if they have the same measure. A **regular polygon** is a polygon with all its sides congruent and all its angles congruent.

example 1

Identify each polygon. List all the diagonals.

a.

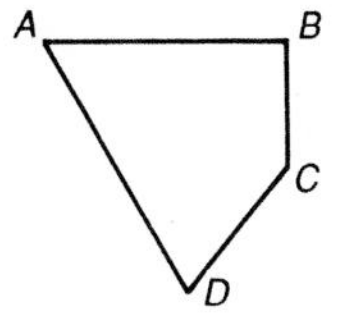

b.

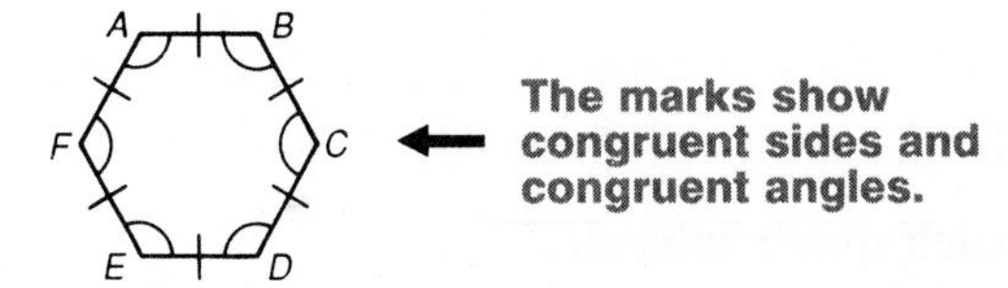

solution

a. quadrilateral;
diagonals: $\overline{AC}$, $\overline{BD}$

b. regular hexagon;
diagonals: $\overline{AC}$, $\overline{AD}$, $\overline{AE}$, $\overline{BD}$, $\overline{BE}$, $\overline{BF}$, $\overline{CE}$, $\overline{CF}$, $\overline{DF}$

your turn

Identify each polygon. List all the diagonals.

1.

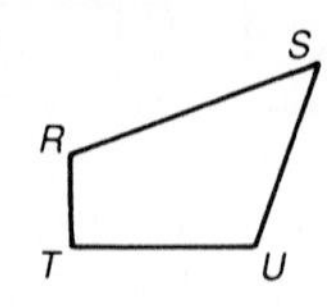

2.

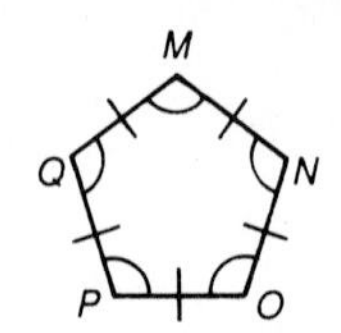

3.

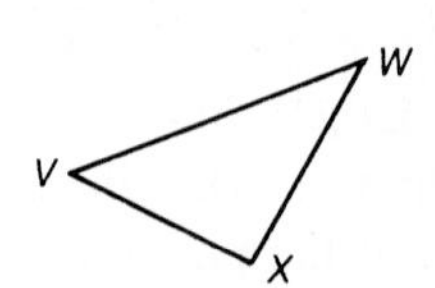

Quadrilaterals can be classified according to their properties.

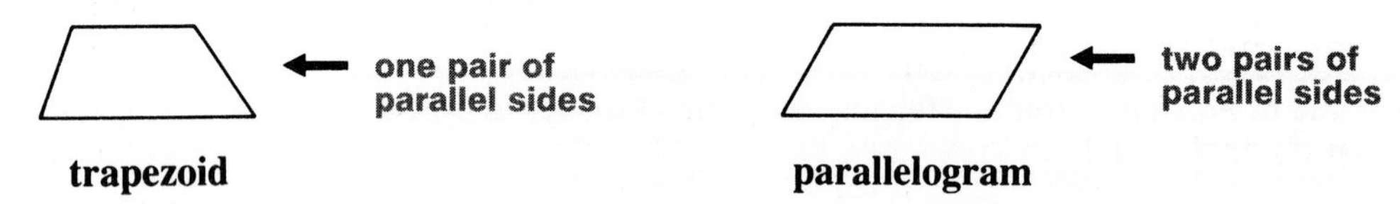

trapezoid **parallelogram**

The opposite sides and the opposite angles of a parallelogram are congruent. Some parallelograms have special names.

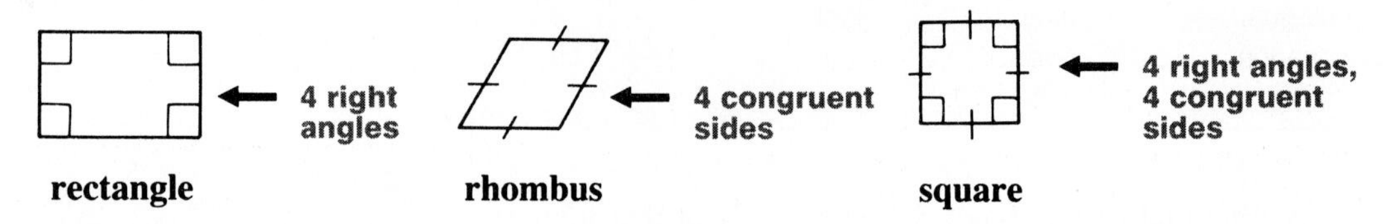

rectangle **rhombus** **square**

example 2

Identify each quadrilateral.

a.

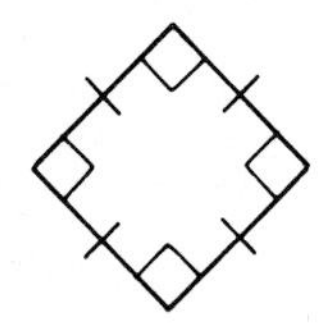

b.

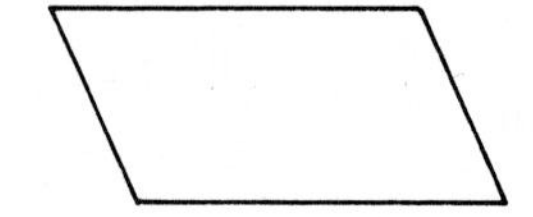

c.

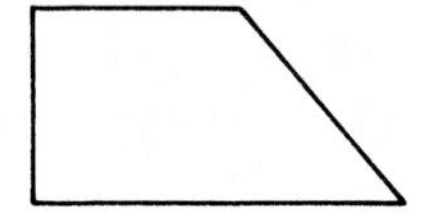

solution

a. square **b.** parallelogram **c.** trapezoid

your turn

Identify each quadrilateral.

4.

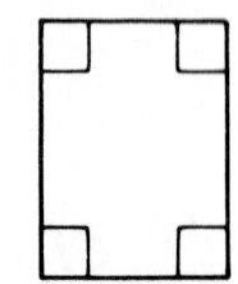

5.

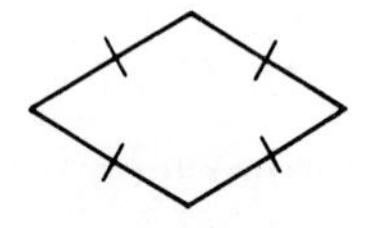

6.

practice exercises

practice for example 1 (pages 487–488)

Identify each polygon. List all the diagonals.

1.

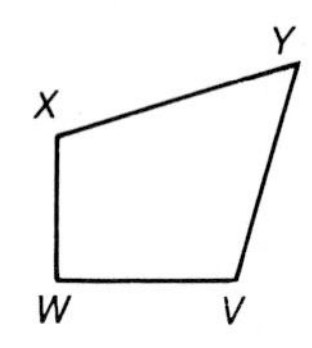

2.

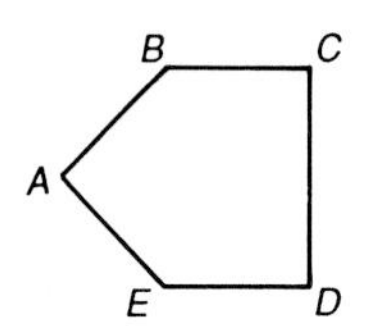

3.

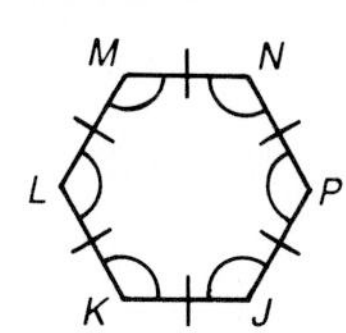

4.

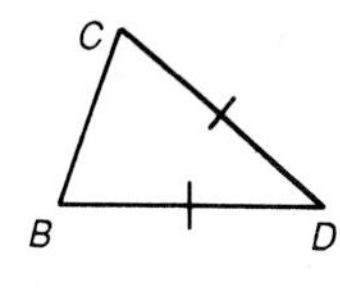

practice for example 2 (page 488)

Identify each quadrilateral.

5.

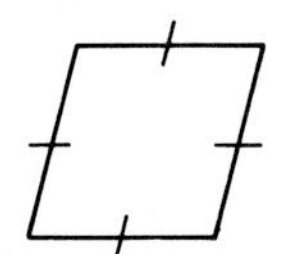

6.

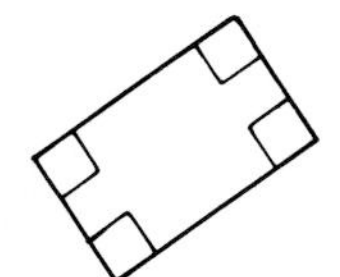

7.

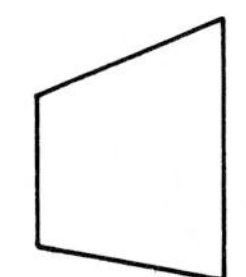

8.

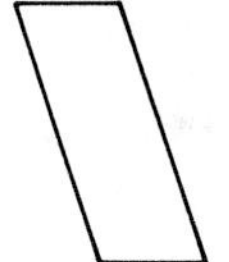

mixed practice (pages 487–488)

Identify each polygon in the figure at the right. Use the word that describes the polygon *most* accurately.

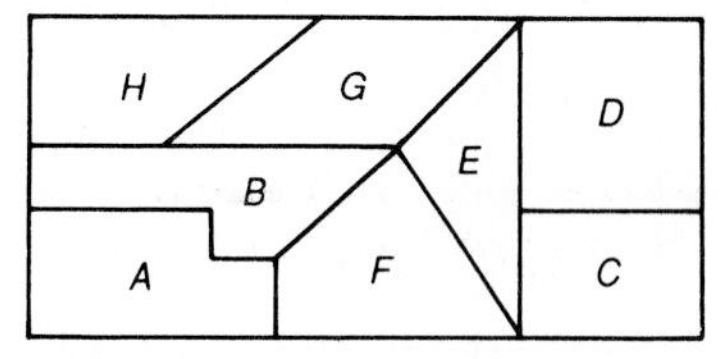

9. *C* **10.** *A* **11.** *E* **12.** *D*

13. *F* **14.** *H* **15.** *B* **16.** *G*

List all the types of quadrilaterals that satisfy the given property.

17. four congruent sides

18. four right angles

19. exactly one pair of parallel sides

20. two pairs of parallel sides

Make a drawing of each polygon.

21. octagon **22.** hexagon **23.** triangle **24.** square

review exercises

Write each answer in simplest form.

1. 3 ft 7 in. + 9 ft 6 in.

2. 24 lb 5 oz + 16 lb 14 oz

3. 12 gal 3 qt − 9 gal 1 qt

4. 75 yd − 34 yd 2 ft

5. 15 c 4 fl oz × 5

6. 3 t 1480 lb × 4

7. 22 gal 3 qt ÷ 7

8. 37 ft 4 in. ÷ 4

22-5 PERIMETER

The distance around a polygon is called its **perimeter** (P). You find the perimeter of a polygon by adding the lengths of all its sides.

example 1

Find the perimeter of the pentagon at the right.

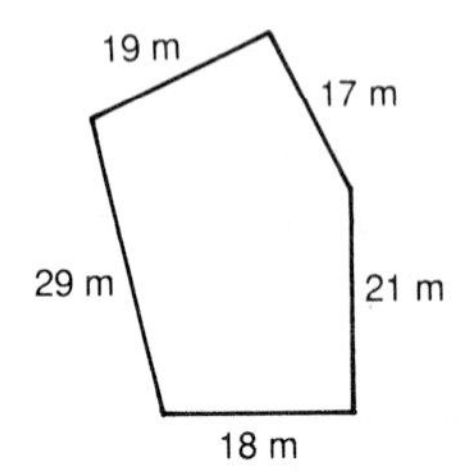

solution

$P = 19 + 17 + 21 + 18 + 29$

$P = 104$ The perimeter is 104 m.

your turn

Find the perimeter of each polygon.

1.

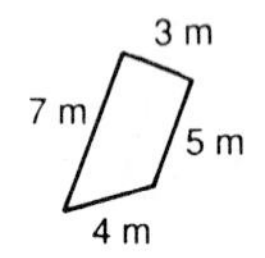

2.

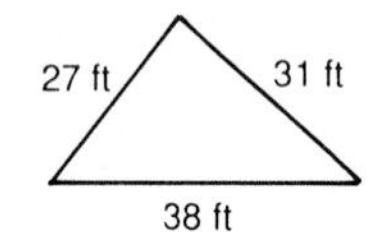

3.

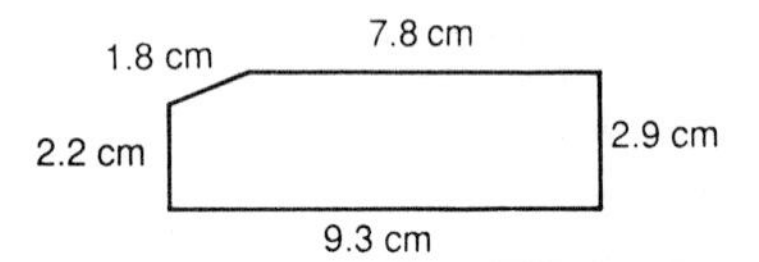

You can find the perimeter of some polygons by using a formula.

rectangle
$P = 2 \times \text{length} + 2 \times \text{width}$
$P = 2l + 2w$

square
$P = 4 \times \text{length of one side}$
$P = 4s$

example 2

Use a formula to find the perimeter of each polygon.

a. rectangle with $l = 9$ m and $w = 6$ m

b. square with $s = \frac{3}{4}$ in.

solution

a. $P = 2l + 2w$

$P = 2(9) + 2(6)$

$P = 18 + 12 = 30$ The perimeter is 30 m.

b. $P = 4s$

$P = 4\left(\frac{3}{4}\right)$

$P = 3$ The perimeter is 3 in.

your turn

Use a formula to find the perimeter of each polygon.

4. rectangle with $l = 5$ cm and $w = 3$ cm

5. square with $s = 25.25$ ft

practice exercises

practice for example 1 (page 490)

Find the perimeter of each polygon.

1.

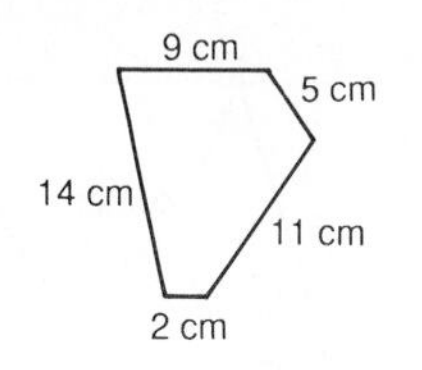

2.

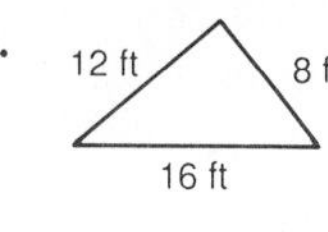

3.

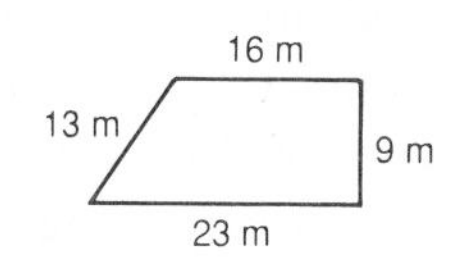

4.

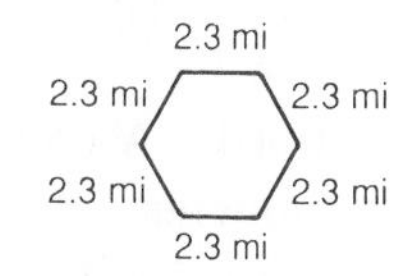

practice for example 2 (page 490)

Use a formula to find the perimeter of each polygon.

5. square with $s = 82$ mm

6. square with $s = 17\frac{1}{4}$ yd

7. rectangle with $l = 16$ ft and $w = 8$ ft

8. rectangle with $l = 7.3$ m and $w = 6$ m

mixed practice (page 490)

Find the perimeter. Use a formula whenever possible.

9. a triangle with sides that measure 8.9 cm, 7.4 cm, and 4.1 cm

10. a rectangle with length $6\frac{1}{2}$ ft and width $3\frac{1}{4}$ ft

11. a rhombus with sides that each measure 17 yd

12. a regular octagon with sides that each measure 3.23 m

13. A role of weather stripping is 180 cm long. How many rolls are needed to go around one square window that measures 90 cm on each side?

14. Fred wants to put new fencing around his rectangular garden. The garden is 18 ft long and 12 ft wide. A 6-ft section of the fencing costs $20. What will be the total cost?

review exercises

Find each answer.

1. $6.4 + 0.2 + 17$

2. $95.71 - 73.37$

3. 3.14×12.5

4. $0.9\overline{)801}$

5. $\frac{1}{2} \times \frac{22}{7}$

6. $\frac{2}{3} \times 9\frac{1}{4}$

7. $\frac{2}{9} \div \frac{1}{3}$

8. $5\frac{1}{2} \div 2\frac{1}{6}$

22-6 CIRCLES AND CIRCUMFERENCE

A **circle** is the set of all points in a plane that are the same distance from a given point in the plane. The given point is the **center** of the circle. A **radius** (plural: *radii*) is a line segment that joins the center to any point on the circle.

A line segment that joins two points on a circle is a **chord.** A chord that contains the center of a circle is a **diameter.** A diameter is twice as long as a radius. A **central angle** is an angle whose vertex is at the center of the circle.

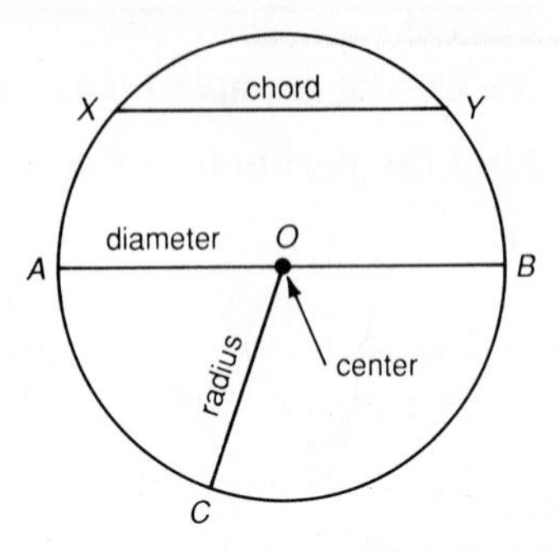

example 1

Use the figure above. Name as many of the following as are shown.

a. the circle **b.** diameters **c.** radii **d.** chords **e.** central angles

solution

a. circle O **b.** $\overline{AB}$ **c.** $\overline{OC}$, $\overline{OA}$, $\overline{OB}$ **d.** $\overline{XY}$, $\overline{AB}$ **e.** $\angle COA$, $\angle COB$, $\angle AOB$

your turn

Use the figure at the right. Name as many of the following as are shown.

1. the circle
2. diameters
3. radii
4. chords
5. central angles

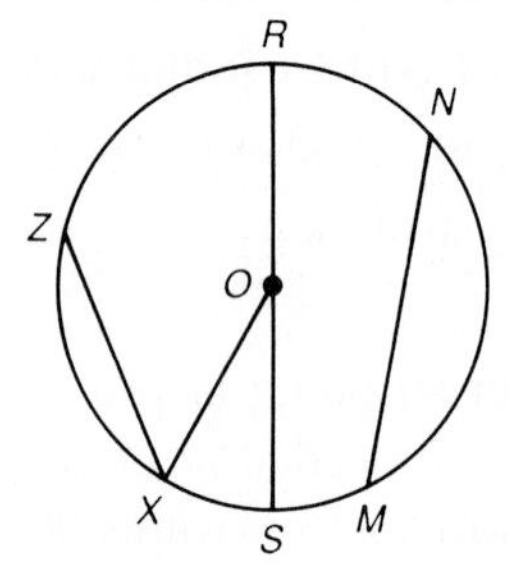

The distance around a circle is called its **circumference** (C). When the circumference of a circle is divided by the length of its diameter (d), the quotient is always the same number. This number is represented by the Greek letter π (read as "pi"). Therefore:

$$C \div d = \pi$$
$$C = \pi d$$

Since a diameter is twice as long as a radius (r), you can also write:

$$C = 2\pi r$$

To find the circumference of a circle you can use either 3.14 or $\frac{22}{7}$ as an *approximate* value for π. The value $\frac{22}{7}$ is often used when the diameter is a multiple of 7.

example 2

Find the circumference of each circle.

a. diameter = 5 cm (Use $\pi \approx 3.14$.)

b. radius = 7 ft (Use $\pi \approx \frac{22}{7}$.)

solution

a. $C = \pi d$

$C \approx 3.14(5)$ ← To use a calculator, enter π × 5 =. Then round.

$C \approx 15.7$

The circumference is approximately 15.7 cm.

b. $C = 2\pi r$

$C \approx (2)\left(\frac{22}{7}\right)(7)$

$C \approx 44$

The circumference is approximately 44 ft.

your turn

Find the circumference of each circle.

6. diameter = 5.5 km (Use $\pi \approx 3.14$.)

7. diameter = 21 in. (Use $\pi \approx \frac{22}{7}$.)

8. radius = 10 m (Use $\pi \approx 3.14$.)

9. radius = $\frac{7}{8}$ yd (Use $\pi \approx \frac{22}{7}$.)

practice exercises

practice for example 1 (page 492)

Use the figure at the right. Name as many of the following as are shown.

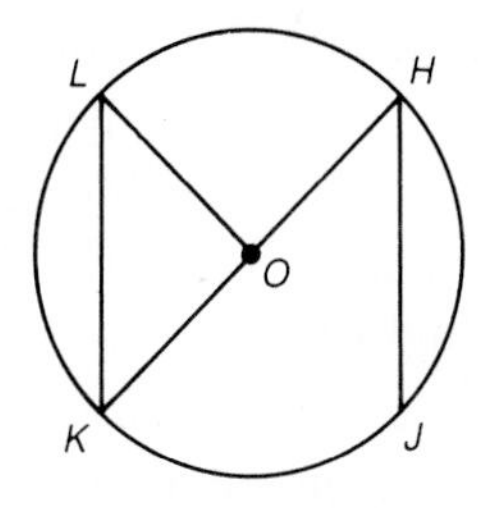

1. the circle

2. diameters

3. radii

4. chords

5. central angles

practice for example 2 (page 493)

Find the circumference of each circle. Use $\pi \approx 3.14$.

6. diameter = 22 yd

7. diameter = 30 in.

8. radius = 72 m

9. radius = 7 cm

10. diameter = 4.5 ft

11. diameter = 6.5 km

Find the circumference of each circle. Use $\pi \approx \frac{22}{7}$.

12. radius = 7 cm

13. radius = 28 ft

14. diameter = 14 in.

15. diameter = 49 m

16. radius = $\frac{1}{2}$ mi

17. radius = $\frac{1}{4}$ yd

mixed practice (pages 492–493)

Use the figure at the right. Is each line segment best described as a *radius, diameter,* or *chord*?

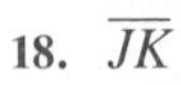

18. $\overline{JK}$ 19. $\overline{OT}$ 20. $\overline{LM}$

21. $\overline{XY}$ 22. $\overline{YO}$ 23. $\overline{RO}$

Find the circumference of each circle. Use $\pi \approx 3.14$.

24.

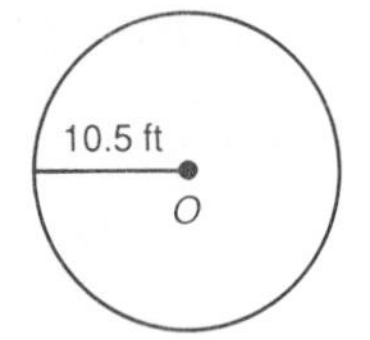

25.

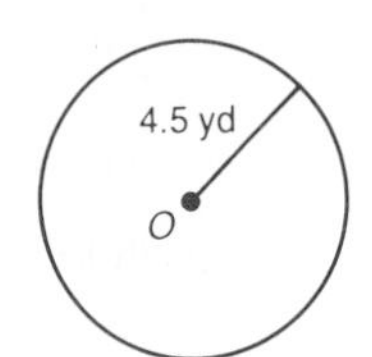

26.

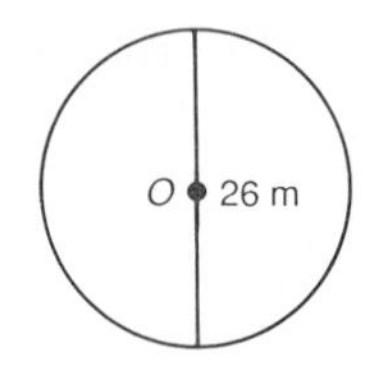

27.

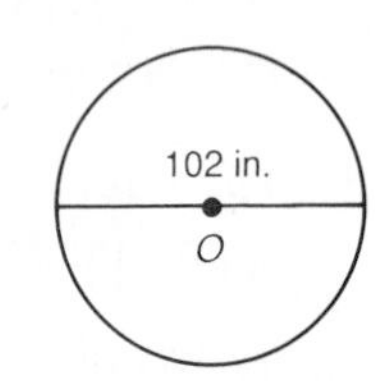

Complete the chart. Use $\pi \approx 3.14$.

	28.	29.	30.	31.	32.	33.
radius	4 cm	?	?	8 cm	?	?
diameter	?	9 ft	12 ft	?	?	?
circumference	?	?	?	?	31.4 mm	40.82 mm

34. The radius of a water wheel is 3.5 m. What is the approximate circumference of the wheel? Use $\pi \approx 3.14$.

35. The diameter of a bicycle wheel is 28 in. Approximately how far does the wheel travel in three complete turns? Use $\pi \approx \frac{22}{7}$.

36. The length of a rectangular park is 80 m, and its width is 30 m. Eliza jogged around the park five times. How far did Eliza jog?

37. Approximately how many feet of fencing are required to enclose a circular garden with a radius of 14 ft? If the fencing is sold in 5-ft sections, how many sections must be bought? Use $\pi \approx \frac{22}{7}$.

review exercises

Find each answer.

1. $24 - 3 \times 8 + 17$
2. $2 \times 5 \times (5 + 7)$
3. $(36 - 4) \div 4 \times 2^3$
4. $3 \times 6 \times (9 + 16)$
5. $\frac{6 \times 3}{3 + 3}$
6. $\frac{24 - 4}{12 \div 2 - 1}$

22-7 AREA OF POLYGONS

The **area** (A) of a figure is the amount of surface it covers. Area is measured in *square units,* such as square centimeters and square feet.

To find the area of the rectangle at the right, you could count the number of square units. The area is twelve square centimeters. Another way to find this area is to use the formula for the area of a rectangle, $A = lw$.

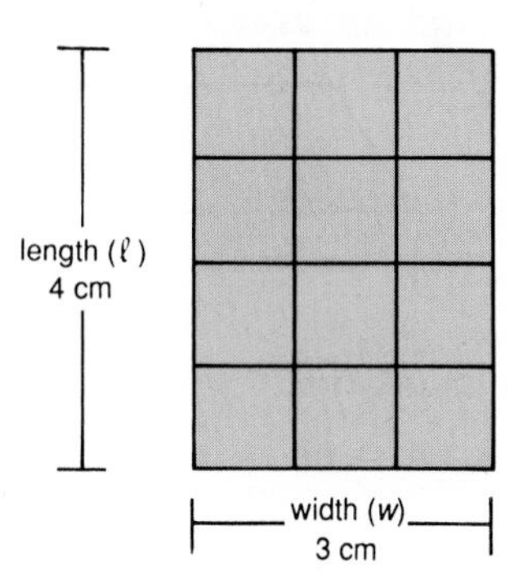

$A = lw$

$A = 4 \text{ cm} \times 3 \text{ cm}$

$A = 12 \text{ cm}^2$ ← **You write cm × cm as cm².**

The area is 12 cm².

You can also use a formula to find the area of a square and of a parallelogram.

square

$A = s \times s$

$A = s^2$

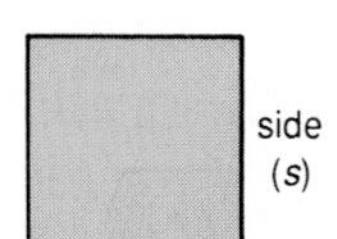

parallelogram

$A = bh$

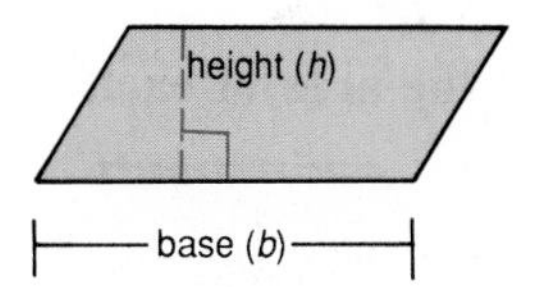

example 1

Find the area of the rectangle and the square.

a.

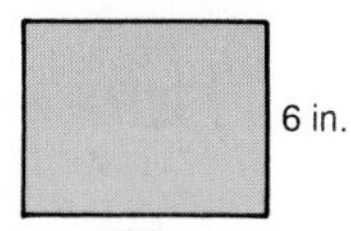

b.

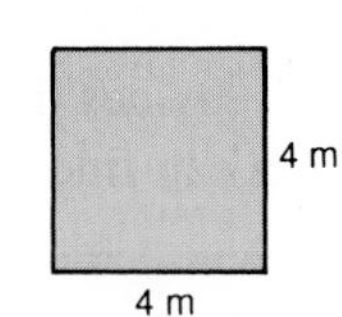

solution

a. $A = lw$

$A = 8(6)$

$A = 48$

The area is 48 in.².

b. $A = s^2$

$A = 4^2$

$A = 16$

The area is 16 m².

your turn

Find the area of each rectangle and each square.

1.

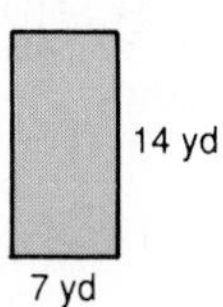

2.

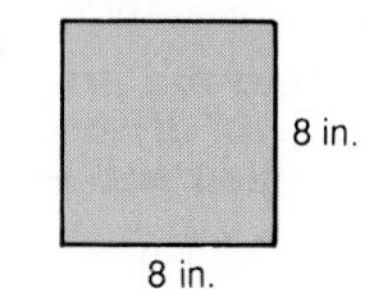

3.

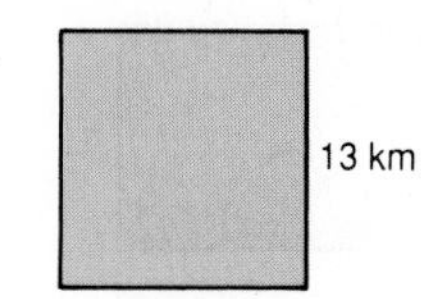

4.

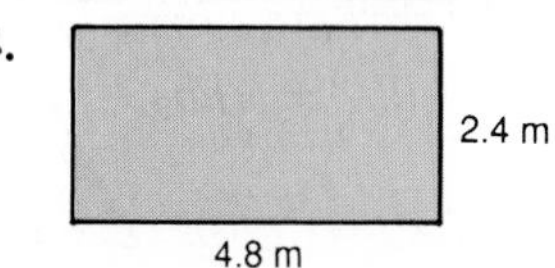

example 2

Find the area of each parallelogram.

a.

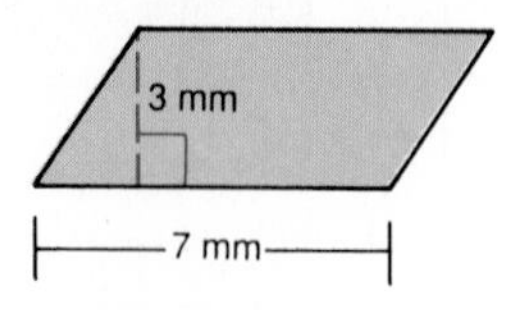

b.

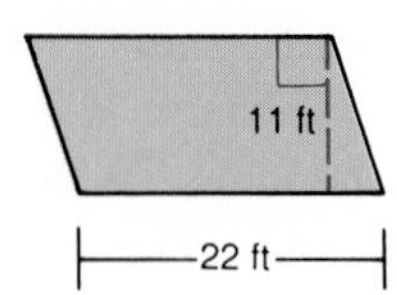

solution

a. $A = bh$

$A = 7(3)$

$A = 21$

The area is 21 mm^2.

b. $A = bh$

$A = 22(11)$

$A = 242$

The area is 242 ft^2.

your turn

Find the area of each parallelogram.

5.

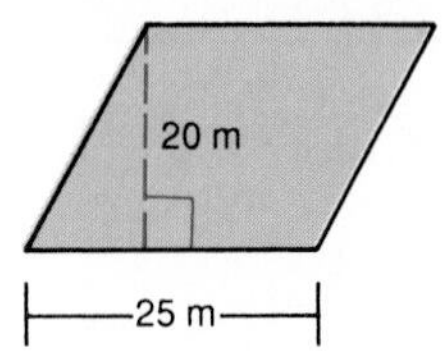

6.

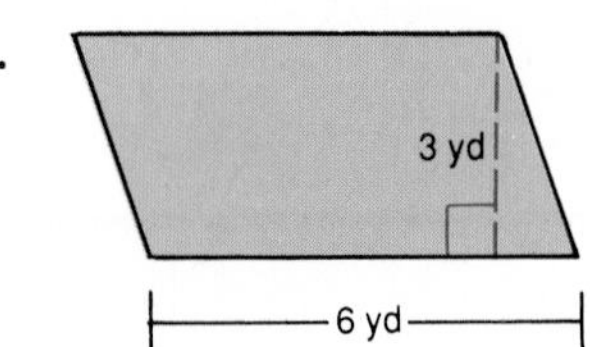

A diagonal of a quadrilateral divides the quadrilateral into two triangles. Knowing this fact, you can find the formula for the area of a triangle.

triangle

$A = \frac{1}{2}bh$

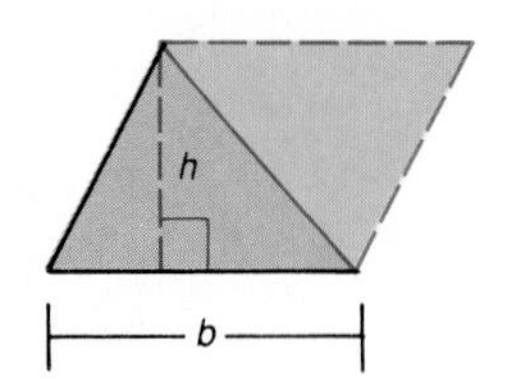

← **The area of one triangle is one half the area of the parallelogram.**

You can also use this fact to find the formula for the area of a trapezoid.

trapezoid

$A = \frac{1}{2}hb_1 + \frac{1}{2}hb_2$

$A = \frac{1}{2}h(b_1 + b_2)$

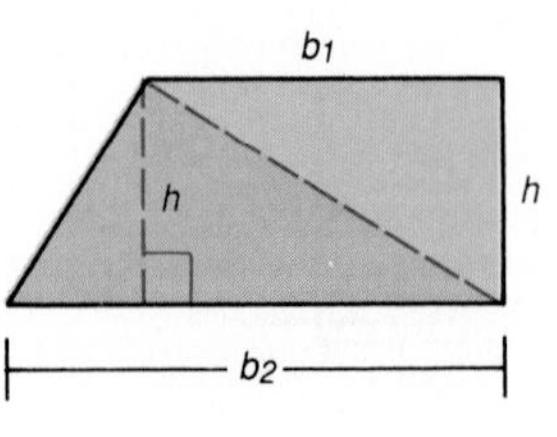

← **To find the area of a trapezoid, you can divide it into two triangles and add the areas.**

example 3

Find the area of the triangle and the trapezoid.

a.

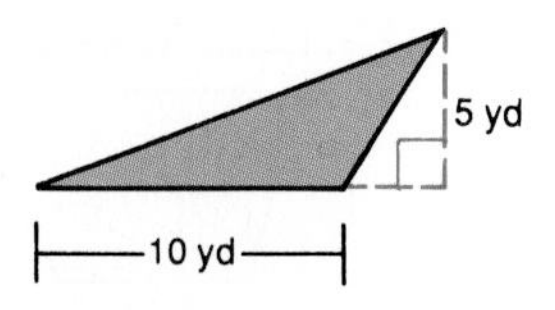

b.

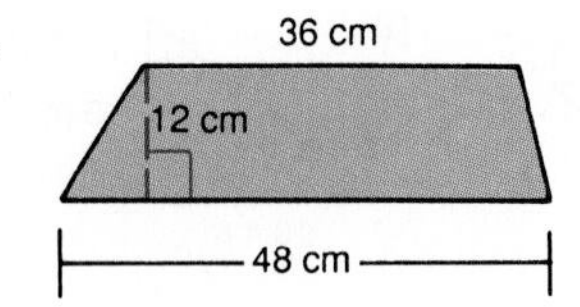

solution

a. $A = \frac{1}{2}bh$

$A = \frac{1}{2}(10)(5)$

$A = 25$

The area is 25 yd^2.

b. $A = \frac{1}{2}h(b_1 + b_2)$

$A = \frac{1}{2}(12)(36 + 48)$

$A = \frac{1}{2}(12)(84) = 504$

The area is 504 cm^2.

your turn

Find the area of the triangle and the trapezoid.

7.

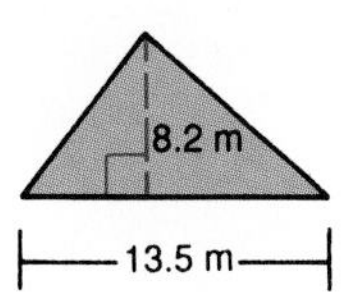

8.

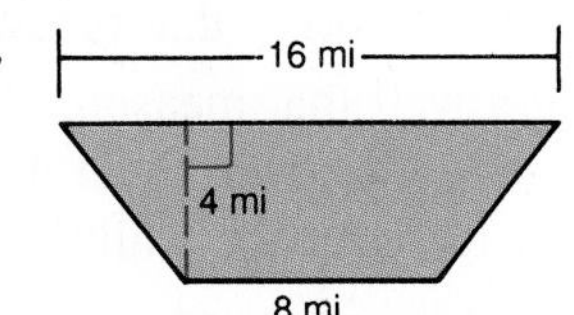

practice exercises

practice for example 1 *(page 495)*

Find the area of each rectangle and each square.

1.

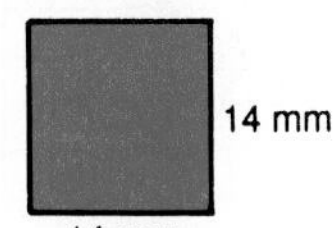

2.

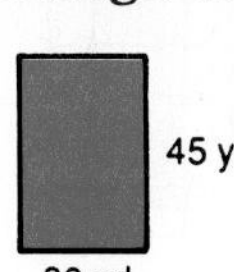

3.

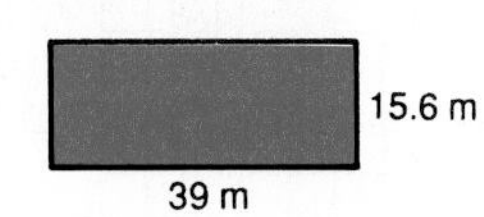

4.

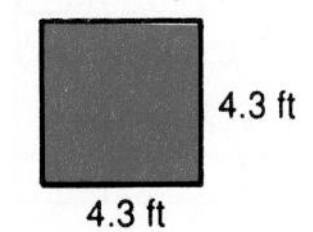

practice for example 2 *(page 496)*

Find the area of each parallelogram.

5.

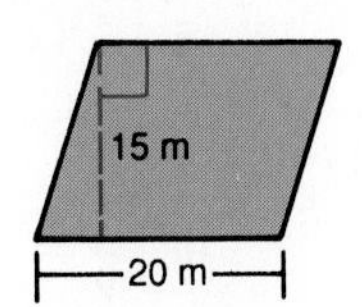

6.

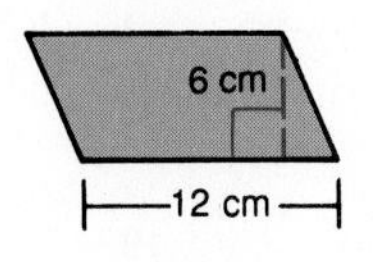

7.

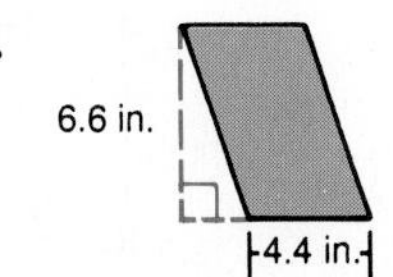

8.

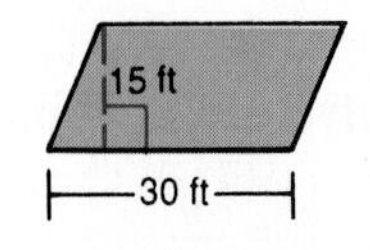

practice for example 3 (page 497)

Find the area of each triangle and each trapezoid.

9. 28 km, 14 km

10. 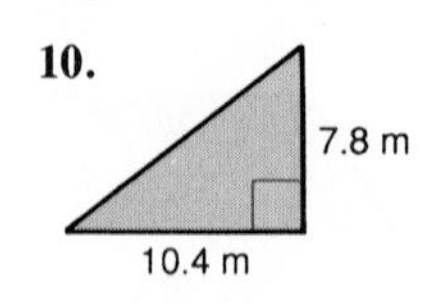

11. 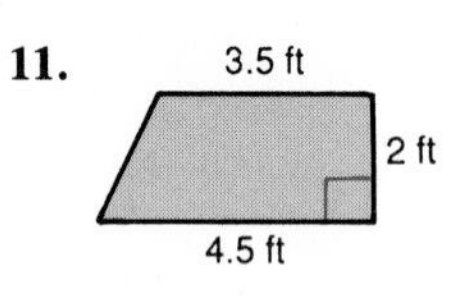

12. 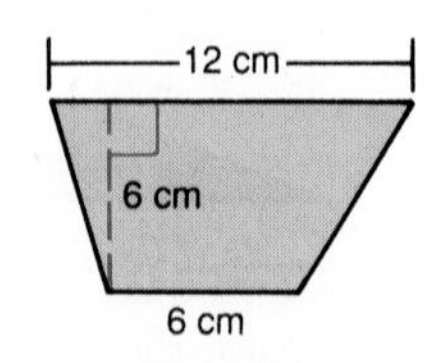

mixed practice (pages 495–497)

Find the area of each figure.

13. a square with sides that each measure 2.6 m
14. a parallelogram with a base that measures 63 mm and height 29 mm
15. a triangle with a base that measures 4 yd and height 3 yd
16. a trapezoid with bases that measure 16 in. and 12 in., and height 8 in.
17. How many square yards of artificial turf are needed to cover a rectangular football field that measures 100 yd by 60 yd?
18. Thrifty Store sells vinyl flooring for $8.39/yd^2. What is the cost of a rectangular piece of flooring that measures 4 yd by 5 yd?
19. Sandra is painting a wall that measures 8 ft by 14 ft. She estimates that one can of paint will cover about 45 ft^2. How many cans of paint must she buy to cover the wall?
20. Ny is buying a fence for his yard. The yard is a rectangular area with length 11 ft and width 15 ft. The fence is sold in 12-ft sections. How many sections must Ny buy to enclose his yard?

Estimate the area in square units of each shaded figure.

21. 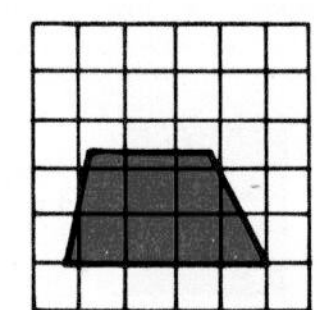

22. 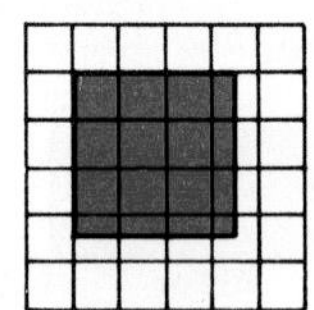

23. 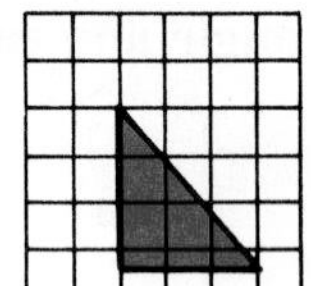

24. 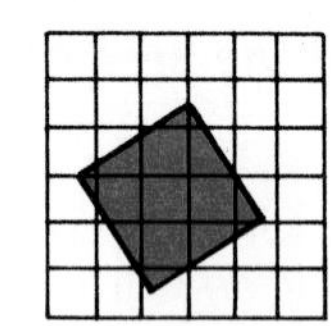

review exercises

Find each answer.

1. 2^4
2. 6^3
3. $3(4^2)$
4. $2(5^3)$
5. $7^2(2^1)(6)$

Solve each proportion.

6. $\frac{x}{4} = \frac{9}{12}$
7. $\frac{15}{6} = \frac{30}{a}$
8. $\frac{9}{72} = \frac{t}{24}$
9. $\frac{25}{m} = \frac{5}{8}$
10. $\frac{2}{9} = \frac{2.8}{c}$

22-8 AREA OF CIRCLES

Suppose you draw a circle divided into equal parts. Now rearrange the parts as shown at the right. The new figure looks very much like a parallelogram.

The base of this "parallelogram" is half the circumference of the circle, or $\frac{1}{2}C$. The radius of the circle is r, so the height of the "parallelogram" must be r.

The area of the "parallelogram" must be equal to the area of the circle. Recall that the formula for the area of a parallelogram is $A = bh$. You can now use this fact to find the formula for the area of a circle.

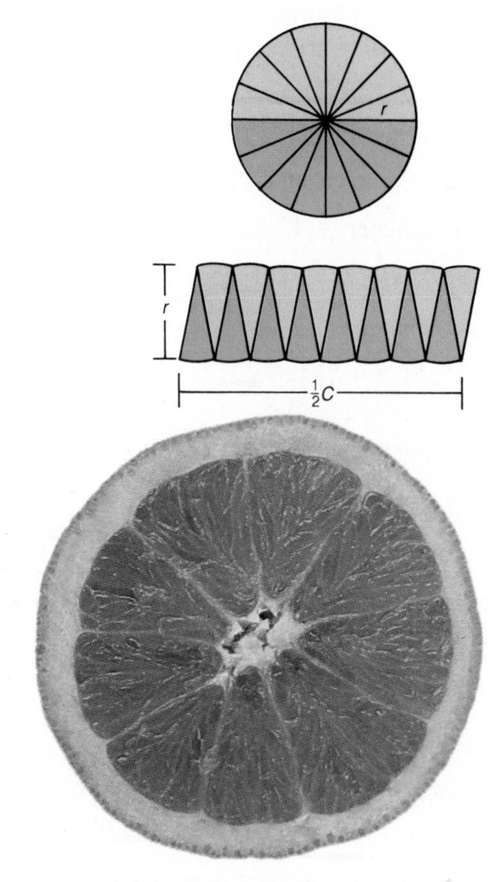

$A = b \times h$

$A = \frac{1}{2}C \times r$ ← **Substitute $\frac{1}{2}C$ for b and r for h.**

$A = \frac{1}{2}(2\pi r) \times r$ ← **Recall that $C = 2\pi r$.**

$A = \pi r^2$

example 1

Find the area of each circle.

a. radius = 3 cm (Use $\pi \approx 3.14$.)

b. radius = 70 yd (Use $\pi \approx \frac{22}{7}$.)

solution

a. $A = \pi r^2$

$A \approx 3.14(3^2)$

$A \approx 3.14(9)$

$A \approx 28.26$

The area is approximately 28.26 cm^2.

b. $A = \pi r^2$

$A \approx \frac{22}{7}(70^2)$

$A \approx \frac{22}{7}(4900)$

$A \approx 15{,}400$

The area is approximately 15,400 yd^2.

your turn

Find the area of each circle.

1. radius = 11 mm (Use $\pi \approx 3.14$.)

2. radius = 28 ft (Use $\pi \approx \frac{22}{7}$.)

To find the area of a circle, you need to know its *radius*. The radius of a circle is one half the diameter. So if you are given the *diameter*, you must first multiply it by one half to find the radius. Then use the formula $A = \pi r^2$.

example 2

Find the area of each circle.

a. diameter = 12 km (Use $\pi \approx 3.14$.)

b. diameter = 42 ft (Use $\pi \approx \frac{22}{7}$.)

solution

a. $A = \pi r^2$

$A \approx 3.14(6^2)$ ← $r = \frac{1}{2} \times 12 = 6$

$A \approx 3.14(36)$

$A \approx 113.04$

The area is approximately 113.04 km^2.

b. $A = \pi r^2$

$A \approx \frac{22}{7}(21^2)$ ← $r = \frac{1}{2} \times 42 = 21$

$A \approx \frac{22}{7}(441)$

$A \approx 1386$

The area is approximately 1386 ft^2.

your turn

Find the area of each circle.

3. diameter = 8 cm (Use $\pi \approx 3.14$.)
4. diameter = 14 ft (Use $\pi \approx \frac{22}{7}$.)

practice exercises

practice for example 1 (page 499)

Find the area of each circle.

1. radius = 9 m (Use $\pi \approx 3.14$.)
2. radius = 13 km (Use $\pi \approx 3.14$.)
3. radius = 7 yd (Use $\pi \approx \frac{22}{7}$.)
4. radius = 14 mi (Use $\pi \approx \frac{22}{7}$.)
5. radius = 17 yd (Use $\pi \approx 3.14$.)
6. radius = 23 cm (Use $\pi \approx 3.14$.)
7. radius = 21 in. (Use $\pi \approx \frac{22}{7}$.)
8. radius = 35 ft (Use $\pi \approx \frac{22}{7}$.)

practice for example 2 (page 500)

Find the area of each circle.

9. diameter = 2 km (Use $\pi \approx 3.14$.)
10. diameter = 16 m (Use $\pi \approx 3.14$.)
11. diameter = 14 yd (Use $\pi \approx \frac{22}{7}$.)
12. diameter = 42 mi (Use $\pi \approx \frac{22}{7}$.)
13. diameter = 38 ft (Use $\pi \approx 3.14$.)
14. diameter = 52 yd (Use $\pi \approx 3.14$.)
15. diameter = 28 in. (Use $\pi \approx \frac{22}{7}$.)
16. diameter = 56 ft (Use $\pi \approx \frac{22}{7}$.)

mixed practice (pages 499–500)

Find the area of each circle. Use $\pi \approx 3.14$.

17.

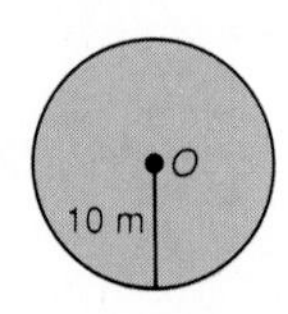

18.

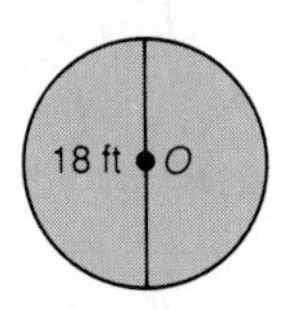

19.

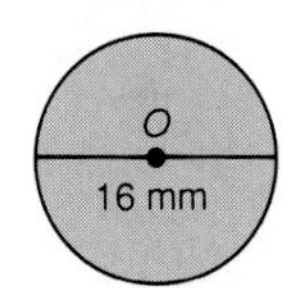

20.

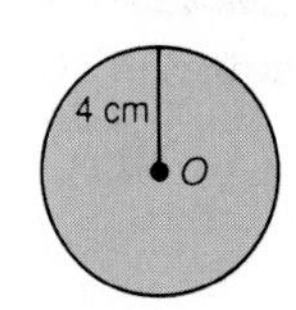

Find the area of each circle. Use $\pi \approx \frac{22}{7}$.

21.

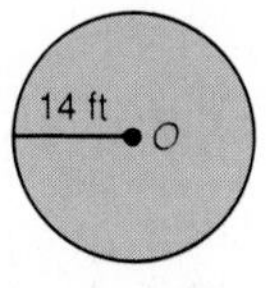

22.

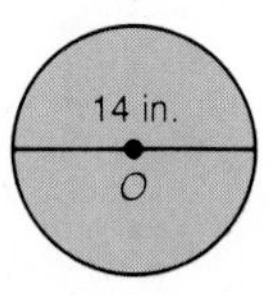

23.

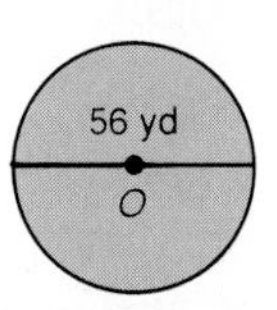

24.

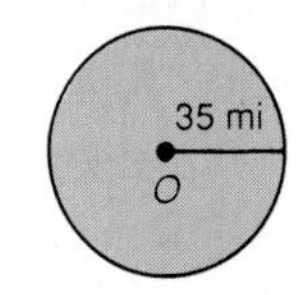

Assume that each circle is separated into equal parts as shown. Find the area of the shaded region. Use $\pi \approx 3.14$.

25.

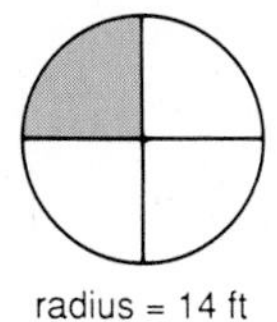

radius = 14 ft

26.

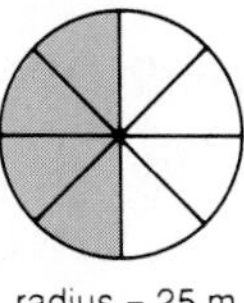

radius = 25 m

27.

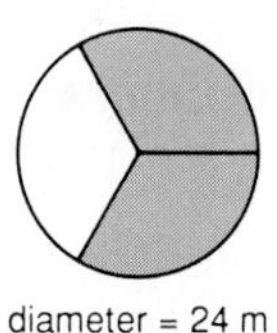

diameter = 24 m

28.

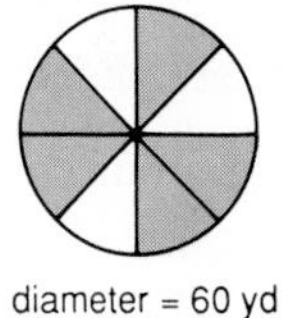

diameter = 60 yd

29. The diameter of a circular dart board is 40 cm. Find the area. Use $\pi \approx 3.14$.

30. Jill jogs three times around a circular track. The radius of the track is 554 ft. How far does Jill jog around the track? Use $\pi \approx 3.14$.

31. A lighthouse beacon can be seen 28 mi in all directions. Over how many square miles can the beacon be seen? Use $\pi \approx \frac{22}{7}$.

32. A baseball diamond is a square with sides that each measure 30 yd. What is the area of the baseball diamond?

review exercises

Write each ratio as a fraction in lowest terms.

1. 12 : 36 **2.** 5 to 40 **3.** 6 : 32 **4.** 4 sec : 2 min **5.** 5 ft to 3 in.

The length and width of a rectangle can only be whole numbers, and the perimeter must be 14. How many different rectangles can be drawn?

SKILL REVIEW

Use the figure at the right. Name as many of the following as are shown. **22-1**

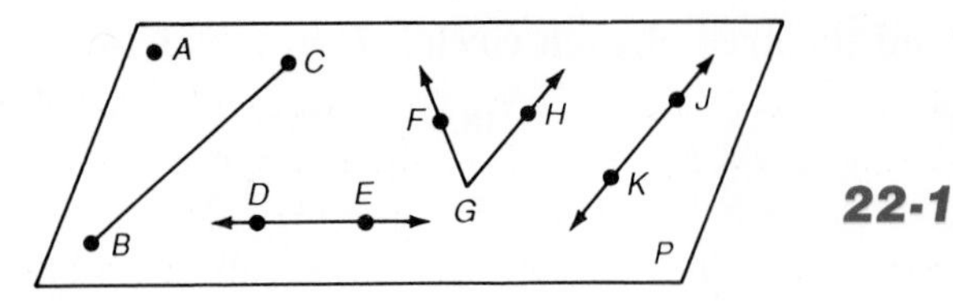

1. points
2. lines
3. line segments
4. angles

Use a protractor to measure the given angle in the figure at the right. **22-2**

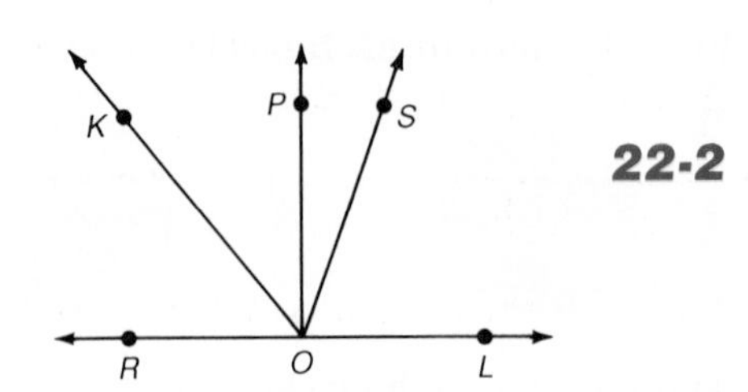

5. $\angle POR$
6. $\angle SOL$
7. $\angle KOL$
8. $\angle SOR$

Classify the given angle as *straight, right, acute,* or *obtuse*. **22-3**

9. 135°
10. 180°
11. 90°
12. 62°

List all the types of quadrilaterals that satisfy the given property. **22-4**

13. four congruent sides
14. exactly one pair of parallel sides

Find the perimeter. Use a formula whenever possible. **22-5**

15. a square with sides that each measure 12 ft
16. a rectangle with length $4\frac{1}{2}$ m and width 3 m
17. a triangle with sides that measure 5 yd, 4 yd, and 1.2 yd
18. a regular hexagon with sides that each measure 23 in.

Use the figure at the right. Name as many of the following as are shown. **22-6**

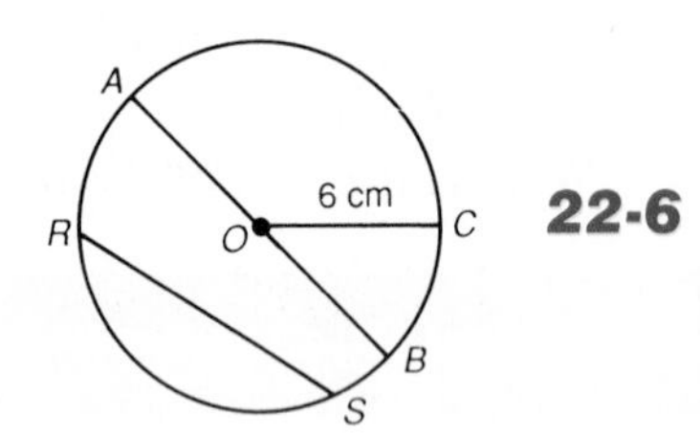

19. the circle
20. diameters
21. radii
22. chords
23. Find the circumference. Use $\pi \approx 3.14$.

Find the area of each figure. **22-7**

24. a parallelogram with a base that measures 4 m and height 2.5 m
25. a triangle with a base that measures 4.5 cm and height 3.6 cm
26. a trapezoid with bases that measure 20 yd and 12 yd, and height 11 yd

22-8

27. a circle with diameter = 28 ft (Use $\pi \approx \frac{22}{7}$.)
28. a circle with radius = 8 in. (Use $\pi \approx 3.14$.)

APPLICATION

22-9 SYMMETRY

If you could fold the triangle at the right along the dashed line $\overleftrightarrow{AB}$, one part of the triangle would fit exactly over the other. The triangle is said to be *symmetric with respect to* $\overleftrightarrow{AB}$. You call $\overleftrightarrow{AB}$ a **line of symmetry.**

A geometric figure may have more than one line of symmetry, or it may have no lines of symmetry.

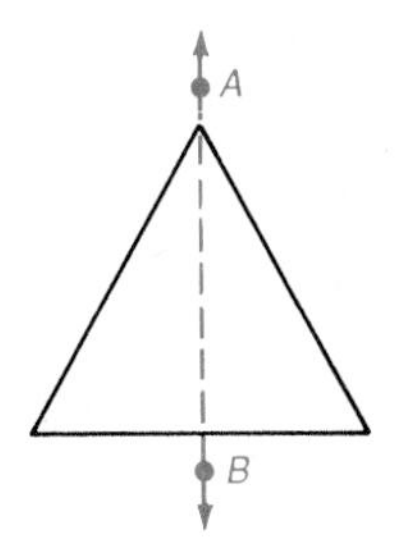

example

Draw all the lines of symmetry in each figure.

a. 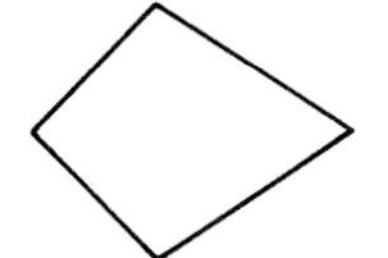

b. 

c. 

solution

a. 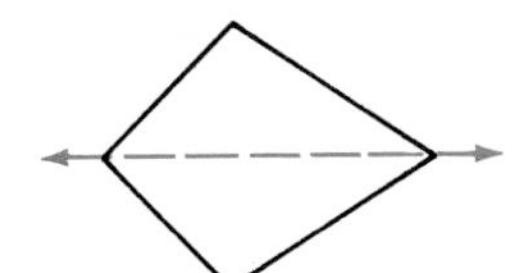

b. 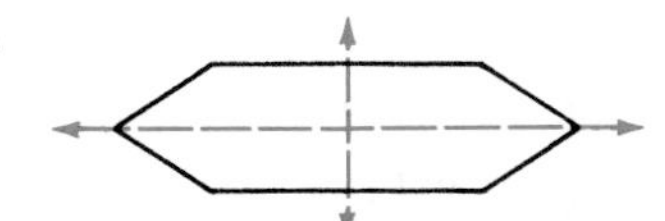

c. There are no lines of symmetry.

exercises

Is the dashed line a line of symmetry? Write *Yes* or *No*.

1. 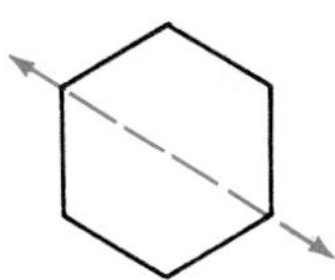

2. 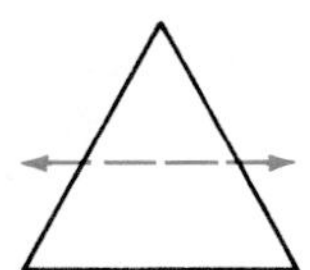

3. 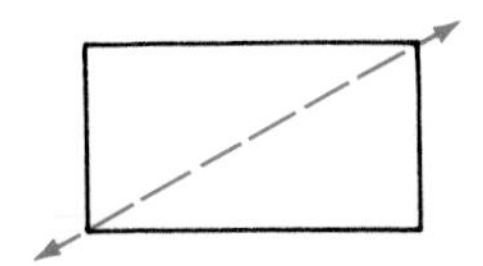

4. 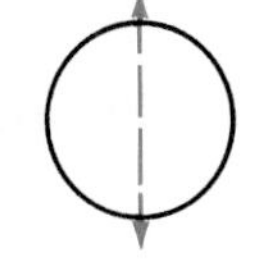

Trace each figure on a piece of paper. Then draw all the lines of symmetry. If the figure has no lines of symmetry, write *None*.

5. 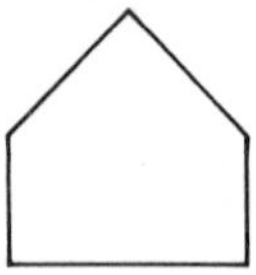 **6.** **7.** 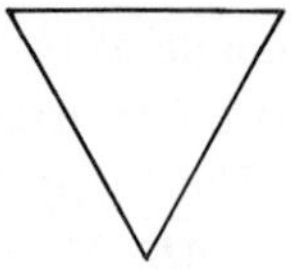 **8.**

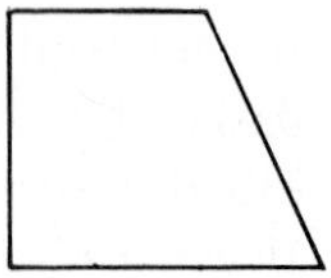

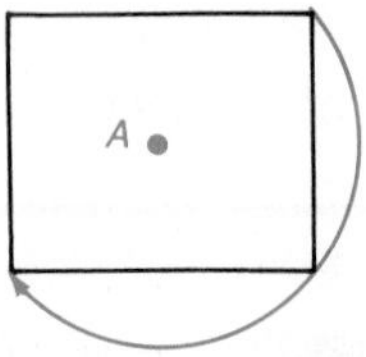

In the rectangle at the right, suppose that you could hold the tip of a pencil at point A, then turn the rectangle. When you have made a *half-turn*, or a turn of 180°, the rectangle will fit exactly over its original position. The rectangle is said to be *symmetric with respect to point A*. You call point A a **point of symmetry.**

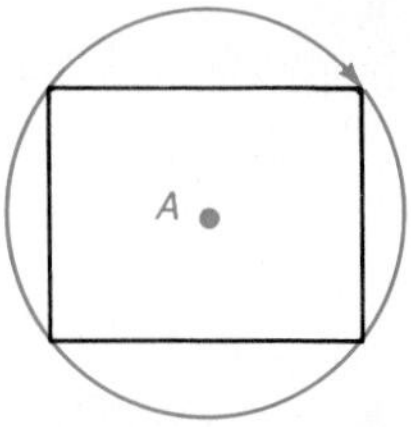

A

Any geometric figure will fit exactly over its original position if you give it a *complete turn,* or a 360° turn. Therefore, there are *two* ways that you can turn the rectangle so that it fits exactly over its original position: a half-turn and a complete turn. If a figure only fits its original position in *one* way, after a complete turn, the figure has no point of symmetry.

A

Is the red point a point of symmetry? Write *Yes* or *No*.

9. 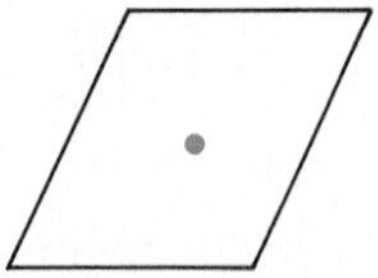 **10.** 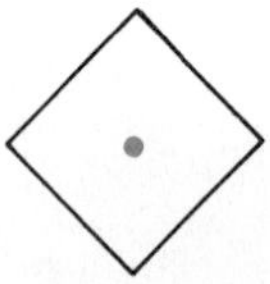 **11.** 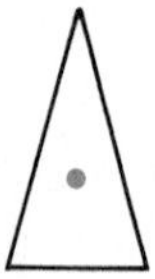 **12.**

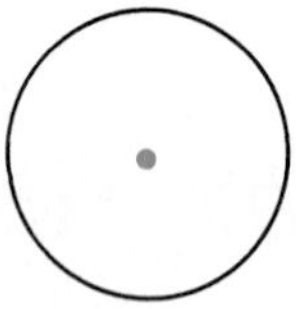

Determine the number of ways that each figure can be turned so that it fits exactly over its original position.

13. 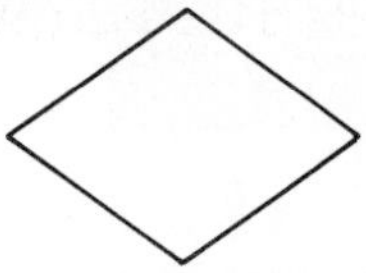 **14.** 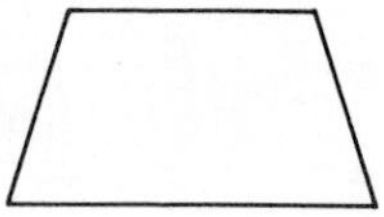 **15.**  **16.**

For each of the following regular polygons, determine the following.

a. **the number of lines of symmetry**

b. **the number of ways it can be turned so that it fits exactly over its original position**

17. 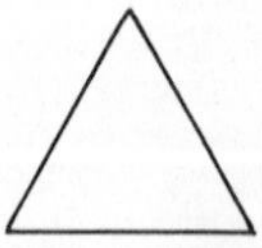 **18.** 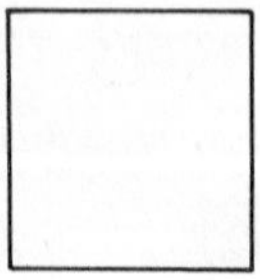 **19.** 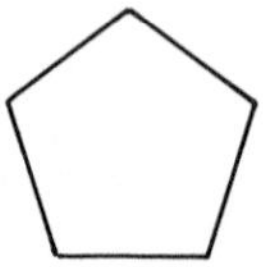 **20.**

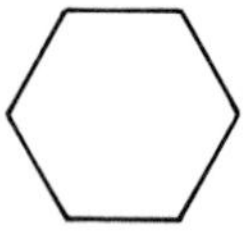

APPLICATION

22-10 DRAWING A CIRCLE GRAPH

When data are expressed as percents of a whole, you can draw a **circle graph** to display them. The whole circle represents 100% of the data. You can divide the circle into **sectors** that represent the parts of the data. The sum of the angles formed by the sectors is 360°.

To draw a circle graph, use the following method.

1. Write each percent as a decimal, then multiply by 360° to find the number of degrees for each sector. When necessary, round the number of degrees to the nearest whole number.
2. Draw a circle and a radius.
3. Use a protractor to draw the central angle for each sector.
4. Label each sector and give the graph a title.

example

Draw a circle graph to display the given data.

Expenses of Urban Residents

Food	15%
Housing	31%
Clothing	5%
Transportation	21%
Medical Care	4%
Entertainment	5%
Other	19%

solution

Food: $360° \times 0.15 = 54°$

Housing: $360° \times 0.31 = 111.6° \approx 112°$

Clothing: $360° \times 0.05 = 18°$

Transportation: $360° \times 0.21 = 75.6° \approx 76°$

Medical Care: $360° \times 0.04 = 14.4° \approx 14°$

Entertainment: $360° \times 0.05 = 18°$

Other: $360° \times 0.19 = 68.4° \approx 68°$

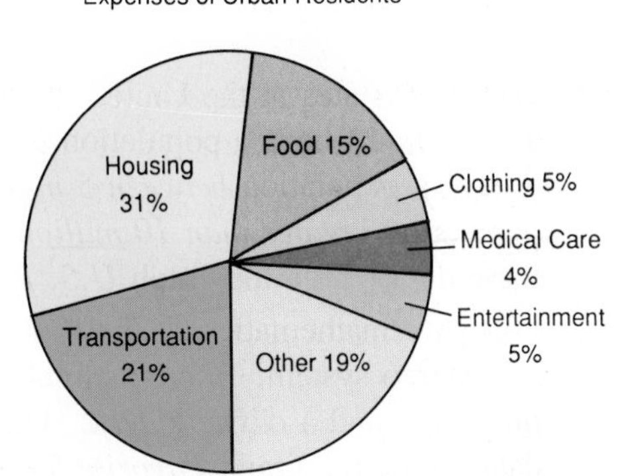

exercises

Draw a circle graph to display the given data.

1. **How High School Students Use Computers at School**

Programming	48%
Word Processing	20%
Drill and Practice	18%
Other	14%

2. **World Population Projections for the Year 2010**

Under 5 Years Old	10%
5 to 14 Years Old	19%
15 to 64 Years Old	64%
65 Years Old and Over	7%

3. **Composition of Earth's Atmosphere**

Nitrogen	78%
Oxygen	21%
Other	1%

4. **Where People Have Checking Accounts**

Commercial Bank	67%
Savings and Loan	23%
Other	10%

5. **Distribution of Medals at the 1988 Winter Olympic Games**

Soviet Union	21%
East Germany	18%
Switzerland	11%
Austria	7%
West Germany	6%
Other	37%

6. **Reasons Why People Work at Two Jobs**

Meeting Regular Expenses	32%
Enjoyment of Second Job	18%
Saving for the Future	10%
Buying Something Special	9%
Gaining Experience	6%
Other	25%

7. Of the 50 states in the United States, 12 have a population *less than 1 million,* 24 have a population *between 1 million and 5 million,* 8 have a population *between 5 million and 10 million,* and 6 have a population *greater than 10 million.* Draw a circle graph to display these data. Title the graph *U.S. Population by State.*

8. Poll your mathematics class about each student's favorite component of a stereo system. Have them choose from a *compact disc player,* a *turntable,* and a *cassette deck.* Draw a circle graph to display these data. Title the graph *Favorite Stereo System Components.*

22-11 COMPOSITE FIGURES

A **composite figure** is a figure that is made up, or *composed,* of other figures. To find the area of a composite figure, you must divide it into figures whose areas you know how to find, and then find the sum of the areas.

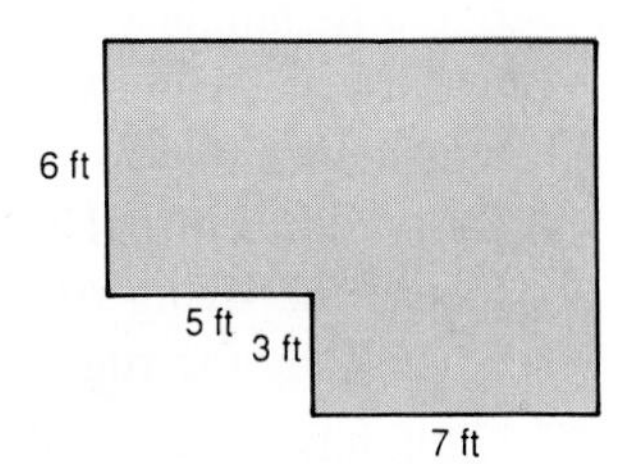

example

Find the area of the figure above.

solution

Divide the figure into two rectangles. Label them I and II.

6 ft I II 9 ft
5 ft 3 ft 7 ft

Add 6 ft + 3 ft to find the unknown length.

$A(\text{I}) = lw$ $\qquad A(\text{II}) = lw$

$= 6(5)$ $\qquad = 9(7)$

$= 30$ $\qquad = 63$

Total Area $= A(\text{I}) + A(\text{II})$

$= 30 + 63 = 93$

The area is 93 ft^2.

exercises

Find the area of each figure. When appropriate, use $\pi \approx 3.14$.

1. 30 m, 45 m, 15 m, 15 m

2. 5 cm, 6 cm, 3 cm, 15 cm

3.

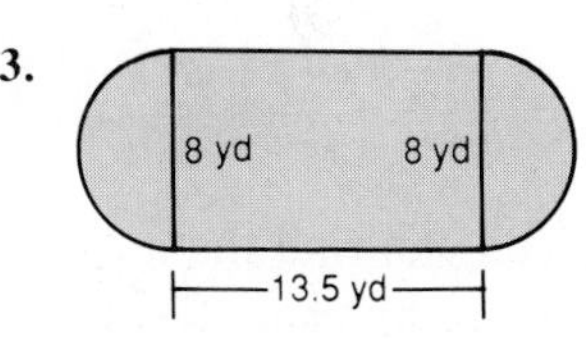

Sometimes you need to find the area of only a *shaded part* of a composite figure. Usually you can find this area by subtracting the areas of known figures.

Find the area of each shaded region. When appropriate, use $\pi \approx 3.14$.

4. 15 yd, 15 yd, 30 yd, 60 yd

5. 2 cm, 8 cm, 8 cm, 12 cm

6.

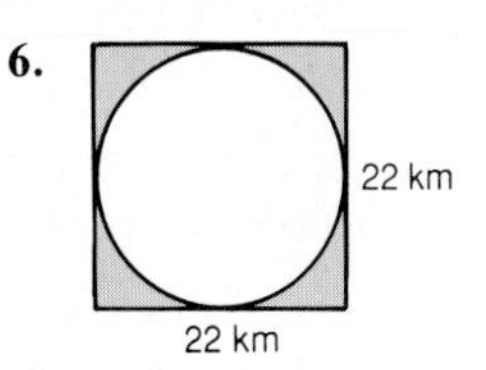

7. Mark Venezia placed a new rug on his living room floor. The rug is shaped like a rectangle with length 6 ft and width 4 ft. His living room floor is a rectangular region with length 18 ft and width 15 ft. How many square feet of the living room floor are not covered by the rug?

8. Carla has a rectangular yard with length 80 yd and width 50 yd. One part of her yard is a swimming pool. The pool is a rectangular area with length 30 yd and width 10 yd. Find the total area of the yard that is not a part of the swimming pool.

9. The top of Ben's kitchen table is shaped like a rectangle. The length of the table is 5 ft and the width is 3 ft. He can enlarge the table top by adding two leaves. The length of each leaf is 5 ft and the width is 2 ft. What is the maximum area of the table top?

10. Nancy DiLoreto wants to put wall-to-wall carpeting in a rectangular area with length 23 ft and width 12 ft. Within this area, there is a circular region where Nancy wants to put ceramic tile. The diameter of the circular region is 10 ft. Find the total area that she would like to be carpeted. Use $\pi \approx 3.14$.

calculator corner

You can find the area of a composite figure by using a calculator.

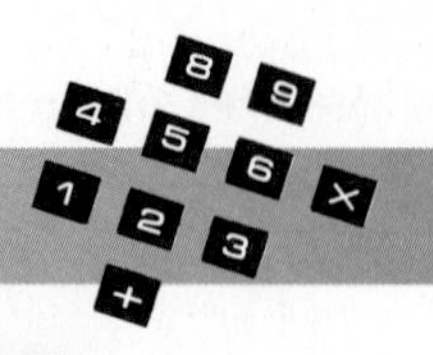

example Find the area of the shaded region.

solution Shaded Area = A(rectangle) − A(square)

Enter 3 × 3 = M+ . Then enter 9 × 6 − MR =.

The result is 45. The area of the shaded region is 45 in.^{2}.

Find the area of each shaded region. When appropriate, use $\pi \approx 3.14$.

1. 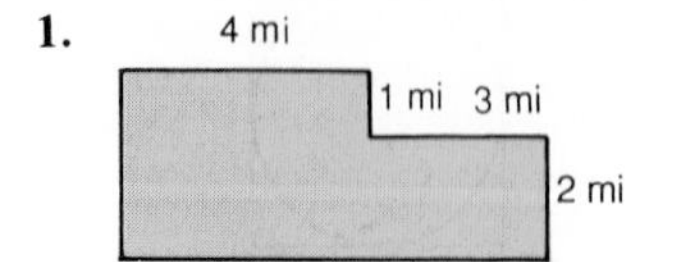

2. 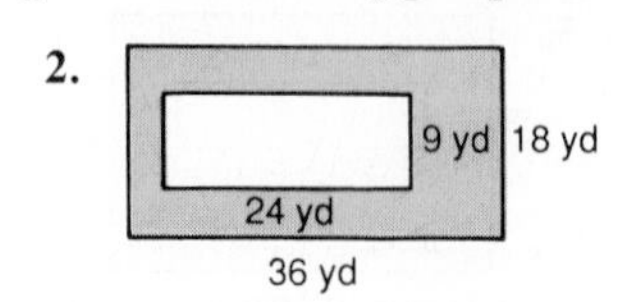

3. 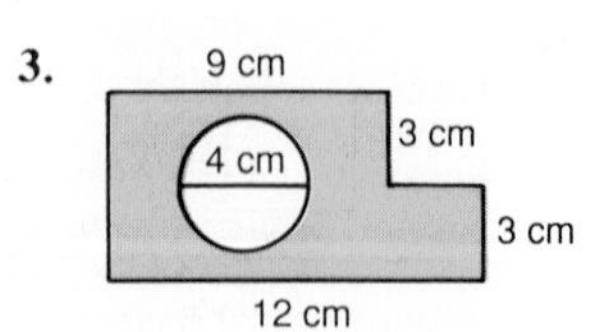

CHAPTER REVIEW

vocabulary vo·cab·u·lar·y

Choose the correct word to complete each sentence.

1. (*Vertical, Adjacent*) angles are angles that share a common side.
2. When the sum of the measures of two angles is 90°, the angles are (*complementary, supplementary*).
3. When two lines intersect to form a right angle, the lines are (*perpendicular, parallel*).
4. The distance around a polygon is called its (*perimeter, area*).

skills

Use the figure at the right.

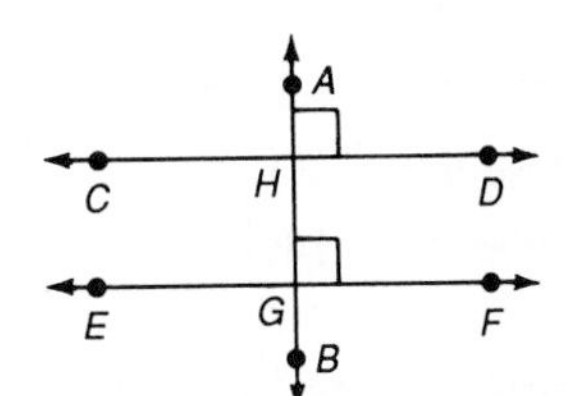

5. Name two pairs of vertical angles.
6. Name a pair of parallel lines.
7. Name a pair of perpendicular lines.

Use a protractor to draw an angle of the given measure.

8. 25°
9. 130°
10. 180°
11. 15°
12. 90°

Tell whether two angles with the given measures are *complementary, supplementary,* or *neither*.

13. 18°, 77°
14. 28°, 62°
15. 52°, 38°
16. 134°, 46°
17. 45°, 139°

Identify each polygon. List all the diagonals.

18.

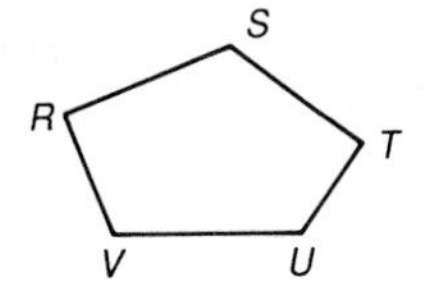

19.

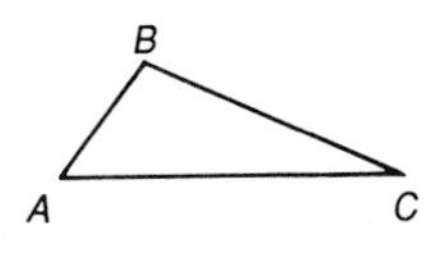

20.

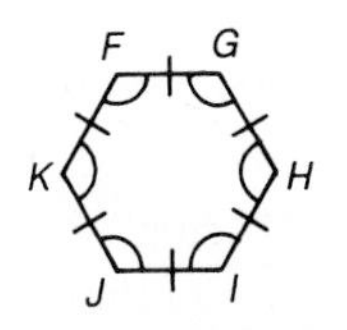

List all the types of quadrilaterals that satisfy the given property.

21. four right angles
22. four right angles and four congruent sides
23. exactly one pair of parallel sides
24. two pairs of parallel sides

Find the perimeter. Use a formula whenever possible.

25. a square with sides that each measure 3 cm
26. a rectangle with length $5\frac{1}{2}$ ft and width 1 ft
27. a regular pentagon with sides that each measure 4.5 m

Find the area of each figure.

28. rectangle

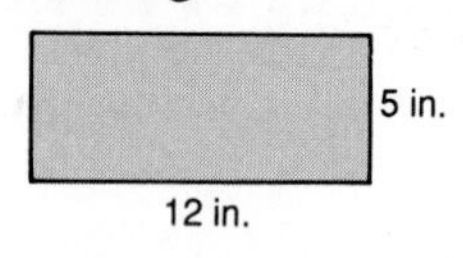

29. square

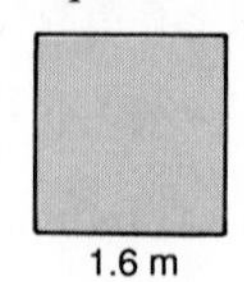

30. triangle

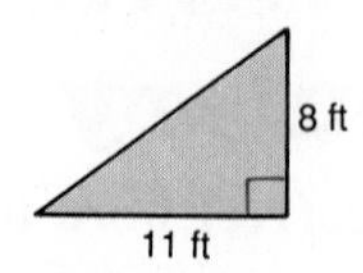

31. trapezoid

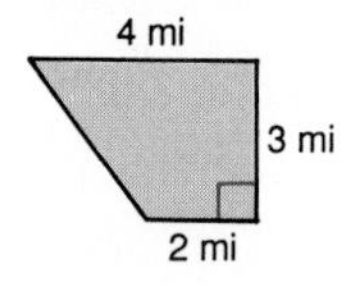

32. triangle

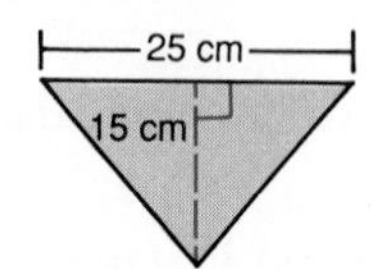

33. trapezoid

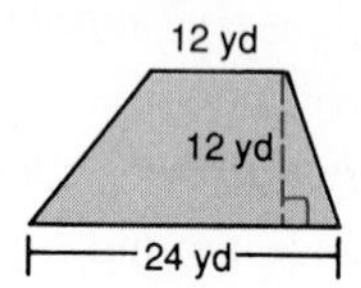

Find the circumference and the area of each circle.

34. diameter = 2 cm (Use $\pi \approx 3.14$.)

35. diameter = 14 ft (Use $\pi \approx \frac{22}{7}$.)

36. diameter = 28 m (Use $\pi \approx \frac{22}{7}$.)

37. radius = 9 km (Use $\pi \approx 3.14$.)

38. radius = 23 mi (Use $\pi \approx 3.14$.)

39. radius = 21 in. (Use $\pi \approx \frac{22}{7}$.)

For each of the following figures, determine the following.

a. the number of lines of symmetry

b. the number of ways it can be turned so that it fits exactly over its original position

40.

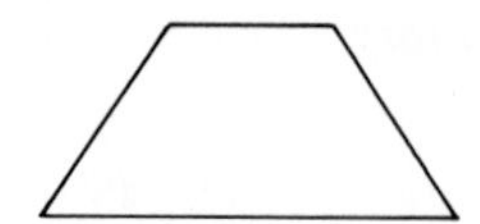

41.

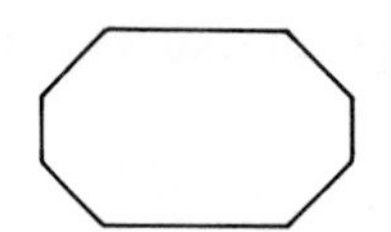

42.

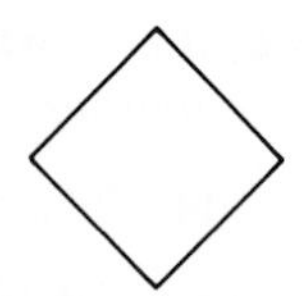

43. Find the area of the figure.

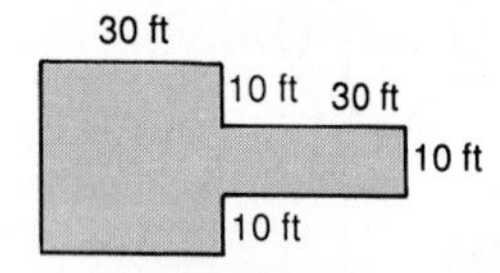

44. Find the area of the shaded region.

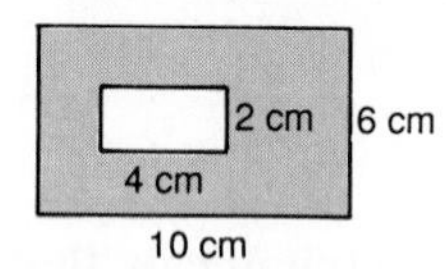

45. Draw a circle graph to display the given data.

Tropical Fish in an Aquarium

Blue	60%
Red	25%
Yellow	15%

CHAPTER TEST

Use the figure at the right. Tell whether each statement is *true* or *false*.

22-1

1. $\angle EJB$ and $\angle FJB$ are a pair of adjacent angles.
2. $\angle CHA$ and $\angle HJF$ are a pair of vertical angles.
3. $\overleftrightarrow{AB}$ and $\overleftrightarrow{EF}$ are intersecting lines.

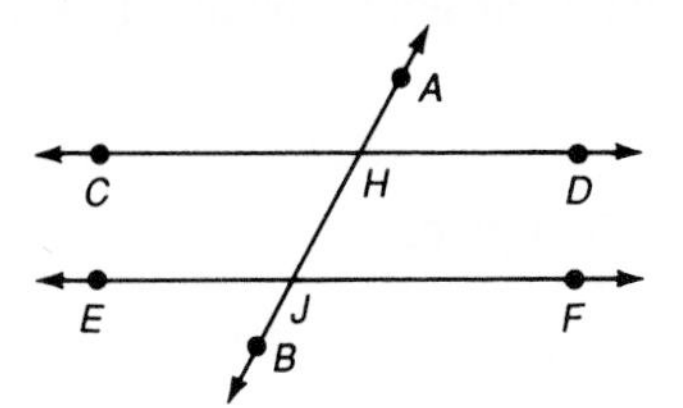

Use a protractor to measure each angle.

22-2

4. 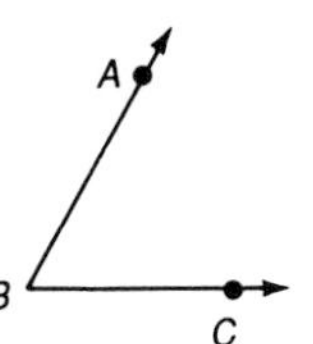

5. 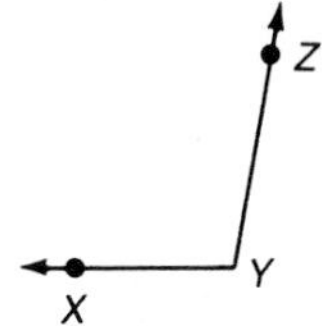

6. 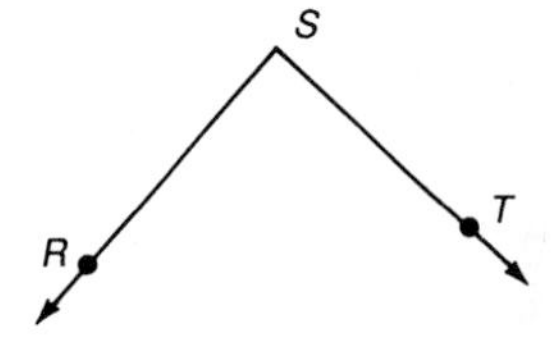

Use the figure at the right. Tell whether each statement is *true* or *false*.

22-3

7. $\overleftrightarrow{AM} \parallel \overleftrightarrow{DK}$
8. $\overleftrightarrow{CF} \parallel \overleftrightarrow{GL}$
9. $\overleftrightarrow{DK} \perp \overleftrightarrow{CF}$
10. $\overleftrightarrow{CF} \perp \overleftrightarrow{GL}$
11. $\angle ABC$ and $\angle LJK$ are supplementary angles.
12. $\angle DEF$ and $\angle BEJ$ are supplementary angles.

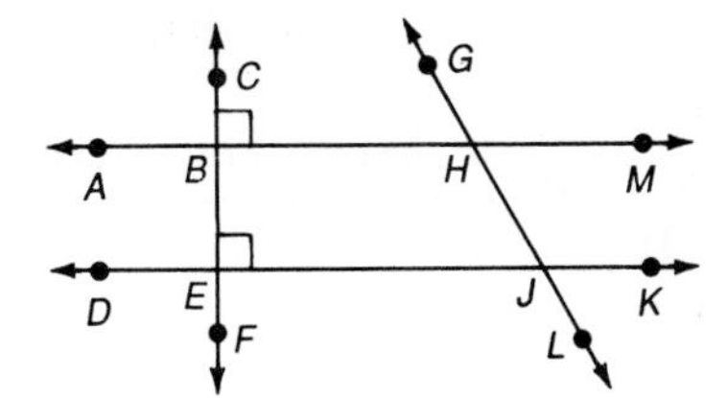

Identify each polygon. List all the diagonals.

22-4

13. 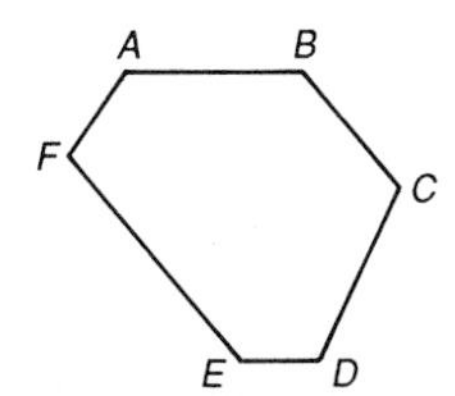

14. 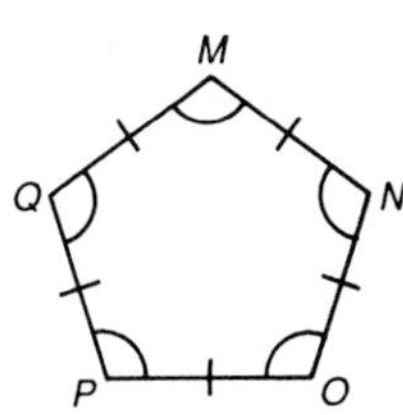

15. 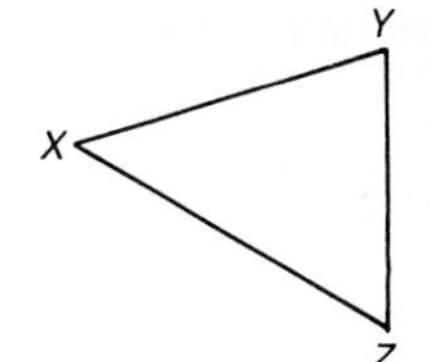

Find the perimeter of each polygon.

22-5

16. 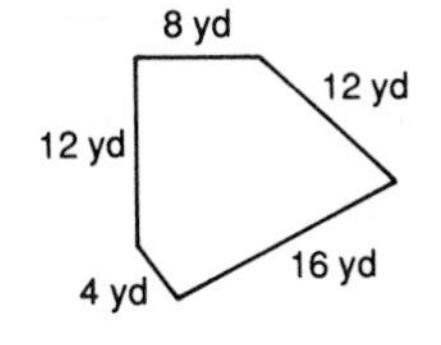

17. a square with sides that each measure 6 mi
18. a rectangle with length 8.4 m and width 3.6 m

Find the circumference of each circle. Use $\pi \approx 3.14$.

19. diameter = 42 mi **20.** radius = 40 km **22-6**

Find the circumference of each circle. Use $\pi \approx \frac{22}{7}$.

21. diameter = 28 cm **22.** radius = 35 yd

Find the area of each figure.

23. parallelogram **24.** triangle **25.** square **22-7**

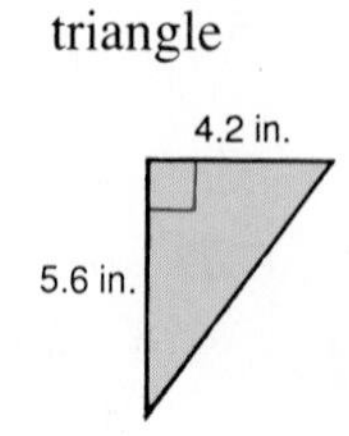

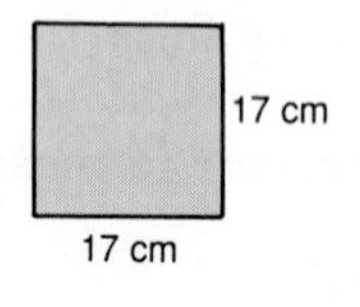

Find the area of each circle. Use $\pi \approx 3.14$.

26. radius = 20 ft **27.** radius = 45 km **22-8**

Find the area of each circle. Use $\pi \approx \frac{22}{7}$.

28. diameter = 14 m **29.** diameter = 42 in.

Is the dashed line a line of symmetry? Write *Yes* or *No*.

30. **31.** **32.** **22-9**

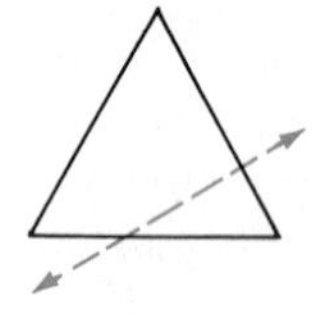

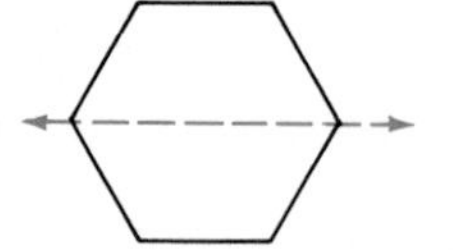

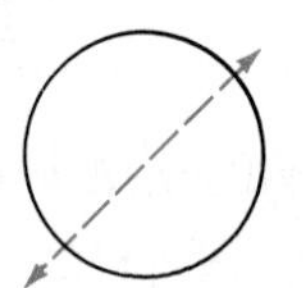

33. Draw a circle graph to display the given data. **22-10**

World Energy Use

Oil	Coal	Natural Gas	Water Power	Nuclear Power
46%	27%	19%	6%	2%

34. Find the area of the figure. **35.** Find the area of the shaded region. Use $\pi \approx 3.14$. **22-11**

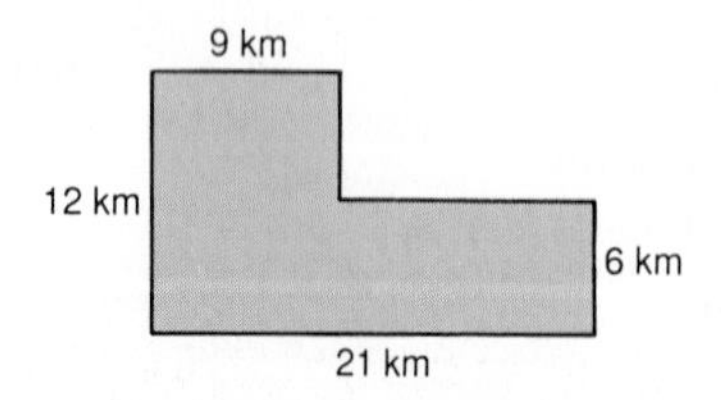

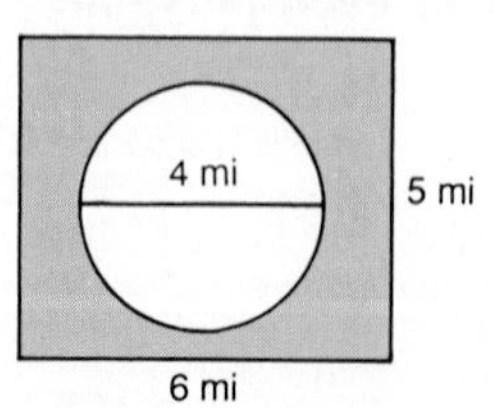

Since the shapes and sizes of these boxes vary considerably, careful planning is needed in order to store them efficiently.

CHAPTER 23

SURFACE AREA AND VOLUME

23-1 SPACE FIGURES

Space figures are three-dimensional figures that enclose part of space.

A **polyhedron** is a space figure with flat surfaces that are called **faces.** Two faces intersect at an **edge.** Three or more edges intersect at a **vertex.**

vertex

face

edge

A **prism** is a polyhedron with two identical, parallel **bases.** A prism is identified by the shape of its bases. A **cube** is a rectangular prism all of whose edges are the same length.

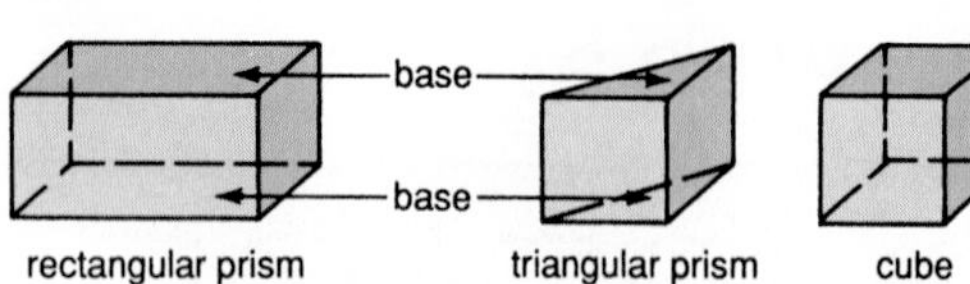

rectangular prism triangular prism cube

A **pyramid** is a polyhedron with one base. The faces other than the base are triangles. A pyramid is identified by the shape of its base.

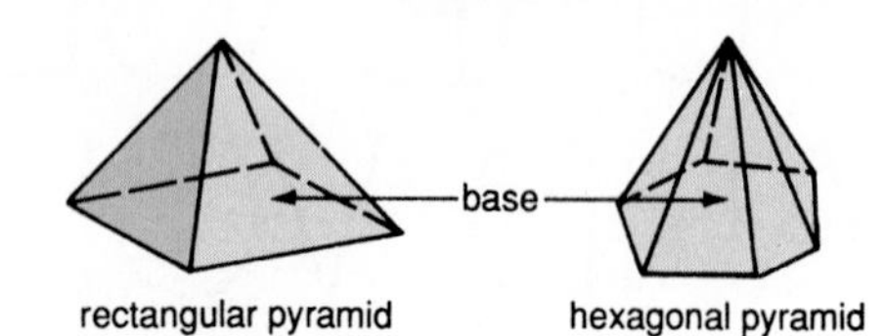

rectangular pyramid hexagonal pyramid

Some space figures are not polyhedrons. They have curved surfaces.

A **cylinder** has two identical, parallel, circular bases.

A **cone** has one circular base and one vertex.

A **sphere** is the set of all points in space that are the same distance from a given point called the **center.**

base

base

cylinder

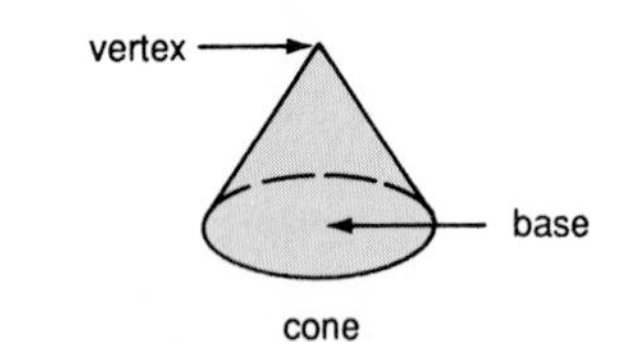

cone

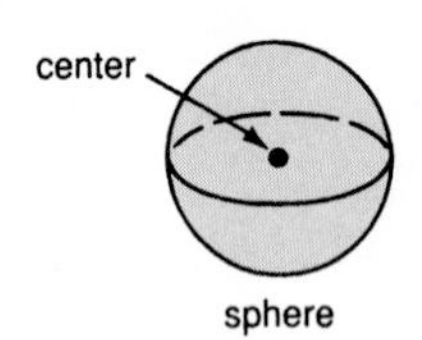

sphere

example 1

Identify each space figure.

a. 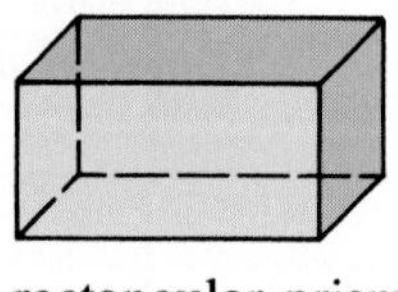
rectangular prism

b. 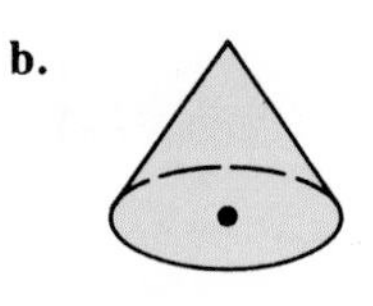
cone

c. 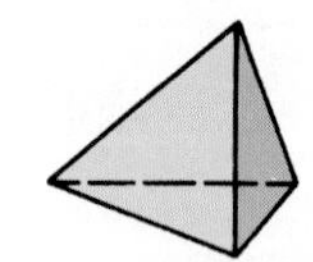
triangular pyramid

d. 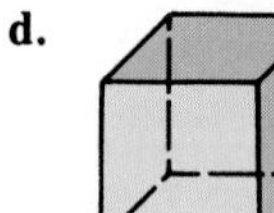
cube

your turn

Identify each space figure.

1. 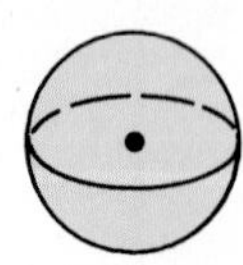

2. 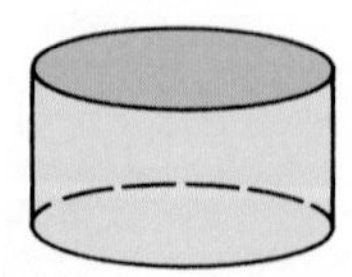

3. 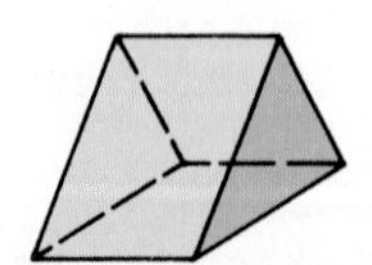

4. 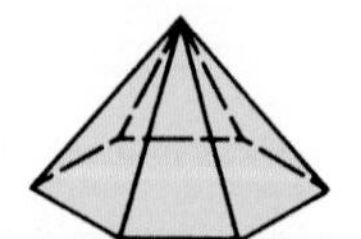

example 2

Identify the space figure that would be formed if you folded the pattern at the right.

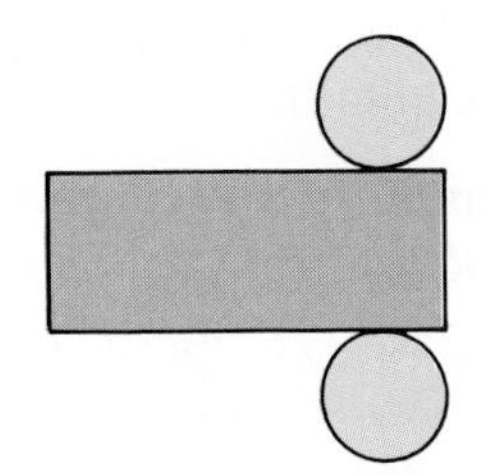

solution

The pattern shows two identical, circular bases. You would form a *cylinder*.

your turn

Identify the space figure that would be formed if you folded each pattern.

5. 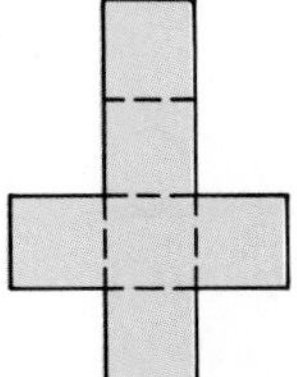

6. 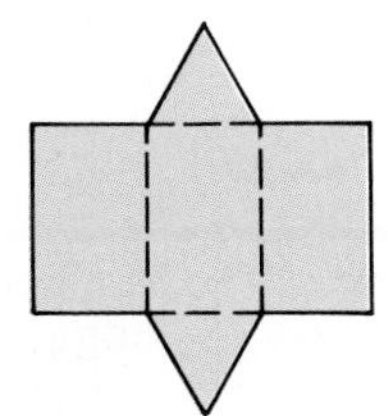

7. 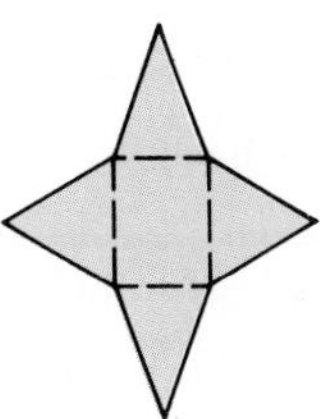

8. 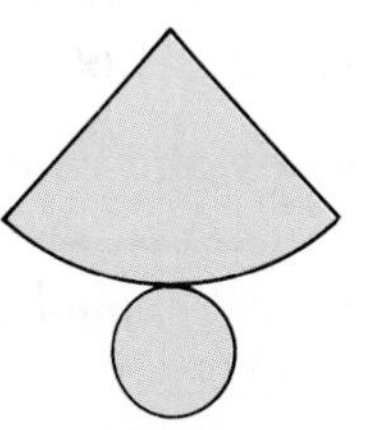

practice exercises

practice for example 1 (page 514)

Identify each space figure.

1. 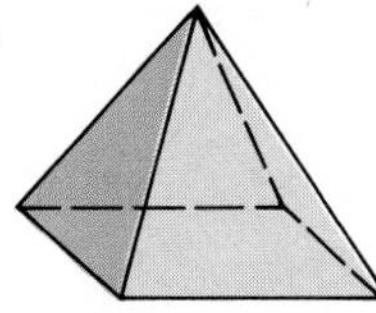

2. 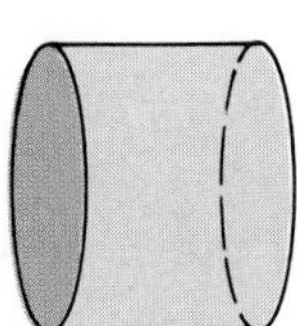

3. 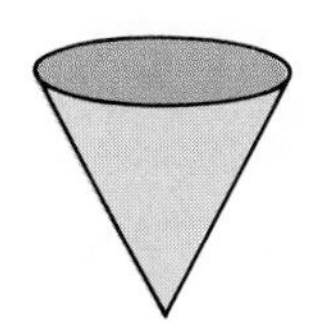

4. 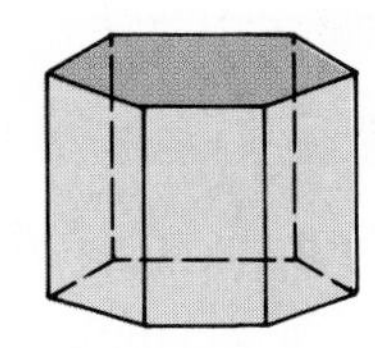

practice for example 2 (page 515)

Identify the space figure that would be formed if you folded each pattern.

5. 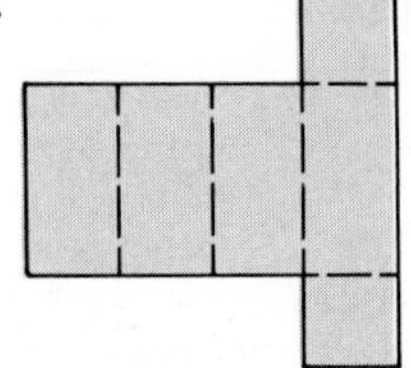

6. 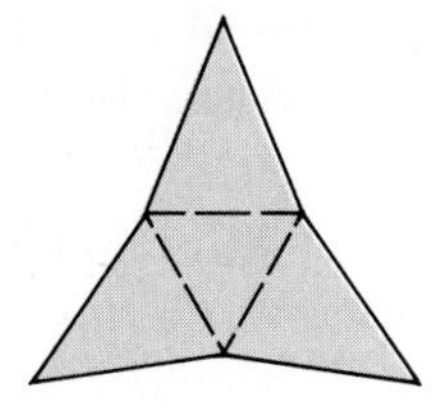

7. 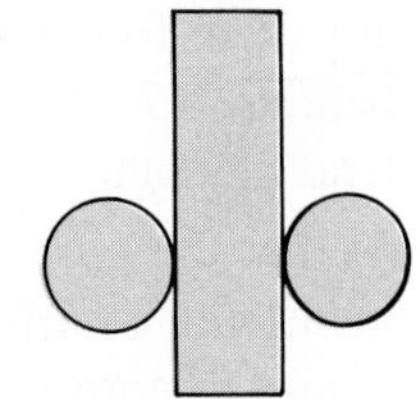

8. 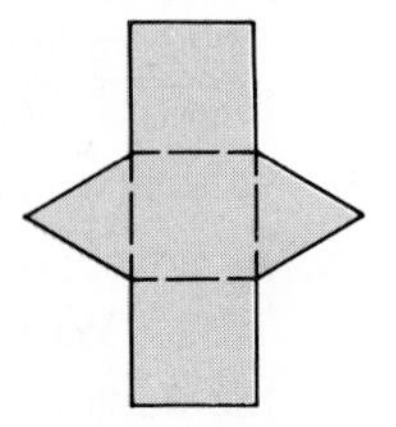

mixed practice *(pages 514–515)*

Imagine that you could fold each pattern below along the dashed lines. Which pattern could you fold to form the space figure shown at the left?

9. 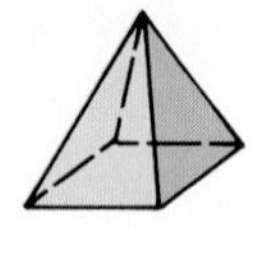 **a.** 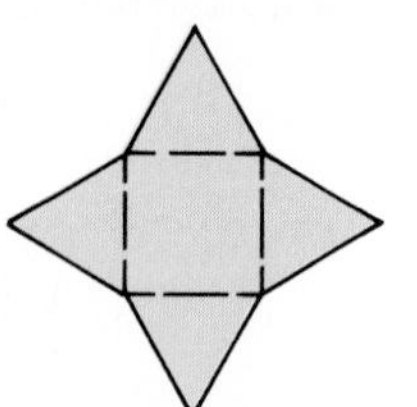 **b.** 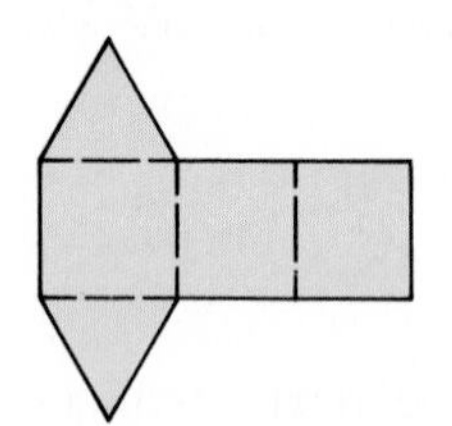 **c.** 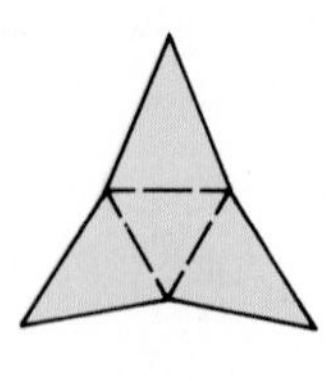

10. 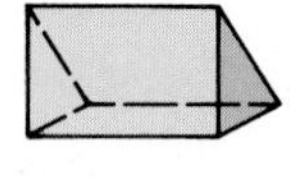 **a.** 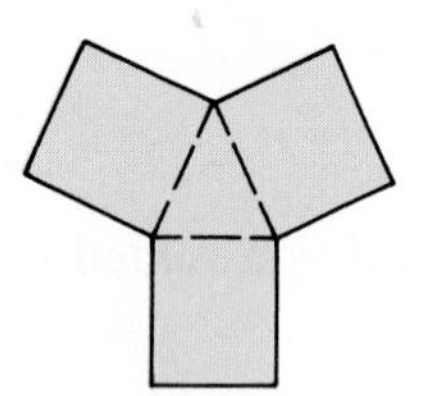 **b.** 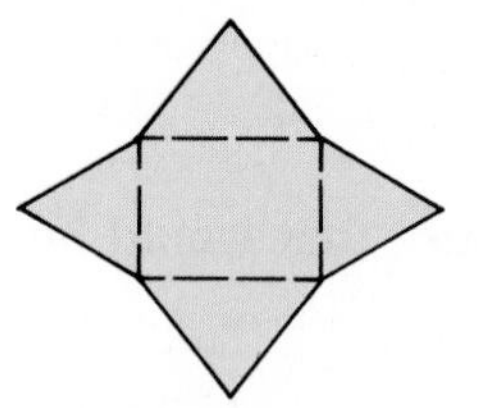 **c.** 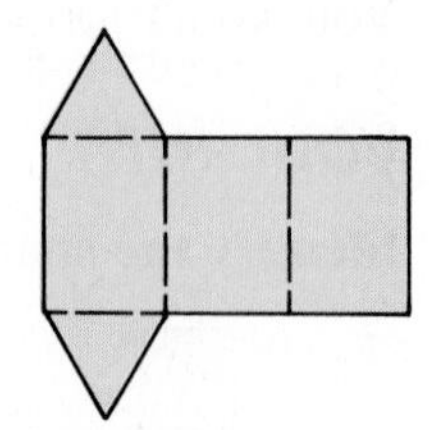

Copy and complete the chart.

	Polyhedron	Number of Faces (*F*)	Number of Vertices (*V*)	Number of Edges (*E*)	*F* + *V* − *E*
11.	rectangular prism	?	?	?	?
12.	triangular prism	?	?	?	?
13.	rectangular pyramid	?	?	?	?
14.	triangular pyramid	?	?	?	?

review exercises

Find the area of each figure. When appropriate, use $\pi \approx 3.14$.

1. a rectangle with length 10 in. and width 9 in.
2. a triangle with length of base 4 m and height 8 m
3. a square with side 7 ft
4. a circle with diameter 6 cm
5. The radius of a circle is 42 in. Use $\pi \approx \frac{22}{7}$ to find the circumference.
6. The Graduation Committee found that four fifths of the class plan to attend a dance. There are 470 students in the class. How many students plan to attend the dance?
7. A newspaper reports that the probability that it will rain tomorrow is 10%. Write this probability as a fraction in lowest terms. Then find the odds in favor of rain tomorrow.

23-2 SURFACE AREA OF PRISMS

The **surface area of a polyhedron** is the sum of the areas of its faces.

example 1

Find the surface area of the rectangular prism at the right.

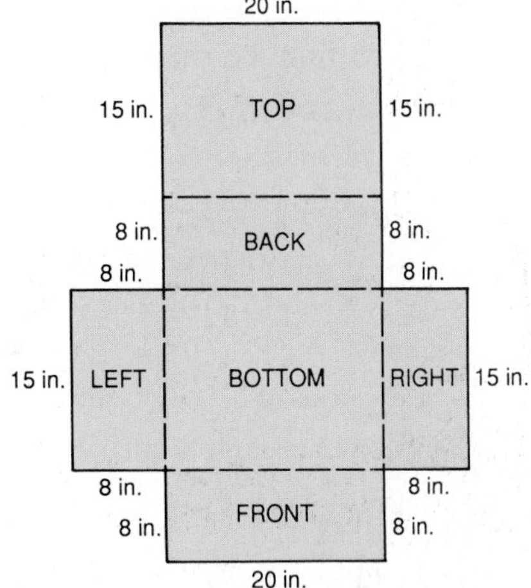

solution

The faces of the rectangular prism are six rectangles. To find the area of each face, use the following formula.

$$A = lw$$

Add the areas of the six rectangular faces to find the surface area.

top:	$A = 20(15)$	=	300
bottom:	$A = 20(15)$	=	300
front:	$A = 20(8)$	=	160
back:	$A = 20(8)$	=	160
left side:	$A = 15(8)$	=	120
right side:	$A = 15(8)$	=	120
Total:			1160

The surface area is 1160 in.^{2}.

your turn

Find the surface area of each rectangular prism.

1. 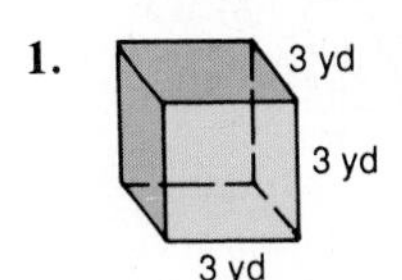

2. 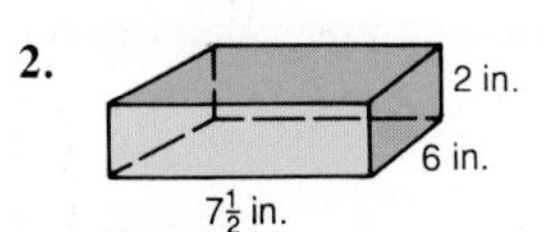

3. 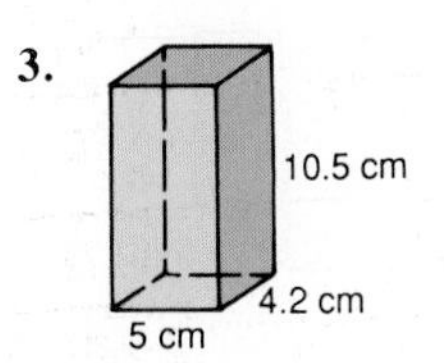

example 2

Find the surface area of the triangular prism at the right.

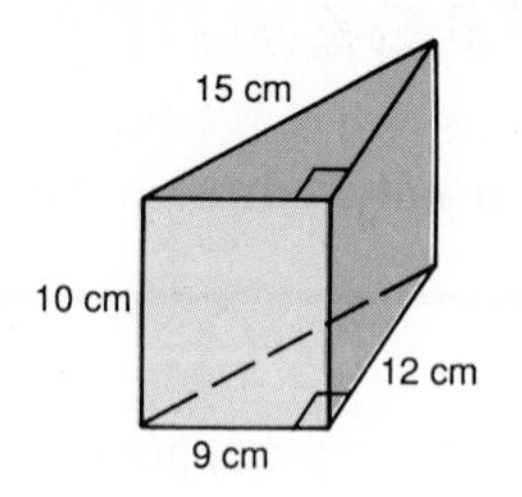

solution

The bases of the triangular prism are two right triangles. To find the area of each base, use the following formula.

$$A = \frac{1}{2}bh$$

The other three faces are rectangles. Add the areas of these three faces to the area of the bases to find the surface area.

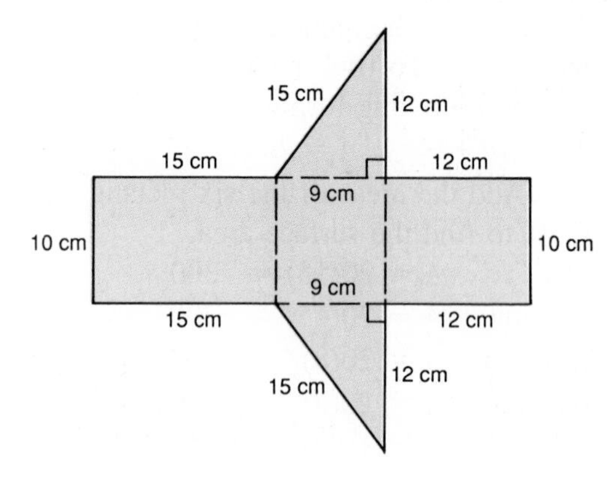

top: $A = \frac{1}{2}(9)(12) = 54$

bottom: $A = \frac{1}{2}(9)(12) = 54$

front: $A = (9)(10) = 90$

left side: $A = (15)(10) = 150$

right side: $A = (12)(10) = 120$

Total: 468

The surface area is 468 cm^2.

your turn

Find the surface area of each triangular prism.

4.

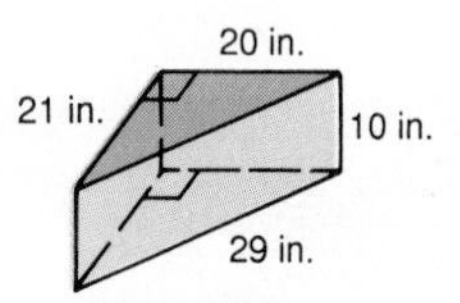

5.

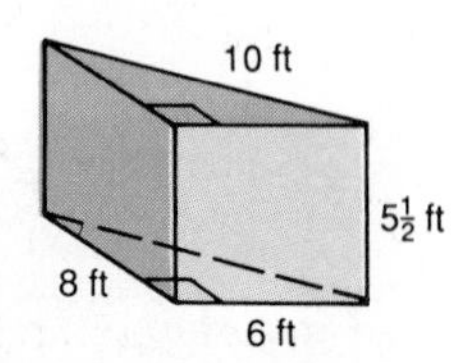

6.

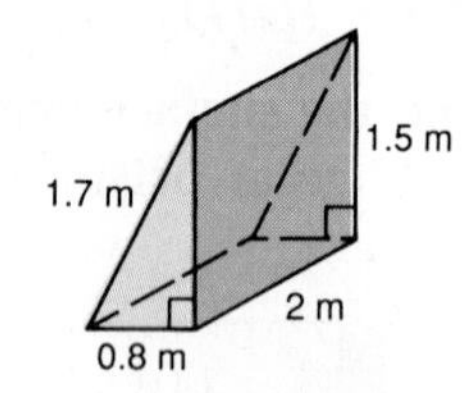

practice exercises

practice for example 1 (page 517)

Find the surface area of each rectangular prism.

1.

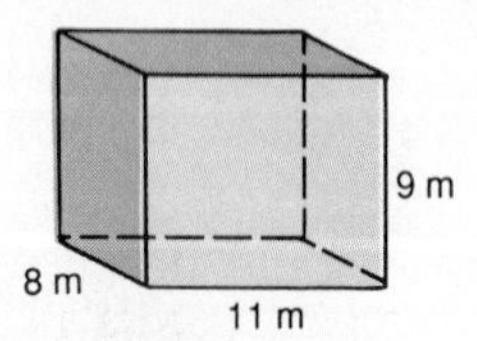

2.

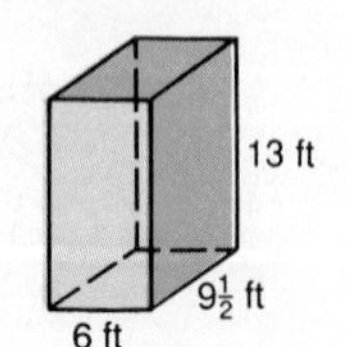

3.

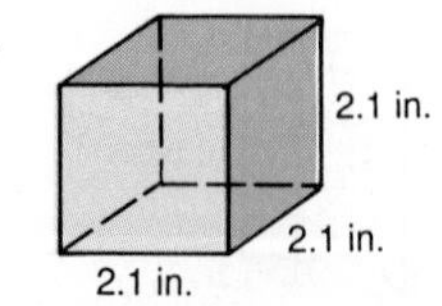

practice for example 2 (page 518)

Find the surface area of each triangular prism.

4. 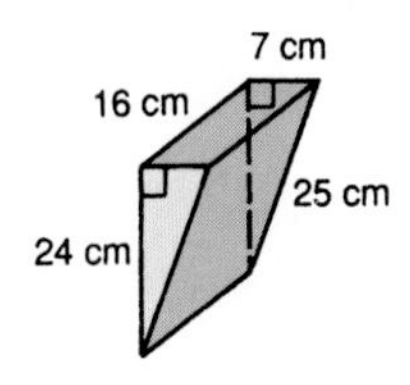

5. 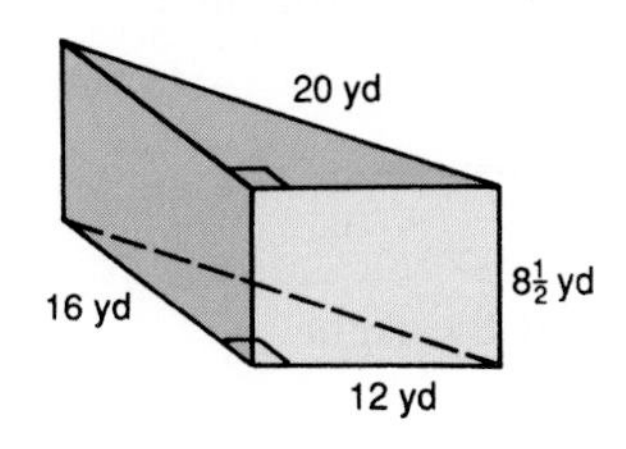

6. 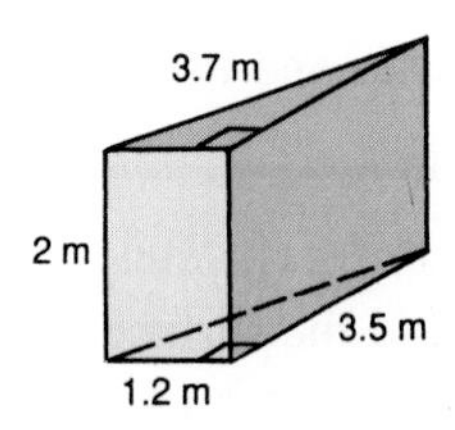

mixed practice (pages 517–518)

Find the surface area of each prism.

7. 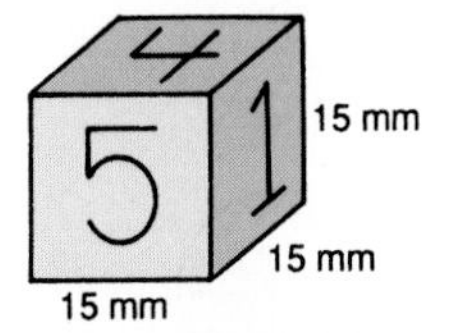

8. 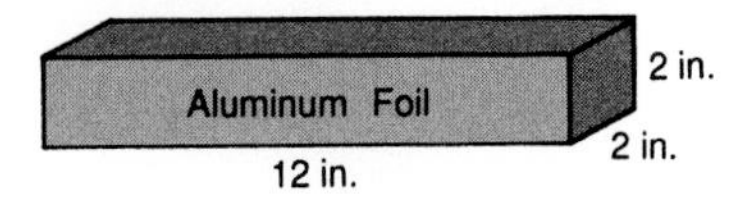

9. 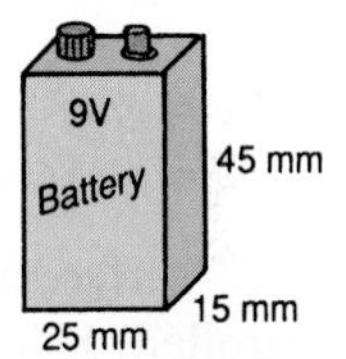

10. 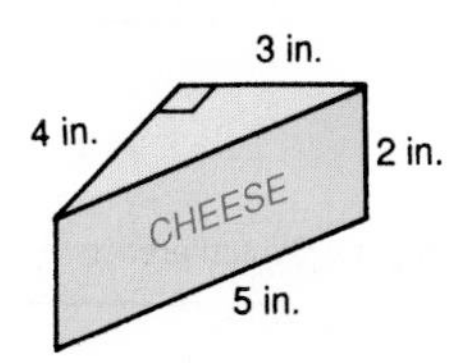

11. 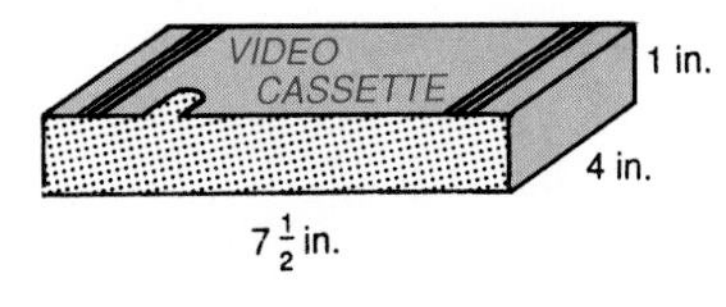

12. 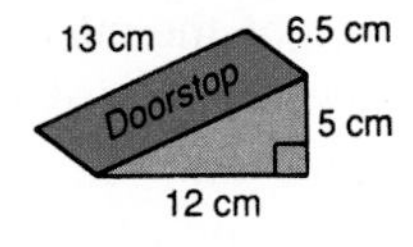

13. What is the surface area of a tool box with length 20 in., width 8 in., and height 9 in.?
14. What is the surface area of a box of facial tissues with length 25 cm, width 12 cm, and height 9 cm?
15. What is the surface area of an album cover with both length and width 313 mm and depth 3 mm?
16. A glass paperweight is shaped like a rectangular prism. The length is 3 in., the width is $1\frac{1}{2}$ in., and the height is $1\frac{1}{2}$ in. What is the surface area of the paperweight?

review exercises

Find each answer.

1. $^-6 + {}^-11$
2. $33.1 - 17.9$
3. $36 \div {}^-4$
4. $(7)(^-8)$
5. $\left(\frac{^-1}{2}\right)\left(\frac{7}{8}\right)$
6. $\frac{3}{5} \div \frac{11}{15}$
7. $^-6 - {}^-6\frac{1}{3}$
8. $9 + \frac{^-1}{4}$
9. $0.9 \div {}^-3$
10. $^-4.8 - 2.3$
11. $6.4 + {}^-3.8$
12. $(^-2)(6.5)(^-3)$

23-3 SURFACE AREA OF CYLINDERS

The **surface area of a cylinder** is the sum of the areas of its two circular bases and its curved surface.

example 1

Find the surface area of the cylinder at the right. Use $\pi \approx 3.14$.

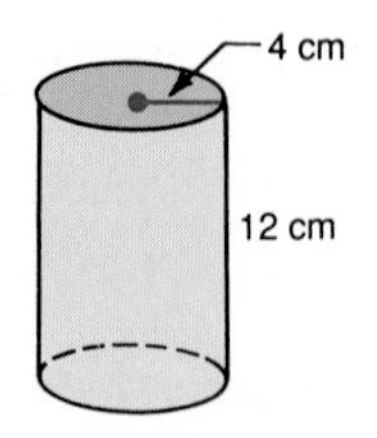

solution

Each base of a cylinder is a circle. To find the area of each base, use the formula for the area of a circle.

$$A = \pi r^2$$
$$A \approx 3.14(4^2) = 50.24$$

The curved surface is like a label on a soup can. When unrolled, it forms a rectangle. To find the area of the curved surface, use the formula for the area of a rectangle ($A = lw$). Then substitute $2\pi r$ (circumference of the base) for l, and h (height of the cylinder) for w.

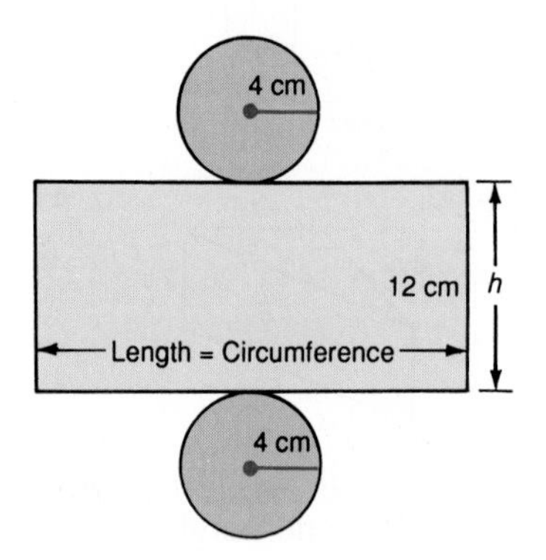

$$A = 2\pi rh$$
$$A \approx 2(3.14)(4)(12) = 301.44$$

Add the three areas to find the surface area.

top base	+	bottom base	+	curved surface	
50.24	+	50.24	+	301.44	= 401.92

The surface area is approximately 401.92 cm^2.

your turn

Find the surface area of each cylinder. Use $\pi \approx 3.14$.

1.

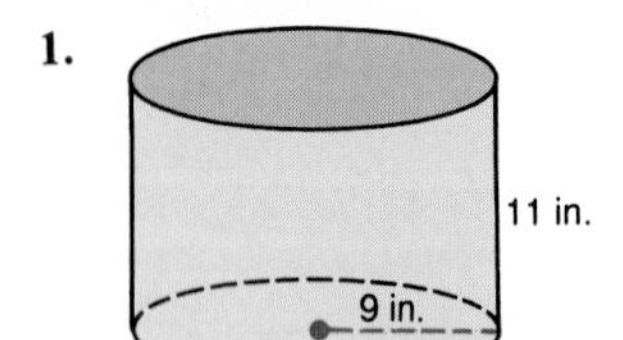

2.

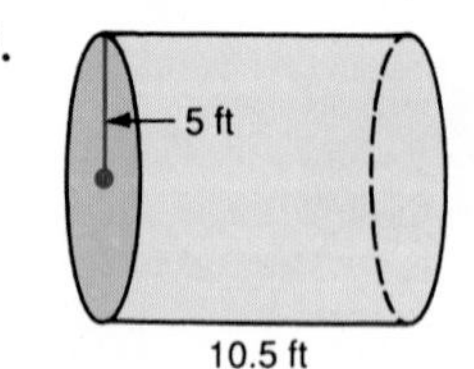

3.

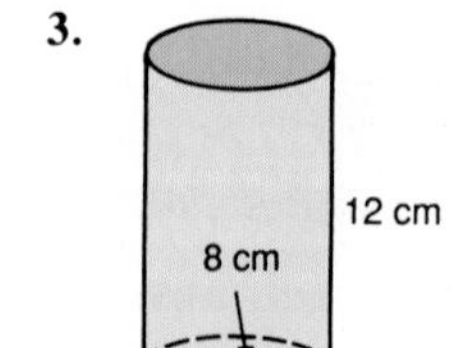

When finding the surface area of a cylinder, you can also use $\frac{22}{7}$ as an approximation for π.

example 2

Find the surface area of the cylinder at the right. Use $\pi \approx \frac{22}{7}$.

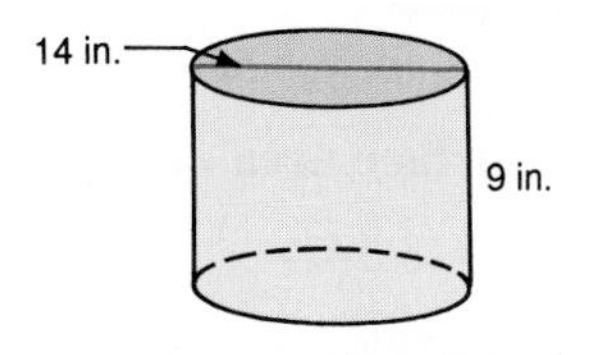

solution

Area of each base: $A = \pi r^2 = \frac{22}{7}(7^2) = 154$ ← **Replace *r* with $\frac{1}{2}$(14), or 7.**

Area of curved surface: $A = 2\pi rh \approx 2\left(\frac{22}{7}\right)(7)(9) = 396$ ← **Replace *r* with 7 and *h* with 9.**

Surface area of cylinder:

top base + bottom base + curved surface
154 + 154 + 396 = 704

The surface area is approximately 704 in.^{2}.

your turn

Find the surface area of each cylinder. Use $\pi \approx \frac{22}{7}$.

4.

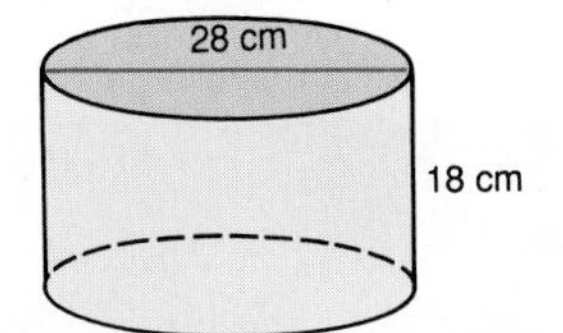

5.

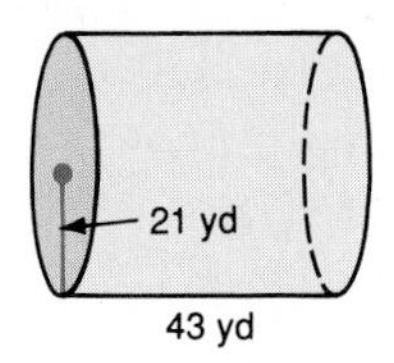

6.

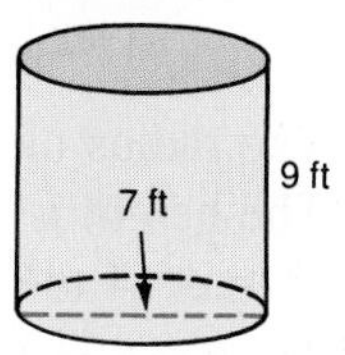

practice exercises

practice for example 1 (page 520)

Find the surface area of each cylinder. Use $\pi \approx 3.14$.

1.

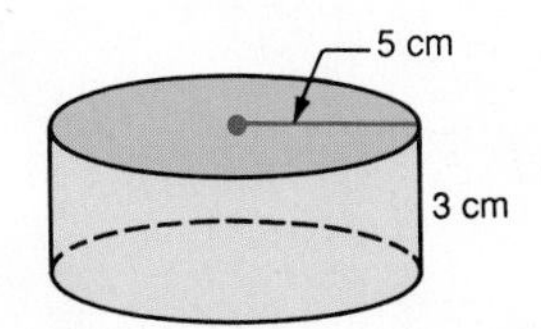

2.

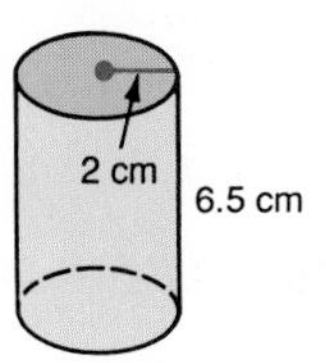

3.

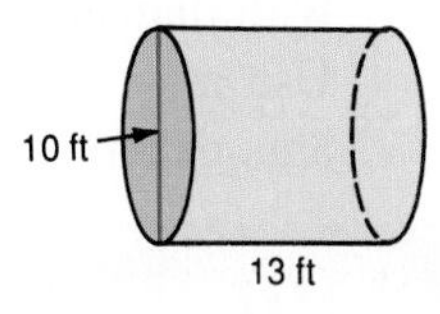

practice for example 2 (page 521)

Find the surface area of each cylinder. Use $\pi \approx \frac{22}{7}$.

4.

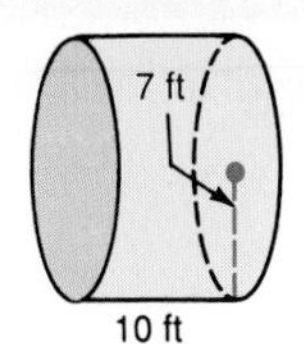

5.

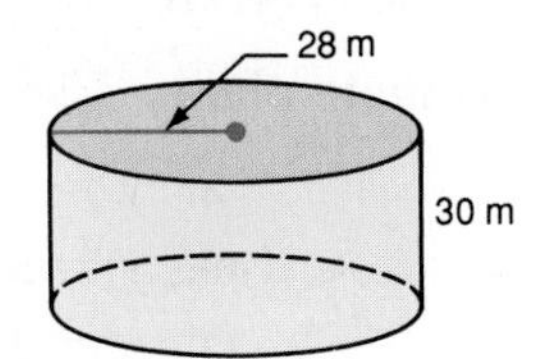

6.

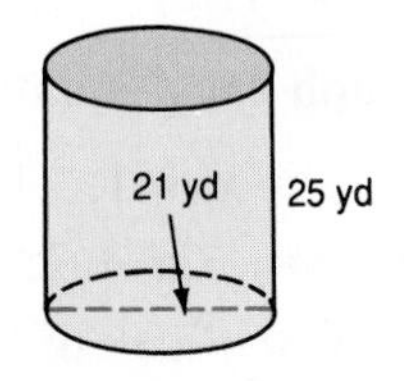

mixed practice (pages 520–521)

Find the surface area of each cylinder.

7. Use $\pi \approx 3.14$.

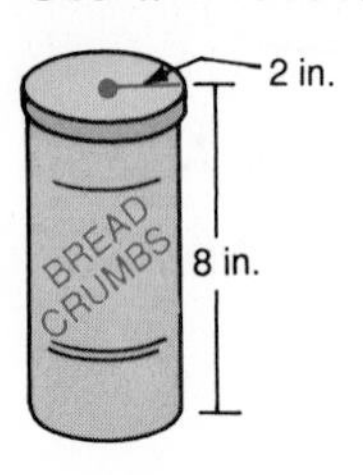

8. Use $\pi \approx 3.14$.

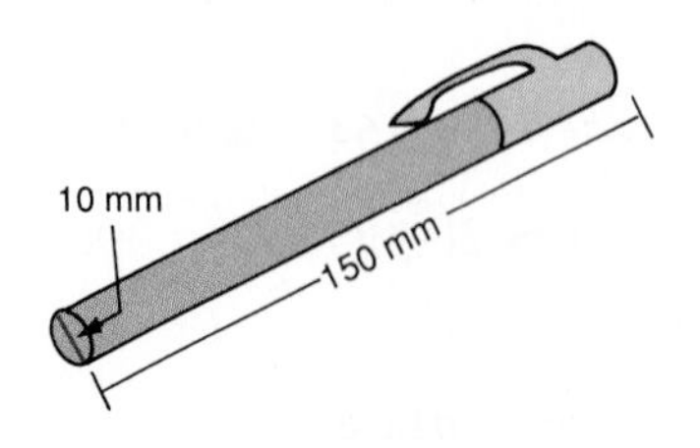

9. Use $\pi \approx \frac{22}{7}$.

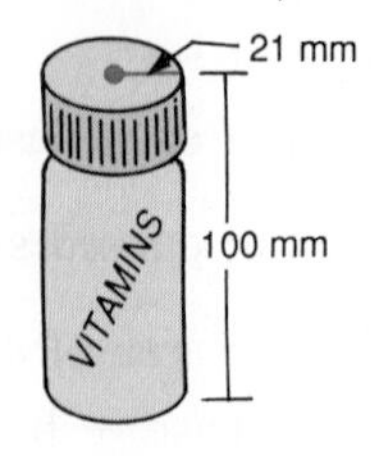

10. Use $\pi \approx \frac{22}{7}$.

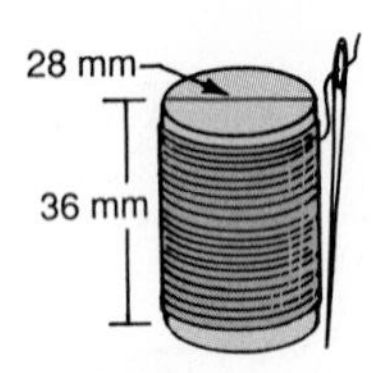

11. Use $\pi \approx \frac{22}{7}$.

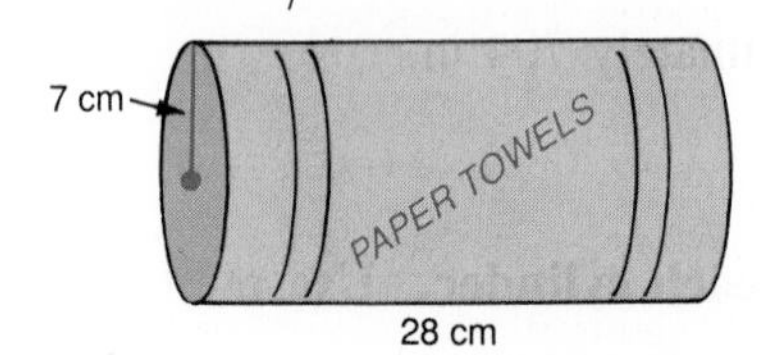

12. Use $\pi \approx 3.14$.

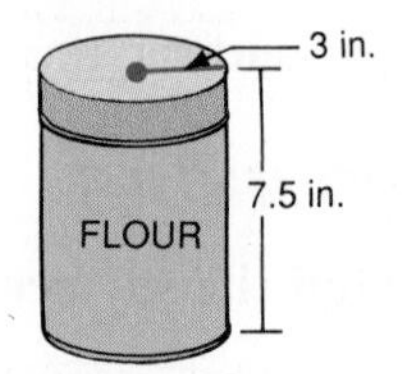

13. The radius of a quarter is 12 mm. The height is 1.5 mm. Use $\pi \approx 3.14$ to find the surface area.
14. The hotel in Detroit's Renaissance Center is shaped like a cylinder with radius 64 ft and height 712 ft. Use $\pi \approx 3.14$ to find the surface area.
15. A salt shaker and the matching pepper shaker are shaped like cylinders, each with diameter 28 mm and height 50 mm. Use $\pi \approx \frac{22}{7}$ to find the total surface area of the two shakers.

review exercises

Graph the given points on a coordinate plane.

1. $P(1, 3)$
2. $Q(^-2, ^-2)$
3. $R(3, ^-1)$
4. $S(^-1, 2)$
5. $T(0, ^-2)$
6. $V(4, 0)$

Graph each equation on a coordinate plane. Use the given values for x.

7. $y = x + 1$; $^-1, 3, 4$
8. $y = -x + 4$; $0, 2, 4$
9. $y = {}^-2x + 2$; $^-1, 2, 3$
10. $y = 3x - 2$; $^-1, 0, 2$
11. $y = -x - 6$; $^-3, 0, 2$
12. $y = 2x$; $^-1, 0, 3$

23-4 VOLUME OF PRISMS AND CYLINDERS

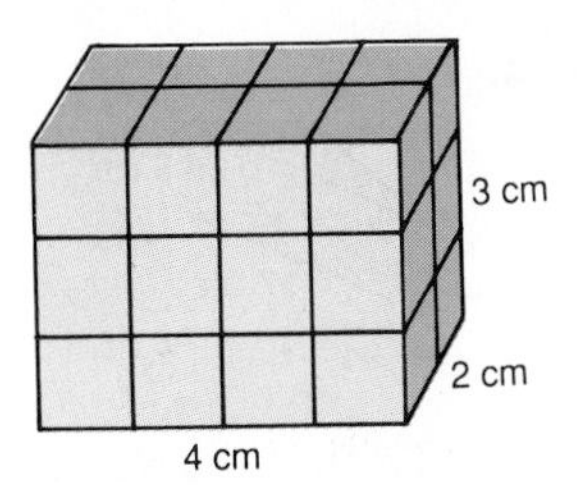

The **volume** (V) of a space figure is the amount of space inside it. Volume is measured in cubic units, such as cubic inches and cubic centimeters.

To find the volume of the prism at the right, you can count the number of cubic units. The volume is 24 cm^3.

You can also find the volume of a prism by finding the product of the area of its base (B) and height (h).

$$V = Bh$$

example 1

Find the volume of the rectangular prism at the right.

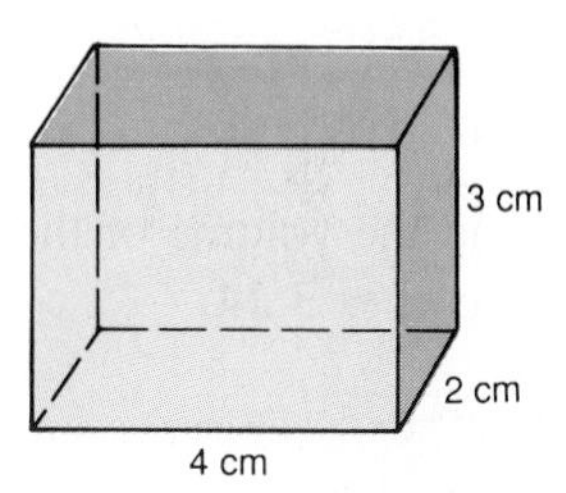

solution

Use the area formula for a rectangle to find B.

$A = lw$
$A = 4(2)$
$A = 8$

Use the volume formula for prisms. Replace B with 8.

$V = Bh$
$V = 8(3)$
$V = 24$

The volume is 24 cm^3.

your turn

Find the volume of each rectangular prism.

1.

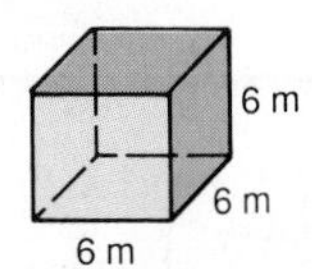

2.

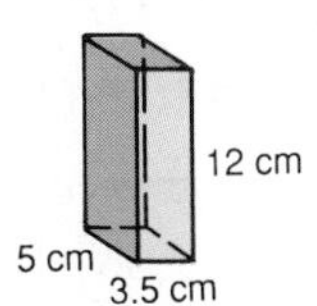

3.

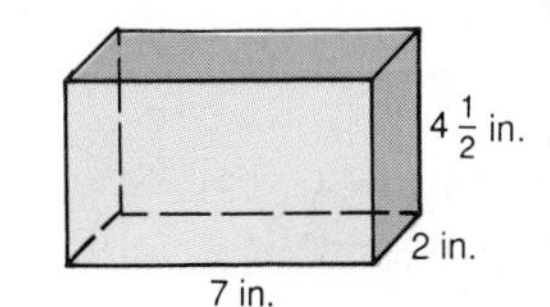

example 2

Find the volume of the triangular prism at the right.

37 ft
8 ft
35 ft
12 ft

solution

Use the area formula for a triangle to find B.

$A = \frac{1}{2}bh$ ← Here h is the height of the *triangle*.
$A = \frac{1}{2}(12)(35)$
$A = 210$

Use the volume formula for prisms. Replace B with 210.

$V = Bh$ ← Here h is the height of the *prism*.
$V = 210(8)$
$V = 1680$

The volume is 1680 ft^3.

your turn

Find the volume of each triangular prism.

4.

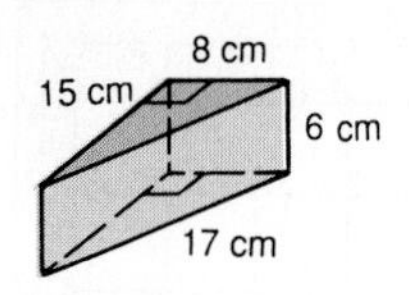

5.

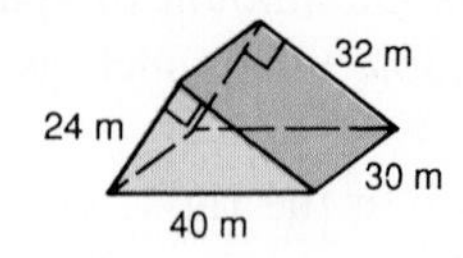

6.

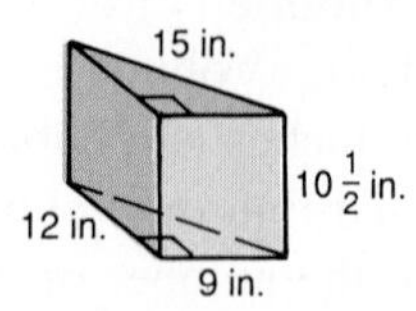

To find the volume of a cylinder, you also use the formula $V = Bh$. Because the base is a circle, you can replace B with πr^2.

$$V = Bh$$
$$V = \pi r^2 h$$

example 3

Find the volume of the cylinder at the right. Use $\pi \approx 3.14$.

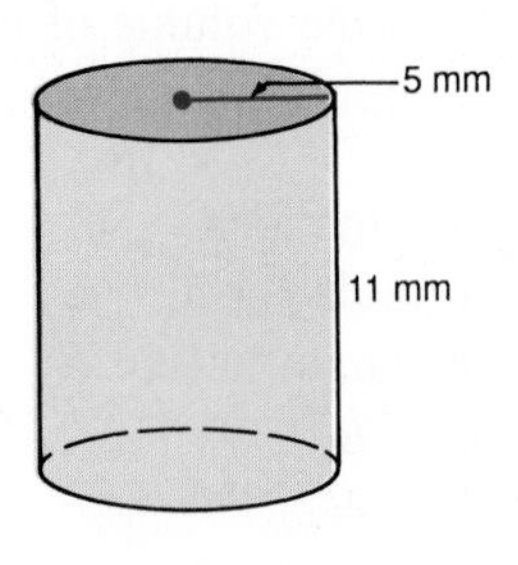

solution

$V = \pi r^2 h$

$V \approx 3.14(5^2)(11)$ ← **Replace r with 5 and h with 11.**

$V \approx 863.5$

The volume is approximately 863.5 mm^3.

your turn

Find the volume of each cylinder.

7. Use $\pi \approx \frac{22}{7}$.

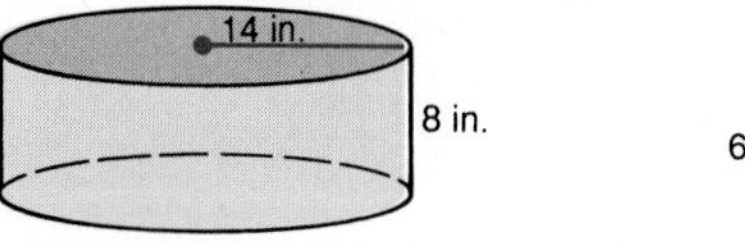

8. Use $\pi \approx 3.14$.

9. Use $\pi \approx \frac{22}{7}$.

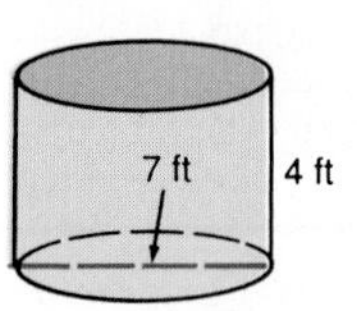

practice exercises

practice for example 1 (page 523)

Find the volume of each rectangular prism.

1.

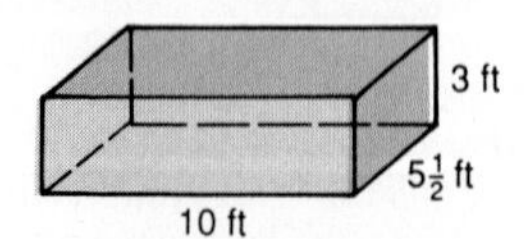

2.

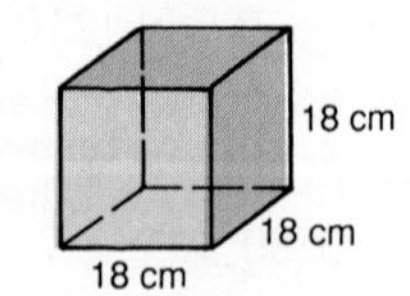

3.

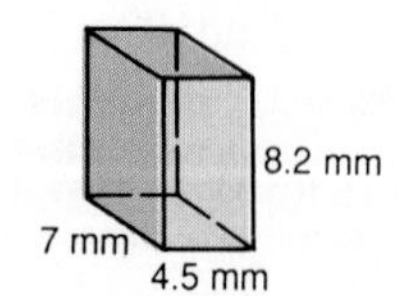

practice for example 2 (pages 523–524)

Find the volume of each triangular prism.

4. 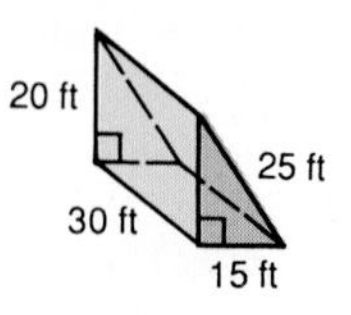

5. 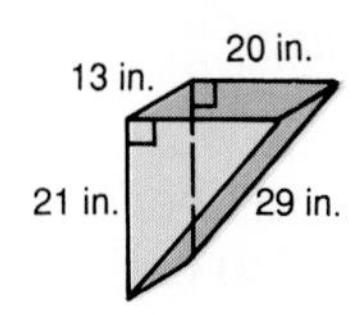

6. 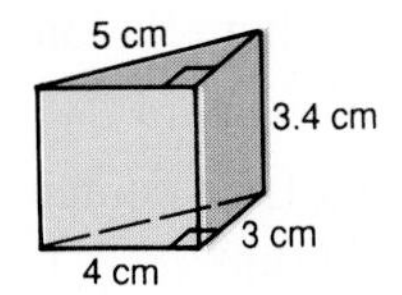

practice for example 3 (page 524)

Find the volume of each cylinder.

7. Use $\pi \approx 3.14$.

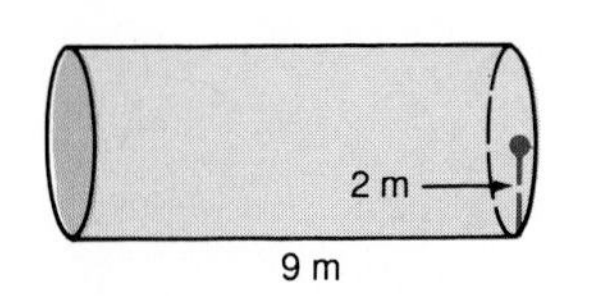

8. Use $\pi \approx \frac{22}{7}$.

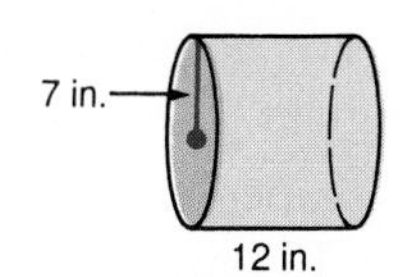

9. Use $\pi \approx 3.14$.

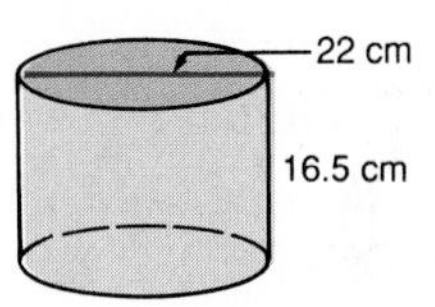

mixed practice (pages 523–524)

Find the volume of each space figure. When appropriate, use $\pi \approx 3.14$.

10. 

11. 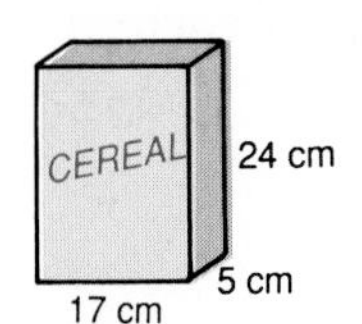

12. 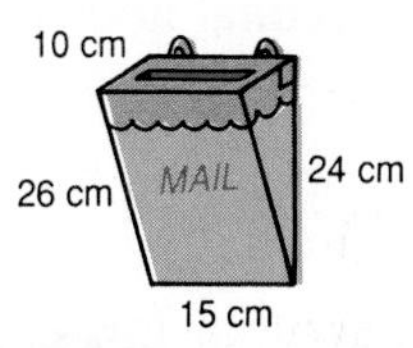

13. 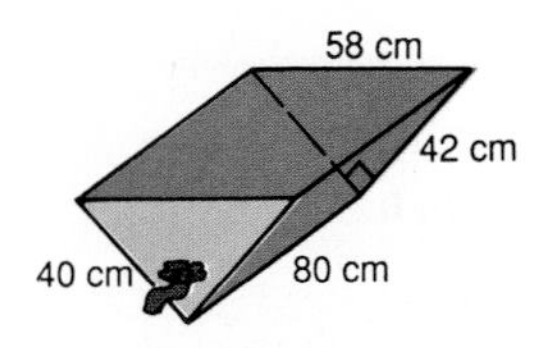

14. 

15. 

16. The length of a box of raisins is 10.5 cm, the width is 4 cm, and the height is 14 cm. What is the volume of the box of raisins?
17. A gas storage tank is shaped like a cylinder with radius 76 ft and height 140 ft. Use $\pi \approx \frac{22}{7}$ to find the volume of the tank.
18. A storage container for one contact lens is shaped like a cylinder. The diameter is 22 mm and the height is 8 mm. Find the total volume of the containers for a pair of contact lenses. Use $\pi \approx 3.14$.
19. The length of a shoe box is 12 in., the width is 4 in., and the height is 7 in. The box contains a pair of size $7\frac{1}{2}$ shoes. What is the volume of the shoe box?

The volume of a container is often called its *capacity*. For containers of liquids, the capacity is often measured in liters or milliliters. You can use the following relationships to find capacity.

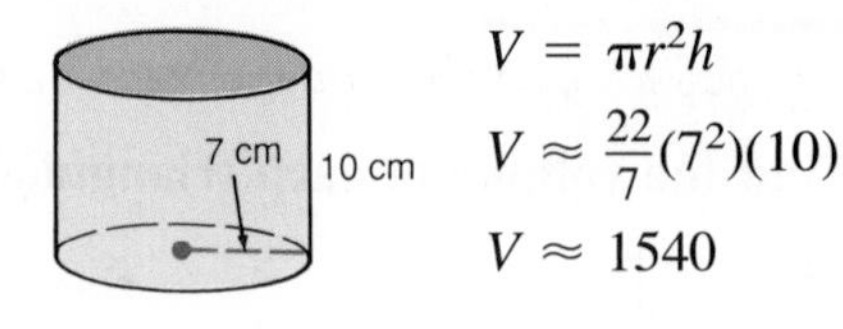

$1 \text{ cm}^3 = 1 \text{ mL} \qquad 1000 \text{ cm}^3 = 1 \text{ L}$

The volume of the container at the right is 1540 cm^3.
Because $1000 \text{ cm}^3 = 1 \text{ L}$ ($\div 1000$), the capacity is $1540 \text{ cm}^3 = 1.54 \text{ L}$ ($\div 1000$).

20. The radius of a can of juice is 5 cm. The height is 18 cm. Find the capacity of the can of juice in milliliters. Use $\pi \approx 3.14$.

21. The length of an aquarium is 92 cm. The width is 45 cm and the height is 45 cm. Find the capacity of the aquarium in liters.

22. The radius of a water glass is 7 cm. The height is 14 cm. How many milliliters of water will fill the glass? Use $\pi \approx \frac{22}{7}$.

23. The lower part of a carton of milk is shaped like a rectangular prism and is half full of milk. The length is 10 cm, the width is 6 cm, and the height is 6 cm. How many milliliters of milk are in the carton?

review exercises

Give the decimal and fraction equivalent of each percent.

1. 25% **2.** 80% **3.** 75% **4.** $37\frac{1}{2}\%$ **5.** $33\frac{1}{3}\%$ **6.** $16\frac{2}{3}\%$

mental math

You can check the reasonableness of a volume by estimating mentally. To check the volume of a rectangular prism, use the formula $V = lwh$, where l is the length of the base and w is the width of the base.

Volume $\underline{?}$ 1776.5 cm^3

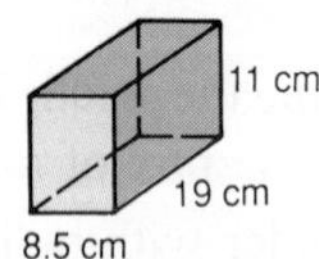

$V = lwh$
$V = 19(8.5)(11)$
Think: $V \approx 20\ (9)\ (10) = 1800$
So 1776.5 cm^3 is *reasonable*.

Estimate mentally to tell if the given volume is *reasonable* or *not reasonable*.

1.

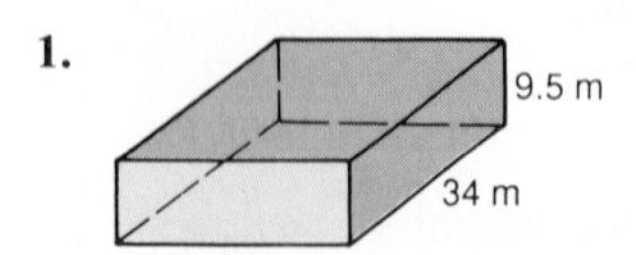

Volume $\underline{?}$ 872.1 m^3

2.

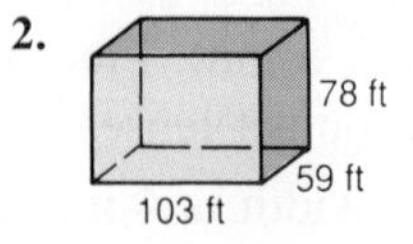

Volume $\underline{?}$ $47{,}400.6 \text{ ft}^3$

3.

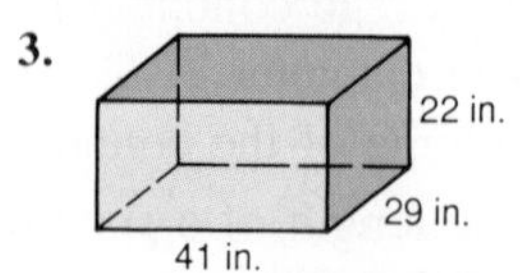

Volume $\underline{?}$ $26{,}158 \text{ in.}^3$

23-5 VOLUME OF RECTANGULAR PYRAMIDS AND CONES

Suppose you empty a pyramid filled with water into a prism with the same base and height. You will find that the prism is only one-third full.

Now suppose you empty a cone filled with water into a cylinder with the same base and height. You will find that the cylinder is only one-third full.

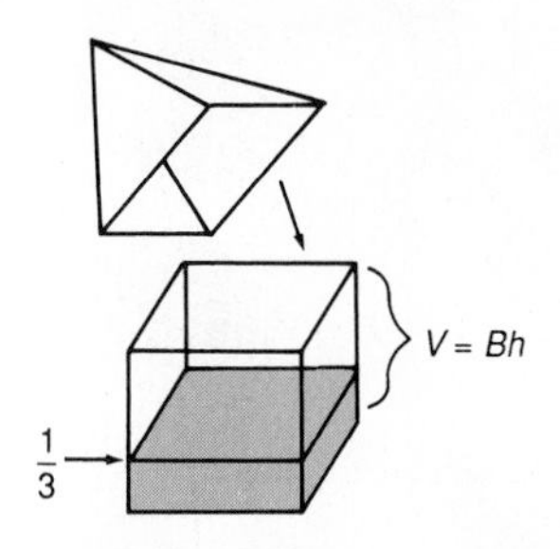

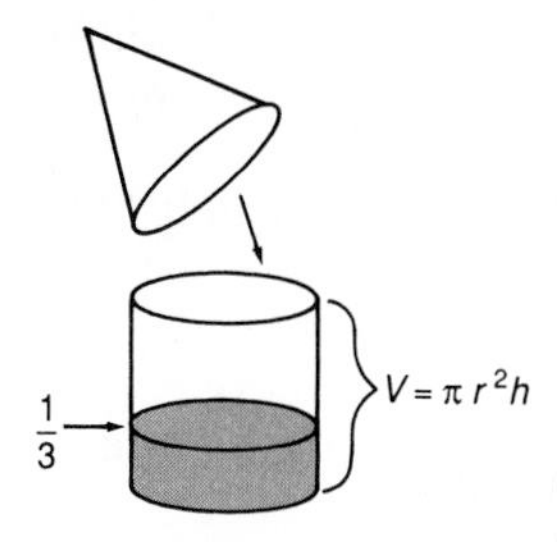

To find the volume of a rectangular pyramid or a cone, you use the following formulas.

rectangular pyramid

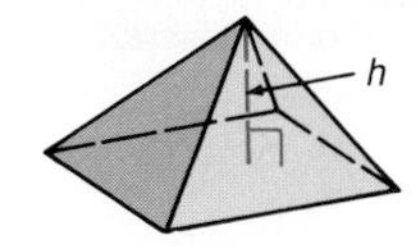

$$V = \frac{1}{3}Bh$$

cone

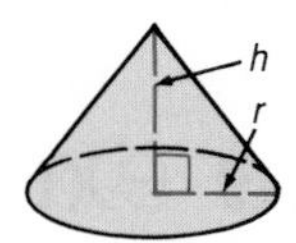

$$V = \frac{1}{3}\pi r^2 h$$

example 1

Find the volume of the rectangular pyramid at the right.

solution

$V = \frac{1}{3}Bh$

$V = \frac{1}{3}(8)(9)(6)$ ⟵ **Replace B with (8)(9), and h with 6.**

$V = 144$

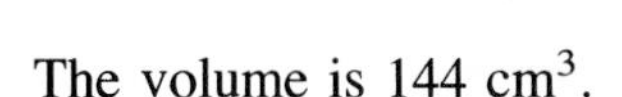

The volume is 144 cm^3.

your turn

Find the volume of each rectangular pyramid.

1.

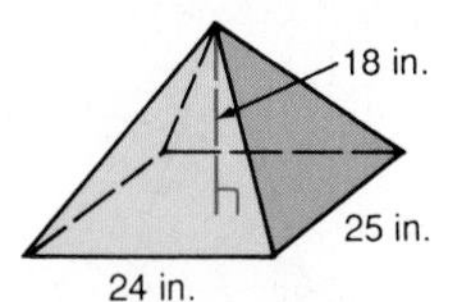

2.

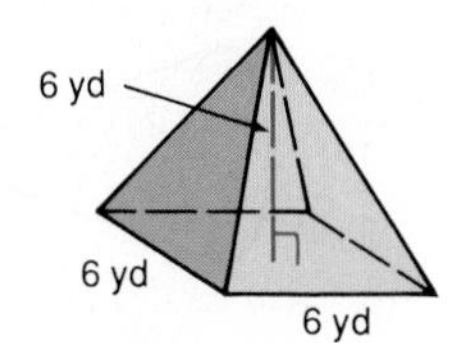

3.

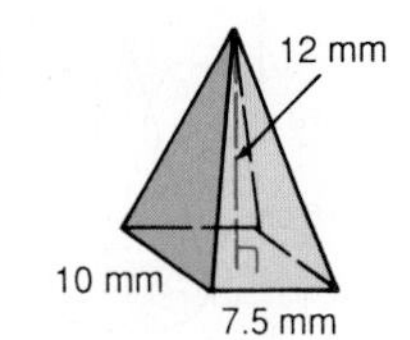

example 2

Find the volume of the cone at the right. Use $\pi \approx \frac{22}{7}$.

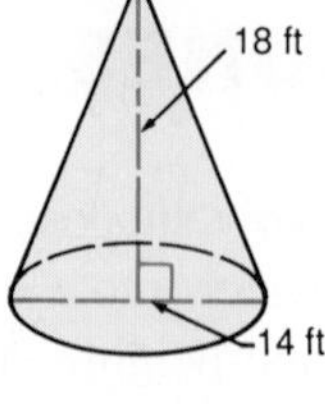

solution

$V = \frac{1}{3}\pi r^2 h$

$V \approx \left(\frac{1}{3}\right)\left(\frac{22}{7}\right)(7^2)(18)$ ← **Replace r with $\frac{1}{2}(14)$, or 7, and h with 18.**

$V \approx 924$

The volume is approximately 924 ft^3.

your turn

Find the volume of each cone.

4. Use $\pi \approx 3.14$.

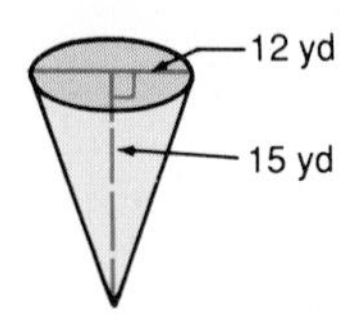

5. Use $\pi \approx \frac{22}{7}$.

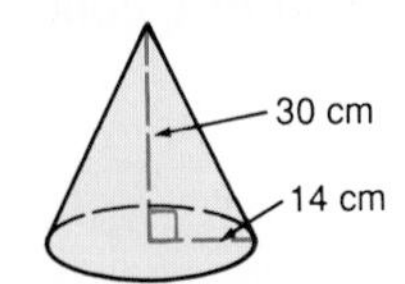

practice exercises

practice for example 1 (page 527)

Find the volume of each rectangular pyramid.

1.

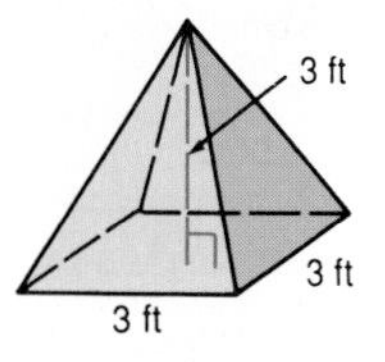

2.

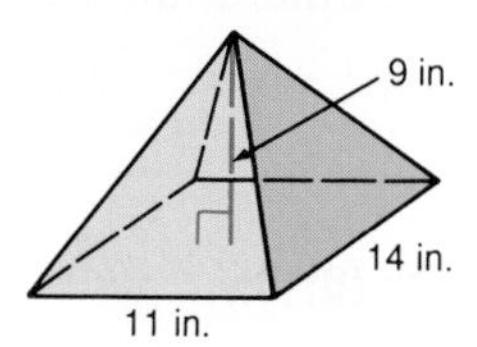

3.

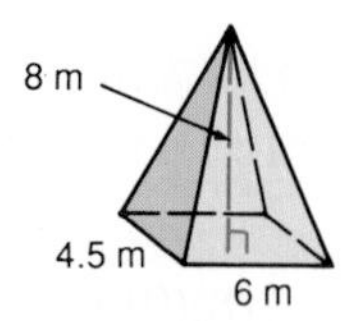

practice for example 2 (page 528)

Find the volume of each cone.

4. Use $\pi \approx 3.14$.

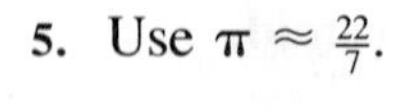

5. Use $\pi \approx \frac{22}{7}$.

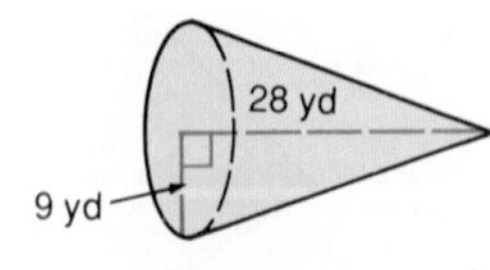

6. Use $\pi \approx \frac{22}{7}$.

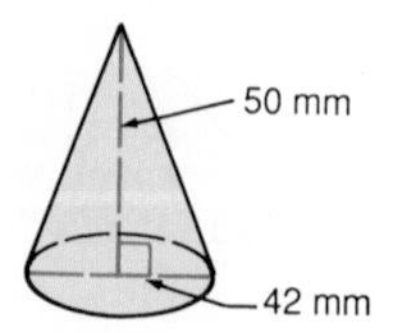

mixed practice *(pages 527–528)*

Find the volume of each space figure. When appropriate, use $\pi \approx 3.14$.

7.

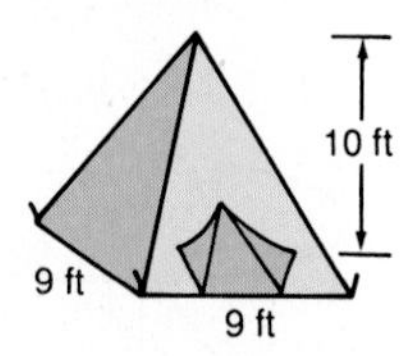

8.

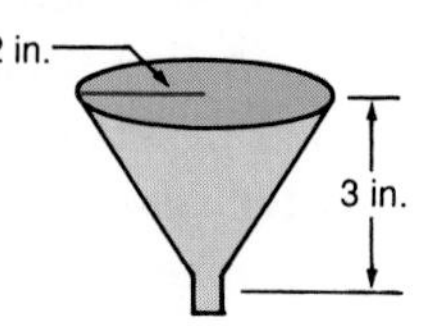

9.

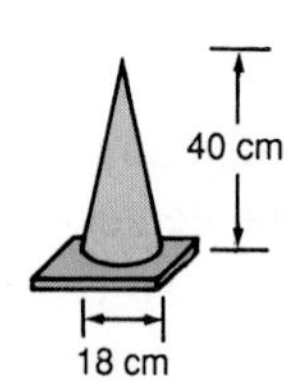

10. A pile of sand is in the shape of a cone. Its radius is 4 in. and its height is 3 in. Find the volume of the pile of sand. Use $\pi \approx 3.14$.

11. A glass paperweight is in the shape of a rectangular pyramid. Its volume is 90 cm^3. The length of its base is 9 cm. The width of its base is 5 cm. What is the height of the paperweight?

12. A paper cup is in the shape of a cone. Its diameter is 6 cm and its height is 7 cm. Find the capacity of the paper cup in milliliters. Use $\pi \approx \frac{22}{7}$. Recall that 1 cm^3 = 1 mL.

review exercises

Solve and graph each inequality.

1. $b - 8 > {}^{-}6$ **2.** $r + 3 \leq 6$ **3.** $4y \geq {}^{-}4$ **4.** $\frac{c}{2} < {}^{-}1$

calculator corner

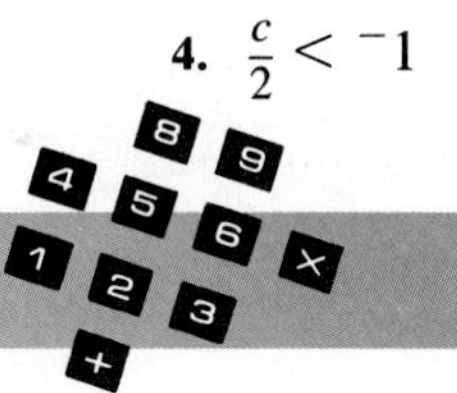

Some calculators have a [π] key and a [x²] key. You can use these keys to find the volume of a space figure.

example Find the volume of the space figure at the right.

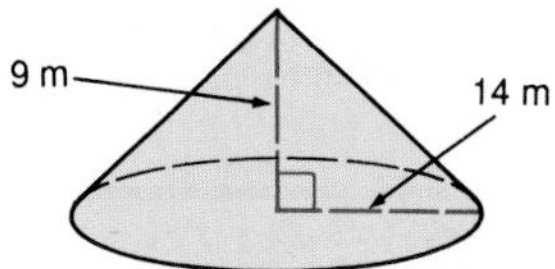

solution $V = \frac{1}{3}\pi r^2 h$

Enter 1 ÷ 3 × [π] × 14 [x²] × 9 =.

The result is 1847.25648.

Rounded to the nearest hundredth, the volume is 1847.26 m^3.

Use a calculator to find the volume of each space figure.

1.

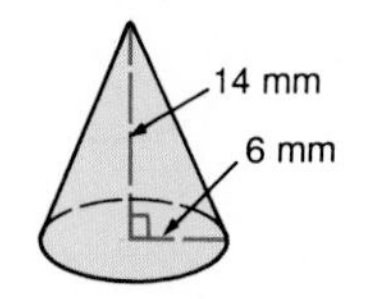

2.

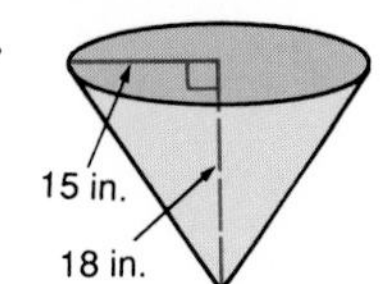

3.

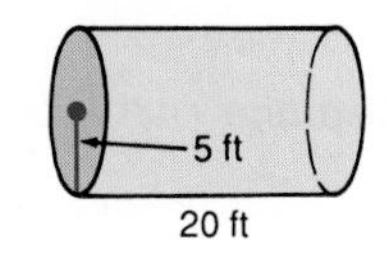

4.

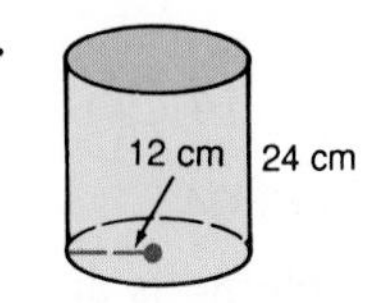

23-6 SPHERES

A line segment that joins the center of a sphere to any point on the sphere is a **radius** (r). A line segment that joins any two points on a sphere and passes through the center of the sphere is a **diameter** (d).

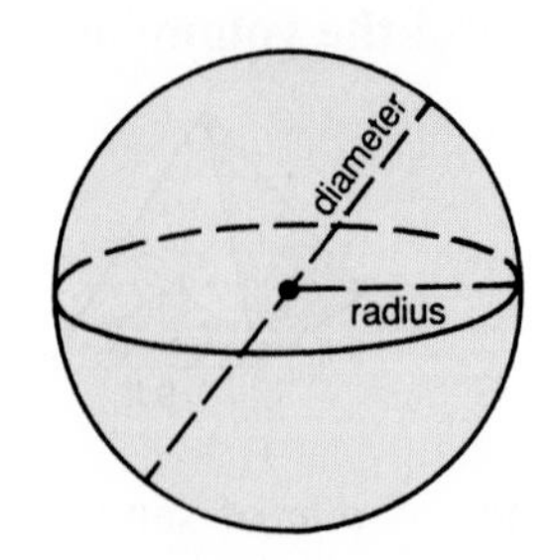

To find the surface area of any sphere, you use the following formula.

$$A = 4\pi r^2$$

example 1

Find the surface area of the sphere at the right. Use $\pi \approx 3.14$.

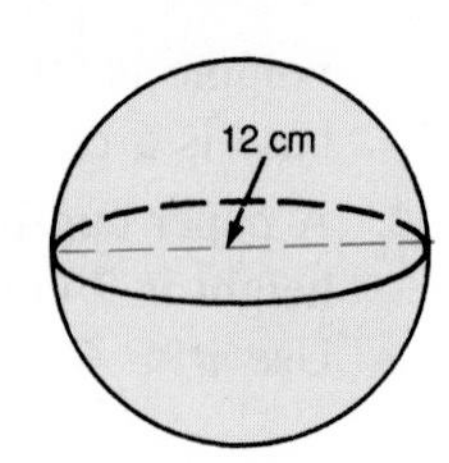

solution

$A = 4\pi r^2$

$A \approx 4(3.14)(6^2)$ ← **Replace r with $\frac{1}{2}(12)$, or 6.**

$A \approx 452.16$

The surface area is approximately 452.16 cm^2.

your turn

Find the surface area of each sphere. Use $\pi \approx 3.14$.

1.

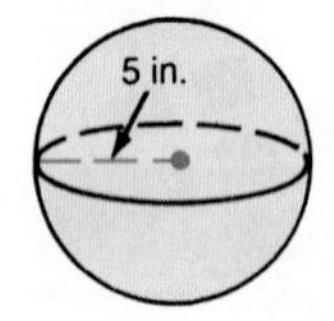

2.

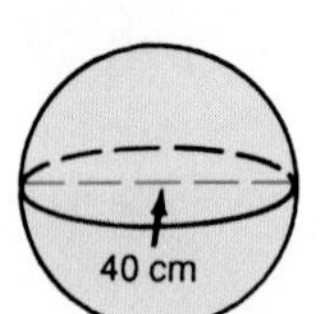

3.

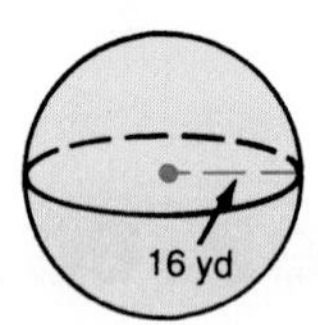

To find the volume of any sphere, you use the following formula.

$$V = \frac{4}{3}\pi r^3$$

example 2

Find the volume of the sphere at the right. Use $\pi \approx 3.14$.

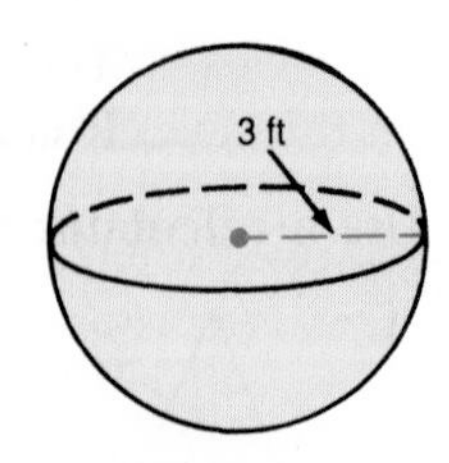

solution

$V = \frac{4}{3}\pi r^3$

$V \approx \frac{4}{3}(3.14)(3^3)$ ← **Replace r with 3.**

$V \approx 113.04$

The volume is approximately 113.04 ft^3.

your turn

Find the volume of each sphere. Use $\pi \approx 3.14$.

4. 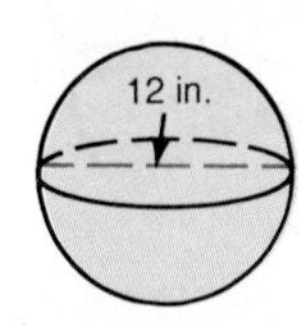

5. 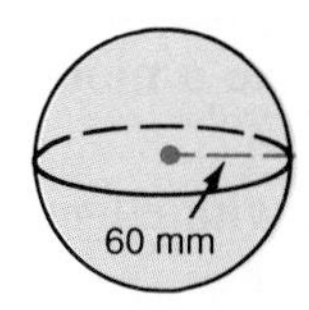

6. 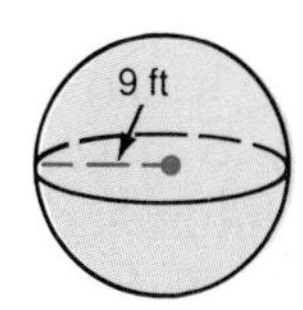

practice exercises

practice for example 1 (page 530)

Find the surface area of each sphere. Use $\pi \approx 3.14$.

1. 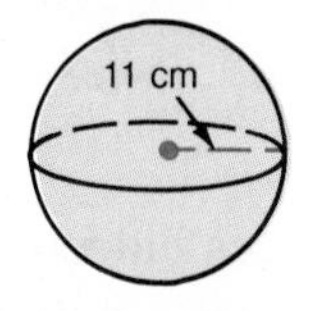

2. 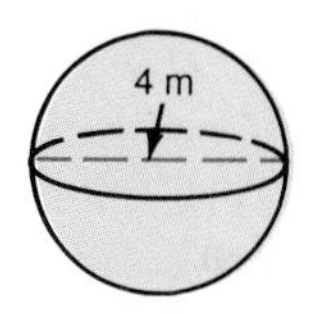

3. 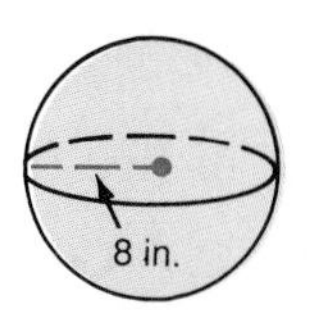

practice for example 2 (pages 530–531)

Find the volume of each sphere. Use $\pi \approx 3.14$.

4. 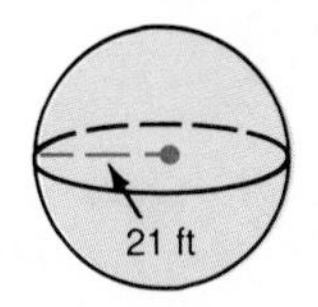

5. 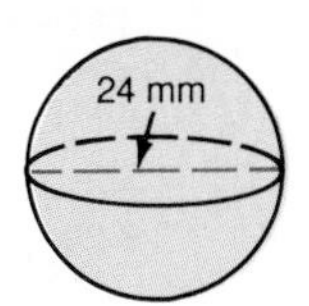

6. 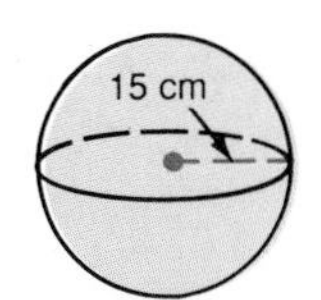

mixed practice (pages 530–531)

Find the surface area and volume of each sphere. Use $\pi \approx 3.14$.

7. 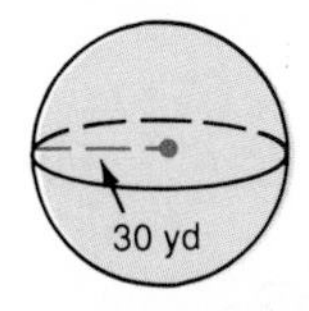

8. 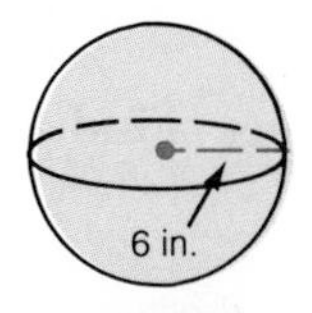

9. 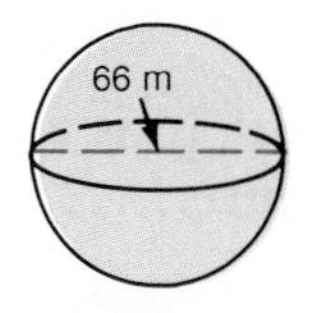

10. The radius of a baseball is 4 cm. Find the surface area. Use $\pi \approx 3.14$.

11. The diameter of a marble is 18 mm. Find the volume. Use $\pi \approx 3.14$.

review exercises

Estimate each decimal as a simple fraction or mixed number.

1. 0.19 **2.** 0.336 **3.** 0.627 **4.** 0.48 **5.** 5.872 **6.** 3.17

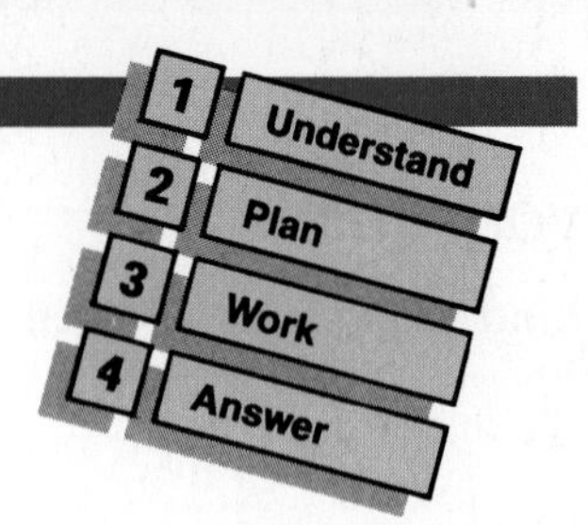

23-7 MAKING A MODEL

Sometimes it may be difficult to visualize a problem that involves a space figure. It may be helpful to make a **model** of the figure. One way to make a model is to use *centimeter cubes*. A **centimeter cube** is a cube with edges that each measure 1 cm. The area of each face of a centimeter cube is 1 cm^2.

example

In the space figure at the right, each is a centimeter cube. Find the surface area of the space figure.

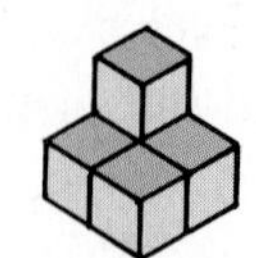

solution

The figure is composed of five centimeter cubes.

Count the faces visible at the "front" of the figure.

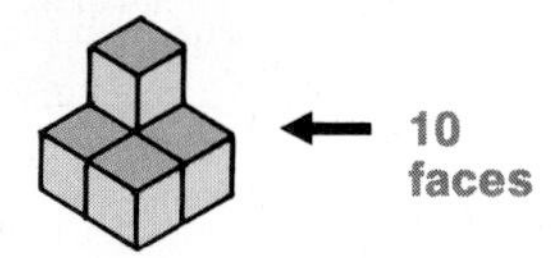

Count the faces hidden from sight at the "back" of the figure.

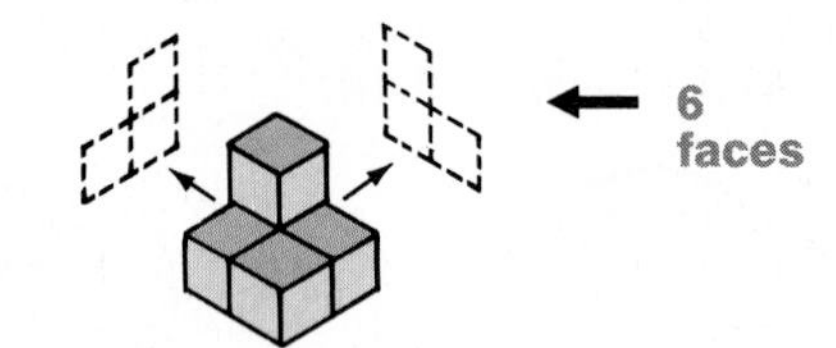

Count the faces hidden from sight at the "bottom" of the figure.

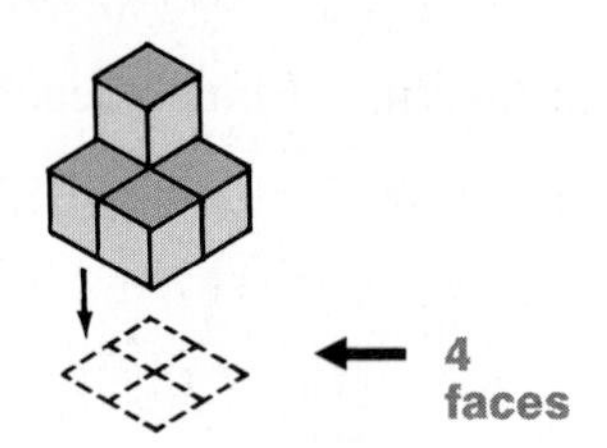

Find the total number of faces: $10 + 6 + 4 = 20$

The surface area of the space figure is 20 cm^2.

exercises

In Exercises 1–8, each is a centimeter cube. Find the surface area of each space figure.

1. 

2. 

3. 

4. 

5. 

6. 

7. 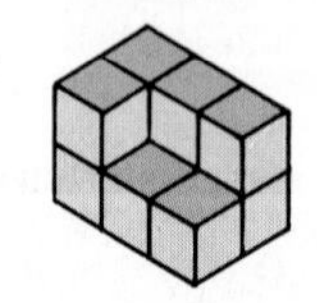

8. 

Find the surface area of each space figure.

9. 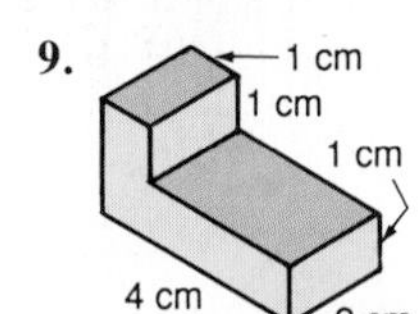

10. 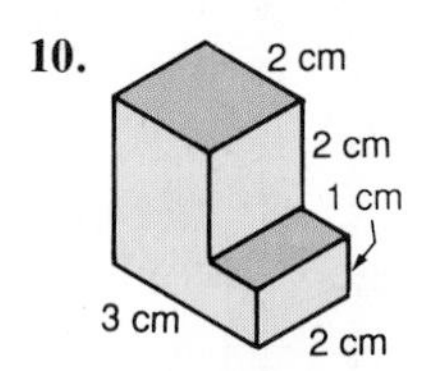

11. 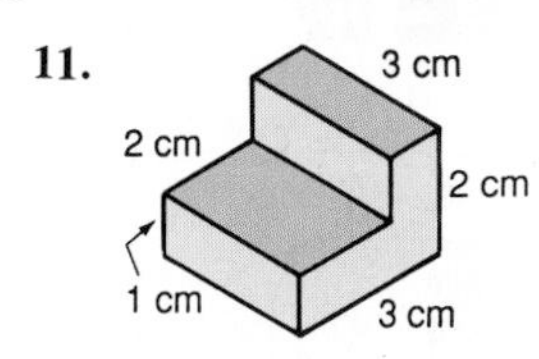

12. 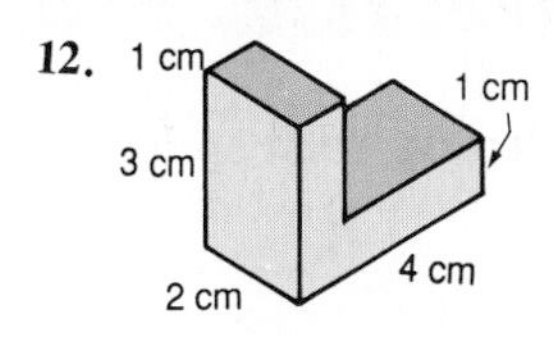

13. The figures below show all possible space figures that can be formed by four centimeter cubes. Which of these space figures has the least surface area?

a. 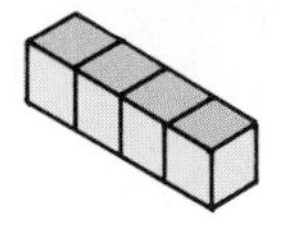

b. 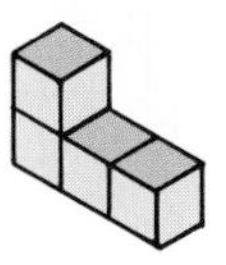

c. 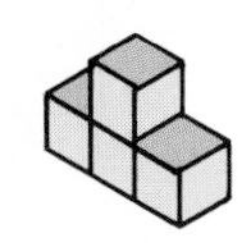

d. 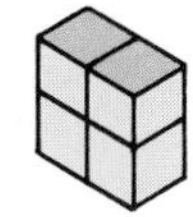

e. 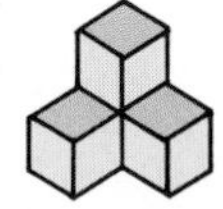

14. A space figure is to be formed by arranging eight centimeter cubes. Make a drawing of the arrangement that you think would have the least surface area.

A large wooden cube is painted red. It is then divided into twenty-seven small cubes as shown at the right.

15. How many small cubes have exactly three faces painted red?
16. How many small cubes have exactly two faces painted red?
17. How many small cubes have exactly one face painted red?
18. How many small cubes have no faces painted red?

review exercises

1. To fill a piñata for 10 people, Mr. Oliva uses 20 toys. How many toys does he use to fill a piñata for 30 people?
2. Kelly Swanson buys a sweater for \$15. She has \$25 left. How much money did she have before buying the sweater?
3. In how many different ways, without regard to order, can you pay \$50 with bills of \$50, \$20, and \$10?
4. Sam weighs 80 lb more than his dog. If their total weight is 140 lb, how much does each weigh?

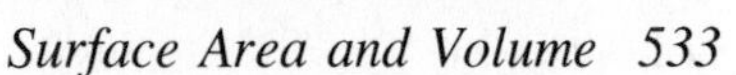

SKILL REVIEW

Identify each space figure.

1. 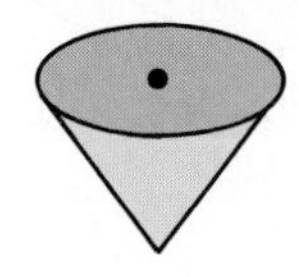

2. 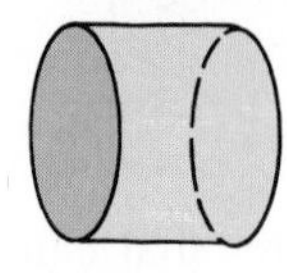

3. 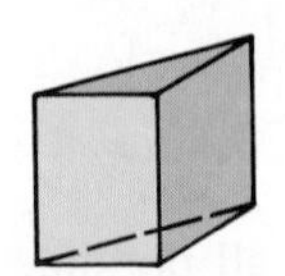

4. 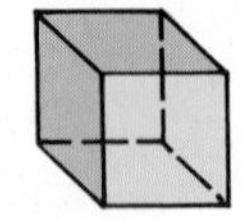

23-1

Find the surface area of each prism.

5. 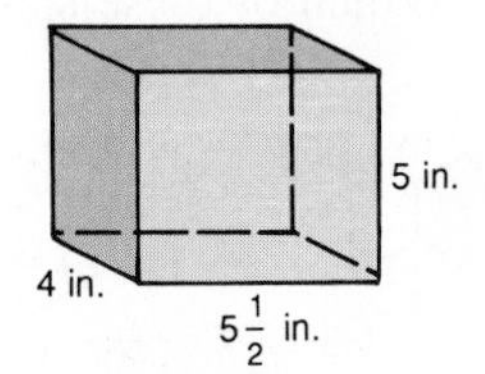

6. 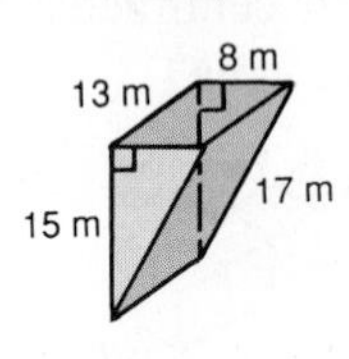

23-2

Find the surface area of each cylinder. Use $\pi \approx 3.14$.

7. 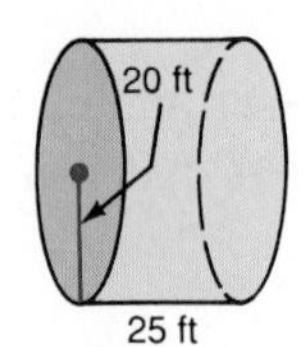

8. 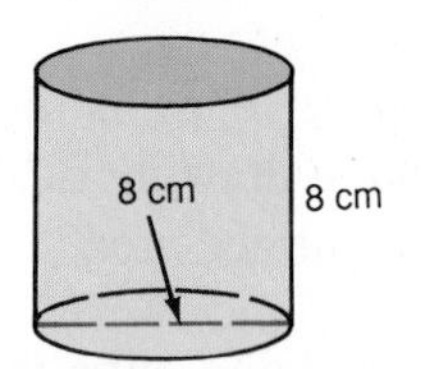

23-3

Find the volume of each space figure. When appropriate, use $\pi \approx \frac{22}{7}$.

9. 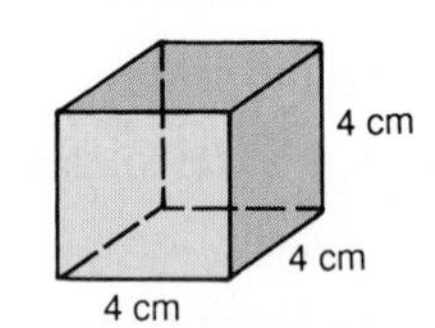

10. 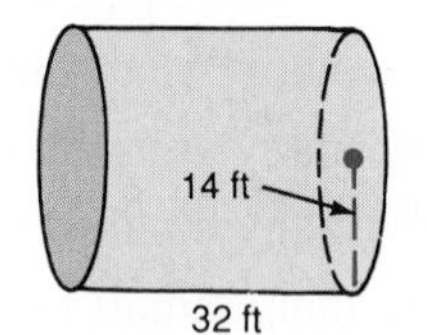

23-4

Find the volume of each space figure. When appropriate, use $\pi \approx \frac{22}{7}$.

11. 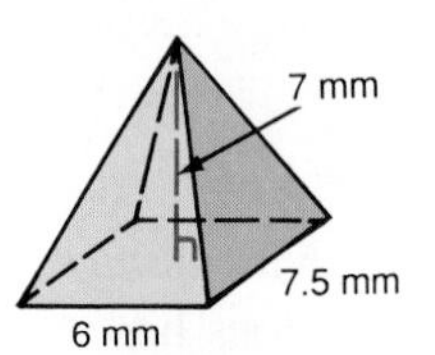

12. 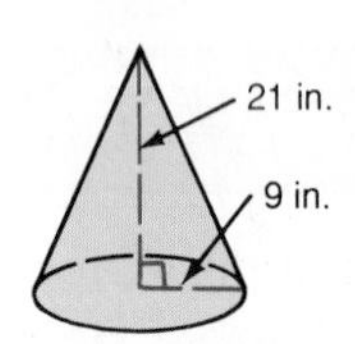

23-5

Find the surface area and volume of each sphere. Use $\pi \approx 3.14$.

13. 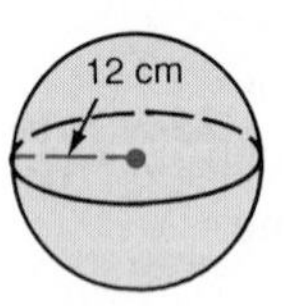

14. 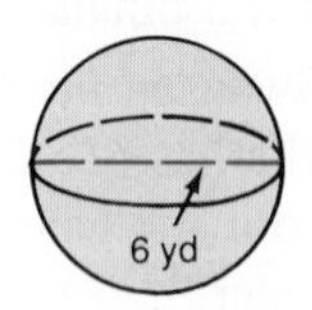

23-6

Each cube is a centimeter cube. Find the surface area of each space figure.

15. 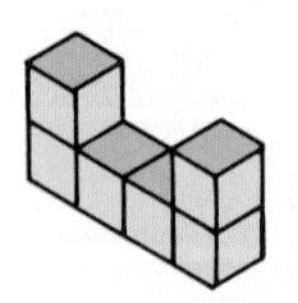

16. 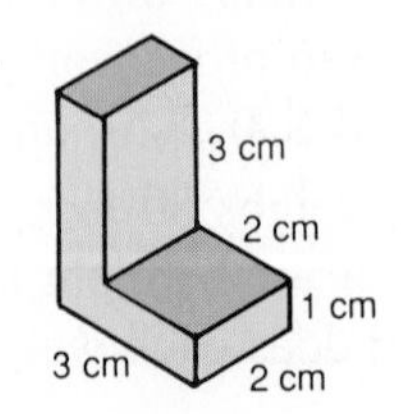

23-7

APPLICATION

23-8 COMPOSITE SPACE FIGURES

A **composite space figure** is a combination of polyhedrons, cylinders, cones, or spheres. To find the volume of a composite space figure, first divide it into space figures whose volumes you know how to find. Then find the sum of the volumes.

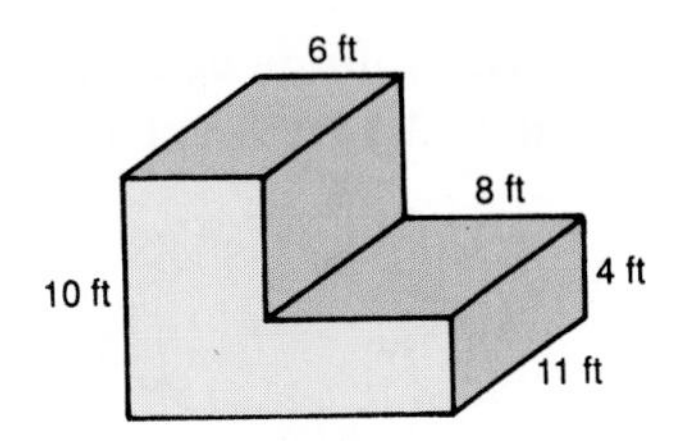

example

Find the volume of the composite space figure above.

solution

Divide the space figure into two rectangular prisms. Label them I and II.

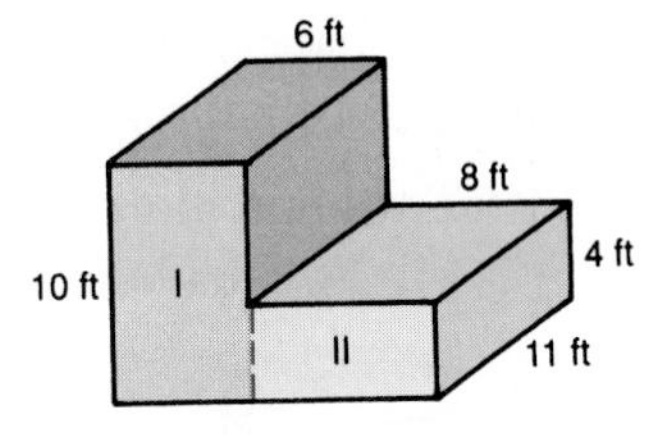

$V(\text{I}) = Bh$
$= 6(11)(10)$
$= 660$

$V(\text{II}) = Bh$
$= 8(11)(4)$
$= 352$

$V(\text{I}) + V(\text{II}) = 660 + 352 = 1012$

The volume is 1012 ft^3.

exercises

Find the volume of each composite space figure. A calculator may be helpful.

1.
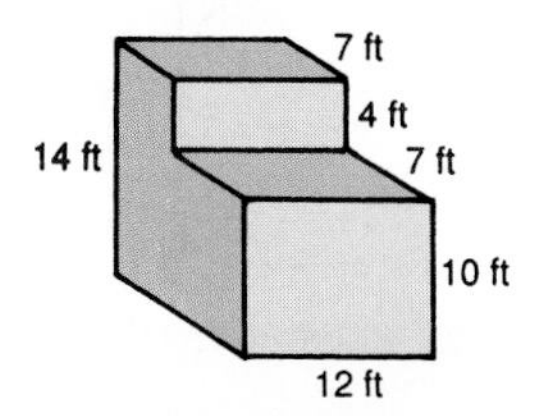

2.
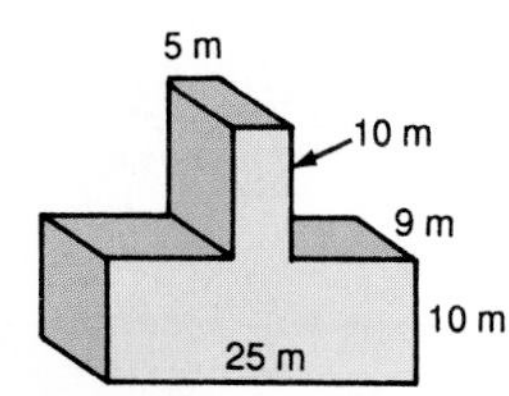

3.
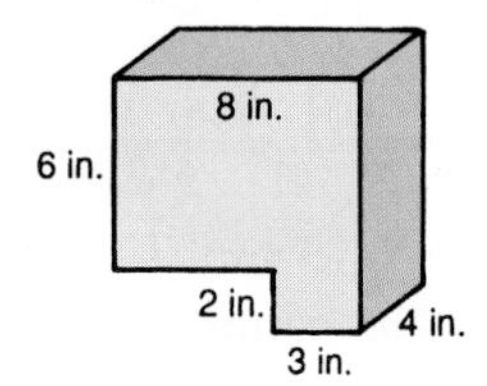

Sometimes a cylinder, pyramid, cone, or sphere is one part of a composite space figure.

Find the volume of each composite space figure. When appropriate, use $\pi \approx 3.14$. A calculator may be helpful.

4.
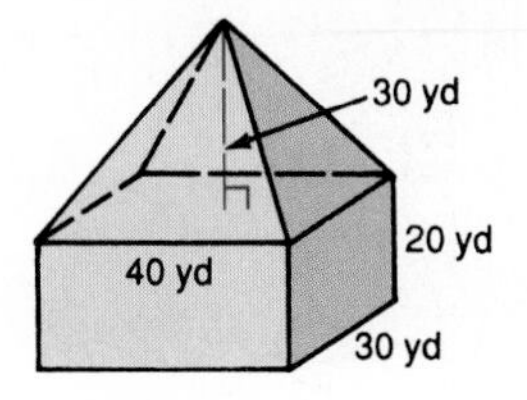

5.
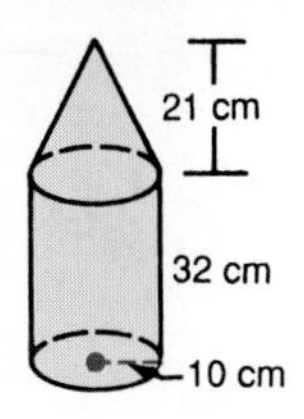

6.
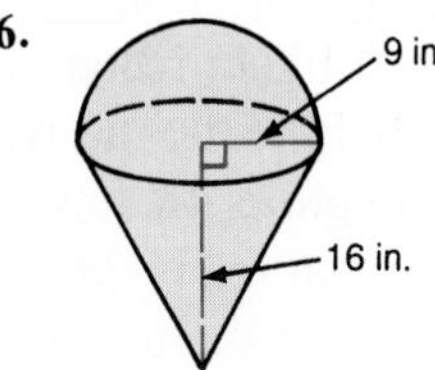

Sometimes composite space figures are composed of three or more space figures.

Find the volume of each composite space figure. When appropriate, use $\pi \approx 3.14$. A calculator may be helpful.

7.

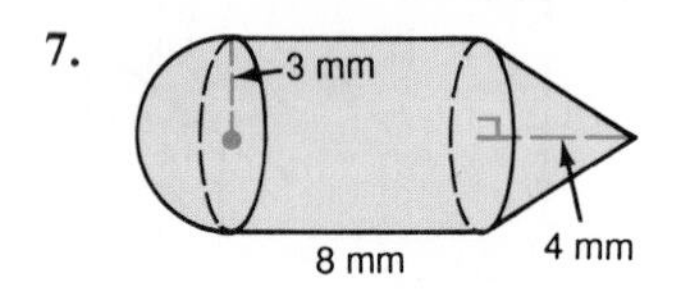

8.

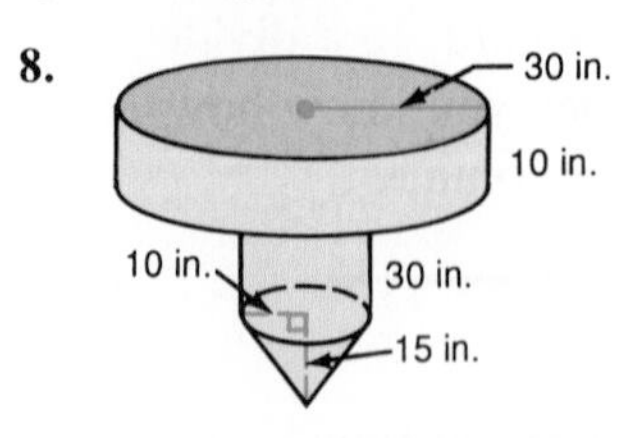

9.

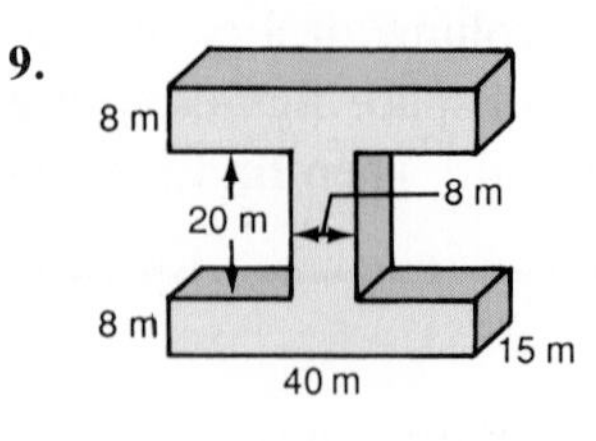

Sometimes you need to find the volume of only a *part* of a composite space figure. Usually you can find this volume by subtracting the volumes of known space figures.

Find the volume of each shaded region. When appropriate, use $\pi \approx 3.14$. A calculator may be helpful.

10.

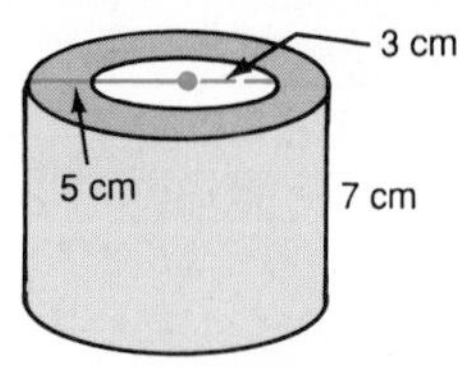

11.

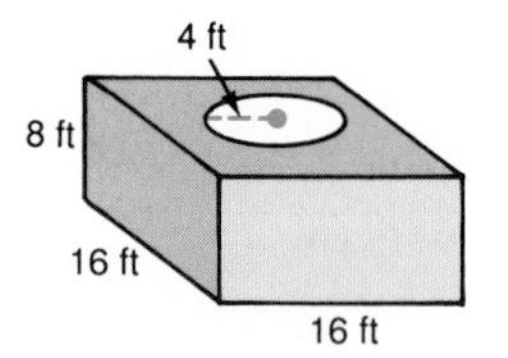

12.

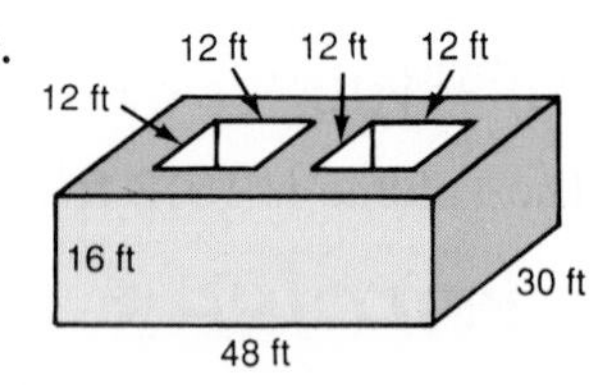

13.

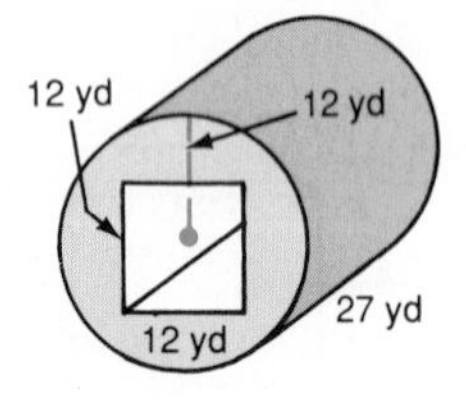

14.

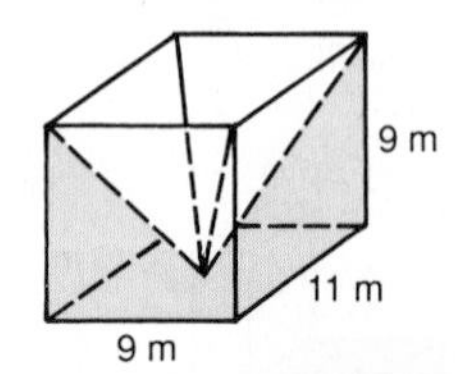

15.

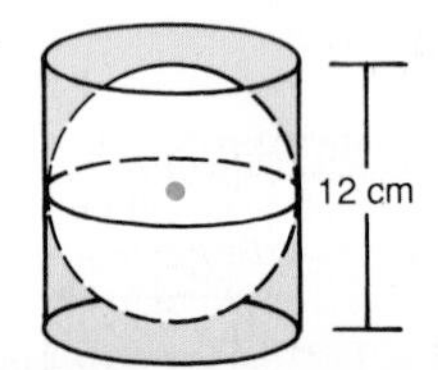

16. The bottom part of an observatory is shaped like a cylinder with diameter 12 m and height 5 m. The top part is shaped like half of a sphere. The radius is 6 m. What is the volume of the observatory? Use $\pi \approx 3.14$.

17. The top part of a garage is shaped like a triangular prism. The bottom part is shaped like a rectangular prism. The length is 15 ft, the width is 12 ft, and the height is 10 ft. The total volume of the garage is 2340 ft^3. What is the volume of the top part?

A draftsperson often makes a drawing of an object that is to be manufactured. The drawing usually includes the front, top, and side views of the object. A solid line indicates an edge of the object. The top, front, and side views of the object at the right are shown below.

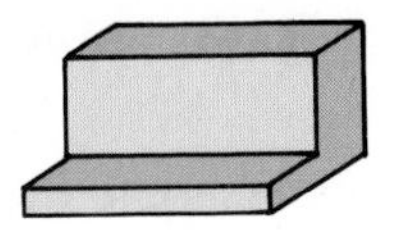

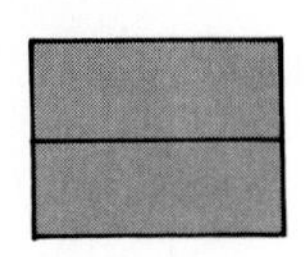

TOP VIEW

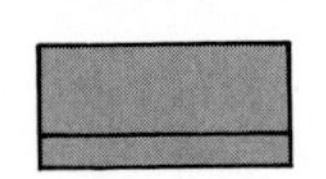

FRONT VIEW

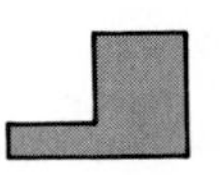

SIDE VIEW

Match each object with the correct view.

18. top view of

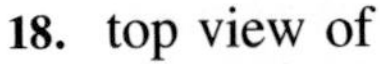

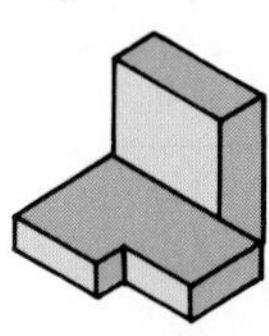

a.

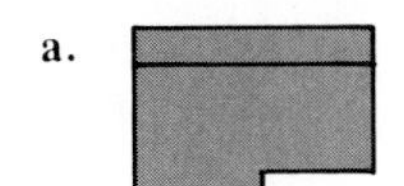

b.

c.

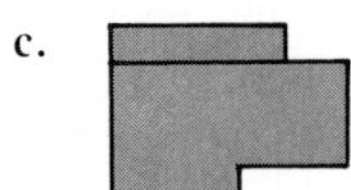

19. front view of

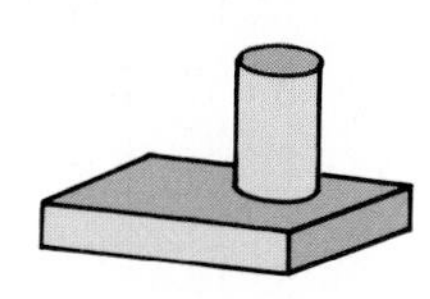

a.

b.

c.

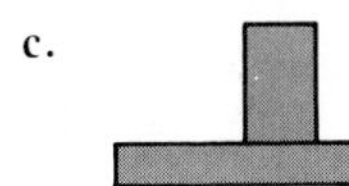

20. side view of

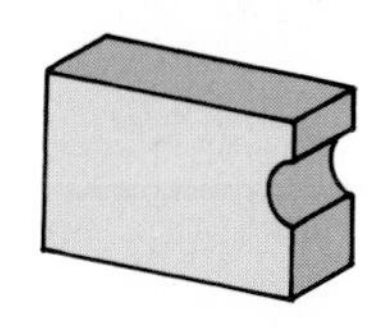

a.

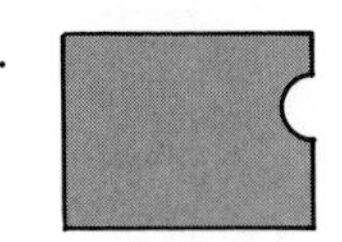

b.

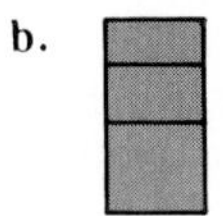

c.

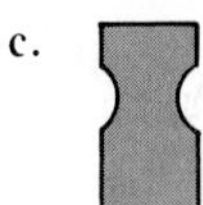

21. Sketch the top, front, and side views of the object at the right.

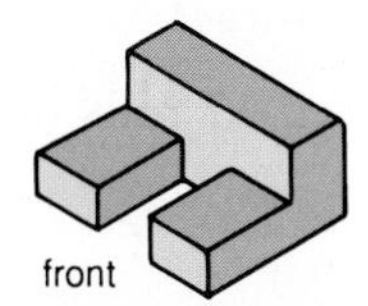

In the composite space figure at the right, each cube is a centimeter cube.

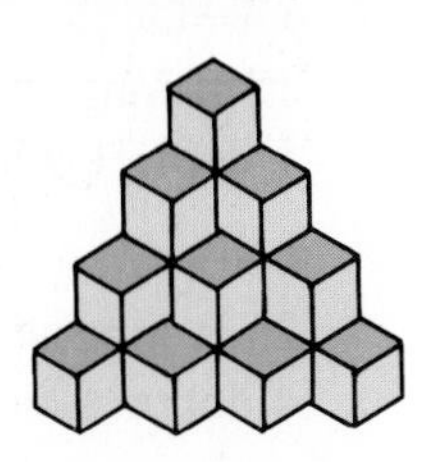

a. How many cubes are not visible?

b. What is the volume of the composite space figure?

c. What is the surface area of the composite space figure?

APPLICATION

23-9 PAPERING AND PAINTING WALLS

Wallpaper is sold in rolls that usually contain about 36 ft^2 of paper each. However, because you need to match patterns and to leave openings for doors and windows, you actually use only about 30 ft^2 from each roll. To estimate the amount of wallpaper that you need, you can use a chart like the one shown below.

example

Sheldon and Samantha plan to wallpaper their living room. The room is 14 ft long, 10 ft wide, and 8 ft high. It has 4 windows and 2 doors. How many rolls of wallpaper do they need?

Perimeter of Room (ft)	Number of Rolls of Wallpaper		
	Height of Ceiling		
	8 ft	9 ft	10 ft
38	10	12	12
40	10	12	12
42	12	12	14
44	12	12	14
46	12	14	14
48	14	14	16
50	14	14	16
52	14	14	16
54	14	16	18
56	14	16	18
58	16	16	18

solution

- Find the perimeter of the room.

  Perimeter $= 2 \times 14 + 2 \times 10$
  $= 48$ (ft)

- Read down the *Perimeter* column to find 48.
- Read across to the *8 ft* column. The number is 14 (rolls).
- Subtract one roll for every two openings (doors and windows).

  $4 + 2 = 6$ ← **There are 4 windows and 2 doors.**
  $6 \div 2 = 3$ ← **Divide by 2.**
  $14 - 3 = 11$ ← **Subtract from 14.**

Sheldon and Samantha need 11 rolls of wallpaper.

exercises

Use the chart above. Find how many rolls of wallpaper are needed to wallpaper the given room.

1. kitchen:
   12 ft long, 8 ft wide, 8 ft high
   1 door, 1 window
2. bedroom:
   15 ft long, 10 ft wide, 8 ft high
   2 doors, 2 windows
3. hallway:
   25 ft long, 4 ft wide, 9 ft high
   4 doors
4. dining room:
   15 ft long, 13 ft wide, 10 ft high
   2 doors, 2 windows

When you are painting walls, you can estimate that one gallon of paint will cover about 440 ft^2. To determine how many gallons of paint you must buy, you find the total surface area of the walls and divide by 440. Then you round this quotient *up* to the next greater whole number.

Find the number of gallons of paint you should buy to paint each room.

5. family room:
   12 ft long, 12 ft wide, 8 ft high
6. living room:
   18 ft long, 15 ft wide, 8 ft high
7. dining room:
   15 ft long, 11 ft wide, 9 ft high
8. kitchen:
   10 ft long, 9 ft wide, 9 ft high
9. Cindy wants to paint the ceiling of her combined living and dining room. The ceiling is 25 ft long and 13 ft wide. Will she be able to paint the ceiling with just one gallon of paint? Explain.
10. Alan wants to paint two rooms the same color. One is a living room that is 15 ft long, 11 ft wide, and 9 ft high. The other is a dining room that is 11 ft long, 10 ft wide, and 9 ft high. Alan decides he must buy three gallons of paint. Is he correct? Explain.
11. Bart plans to paint a room that is 18 ft long, 12 ft wide, and 10 ft high. The paint costs $14.99 per gallon, and there is a 5% sales tax. What is the total cost of the paint?
12. Kathy wants to wallpaper a room that is 18 ft long, 11 ft wide, and 10 ft high. The room has one door and three windows. The wallpaper costs $15 per roll, and there is a 4% sales tax. What is the total cost of the wallpaper?
13. Marie's living room is 14 ft long, 12 ft wide, and 8 ft high. It has two doors and two windows. She is considering paint that costs $18 per gallon and wallpaper that costs $13 per roll. Would it cost her less to paint or wallpaper the room? About how much less?
14. Assume that the cost of a gallon of paint is $12. Work together with a classmate to find the cost of painting the walls of your classroom.

CAREER

FARMER

Mike Mihalko is a farmer. He stores his animal feed in a *tower silo,* an upright structure that is shaped like a cylinder. The silo protects the feed from spoiling. Mike uses the following relationship to determine the amount of feed that he can store in a silo.

$$1 \text{ ft}^3 \text{ (approximately)} = 0.8035 \text{ bushel}$$

example

Mike is planning to build a tower silo with diameter 20 ft and height 55 ft. How many bushels of feed will he be able to store in this silo?

solution

Find the volume of the silo. Use $\pi \approx 3.14$.

$V = \pi r^2 h$

$V \approx 3.14(10^2)(55)$ ← **Replace r with $\frac{1}{2} \times 20$, or 10, and h with 55.**

$V \approx 17{,}270 \text{ (ft}^3\text{)}$

Since $1 \text{ ft}^3 = 0.8035$ bushel, $17{,}270 \text{ ft}^3 = 13{,}876.445$ bushels.

(× 0.8035) (× 0.8035)

Mike will be able to store about 13,876 bushels in this silo.

exercises

1. Mike has a tower silo with diameter 10 ft and height 40 ft. What is the volume of this silo?
2. A tower silo is built with diameter 30 ft and height 70 ft. How many bushels of animal feed can be stored in this silo?
3. Mike plans to build a tower silo with radius 10 ft and height 45 ft. Then he realizes that he can afford to build the same silo with height 48 ft. How many more bushels of animal feed would he be able to store in the taller silo?
4. Roger Parkinson has a tower silo with radius 15 ft and height 40 ft. Floyd Prentice has a tower silo with radius 10 ft and height 80 ft. Whose silo can store more animal feed? How many bushels more?

CHAPTER REVIEW

vocabulary vo·cab·u·lar·y

Choose the correct word to complete each sentence.

1. The (*volume, surface area*) of a space figure is the amount of space inside it.
2. A (*sphere, cone*) is the set of all points in space that are the same distance from a given point called the center.

skills

Identify the space figure that would be formed if you folded each pattern.

3. 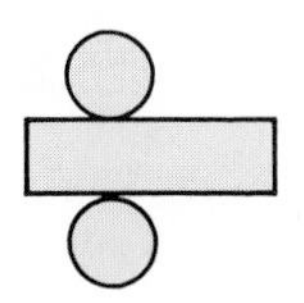

4. 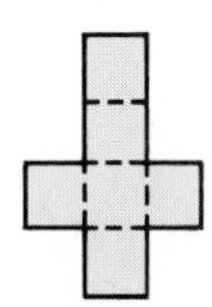

5. 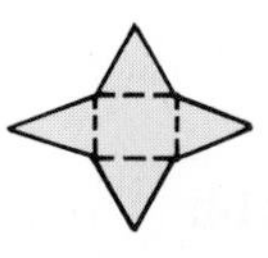

6. 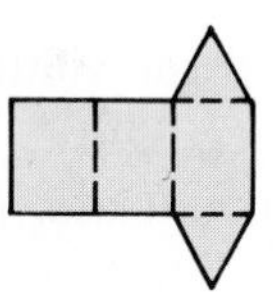

Find the surface area of each space figure. When appropriate, use $\pi \approx \frac{22}{7}$.

7. 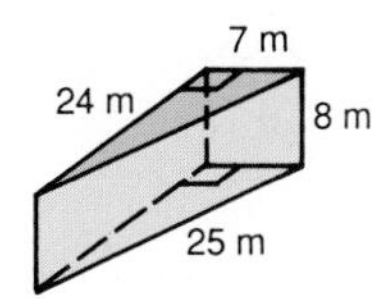

8. 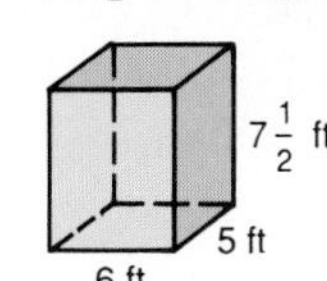

9. 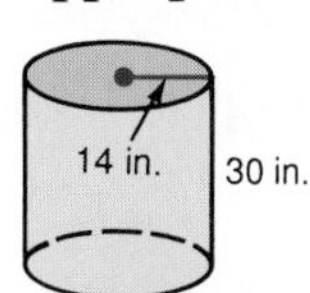

10. 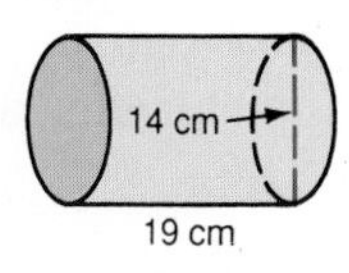

Find the volume of each space figure. When appropriate, use $\pi \approx 3.14$.

11. 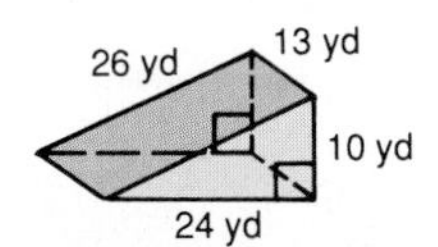

12. 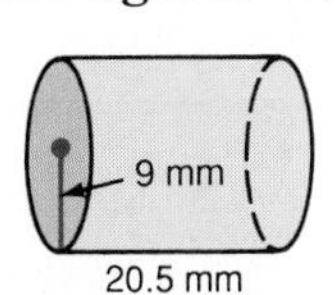

13. 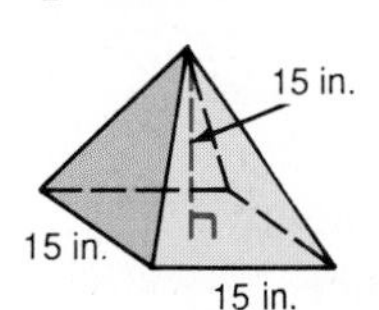

14. 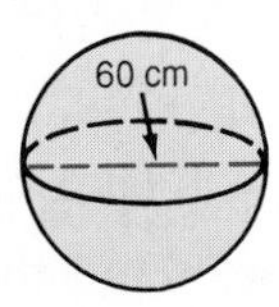

Each (cube) is a centimeter cube. Find the surface area of each space figure.

15. 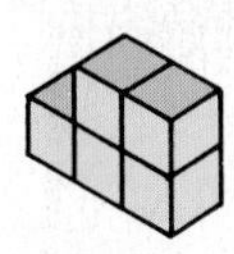

16. 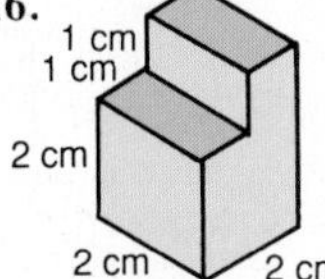

Find the volume of each shaded region. When appropriate, use $\pi \approx 3.14$.

17. 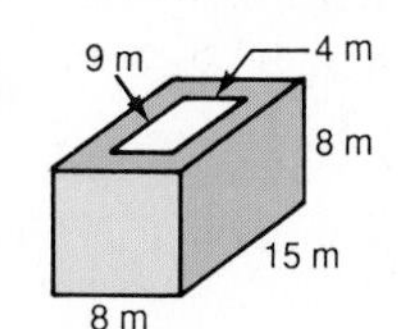

18. 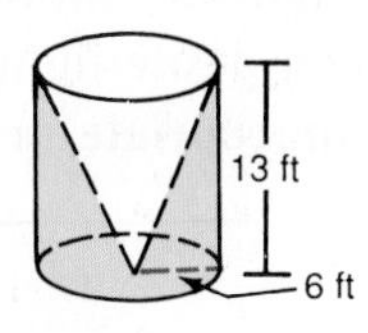

19. Jill decides to paint the walls of a room that is 18 ft long, 14 ft wide, and 8 ft high. One gallon of paint covers 440 ft^2. How many gallons of paint does she need?

CHAPTER TEST

Identify each space figure. **23-1**

1.

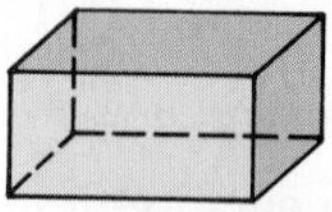

2.

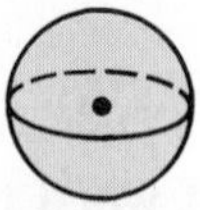

Find the surface area of each prism. **23-2**

3. 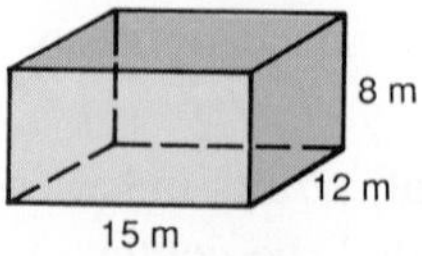 8 m, 12 m, 15 m

4. 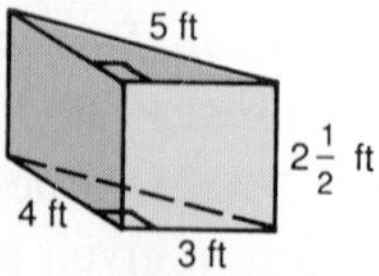 5 ft, $2\frac{1}{2}$ ft, 4 ft, 3 ft

Find the surface area of each cylinder. Use $\pi \approx \frac{22}{7}$. **23-3**

5. 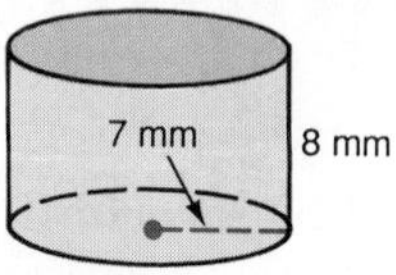 7 mm, 8 mm

6. 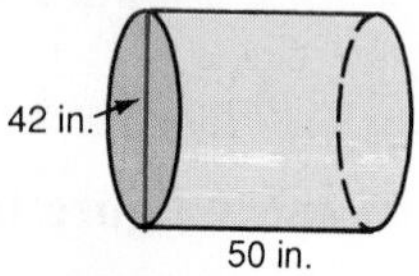 42 in., 50 in.

Find the volume of each space figure. When appropriate, use $\pi \approx 3.14$. **23-4**

7. 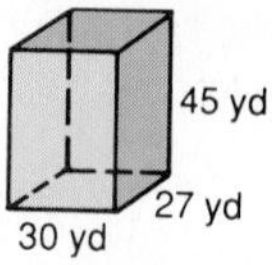 45 yd, 27 yd, 30 yd

8. 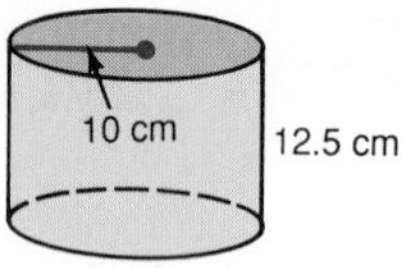 10 cm, 12.5 cm

Find the volume of each space figure. When appropriate, use $\pi \approx 3.14$. **23-5**

9. 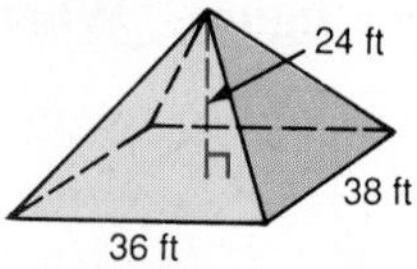 24 ft, 38 ft, 36 ft

10. 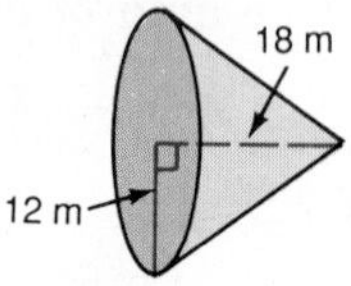 18 m, 12 m

Find the surface area and volume of each sphere. Use $\pi \approx 3.14$. **23-6**

11. 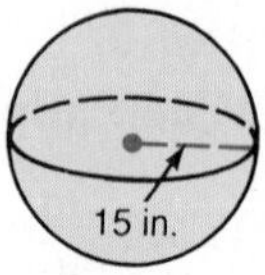 15 in.

12. 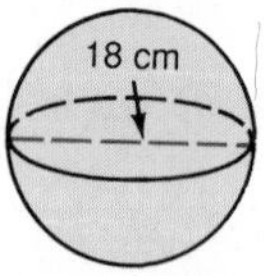 18 cm

Each [cube] is a centimeter cube. Find the surface area of each space figure. **23-7**

13.

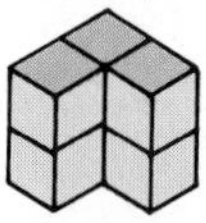

14. 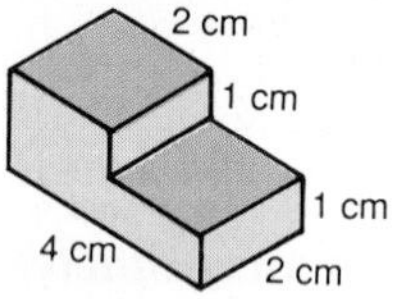 2 cm, 1 cm, 1 cm, 4 cm, 2 cm

Find the volume of each composite figure. When appropriate, use $\pi \approx 3.14$. **23-8**

15. 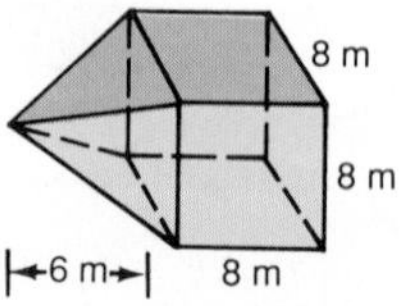 8 m, 8 m, 6 m, 8 m

16. 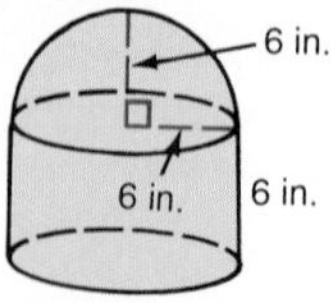 6 in., 6 in., 6 in.

17. Use the chart on page 538. Ed plans to wallpaper a room with perimeter 54 ft and height 9 ft. The room has one door and three windows. How many rolls of wallpaper does he need? **23-9**

A hang glider has to be strong but light in weight. Triangles built into the frame help to provide the needed strength.

CHAPTER 24

TRIANGLES

24-1 FACTS ABOUT TRIANGLES

There are many types of triangles. One way to *classify* triangles is by the number of congruent sides.

equilateral triangle	**isosceles triangle**	**scalene triangle**
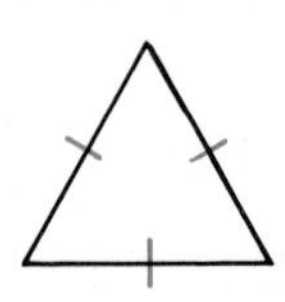	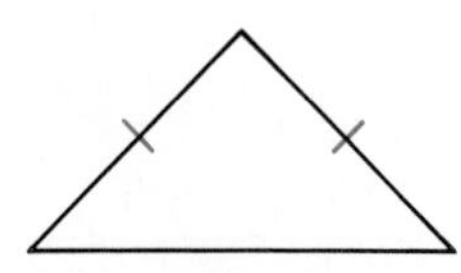	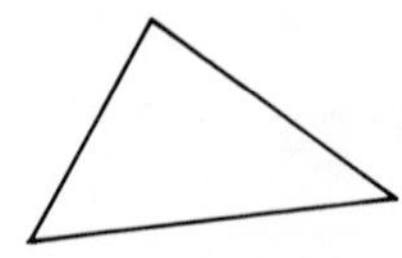
three congruent sides	two congruent sides	no congruent sides

Another way to classify triangles is by their angles.

acute triangle	**right triangle**	**obtuse triangle**
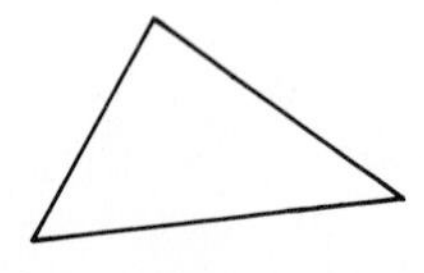	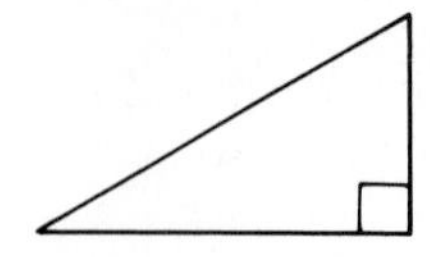	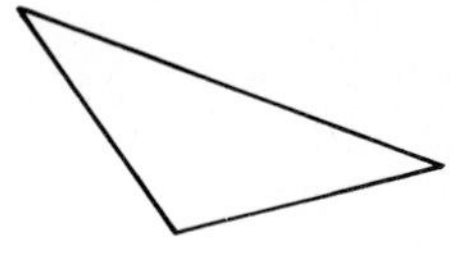
three acute angles	one right angle	one obtuse angle

example 1

Classify each triangle first by its sides and then by its angles.

a.

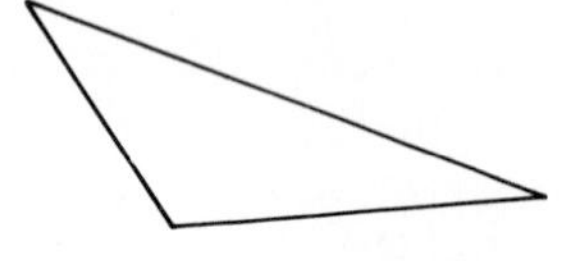

no congruent sides → *scalene*
one obtuse angle → *obtuse*

b.

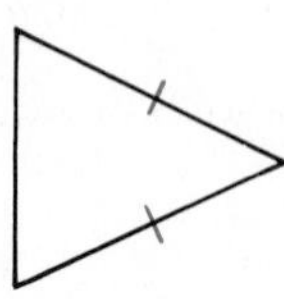

two congruent sides → *isosceles*
three acute angles → *acute*

your turn

Classify each triangle first by its sides and then by its angles.

1.

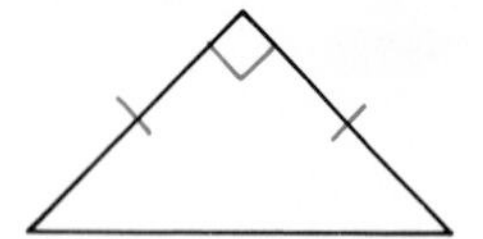

2.

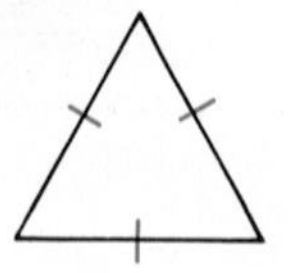

3.

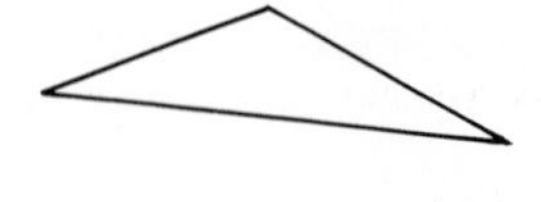

In any triangle, the sum of the lengths of any two sides is *greater than* the length of the third side.

example 2

Can the given lengths be the lengths of the sides of a triangle? Write *Yes* or *No*.

a. 8 ft, 6 ft, 5 ft **b.** 10 cm, 4 cm, 3 cm

solution

a. $8 + 6 \ \underline{?}\ 5$ $8 + 5 \ \underline{?}\ 6$ $6 + 5 \ \underline{?}\ 8$

$14 > 5$ $13 > 6$ $11 > 8$

Each sum of two lengths is greater than the third length.

Yes, 8 ft, 6 ft, and 5 ft can be the lengths of the three sides of a triangle.

6 ft, 5 ft, 8 ft

b. $10 + 4 \ \underline{?}\ 3$ $10 + 3 \ \underline{?}\ 4$ $4 + 3 \ \underline{?}\ 10$

$14 > 3$ $13 > 4$ $7 < 10$

The sum 4 + 3 is *not* greater than the third length, 10.

No, 10 cm, 4 cm, and 3 cm cannot be the lengths of the three sides of a triangle.

3 cm, 4 cm, 10 cm

your turn

Can the given lengths be the lengths of the sides of a triangle? Write *Yes* or *No*.

4. 4 ft, 8 ft, 3 ft **5.** 5 cm, 5 cm, 6 cm **6.** 1 in., 2 in., 3 in.

In any triangle, the sum of the measures of the angles is 180°.

example 3

The measures of two angles of a triangle are 30° and 100°. Find the measure of the third angle.

solution

Add the two known measures.

$30° + 100° = 130°$

Subtract the sum from 180°.

$180° - 130° = 50°$

The measure of the third angle is 50°.

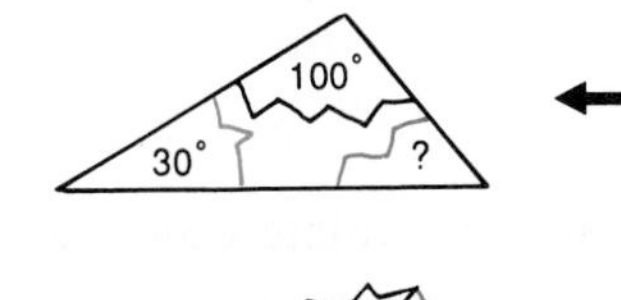

If you could tear off the three angles, they would combine to form a straight angle.

your turn

The measures of two angles of a triangle are given. Find the measure of the third angle.

7. 36°, 110° **8.** 90°, 57° **9.** 37°, 26°

practice exercises

practice for example 1 (page 544)

Classify each triangle first by its sides and then by its angles.

1. 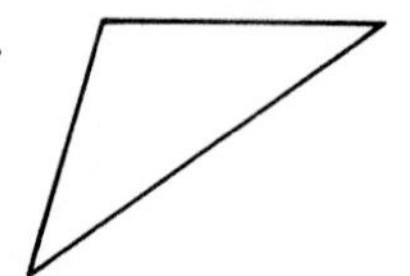
2. 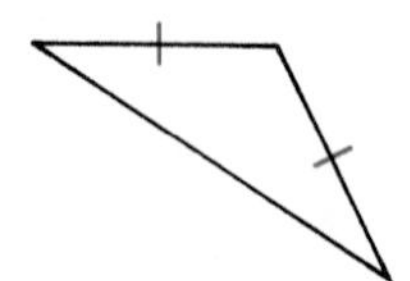
3. 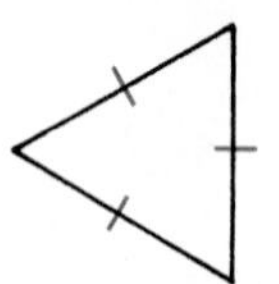
4. 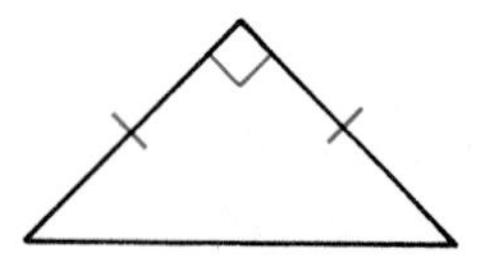

practice for example 2 (page 545)

Can the given lengths be the lengths of the sides of a triangle? Write *Yes* or *No*.

5. 5 m, 10 m, 13 m
6. 8 in., 21 in., 13 in.
7. 24 km, 13 km, 5 km
8. 16 cm, 8 cm, 11 cm
9. 6 ft, 9 ft, 18 ft
10. 20 yd, 16 yd, 12 yd

practice for example 3 (page 545)

The measures of two angles of a triangle are given. Find the measure of the third angle.

11. 69°, 43°
12. 24°, 26°
13. 90°, 52°
14. 47°, 38°

mixed practice (pages 544–545)

Can the given lengths be the lengths of the sides of a triangle? Write *Yes* or *No*. If *Yes*, classify the triangle by its sides.

15. 1 ft, 1 ft, 1 ft
16. 3 km, 1 km, 4 km
17. 7 mi, 9 mi, 17 mi
18. 21 yd, 28 yd, 35 yd
19. 1.4 m, 2.1 m, 3.5 m
20. 7.1 cm, 6.4 cm, 7.1 cm

The measures of two angles of a triangle are given. Find the measure of the third angle. Then classify each triangle by its angles.

21. 27°, 141°
22. 14°, 90°
23. 108°, 49°
24. 39°, 66°

Choose the correct word to complete each statement.

25. A triangle that has (*two, three*) congruent sides is called an equilateral triangle.
26. A(n) (*acute, right*) triangle has one right angle.
27. In any triangle, the sum of the lengths of any two sides is (*greater, less*) than the length of the third side.
28. In any triangle, the sum of the measures of the angles is (*360, 180*) degrees.

29. A triangle with three angles of equal measure is called an **equiangular triangle.** Find the measures of the three angles of an equiangular triangle.

30. One property of an isosceles triangle is that it has two congruent angles. Suppose that the measure of one angle of an isosceles triangle is 120°. Find the measures of the other two angles.

review exercises

Evaluate each expression when $a = {}^{-}2$, $b = 5$, and $c = 7$.

1. $a - 1$	**2.** $7c$	**3.** ac^2	**4.** $4b^2$
5. $3a + 6$	**6.** $2c + 5b$	**7.** $10c - 13$	**8.** $7b - 20$
9. $c + 4 + c$	**10.** $5 + a + c$	**11.** $b^2 - 4$	**12.** a^2b

mental math

To find the sum of the measures of the angles of a polygon mentally, use the fact that the sum of the measures of the angles of a triangle is 180°.

Think: A pentagon has 2 diagonals from one vertex. The diagonals form 3 triangles.

The sum of the measures of the angles is

$$3(180°) = 540°$$

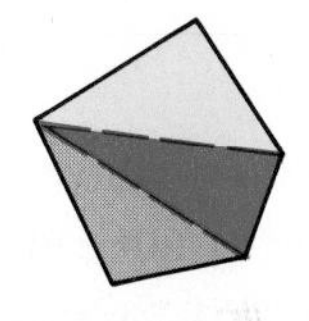

Mentally find the sum of the measures of the angles of each polygon.

1.

2. 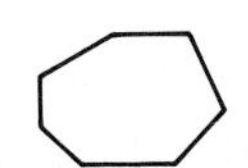

3. 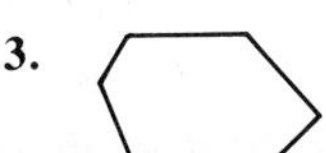

4. 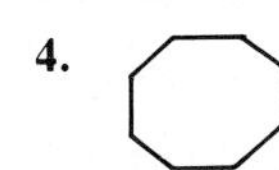

puzzle corner

Fold a piece of paper two times as shown. By making one more fold, you can create an equilateral triangle. Describe how to make the third fold.

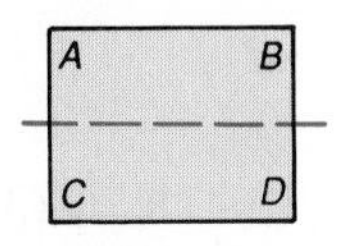

Fold the paper in half.

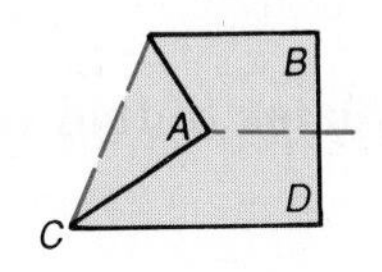

Match the left corner to the center fold.

24-2 SQUARES AND SQUARE ROOTS

The length of each side of the square at the right is 5 units. The area of this square is 25 square units. For this reason, you can say that the **square** of 5 is 25. You write $5^2 = 25$.

If the area of a square is 25 square units, you know that the length of each side must be 5 units. So you can say that a **square root** of 25 is 5. You write $\sqrt{25} = 5$. The symbol $\sqrt{\ }$ is called the **radical sign.**

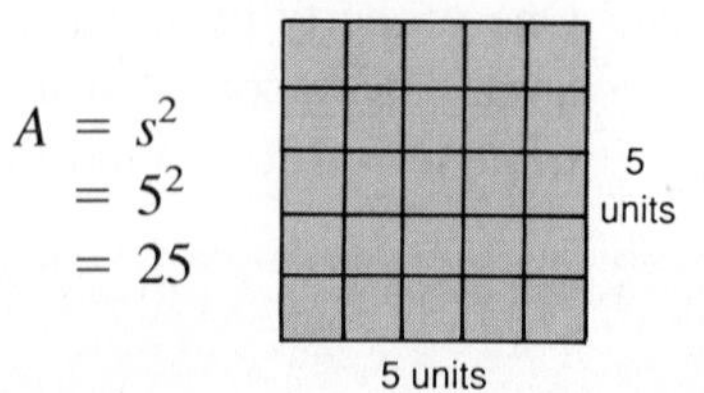

example 1

Find each square root: **a.** $\sqrt{36}$ **b.** $\sqrt{4900}$

solution

a. Since $6^2 = 36$, $\sqrt{36} = 6$.

b. Since $70^2 = 4900$, $\sqrt{4900} = 70$.

your turn

Find each square root.

1. $\sqrt{4}$ **2.** $\sqrt{64}$ **3.** $\sqrt{900}$ **4.** $\sqrt{8100}$

A number whose square root is a whole number is a **perfect square.** If a number is not a perfect square, you can use a calculator or a table like the one on page 549 to find its *approximate* square root. You can also use the table to find square roots of some greater numbers.

example 2

Use the table on page 549 to find each square root: **a.** $\sqrt{50}$ **b.** $\sqrt{9604}$

solution

a. Find 50 in the *No.* (number) column. Read the number across from 50 in the *Square Root* column.
$\sqrt{50} \approx 7.071$ ← **On a calculator enter 50 [√].**

b. Find 9604 in the *Square* column. Read the number across from 9604 in the *No.* column.
$\sqrt{9604} = 98$

your turn

Use the table on page 549 or a calculator to find each square root.

5. $\sqrt{33}$ **6.** $\sqrt{74}$ **7.** $\sqrt{784}$ **8.** $\sqrt{13{,}456}$

Table of Squares and Square Roots

NO.	SQUARE	SQUARE ROOT	NO.	SQUARE	SQUARE ROOT	NO.	SQUARE	SQUARE ROOT
1	1	1.000	**51**	2,601	7.141	**101**	10,201	10.050
2	4	1.414	**52**	2,704	7.211	**102**	10,404	10.100
3	9	1.732	**53**	2,809	7.280	**103**	10,609	10.149
4	16	2.000	**54**	2,916	7.348	**104**	10,816	10.198
5	25	2.236	**55**	3,025	7.416	**105**	11,025	10.247
6	36	2.449	**56**	3,136	7.483	**106**	11,236	10.296
7	49	2.646	**57**	3,249	7.550	**107**	11,449	10.344
8	64	2.828	**58**	3,364	7.616	**108**	11,664	10.392
9	81	3.000	**59**	3,481	7.681	**109**	11,881	10.440
10	100	3.162	**60**	3,600	7.746	**110**	12,100	10.488
11	121	3.317	**61**	3,721	7.810	**111**	12,321	10.536
12	144	3.464	**62**	3,844	7.874	**112**	12,544	10.583
13	169	3.606	**63**	3,969	7.937	**113**	12,769	10.630
14	196	3.742	**64**	4,096	8.000	**114**	12,996	10.677
15	225	3.873	**65**	4,225	8.062	**115**	13,225	10.724
16	256	4.000	**66**	4,356	8.124	**116**	13,456	10.770
17	289	4.123	**67**	4,489	8.185	**117**	13,689	10.817
18	324	4.243	**68**	4,624	8.246	**118**	13,924	10.863
19	361	4.359	**69**	4,761	8.307	**119**	14,161	10.909
20	400	4.472	**70**	4,900	8.367	**120**	14,400	10.954
21	441	4.583	**71**	5,041	8.426	**121**	14,641	11.000
22	484	4.690	**72**	5,184	8.485	**122**	14,884	11.045
23	529	4.796	**73**	5,329	8.544	**123**	15,129	11.091
24	576	4.899	**74**	5,476	8.602	**124**	15,376	11.136
25	625	5.000	**75**	5,625	8.660	**125**	15,625	11.180
26	676	5.099	**76**	5,776	8.718	**126**	15,876	11.225
27	729	5.196	**77**	5,929	8.775	**127**	16,129	11.269
28	784	5.292	**78**	6,084	8.832	**128**	16,384	11.314
29	841	5.385	**79**	6,241	8.888	**129**	16,641	11.358
30	900	5.477	**80**	6,400	8.944	**130**	16,900	11.402
31	961	5.568	**81**	6,561	9.000	**131**	17,161	11.446
32	1,024	5.657	**82**	6,724	9.055	**132**	17,424	11.489
33	1,089	5.745	**83**	6,889	9.110	**133**	17,689	11.533
34	1,156	5.831	**84**	7,056	9.165	**134**	17,956	11.576
35	1,225	5.916	**85**	7,225	9.220	**135**	18,225	11.619
36	1,296	6.000	**86**	7,396	9.274	**136**	18,496	11.662
37	1,369	6.083	**87**	7,569	9.327	**137**	18,769	11.705
38	1,444	6.164	**88**	7,744	9.381	**138**	19,044	11.747
39	1,521	6.245	**89**	7,921	9.434	**139**	19,321	11.790
40	1,600	6.325	**90**	8,100	9.487	**140**	19,600	11.832
41	1,681	6.403	**91**	8,281	9.539	**141**	19,881	11.874
42	1,764	6.481	**92**	8,464	9.592	**142**	20,164	11.916
43	1,849	6.557	**93**	8,649	9.644	**143**	20,449	11.958
44	1,936	6.633	**94**	8,836	9.695	**144**	20,736	12.000
45	2,025	6.708	**95**	9,025	9.747	**145**	21,025	12.042
46	2,116	6.782	**96**	9,216	9.798	**146**	21,316	12.083
47	2,209	6.856	**97**	9,409	9.849	**147**	21,609	12.124
48	2,304	6.928	**98**	9,604	9.899	**148**	21,904	12.166
49	2,401	7.000	**99**	9,801	9.950	**149**	22,201	12.207
50	2,500	7.071	**100**	10,000	10.000	**150**	22,500	12.247

practice exercises

practice for example 1 (page 548)

Find each square root.

1. $\sqrt{9}$ 2. $\sqrt{81}$ 3. $\sqrt{1600}$ 4. $\sqrt{6400}$ 5. $\sqrt{0}$

practice for example 2 (page 548)

Use the table on page 549 or a calculator to find each square root.

6. $\sqrt{17}$ 7. $\sqrt{40}$ 8. $\sqrt{92}$ 9. $\sqrt{529}$ 10. $\sqrt{4356}$

mixed practice (page 548)

Find each square root. Use the table on page 549 or a calculator if necessary.

11. $\sqrt{1}$ 12. $\sqrt{16}$ 13. $\sqrt{90}$ 14. $\sqrt{130}$ 15. $\sqrt{10{,}000}$
16. $\sqrt{14{,}400}$ 17. $\sqrt{141}$ 18. $\sqrt{150}$ 19. $\sqrt{11{,}236}$ 20. $\sqrt{20{,}736}$

Tell whether each statement is *True* or *False*.

21. $\sqrt{6}$ is between 2 and 3.
22. $\sqrt{42}$ is between 5 and 6.
23. $\sqrt{3}$ is between 1.7 and 1.8.
24. $\sqrt{58}$ is between 7.5 and 7.6.

Compare. Replace each ? with >, <, or = .

25. $11 \ \underline{?}\ \sqrt{115}$ 26. $\sqrt{5041}\ \underline{?}\ 63$ 27. $\sqrt{3249}\ \underline{?}\ 8^2$ 28. $4^2\ \underline{?}\ \sqrt{256}$

29. The perimeter of a square is 84 cm. What is its area?
30. The area of a square is 49 m^2. What is its perimeter?
31. The area of the square base of a building is 2500 ft^2. Find the length of a side of the base.
32. The target for a parachute jump is in a circular region with a diameter of 100 m. Use $\pi \approx 3.14$ to find the area of the region.

review exercises

Solve each equation.

1. $x + 6 = 21$ 2. $y - 9 = 45$
3. $13 + b = {}^-24$ 4. $z - 75 = 64$
5. ${}^-132 = 6m$ 6. $\frac{1}{2}p = 50$

24-3 THE PYTHAGOREAN THEOREM

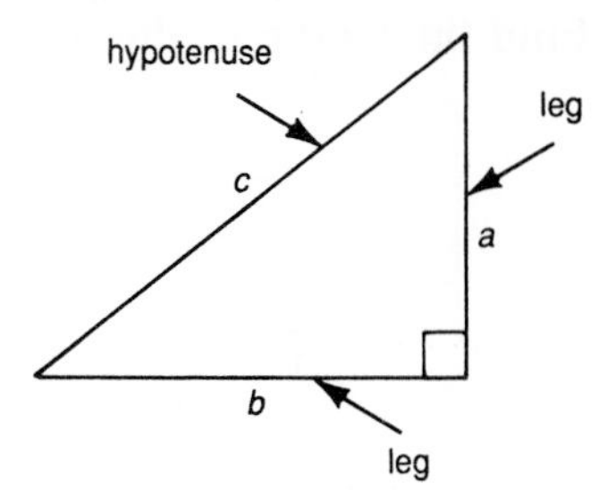

In a right triangle, the side opposite the right angle is called the **hypotenuse** (c). The hypotenuse is always the longest side. The other sides (a and b) are called **legs.**

The **Pythagorean Theorem** is an important property of right triangles. The theorem was named in honor of the Greek mathematician Pythagoras. This property of right triangles can be stated in the following way.

> For any right triangle, the square of the length of the hypotenuse is equal to the sum of the squares of the lengths of the two legs.
>
> $$c^2 = a^2 + b^2$$

The *converse* of the Pythagorean Theorem is also true.

> If the sides of a triangle have lengths a, b, and c such that $c^2 = a^2 + b^2$, then the triangle is a right triangle.

example 1

Can the given lengths be the lengths of the sides of a right triangle? Write *Yes* or *No*.

a. 2, 4, 5

b. 5, 12, 13

solution

a.
$$c^2 = a^2 + b^2$$
$$5^2 \stackrel{?}{=} 2^2 + 4^2$$ ← **Substitute the greatest length for c.**
$$25 \stackrel{?}{=} 4 + 16$$
$$25 \neq 20$$

No, 2, 4, and 5 cannot be the lengths of the sides of a right triangle.

b.
$$c^2 = a^2 + b^2$$
$$13^2 \stackrel{?}{=} 5^2 + 12^2$$
$$169 \stackrel{?}{=} 25 + 144$$
$$169 = 169$$

Yes, 5, 12, and 13 can be the lengths of the sides of a right triangle.

your turn

Can the given lengths be the lengths of the sides of a right triangle? Write *Yes* or *No*.

1. 3, 4, 5 **2.** 6, 11, 12 **3.** 15, 8, 17 **4.** 9, 9, 12

If you know the lengths of two sides of a right triangle, you can use the Pythagorean Theorem to find the length of the third side.

example 2

Find the length of the third side. If necessary, round to the nearest tenth.

a.

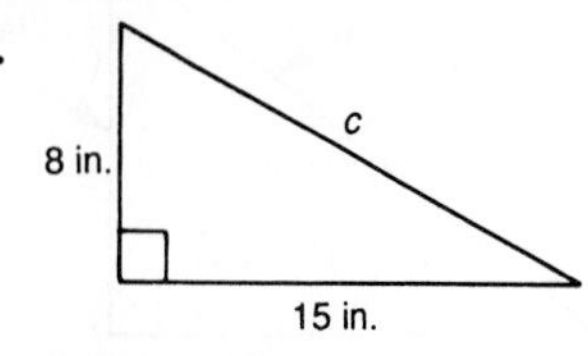

b.

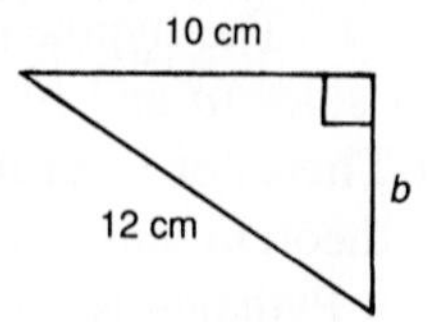

solution

a.

$c^2 = a^2 + b^2$

$c^2 = 8^2 + 15^2$

$c^2 = 64 + 225$

$c^2 = 289$

$c = \sqrt{289} = 17$ ← **Use the table on page 549.**

The length of the hypotenuse is 17 in.

b.

$c^2 = a^2 + b^2$

$12^2 = 10^2 + b^2$

$144 = 100 + b^2$ ← **Subtract 100 from both sides.**

$44 = b^2$

$b = \sqrt{44} \approx 6.633$

The length of the leg is about 6.6 cm.

your turn

Find the length of the third side. If necessary, round to the nearest tenth.

5.

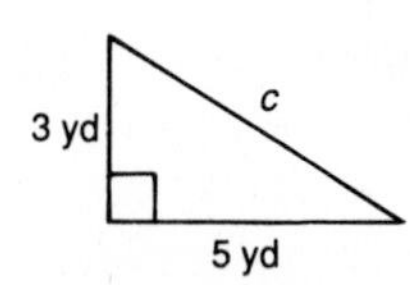

6.

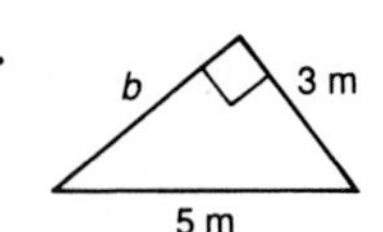

7.

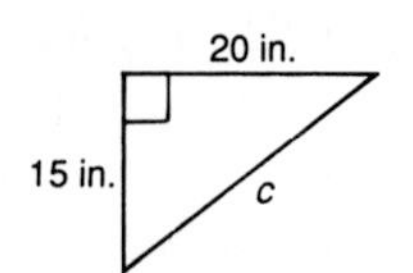

8.

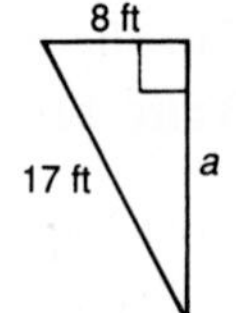

practice exercises

practice for example 1 *(page 551)*

Can the given lengths be the lengths of the sides of a right triangle? Write *Yes* or *No*.

1. 4, 5, 7 **2.** 3, 9, 12 **3.** 15, 9, 12 **4.** 8, 6, 10

5. 7, 24, 25 **6.** 17, 21, 14 **7.** 20, 12, 16 **8.** 37, 35, 12

practice for example 2 *(page 552)*

Find the length of the third side. If necessary, round to the nearest tenth.

9.

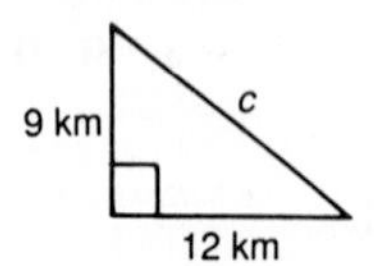

10.

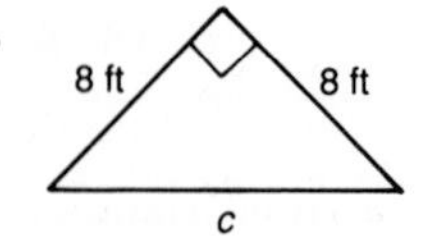

11.

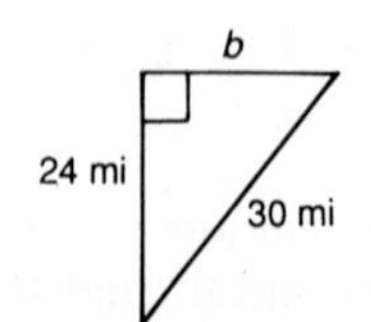

12.

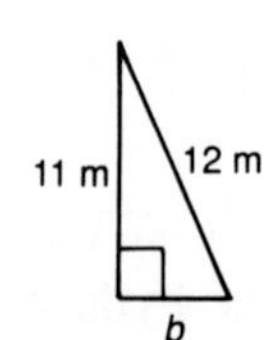

mixed practice (pages 551–552)

The lengths of two sides of a right triangle are given. Assume that c is the length of the hypotenuse and a and b are the lengths of the legs. Sketch the triangle and then find the length of the third side.

13. $a = 8, b = 6$
14. $a = 12, b = 16$
15. $a = 7, c = 25$
16. $a = 27, c = 45$
17. $c = 29, b = 21$
18. $c = 26, b = 10$

Find the unknown length. If necessary, round to the nearest tenth.

19.

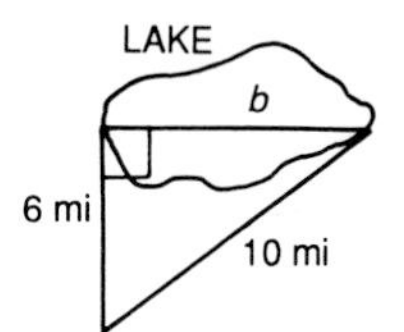

20.

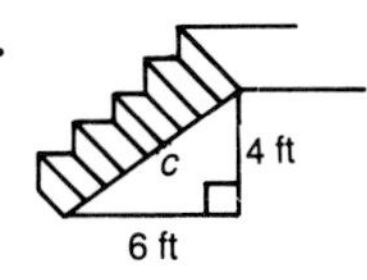

21.

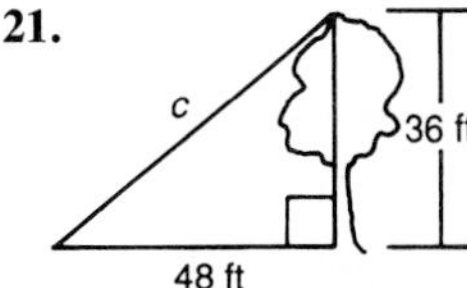

22.

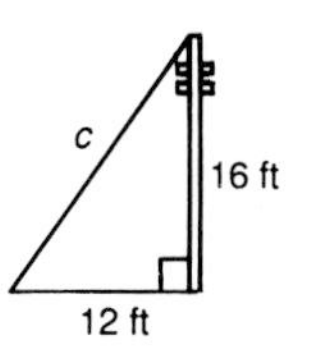

23.

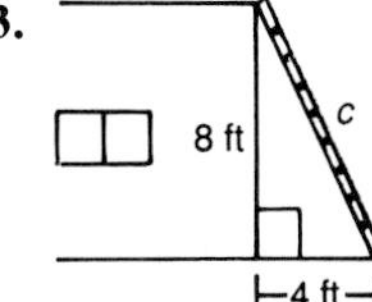

24.

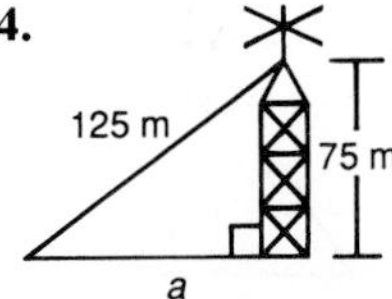

Solve. Making a drawing may be helpful.

25. Elizabeth Calo left her house and walked 1 mi due east and then 3 mi due north. To the nearest tenth of a mile, how far is she from her house?

26. The size of a television screen is indicated by the length of the diagonal across the screen. What is the size of a television screen that measures approximately 16 in. by 12 in.?

review exercises

Solve each proportion.

1. $\frac{3}{33} = \frac{12}{n}$
2. $\frac{8}{y} = \frac{4}{9}$
3. $\frac{h}{28} = \frac{3}{21}$
4. $\frac{c}{0.2} = \frac{45}{3}$

Find each answer using a proportion.

5. What number is 20% of 35?
6. 60% of 185 is what number?
7. 2 is what percent of 40?
8. What percent of 384 is 288?
9. 119 is 17% of what number?
10. 150% of what number is 120?

24-4 SIMILAR TRIANGLES

Triangles that have the same shape are called **similar** triangles. Similar triangles do not necessarily have the same size. When two triangles are similar, their *corresponding angles* are congruent. The lengths of their *corresponding sides* are **proportional.** This means that the ratios of the lengths of the corresponding sides are equal.

If the lengths of their corresponding sides are in a 1 to 1 ratio, then the triangles are **congruent.** Congruent triangles have the same shape *and* the same size.

example 1

In each pair, tell whether the triangles are *similar* or *not similar.* If the triangles are similar, tell whether they are *congruent* or *not congruent.*

a.

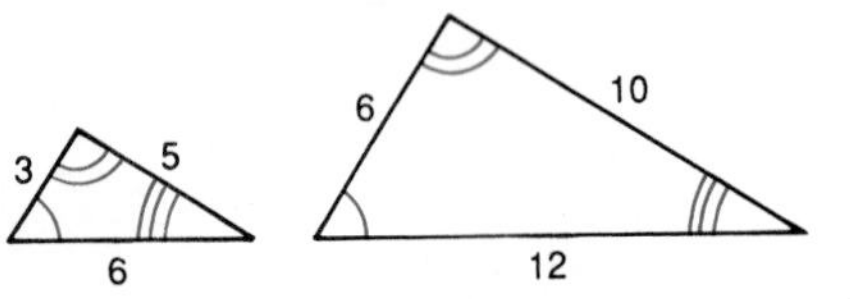

b.

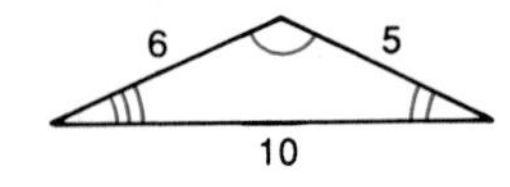

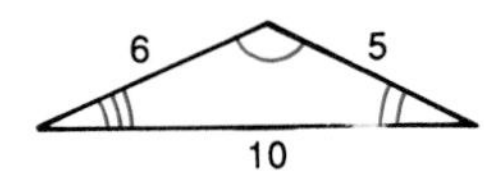

solution

a. Corresponding angles are congruent.

corresponding sides: $\frac{\text{first triangle}}{\text{second triangle}} \begin{matrix}\to\\ \to\end{matrix} \frac{3}{6} = \frac{1}{2} \qquad \frac{5}{10} = \frac{1}{2} \qquad \frac{6}{12} = \frac{1}{2}$

The ratios of the lengths of corresponding sides are equal, so the triangles are *similar*. The ratios are not 1 to 1, so the triangles are *not congruent.*

b. Corresponding angles are congruent.

corresponding sides: $\frac{\text{first triangle}}{\text{second triangle}} \begin{matrix}\to\\ \to\end{matrix} \frac{6}{6} = \frac{1}{1} \qquad \frac{5}{5} = \frac{1}{1} \qquad \frac{10}{10} = \frac{1}{1}$

The ratios of the lengths of corresponding sides are equal, so the triangles are *similar*. The ratios are 1 to 1, so the triangles are *congruent*.

your turn

In each pair, tell whether the triangles are *similar* or *not similar.* If the triangles are similar, tell whether they are *congruent* or *not congruent.*

1.

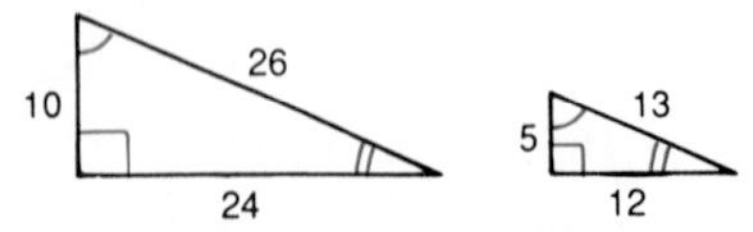

2.

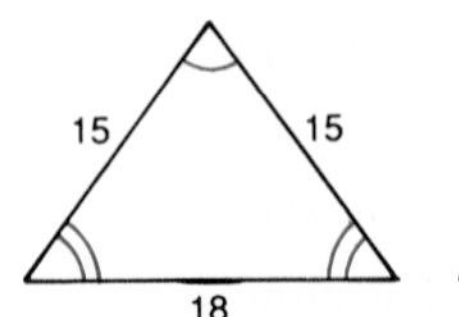

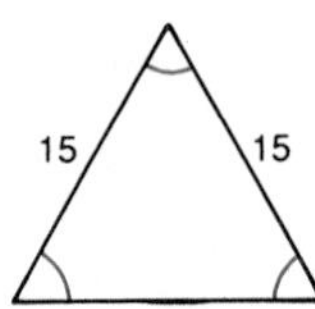

example 2

The triangles are similar.
Find the unknown length *x*.

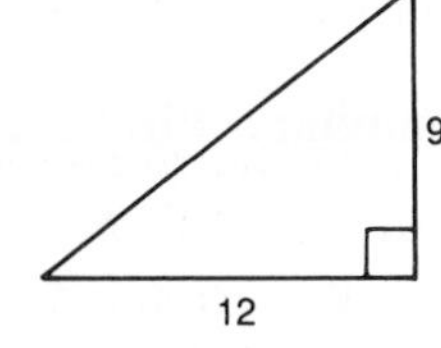

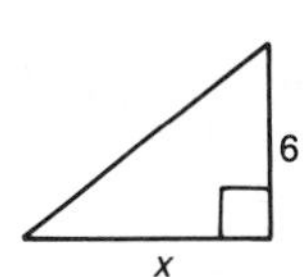

solution

Find the ratio of the known lengths of corresponding sides.

$$\frac{\text{first triangle}}{\text{second triangle}} \rightarrow \frac{9}{6} = \frac{3}{2} \quad \leftarrow \text{Write the ratio in lowest terms.}$$

Set up and solve a proportion involving the unknown length.

$$\frac{\text{first triangle}}{\text{second triangle}} \rightarrow \frac{3}{2} = \frac{12}{x}$$

$$3x = 24$$

$$x = 8$$

The unknown length is 8.

your turn

In each pair, the triangles are similar. Find the unknown length *x*.

3.

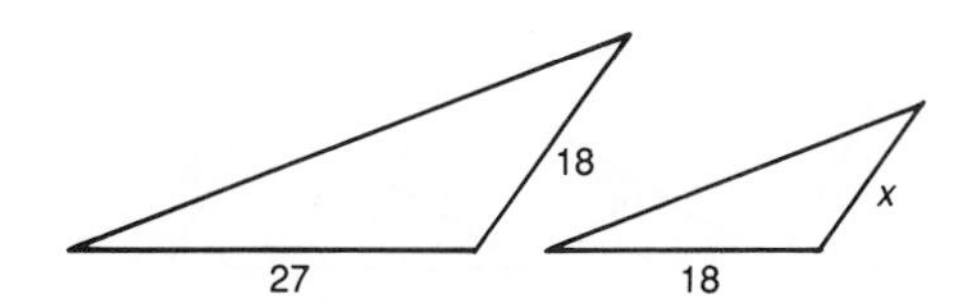

4.

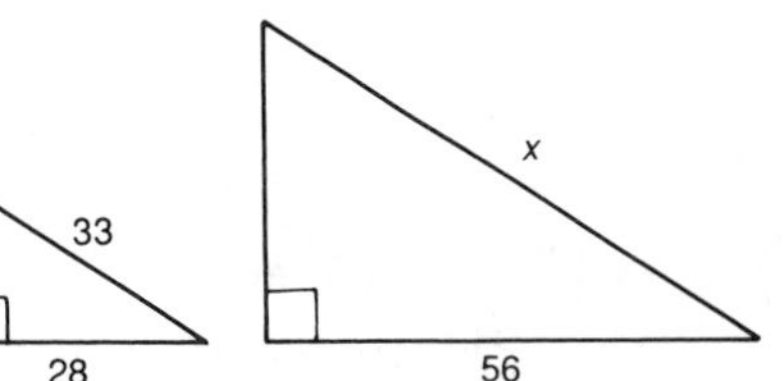

practice exercises

practice for example 1 (page 554)

In each pair, tell whether the triangles are *similar* or *not similar*. If the triangles are similar, tell whether they are *congruent* or *not congruent*.

1.

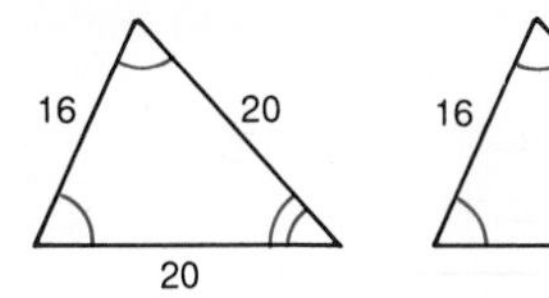

2.

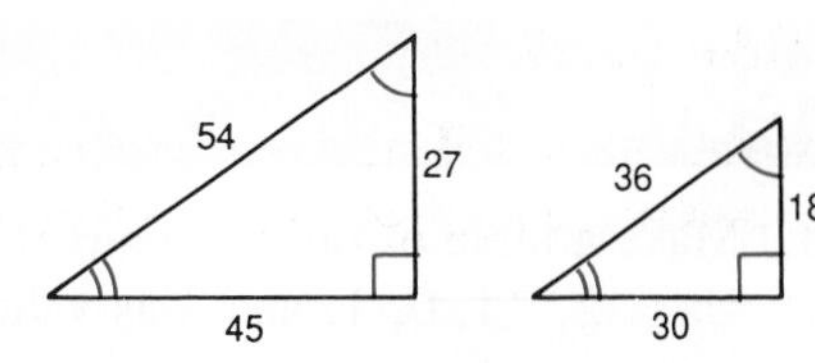

3.

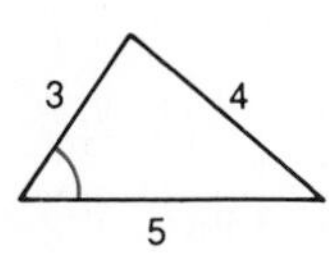

3 4 6

4.

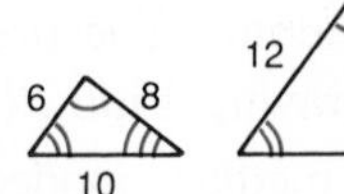

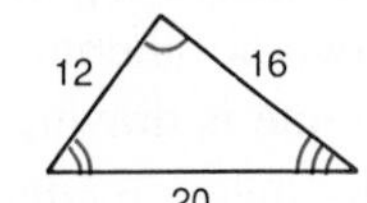

practice for example 2 (page 555)

In each pair, the triangles are similar. Find the unknown length *x*.

5.

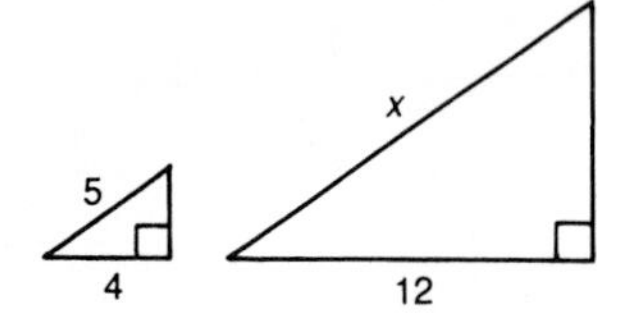

6.

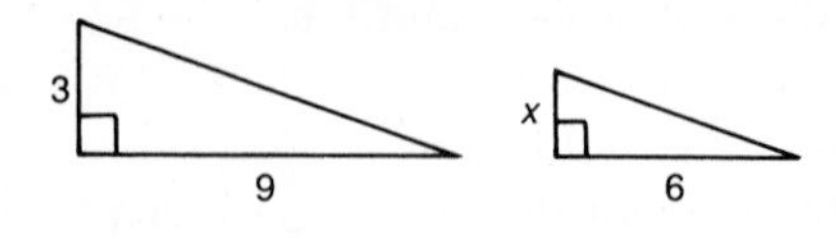

7.

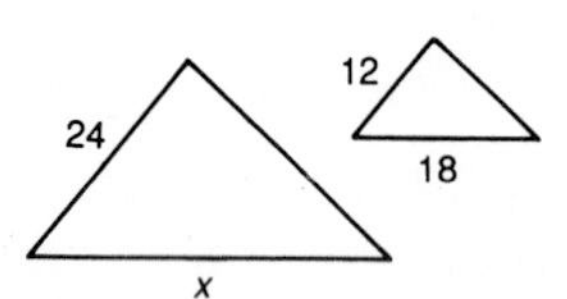

8.

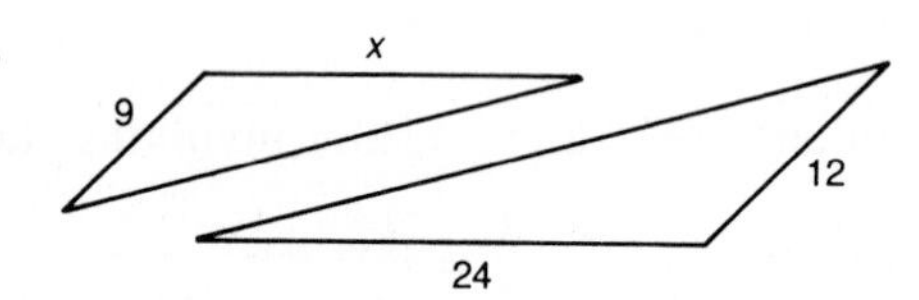

mixed practice (pages 554–555)

In each pair, the triangles are similar. Find the unknown lengths. Then tell whether the triangles are *congruent* or *not congruent*.

9.

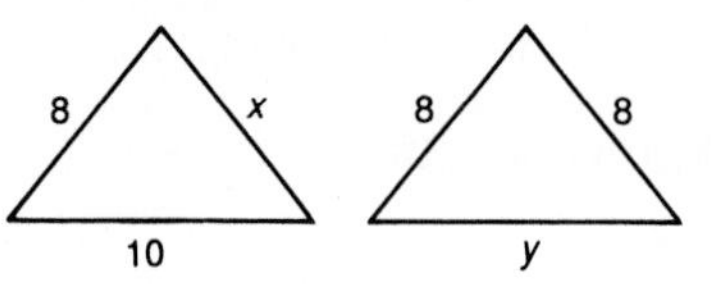

10.

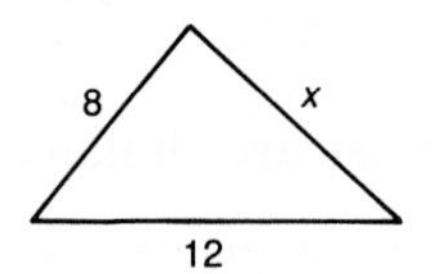

11.

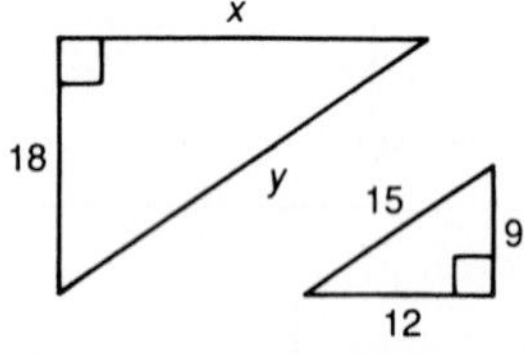

12.

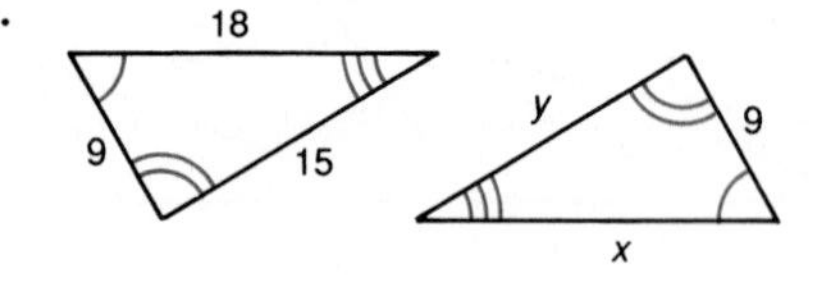

13. Draw any scalene triangle. Label it *ABC*. Now use a protractor and a ruler to draw a triangle *DEF* that is congruent to triangle *ABC*.
14. Draw an isosceles triangle with angles that measure 90°, 45°, and 45°. Label it *XYZ*. Now use a protractor and a ruler to draw a triangle *RST* that is similar, but not congruent, to triangle *XYZ*.

review exercises

1. Make a table of ordered pairs that are solutions of $y = 3x - 5$. Use $^{-}2$, $^{-}1$, 0, 1, and 2 as values for *x*.
2. A bag contains 4 green, 1 blue, and 7 red marbles. Two marbles are drawn at random. The first marble is not replaced before the second one is drawn. Find *P*(red, then green).
3. Find the mean, median, mode(s), and range: 24, 36, 21, 41, 33, 36, 33

24-5 TRIGONOMETRIC RATIOS

A leg of a right triangle is sometimes identified by its relationship to one of the acute angles. For instance, in triangle ABC at the right, $\overline{AC}$ is called the side **adjacent** to $\angle A$. $\overline{BC}$ is the side **opposite** $\angle A$. These names are used in defining the following **trigonometric ratios** for $\angle A$.

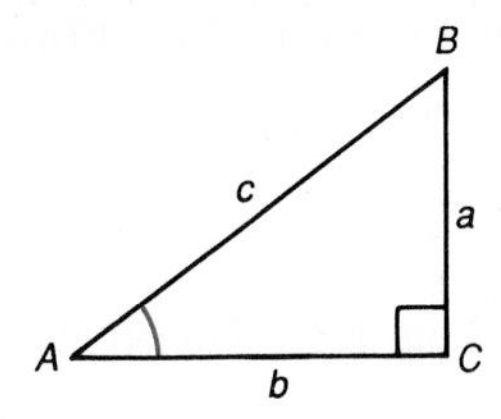

tangent of $\angle A$: $\tan A = \dfrac{\text{length of the side opposite } \angle A}{\text{length of the side adjacent to } \angle A} = \dfrac{a}{b}$

sine of $\angle A$: $\sin A = \dfrac{\text{length of the side opposite } \angle A}{\text{length of the hypotenuse}} = \dfrac{a}{c}$

cosine of $\angle A$: $\cos A = \dfrac{\text{length of the side adjacent to } \angle A}{\text{length of the hypotenuse}} = \dfrac{b}{c}$

In working with the trigonometric ratios, you may find it helpful to use these shortened forms of the definitions.

$\tan A = \dfrac{\text{opposite}}{\text{adjacent}}$ $\sin A = \dfrac{\text{opposite}}{\text{hypotenuse}}$ $\cos A = \dfrac{\text{adjacent}}{\text{hypotenuse}}$

example 1

Use the triangle at the right. Find each ratio in lowest terms.

a. $\tan A$ **b.** $\sin A$ **c.** $\cos A$

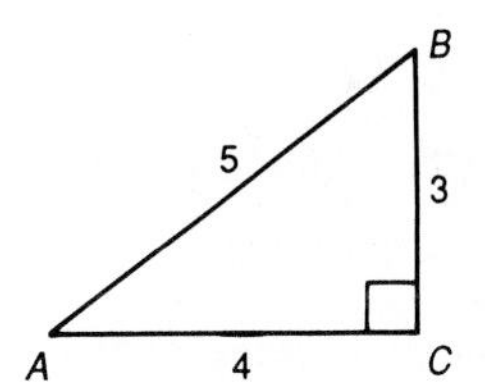

solution

a. $\tan A = \dfrac{\text{opposite}}{\text{adjacent}} = \dfrac{3}{4}$

b. $\sin A = \dfrac{\text{opposite}}{\text{hypotenuse}} = \dfrac{3}{5}$

c. $\cos A = \dfrac{\text{adjacent}}{\text{hypotenuse}} = \dfrac{4}{5}$

your turn

Use the triangle above. Find each ratio in lowest terms.

1. $\tan B$ **2.** $\sin B$ **3.** $\cos B$

Sometimes the length of one of the sides is unknown and is represented by a variable. Before finding the trigonometric ratios, you can use the Pythagorean Theorem to find this unknown length.

example 2

Use the triangle at the right. Find each ratio in lowest terms.

a. $\sin M$ **b.** $\cos M$ **c.** $\tan M$

solution

First use the Pythagorean Theorem to find the length of $\overline{NP}$.

$$\begin{aligned} 13^2 &= a^2 + 5^2 \\ 169 &= a^2 + 25 \\ 144 &= a^2 \\ 12 &= a \end{aligned}$$

a. $\sin M = \dfrac{\text{opposite}}{\text{hypotenuse}} = \dfrac{12}{13}$

b. $\cos M = \dfrac{\text{adjacent}}{\text{hypotenuse}} = \dfrac{5}{13}$

c. $\tan M = \dfrac{\text{opposite}}{\text{adjacent}} = \dfrac{12}{5}$

your turn

Use the triangle at the right. Find each ratio in lowest terms.

4. $\sin T$
5. $\cos T$
6. $\tan T$

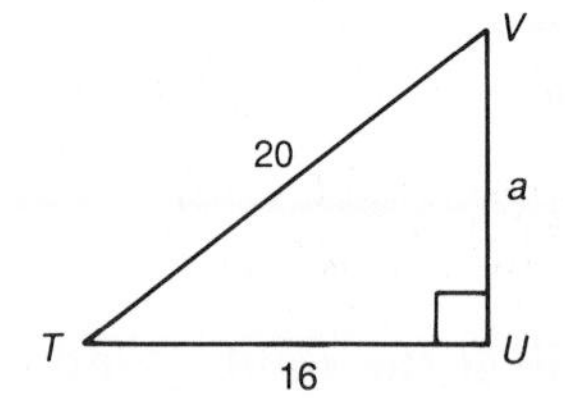

practice exercises

practice for example 1 (page 557)

Use the triangle at the right. Find each ratio in lowest terms.

1. $\tan X$ 2. $\tan Y$
3. $\cos Y$ 4. $\cos X$
5. $\sin X$ 6. $\sin Y$

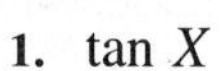

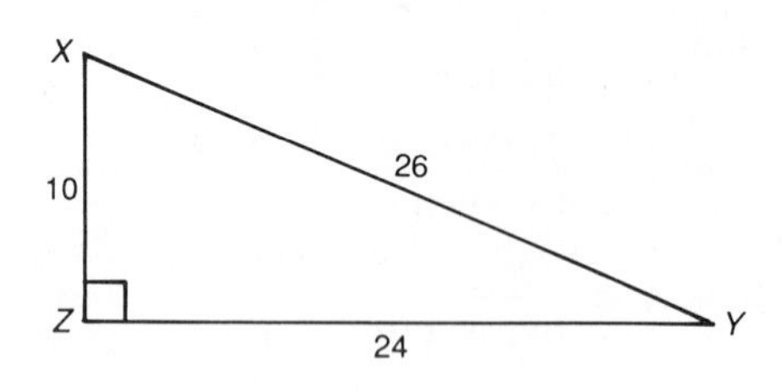

practice for example 2 (page 558)

Use the triangle at the right. Find each ratio in lowest terms.

7. $\sin A$ 8. $\sin B$
9. $\tan A$ 10. $\tan B$
11. $\cos A$ 12. $\cos B$

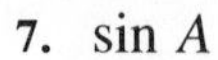

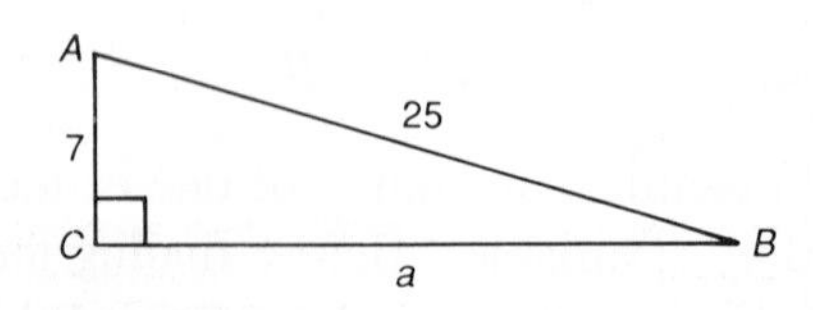

mixed practice (pages 557–558)

Use triangle *JKL* at the right. Find each ratio in lowest terms.

13. sin J **14.** cos L **15.** tan L

16. sin L **17.** cos J **18.** tan J

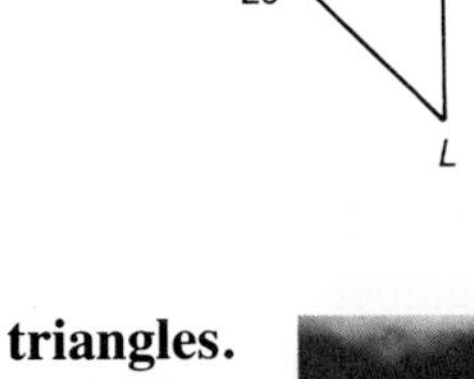

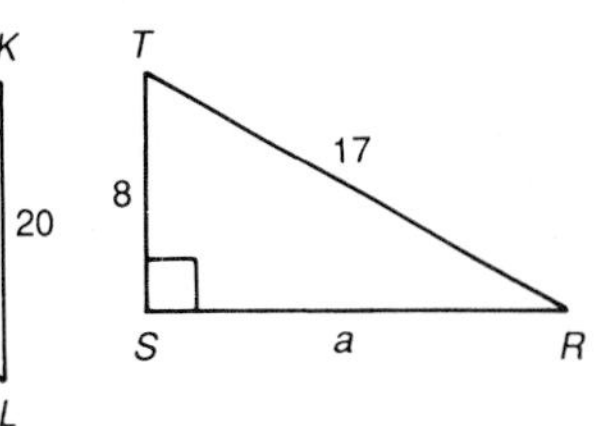

Use triangle *RST* at the right. Find each ratio in lowest terms.

19. sin T **20.** cos R **21.** tan R

22. sin R **23.** cos T **24.** tan T

Triangles *CDE* and *MNO* are similar right triangles. Angles *D* and *N* are right angles. In triangle *CDE*, the length of side $\overline{CD}$ is 15 m and the length of side $\overline{DE}$ is 20 m. In triangle *MNO*, the length of side $\overline{MN}$ is 30 m and the length of side $\overline{NO}$ is 40 m.

25. Sketch triangle *CDE*.

26. Sketch triangle *MNO*.

27. Find the length of the hypotenuse of triangle *CDE*.

28. Find the length of the hypotenuse of triangle *MNO*.

29. Find sin O in lowest terms.

30. Find sin E in lowest terms.

31. Find cos C in lowest terms.

32. Find cos M in lowest terms.

33. Find tan E in lowest terms.

34. Find tan O in lowest terms.

review exercises

Estimate.

1. $212 + 197$ **2.** $\$798 + \$290$ **3.** $498 - 292$ **4.** $1774 - 414$

5. $4.34 + 5.63$ **6.** $287.3 - 46.7$ **7.** 57×21 **8.** 98×19

9. $5\frac{3}{4} + 8\frac{7}{9}$ **10.** $13\frac{2}{15} + 21\frac{4}{5}$ **11.** $18\frac{5}{6} - 13\frac{2}{3}$ **12.** $25\frac{3}{11} - 6\frac{9}{10}$

13. 8.9×5.9 **14.** 0.781×0.93 **15.** $3\overline{)239}$ **16.** $9\overline{)2904}$

17. $7\overline{)4723}$ **18.** $90\overline{)19{,}063}$ **19.** 26% of 82 **20.** 61% of 407

PROBLEM SOLVING STRATEGY

24-6 SOLVING A SIMPLER PROBLEM

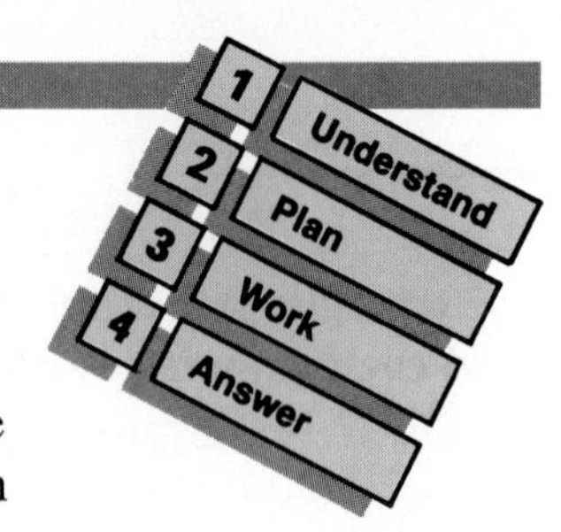

Sometimes a complicated problem can be broken down into one or more simpler problems. The solution to the original problem can then be found through the solutions of the simpler problems.

example

Korbo Electronics Corporation has hired a contractor to number all the office doors in its building. The office doors will be numbered consecutively starting with the number 1. Some doors will require more digits than others, so the corporation agrees to pay only for the number of digits that are actually used. If the contractor used 513 digits, how many different office doors are there?

solution

Step 1 Given: 513 digits used
consecutive numbering starting with 1
Find: the number of office doors

Step 2 Solve a *series* of simpler problems.

- Find the number of 1-digit numbers used.
- Find the number of 2-digit numbers used.
- Find the number of 3-digit numbers used.

Organize the simpler problems in a table.

Step 3

	Office Number	Number of Digits	Digits Remaining
1-digit numbers	1 to 9	$9 \times 1 = 9$	$513 - 9 = 504$
2-digit numbers	10 to 99	$90 \times 2 = 180$	$504 - 180 = 324$
3-digit numbers	100 to 199	$100 \times 3 = 300$	$324 - 300 = 24$

There are 24 digits remaining. Each number must have 3 digits.

$$24 \div 3 = 8 \qquad 199 + 8 = 207$$

Step 4 There are 207 office doors.

problems

1. The doors of an office building are numbered consecutively starting at 1. If 222 digits are used, how many doors are there?

2. Gavin Laidley uses a printing press that requires one piece of type for each digit that he prints. He used 801 pieces of type to number the pages of a certain book. If the first page number was 1, how many numbered pages did the book contain?

3. Kelley Rivers runs a printing press that requires one piece of type for each digit that she prints. The pages of a book are numbered from 1 to 333. How many pieces of type will be needed to number the pages of this book?

4. The houses on Lyons Drive are numbered consecutively from 1 to 124.
   a. How many house numbers contain at least one digit 4?
   b. How many house numbers contain at least one digit 9?
   c. How many house numbers contain at least one digit 1?

5. The houses on Maple Avenue are numbered consecutively from 10 to 899. How many digits are needed to form all the house numbers?

6. The doors of an office building are numbered consecutively starting at 100. If 2800 digits are used, how many doors are numbered?

7. Maura has a board that looks like the one at the right. There are 3 squares along each side. What is the total number of squares on the board?

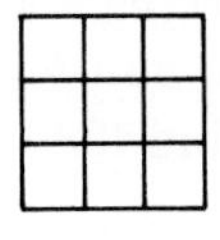

8. Emilio Ramerez drew the triangular figure at the right. What is the total number of triangles in the figure?

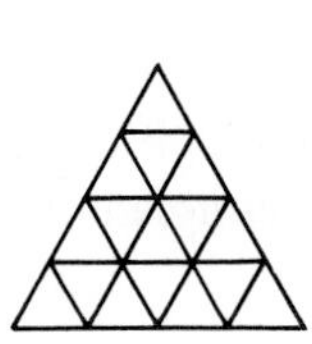

review exercises

In Exercises 1–3, each [cube] is a centimeter cube. Find the surface area of each space figure.

1. 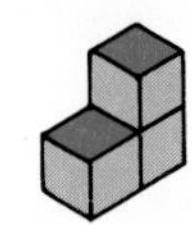

2. 

3. 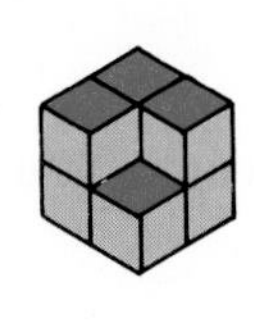

4. Eight people are attending a holiday party. Each of the eight people exchanges a present with each of the others. What is the total number of presents that are exchanged?

SKILL REVIEW

Classify each triangle first by its sides and then by its angles.

24-1

1. 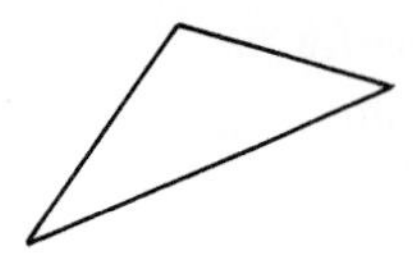

2. 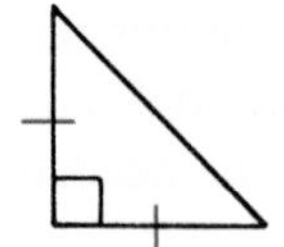

3. 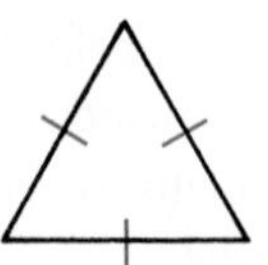

Can the given lengths be the lengths of the sides of a triangle? Write *Yes* or *No*.

4. 9 cm, 10 cm, 11 cm **5.** 8 ft, 3 ft, 5 ft **6.** 4 m, 10 m, 12 m

Find each square root. Use the table on page 549 or a calculator if necessary.

24-2

7. $\sqrt{49}$ **8.** $\sqrt{100}$ **9.** $\sqrt{400}$ **10.** $\sqrt{12{,}100}$

11. $\sqrt{102}$ **12.** $\sqrt{128}$ **13.** $\sqrt{1296}$ **14.** $\sqrt{21{,}316}$

The lengths of two sides of a right triangle are given. Assume that *c* is the length of the hypotenuse and *a* and *b* are the lengths of the legs. Sketch the triangle and then find the length of the third side.

24-3

15. $a = 5, b = 12$ **16.** $a = 20, b = 21$ **17.** $a = 15, c = 25$

18. $a = 35, c = 37$ **19.** $b = 8, c = 17$ **20.** $b = 14, c = 50$

In each pair, the triangles are similar. Find the unknown lengths. Then tell whether the triangles are *congruent* or *not congruent*.

24-4

21. 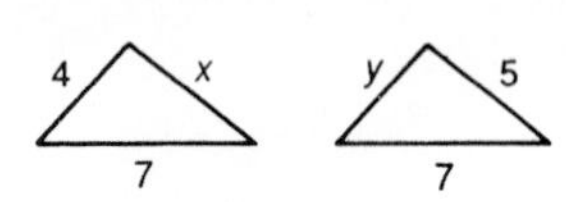

22. 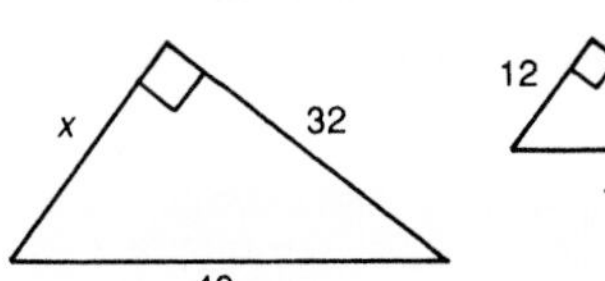

23. 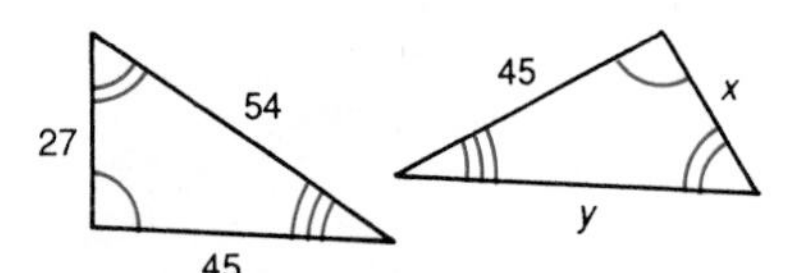

24. 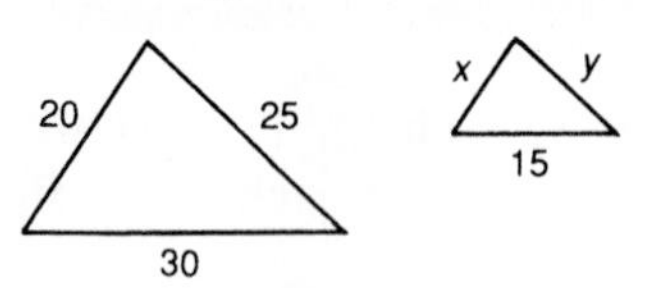

Use the triangle at the right. Find each ratio in lowest terms.

24-5

25. sin *B* **26.** cos *A*

27. cos *B* **28.** tan *A*

29. sin *A* **30.** tan *B*

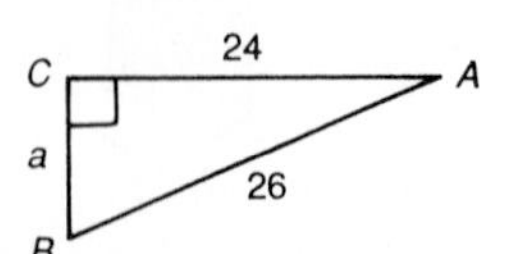

24-6

31. The pages of a certain book are numbered consecutively from 1 to 248. How many page numbers contain at least one digit 2?

APPLICATION

24-7 INDIRECT MEASUREMENT

If you cannot measure a length directly, you sometimes can use similar triangles to make an **indirect measurement.**

example

Mary Campo was standing near the school flagpole on a sunny day. She cast a shadow of 10 ft when the flagpole cast a shadow of 30 ft. Mary knows that she is 5 ft tall. How tall is the flagpole?

solution

Make a drawing.

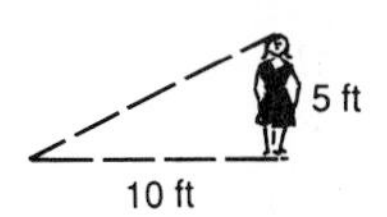

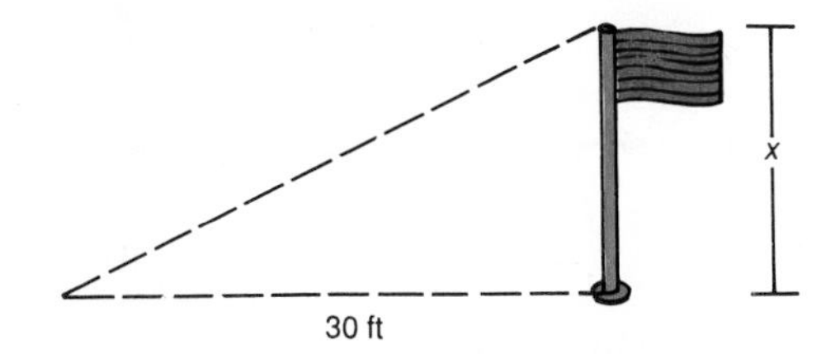

Find the ratio of the known lengths of corresponding sides.

$$\frac{\text{first triangle}}{\text{second triangle}} \rightarrow \frac{10}{30} = \frac{1}{3}$$ ← **Write the ratio in lowest terms.**

Set up and solve a proportion involving the unknown length.

$$\frac{\text{first triangle}}{\text{second triangle}} \rightarrow \frac{1}{3} = \frac{5}{x} \rightarrow x = 15$$

The flagpole is 15 ft tall.

exercises

In each pair, the triangles are similar. Find the unknown length *x*.

1.

2.

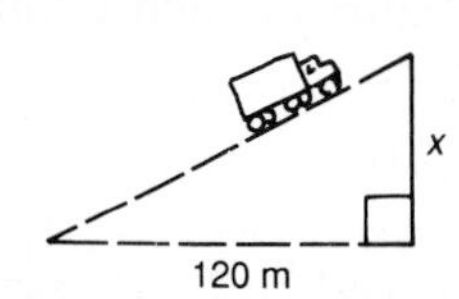

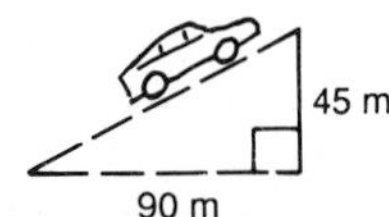

3. Lawrence Vozella casts a shadow of 8 ft at the same time that an elm tree casts a shadow of 24 ft. Lawrence Vozella is 6 ft tall. How tall is the elm tree?
4. A 7-ft fence post casts a shadow that is 15 ft long. At the same time, a building casts a shadow that is 60 ft long. Find the height of the building.

APPLICATION

24-8 USING A TRIGONOMETRIC TABLE

The trigonometric table on page 565 lists the values of the sine, cosine, and tangent of some angles with measures between 0° and 90°. You can use this trigonometric table to find the unknown lengths in a right triangle.

example

Find the unknown lengths in the triangle at the right. Round to the nearest tenth.

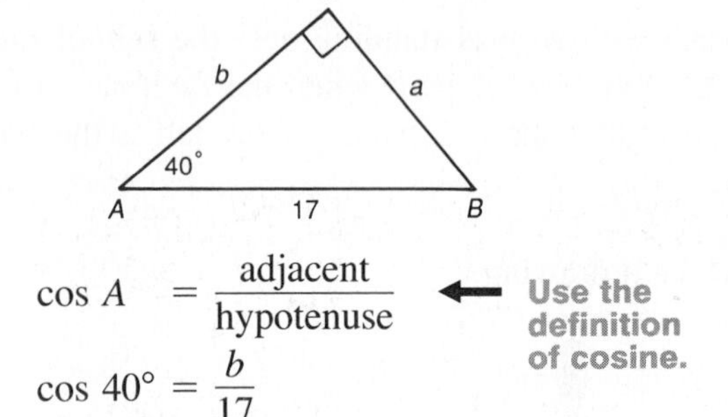

solution

$\sin A = \dfrac{\text{opposite}}{\text{hypotenuse}}$ ← Use the definition of sine.

$\sin 40° = \dfrac{a}{17}$

Find 40° in the *Angle* column of the table. Read the number across from 40° in the *Sine* column: .6428

$$0.6428 \approx \frac{a}{17}$$

$$17(0.6428) \approx 17\left(\frac{a}{17}\right)$$

$$10.9276 \approx a$$

$$a \approx 10.9$$

$\cos A = \dfrac{\text{adjacent}}{\text{hypotenuse}}$ ← Use the definition of cosine.

$\cos 40° = \dfrac{b}{17}$

Find 40° in the *Angle* column of the table. Read the number across from 40° in the *Cosine* column: .7660

$$0.7660 \approx \frac{b}{17}$$

$$17(0.7660) \approx 17\left(\frac{b}{17}\right)$$

$$13.022 \approx b$$

$$b \approx 13.0$$

exercises

Use the table on page 565 to find each value.

1. sin 30° **2.** sin 17° **3.** cos 82° **4.** cos 56° **5.** tan 45° **6.** tan 26°

Find the unknown lengths in each triangle. Round to the nearest tenth.

7.

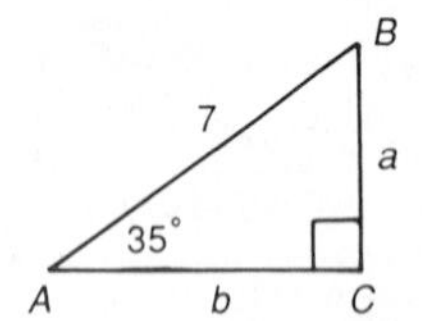

8.

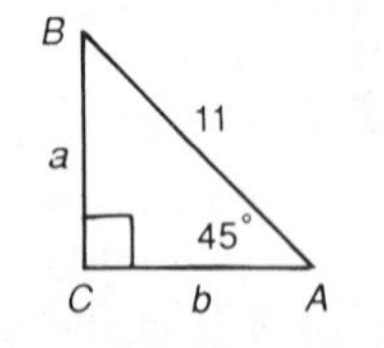

9.

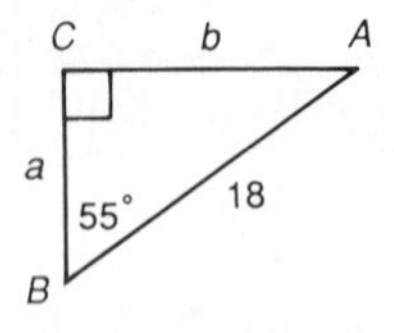

10.

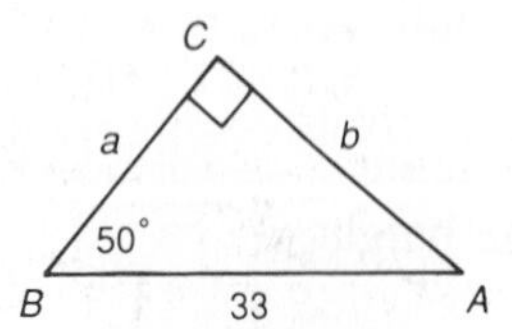

11.

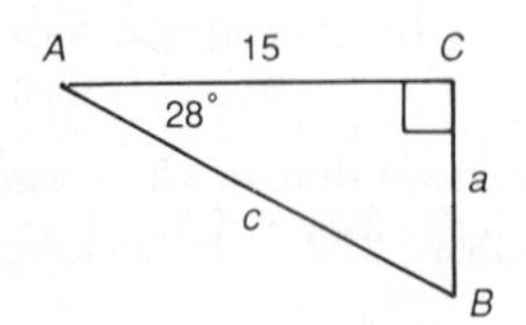

12.

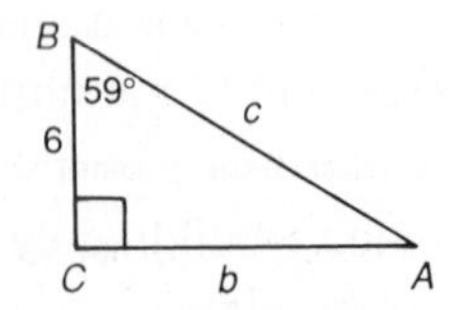

Table of Trigonometric Ratios

ANGLE	SINE	COSINE	TANGENT
1°	.0175	.9998	0.175
2°	.0349	.9994	.0349
3°	.0523	.9986	.0524
4°	.0698	.9976	.0699
5°	.0872	.9962	.0875
6°	.1045	.9945	.1051
7°	.1219	.9925	.1228
8°	.1392	.9903	.1405
9°	.1564	.9877	.1584
10°	.1736	.9848	.1763
11°	.1908	.9816	.1944
12°	.2079	.9781	.2126
13°	.2250	.9744	.2309
14°	.2419	.9703	.2493
15°	.2588	.9659	.2679
16°	.2756	.9613	.2867
17°	.2924	.9563	.3057
18°	.3090	.9511	.3249
19°	.3256	.9455	.3443
20°	.3420	.9397	.3640
21°	.3584	.9336	.3839
22°	.3746	.9272	.4040
23°	.3907	.9205	.4245
24°	.4067	.9135	.4452
25°	.4226	.9063	.4663
26°	.4384	.8988	.4877
27°	.4540	.8910	.5095
28°	.4695	.8829	.5317
29°	.4848	.8746	.5543
30°	.5000	.8660	.5774
31°	.5150	.8572	.6009
32°	.5299	.8480	.6249
33°	.5446	.8387	.6494
34°	.5592	.8290	.6745
35°	.5736	.8192	.7002
36°	.5878	.8090	.7265
37°	.6018	.7986	.7536
38°	.6157	.7880	.7813
39°	.6293	.7771	.8098
40°	.6428	.7660	.8391
41°	.6561	.7547	.8693
42°	.6691	.7431	.9004
43°	.6820	.7314	.9325
44°	.6947	.7193	.9657
45°	.7071	.7071	1.0000

ANGLE	SINE	COSINE	TANGENT
46°	.7193	.6947	1.0355
47°	.7314	.6820	1.0724
48°	.7431	.6691	1.1106
49°	.7547	.6561	1.1504
50°	.7660	.6428	1.1918
51°	.7771	.6293	1.2349
52°	.7880	.6157	1.2799
53°	.7986	.6018	1.3270
54°	.8090	.5878	1.3764
55°	.8192	.5736	1.4281
56°	.8290	.5592	1.4826
57°	.8387	.5446	1.5399
58°	.8480	.5299	1.6003
59°	.8572	.5150	1.6643
60°	.8660	.5000	1.7321
61°	.8746	.4848	1.8040
62°	.8829	.4695	1.8807
63°	.8910	.4540	1.9626
64°	.8988	.4384	2.0503
65°	.9063	.4226	2.1445
66°	.9135	.4067	2.2460
67°	.9205	.3907	2.3559
68°	.9272	.3746	2.4751
69°	.9336	.3584	2.6051
70°	.9397	.3420	2.7475
71°	.9455	.3256	2.9042
72°	.9511	.3090	3.0777
73°	.9563	.2924	3.2709
74°	.9613	.2756	3.4874
75°	.9659	.2588	3.7321
76°	.9703	.2419	4.0108
77°	.9744	.2250	4.3315
78°	.9781	.2079	4.7046
79°	.9816	.1908	5.1446
80°	.9848	.1736	5.6713
81°	.9877	.1564	6.3138
82°	.9903	.1392	7.1154
83°	.9925	.1219	8.1443
84°	.9945	.1045	9.5144
85°	.9962	.0872	11.4301
86°	.9976	.0698	14.3007
87°	.9986	.0523	19.0811
88°	.9994	.0349	28.6363
89°	.9998	.0175	57.2900

To *solve* a right triangle, you find the measures of all the sides and all the angles of the triangle.

Solve each triangle. Use the table on page 565. If necessary, round to the nearest tenth.

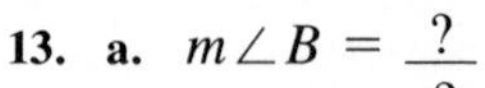

13. **a.** $m\angle B = \underline{?}$
b. $m\angle C = \underline{?}$
c. $a = \underline{?}$
d. $b \approx \underline{?}$

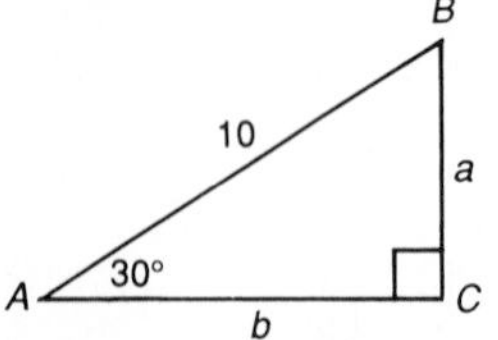

14. **a.** $m\angle A = \underline{?}$
b. $m\angle C = \underline{?}$
c. $a \approx \underline{?}$
d. $b \approx \underline{?}$

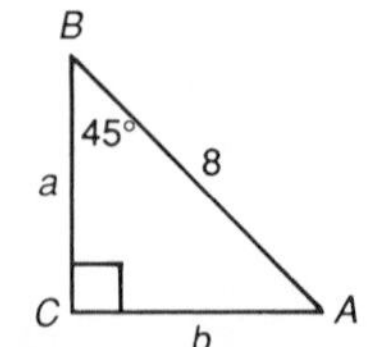

15. A wire that is attached to the top of a tent is 15 ft long. The wire forms a 60° angle with the ground. What is the height of the tent? Round to the nearest foot.

16. How high is a kite when 80 ft of string is let out and the string is secured so that it makes an angle of 30° with the ground?

17. Herman and Cal secured a 50-m rope to the top of a building. The rope forms a 50° angle with the ground. What is the height of the building? Round to the nearest meter.

18. Robert Lloyd needs to place his ladder so that it reaches 24 ft up the side of his house. When the ladder is fully extended, its length is 25 ft. How far from the house should Robert place the base of the ladder?

calculator corner

Some calculators have sine SIN, cosine COS, and tangent TAN keys.

example Use a calculator to find sin 14°, cos 14°, and tan 14°.

solution

Enter 14 SIN.	The result is 0.241921895.	sin 14° ≈ 0.2419
Enter 14 COS.	The result is 0.970295726.	cos 14° ≈ 0.9703
Enter 14 TAN.	The result is 0.249328002.	tan 14° ≈ 0.2493

Use a calculator to find the sine, cosine, and tangent of each angle.

1. 21° **2.** 35° **3.** 45° **4.** 72° **5.** 83° **6.** 90°

APPLICATION

24-9 CONSTRUCTING CONGRUENT LINE SEGMENTS AND ANGLES

A **construction** is made using only a *straightedge* and a *compass*. The **straightedge** is used to draw a line or a part of a line. The **compass** is used to draw a circle or a part of a circle. A part of a circle is called an **arc.**

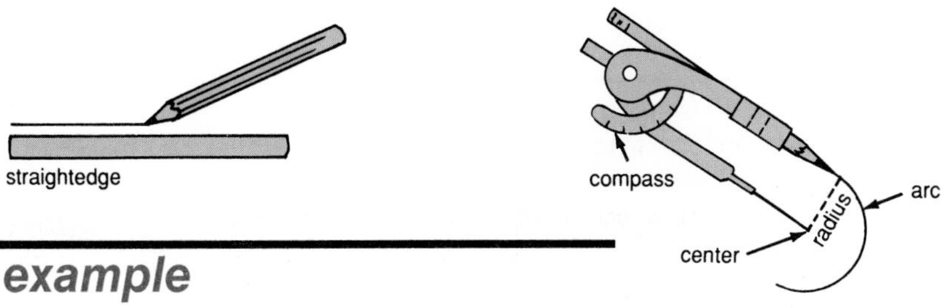

example

Construct $\overline{PQ}$ so that $\overline{PQ} \cong \overline{MN}$. ← **The symbol ≅ means *is congruent to.***

M N

solution

First trace $\overline{MN}$ onto a piece of paper.

Draw a ray with endpoint P.

P

Set the compass to a radius equal to the length of $\overline{MN}$.

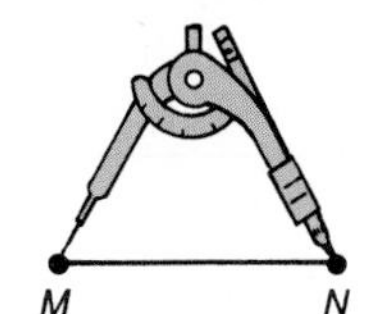

Using P as center, draw an arc intersecting the ray. Label point Q.

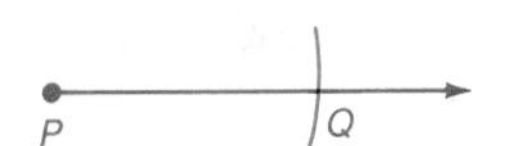

exercises

Trace each line segment onto a piece of paper. Then construct a line segment congruent to it.

1. 3.

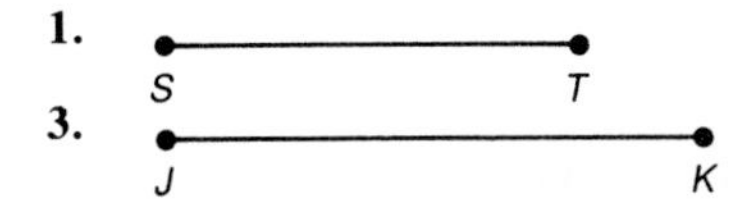

2. 4.

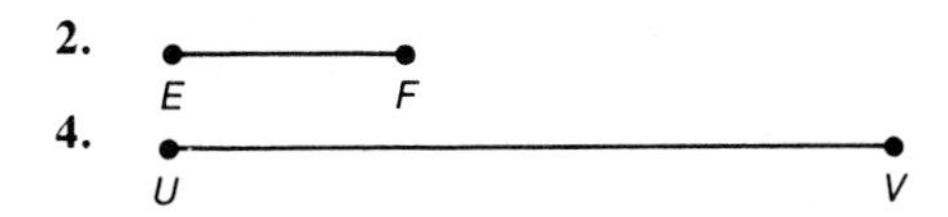

Trace each line segment onto a piece of paper. Then construct a line segment that is twice as long as the given segment.

5. 7.

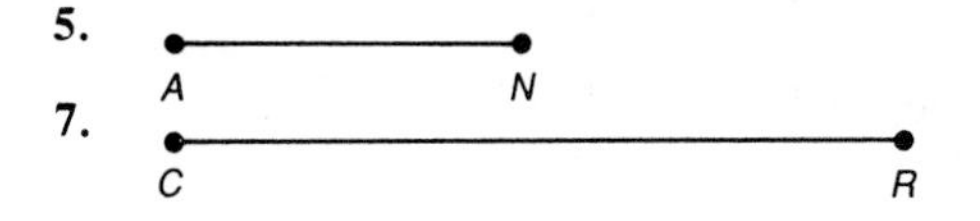

6. 8.

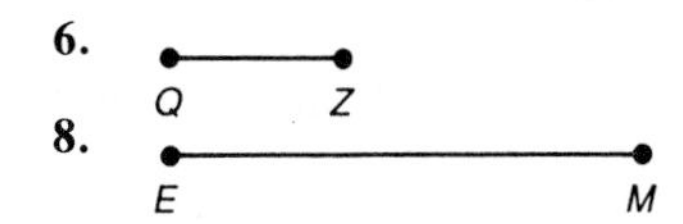

Given $\angle A$, you can use a straightedge and compass to construct an $\angle Q$ congruent to it. You write $\angle Q \cong \angle A$.

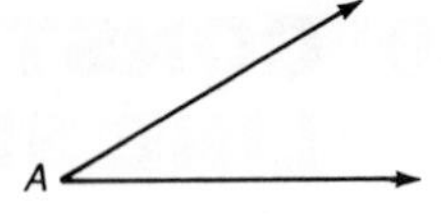

Begin by tracing $\angle A$ onto your paper.

Draw a ray with endpoint Q.

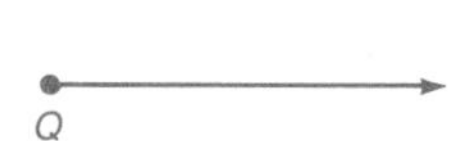

Using A as center, draw an arc. Label points X and Y.

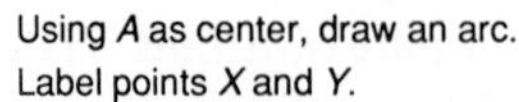

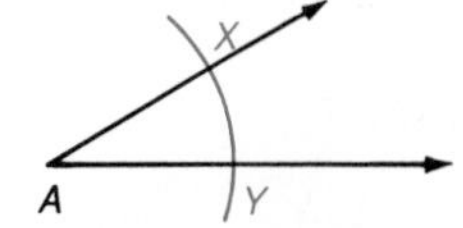

Using the same radius and Q as center, draw an arc. Label point R.

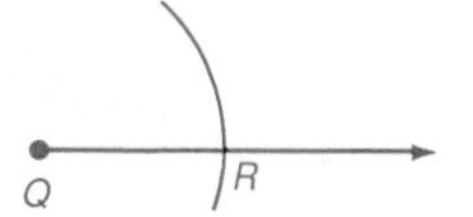

Set the compass to a radius equal to the length of $\overline{XY}$.

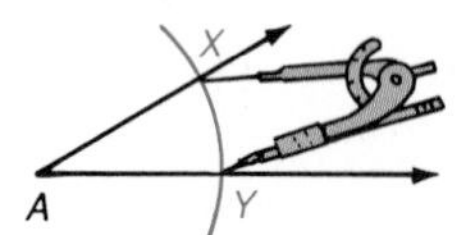

Using R as center, draw an arc. Label point P.

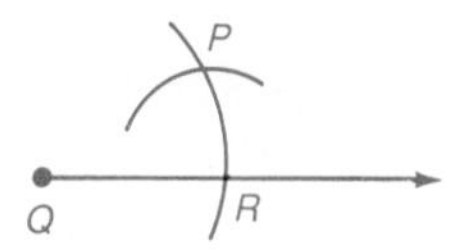

Draw $\overrightarrow{QP}$.

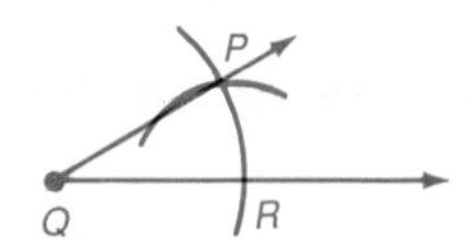

Trace each angle onto a piece of paper. Then construct an angle congruent to it.

9.

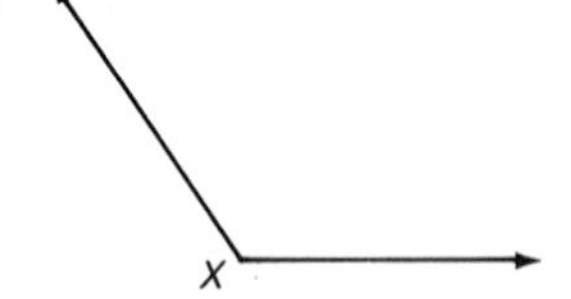

10.

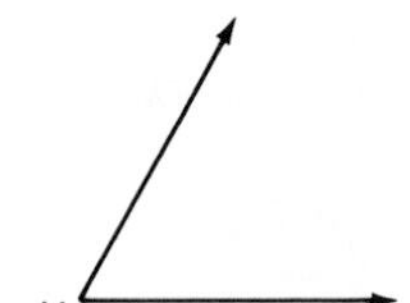

11.

Trace each angle onto a piece of paper. Then construct an angle with a measure that is twice the measure of the given angle.

12.

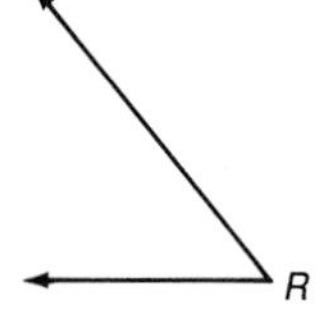

13.

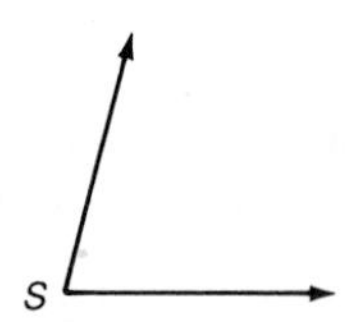

14.

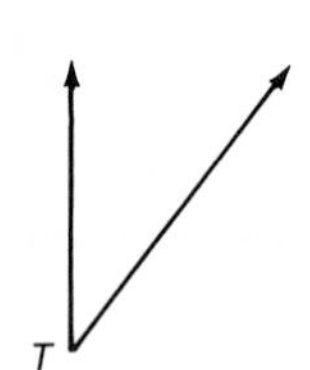

Trace each set of line segments and angles onto a piece of paper. Then use them to construct a triangle.

15.

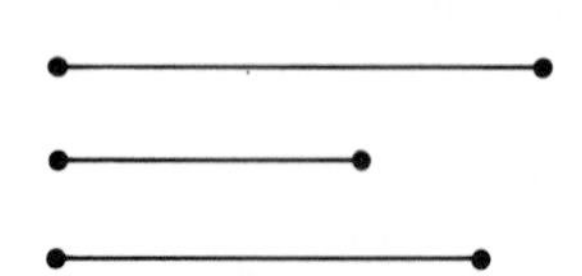

16.

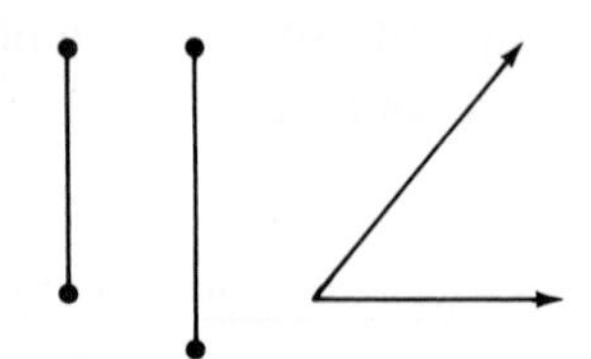

APPLICATION

24-10 BISECTING LINE SEGMENTS AND ANGLES

In the figure at the right, $\overline{AB}$ separates $\overline{CD}$ into two congruent segments. You say that $\overline{AB}$ **bisects** $\overline{CD}$. Any ray, segment, or line that bisects $\overline{CD}$ is called a **bisector** of $\overline{CD}$. You can construct a bisector using a compass and straightedge.

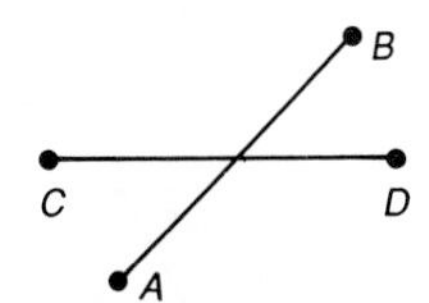

example

Bisect $\overline{XY}$.

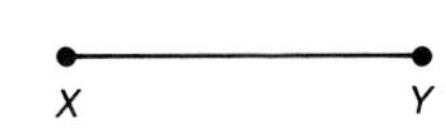

solution

First trace $\overline{XY}$ onto a piece of paper.

Choose a point Z on $\overline{XY}$ that is clearly closer to Y than to X. Set the compass to a radius equal to the length of $\overline{XZ}$.

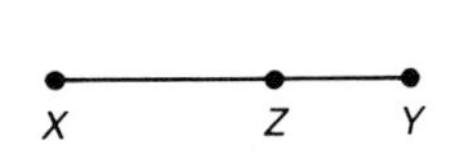

Using X as center, draw an arc.

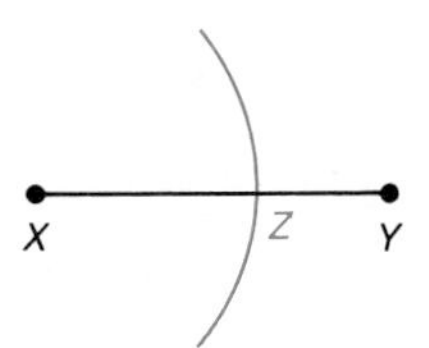

Using the same radius and Y as center, draw an arc intersecting the first in two points. Label them V and W.

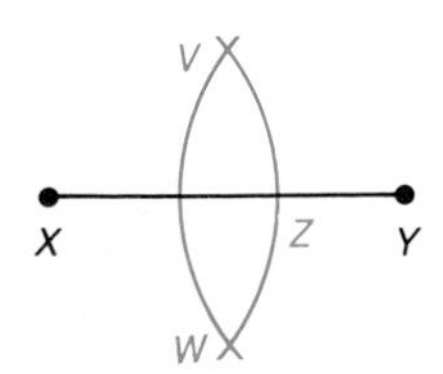

Draw $\overleftrightarrow{VW}$.

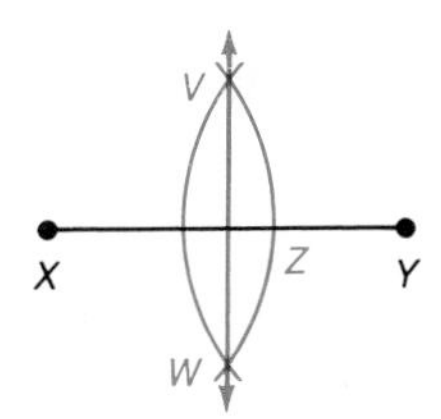

exercises

Trace each line segment onto a piece of paper. Then bisect it.

1. M ———— N

2. 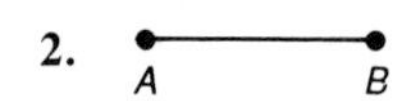

3. X ———— Y

4. J ———— K

5. Trace the line segment given in Exercise 1 onto a piece of paper. Construct a line segment with a length equal to one half the length of the given segment.
6. Trace the line segment given in Exercise 4 onto a piece of paper. Construct a line segment with a length equal to one fourth the length of the given segment.

An **angle bisector** is a ray that separates the angle into two congruent angles. To bisect $\angle B$, begin by tracing $\angle B$ onto a piece of paper.

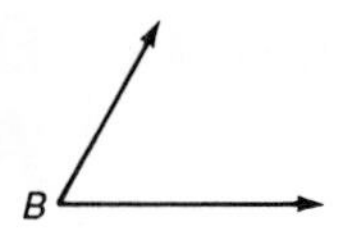

Using *B* as center, draw an arc. Label points *E* and *F*.

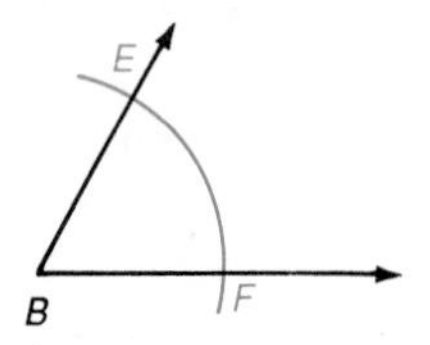

Using *E* as center, draw an arc.

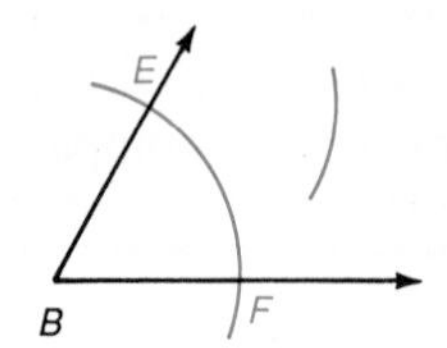

Using the same radius and *F* as center, draw an arc. Label point *X*.

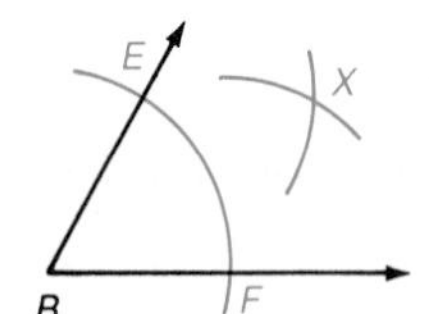

Draw $\overrightarrow{BX}$.

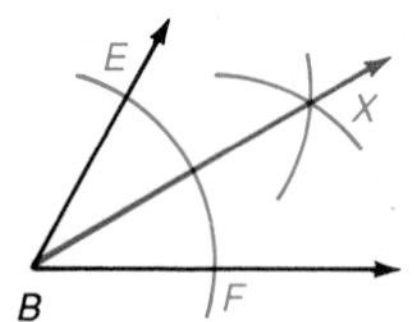

Trace each angle onto a piece of paper. Then bisect it.

7. 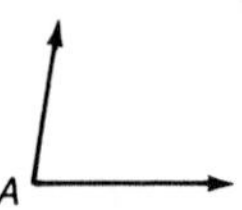

8. 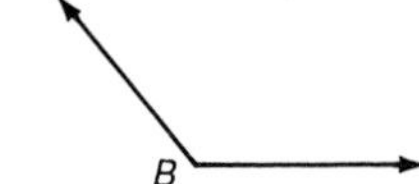

9. 

Trace each angle onto a piece of paper. Then construct an angle with a measure that is half the measure of the given angle.

10. 

11. 

12. 

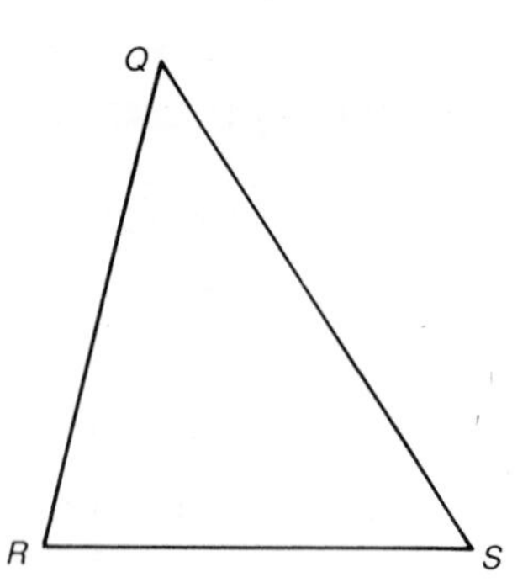

13. Trace the triangle at the right onto a piece of paper. Bisect $\angle Q$, $\angle R$, and $\angle S$. Do the bisectors intersect in one point?

14. Trace the triangle at the right onto a piece of paper. Bisect $\overline{QR}$, $\overline{QS}$, and $\overline{RS}$.

15. Refer to your work in Exercise 14. Notice that the bisectors intersect in one point. Label the point *P*. Draw a circle with center *P* and radius equal to the length of $\overline{PQ}$. What do you notice about points *R* and *S*?

16. Draw a line segment on a piece of paper. Explain how you could fold the paper to find the bisector of the line segment.

17. Draw an angle on a piece of paper. Explain how you could fold the paper to find the bisector of the angle.

24-11 CONSTRUCTING PERPENDICULAR LINES

In Chapter 22 you learned that two lines that intersect to form right angles are called perpendicular lines. Given any point P on a line, you can construct a line perpendicular to the given line at P.

example

Construct a line perpendicular to $\overleftrightarrow{AB}$ at P.

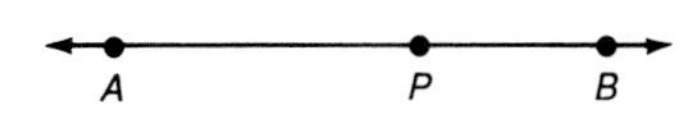

solution

First trace $\overleftrightarrow{AB}$ with point P onto a piece of paper.

Using P as center, draw an arc intersecting $\overleftrightarrow{AB}$ at two points. Label points R and S.

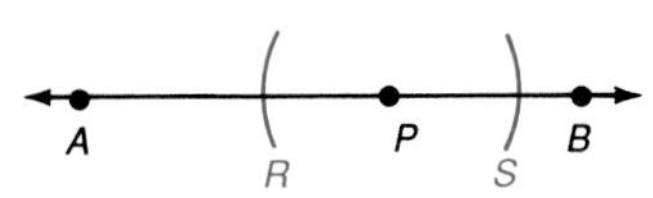

Set the compass to a radius greater than the length of $\overline{PR}$. Using R as center, draw an arc.

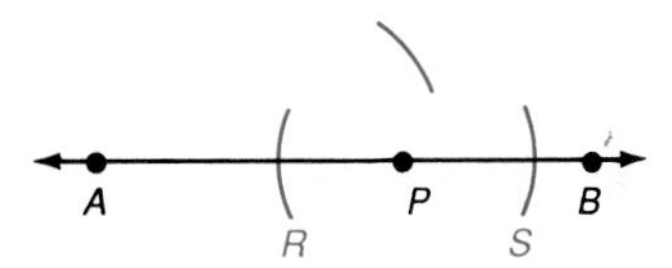

Using the same radius and S as center, draw an arc. Label point Q.

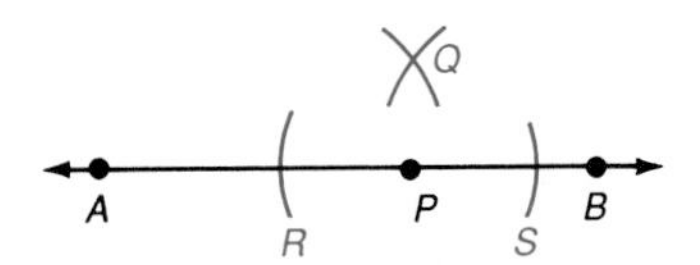

Draw $\overleftrightarrow{QP}$.

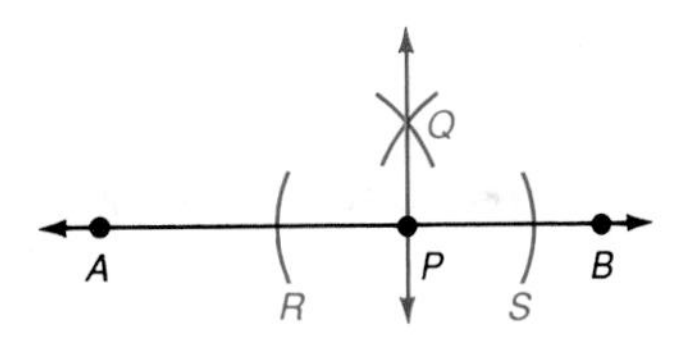

exercises

Trace each line onto a piece of paper. Then construct a line that is perpendicular to the given line at P.

1. D P F

2. O P Q

3. W P Y

4. R P T

For Exercises 5–7, first trace the given line segments onto a piece of paper.

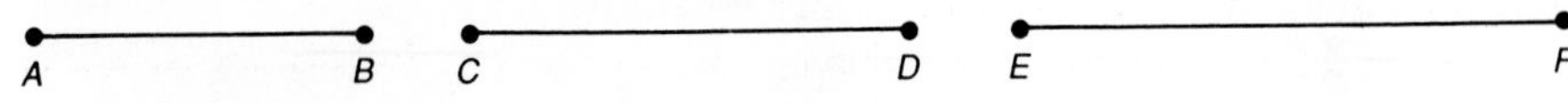

5. Construct a triangle XYZ with $\overline{XY} \cong \overline{AB}$, $\overline{YZ} \cong \overline{CD}$, and $\overline{XY} \perp \overline{YZ}$.
6. Construct a square with sides congruent to $\overline{CD}$.
7. Construct rectangle $PQRS$ with $\overline{PQ}$ and $\overline{RS}$ congruent to $\overline{CD}$, and with $\overline{QR}$ and $\overline{PS}$ congruent to $\overline{EF}$.

Given any line and a point P not on the line, you can also construct a line through P that is perpendicular to the given line. Begin by tracing $\overleftrightarrow{AB}$ and point P onto a piece of paper.

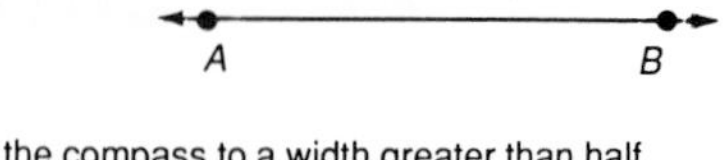

Using P as center, draw an arc intersecting $\overleftrightarrow{AB}$ in two points. Label the points R and S.

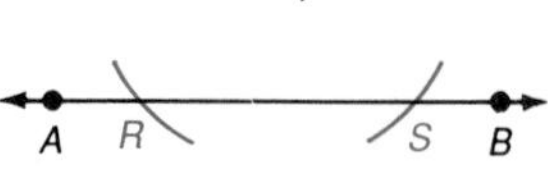

Set the compass to a width greater than half the length of $\overline{RS}$. Using R as center, draw an arc.

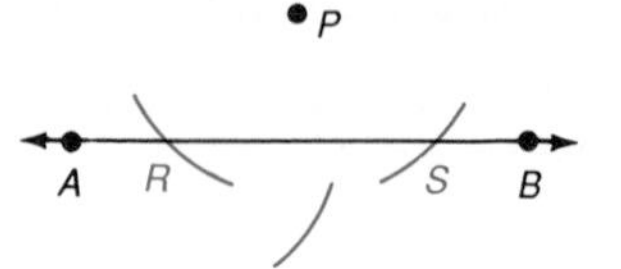

Using the same radius and S as center, draw an arc. Label point Q.

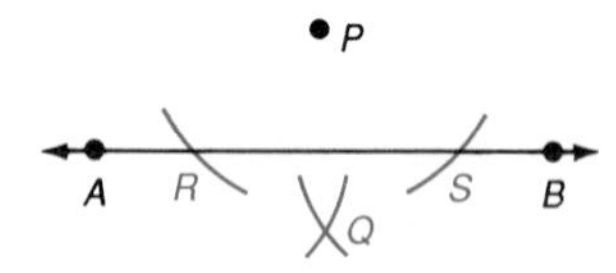

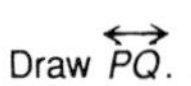

Draw $\overleftrightarrow{PQ}$.

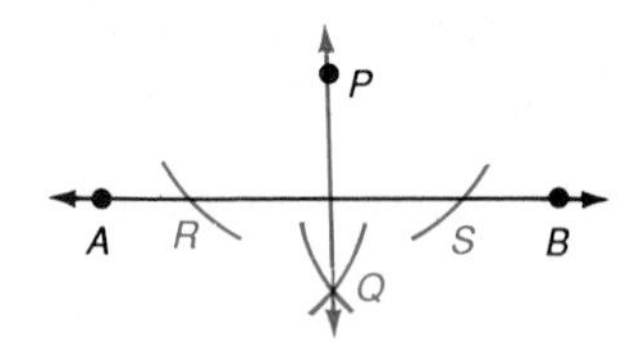

Trace each figure onto a piece of paper. Then construct a line through *P* that is perpendicular to the given line.

8.

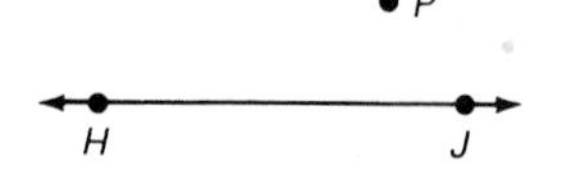

9.

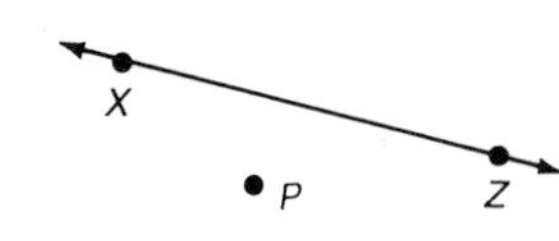

10.

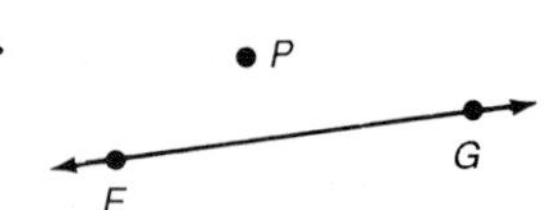

11. Trace this figure onto a piece of paper.

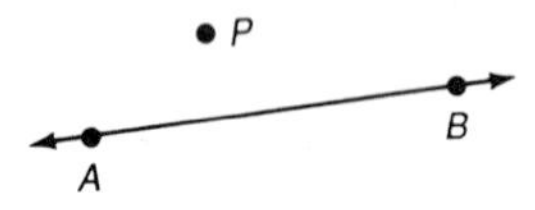

Through P, construct $\overleftrightarrow{PQ}$ perpendicular to $\overleftrightarrow{AB}$. At P, construct $\overleftrightarrow{PN}$ perpendicular to $\overleftrightarrow{PQ}$. What appears to be true of $\overleftrightarrow{PN}$ and $\overleftrightarrow{AB}$?

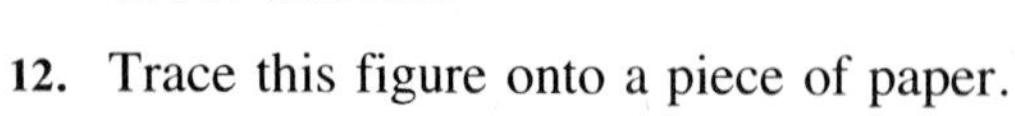

12. Trace this figure onto a piece of paper.

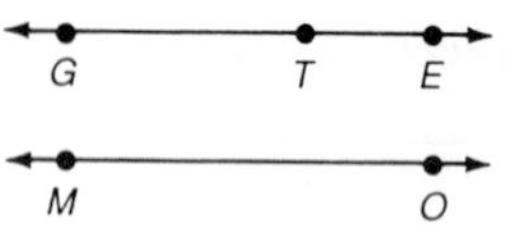

At T, construct $\overleftrightarrow{TR}$ perpendicular to $\overleftrightarrow{GE}$. What appears to be true of $\overleftrightarrow{TR}$ and $\overleftrightarrow{MO}$?

CHAPTER REVIEW

vocabulary vo·cab·u·lar·y

Choose the correct word to complete each sentence.

1. A triangle with a 95° angle is called an (*acute, obtuse*) triangle.
2. In a right triangle, the side opposite the right angle is called the (*hypotenuse, leg*).

skills

The measures of two angles of a triangle are given. Find the measure of the third angle. Then classify each triangle by its angles.

3. 32°, 77° 4. 105°, 20° 5. 90°, 53° 6. 67°, 15°

Find each square root. Use the table on page 549 or a calculator if necessary.

7. $\sqrt{144}$ 8. $\sqrt{225}$ 9. $\sqrt{3600}$ 10. $\sqrt{104}$ 11. $\sqrt{17{,}424}$

12. The triangles are similar. Find the unknown lengths.

Use the triangle at the right. Find each ratio in lowest terms.

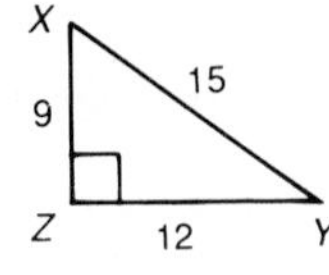

13. tan X 14. sin Y 15. cos Y
16. sin X 17. cos X 18. tan Y

19. The houses on Richdale Avenue are numbered consecutively from 1 to 150. How many house numbers contain at least one digit 4?
20. A wire from the top of a tower is 50 m long. The wire forms a 54° angle with the ground. What is the height of the tower? Round to the nearest meter.

For Exercises 21–23, first trace the figures below onto a piece of paper.

21. Construct a line segment $\overline{RS}$ that is congruent to $\overline{MN}$.
22. Bisect $\angle B$.
23. Construct a line perpendicular to $\overleftrightarrow{XY}$ at P.

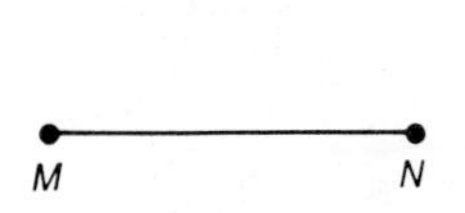

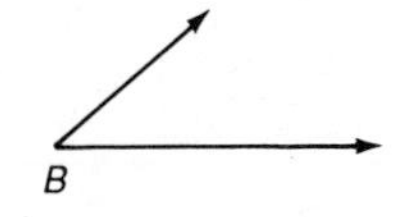

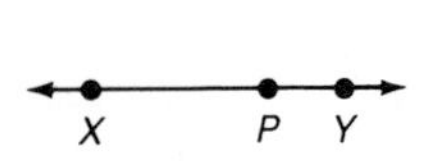

CHAPTER TEST

Classify each triangle by its sides and then by its angles. **24-1**

1.

2. 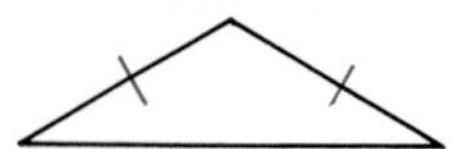

3. 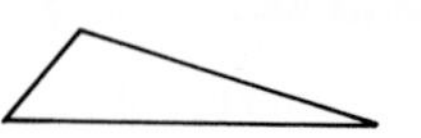

Find each square root. Use the table on page 549 or a calculator if necessary. **24-2**

4. $\sqrt{39}$ 5. $\sqrt{64}$ 6. $\sqrt{2116}$ 7. $\sqrt{15,376}$

The lengths of two sides of a right triangle are given. Assume that c is the length of the hypotenuse and a and b are the lengths of the legs. Sketch the triangle and then find the length of the third side. **24-3**

8. $a = 16,\ b = 30$ 9. $a = 24,\ c = 25$ 10. $b = 21,\ c = 29$

In each pair, the triangles are similar. Find the unknown lengths. **24-4**

11. 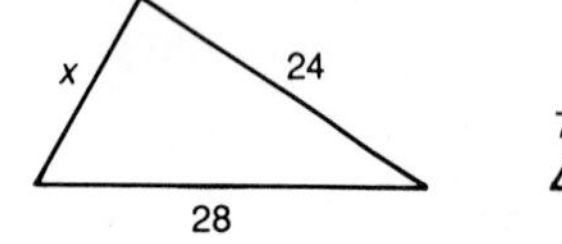

12. 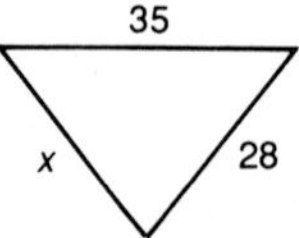

Use the triangle at right. Find each ratio in lowest terms. **24-5**

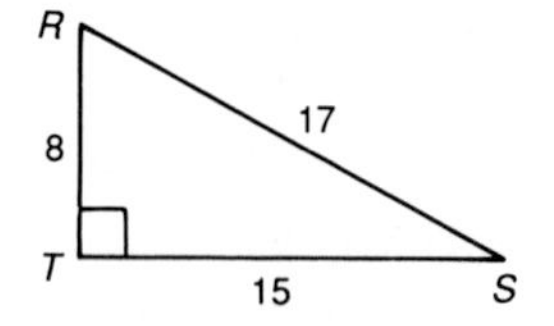

13. $\sin R$ 14. $\tan S$ 15. $\cos R$

16. $\sin S$ 17. $\tan R$ 18. $\cos S$

19. The classrooms in Dey School are numbered consecutively from 1 to 85. How many classroom numbers contain at least one digit 1? **24-6**

20. Juan casts a shadow of 12 ft at the same time that a tree casts a shadow of 32 ft. The tree is 16 ft tall. How tall is Juan? **24-7**

21. A 28-m rope is attached to the top of a post. The rope forms a 37° angle with the ground. What is the height of the post? Round to the nearest meter. **24-8**

For Exercises 22–24, first trace the figures below onto a piece of paper.

22. Construct an $\angle S$ that is congruent to $\angle K$. **24-9**

23. Bisect $\overline{VW}$. **24-10**

24. Construct a line perpendicular to $\overleftrightarrow{AB}$ through P. **24-11**

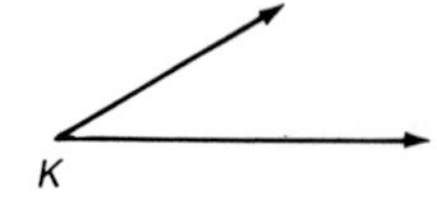

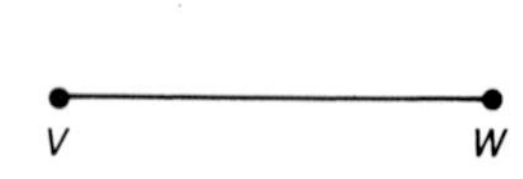

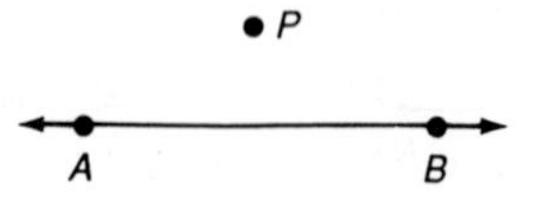

COMPUTER

Logo

The computer language **Logo** is used to draw pictures. To create a picture, you use a small triangle, called a **turtle,** that moves around the screen. Many turtle instructions, or **commands,** are familiar words that describe exactly what you want the turtle to do. For instance, the turtle can move *forward* or *back* a specified distance, or turn *left* or *right* a given number of degrees. The chart at the right gives the short form of some commands.

Command	Short Form
forward	FD
back	BK
right	RT
left	LT

The HOME command brings the turtle to home position, as shown at the right. Notice that the turtle points to the top of the screen.

Home Position

Any number of commands can be entered on one line. However, you may find it easier to read if you put no more than four commands on one line. You can use the following commands to create a square.

```
FD 35 RT 90   ← The turtle moves forward 35 "steps."
FD 35 RT 90
FD 35 RT 90   ← The turtle turns right 90°.
FD 35 RT 90
```

After completing these commands, a figure such as the one shown at the right should be on the screen.

The REPEAT command allows you to write fewer commands. When you created a square using the commands above, you entered FD 35 RT 90 four times. You can also enter the following.

REPEAT 4 [FD 35 RT 90]

exercises

Use grid paper to draw the path of the turtle and its position after completing each set of commands. Assume that the turtle starts in the home position.

1. FD 40 RT 90
FD 20 RT 90
FD 40 RT 90
FD 20 RT 90

2. BK 30 RT 90
BK 30 LT 90
FD 30 RT 90
FD 30 LT 90

3. REPEAT 3 [FD 50 RT 120]

4. Using the REPEAT command, write a set of Logo commands that will produce a rectangle.

COMPETENCY TEST

Choose the letter of the correct answer.

1. Which describes the triangle?

A. right, isosceles
B. acute, scalene
C. equilateral
D. right, scalene

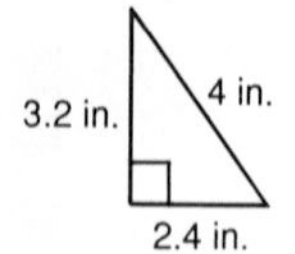

2. How many digits are used in writing the numbers from 1 to 385?

A. 385 B. 1068
C. 1155 D. 1047

3. Find the surface area. Use $\pi \approx \frac{22}{7}$.

A. 550 ft^2
B. 704 ft^2
C. 63 ft^2
D. 371 ft^2

4. Refer to Exercise 3. Find the volume. Use $\pi \approx 3.14$.

A. 1384.74 ft^3
B. 395.64 ft^3
C. 5538.96 ft^3
D. 461.58 ft^3

5. The triangles are similar. Find *x*.

A. $x = 7$
B. $x = 60$
C. $x = 63$
D. $x = 126.15$

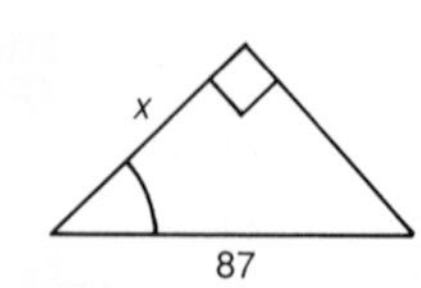

6. Find the surface area.

A. 240 m^2
B. 20 m^2
C. 124 m^2
D. 248 m^2

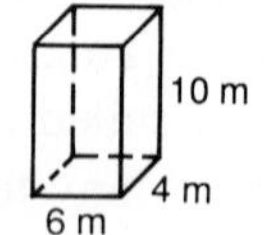

7. The lengths of the legs of a right triangle are 7 mm and 8 mm. Find the length of the hypotenuse.

A. 113 mm B. $\sqrt{113}$ mm
C. 15 mm D. $\sqrt{15}$ mm

8. Which three lengths *cannot* be the lengths of the sides of a triangle?

A. 3 m, 3 m, 3 m
B. 3 m, 3 m, 4 m
C. 3 m, 3 m, 5 m
D. 3 m, 3 m, 6 m

9. Find the circumference of a circle with diameter 8 in. Use $\pi \approx 3.14$.

A. 25.12 in.
B. 12.56 in.
C. 50.24 in.
D. 11.14 in.

10. Choose the true statement.

A. $\overleftrightarrow{AB} \parallel \overleftrightarrow{CD}$
B. $\overleftrightarrow{AB} \perp \overleftrightarrow{CD}$
C. $\angle 1$ and $\angle 2$ are supplementary angles.
D. $\angle 3$ and $\angle 4$ are complementary angles.

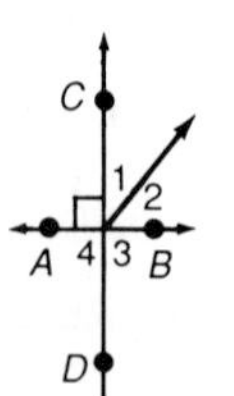

GEOMETRY

11. Choose the best estimate of the measure of the angle.

A. 90°
B. 45°
C. 135°
D. 60°

12. Find the area of a parallelogram with a base that measures 9 cm and height 4 cm.

A. 18 cm^2
B. 26 cm^2
C. 36 cm^2
D. 13 cm^2

13. Choose the word that best describes the figure.

A. prism
B. polygon
C. polyhedron
D. pyramid

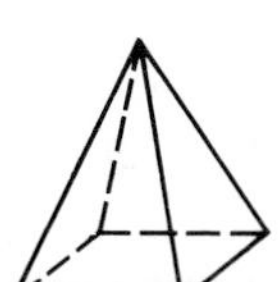

14. Find the perimeter of a regular hexagon with sides that each measure 5.4 cm.

A. 43.2 cm B. 32.4 cm
C. 27 cm D. 21.6 cm

15. Find the volume.

A. 32 m^3
B. 48 m^3
C. 8 m^3
D. 96 m^3

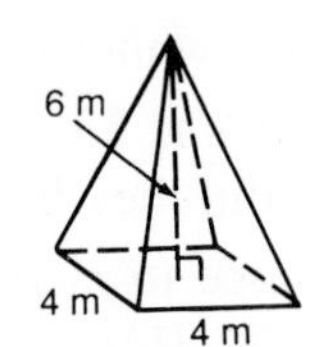

16. $\angle A$ and $\angle B$ are supplementary angles. If the measure of $\angle A$ is 46°, find the measure of $\angle B$.

A. 46° B. 44°
C. 144° D. 134°

17. Find the volume of a rectangular prism with length 8 in., width 9 in., and height 13 in.

A. 936 $in.^3$
B. 586 $in.^3$
C. 312 $in.^3$
D. 27,000 $in.^3$

18. Which ratio is greatest?

A. $\sin R$
B. $\cos R$
C. $\tan P$
D. $\sin P$

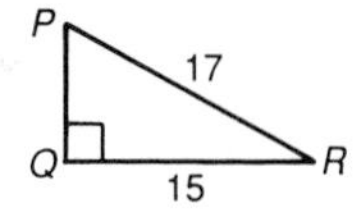

19. The measures of two angles of a triangle are 54° and 72°. Find the measure of the third angle.

A. 54° B. 72°
C. 126° D. 108°

20. Find the area. Use $\pi \approx \frac{22}{7}$.

A. 252 ft^2
B. 560 ft^2
C. 406 ft^2
D. 329 ft^2

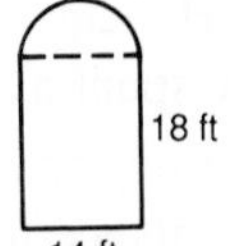

CUMULATIVE REVIEW

CHAPTERS 1–24

Find each answer. Write fractions in lowest terms.

1. $\sqrt{81}$ **2.** $0 \div {}^{-}7$ **3.** $9 + 6^2 \div 3$ **4.** 0.35×81.4

5. $({}^{-}14)({}^{-}8)$ **6.** $\$73.50 - \$15$ **7.** $({}^{-}2)(3)({}^{-}9)$ **8.** $0.65\overline{)159.9}$

9. ${}^{-}17\frac{1}{3} - {}^{-}8\frac{3}{4}$ **10.** $5\frac{3}{4} \div \frac{{}^{-}2}{3}$ **11.** $11\frac{3}{5} + 2\frac{1}{2}$ **12.** $412(381)$

13. 16 gal − 3 gal 3 qt

14. 5 ft 8 in. + 12 ft 10 in.

15. Write $7\frac{5}{8}$ as a decimal.

16. Write 0.3125 as a fraction.

17. Complete: **a.** 5.4 L = _?_ mL **b.** 23,000 g = _?_ kg

18. Write $127.53 in words as it would appear on a check.

19. Find the circumference of a circle with diameter 10 cm. Use $\pi \approx 3.14$.

20. Find the volume of a cone with radius 6 ft and height 7 ft. Use $\pi \approx \frac{22}{7}$.

21. Find the perimeter of a regular octagon with sides that each measure 2 in.

22. Graph $y = {}^{-}3x + 1$ on a coordinate plane. Use ${}^{-}2$, 0, 2 as values for x.

Estimate.

23. $427 - 336$ **24.** 0.738×4125 **25.** $\$56.51 + \$23.79$

26. $\frac{7}{34} \times 248$ **27.** $0.82\overline{)334}$ **28.** $4301 \div 597$

Solve.

29. $t - 3 < {}^{-}4$ **30.** ${}^{-}7y = 105$ **31.** $5z + 2 = 32$ **32.** $\frac{2}{3}y = {}^{-}22$

33. Louanne left her house and rode her bicycle 8 km due north and then 6 km due east. How far was she from her house?

34. Teresa decides to paint the walls of a room that is 16 ft long, 15 ft wide, and 9 ft high. One gallon of paint covers 440 ft^2. How many gallons of paint does she need?

35. Ramon bought two albums at $7.99 each. The sales tax rate is 6%. He paid with a $20 bill. How much change should Ramon receive?

36. A drawer contains 8 black, 6 brown, and 6 white socks. Two socks are drawn at random. The first sock is not replaced before the second one is drawn. Find the probability of drawing 2 white socks.

37. The Hamans bought sandwiches at $5 each and salads at $3 each. They spent a total of $19. How many sandwiches did they buy?

38. The Booster Club has $580 in its treasury. They hope to increase that balance to $1000 by charging annual dues of $2. How many members must they enroll to reach their goal?

table of measures

Time

60 seconds (s)	= 1 minute (min)	365 days, 52 weeks (approx.), 12 months	= 1 year
60 minutes	= 1 hour (h)	10 years	= 1 decade
24 hours	= 1 day	100 years	= 1 century
7 days	= 1 week		
4 weeks (approx.)	= 1 month		

Metric

Length

10 millimeters (mm)	= 1 centimeter (cm)
100 cm, 1000 mm	= 1 meter (m)
1000 m	= 1 kilometer (km)

Area

100 square millimeters (mm^2)	= 1 square centimeter (cm^2)
10,000 cm^2	= 1 square meter (m^2)
10,000 m^2	= 1 hectare (ha)

Volume

1000 cubic millimeters (mm^3)	= 1 cubic centimeter (cm^3)
1,000,000 cm^3	= 1 cubic meter (m^3)

Liquid Capacity

1000 milliliters (mL)	= 1 liter (L)
1000 L	= 1 kiloliter (kL)

Mass

1000 milligrams (mg)	= 1 gram (g)
1000 g	= 1 kilogram (kg)
1000 kg	= 1 metric ton (t)

Temperature–Degrees Celsius (°C)

0°C	= freezing point of water
37°C	= normal body temperature
100°C	= boiling point of water

United States Customary

Length

12 inches (in.)	= 1 foot (ft)
36 in., 3 ft	= 1 yard (yd)
5280 ft, 1760 yd	= 1 mile (mi)

Area

144 square inches (in.^{2})	= 1 square foot (ft^2)
9 ft^2	= 1 square yard (yd^2)
43,560 ft^2, 4840 yd^2	= 1 acre (A)

Volume

1728 cubic inches (in.^{3})	= 1 cubic foot (ft^3)
27 ft^3	= 1 cubic yard (yd^3)

Liquid Capacity

8 fluid ounces (fl oz)	= 1 cup (c)
2 c	= 1 pint (pt)
2 pt	= 1 quart (qt)
4 qt	= 1 gallon (gal)

Weight

16 ounces (oz)	= 1 pound (lb)
2000 lb	= 1 ton (t)

Temperature–Degrees Fahrenheit (°F)

32°F	= freezing point of water
98.6°F	= normal body temperature
212°F	= boiling point of water

Write the short word form of each number. 1-1

1. 278.361
2. 42,104,000
3. 11,000,320,434
4. 0.67
5. 0.5302
6. 210.3
7. 75.04
8. 9.008

Write the numeral form of each number.

9. 24 billion, 34 million, 129
10. 86 and 16 thousandths
11. nine million, nineteen thousand, two hundred eleven
12. three hundred fifty-two ten-thousandths
13. The area of Texas is about 262,017 mi^2 (square miles). Write the word form of this number.

Write the numeral form of each number. 1-2

14. 37.7 million
15. 129.4 thousand
16. 4.15 billion
17. 200.6 thousand
18. 7.022 billion
19. 60.405 million

Complete. Replace each __?__ with *thousand, million,* or *billion.*

20. 42,600,000 = 42.6 __?__
21. 3300 = 3.3 __?__
22. 5,020,000,000 = 5.02 __?__
23. 600,280,000 = 600.28 __?__
24. 117,200 = 117.2 __?__
25. 888,800,000,000 = 888.8 __?__
26. The population of Mexico is about 81.9 million. Write the numeral form of this number.

Round to the place of the underlined digit. 1-3

27. $\underline{7}48.2$
28. $0.3\underline{0}6$
29. $\$2\underline{5}.23$
30. $0.00\underline{4}3$
31. $\$14.7\underline{9}8$
32. $\underline{2}704$
33. $1\underline{1}.363$
34. $\$\underline{8}.64$
35. $\underline{9}5$
36. $0.0\underline{9}19$
37. $\$87.5\underline{2}6$
38. $21.\underline{9}9$
39. The weight of the ocean liner *Queen Elizabeth 2* is 65,983 t (tons). Round this number to the nearest thousand.

Compare. Replace each __?__ with >, <, or =. 1-4

40. 3651 __?__ 365
41. $938 __?__ $9380
42. 17.327 __?__ 17.272
43. 0.002 __?__ 0.020
44. 0.34 __?__ 0.342
45. 0.1 __?__ 0.011
46. 10.4 __?__ 10.40
47. 8.64 __?__ 86.4
48. $36.00 __?__ $36
49. 1.73 __?__ 1.094
50. 29.73 __?__ 29.7
51. $44.04 __?__ $44.40
52. At the hardware store, Al bought items that were priced at $3.69, $3.50, and $3.75. Write the prices in order from least to greatest.

Find each sum. **2-1**

1. 86 + 72 + 38
2. 3555 + 19 + 258
3. \$23 + \$1.16 + \$.09
4. 7.05 + 18 + 18.6
5. 11,047 + 983 + 11
6. \$14 + \$7987 + \$605
7. \$25 + \$52 + \$5
8. \$4 + \$50 + \$836
9. \$4.18 + \$2 + \$1.02
10. \$11.38 + \$18.71
11. 2.71 + 0.6 + 13.98
12. 0.093 + 16.41 + 0.04

Find each difference. **2-2**

13. 463 − 121
14. 11.55 − 0.26
15. \$1174 − \$73
16. \$28.35 − \$19.61
17. 9000 − 6342
18. 2677 − 1395
19. 80 − 65.6
20. 25.33 − 19.1
21. \$160 − \$94.25
22. \$198 − \$1.56
23. 1111 − 199
24. 34.81 − 5.006
25. Jesse saved \$80 for his January heating bill. The actual bill was \$73.24. How much money was left after Jesse paid his bill?

Estimate by rounding. **2-3**

26. 414 + 129
27. \$881 + \$712 + \$493
28. 8.73 + 6.4 + 3.25
29. 598 − 306
30. 4294 − 1973
31. \$73.12 − \$28.04

Estimate by using the front-end digits.

32. 641 + 862 + 395
33. 16.8 + 49.52 + 3.4
34. \$203 + \$81 + \$617
35. 82.6 − 17.3
36. \$3924 − \$3305
37. 194.2 − 76.8
38. At the Kite Festival, there were 198 boys and 167 girls. About how many more boys than girls were at the festival?

If possible, solve. If it is not possible to solve, tell what additional information is needed. **2-4**

39. Elizabeth bought a history book for \$14.50, a music book for \$12.95, and a Spanish book for \$13.95. How much change did Elizabeth receive?
40. Pedro earned \$360 delivering groceries and \$249 delivering newspapers. How much did he earn from these two jobs?
41. On Saturday, 1485 people visited the zoo. On Sunday, 1298 people visited the zoo. The cost of admission is \$5.50. How many more people visited on Saturday than on Sunday?
42. In January, 17 new members joined the Milville Gym. In February, 35 new members joined. What was the total membership of the Milville Gym at the end of February?

Estimate by finding a range of reasonable numbers. **3-1**

1. $2 \times \$54$ 2. $4 \times \$880$ 3. 27×85 4. 66×19
5. $3 \times \$179$ 6. $7 \times \$28$ 7. 33×428 8. 54×318
9. Sarah types 43 words per minute. Find a range of numbers to describe about how many words she types in 25 min (minutes).

Estimate by rounding to the place of the leading digit.

10. 3×29 11. $\$6 \times 93$ 12. 28×16 13. 42×74
14. $55 \times \$81$ 15. 21×17 16. $\$54 \times 815$ 17. 67×329
18. 76×582 19. $32 \times \$991$ 20. 63×458 21. 99×745

Find each product. Check the reasonableness of your answer. **3-2**

22. 5×16 23. $24 \times \$88$ 24. 32×93 25. 49×31
26. 19×337 27. 9×409 28. 7×67 29. $4 \times \$555$
30. 2×606 31. 36×35 32. 8×77 33. 62×92
34. $6 \times \$980$ 35. 393×610 36. $96 \times \$228$ 37. 114×835
38. 49×855 39. 571×380 40. $200 \times \$407$ 41. $95 \times \$118$
42. On his trip to Boston, Carl drove 265 mi (miles) each day. How many miles did he drive in eight days?

Use the properties of multiplication to find each product. **3-3**

43. 83×66 44. 478×363 45. $50 \times 32 \times 2$ 46. 25×77
47. 7×42 48. 3×710 49. 4×79 50. 6×594
51. 624×552 52. 365×144 53. $2 \times 37 \times 5$ 54. $5 \times 17 \times 20$
55. 9×37 56. 6×412 57. 3×98 58. $25 \times 26 \times 4$
59. 642×722 60. $4 \times 28 \times 25$ 61. $2 \times 43 \times 50$ 62. 8×797
63. The Music Club reserved five rows of seats at Symphony Hall. Each row has 38 seats. How many seats did the club reserve?

Find each answer. **3-4**

64. 6^2 65. 3^3 66. 2^4 67. 5^3
68. 1^8 69. 64^1 70. 18^0 71. 10^4
72. 10^9 73. 3×4^3 74. $2^5 \times 5^2$ 75. $7^3 \times 1^6$
76. $6^2 \times 4 \times 2^3$ 77. $8^2 \times 1^4 \times 7^1$ 78. $10^2 \times 5^0 \times 9^2$ 79. $2^0 \times 3^4 \times 4^2$
80. The distance from Earth to the moon is approximately 2×10^5 mi. Write the numeral form of this number.

extra practice

CHAPTER 4

Estimate using compatible numbers.

4-1

1. $9\overline{)8230}$
2. $5\overline{)1447}$
3. $4\overline{)3922}$
4. $11{,}871 \div 3$
5. $73{,}658 \div 8$
6. $35{,}221 \div 62$
7. $19\overline{)21{,}453}$
8. $44\overline{)15{,}274}$
9. $20\overline{)41{,}098}$
10. $297{,}841 \div 58$
11. $622{,}731 \div 71$
12. $414{,}998 \div 37$
13. $316\overline{)593{,}024}$
14. $163\overline{)194{,}351}$
15. $723\overline{)550{,}399}$
16. Ana works after school for a total of 12 h (hours) each week. She earns \$59 per week. About how much does she earn per hour?

Find each quotient. Check the reasonableness of your answer.

4-2

17. $9164 \div 4$
18. $8412 \div 7$
19. $70{,}528 \div 3$
20. $17\overline{)2405}$
21. $61\overline{)1121}$
22. $32\overline{)1396}$
23. $21{,}585 \div 15$
24. $38{,}854 \div 64$
25. $49{,}497 \div 93$
26. $42\overline{)44{,}030}$
27. $21\overline{)394{,}608}$
28. $12\overline{)292{,}428}$
29. $998{,}135 \div 34$
30. $429{,}213 \div 173$
31. $637{,}132 \div 229$
32. Sam Chin has 375 stamps to put in his album. There is room for 25 stamps on each page. How many pages can Sam fill?

Find each answer.

4-3

33. $12 \div 3 + 4 \times 2$
34. $11 + 3 \times 8 - 5$
35. $90 - 3 + 2^3$
36. $(2 - 1) \times (7 + 3)$
37. $(6^2 + 14) \div 25$
38. $3 \times (7^2 + 1) \div 5$
39. $\frac{19 + 9}{2 \times 7}$
40. $\frac{15 \div 3}{8 - 3}$
41. $\frac{9 + 15}{36 - 3 \times 10}$
42. $26 + 3^2 \div 3 - 2$
43. $4^2 \times 5 \div 2 + 17$
44. $4 \times (2^4 - 6) \div 4$
45. $5^2 + 6 \times 12 - 2^3$
46. $\frac{3 \times 15}{7 + 16 \div 2}$
47. $\frac{24 + 64 \div 4}{5 \times 2}$
48. Jackson bought four tapes that cost \$9 each and a tape player that cost \$30. What was the total amount of his purchase?

Solve.

4-4

49. Manuel bought a calendar that cost \$11.50 and a book that cost \$8.75. He gave the clerk \$40. How much change did he receive?

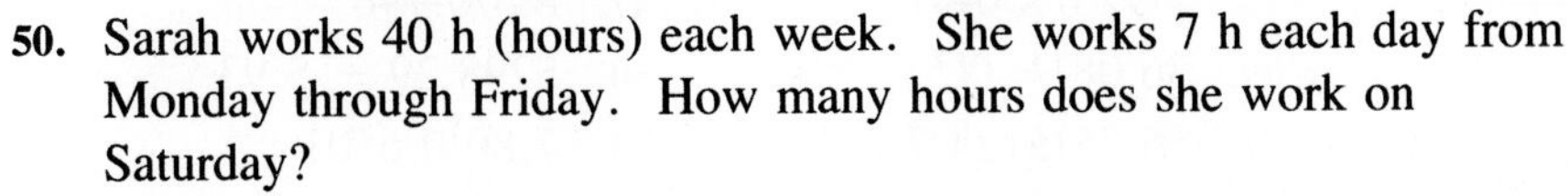

50. Sarah works 40 h (hours) each week. She works 7 h each day from Monday through Friday. How many hours does she work on Saturday?
51. A factory makes 68,400 nails each day. One box holds 150 nails. How many boxes of nails does the factory make each day?

extra practice CHAPTER 5

Find each product. If necessary, round to the nearest cent. **5-1**

1. 6×32.7
2. 8.14×25
3. 3×0.635
4. 0.05×0.38
5. 1.9×7.42
6. 0.651×0.7
7. $4 \times \$2.70$
8. $0.18 \times \$5.29$
9. $\$7.86 \times 63$
10. $10 \times \$3.03$
11. $0.01 \times \$14.79$
12. $\$.98 \times 0.52$
13. Each day Neil Gillis earns money by delivering 96 newspapers in his neighborhood. Each newspaper costs $.35. What is the total cost of the newspapers that Neil delivers?

Estimate by rounding to the place of the leading digit. **5-2**

14. $2.7 \times \$49.63$
15. 4.63×13.45
16. $\$.23 \times 0.34$
17. 0.606×986
18. $0.69 \times \$8.55$
19. 814×0.52
20. 0.94×0.492
21. 0.872×3.59
22. 0.13×3.041
23. $0.87 \times \$64.13$
24. 6.99×7.6
25. 0.41×5203
26. Red cabbage costs $.59/lb (per pound). About how much does a 3.2-lb head of red cabbage cost?
27. The senior class sold 276 tickets that cost $2.25 each. About how much money did the senior class earn from ticket sales?

Find the quotient. If necessary, round to the nearest thousandth. **5-3**

28. $6\overline{)\$65.82}$
29. $4\overline{)\$28.12}$
30. $18\overline{)36.63}$
31. $68 \div 23$
32. $0.72 \div 0.9$
33. $55 \div 0.25$
34. $1.42\overline{)8.591}$
35. $0.15\overline{)1.8}$
36. $53\overline{)45}$
37. $6.53 \div 0.36$
38. $32.488 \div 6.2$
39. $9.213 \div 0.003$
40. The total cost of a skiing trip for 20 students was $350. What was the cost for each student?
41. Alisa bought 7 qt (quarts) of strawberries for $13.23. How much did one quart of strawberries cost?

Estimate using compatible numbers. **5-4**

42. $\$18.52 \div \$4.98$
43. $1139.4 \div 6.2$
44. $544 \div 0.7$
45. $0.37\overline{)0.152}$
46. $\$7.23\overline{)\$50.31}$
47. $8.5\overline{)0.796}$
48. $0.287 \div 0.9$
49. $36.08 \div 0.51$
50. $\$298.29 \div \$.93$
51. $3.21\overline{)22.09}$
52. $\$5.35\overline{)\$124.73}$
53. $7.89\overline{)0.6501}$
54. An 18.2-oz (ounce) box of oatmeal costs $1.19. About how much does one ounce of oatmeal cost?

Select the most reasonable unit. Choose mm, cm, m, or km. 6-1

1. length of a comb: 17 _?_
2. height of a building: 10.5 _?_
3. width of a button: 12 _?_
4. length of a race: 6 _?_

Measure each line segment in: a. millimeters b. centimeters

5. ├──────────────┤
6. ├────────────────────┤

Complete.

7. 7.2 cm = _?_ mm
8. 1500 m = _?_ km
9. 1230 cm = _?_ m
10. 2.65 km = _?_ m
11. 19 m = _?_ cm
12. 384 mm = _?_ cm
13. Anthony is 1.8 m tall. Theodore is 1.69 m tall. How many centimeters taller is Anthony?

Select the more reasonable unit. Choose mL or L. 6-2

14. capacity of a soup bowl: 354 _?_
15. capacity of a pitcher: 2 _?_
16. capacity of a tea cup: 0.24 _?_
17. capacity of a bath tub: 94.6 _?_

Complete.

18. 17 L = _?_ mL
19. 92,371 mL = _?_ L
20. 8600 mL = _?_ L
21. 0.295 L = _?_ mL
22. 565 mL = _?_ L
23. 4.3 L = _?_ mL

Select the most reasonable unit. Choose mg, g, or kg. 6-3

24. mass of a block of cheese: 400 _?_
25. mass of a radio: 2270 _?_
26. mass of a bag of potatoes: 3 _?_
27. mass of a butterfly: 2 _?_

Complete.

28. 2.9 kg = _?_ g
29. 163 mg = _?_ g
30. 5940 g = _?_ kg
31. 0.3 g = _?_ mg
32. 867 g = _?_ kg
33. 17 kg = _?_ g

Solve. 6-4

34. At 4:00 P.M., 35 students are studying in the library. At 4:30 P.M., 7 students leave. At 5:00 P.M., 3 students arrive. At 5:30 P.M., 7 more students leave. If this pattern continues, how many students will be studying in the library at 9:05 P.M.?
35. One pint of blueberries costs \$1.75. Two pints cost \$3.25, three pints cost \$4.75, and four pints cost \$6.25. If this pattern continues, how much will seven pints of blueberries cost?

Is the first number divisible by the second number? Write *Yes* or *No*. 7-1

1. 51,733; 3
2. 8665; 5
3. 112,842; 9
4. 24,891; 2
5. 74,816; 8
6. 439,654; 10
7. 45,102; 3
8. 9,024,874; 4

Write the prime factorization of each number. 7-2

9. 195
10. 40
11. 104
12. 170
13. 147
14. 90

Find the GCF and the LCM.

15. 14 and 28
16. 9 and 24
17. 15 and 35
18. 11 and 54
19. 6 and 44
20. 18, 21, and 45
21. 20, 65, and 80
22. 8, 16, and 56
23. List all the prime numbers between 22 and 32.

Replace each __?__ with the number that will make the fractions equivalent. 7-3

24. $\frac{3}{4} = \frac{?}{16}$
25. $\frac{18}{27} = \frac{2}{?}$
26. $\frac{15}{?} = \frac{3}{10}$
27. $\frac{?}{2} = \frac{7}{14}$

Write each fraction in lowest terms.

28. $\frac{3}{18}$
29. $\frac{6}{15}$
30. $\frac{25}{30}$
31. $\frac{40}{80}$
32. $\frac{32}{76}$
33. $\frac{93}{105}$

Compare. Replace each __?__ with >, <, or =. 7-4

34. $\frac{7}{12}$ __?__ $\frac{13}{24}$
35. $\frac{3}{4}$ __?__ $\frac{4}{5}$
36. $2\frac{5}{6}$ __?__ $\frac{17}{6}$
37. $\frac{13}{10}$ __?__ $1\frac{3}{5}$

Write each set of numbers in order from least to greatest.

38. $\frac{5}{6}, \frac{11}{12}, \frac{1}{6}$
39. $\frac{2}{3}, \frac{7}{18}, \frac{4}{9}$
40. $\frac{3}{5}, \frac{2}{3}, \frac{7}{10}$
41. $1\frac{2}{7}, 1\frac{3}{14}, 1\frac{1}{4}$

Write a simpler fraction as an estimate of each fraction. 7-5

42. $\frac{8}{15}$
43. $\frac{11}{30}$
44. $\frac{29}{39}$
45. $\frac{20}{98}$
46. $\frac{41}{63}$
47. $\frac{202}{500}$

Tell whether each fraction is closest to 0, $\frac{1}{2}$, or 1.

48. $\frac{9}{17}$
49. $\frac{85}{87}$
50. $\frac{16}{300}$
51. $\frac{33}{68}$
52. $\frac{2}{95}$
53. $\frac{144}{149}$
54. About what part of a day is nine hours?
55. There are 197 sophomores at Valley Regional High School. The number of sophomores who come to school on the bus is 153. About what part of the sophomores come to school on the bus?

Write each product in lowest terms. **8-1**

1. $\frac{1}{2} \times \frac{4}{7}$
2. $\frac{13}{20} \times \frac{5}{26}$
3. $5 \times \frac{3}{8}$
4. $\frac{3}{14} \times 16$
5. $\frac{2}{5} \times 1\frac{7}{8}$
6. $2\frac{5}{6} \times 12$
7. $3\frac{1}{9} \times 1\frac{3}{7}$
8. $2\frac{2}{15} \times 3\frac{3}{4}$

Estimate. **8-2**

9. $6\frac{1}{2} \times 3\frac{4}{5}$
10. $4\frac{5}{8} \times 8\frac{1}{3}$
11. $5\frac{2}{5} \times 11\frac{3}{10}$
12. $7\frac{2}{9} \times 9\frac{5}{7}$
13. $\frac{1}{8} \times 239$
14. $49 \times \frac{5}{32}$
15. $\frac{7}{20} \times 8\frac{3}{4}$
16. $\frac{62}{81} \times 15\frac{2}{9}$
17. Julie Yee works at a savings bank. She works $5\frac{1}{2}$ days each week. Each day Julie works $6\frac{1}{4}$ h. About how many hours does Julie work each week?

Write the reciprocal of each number, if possible. **8-3**

18. $\frac{1}{5}$
19. $\frac{7}{8}$
20. 0
21. 10
22. $1\frac{4}{9}$
23. $3\frac{1}{2}$

Write each quotient in lowest terms.

24. $\frac{2}{5} \div \frac{3}{4}$
25. $\frac{6}{7} \div \frac{15}{28}$
26. $20 \div \frac{10}{19}$
27. $\frac{7}{9} \div 7$
28. $2\frac{2}{3} \div 1\frac{1}{15}$
29. $3\frac{1}{2} \div \frac{5}{12}$
30. $35 \div 1\frac{2}{5}$
31. $1\frac{3}{8} \div 2\frac{3}{4}$
32. A recipe for one loaf of raisin bread requires $\frac{1}{4}$ c of raisins. How many loaves of raisin bread can be made using 3 c of raisins?

Estimate. **8-4**

33. $20\frac{2}{3} \div 6\frac{4}{7}$
34. $12\frac{2}{5} \div 5\frac{1}{3}$
35. $40\frac{2}{3} \div 9\frac{5}{8}$
36. $36\frac{1}{4} \div 5\frac{2}{9}$
37. $3\frac{5}{6} \div \frac{8}{17}$
38. $23\frac{4}{5} \div \frac{11}{40}$
39. $9\frac{1}{8} \div \frac{12}{59}$
40. $11\frac{2}{7} \div \frac{6}{21}$

8-5

41. A skirt pattern requires $3\frac{1}{2}$ yd (yards) of material. How many skirts can be made from 11 yd of material?
42. Doreen can read 40 pages in one hour. How many hours should it take her to read a book that has 180 pages?
43. Stanley Warner paid $7 for four jars of tomato sauce. What was the cost of each jar of sauce?

Write each sum or difference in lowest terms. 9-1

1. $\frac{11}{13} + \frac{2}{13}$
2. $\frac{9}{10} - \frac{3}{10}$
3. $\frac{1}{3} + \frac{4}{9}$
4. $\frac{3}{14} + \frac{6}{7}$
5. $\frac{2}{3} - \frac{8}{15}$
6. $\frac{5}{6} - \frac{11}{24}$
7. $\frac{4}{5} + \frac{3}{4}$
8. $\frac{8}{9} + \frac{1}{2}$
9. $\frac{5}{6} - \frac{5}{8}$
10. $\frac{7}{9} - \frac{1}{6}$
11. Tina planted potatoes in $\frac{1}{4}$ of her garden and corn in $\frac{3}{10}$ of her garden. What fraction of her garden contains the corn and the potatoes?

Estimate by rounding. 9-2

12. $11\frac{4}{7} + 8\frac{2}{3}$
13. $18\frac{5}{6} + 4\frac{3}{11}$
14. $15\frac{1}{4} - 6\frac{3}{7}$
15. $8\frac{2}{7} - 2\frac{1}{2}$
16. $7\frac{2}{15} + 9\frac{3}{8}$
17. $21\frac{1}{3} + 34\frac{3}{5}$
18. $10\frac{5}{8} - 3\frac{6}{11}$
19. $26\frac{5}{9} - 13\frac{1}{12}$

Estimate by adding the whole numbers and then adjusting.

20. $\frac{4}{9} + 3\frac{5}{8} + 7\frac{1}{10}$
21. $6\frac{10}{11} + 1\frac{2}{3} + 4\frac{5}{18}$
22. $2\frac{3}{4} + 12 + 5\frac{3}{16}$
23. $9\frac{1}{6} + 4\frac{3}{8} + 1\frac{11}{24}$
24. $10\frac{5}{6} + \frac{9}{10} + 5\frac{1}{3}$
25. $\frac{4}{19} + 16 + 4\frac{4}{5}$
26. $2\frac{7}{18} + 7\frac{3}{10} + 8\frac{5}{14}$
27. $6\frac{14}{15} + 3\frac{8}{9} + \frac{1}{8}$
28. $1\frac{6}{17} + 9\frac{21}{30} + \frac{6}{7}$

Write each sum in lowest terms. 9-3

29. $6\frac{3}{8} + 5\frac{1}{8}$
30. $8 + 10\frac{4}{9}$
31. $9\frac{2}{5} + 4\frac{3}{10}$
32. $15\frac{3}{4} + 4\frac{1}{12}$
33. $7\frac{1}{6} + 2\frac{5}{8}$
34. $25\frac{2}{15} + \frac{7}{10}$
35. $7\frac{6}{11} + 6\frac{5}{11}$
36. $20\frac{4}{9} + 5\frac{20}{27}$
37. $8\frac{5}{6} + 4\frac{2}{3}$
38. $1\frac{3}{5} + 2\frac{5}{9}$
39. $5\frac{2}{3} + 3\frac{4}{7}$
40. $13\frac{1}{3} + 10\frac{9}{11}$
41. Jerry hiked $6\frac{3}{4}$ mi (miles) on Saturday and $5\frac{1}{2}$ mi on Sunday. How many miles did he hike in all on Saturday and Sunday?

Write each difference in lowest terms. 9-4

42. $10\frac{5}{6} - 8\frac{1}{6}$
43. $9\frac{11}{12} - 4\frac{3}{4}$
44. $21\frac{7}{9} - 6\frac{5}{18}$
45. $17\frac{2}{3} - 15\frac{5}{8}$
46. $35\frac{3}{8} - 25\frac{3}{10}$
47. $48\frac{5}{7} - 23\frac{1}{4}$
48. $6\frac{1}{4} - 1\frac{3}{4}$
49. $16\frac{1}{8} - 4\frac{3}{16}$
50. $9 - \frac{7}{10}$
51. $12 - 5\frac{2}{3}$
52. $27\frac{3}{4} - 12\frac{5}{6}$
53. $40\frac{2}{7} - 32\frac{4}{5}$

Write each decimal as a fraction or a mixed number in lowest terms. **10-1**

1. 0.8 2. 0.36 3. 0.325 4. 0.04 5. 0.002
6. 6.5 7. 2.14 8. 9.55 9. 7.025 10. 11.625
11. $0.28\frac{4}{7}$ 12. $0.47\frac{1}{2}$ 13. $0.33\frac{1}{3}$ 14. $0.87\frac{1}{2}$ 15. $0.16\frac{2}{3}$
16. $0.66\frac{2}{3}$ 17. $0.51\frac{1}{9}$ 18. $0.83\frac{1}{3}$ 19. $0.03\frac{3}{4}$ 20. $0.09\frac{1}{6}$

Write each fraction or mixed number as a decimal. If the decimal is a repeating decimal, use a bar to show the repeating part of the decimal. **10-2**

21. $\frac{3}{5}$ 22. $\frac{7}{8}$ 23. $\frac{13}{25}$ 24. $6\frac{9}{20}$ 25. $13\frac{3}{8}$
26. $\frac{1}{9}$ 27. $\frac{1}{6}$ 28. $\frac{7}{11}$ 29. $7\frac{2}{15}$ 30. $1\frac{2}{3}$

Compare. Replace each __?__ with >, <, or =.

31. 0.65 __?__ $\frac{7}{10}$ 32. 0.38 __?__ $\frac{9}{25}$ 33. $\frac{1}{6}$ __?__ 0.2 34. $\frac{7}{8}$ __?__ 0.875
35. $3\frac{4}{9}$ __?__ 3.4 36. $7\frac{1}{2}$ __?__ 7.51 37. 2.25 __?__ $2\frac{1}{4}$ 38. 8.9 __?__ $8\frac{4}{5}$

39. One glass holds $6\frac{3}{4}$ oz (ounces) of milk. Another glass holds 6.5 oz of milk. Which glass holds more? Explain.

Estimate each decimal as a simple fraction or mixed number. **10-3**

40. 0.13 41. 0.99 42. 0.874 43. 0.591 44. 0.512
45. 0.7482 46. 9.21 47. 24.67 48. 5.167 49. 18.483

Estimate using compatible numbers.

50. 0.19×398 51. $0.37 \times \$465$ 52. 69.54×0.517 53. $0.249 \times \$25.16$
54. 217×0.79 55. 713×0.1246 56. 0.657×1462 57. $0.3324 \times \$9106$

10-4

58. Vanessa bought some apples for $.45 each and some pears for $.59 each. She spent a total of $3.71. How many of each fruit did she buy?
59. For the festival, Yolanda made a total of 35 potholders and aprons. She made 9 more potholders than aprons. How many of each item did Yolanda make?
60. Marvin sold shirts for $17 each and sweaters for $21 each. His sales totaled $110. How many of each type of clothing did he sell?

Complete. **11-1**

1. 60 in. = ? ft
2. 42 ft = ? yd
3. $4\frac{1}{3}$ yd = ? in.
4. $2\frac{1}{4}$ mi = ? ft
5. $5\frac{3}{4}$ ft = ? in.
6. 10,560 yd = ? mi
7. 12 ft 8 in. = ? in.
8. 549 in. = ? yd ? in.
9. 7 yd 2 ft = ? ft

Measure each line segment to the nearest $\frac{1}{16}$ inch.

10. ├──────────────┤
11. ├──────────────┤

Complete. **11-2**

12. 128 oz = ? lb
13. 7 lb = ? oz
14. 26,000 lb = ? t
15. 6 t = ? lb
16. $9\frac{1}{4}$ lb = ? oz
17. $7\frac{1}{5}$ t = ? lb
18. 8 lb 11 oz = ? oz
19. 2 t 584 lb = ? lb
20. 373 oz = ? lb ? oz

Complete. **11-3**

21. 64 fl oz = ? c
22. 9 gal = ? pt
23. 44 qt = ? gal
24. 14 c = ? pt
25. $23\frac{1}{4}$ gal = ? qt
26. $7\frac{1}{2}$ qt = ? pt
27. 8 gal 2 qt = ? qt
28. 35 pt = ? qt ? pt
29. 6 c 5 fl oz = ? oz
30. 19 qt = ? gal ? qt

Write each answer in simplest form. **11-4**

31. 23 ft 8 in. + 15 ft 7 in.
32. 11 mi 482 yd + 12 mi 1365 yd
33. 7 lb 6 oz × 3
34. 12 c 5 fl oz × 4
35. 8 t 530 lb − 3 t 980 lb
36. 32 qt − 17 qt 1 pt
37. 16 yd 4 in. ÷ 5
38. 46 gal 2 qt ÷ 6
39. Leslie bought an 18-ft roll of red ribbon. She needs to cut six pieces of red ribbon that each measure 2 ft 8 in. Did Leslie buy enough red ribbon? Explain.

Give the number of significant digits and the unit of precision. **11-5**

40. 205 m
41. $60{,}0\underline{0}0$ mi
42. 0.074 L
43. 1470 lb
44. 13.90 gal

Find each answer. Round to the correct number of significant digits.

45. 2817 ft + 460 ft
46. 57.21 km + 6.3 km
47. 19,000 mi − 3200 mi
48. 14.04 kg − 0.018 kg

Write each ratio as a fraction in lowest terms. 12-1

1. 9 to 25
2. 27:9
3. $\frac{65}{15}$
4. 14 to 28
5. 10:35
6. $\frac{24}{54}$
7. 48 to 21
8. $\frac{64}{8}$
9. $4:30¢
10. 15 s to 5 min
11. 6 ft to 10 in.
12. 1 m:20 cm

Write the unit rate.

13. 220 mi in 4 h
14. $14 for 8 lb
15. 168 books for 8 shelves
16. $23.15 in 5 h
17. 150.5 mi on 7 gal
18. 21.6 m in 2.4 s
19. Connie Marston earned $38.10 for working 6 h. How much did she earn per hour?

Tell whether each proportion is *true* or *false*. 12-2

20. $\frac{3}{2} = \frac{9}{6}$
21. $\frac{3}{10} = \frac{9}{90}$
22. $\frac{10}{6} = \frac{0.5}{0.3}$
23. $\frac{13}{16} = \frac{39}{18}$
24. $\frac{16}{46} = \frac{4}{12}$
25. $\frac{25}{1.3} = \frac{30}{1.2}$
26. $\frac{25}{100} = \frac{5}{20}$
27. $\frac{0.6}{75} = \frac{0.4}{50}$

Solve each proportion.

28. $\frac{x}{4} = \frac{20}{5}$
29. $\frac{3.5}{7} = \frac{h}{6}$
30. $\frac{9}{63} = \frac{2}{k}$
31. $\frac{32}{c} = \frac{16}{1}$
32. $\frac{3}{n} = \frac{20}{160}$
33. $\frac{0.6}{5} = \frac{1.2}{w}$
34. $\frac{45}{12} = \frac{y}{24}$
35. $\frac{m}{14} = \frac{2}{7}$
36. $\frac{b}{1} = \frac{27}{2.7}$
37. $\frac{16}{3} = \frac{p}{9}$
38. $\frac{55}{11} = \frac{5}{x}$
39. $\frac{4.8}{a} = \frac{2.4}{2}$

Solve using a proportion. 12-3

40. The ratio of boys to girls in Mr. Russell's class is 2:3. There are 10 boys in the class. How many girls are in the class?
41. Derrick Slattery earns $14.25 for working 3 h. How much money does he earn for working 5 h?
42. Joe drove 102 mi on 3 gal of gasoline. At that rate, how much farther can he drive on 18 more gallons of gasoline?
43. The cost of an 8-oz block of cheese is $1.60. How much would a 12-oz block of the same type of cheese cost?
44. The cost of three pens is $2.10. How much do seven pens cost?

Write each percent as a decimal. Write each decimal as a percent.

13-1

1. 27%
2. 63%
3. 9%
4. 5.4%
5. 0.85
6. 0.07
7. 0.5
8. $0.45\frac{3}{8}$
9. $0.03\frac{3}{4}$
10. 1.14
11. $15\frac{1}{3}\%$
12. $6\frac{1}{2}\%$
13. 1.4
14. 2.7
15. 0.006
16. 0.987
17. 8.9%
18. 1.15%
19. 0.4%
20. 0.35%

Write each percent as a fraction or a mixed number in lowest terms.

13-2

21. 22%
22. 8%
23. 175%
24. 235%
25. 4.6%
26. 0.05%
27. $\frac{3}{5}\%$
28. $6\frac{1}{4}\%$
29. $56\frac{1}{2}\%$
30. $83\frac{1}{3}\%$

Write each fraction or mixed number as a percent.

31. $\frac{1}{20}$
32. $\frac{2}{5}$
33. $\frac{1}{8}$
34. $\frac{7}{9}$
35. $\frac{4}{25}$
36. $\frac{7}{10}$
37. $\frac{3}{16}$
38. $1\frac{2}{3}$
39. $2\frac{1}{2}$
40. $1\frac{1}{16}$

Find each answer.

13-3

41. What number is 80% of 205?
42. What number is 25% of 143?
43. $16\frac{2}{3}\%$ of 96 is what number?
44. $33\frac{1}{3}\%$ of 127 is what number?
45. What number is 42% of 350?
46. What number is 8% of 64.7?
47. 13.2% of 164 is what number?
48. 7.5% of 29 is what number?
49. What number is 0.1% of 67.1?
50. What number is 200% of 48.5?
51. The Samuelsons pay $700 per month to rent their apartment. On January 1, the landlord is raising the rent 4%. How much will the Samuelsons pay each month for rent after January 1?

Estimate.

13-4

52. 37% of $637
53. 41% of 151
54. 16% of 17
55. $4\frac{1}{2}\%$ of $3972
56. $7\frac{1}{3}\%$ of $19.55
57. 9.7% of $103.16
58. 32.8% of 22
59. 49.7% of $79.43
60. 2.25% of 411
61. A magazine mailed a survey to 2950 subscribers. Thirty-four percent of the subscribers answered the survey. About how many people answered the survey?

Find each answer.

14-1

1. What percent of 30 is 24?
2. What percent of $150 is $39?
3. $18 is what percent of $48?
4. 132 is what percent of 72?
5. What percent of 400 is 56?
6. What percent of 5 is 17?
7. Nadia earns $350 each week. She pays $20 each week for transportation. What percent of her weekly earnings does Nadia pay for transportation?

Tell whether there is an *increase* or *decrease*. Then find the percent of increase or decrease.

14-2

8. original number of subscriptions: 125
   new number of subscriptions: 155
9. original price: $240
   new price: $440
10. original cost: $72
    new cost: $54
11. original number of pages: 590
    new number of pages: 354
12. original width: 56 in.
    new width: 63 in.
13. original hours: 48
    new hours: 16

Find each answer.

14-3

14. 7% of what number is 203?
15. 350% of what number is 49?
16. 92 is 9.2% of what number?
17. 481 is 14.8% of what number?
18. $16\frac{2}{3}$% of what number is 13?
19. 75% of what number is 129?

Find each answer using a proportion.

14-4

20. What number is 36% of 175?
21. What number is 5.8% of 650?
22. 160% of 98 is what number?
23. 2 is what percent of 25?
24. What percent of 40 is 9.2?
25. What percent of 54 is 45?
26. 95% of what number is 76?
27. 2.5% of what number is 16?

14-5

28. Alex must be at the school bus stop at 7:25 A.M. It takes him 10 min to walk to the bus stop, and he wants to allow 45 min to dress and have breakfast. At what time should Alex get up?
29. Rosa must work 9 h in order to earn enough money to buy a pair of sneakers and a book bag. The sneakers cost $42. The book bag cost $21. How much money does Rosa earn per hour?
30. Kenneth's lunch cost $6.40 plus 5% meals tax. He received $3.30 change from the cashier. How much money did Kenneth give the cashier?

Find the interest. If necessary, round to the nearest cent.

15-1

1. \$700 at 6% per year for 1 year
2. \$425 at 5% per year for 1 year
3. \$7055 at 10.6% per year for 2 years
4. \$2500 at 7.3% per year for 3 years
5. \$6750 at $8\frac{1}{2}$% per year for 1 year
6. \$8550 at $4\frac{1}{2}$% per year for 1 year
7. \$9005 at 13% per year for 6 months
8. \$3175 at 12% per year for 6 months

Find the amount due. If necessary, round to the nearest cent.

9. \$500 at 14% per year for 1 year
10. \$250 at 12.3% per year for 1 year
11. \$630 at $9\frac{1}{2}$% per year for 2 years
12. \$1755 at $11\frac{1}{2}$% per year for 5 years
13. \$2120 at 7% per year for 6 months
14. \$4575 at 8.2% per year for 6 months
15. Lisa Reinhold borrowed \$2000 for 1 year at an interest rate of 7.5% per year. How much interest did she pay? What was the total amount due?
16. Ralph Hill bought a couch for \$980. He made a down payment of \$100. Ralph borrowed \$880 for 6 months at an interest rate of $6\frac{1}{2}$% per year. What was the total amount due on his loan?

Find the balance. If possible, use the table on page 296. If necessary, round to the nearest cent.

15-2

17. \$500 at 5% compounded semiannually for 1 year
18. \$2000 at 7.5% compounded semiannually for 1 year
19. \$1500 at 4% compounded semiannually for 2 years
20. \$1850 at 6% compounded semiannually for 3 years
21. \$4000 at 7% compounded annually for 2 years
22. \$900 at 5.5% compounded annually for 2 years
23. \$3600 at 4% compounded quarterly for 1 year
24. \$4350 at 6% compounded quarterly for 2 years
25. \$650 at 8% compounded semiannually for 4 years
26. \$700 at 4% compounded semiannually for 1 year
27. \$2050 at 6% compounded quarterly for 1 year
28. \$1000 at 8% compounded quarterly for 2 years
29. Irene Pollet has \$1500 in a savings account that earns 5.5% annual interest compounded semiannually. She makes no deposits or withdrawals for one year. Find the balance at the end of one year.
30. Sherwood Benson has \$400 in an account that earns 6% annual interest compounded quarterly. He makes no deposits or withdrawals for two years. Find the balance at the end of two years.

Use the double bar graph at the right.

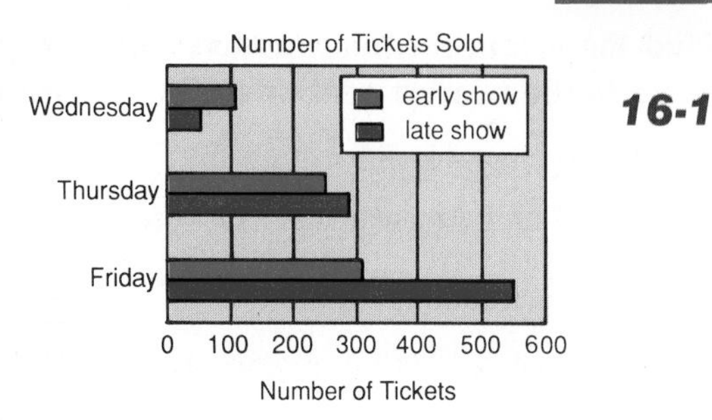

16-1

1. Estimate the number of tickets sold for the early show on Wednesday.
2. On what day were about 300 tickets sold for the late show?
3. On Friday, about how many more tickets were sold for the late show than for the early show?

16-2

4. Draw a line graph to display the given data.

Monthly Rainfall

Month	May	June	July	August
Rainfall (cm)	35	25	30	15

16-3

5. Draw a pictograph to display the given data.

Magazine Sales

Type	Weekly	Monthly	Bimonthly	Quarterly
Number of Magazines Sold	350	300	425	375

The circle graph at the right shows how the Johnsons budget their monthly net income.

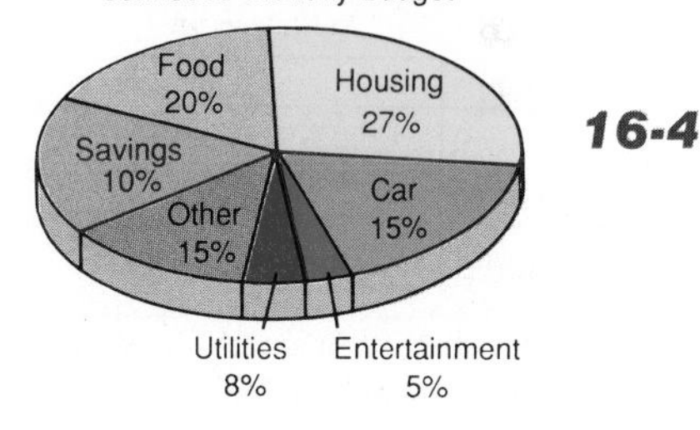

16-4

6. If their monthly income is $3000, find the amount that they budget monthly for food.
7. If they budget $200 for utilities, find their monthly net income.
8. If they budget $594 for housing, how much do they budget for entertainment?
9. Estimate the fraction of their monthly income that is budgeted for housing.

Use the line graph at the right. Estimate the median income for each year.

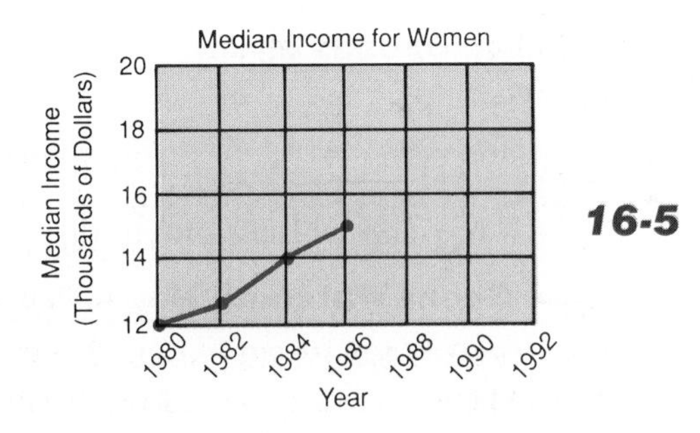

16-5

10. 1981   11. 1983   12. 1985

Assume that the trend will continue. Predict the median income for each year.

13. 1988   14. 1990   15. 1992

extra practice CHAPTER 17

Find the mean, median, mode(s), and range. If necessary, round the mean to the nearest cent or to the nearest tenth.

17-1

1. 52, 31, 10, 45, 52
2. 8.7, 4.5, 6.8, 3.3
3. 2.7, 9.3, 5.8, 2.7, 1.4, 7.5
4. \$6.32, \$5.48, \$9.15, \$8.56

17-2

5. Make a frequency table for the given data.

Number of Hours Worked

11	8	9	10	9	11	8	11
9	12	10	9	11	10	12	9

Exercises 6–9 refer to the frequency polygon below.

17-3

6. Find the mode(s) of the daily miles traveled.
7. What is the mean number of miles traveled?
8. For how many days are there less than 50 mi traveled?
9. Draw a histogram for the data in the frequency polygon.

Exercises 10–12 refer to the scattergram below.

17-4

10. What is the range of the test scores?
11. Find the mode(s) of the test scores.
12. When Barbara's test score was 70, how many hours did she spend studying?

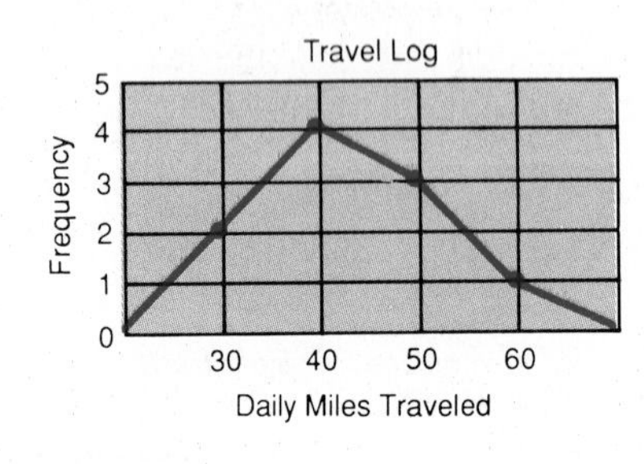

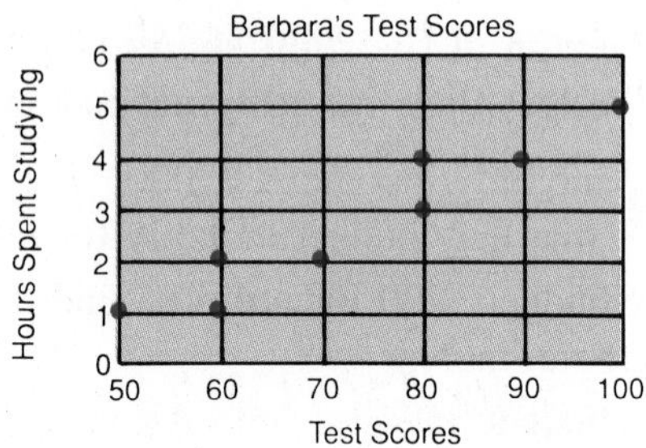

17-5

13. Make a stem-and-leaf plot to display the given data.

Average Speeds (mi/h)

15	35	25	32	53	33	24	60	33	40	45	30
28	50	30	45	47	55	35	42	26	20	35	10

17-6

14. Make a box-and-whisker plot to display the given data.

Time Spent Watching News Programs (h)

1.0	7.0	8.0	2.0	4.5	5.5	1.0
0.0	4.0	3.5	1.0	4.0	3.0	

A number cube is rolled. Find each probability. 18-1

1. $P(5)$
2. $P(\text{odd})$
3. $P(\text{number} < 7)$
4. $P(1)$
5. $P(6 \text{ or } 1)$
6. $P(\text{not } 4)$
7. $P(2 \text{ or } 3)$
8. $P(\text{number} > 2)$

A number cube is rolled. Find the odds in favor of each event.

9. 1 or 5
10. not 3
11. number < 6
12. 2 or 4
13. 3
14. odd number
15. number > 5
16. not 2 or not 6

18-2

17. Use a tree diagram to find the number of possible outcomes when rolling a number cube and selecting a letter from B through E.
18. A number cube is rolled twice. Use a tree diagram to find the probability of rolling a 4 and then a 5.

Use the counting principle to find each probability when choosing a letter from D to M and a digit from 1 to 8 at random. 18-3

19. $P(\text{D}, 1)$
20. $P(\text{F}, \text{odd digit})$
21. $P(\text{L}, \text{not } 8)$
22. $P(\text{H or J}, 3)$
23. $P(\text{M}, \text{digit} < 6)$
24. $P(\text{not K}, 4 \text{ or } 7)$

A number cube is rolled and a marble is drawn at random from those at the right. Find each probability. 18-4

25. $P(5, \text{then green})$
26. $P(\text{odd number, then red})$
27. $P(2 \text{ or } 3, \text{then black})$
28. $P(\text{not } 4, \text{then not green})$

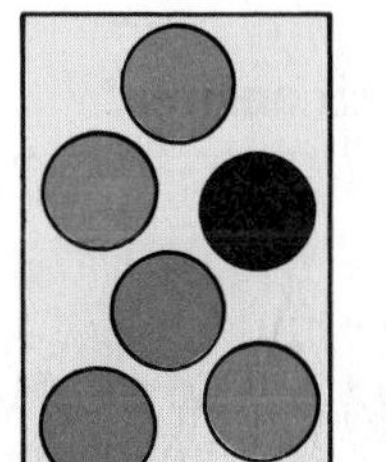

Two marbles are drawn at random from those at the right. The first marble is not replaced before the second one is drawn. Find each probability.

29. $P(\text{red, then black})$
30. $P(\text{not green, then green})$
31. $P(\text{two red})$
32. $P(\text{two green})$

Find the number of permutations and the number of combinations. 18-5

33. 2 out of 5 students
34. 3 out of 6 chairs
35. 4 out of 7 horses
36. 6 out of 6 courses

18-6

37. Mimi has only pennies, nickels, and dimes. In how many different ways, without regard to order, can she make 30¢ in change?
38. Ed has a bottle that measures one cup, a bottle that measures two cups, and a bottle that measures three cups. In how many different ways, without regard to order, can he measure exactly nine cups?

extra practice — CHAPTER 19

Rewrite each rational number as a quotient of two integers. *19-1*

1. 7 2. $^{-}46$ 3. 123 4. 0 5. $5\frac{1}{4}$ 6. $^{-}2\frac{1}{8}$

7. $^{-}0.9$ 8. 0.83 9. 2.14 10. $^{-}1.39$ 11. $0.\overline{3}$ 12. $^{-}0.\overline{7}$

Write the opposite and the absolute value of each number. *19-2*

13. 15 14. $^{-}37$ 15. 0 16. 9.1 17. $^{-}23.8$

18. 0.2 19. $\frac{5}{8}$ 20. $\frac{^{-}1}{16}$ 21. $^{-}19\frac{4}{11}$ 22. $4\frac{3}{5}$

Compare. Replace each __?__ with > or <. *19-3*

23. $^{-}7$ __?__ $^{-}6$ 24. 5 __?__ $^{-}4$ 25. $^{-}10$ __?__ $^{-}12$ 26. 1.6 __?__ $^{-}3.1$

27. $^{-}8.2$ __?__ 4.5 28. $^{-}2$ __?__ $^{-}8\frac{1}{2}$ 29. $\frac{^{-}1}{4}$ __?__ $^{-}1$ 30. $^{-}9$ __?__ 0

Write each set of numbers in order from least to greatest.

31. $^{-}7$; 7; $^{-}8$; 6 32. $^{-}5.2$; $^{-}2.7$; $^{-}5.6$; 2.3 33. $2\frac{1}{6}$; $^{-}4\frac{1}{6}$; $1\frac{1}{6}$; $\frac{^{-}5}{6}$

34. Monday night the recorded temperature was $^{-}3$°F. Tuesday night the recorded temperature was $^{-}9$°F. On which night was the recorded temperature higher?

Find each answer.

19-4

35. $12 + 16$ 36. $^{-}10 + {}^{-}7$ 37. $^{-}9 + {}^{-}21$ 38. $^{-}15 + 6$

39. $39 + {}^{-}17$ 40. $^{-}52 + 52$ 41. $^{-}3.7 + {}^{-}1.2$ 42. $^{-}5 + 6.4$

43. $9.1 + {}^{-}12.4$ 44. $\frac{^{-}1}{4} + {}^{-}3\frac{3}{4}$ 45. $^{-}1\frac{1}{8} + {}^{-}2\frac{3}{8}$ 46. $^{-}3\frac{2}{5} + {}^{-}2\frac{1}{5}$

19-5

47. $7 - 14$ 48. $25 - 32$ 49. $12 - {}^{-}41$ 50. $^{-}60 - {}^{-}56$

51. $^{-}3 - 24$ 52. $^{-}45 - 15$ 53. $79.5 - 31.6$ 54. $11.4 - {}^{-}2.8$

55. $^{-}4.2 - {}^{-}6.5$ 56. $^{-}18.4 - 5.3$ 57. $\frac{^{-}1}{2} - {}^{-}1\frac{1}{2}$ 58. $^{-}2\frac{3}{10} - 4\frac{1}{10}$

19-6

59. $8(3)$ 60. $(^{-}5)(^{-}9)$ 61. $6(^{-}7)$ 62. $(^{-}4.1)(2.3)$

63. $^{-}36 \div 6$ 64. $72 \div {}^{-}8$ 65. $^{-}48 \div {}^{-}12$ 66. $^{-}24.6 \div 3$

67. $\left(\frac{^{-}3}{5}\right)\left(\frac{^{-}2}{7}\right)$ 68. $0 \div {}^{-}1$ 69. $2(^{-}6)(4)$ 70. $\left(\frac{^{-}1}{2}\right)\left(\frac{3}{10}\right)\left(\frac{^{-}2}{9}\right)$

71. The temperature dropped 10°F in five hours. What was the average change in temperature per hour?

Evaluate each expression when $w = {}^-4$, $x = 0$, $y = 8$, and $z = 1$.

20-1

1. $y - z$
2. $z + 4 + z$
3. $w + y$
4. $x - y$
5. $3wz$
6. ${}^-4x$
7. y^2
8. w^2y
9. $\frac{0}{z}$
10. $\frac{y}{w} + x$
11. $2w + 3z$
12. $4y - 10w$

Solve each equation. Check your answers.

20-2

13. $a + 7 = 5$
14. ${}^-2 = {}^-6 - r$
15. $m + \frac{3}{7} = \frac{1}{7}$
16. $1\frac{2}{5} = p - \frac{1}{5}$
17. $w - 7 = {}^-12$
18. $20 = s - 0.7$

20-3

19. $5q = {}^-20$
20. $24.3 = 9y$
21. $\frac{r}{{}^-0.9} = 4$
22. ${}^-7 = \frac{h}{2}$
23. $\frac{2}{9}x = 42$
24. ${}^-18 = \frac{{}^-3}{10}t$

20-4

25. ${}^-4x + 22 = 10$
26. $5.6 = 2a - 4$
27. $\frac{y}{3} + 10 = {}^-11$
28. ${}^-40 = \frac{d}{9} - 35$
29. ${}^-7 = \frac{5}{6}b + 8$
30. $\frac{{}^-1}{4}w - 21 = {}^-23$

Solve and graph each inequality.

20-5

31. $y + 9 > 13$
32. $t - 7 \geq {}^-9$
33. $\frac{a}{5} \leq {}^-1$
34. $6x \leq 18$
35. $\frac{m}{4} > 0$
36. $11h \leq {}^-33$

Solve using an equation.

20-6

37. Malika paid \$172 for an eight-performance symphony concert series. What is the price for each performance?
38. Aretha Wilkes used $6\frac{1}{2}$ yd of material to make curtains. She has $2\frac{1}{2}$ yd of material left. How much material did she have before she made the curtains?
39. Damon rented a car for \$18.95 per day plus \$.17 per mile. His total rental expense for one day came to \$52.95. How many miles did Damon drive?
40. Shirley Weingarden subscribes to a weekly news magazine that costs \$28.84 per year. She makes an initial payment of \$16. She then must make three equal payments. What is the amount of each of these three payments?

Use the coordinate plane at the right. Write the letter of the point with the given coordinates.

21-1

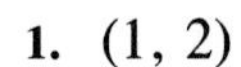

1. $(1, 2)$	**2.** $(^-3, ^-1)$	**3.** $(^-3, 4)$
4. $(^-2, 2)$	**5.** $(^-4, ^-4)$	**6.** $(2, ^-3)$
7. $(0, 0)$	**8.** $(0, 1)$	**9.** $(4, ^-2)$
10. $(^-4, 0)$	**11.** $(3, 0)$	**12.** $(0, ^-4)$

Graph the given points on a coordinate plane.

13. $A(4, 3)$	**14.** $B(^-2, ^-1)$	**15.** $C(0, 4)$
16. $D(^-2, 0)$	**17.** $E(1, ^-4)$	**18.** $F(^-2, ^-3)$
19. $G(^-3, 2)$	**20.** $H(3, 2)$	**21.** $J(3, ^-2)$

Is each ordered pair a solution of $y = {}^-4x + 1$? Write *Yes* or *No*.

21-2

22. $(1, 5)$	**23.** $(^-7, 2)$	**24.** $(^-2, 9)$	**25.** $(^-1, ^-5)$
26. $(2, ^-6)$	**27.** $(^-4, 17)$	**28.** $(\frac{1}{4}, 0)$	**29.** $(^-\frac{1}{2}, 3)$

For each equation, make a table of ordered pairs that are solutions. Use $^-2$, $^-1$, 0, 1, and 2 as values for x.

30. $y = x + 11$	**31.** $y = x - 6$	**32.** $y = 4x$
33. $y = {}^-5x$	**34.** $y = 3x - 8$	**35.** $y = {}^-2x + 9$

Graph each equation on a coordinate plane. Use $^-2$, 0, 2 as values for x.

21-3

36. $y = x + 4$	**37.** $y = x - 3$	**38.** $y = -x - 5$
39. $y = -x + 2$	**40.** $y = -x$	**41.** $y = x$
42. $y = 2x$	**43.** $y = {}^-3x$	**44.** $y = {}^-2x + 1$
45. $y = 3x + 4$	**46.** $y = 2x - 5$	**47.** $y = {}^-3x - 2$

Is the given ordered pair a solution of the system? Write *Yes* or *No*.

21-4

48. $(^-1, 1)$
$y = 6x - 7$
$y = x + 2$

49. $(3, ^-1)$
$y = x - 4$
$y = {}^-2x + 5$

50. $(0, ^-3)$
$y = {}^-4x - 3$
$y = x - 3$

Solve each system by graphing.

51. $y = 2x + 3$
$y = -x - 3$

52. $y = {}^-3x + 2$
$y = -x - 2$

53. $y = 2x + 1$
$y = 2x - 3$

54. $y = x + 5$
$y = -x + 3$

55. $y = 2x - 4$
$y = x - 4$

56. $y = {}^-2x + 7$
$y = x + 1$

extra practice

Use the figure at the right to name the following.

22-1

1. three points
2. three lines
3. three line segments
4. four rays
5. two pairs of vertical angles

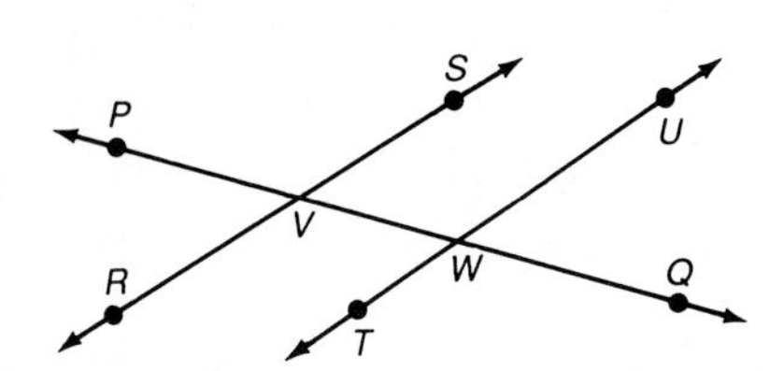

Use a protractor to draw an angle of the given measure.

22-2

6. 30° 7. 120° 8. 65° 9. 155° 10. 90°

Tell whether two angles with the given measures are *complementary*, *supplementary*, or *neither*.

22-3

11. 24°, 66° 12. 105°, 85° 13. 32°, 148° 14. 72°, 8°

Identify each quadrilateral.

22-4

15.

16. 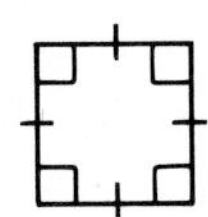

17. 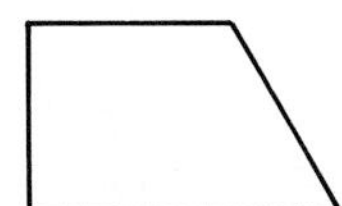

18. 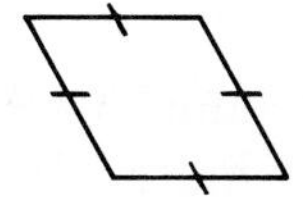

Find the perimeter. Use a formula whenever possible.

22-5

19. a rhombus with sides that each measure 14 m
20. a rectangle with length $5\frac{1}{2}$ yd and width 4 yd
21. a triangle with sides that measure 2.3 cm, 4.1 cm, and 4.9 cm
22. a regular pentagon with sides that each measure 34 in.

Find the circumference of each circle.

22-6

23. diameter = 15 cm (Use $\pi \approx 3.14$.)
24. radius = 9.5 mi (Use $\pi \approx 3.14$.)
25. radius = 63 m (Use $\pi \approx \frac{22}{7}$.)
26. diameter = 21 yd (Use $\pi \approx \frac{22}{7}$.)

Find the area of each figure.

22-7

27. a rectangle with length 4.8 m and width 3.4 m
28. a parallelogram with a base that measures 56 in. and height 10.5 in.
29. a triangle with a base that measures 17 yd and height 22 yd
30. a trapezoid with bases that measure 4 km and 10 km, and height 12 km

22-8

31. a circle with radius = 35 m (Use $\pi \approx \frac{22}{7}$.)
32. a circle with diameter = 26 ft (Use $\pi \approx 3.14$.)

Identify each space figure. **23-1**

1. 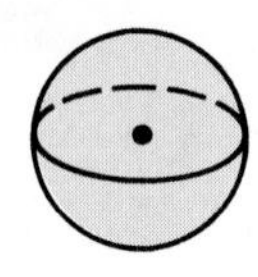

2. 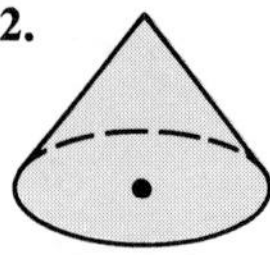

3. 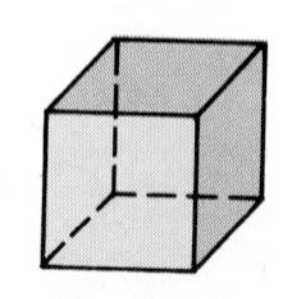

4. 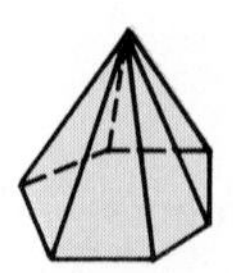

Find the surface area of each prism. **23-2**

5. 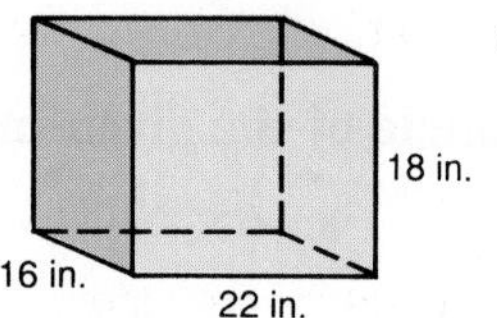

6. 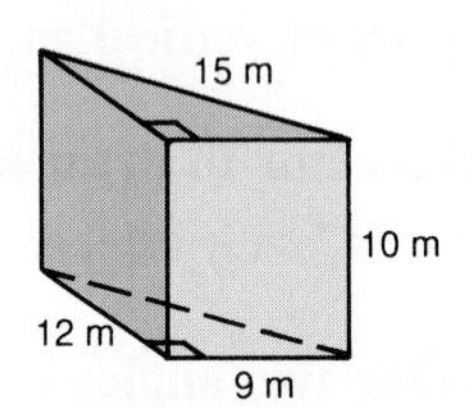

Find the surface area of each cylinder. Use $\pi \approx 3.14$. **23-3**

7. 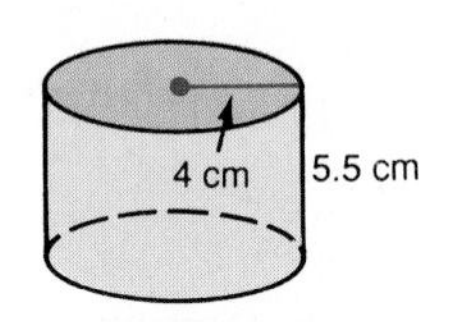

8. 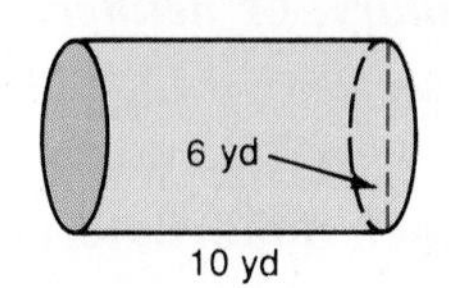

Find the volume of each prism or cylinder. When appropriate, use $\pi \approx \frac{22}{7}$. **23-4**

9. 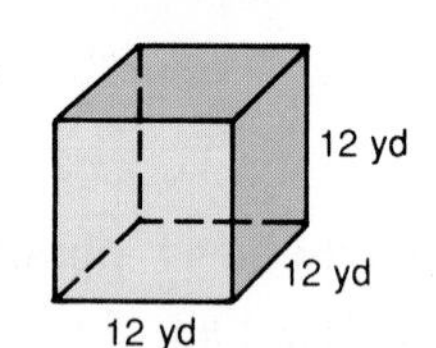

10. 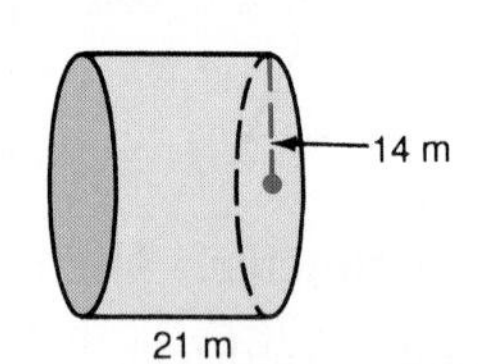

Find the volume of each space figure. When appropriate, use $\pi \approx \frac{22}{7}$. **23-5**

11. 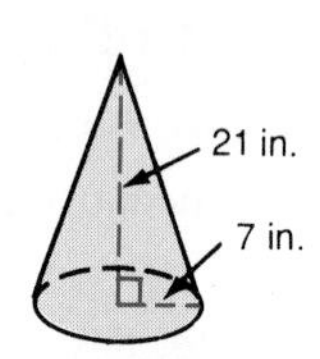

12. 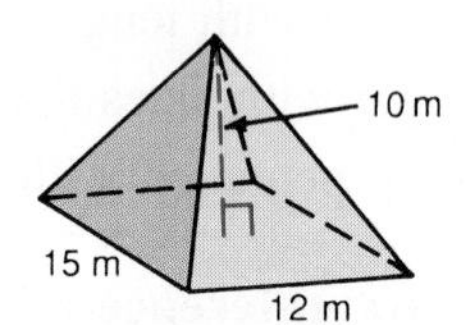

Find the surface area and volume of each sphere. Use $\pi \approx 3.14$. **23-6**

13. 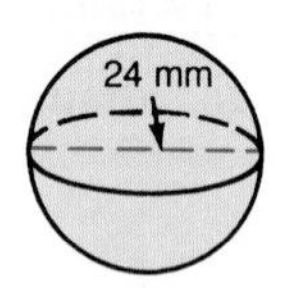

14. 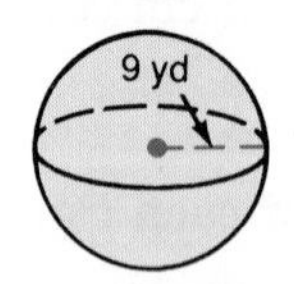

Each ▢ **is a centimeter cube. Find the surface area of each space figure.** **23-7**

15. 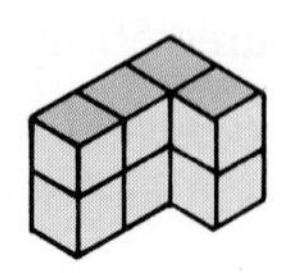

16. 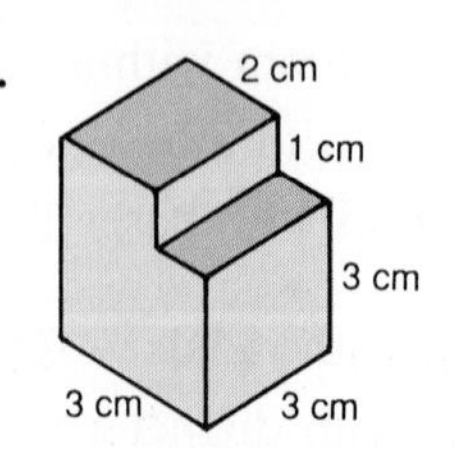

Classify each triangle first by its sides and then by its angles.

24-1

1\. 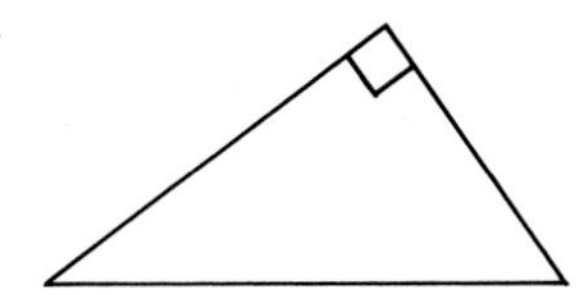

2\. 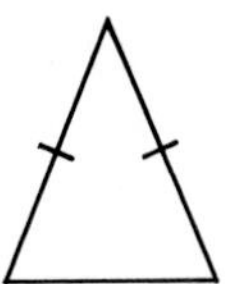

3\. 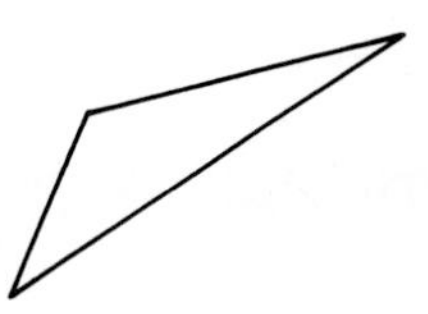

Can the given lengths be the lengths of the sides of a triangle? Write *Yes* or *No*.

4. 5 in., 11 in., 6 in. **5.** 9 m, 23 m, 21 m **6.** 8.3 mi, 14.5 mi, 5.2 mi

Find each square root. Use the table on page 549 or a calculator if necessary.

24-2

7. $\sqrt{25}$ **8.** $\sqrt{121}$ **9.** $\sqrt{4900}$ **10.** $\sqrt{10{,}000}$

11. $\sqrt{95}$ **12.** $\sqrt{138}$ **13.** $\sqrt{3844}$ **14.** $\sqrt{22{,}201}$

The lengths of two sides of a right triangle are given. Assume that *c* is the length of the hypotenuse and *a* and *b* are the lengths of the legs. Sketch the triangle and then find the length of the third side.

24-3

15. $a = 7, b = 24$ **16.** $a = 24, b = 18$ **17.** $a = 9, c = 15$

18. $a = 16, c = 34$ **19.** $b = 36, c = 39$ **20.** $b = 42, c = 58$

In each pair, the triangles are similar. Find the unknown lengths. Then tell whether the triangles are *congruent* or *not congruent*.

24-4

21. 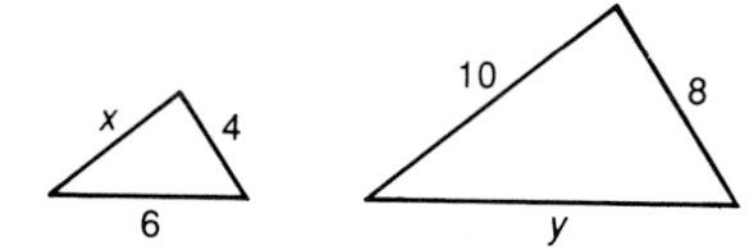

22. 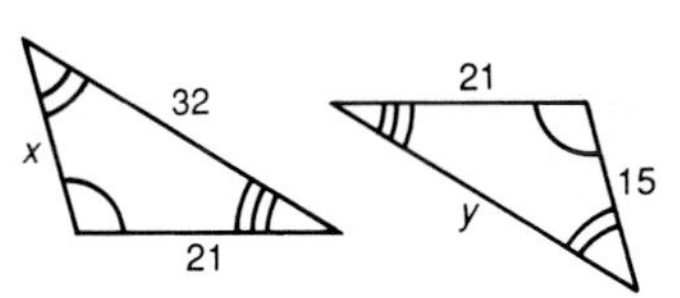

23. 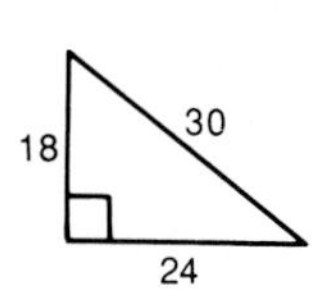

x y 12

24. 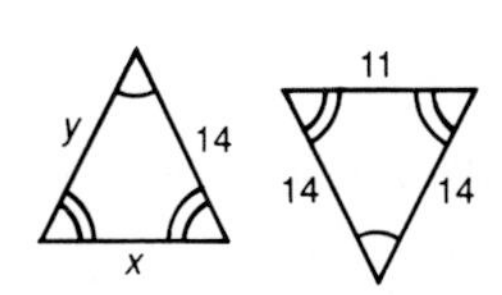

Use the triangle at the right. Find each ratio in lowest terms.

24-5

25. sin *B* **26.** cos *A*

27. cos *B* **28.** tan *A*

29. sin *A* **30.** tan *B*

A 26 10 C B a

24-6

31. The houses on Pine Street are numbered consecutively from 1 to 115. How many house numbers contain at least one digit 3?

appendix a

DEDUCTIVE AND INDUCTIVE REASONING

CONDITIONAL STATEMENTS

On page 551, the Pythagorean Theorem was stated as follows.

> For any right triangle, the square of the length of the hypotenuse is equal to the sum of the squares of the lengths of the two legs.

Another way to state this theorem is to write a sentence in *if-then* form.

> *If* a triangle is a right triangle, *then* the square of the length of the hypotenuse is equal to the sum of squares of the lengths of the two legs.

A sentence that can be written in if-then form is called a **conditional statement.** Every conditional statement has two parts. The "if" part is called the **condition.** The "then" part is called the **conclusion.**

example 1

Rewrite each conditional statement in if-then form. Then identify the condition and the conclusion.

a. I will rake the leaves if you mow the lawn.

b. All rectangles have four right angles.

solution

a. If you mow the lawn,
then I will rake the leaves.

Condition: You mow the lawn.

Conclusion: I will rake the leaves.

b. If a figure is a rectangle,
then it has four right angles.

Condition: A figure is a rectangle.

Conclusion: It has four right angles.

your turn

Rewrite each conditional statement in if-then form. Then identify the condition and the conclusion.

1. I would read the examples first if I were you.
2. The game will be postponed if it rains.
3. Every positive number has a square root.
4. In case of fire, you should call the fire department.

CONVERSES

When you switch the condition and the conclusion of a conditional statement, the result is another conditional statement that is called the **converse** of the original. For instance, on page 551 the converse of the Pythagorean Theorem was stated as follows.

> If the sides of a triangle have lengths a, b, and c such that $c^2 = a^2 + b^2$, then the triangle is a right triangle.

Both the Pythagorean Theorem and its converse are true statements. However, whether the original statement is true or false does not necessarily determine whether its converse is true or false. The converse of a *true* statement may be either true or false, and the converse of a *false* statement may be either true or false.

example 2

Tell whether the given statement is *True* or *False*. Then write its converse and tell whether the converse is *True* or *False*.

a. If $2y = 10$, then $y = 5$.

b. Every rectangle has four sides.

solution

a. Statement: If $2y = 10$, then $y = 5$. *True*

Converse: If $y = 5$, then $2y = 10$. *True*

b. First rewrite the statement in if-then form.

Statement: If a figure is a rectangle, then it has four sides. *True*

Converse: If a figure has four sides, then it is a rectangle. *False*

your turn

Tell whether the given statement is *True* or *False*. Then write its converse and tell whether the converse is *True* or *False*.

5. If there is ice on the road, then the road is slippery.

6. If a clock shows 12:00 all day, then the clock is broken.

7. If $x = {}^{-}2$, then $3x + 7 = 1$.

8. If $z > 3$, then $z > 7$.

9. Every right angle has a measure of 90°.

10. Every trapezoid has a pair of parallel sides.

11. If $a = 0$ or $b = 0$, then $ab = 0$.

12. If $a = 0$ and $b = 0$, then $ab = 0$.

DEDUCTIVE REASONING

Conditional statements play an important role in *deductive reasoning*. **Deductive reasoning** is a process of reaching conclusions based on true statements. Given a true conditional statement and a *case* in which the condition is true, the conclusion must be true.

example 3

Use deductive reasoning to complete the conclusion.

Statement: If a person is 18 years old or older, then he or she can register to vote.
Case: Ian is 21 years old.
Conclusion: Ian ______?______.

solution

Ian is 21, so he is older than 18. For Ian's case, the statement is true.
Conclusion: Ian can register to vote.

your turn

Use deductive reasoning to complete the conclusion.

13. Statement: If Ana finishes her homework, then she can watch television.
Case: Ana has finished her homework.
Conclusion: Ana ______?______.

14. Statement: Every parallelogram has two pairs of parallel sides.
Case: A square is a parallelogram.
Conclusion: A square ______?______.

15. Statement: If the product of two numbers is positive, then either both numbers are positive or both are negative.
Case: The product $5x$ is positive.
Conclusion: x ______?______.

16. Statement: All wombats are marsupials.
Case: Wally is a wombat.
Conclusion: Wally ______?______.

INDUCTIVE REASONING

Inductive reasoning is a process of reaching a conclusion based on a series of individual examples. The conclusion *may* be true, but it is not *necessarily* true.

For instance, suppose that the school cafeteria served hamburgers every Wednesday for the first six weeks of the school year. You might then predict that the school cafeteria will serve hamburgers on the seventh Wednesday. In reaching this conclusion, you would be using inductive reasoning.

example 4

Use inductive reasoning to predict the next three numbers in the pattern:

1, 3, 6, 10, 15, 21, ?, ?, ?

solution

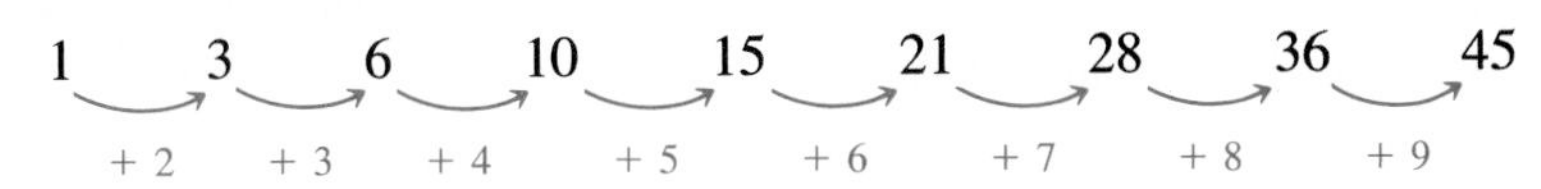

The next three numbers in the pattern are 28, 36, and 45.

your turn

Use inductive reasoning to predict the next three numbers in the pattern.

17. 1, 5, 9, 13, 17, 21, ?, ?, ?

18. $6, 3, \frac{3}{2}, \frac{3}{4}, \frac{3}{8}, \frac{3}{16}$, ?, ?, ?

19. 1, 2, 6, 24, 120, 720, ?, ?, ?

20. 1, 3, 6, 18, 21, 63, 66, ?, ?, ?

In each situation, tell whether the reasoning used is *inductive* or *deductive*.

21. Every member of the school band plays a musical instrument. Jenny is a member of the school band. Therefore, Jenny plays a musical instrument.

22. Gerard draws six pentagons. He measures each angle carefully and finds that the sum of the measures of the angles of each pentagon is 540°. He concludes that the sum of the measures of the angles of any pentagon is 540°.

23. Mimi notices the following: a triangle has no diagonals; a square has two diagonals; a pentagon has three diagonals; and a hexagon has four diagonals. She concludes that the number of diagonals of a polygon is two less than the number of sides.

24. Glen knows that all gymnasts are agile and strong. Glen knows that Francine is a gymnast. Glen concludes that Francine is agile and strong.

appendix b

TRANSFORMATIONS

TRANSLATIONS

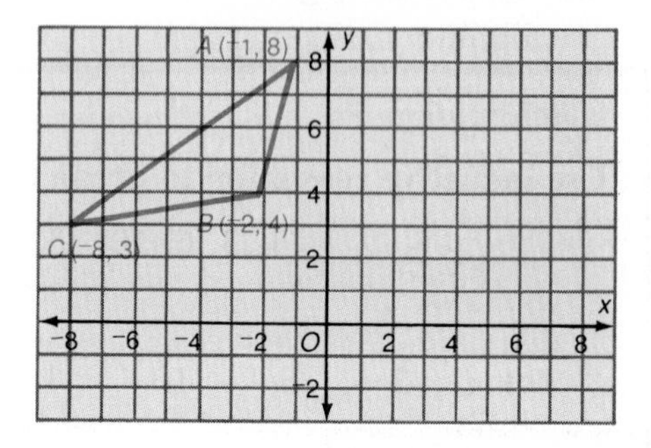

At the right you see triangle ABC on a coordinate plane. Its vertices are the points $A(^-1, 8)$, $B(^-2, 4)$, and $C(^-8, 3)$.

There are many ways that you can move this triangle on the plane. For instance, a **translation** is the result when you *slide* it along a straight line. A move like a translation is called a **tranformation** of the triangle.

example 1

Use triangle *ABC* above. Write the coordinates of the vertices after the given translation.

a. 9 units to the right **b.** 5 units down

solution

a. Add 9 to the x-coordinate of each vertex.

$A(^-1, 8) \rightarrow D(^-1 + 9, 8)$, or $D(8, 8)$
$B(^-2, 4) \rightarrow E(^-2 + 9, 4)$, or $E(7, 4)$
$C(^-8, 3) \rightarrow F(^-8 + 9, 3)$, or $F(1, 3)$

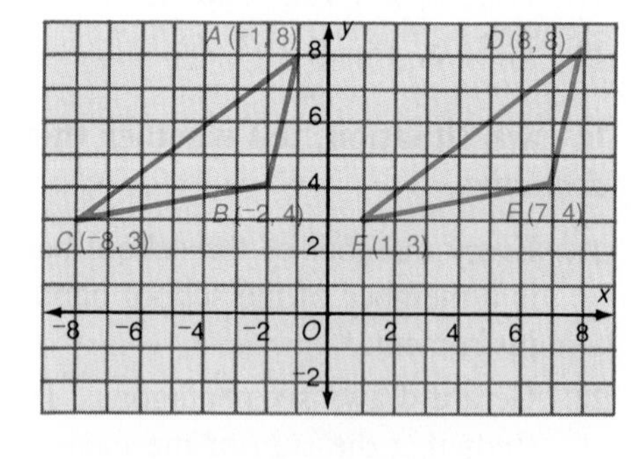

b. Add $^-5$ to the y-coordinate of each vertex.

$A(^-1, 8) \rightarrow G(^-1, 8 + {^-5})$, or $G(^-1, 3)$
$B(^-2, 4) \rightarrow H(^-2, 4 + {^-5})$, or $H(^-2, ^-1)$
$C(^-8, 3) \rightarrow J(^-8, 3 + {^-5})$, or $J(^-8, ^-2)$

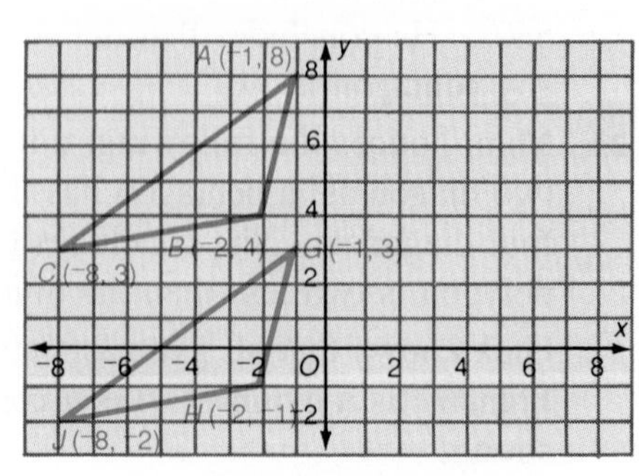

your turn

The given points are the vertices of a triangle on a coordinate plane. Write the coordinates of the vertices after the given translation.

1. $A(^-9, 4)$; $B(^-4, 6)$; $C(^-1, 3)$
   10 units to the right
2. $A(3, 4)$; $B(8, 2)$; $C(7, 7)$
   9 units to the left
3. $A(2, 5)$; $B(4, 1)$; $C(8, 6)$
   7 units down
4. $A(^-7, ^-2)$; $B(^-2, ^-3)$; $C(^-5, ^-6)$
   8 units up
5. $A(^-4, 5)$; $B(0, 7)$; $C(5, 3)$
   2 units to the left
6. $A(1, 3)$; $B(^-3, ^-2)$; $C(4, ^-4)$
   3 units down

example 2

Use triangle ABC at the top of page 608. Write the coordinates of the vertices after a translation 9 units to the right and 5 units down.

solution

Add 9 to the x-coordinate of each vertex.
Add $^-5$ to the y-coordinate of each vertex.

$A(^-1, 8) \rightarrow K(^-1 + 9, 8 + {^-5})$, or $K(8, 3)$
$B(^-2, 4) \rightarrow L(^-2 + 9, 4 + {^-5})$, or $L(7, ^-1)$
$C(^-8, 3) \rightarrow M(^-8 + 9, 3 + {^-5})$, or $M(1, ^-2)$

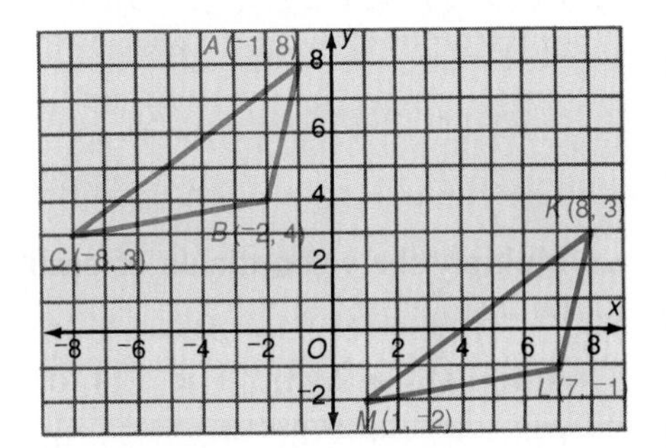

your turn

The given points are the vertices of a triangle on a coordinate plane. Write the coordinates of the vertices after the given translation.

7. $A(^-6, ^-1)$; $B(^-2, ^-3)$; $C(^-5, ^-6)$
   7 units to the right
   9 units up
8. $A(1, 1)$; $B(4, 4)$; $C(7, 2)$
   8 units to the left
   6 units down
9. $A(4, ^-2)$; $B(7, ^-4)$; $C(1, ^-6)$
   9 units to the left
   7 units up
10. $A(^-2, ^-3)$; $B(^-8, ^-5)$; $C(^-5, ^-1)$
    10 units to the right
    8 units down
11. $A(5, 0)$; $B(0, 4)$; $C(^-3, ^-2)$
    2 units to the left
    3 units down
12. $A(^-5, 2)$; $B(2, 2)$; $C(^-5, ^-4)$
    5 units to the right
    4 units up

REFLECTIONS

A **reflection** is the transformation that results when you *flip* a figure across a line.

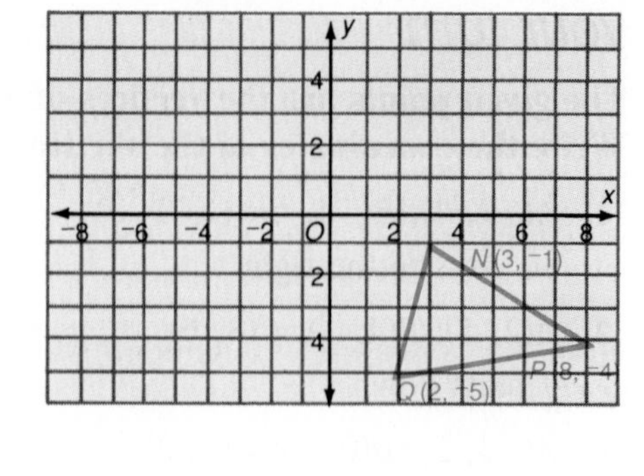

Use triangle *NPQ* at the right. Write the coordinates of the vertices after each transformation.

a. a reflection across the *y*-axis

b. a reflection across the *x*-axis

solution

a. Multiply the *x*-coordinate of each vertex by $^-1$.

$N(3, ^-1) \rightarrow R(3 \times {^-1}, ^-1)$, or $R(^-3, ^-1)$

$P(8, ^-4) \rightarrow S(8 \times {^-1}, ^-4)$, or $S(^-8, ^-4)$

$Q(2, ^-5) \rightarrow T(2 \times {^-1}, ^-5)$, or $T(^-2, ^-5)$

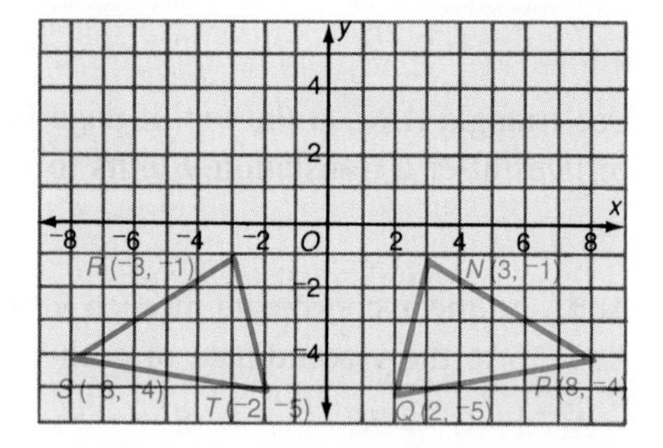

b. Multiply the *y*-coordinate of each vertex by $^-1$.

$N(3, ^-1) \rightarrow U(3, ^-1 \times {^-1})$, or $U(3, 1)$

$P(8, ^-4) \rightarrow V(8, ^-4 \times {^-1})$, or $V(8, 4)$

$Q(2, ^-5) \rightarrow W(2, ^-5 \times {^-1})$, or $W(2, 5)$

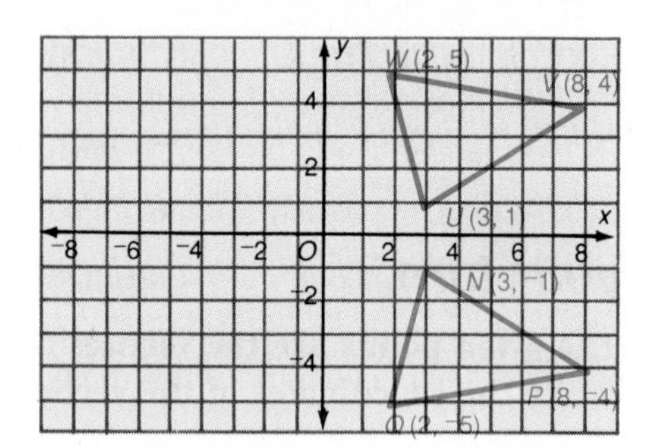

your turn

The given points are the vertices of a triangle on a coordinate plane. Write the coordinates of the vertices after each transformation.

a. a reflection across the *y*-axis

b. a reflection across the *x*-axis

13. $N(3,1)$; $P(4, 5)$; $Q(8, 2)$

14. $N(^-1, 1)$; $P(^-5, 2)$; $Q(^-7, 7)$

15. $N(3, ^-5)$; $P(6, ^-5)$; $Q(3, ^-1)$

16. $N(^-1, ^-2)$; $P(^-3, ^-7)$; $Q(^-5, ^-4)$

17. $N(3, 0)$; $P(7, ^-2)$; $Q(3, ^-7)$

18. $N(^-7, 4)$; $P(^-7, ^-4)$; $Q(0, 0)$

ROTATIONS

A third type of transformation is a *rotation*. A **rotation** is the result when you *turn* a figure about a fixed point. A rotation of 180° is sometimes called a **half-turn** rotation.

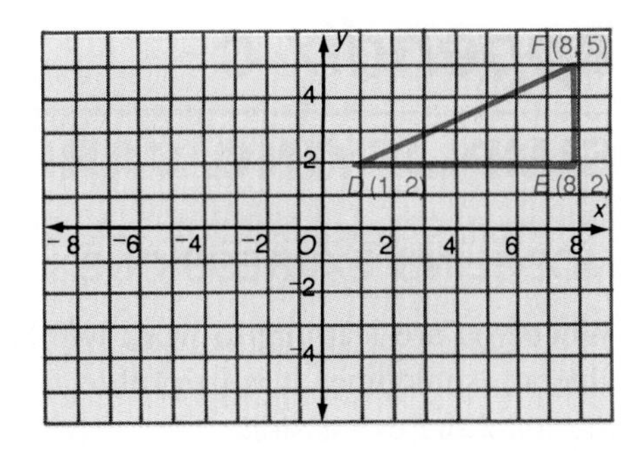

example 4

Use triangle *DEF* at the right. Write the coordinates of the vertices after a half-turn rotation about the origin.

solution

Multiply both the *x*-coordinate and the *y*-coordinate of each vertex by $^-1$.

$D(1, 2) \rightarrow P(1 \times {^-1}, 2 \times {^-1})$, or $P(^-1, ^-2)$
$E(8, 2) \rightarrow Q(8 \times {^-1}, 2 \times {^-1})$, or $Q(^-8, ^-2)$
$F(8, 5) \rightarrow R(8 \times {^-1}, 5 \times {^-1})$, or $R(^-8, ^-5)$

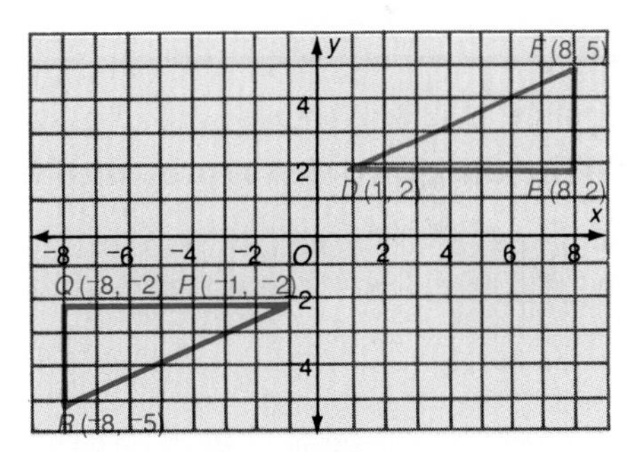

your turn

The given points are the vertices of a triangle on a coordinate plane. Write the coordinates of the vertices after a half-turn rotation about the origin.

19. $D(3, 3)$; $E(5, 1)$; $F(7, 7)$

20. $D(^-4, ^-2)$; $E(^-1, ^-5)$; $F(^-6, ^-8)$

21. $D(^-7, 3)$; $E(^-4, 7)$; $F(^-1, 6)$

22. $D(6, ^-3)$; $E(7, ^-5)$; $F(1, ^-4)$

23. $D(^-4, 0)$; $E(0, 0)$; $F(6, 0)$

24. $D(2, 4)$; $E(2, ^-3)$; $F(^-5, ^-3)$

Points *A*, *B*, and *C* are the vertices of a triangle on a coordinate plane. Points *R*, *S*, and *T* are the coordinates of the vertices after a transformation. Identify the transformation.

25. $A(^-2, 4) \rightarrow R(2, 4)$
$B(^-6, 1) \rightarrow S(6, 1)$
$C(^-4, 0) \rightarrow T(4, 0)$

26. $A(1, 5) \rightarrow R(0, 5)$
$B(7, 2) \rightarrow S(6, 2)$
$C(4, 6) \rightarrow T(3, 6)$

27. $A(^-2, ^-1) \rightarrow R(2, 1)$
$B(^-5, ^-8) \rightarrow S(5, 8)$
$C(^-9, 0) \rightarrow T(9, 0)$

28. $A(7, ^-2) \rightarrow R(7, 3)$
$B(8, ^-4) \rightarrow S(8, 1)$
$C(3, ^-5) \rightarrow T(3, 0)$

29. $A(^-1, ^-9) \rightarrow R(^-1, 9)$
$B(^-4, ^-3) \rightarrow S(^-4, 3)$
$C(^-2, ^-1) \rightarrow T(^-2, 1)$

30. $A(1, 5) \rightarrow R(2, 7)$
$B(^-4, ^-1) \rightarrow S(^-3, 1)$
$C(4, ^-8) \rightarrow T(5, ^-6)$

appendix c

SIMPLIFYING VARIABLE EXPRESSIONS

MODELING EXPRESSIONS

When you are learning to work with expressions in algebra, sometimes it is helpful to use physical objects to model the expressions. One way to do this is by using algebra tiles. A standard set of algebra tiles contains the three different shapes shown at the right. You can model many expressions by using combinations of these shapes.

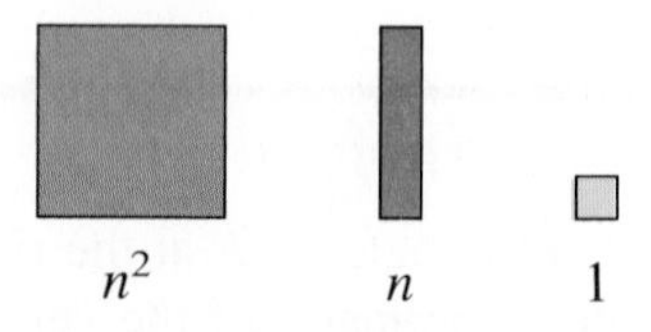

example 1

Write the expression represented by each set of algebra tiles.

a.

b.

c.

d. 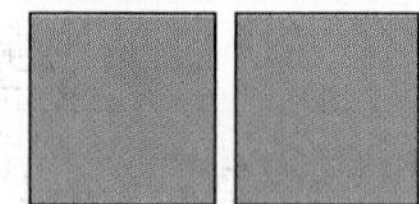

solution

a. There are three n tiles, so the expression is $n + n + n = 3 \cdot n$, or $3n$.

b. There are eight 1 tiles, so the expression is $8 \cdot 1$, or 8.

c. There are four n tiles and eleven 1 tiles. The expression is $4n + 11$.

d. There are two n^2 tiles, nine n tiles, and three 1 tiles. The expression is $2n^2 + 9n + 3$.

your turn

Write the expression represented by each set of algebra tiles.

1. 

2. 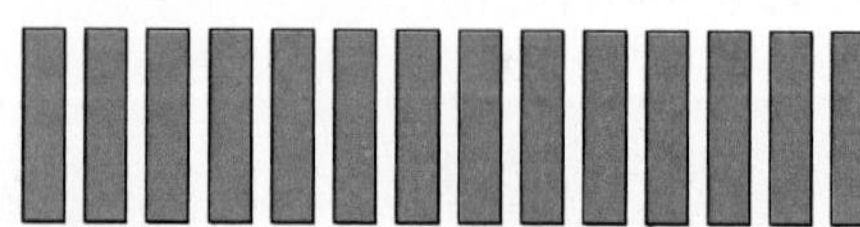

3. 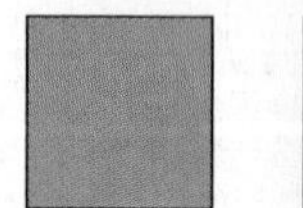 

4. 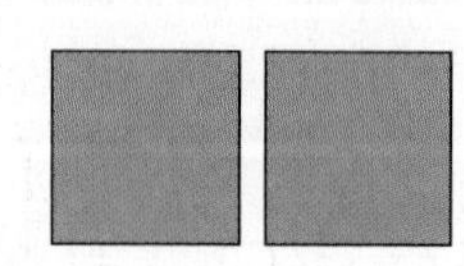

example 2

Show how to use algebra tiles to model the expression $2n^2 + 6n + 3$.

solution

The model requires two n^2 tiles, six n tiles, and three 1 tiles. The solution is shown at the right.

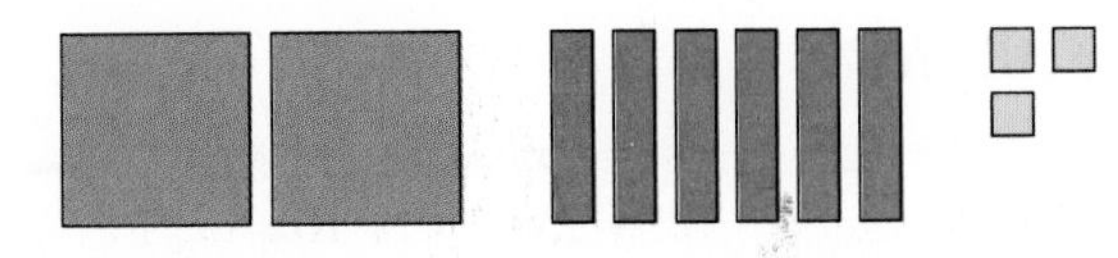

your turn

Show how to use algebra tiles to model each expression.

5. $4n + 9$ **6.** $2n^2 + 7$ **7.** $n^2 + 5n$ **8.** $2n^2 + n + 5$

9. $6 + n$ **10.** $3n + 2n^2$ **11.** $12 + n^2$ **12.** $8 + 2n^2 + n$

COMBINING LIKE TERMS

When you are given a variable expression, sometimes it is possible to write it in a simpler form. There are many ways that you can do this. One method that is used frequently is called *combining like terms.*

For instance, the expression $3n + 7 + 2n$ contains three **terms:** $3n$, 7, and $2n$. When you model this expression with algebra tiles, you can see that the terms $3n$ and $2n$ are **like terms** because they are represented by the same shape of tile.

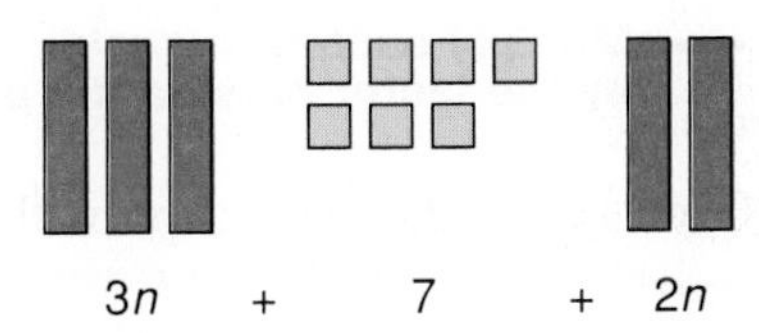

Now, if you were to rearrange the tiles so that all the like shapes were together, you could calculate a new total for the number of n tiles.

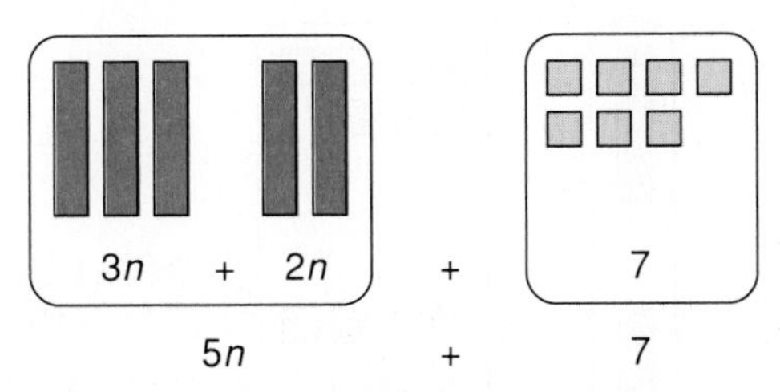

So it is possible to write $3n + 7 + 2n$ in a simpler form as $5n + 7$.

In any variable expression, you can recognize the like terms because they have identical variable parts. Terms that have *different* variable parts are called **unlike terms.**

example 3

Rewrite each expression by combining like terms.

a. $7x + 2x^2 + 3x$ **b.** $c^2 + 9 + c^2$

solution

a. $7x + 2x^2 + 3x = 2x^2 + (7x + 3x)$ ← **The only like terms are 7*x* and 3*x*.**
$= 2x^2 + 10x$ **$2x^2$ has a different variable part.**

b. $c^2 + 9 + c^2 = (c^2 + c^2) + 9$
$= (1c^2 + 1c^2) + 9$ ← **Think of each c^2 as $1c^2$.**
$= 2c^2 + 9$

your turn

Rewrite each expression by combining like terms.

13. $3a + 8 + 2a$ **14.** $2z^2 + 7z + 4z^2$ **15.** $m^2 + 5m + m^2$

16. $10 + t + 6t$ **17.** $j^2 + j^2 + 3$ **18.** $1 + 3r + 5r + 1$

USING THE DISTRIBUTIVE PROPERTY

If you are given a variable expression that contains parentheses, you might be able to make it simpler by using the distributive property. For instance, here is a way to rewrite the expression $2(n + 7)$.

$2(n + 7) = 2(n) + 2(7)$ ← **Multiply each term *inside* the parentheses by the factor *outside* the parentheses.**
$= 2n + 14$

You can see why this method works if you model the expression with algebra tiles.

Model $2(n + 7)$ as two groups of $(n + 7)$.

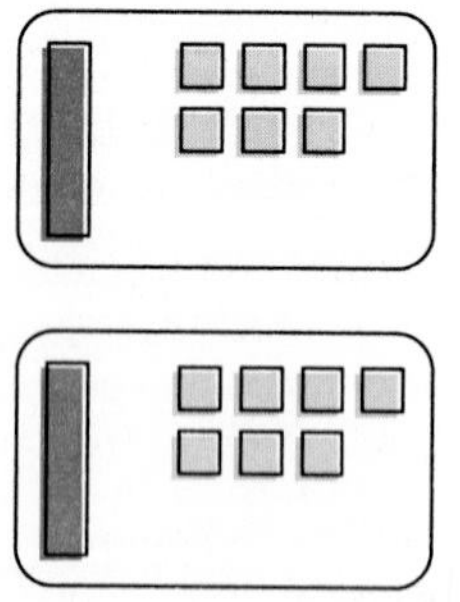

Rearrange the tiles to form one group of $2n$ and one group of 14.

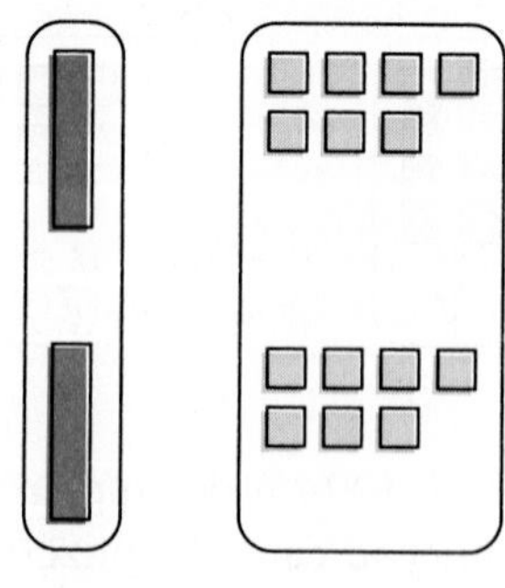

$2(n + 7) = 2n + 14$

example 4

Rewrite $6(z^2 + z + 7)$ by using the distributive property.

solution

$$6(z^2 + z + 7) = 6(z^2) + 6(z) + 6(7)$$ ← **Multiply each term inside the parentheses by 6.**

$$= 6z^2 + 6z + 42$$

your turn

Rewrite each expression by using the distributive property.

19. $3(y + 5)$ **20.** $7(5 + c)$ **21.** $6(w + 1)$ **22.** $4(g^2 + g)$

23. $(n^2 + n)2$ **24.** $(z + 9)3$ **25.** $7(s^2 + s + 3)$ **26.** $(1 + p + p^2)12$

Combining like terms and using the distributive property are two methods used to *simplify* an expression. You **simplify** an expression when you perform as many of the indicated operations as possible. Sometimes you can simplify an expression by combining methods.

example 5

Simplify $3(6 + t) + 11$.

solution

$$3(6 + t) + 11 = 3(6) + 3(t) + 11$$ ← **First use the distributive property.**

$$= 18 + 3t + 11$$

$$= 3t + (18 + 11)$$ ← **Then combine like terms.**

$$= 3t + 29$$

your turn

Simplify each expression.

27. $4(v + 2) + 1$ **28.** $6(b^2 + 3) + 9b^2$ **29.** $12m + 3(m + 8)$

30. $3(1 + r^2) + 14$ **31.** $8(a^2 + a) + 2a$ **32.** $9 + 2(k^2 + k)$

Simplify each expression using any method. If you are not sure how to proceed, try to model the expression with algebra tiles.

33. $2(v + 5)$ **34.** $3j + 11j$ **35.** $q^2 + 9 + 10q^2$

36. $(h + 1)8$ **37.** $6 + 4(n^2 + 4)$ **38.** $3(c^2 + c) + c$

39. $4(b + 1) + 2b + 7$ **40.** $7(x + 2) + 3(x + 4)$ **41.** $2(m^2 + m + 3) + 5$

42. $7s^2 + 5(s + 2)$ **43.** $3(4n)$ **44.** $2(5n + 3)$

glossary

a

absolute value (p. 405): The distance of a number from zero on a number line.

acute angle (p. 483): An angle that has a measure between 0° and 90°.

acute triangle (p. 544): A triangle with three acute angles.

addends (p. 20): Numbers that are combined by the operation of addition. In $13 + 7 = 20$, 13 and 7 are addends.

adjacent angles (p. 479): Two angles in a plane having a common vertex and side but no common interior points.

angle (p. 478): A figure formed by two rays that have the same endpoint.

angle bisector (p. 570): Any ray that separates the angle into two congruent angles.

area (p. 495): The amount of surface covered by a plane figure.

associative property of addition (p. 21): Changing the grouping of the addends does not change the sum.

$$(8 + 6) + 9 = 8 + (6 + 9)$$

associative property of multiplication (p. 45): Changing the grouping of the factors does not change the product.

$$(2 \times 4) \times 5 = 2 \times (4 \times 5)$$

average (p. 346): *See* mean.

axes (p. 318): Two intersecting number lines, one horizontal and one vertical, used in graphing.

b

bar graph (p. 318): A display of numerical facts in which the lengths of bars are used to compare numbers.

base of a power (p. 48): A number that is used as a factor a given number of times. In 8^3, 8 is the base.

base pay (p. 284): A worker's regular salary.

bisector of a line segment (p. 569): Any ray, line segment, or line that bisects a line segment.

box-and-whisker plots (p. 361): A way of displaying data that shows the quartiles.

budget (p. 339): A plan for managing money.

c

capacity (p. 526): Another name for the volume of a container.

Celsius scale (p. 419): A system of measuring temperature in which the boiling point of water is 100° and the freezing point of water is 0°.

central angle of a circle (p. 492): An angle whose vertex is at the center of the circle.

certain event (p. 374): An event that will definitely occur.

chord (p. 492): A line segment that joins two points on a circle.

circle (p. 492): The set of all points in a plane that are the same distance from a given point called the *center* of the circle.

circle graph (p. 330): A graph that shows the relationship of data expressed as percents or parts of a whole.

circumference (p. 492): The distance around a circle.

cluster (p. 351): To be grouped about a central number or item.

combination (p. 386): A group formed by part or all of the objects in a given group without regard to order.

commission (p. 284): An amount of money paid to a salesperson based on a percent of total sales.

commutative property of addition (p. 21): Numbers may be added in any order.

$$8 + 3 = 3 + 8$$

commutative property of multiplication (p. 45): Numbers may be multiplied in any order.

$$4 \times 6 = 6 \times 4$$

compass (p. 567): An instrument used to draw a circle or a part of a circle.

compatible numbers (pp. 60, 147): Numbers that multiply or divide easily, used in estimating.

complementary angles (p. 483): Two angles whose measures have a sum of 90°.

composite figure (pp. 507, 535): An irregularly shaped figure that can be divided into simpler figures.

composite number (p. 127): A number that has more than two factors.

compound interest (p. 295): Interest paid on the principal and on any interest previously earned.

conditional statement (p. 604): A sentence that contains two specific parts, a condition (the *if* part) and a conclusion (the *then* part).

cone (p. 514): A space figure with one circular base and one vertex.

congruent angles (p. 487): Two or more angles that have the same measure.

congruent line segments (p. 487): Two or more line segments that have the same length.

congruent triangles (p. 554): Triangles that have the same shape and size. The lengths of their corresponding sides are in a 1 to 1 ratio.

construction (p. 567): A drawing that is made using only a straightedge and a compass.

converse (p. 605): A statement in which the condition and conclusion of a conditional statement are interchanged.

coordinate plane (p. 450): A plane divided into sections by a horizontal and a vertical number line.

coordinates of a point (p. 450): The numbers in the ordered pair that locate a point on the coordinate plane.

correlation (p. 355): The relationship between two sets of data as indicated by a trend line.

cosine (p. 557): In a right triangle, the cosine of an acute angle is the ratio of the length of the adjacent leg to the length of the hypotenuse.

counting principle (p. 380): In probability, the total number of possible outcomes of an event with multiple steps is the product of the number of choices for each step.

cross products (p. 231): In the proportion $\frac{2}{5} = \frac{6}{15}$, 2×15 and 5×6 are cross products.

cube (p. 514): A rectangular prism all of whose edges are the same length.

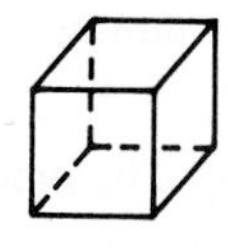

cylinder (p. 514): A space figure with two identical, parallel circular faces.

d

data (p. 318): A collection of numerical facts.

decimal (p. 2): A number written with a decimal point and digits to the right of the ones' place. 0.23 and 14.5 are decimals.

deductions (p. 92): Amounts of money subtracted from a worker's earnings by an employer.

deductive reasoning (p. 606): A process of reaching conclusions based on true statements.

degree (p. 481): A unit of angle measure.

denominator (p. 128): In the fraction $\frac{5}{16}$, 16 is the denominator.

dependent events (p. 382): Events in which the occurrence of one event affects the occurrence of another event.

diagonal of a polygon (p. 487): A line segment that joins two vertices of the polygon and is not a side.

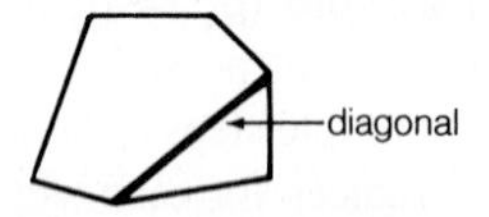

diameter of a circle (p. 492): A chord that passes through the center of the circle.

difference (p. 23): The result of subtracting one number from another. In $17 - 9 = 8$, 8 is the difference.

digit (p. 2): In our number system, one of the symbols 0, 1, 2, 3, 4, 5, 6, 7, 8, 9.

discount (p. 286): The amount by which the list price of an item is decreased.

distributive property (p. 45): Numbers inside parentheses may be multiplied by a factor outside the parentheses as follows.

$$4 \times (6 + 2) = (4 \times 6) + (4 \times 2) = 24 + 8 = 32$$

dividend (p. 60): The number that is divided by another number.

divisible (p. 122): A number is divisible by a second number if the remainder is zero when the first number is divided by the second.

divisor (p. 60): The number by which another number is divided.

down payment (p. 306): The initial amount of money paid by the purchaser of a house, automobile, or other item.

e

edge of a polyhedron (p. 514): The intersection of two faces of the polyhedron.

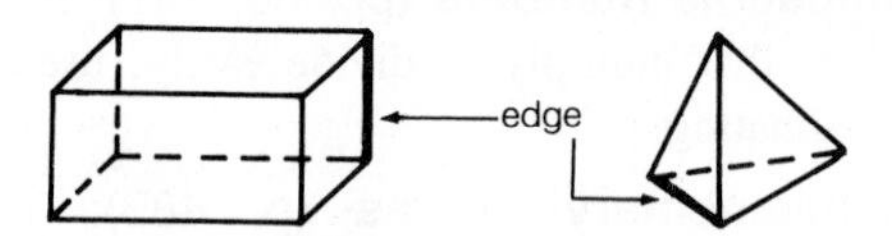

elapsed time (p. 71): The amount of time that passes between the start and the end of an event.

equation (p. 429): A statement that two numbers or quantities are equal.

equivalent fractions (p. 128): Fractions that represent the same amount.

equilateral triangle (p. 544): A triangle with three congruent sides.

evaluate (p. 426): To find the value of a variable expression when the variables are replaced by given values.

event (p. 374): In probability, an activity that has a set of outcomes.

exponent (p. 48): A number used to show how many times another number is to be used as a factor. In 5^4, 4 is the exponent.

f

face of a polyhedron (p. 514): Any one of the surfaces of the polyhedron.

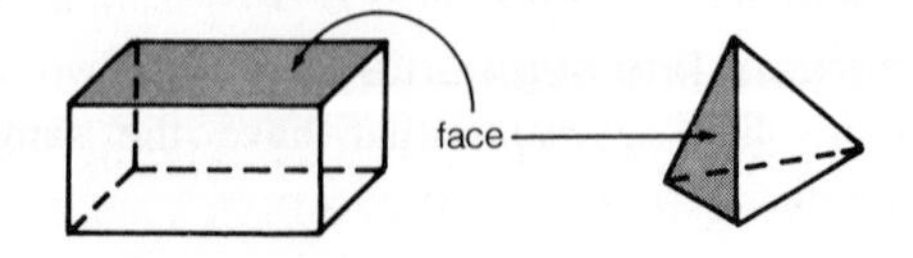

factor (p. 125): When one number is divisible by a second number, the second number is a factor of the first. 7 is a factor of 21.

factorial (p. 385): A notation for the product of a number and all nonzero whole numbers less than the given number.

$5! = 5 \times 4 \times 3 \times 2 \times 1 = 120$

factors (p. 40): Numbers that are multiplied to give a product. In 3×8, 3 and 8 are factors.

Fahrenheit scale (p. 419): A system of measuring temperature in which the boiling point of water is 212° and the freezing point of water is 32°.

formula (p. 443): An equation that describes a relationship between two or more quantities.

fraction (p. 128): A number written in the form $\frac{a}{b}$, where $b \neq 0$. $\frac{9}{16}$ and $\frac{7}{2}$ are fractions.

frequency (p. 349): The number of times an item appears in a given set of data.

frequency polygon (p. 353): A line graph used to show frequencies.

function (p. 464): A relationship between two quantities in which one quantity depends on the other. A function can be described by a rule, a table, or a graph.

g

greatest common factor (GCF) (p. 125): The greatest number that is a factor of two or more numbers. 13 is the greatest common factor of 26 and 65.

gross pay (p. 90): The sum of regular and overtime earnings.

h

histogram (p. 352): A type of bar graph used to show frequencies.

hypotenuse (p. 551): The side opposite the right angle in a right triangle.

i

impossible event (p. 374): An event that will never occur.

independent events (p. 382): Events for which the occurrence of one event does not affect the occurrence of another event.

inductive reasoning (p. 606): A process of reaching conclusions based on a series of individual examples.

inequality (p. 437): A sentence that has an inequality symbol between two numbers or quantities.

integers (p. 402): The numbers . . . , $^-3$, $^-2$, $^-1$, 0, 1, 2, ...

interest (p. 292): Money paid for the use of money.

intersecting lines (p. 479): Lines that meet in one point.

inverse operations (pp. 23, 62): Operations that "undo" each other. Addition and subtraction are inverse operations.

isosceles triangle (p. 544): A triangle with two congruent sides.

l

leading digit (p. 7): The first nonzero digit of a whole number or a decimal.

least common denominator (LCD) (p. 131): The least common multiple of the denominators of two or more fractions.

least common multiple (LCM) (p. 126): The least positive number that is a multiple of two or more nonzero numbers. 60 is the LCM of 15 and 20.

legs (p. 551): The sides of a right triangle that form the right angle.

line (p. 478): A set of points that extends without end in opposite directions.

line graph (p. 322): A display of numerical facts that uses line segments to show both amount and direction of change.

line of symmetry (p. 503): A line through a figure so that if the figure were folded on the line, one side would fit exactly on the other.

line segment (p. 478): A part of a line that consists of two endpoints and all the points between them.

list price (p. 286): The manufacturer's suggested retail price for an item.

lowest terms (p. 128): A fraction is in lowest terms if the greatest common factor of the numerator and the denominator is 1.

m

markup (p. 286): The amount added to the wholesale price of an item to establish the selling price.

mean (average) (p. 346): The sum of the items in a set of data divided by the number of items.

median (p. 346): The middle number (or the mean of the two middle numbers) of a set of data arranged in numerical order.

mixed number (p. 131): A number that consists of both a whole number and a fraction. $7\frac{2}{3}$ is a mixed number.

mode (p. 346): The item that appears most often in a set of data.

multiple (p. 126): When a number is multiplied by a nonzero whole number, the product is a multiple of the given number. Multiples of 3 are 3, 6, 9, 12, 15, ...

n

negative numbers (p. 402): Numbers less than zero.

net pay (take-home pay) (p. 92): A worker's earnings after deductions have been made from gross pay.

numerator (p. 128): In the fraction $\frac{5}{16}$, 5 is the numerator.

o

obtuse angle (p. 483): An angle that has a measure between 90° and 180°.

obtuse triangle (p. 544): A triangle with one obtuse angle.

odds (p. 374): The odds in favor of an event are the ratio of the number of favorable outcomes to the number of unfavorable outcomes of the event.

opposites (p. 404): Numbers that are the same distance from zero on a number line but on different sides of zero. The opposite of zero is zero.

ordered pair (p. 450): A pair of numbers such as $(4, {}^{-}6)$ in which the order of the numbers is important.

origin (p. 450): The point on a coordinate plane where the x-axis and the y-axis meet.

outcome (p. 374): A possible result. One outcome of tossing a coin is *tails*.

p

parallel lines (p. 484): Lines in a plane that do not intersect.

parallelogram (p. 488): A quadrilateral with two pairs of parallel sides.

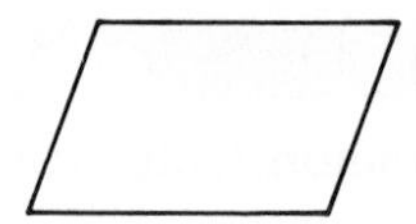

percent (p. 246): A notation that represents a ratio that compares a number to 100. The symbol % is read "percent."

$$4\% = \frac{4}{100} = 0.04$$

perfect square (p. 548): A number whose square root is a whole number.

perimeter (p. 490): The distance around a polygon.

permutation (p. 385): An arrangement of objects in a particular order.

perpendicular lines (p. 484): Lines that intersect to form right angles.

pi (π) (p. 492): A symbol representing the quotient when the circumference of a circle is divided by the length of the diameter, approximated by either $\frac{22}{7}$ or 3.14.

pictograph (p. 326): A way of displaying data in which a picture symbol represents a specified number of items.

place value (p. 2): The value given to the place where a digit appears in a number.

plane (p. 478): A set of points that forms a flat surface that extends without end.

point (p. 478): An exact location in space.

point of symmetry (p. 504): A point about which a geometric figure can be turned to a position that exactly matches its original position.

polygon (p. 487): A closed plane figure formed by joining three or more line segments at their endpoints.

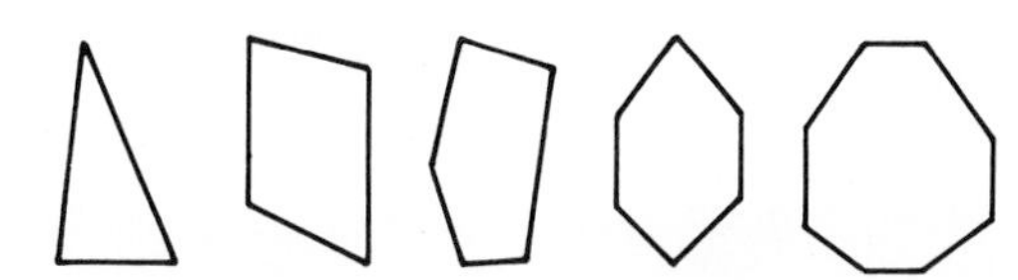

polyhedron (p. 514): A space figure with flat surfaces.

positive numbers (p. 402): Numbers greater than zero.

power of a number (p. 48): The result of multiplying a number by itself a given number of times. 16, or 2^4, is the fourth power of 2.

precision of a measurement (p. 214): Precision is determined by the smallest unit on a measuring instrument.

prime factorization (p. 125): The process of writing a whole number as the product of prime numbers.

prime number (p. 125): A whole number greater than 1 that has exactly two factors, 1 and the number itself.

principal (p. 292): Money on which interest is paid.

prism (p. 514): A polyhedron with two identical and parallel polygons as its bases and parallelograms as its other faces.

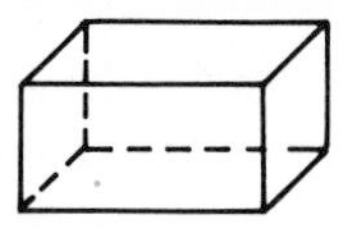

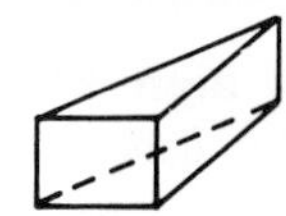

probability (p. 374): The chance that an event will occur, calculated by finding the ratio of the number of favorable outcomes to the number of possible outcomes of the event.

product (p. 40): The result of multiplying two or more numbers. The product of 6 and 8 is 48.

proportion (p. 231): A statement that two ratios are equal.

protractor (p. 481): An instrument used in measuring and drawing angles.

pyramid (p. 514): A polyhedron with any polygon as its base and triangles as its other faces.

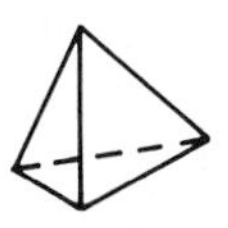

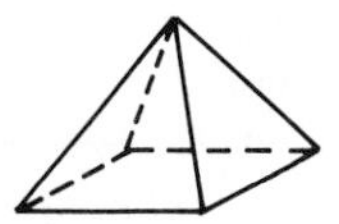

Pythagorean Theorem (p. 551): For any right triangle, the square of the length of the hypotenuse is equal to the sum of the squares of the lengths of the two legs.

q

quadrilateral (p. 487): A four-sided polygon.

quartiles (p. 361): The *first quartile* is the median of the lower half of a set of data. The *third quartile* is the median of the upper half of a set of data.

quotient (p. 60): The result of dividing one number by another. In $72 \div 9 = 8$, 8 is the quotient.

r

radical sign (p. 548): The symbol, $\sqrt{\ }$, used to denote the square root of a number. $\sqrt{36} = 6$

radius of a circle (p. 492): A line segment that joins the center of the circle with any point on the circle.

range of a set of data (p. 346): The difference between the greatest and least values of data in a given set.

rate (p. 228): A ratio of two different types of measurement.

ratio (p. 228): A comparison of two numbers by division.

rational number (p. 402): Any number that can be written as the quotient of two integers, $\frac{a}{b}$, where b does not equal 0.

ray (p. 478): A part of a line that has one endpoint and extends without end in one direction.

reciprocals (p. 149): Two numbers whose product is 1. $\frac{5}{7}$ and $\frac{7}{5}$ are reciprocals.

rectangle (p. 488): A parallelogram with four right angles.

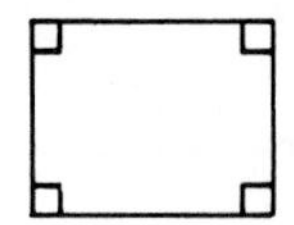

reflection (p. 610): The transformation that results from flipping a figure across a line.

regular polygon (p. 487): A polygon with all sides congruent and all angles congruent.

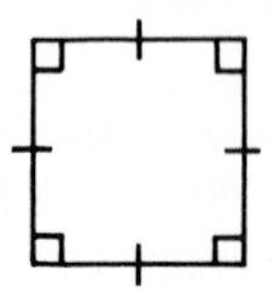

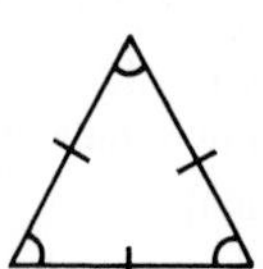

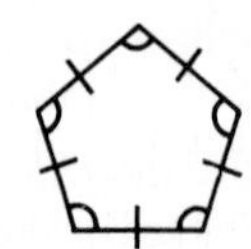

remainder (p. 60): The whole number left after one number is divided by another number.

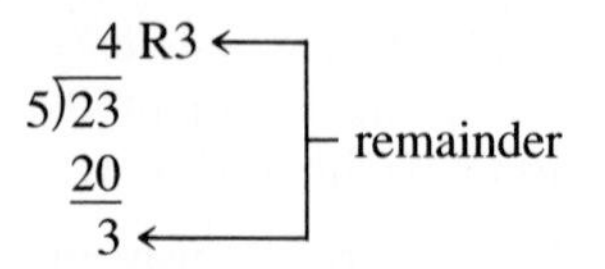

repeating decimal (p. 188): A decimal in which a digit or block of digits repeats endlessly. $0.8\overline{3}$ is a repeating decimal.

rhombus (p. 488): A parallelogram with all sides congruent.

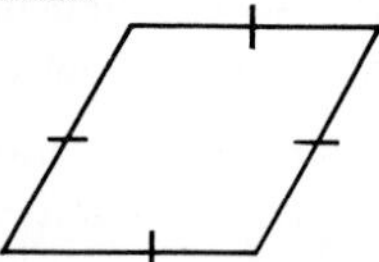

right angle (p. 483): An angle that has a measure of 90°.

right triangle (p. 544): A triangle with one right angle.

rotation (p. 611): The transformation that results from turning a figure about a fixed point.

s

sales tax (p. 258): A tax on the selling price of an item or service that you purchase.

sample (p. 392): In statistics, part of a larger group.

sample space (p. 377): A list of all possible outcomes.

scale drawing (p. 240): A drawing that shows the correct shape of an object but usually is different in size.

scale model (p. 242): An accurate representation of an actual object in a smaller size.

scalene triangle (p. 544): A triangle with no congruent sides.

scattergram (p. 355): A graph that shows the relationships between two different sets of data.

scientific notation (p. 420): A method of writing a number as the product of a power of 10 and a number that is at least 1 but less than 10. 2,300,000 in scientific notation is 2.3×10^6.

sides of an angle (p. 478): The two rays that form the angle.

sides of a polygon (p. 487): The line segments that form the polygon.

similar triangles (p. 554): Triangles that have the same shape.

simple interest (p. 292): Interest paid only on the principal.

sine (p. 557): In a right triangle, the sine of an acute angle is the ratio of the length of the opposite leg to the length of the hypotenuse.

solution of an equation (p. 429): The values of the variables that make the equation true.

solution of an inequality (p. 437): All values of the variable that make the inequality true.

solution of a system of equations (p. 459): An ordered pair that is a solution of all equations in the system.

space figure (p. 514): A three-dimensional figure that encloses part of space.

sphere (p. 514): The set of all points in space that are the same distance from a given point. The given point is called the *center* of the sphere.

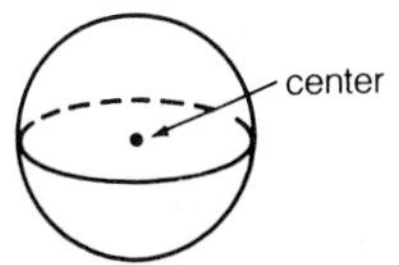

square (p. 488): A parallelogram with four right angles and all sides congruent.

square of a number (p. 548): The result when a number is used as a factor twice. 25 is the square of 5.

square root of a number (p. 548): One of the two equal factors of a number. 7 is the square root of 49.

statistics (p. 346): The branch of mathematics that deals with organizing and analyzing data.

stem-and-leaf plot (p. 358): A way of organizing data to show their frequency.

straight angle (p. 483): An angle that has a measure of 180°.

sum (p. 20): The result of adding two or more numbers. The sum of 5 and 9 is 14.

supplementary angles (p. 483): Two angles whose measures have a sum of 180°.

surface area of a space figure (p. 517): The sum of the areas of all surfaces of the figure.

system of equations (p. 459): Two or more equations that have the same variables.

t

tangent (p. 557): In a right triangle, the tangent of an acute angle is the ratio of the length of the opposite leg to the length of the adjacent leg.

terms of a proportion (p. 231): The numbers used in forming the proportion. In $\frac{3}{20} = \frac{15}{100}$, the terms are 3, 20, 15, and 100.

transform a formula (p. 445): To use inverse operations to get a particular variable alone on one side of the equals sign.

transformation (p. 608): The moving of a shape on a plane.

translation (p. 608): The transformation that results from sliding a shape on a plane.

trapezoid (p. 488): A quadrilateral with exactly one pair of parallel sides.

tree diagram (p. 377): A way of representing a sample space.

trend line (p. 355): The line around which the data in some scattergrams tend to cluster.

triangle (p. 487): A three-sided polygon.

u

unit of precision (p. 214): The smallest unit shown on a measuring instrument.

unit price (p. 237): The cost of one unit of a particular item, expressed in terms of the unit in which the product is generally measured.

unit rate (p. 228): A rate for one unit of a given quantity. Feet per second is a unit rate.

v

value of a variable (pp. 231, 426): Any of the numbers represented by the variable.

variable (pp. 231, 426): A symbol that represents an unknown number.

variable expression (p. 426): An expression that contains one or more variables.

vertex of an angle (p. 478): The endpoint of the rays that form the angle.

vertex of a polygon (p. 487): The point where two sides of the polygon intersect.

vertex of a polyhedron (p. 514): The point where three or more edges of the polyhedron intersect.

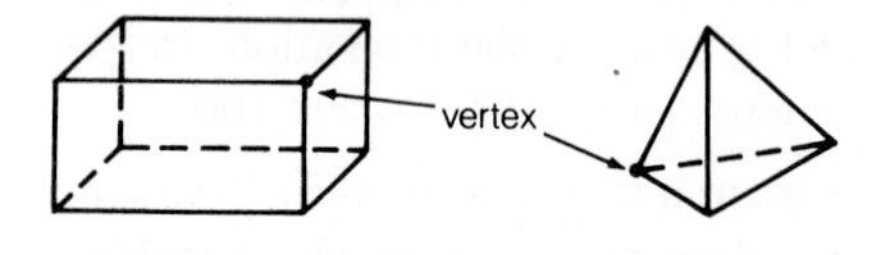

vertical angles (p. 479): The pairs of nonadjacent angles formed by two intersecting lines. $\angle 2$ and $\angle 4$ are vertical angles.

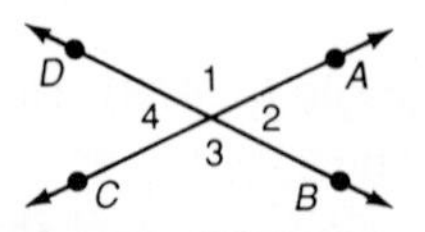

volume (p. 523): The amount of space inside a space figure.

w

whole numbers (p. 2): The numbers 0, 1, 2, 3, and so on.

x

x-axis (p. 450): The horizontal number line on a coordinate plane.

x-coordinate (p. 450): The first number of an ordered pair.

y

y-axis (p. 450): The vertical number line on a coordinate plane.

y-coordinate (p. 450): The second number of an ordered pair.

index

a

b

c

d

e

f

g

h

i

n

o

p

q

r

s

t

u

v

w

x y z

credits

Cover Design — Group Four Design
Cover Photography — John Chomitz
Mechanical Art — Anco/Boston
Illustrations — Linda Phinney

Photographs
1 Craig Wells/After Image **5** Mike J. Howell/ Stock Shop **9** Arthur Sirdofsky/Stock Shop **13–14** Peter Chapman **15** Everett Johnson/Leo de Wys, Inc. **19** John Chomitz **24** Steve Elmore/ Tom Stack and Associates **27–29** Peter Chapman **31** Richard Alcorn/FPG **32** Dave Bartruff/FPG **36** Sybil Shelton/Peter Arnold, Inc. **39** Walter Bibikow/Image Bank **43** Peter Chapman **47** J. Gleiter/H. Armstrong Roberts **51** Michal Heron **52** Brent Peterson/Stock Market **55** Reed Kaestman/Nawrocki **56** Jim Erickson/Stock Market **59** Mark Antman/Image Works **64** E.R. Degginger (cow)/Bruce Coleman, Inc. **64** Mike Blake **66–67** Peter Chapman **71** Ken Baumel **72** Tom Tracy (train)/Stock Shop **72** Ralph Ragsdale **77** © Henley and Savitch/Uniphoto **80** Peter Chapman **85** Hans Reinhard/Bruce Coleman, Inc. **89** Michal Heron/Woodfin Camp and Associates **91** Randy Duchaine/Stock Market **94** Alex S. Maclean (car lot)/Landslides **94** Dick Luria/FPG **97** John Bechtold/Stock Shop **102** James Ballard **104** Camerique **107** Index Stock, Inc. **109** Harry Harttman/Bruce Coleman, Inc. **112** Ed Lettau/FPG **113** Index Stock **117** Penn Archives **117** Dan Esgrow (microchip)/Image Bank **121** Harvey Lloyd/Stock Market **130** Brian Parker/Tom Stack and Associates **137** Russ Lappa/Picture Cube **138–139** Peter Chapman **140** Tad Goodale (tool kit)/Tad Goodale **140** Gabe Palmer/Stock Market **143** Gabe Palmer/Stock Market **148** R. Pleasant/FPG **165** David Witbeck/Picture Cube **171** Focus on Sports **173** Camerique **174** Index Stock **177** Gabe Palmer/After Image **179** Index Stock **180** Peter Vadnai/Stock Market **181** Peter Chapman **182** Stuart Cohen/Stock Market **190** Ballard **193** C. Delbert **194** John P. Kelley/ Image Bank **198** Michael Tamborrino/Stock Market **199** Paul Light (girls)/Paul Light **199** Index Stock **200** Peter Chapman **203** Steve Dunwell/Image Bank **210** Index Stock **211** Index Stock **213** Carlos Fergara/ Nawrocki **217** Jamie Tanaka (girl)/Bruce Coleman, Inc. **217** Eric Roth/Picture Cube **220** Dick Luria/FPG **223** Lawrence Migdale/Photo Researchers **227** Murray and Assoc./After Image **234** Four By Five **235** Peter Chapman **237** Frank Oberle/After Image **238–239** Jim Arndt/Visual Images **239** Tad Goodale **241** Henley and Savage/Stock market **242** Jon Riley/After Image **245** Harold Sund **253–261** Peter Chapman **261** Stacy Pick (exterior)/Uniphoto **262** Shostal **262** D.P. Hershkowitz (house)/Bruce Coleman, Inc. **265** Magna Graphics **268** Michael Furman/ Stock Market **271** Ron Kimball **274** Thomas Kitchin/Tom Stack and Associates **279** A. Brilliant (phone store) **279** Index Stock **281** Janeart Ltd./Image Bank **283** Peter Chapman **285** Laffont/Sygma **287** Peter Chapman **291** David Bentley/Nawrocki **292** John Martucci **299** Chris Jones/Stock Market **302** Gordon/Traub/Wheeler Pictures **305** Gabe Palmer/After Image **307** Joan Sloane/Berg and Associates **307** Steve Smith (steering wheel)/Wheeler Pictures **309** Henley and Savage/Stock Market **310** Tom Campbell/ Click/Chicago **313** Michal Heron **317** Mark Newman/Tom Stack and Associates **319** Ron Kimball/Ron Kimball **320–321** Kunio Owaki/ Stock Market **325** Bill Brooks/Masterfile **326** H. Armstrong Roberts **329** Joel Elkins/FPG **331** Stock Shop **337** Sam Sweezy **339** Carol Lee/ Click/Chicago **345** John Chomitz **348** JVC **353** David Madison/Bruce Coleman, Inc. **358** Pavloff, N./Image Bank **360** Harold Sund/ Image Bank **365** Chris Jones/Stock Market **366** McCoy, Dan/Rainbow **368** Tom Tracy/FPG **369** Bertsch/Stock Shop **376** Peter Chapman **377** John Martucci **379** Ralph B. Pleasant/FPG **382** Peter Chapman **384** Peter Chapman **387** Larry Allan/Image Bank **392** Eisman **394** L.C.T. Rhodes/Taurus **394** James Ballard **397** James Simon/Picture Cube **406** Eisman **408** Ken Stepwell/Bruce Coleman, Inc. **411** Stephen Frink/Stock Market **414** Gabe Palmer/Stock Market **419** Dave Bartruff/FPG **419** Will and Denni MacIntyre/Photo Researchers **421** Mark Segal/Click/Chicago **431** Bob Daemmrich/Stock Boston **439** Barbara Alper/Stock Boston **440** M. Thonig/H. Armstrong Roberts **443** Eunice Harris/Photo Researchers **443** Donald Dietz/Stock Boston **445** Gene Dekovic (car) **445** Mike Blake **452** P. E. Ken/ Uniphoto **458** Jeffrey Muir Hamilton/Stock Boston **463** Ralph Oberlander/Stock Boston **465** Carol Lee/Uniphoto **467** John Deere **470** Gabe Palmer/Stock Market **473** David Jeffery/The Image Bank **474** B. Taylor/ALPHA **480** Ballard **486** Peter Chapman/Pencils **491** James R. Holland/Stock Boston **495** Steven Grohe/Stock Boston **499** Everett Johnson C./Folio **503** Dallas and John Heaton/Stock Boston **505** G. Cloyd/Taurus **508** Max Hirshfeld/Folio **517** Ron Solomon/Uniphoto **522** Shepard Sherbell/Stock Boston **528** Cary Wolinsky/Stock Boston **531** Mike Blake **533** Bob Daemmrich/Stock Boston **536** Thomas Russell/Picture Cube **539** Lemau/FPG **540** Robert Bounemann/Photo Researchers **540** Bob Daemmrich/Uniphoto **543** William Carter/Bruce Coleman **550** Jerry Irwin/ Uniphoto **553** D. and I. McDonald/Picture Cube **559** Everett Johnson C./Leo de Wys **561** R. Laird/FPG **572** Peter Chapman

answers to your turn exercises

CHAPTER 1

Page 3 **1.** fifty-eight thousand, five hundred ninety-three **2.** eighty-one billion, eight hundred eighteen thousand, one hundred **3.** four hundred twenty-nine and six tenths **4.** two hundred thirty-five ten-thousandths **5.** 872 million, 301 thousand, 246 **6.** 1 billion, 700 million, 17 **7.** 92 hundredths **8.** 3 thousand and 3 thousandths **9.** 28,990,000 **10.** 157,064.3 **11.** 49,803,000,271 **12.** 0.5013

Page 5 **1.** 38,100,000 **2.** 7,250,000,000 **3.** 460,900 **4.** 8,504,000 **5.** thousand **6.** billion **7.** 4.942 **8.** 9.3

Page 7 **1.** 15.8 **2.** 45 **3.** 0.10 **4.** 1000 **5.** 400 **6.** 10,000 **7.** 0.7 **8.** 0.09

Page 8 **9.** $15 **10.** $212.37 **11.** $1 **12.** $110.00

Page 10 **1.** $>$ **2.** $=$ **3.** $<$ **4.** 6571; 65,710; 65,791 **5.** 0.8; 0.8192; 0.892 **6.** $14.44; $14.50; $14.54

CHAPTER 2

Page 20 **1.** 1194 **2.** 897 **3.** $2448 **4.** 174.57 **5.** 25.99 **6.** $57.76

Page 21 **7.** 94 **8.** 106 **9.** 16.9

Page 23 **1.** 201 **2.** $91 **3.** 1266 **4.** $7489 **5.** 17.11 **6.** 12.83 **7.** 203.9 **8.** $65.47

Page 25 **1.** about 700 **2.** about $400 **3.** about $12 **4.** about 20,000 **5.** about 70 **6.** about 12,000 **7.** about $22,000 **8.** about $500

Page 26 **9.** about 50 **10.** about $1000 **11.** about 41

CHAPTER 3

Page 40 **1.** between 240 and 270 **2.** between 800 and 1200 **3.** between $2700 and $4000 **4.** between $10,000 and $18,000

Page 41 **5.** about 720 **6.** about $2000 **7.** about $1200 **8.** about 8000

Page 42 **1.** 2112 **2.** 22,857 **3.** $312,872 **4.** 77,982 **5.** 60,030 **6.** $13,230 **7.** 315,656 **8.** 79,426

Page 45 **1.** 154,154 **2.** 270,510 **3.** 800 **4.** 100

Page 46 **5.** 432 **6.** 2170 **7.** 4985 **8.** 1755

Page 48 **1.** 9 **2.** 16 **3.** 125 **4.** 625 **5.** 1 **6.** 9 **7.** 768 **8.** 288 **9.** 96 **10.** 1600

CHAPTER 4

Page 60 **1.** about 400 **2.** about 7000 **3.** about 900 **4.** about 600

Page 61 **5.** about 600; under **6.** about 80; under **7.** about 40; over **8.** about 700; over

Page 62 **1.** 97 R2 **2.** 213 R1 **3.** 1099 R1 **4.** 5803

Page 63 **5.** 57 R9 **6.** 107 R2 **7.** 52 R42 **8.** 392 R10 **9.** 1116 R36 **10.** 1445 **11.** 3012 **12.** 963 R40

Page 65 **1.** 24 **2.** 13 **3.** 35 **4.** 4 **5.** 15 **6.** 40

Page 66 **7.** 4 **8.** 2 **9.** 3 **10.** 5

CHAPTER 5

Page 78 **1.** 44.401 **2.** 2352.2 **3.** 79.3 **4.** 42 **5.** 30.15 **6.** 0.08 **7.** 0.0072 **8.** 0.03

Page 79 **9.** $34.56 **10.** $5.40 **11.** $2.23 **12.** $.58

Page 81 **1.** about $200 **2.** about 7.2 **3.** about 0.15 **4.** about $3 **5.** about $90; overestimate **6.** about 500; underestimate **7.** about 5600; underestimate **8.** about 2; can't tell

Page 83 **1.** $2.33 **2.** 3.45 **3.** 2.75 **4.** 1.007 **5.** 1.4 **6.** 30 **7.** 900 **8.** 200

Page 84 **9.** 2.5 **10.** 0.8 **11.** 0.86 **12.** 0.03

Page 86 **1.** about 40 **2.** about 9 **3.** about 500 **4.** about 0.06 **5.** about 2000 **6.** about 800 **7.** about 0.5 **8.** about 0.8

CHAPTER 6

Page 98 **1.** m **2.** mm **3. a.** 32 mm **b.** 3.2 cm **4. a.** 56 mm **b.** 5.6 cm

Page 99 **5.** 83 **6.** 6 **7.** 0.1172

Page 101 **1.** L **2.** mL **3.** 3700 **4.** 80 **5.** 7.61 **6.** 0.468

Page 103 **1.** kg **2.** g **3.** 500 **4.** 0.375 **5.** 6.295

Page 104 **6.** a **7.** b

CHAPTER 7

Page 122 **1.** Yes; No; No **2.** No; No; No **3.** Yes; Yes; Yes **4.** No; No; Yes

Page 123 **5.** Yes; Yes **6.** Yes; No **7.** No; No **8.** Yes; Yes **9.** Yes; No **10.** No; No **11.** Yes; Yes **12.** No; No

Page 125 **1.** $2 \times 5 \times 7$ **2.** $2 \times 3 \times 11$ **3.** $2^2 \times 3^2$ **4.** $2^3 \times 3$ **5.** 3×17 **6.** $3 \times 5 \times 7$ **7.** 5 **8.** 8 **9.** 1 **10.** 4

Page 126 **11.** 12 **12.** 36 **13.** 20 **14.** 30

Page 128 **1.** 18 **2.** 64 **3.** 5 **4.** 6

Page 129 **5.** $\frac{1}{2}$ **6.** $\frac{3}{7}$ **7.** $\frac{3}{11}$ **8.** $\frac{3}{5}$ **9.** $\frac{4}{5}$ **10.** $\frac{9}{16}$

Page 131 **1.** $<$ **2.** $=$ **3.** $<$ **4.** $>$ **5.** $<$ **6.** $<$ **7.** $>$ **8.** $>$

Page 132 **9.** $\frac{1}{7}, \frac{3}{7}, \frac{5}{7}$ **10.** $\frac{1}{2}, \frac{3}{5}, \frac{5}{6}$ **11.** $\frac{1}{6}, \frac{5}{9}, \frac{2}{3}$ **12.** $1\frac{4}{13}, 1\frac{1}{3}, 1\frac{29}{39}$

Page 134 **1.** about $\frac{1}{4}$ **2.** about $\frac{1}{7}$ **3.** about $\frac{1}{2}$ **4.** about $\frac{3}{4}$ **5.** about $\frac{1}{3}$ **6.** about $\frac{4}{5}$ **7.** 1 **8.** 0 **9.** 1 **10.** $\frac{1}{2}$ **11.** $\frac{1}{2}$ **12.** 0

CHAPTER 8

Page 144 **1.** $\frac{20}{63}$ **2.** $\frac{2}{9}$ **3.** $3\frac{1}{3}$ **4.** 1

Page 145 **5.** $2\frac{2}{3}$ **6.** $5\frac{1}{2}$ **7.** $11\frac{1}{5}$ **8.** 7

Page 147 **1.** about 8 **2.** about 35 **3.** about 12 **4.** about 48 **5.** about 7 **6.** about 50 **7.** about 16 **8.** about 8

Page 149 **1.** 8 **2.** $\frac{10}{3}$ **3.** no reciprocal **4.** $\frac{1}{9}$ **5.** $\frac{8}{13}$ **6.** $\frac{3}{17}$ **7.** $\frac{5}{18}$ **8.** $1\frac{1}{2}$ **9.** $\frac{3}{44}$ **10.** 20

Page 150 **11.** $3\frac{1}{3}$ **12.** $3\frac{1}{3}$ **13.** $2\frac{14}{15}$ **14.** $1\frac{5}{9}$

Page 152 **1.** about 4 **2.** about 2 **3.** about 6 **4.** about 7 **5.** about 15 **6.** about 60 **7.** about 12 **8.** about 180

CHAPTER 9

Page 166 **1.** $\frac{13}{17}$ **2.** $\frac{2}{9}$ **3.** 1 **4.** $\frac{1}{2}$

Page 167 **5.** $\frac{2}{3}$ **6.** $\frac{1}{8}$ **7.** $1\frac{1}{14}$ **8.** $\frac{11}{28}$

Page 169 **1.** about 20 **2.** about 12 **3.** about 11 **4.** about 2 **5.** about 10 **6.** about 9 **7.** about 13

Page 172 **1.** $5\frac{3}{5}$ **2.** $15\frac{7}{9}$ **3.** $6\frac{4}{5}$ **4.** $14\frac{3}{10}$ **5.** $11\frac{3}{14}$ **6.** $6\frac{5}{18}$ **7.** $3\frac{1}{6}$ **8.** 14

Page 175 **1.** $\frac{1}{3}$ **2.** $2\frac{5}{18}$ **3.** $2\frac{11}{24}$ **4.** $6\frac{1}{2}$ **5.** $\frac{3}{5}$ **6.** $2\frac{7}{8}$ **7.** $6\frac{11}{16}$ **8.** $2\frac{13}{20}$

CHAPTER 10

Page 186 **1.** $\frac{3}{10}$ **2.** $\frac{9}{20}$ **3.** $\frac{3}{500}$ **4.** $1\frac{1}{25}$ **5.** $2\frac{1}{8}$ **6.** $\frac{1}{3}$ **7.** $\frac{2}{3}$ **8.** $\frac{17}{200}$ **9.** $\frac{11}{18}$ **10.** $\frac{4}{15}$

Page 188 **1.** 0.4 **2.** 0.15 **3.** 0.375 **4.** 1.5 **5.** 2.75 **6.** $0.\overline{3}$ **7.** $1.\overline{2}$ **8.** $0.4\overline{6}$ **9.** $2.1\overline{6}$ **10.** $0.\overline{27}$

Page 189 **11.** $>$ **12.** $>$ **13.** $>$ **14.** $=$

Page 191 **1.** about $\frac{1}{2}$ **2.** about $\frac{3}{4}$ **3.** about $\frac{1}{5}$ **4.** about $\frac{1}{6}$ **5.** about $1\frac{1}{3}$

Page 192 **6.** about 200 **7.** about 300 **8.** about $20,000 **9.** about 490

CHAPTER 11

Page 204 **1.** 13 **2.** 15,840 **3.** 39 **4.** 68 **5.** 6; 3 **6.** 3; 60

Page 205 **7.** $1\frac{3}{8}$ in. **8.** $2\frac{1}{16}$ in.

Page 207 **1.** 4 **2.** 3 **3.** 116 **4.** 169 **5.** 10,350 **6.** 28; 6

Page 209 **1.** 9 **2.** 208 **3.** 29 **4.** 33 **5.** 8; 2 **6.** 10; 1

Page 210 **7.** c **8.** a

Page 212 **1.** 41 ft 10 in. **2.** 6 t 753 lb **3.** 54 qt **4.** 50 ft 8 in. **5.** 13 t 1600 lb **6.** 61 c 2 fl oz **7.** 9 gal 2 qt **8.** 43 ft 5 in. **9.** 12 lb 3 oz **10.** 3 yd 1 ft **11.** 12 gal 3 qt **12.** 3 lb 10 oz

Page 214 **1.** 3; 10 mi **2.** 4; 10 lb **3.** 1; 0.01 km **4.** 4; 0.01 L

Page 215 **5.** 5200 yd **6.** 23.7 kg **7.** 218.7 L

CHAPTER 12

Page 228 **1.** $\frac{12}{35}$ **2.** $\frac{1}{4}$ **3.** $\frac{5}{4}$ **4.** $\frac{7}{1}$ **5.** $\frac{1}{8}$ **6.** $\frac{16}{49}$ **7.** $\frac{12}{1}$ **8.** $\frac{3}{1}$

Page 229 **9.** 55 mi/h **10.** 10 ft/s **11.** $.50/apple **12.** 4.8 m/s

Page 231 **1.** True **2.** True **3.** False **4.** True

Page 232 **5.** 54 **6.** 27 **7.** 3 **8.** 6

CHAPTER 13

Page 246 **1.** 0.85 **2.** 0.04 **3.** 0.027 **4.** 2.7 **5.** $0.27\frac{2}{3}$ **6.** 80% **7.** 0.2% **8.** 64% **9.** $5\frac{1}{2}$% **10.** 325%

Page 248 **1.** $\frac{9}{100}$ **2.** $\frac{3}{20}$ **3.** $2\frac{1}{2}$ **4.** $\frac{221}{250}$ **5.** $\frac{1}{2500}$ **6.** $\frac{7}{800}$ **7.** $\frac{1}{250}$ **8.** $\frac{1}{3}$ **9.** $\frac{43}{400}$ **10.** $\frac{37}{125}$

Page 249 **11.** 42% **12.** $66\frac{2}{3}$% **13.** $62\frac{1}{2}$% **14.** 270% **15.** $144\frac{4}{9}$%

Page 251 **1.** 30 **2.** 86 **3.** $116\frac{2}{3}$ **4.** $58\frac{1}{2}$

Page 252 **5.** 103.14 **6.** 28.7 **7.** 45 **8.** 1.314

Page 254 **1.** about 16 **2.** about 300 **3.** about $180 **4.** about $70 **5.** about 45 **6.** about 480 **7.** about $20 **8.** about $800

CHAPTER 14

Page 266 **1.** 75% **2.** 60% **3.** 28% **4.** 125%

Page 267 **5.** $87\frac{1}{2}\%$ **6.** $83\frac{1}{3}\%$ **7.** $162\frac{1}{2}\%$ **8.** $137\frac{1}{2}\%$ **9.** $7\frac{1}{2}\%$ **10.** $27\frac{1}{4}\%$ **11.** $86\frac{2}{3}\%$ **12.** $316\frac{1}{3}\%$

Page 269 **1.** 30% **2.** 120% **3.** $166\frac{2}{3}\%$ **4.** $83\frac{1}{3}\%$

Page 270 **5.** 80% **6.** 50% **7.** $83\frac{1}{3}\%$ **8.** $37\frac{1}{2}\%$

Page 272 **1.** 50 **2.** 6 **3.** 75 **4.** 23 **5.** 900 **6.** 3000

Page 273 **7.** 75 **8.** 96 **9.** 40 **10.** 24 **11.** 420 **12.** 120

Page 275 **1.** 66 **2.** 14.7 **3.** 12.15 **4.** 105 **5.** 8% **6.** 23% **7.** $33\frac{1}{3}\%$ **8.** 125%

Page 276 **9.** 650 **10.** 84 **11.** 123 **12.** 400

CHAPTER 15

Page 292 **1.** $78 **2.** $4.40 **3.** $142.50 **4.** $13.13

Page 293 **5.** $2240 **6.** $1260 **7.** $5950 **8.** $549.38

Page 295 **1.** $551.25 **2.** $1076.41

Page 296 **3.** $578.80 **4.** $33.77

CHAPTER 16

Page 318 **1.** about 400 m **2.** about 300 m **3.** about 250 m **4.** 9 **5.** 3 **6.** country, folk, jazz

Page 319 **7.**

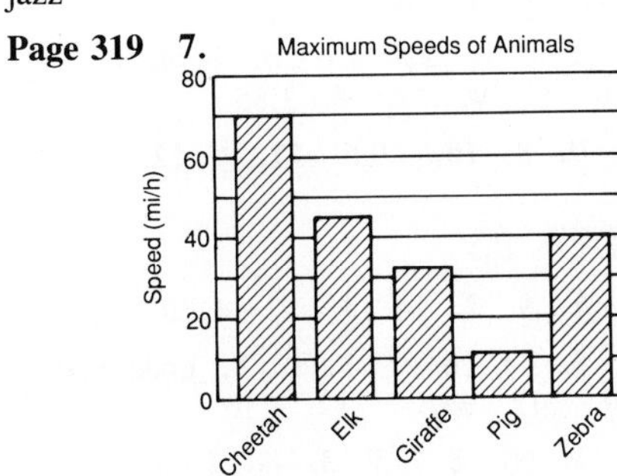

Page 322 **1.** 39°F **2.** 2:00 P.M. and 6:00 P.M. **3.** about 40 million **4.** 1940 and 1960 **5.** 1920 **6.** about 230 million

Page 323 **7.** Telephones in the United States

Page 326 **1.** about 15 thousand **2.** about 42.5 thousand **3.** about 30 thousand **4.** 1960

Page 327 **5.**

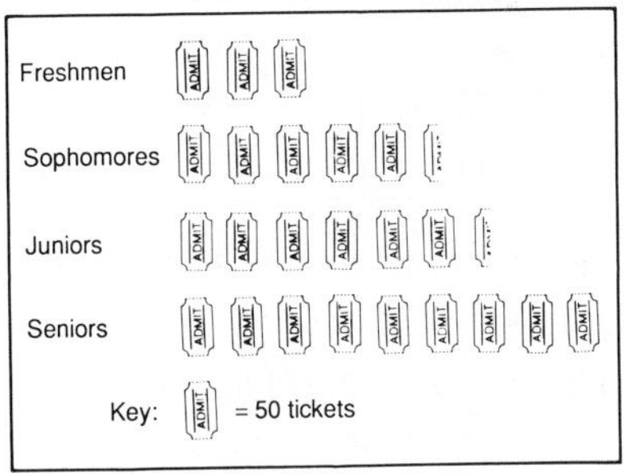

Page 330 **1.** about 48 million **2.** about 84 million **3.** about 57.6 million **4.** about $\frac{1}{4}$

Page 331 **5.** $15,000 **6.** $24,000 **7.** $22,500 **8.** $2500 **9.** $17,600

CHAPTER 17

Page 346 **1.** 77; 78; 78; 33 **2.** 5.2; 4.5; none; 4.4 **3.** 66; 66.5; 57; 18 **4.** 7.9; 8.6; 8.6 and 9.0; 4.3

Page 347 **5.** $92.80; $91; $91; $18 **6.** $8.95; $8.83; none; $2.23 **7.** 41.7; 41.5; none; 13 **8.** 16.4; 17.55; 19.2; 9.3

Page 349

1. Number of Games Won

Number of Games	Tally	Frequency
8	II	2
9	~~IIII~~ II	7
10	~~IIII~~ II	7
11	~~IIII~~	5
12	~~IIII~~ II	7

2. **Typing Rates (Words/Min)**

Number of Words	Tally	Frequency
45	I	1
50	II	2
55	~~IIII~~ II	7
60	I	1
65	~~IIII~~ III	8
70	~~IIII~~ IIII	9

Page 350

3. **Scores on a Spanish Test (%)**

Score	Tally	Frequency
61–70	II	2
71–80	~~IIII~~ III	8
81–90	IIII	4
91–100	II	2

4. **Ages of Baseball Players**

Age (Years)	Tally	Frequency
20–24	II	2
25–29	~~IIII~~	5
30–34	IIII	4
35–39	I	1
40–44	II	2

Page 352 **1.** 7 **2.** 61–70 **3.** 3 **4.** 9

Page 353 **5.** 19 **6.** 15 **7.** 15 **8.** 14

Page 355 **1.** 16 lb **2.** 72 in. **3.** 3
4. 65 in. and 70 in.

Page 356 **5.** about 4 points **6.** about 1 point
7. increase **8.** positive

Page 358

1. **Scores on a Science Test**

6	1, 4, 8, 8
7	0, 0, 2, 3, 3, 7, 7
8	1, 2, 2, 4, 4, 5
9	0, 0, 3

2. **Number of Touchdowns**

9	0, 0, 1, 3
10	0, 5
11	3, 6
12	0, 6

Page 359 **3.** 2 **4.** 15 **5.** none **6.** none

Page 362

1.

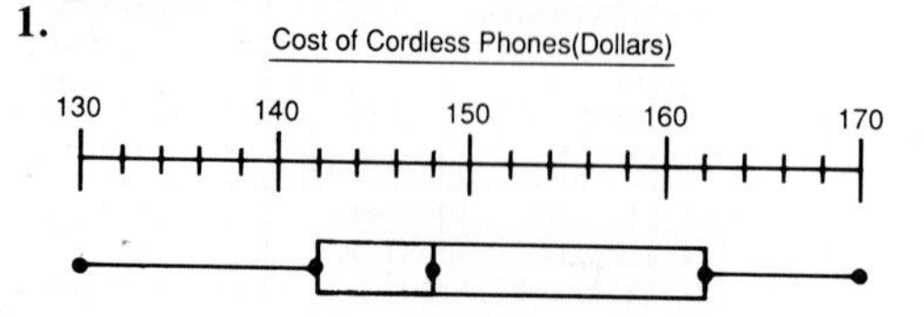

2.

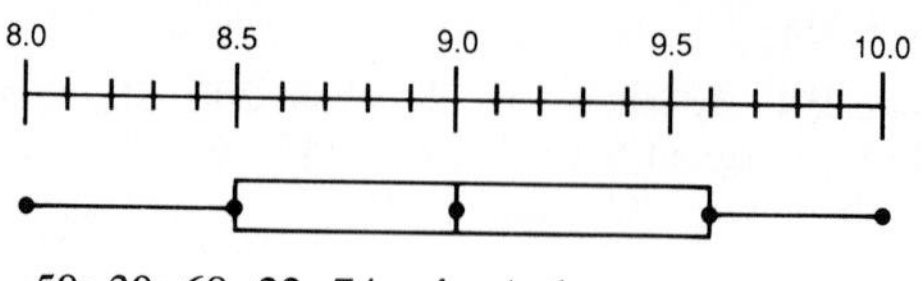

3. 50; 30; 68; 22; 74 **4.** Arden

CHAPTER 18

Page 374 **1.** $\frac{1}{6}$ **2.** $\frac{5}{6}$ **3.** $\frac{1}{3}$ **4.** 1

Page 375 **5.** 1 to 2 **6.** 2 to 1 **7.** 5 to 1
8. 1 to 1

Page 377 **1.** 9 **2.** 6 **3.** $\frac{1}{8}$ **4.** $\frac{3}{8}$ **5.** $\frac{1}{8}$
6. $\frac{1}{2}$

Page 380 **1.** 70 **2.** 1200 **3.** 128 **4.** 1296
5. $\frac{1}{260}$ **6.** $\frac{1}{52}$ **7.** $\frac{9}{130}$ **8.** $\frac{15}{52}$

Page 382 **1.** $\frac{1}{20}$ **2.** $\frac{1}{5}$ **3.** $\frac{4}{25}$ **4.** $\frac{2}{9}$ **5.** $\frac{2}{9}$
6. 0

Page 385 **1.** 24 **2.** 40,320 **3.** 11,880
4. 462

Page 386 **5.** 15 **6.** 56

CHAPTER 19

Page 402 **1.** 2 **2.** $^{-}6$ **3.** 47

Page 403 **4.** $^{-}\frac{8}{1}$ **5.** $\frac{9}{4}$ **6.** $\frac{5}{6}$ **7.** $\frac{399}{100}$ **8.** $\frac{0}{1}$

Page 404 **1.** *C* **2.** *B* **3.** *D* **4.** *A*

Page 405 **5.** $^{-}6$ **6.** $^{-}1\frac{1}{2}$ **7.** 3.2 **8.** 0
9. 8 **10.** 9 **11.** 1.4 **12.** $6\frac{3}{5}$

Page 407 **1.** $<$ **2.** $>$ **3.** $<$ **4.** $>$ **5.** $^{-}9$; $^{-}5$; 0; 9 **6.** $^{-}3.4$; $^{-}3.3$; 1.7; 1.8 **7.** $^{-}5\frac{1}{2}$; $^{-}1\frac{1}{2}$; $1\frac{1}{2}$; $2\frac{3}{4}$

Page 409 **1.** 9 **2.** 2 **3.** $^{-}2.5$ **4.** $^{-}5$
5. 9 **6.** $^{-}50$ **7.** $^{-}2\frac{2}{3}$ **8.** 13.1

Page 410 **9.** $^{-}6$ **10.** 0 **11.** 4.1 **12.** $^{-}3\frac{1}{3}$

Page 412 **1.** $36 + {}^{-}63$ **2.** $14 + 20$
3. $^{-}29.1 + {}^{-}17.5$ **4.** $^{-}6\frac{1}{3} + 3\frac{1}{3}$ **5.** $^{-}23$
6. $^{-}86$ **7.** $7\frac{5}{9}$ **8.** $^{-}4.2$

Page 415 **1.** $^{-}14$ **2.** $^{-}2.6$ **3.** 4 **4.** 0
5. $^{-}36$ **6.** 80 **7.** $^{-}\frac{1}{16}$ **8.** 3.85

Page 416 **9.** 8 **10.** $^{-}0.5$ **11.** $\frac{4}{9}$ **12.** 0

CHAPTER 20

Page 426 **1.** 9 **2.** 14 **3.** $^{-}2$ **4.** 9

Page 427 **5.** $^{-}6$ **6.** 0 **7.** 0 **8.** undefined
9. 13 **10.** $^{-}12$ **11.** 7 **12.** $^{-}10$

Page 429 **1.** 60 **2.** $^{-}5$ **3.** $\frac{2}{11}$ **4.** 3.6

Page 430 **5.** −3 **6.** 24 **7.** 24.3 **8.** $1\frac{5}{9}$

Page 432 **1.** 20 **2.** −4 **3.** −1.3 **4.** 0.5

Page 433 **5.** −40 **6.** 63 **7.** −12.6 **8.** 4.5 **9.** 20 **10.** −60 **11.** −49 **12.** −21

Page 435 **1.** −3 **2.** −16 **3.** 1.3 **4.** 25 **5.** −18 **6.** −28

Page 437 **1.** $y < 2$ **2.** $s \leq -6$ **3.** $a \geq 20$ **4.** $x > -16$ **5.** $a \geq -2$ **6.** $y \leq 3$

Page 438

7.

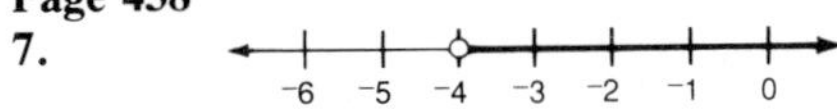

8.

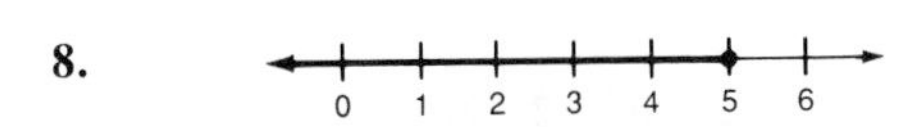

9.

10.

CHAPTER 21

Page 450 **1.** (−2, −4) **2.** (1, 2) **3.** (1, −2) **4.** (−4, 1) **5.** (3, 0) **6.** (0, 3)

7.—12.

Page 453 **1.** Yes **2.** No **3.** Yes **4.** No **5.** Yes **6.** (5, 7) **7.** (1, −1) **8.** (−4, −11) **9.** (0, −3) **10.** ($\frac{1}{2}$, −2)

Page 454

11.

$y = x + 5$		
x	y	(x, y)
−2	3	(−2, 3)
−1	4	(−1, 4)
0	5	(0, 5)
1	6	(1, 6)
2	7	(2, 7)

12.

$y = -2x$		
x	y	(x, y)
−2	4	(−2, 4)
−1	2	(−1, 2)
0	0	(0, 0)
1	−2	(1, −2)
2	−4	(2, −4)

13.

$y = 5x - 2$		
x	y	(x, y)
−2	−12	(−2, −12)
−1	−7	(−1, −7)
0	−2	(0, −2)
1	3	(1, 3)
2	8	(2, 8)

14.

$y = -3x + 1$		
x	y	(x, y)
−2	7	(−2, 7)
−1	4	(−1, 4)
0	1	(0, 1)
1	−2	(1, −2)
2	−5	(2, −5)

Page 456

1.

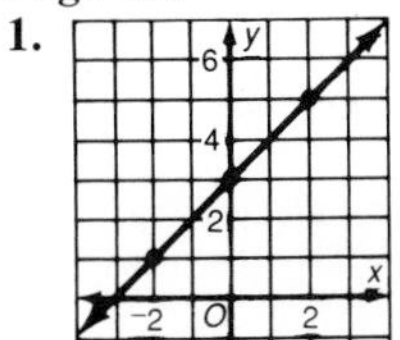

2.

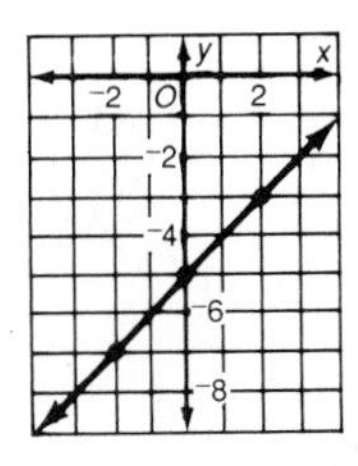

3.

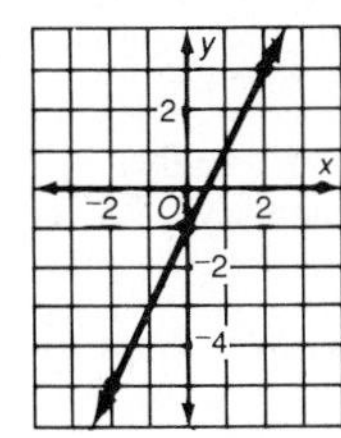

4.

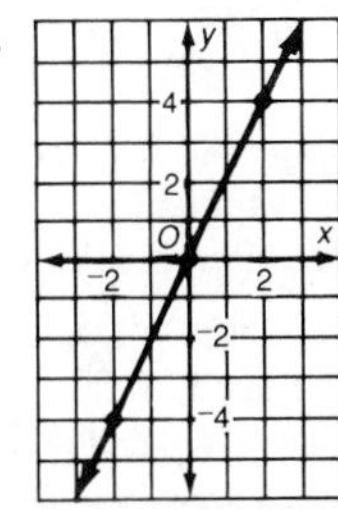

5.

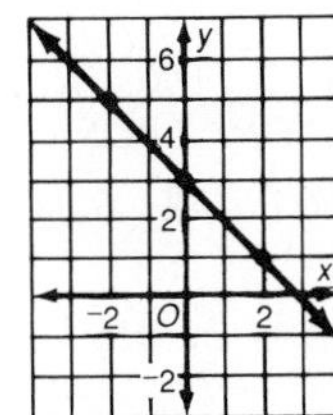

6.

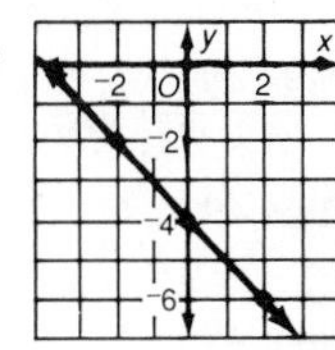

7.

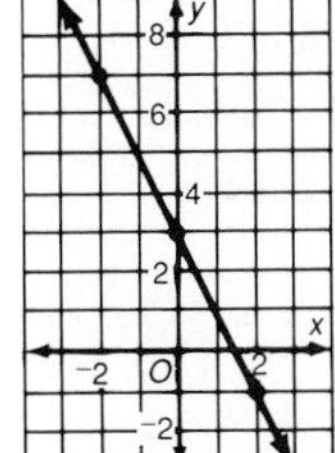

8.

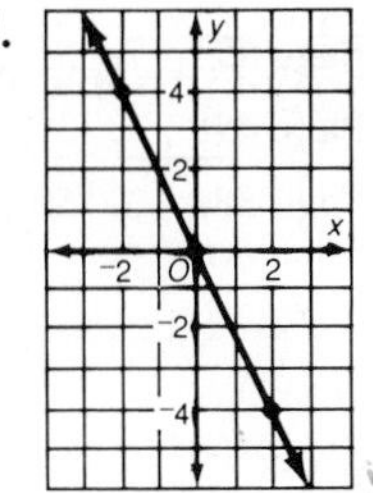

Page 459 **1.** Yes **2.** No **3.** No **4.** Yes

Page 460 **5.** (2, 1) **6.** No solution **7.** $(^-2, ^-5)$ **8.** $(0, ^-2)$

CHAPTER 22

Page 479 **1.** $C, D, E, F, G, N, W, X, Y, Z$ **2.** A **3.** $\overleftrightarrow{FG}, \overleftrightarrow{YZ}$ **4.** $\angle CDE$ **5.** $\overline{YZ}, \overline{FG}, \overline{WX}, \overline{DC}, \overline{DE}$ **6.** adjacent **7.** vertical **8.** adjacent **9.** adjacent

Page 481 **1.** 60° **2.** 135° **3.** 90°

4.

20°

5.

140°

6.

90°

Page 483 **1.** acute **2.** straight **3.** obtuse **4.** right

Page 484 **5.** complementary **6.** neither **7.** supplementary **8.** complementary **9.** False **10.** True **11.** True **12.** False

Page 488 **1.** quadrilateral; $\overline{RU}, \overline{ST}$ **2.** regular pentagon; $\overline{MO}, \overline{MP}, \overline{NP}, \overline{NQ}, \overline{OQ}$ **3.** triangle; none **4.** rectangle **5.** rhombus **6.** trapezoid

Page 490 **1.** 19 m **2.** 96 ft **3.** 24 cm **4.** 16 cm **5.** 101 ft

Page 492 **1.** circle O **2.** $\overline{RS}$ **3.** $\overline{OX}, \overline{OR}, \overline{OS}$ **4.** $\overline{RS}, \overline{XZ}, \overline{MN}$ **5.** $\angle XOS, \angle XOR, \angle ROS$

Page 493 **6.** 17.27 km **7.** 66 in. **8.** 62.8 m **9.** $5\frac{1}{2}$ yd

Page 495 **1.** 98 yd^2 **2.** 64 $in.^2$ **3.** 169 km^2 **4.** 11.52 m^2

Page 496 **5.** 500 m^2 **6.** 18 yd^2

Page 497 **7.** 55.35 m^2 **8.** 48 mi^2

Page 499 **1.** 379.94 mm^2 **2.** 2464 ft^2

Page 500 **3.** 50.24 cm^2 **4.** 154 ft^2

CHAPTER 23

Page 514 **1.** sphere **2.** cylinder **3.** triangular prism **4.** hexagonal pyramid

Page 515 **5.** cube **6.** triangular prism **7.** rectangular pyramid **8.** cone

Page 517 **1.** 54 yd^2 **2.** 144 $in.^2$ **3.** 235.2 cm^2

Page 518 **4.** 1120 $in.^2$ **5.** 180 ft^2 **6.** 9.2 m^2

Page 520 **1.** 1130.4 $in.^2$ **2.** 486.7 ft^2 **3.** 401.92 cm^2

Page 521 **4.** 2816 cm^2 **5.** 8448 yd^2 **6.** 275 ft^2

Page 523 **1.** 216 m^3 **2.** 210 m^3 **3.** 63 $in.^3$

Page 524 **4.** 360 cm^3 **5.** 11,520 m^3 **6.** 567 $in.^3$ **7.** 4928 $in.^3$ **8.** 1469.52 cm^3 **9.** 154 ft^3

Page 527 **1.** 3600 $in.^3$ **2.** 72 yd^3 **3.** 300 mm^3

Page 528 **4.** 565.2 yd^3 **5.** 6160 cm^3

Page 530 **1.** 314 $in.^2$ **2.** 5024 cm^2 **3.** 3215.36 yd^2

Page 531 **4.** 904.32 $in.^3$ **5.** 904,320 mm^3 **6.** 3052.08 ft^3

CHAPTER 24

Page 544 **1.** isosceles; right **2.** equilateral; acute **3.** scalene; obtuse

Page 545 **4.** No **5.** Yes **6.** No **7.** 34° **8.** 33° **9.** 117°

Page 548 **1.** 2 **2.** 8 **3.** 30 **4.** 90 **5.** 5.745 **6.** 8.602 **7.** 28 **8.** 116

Page 551 **1.** Yes **2.** No **3.** Yes **4.** No

Page 552 **5.** 5.8 yd **6.** 4 m **7.** 25 in. **8.** 15 ft

Page 554 **1.** similar; not congruent **2.** not similar

Page 555 **3.** $x = 12$ **4.** $x = 66$

Page 557 **1.** $\frac{4}{3}$ **2.** $\frac{4}{5}$ **3.** $\frac{3}{5}$

Page 558 **4.** $\frac{3}{5}$ **5.** $\frac{4}{5}$ **6.** $\frac{3}{4}$

answers to selected exercises

CHAPTER 1

Pages 3–4 Practice Exercises **1.** fourteen thousand, six hundred twelve **3.** twenty-three hundredths **5.** seventeen ten-thousandths **7.** four thousand, two hundred fifty and one tenth **9.** 7 thousand, 525 **11.** 7 tenths **13.** 8401 ten-thousandths **15.** 494 and 49 hundredths **17.** 934,276 **19.** 7,000,000,300 **21.** 10.092 **23.** ten millions **25.** hundredths **27.** 81.75 **29.** one hundred eighty-six thousand, two hundred eighty-one and seven tenths

Page 4 Review Exercises **1.** 16 **3.** 42 **5.** 4 **7.** 9 **9.** 21 **11.** 7

Page 6 Practice Exercises **1.** 4,300,000 **3.** 45,900 **5.** 9,125,000,000 **7.** 700,160,000 **9.** billion **11.** 14.15 **13.** True **15.** False **17.** 3,900,000,000 **19.** $464.9 million **21.** 1,902,000; $11,800,000

Page 6 Review Exercises **1.** 45,012,768 **3.** 280,010

Pages 8–9 Practice Exercises **1.** 804,000 **3.** 5842 **5.** 0.40 **7.** 915.7 **9.** 40 **11.** 500 **13.** 10 **15.** 0.09 **17.** $6 **19.** $501 **21.** $2.13 **23.** $798.09 **25.** 440 **27.** 16,480 **29.** 50,000,000 **31.** 1 **33.** 0.70 **35.** $194 **37.** $6.08 **39.** 72.6 **43.** 9091 **45.** $324

Page 9 Review Exercises **1.** 4,500,000 **3.** 16,970,000,000 **5.** 9,921,000,000 **7.** 243,200

Page 11 Practice Exercises **1.** > **3.** > **5.** < **7.** $285; $2385; $2583 **9.** 0.0001; 0.001; 0.01 **11.** 5.067; 7.506; 7.605 **13.** True **15.** False **17.** False **19.** True **21.** False **23.** True **25.** Janine **27.** 4081; 4745; 5083; 5098; 5246; 5469

Page 11 Review Exercises **1.** twenty-six and one tenth **3.** one thousand, eight hundred forty-four **5.** two hundred six million, five hundred seventy thousand, three hundred thirteen **7.** four hundred two and thirty-nine thousandths

Page 12 Skill Review **1.** ten thousand, eight hundred thirty-three **3.** two hundred two thousandths **5.** four hundred eighty-one million, six hundred twenty thousand, four hundred seventy **7.** three thousand, ninety-seven and eighteen hundredths **9.** 298,000 **11.** 47.5 **13.** 6,100,000 **15.** 85,500,000,000 **17.** 290,700 **19.** thousand **21.** million **23.** 656,000 **25.** $10 **27.** 800,000 **29.** 0.002 **31.** 60,000 **33.** < **35.** < **37.** < **39.** 0.0038; 0.0388; 0.338

Pages 13–14 Exercises **1.** $6.60 **3.** $118.53 **5.** Eight and $\frac{75}{100}$ dollars **7.** Two hundred fifteen and $\frac{06}{100}$ dollars **9.** John F. Ramirez **11.** River-Bank, Missouri **13.** clothes **17.** Yes **19.** word form

Pages 15–16 Exercises **1.** 2561 **3.** 974 **5.** 8035 **7.** 7286 **9.** the first month; 1390 > 1345 **11.** 43,901.3 **13.** 51,437.0 **15.** 8354.2 **17.** 25.8 **19.** car A **21.** The odometer shows six zeros.

Page 17 Chapter Review **1.** is less than **3.** 54,280,008 **5.** 0.065 **7.** 453,000,213,089 **9.** 12.76 **11.** 81,700,000 **13.** 20 **15.** $631.57 **17.** True **19.** False **21.** $209.35 **23.** 5047

CHAPTER 2

Pages 21–22 Practice Exercises **1.** $231 **3.** 11,468 **5.** $4366 **7.** 7.49 **9.** 3.7892 **11.** $38.44 **13.** 45 **15.** 139 **17.** 13.7 **19.** 1225 **21.** 394.1 **23.** 214.87 **25.** 80 **27.** 250 **29.** 24.9 **31.** associative property of addition, commutative property of addition, associative property of addition; 338 **33.** 6,713,000 **35.** 19,250,000 **37.** 1985

Page 22 Review Exercises **1.** 3 **3.** 6 **5.** 10 **7.** 13 **9.** 18

Page 24 Practice Exercises **1.** 2122 **3.** $4071 **5.** 89,083 **7.** 22.02 **9.** 3.04 **11.** $3901.75 **13.** 511 **15.** $31.87 **17.** $469 **19.** 338.51 **21.** 481.147 **23.** 601 ft **25.** $.10

Page 24 Review Exercises **1.** 30 **3.** 3 **5.** 10,000 **7.** 0.2 **9.** 7.119 **11.** 4000

Pages 26–27 Practice Exercises **1.** about 6000 **3.** about $7 **5.** about $300 **7.** about 11,000 **9.** about 190 **11.** about 500 **13.** about 2000 **15.** about 100 **17.** about $11 **19.** about $80 **21.** about 300 **23.** about 1200 **25.** about $700 **27.** about 2400 **29.** about 2500 **31.** a; c **33.** about 40,000

Page 27 Review Exercises **1.** 2059 **3.** $644 **5.** 29.058 **7.** 11.8085 **9.** $64.44

Page 29 Problems **1.** 1141 pennants **3.** You need to know the number of gallons of gasoline his tank holds. **5.** You need to know what the odometer showed when they started. **7.** 3.75 points

Page 29 Review Exercises 1. $<$ **3.** $>$ **5.** $<$

Page 30 Skill Review 1. 6173 **3.** 20,742 **5.** 46.604 **7.** 47 **9.** 14.5 **11.** 90 **13.** 233 **15.** 91.13 **17.** $3.98 **19.** 11,175 fans **21.** about $1200 **23.** about $500 **25.** about $20 **27.** about 1600 **29.** about 300 **31.** about $700 **33.** You need to know how much the items cost originally. **35.** 13,324

Page 31 Exercises 1. one $1 bill, one quarter, two pennies **3.** one $5 bill, one nickel, one penny **5.** one $10 bill, two $1 bills, one dime, one nickel **7.** one $5 bill, two $1 bills, one quarter, two dimes, three pennies **9.** three $1 bills, three dimes, two pennies **11.** one $5 bill, three $1 bills, three dimes, three pennies **13.** the $5 bill and three pennies

Page 33 Exercises 1. $1198.86 **3.** $1079.66 **5.** No check was written. **7.** $140.56 **9.** $611.01 **11.** $1659.03

Page 35 Exercises 1. $270.27 **3.** $647.14

Page 36 Exercises 1. $1.45 **3.** one $5 bill, four $1 bills, one quarter, two dimes

Page 37 Chapter Review 1. commutative **3.** 9632 **5.** 11 **7.** 51.74 **9.** 41,530 **11.** 160 **13.** about $20 **15.** about 700 **17.** about $400 **19.** You need to know how many tickets they sold. **21.** $877.53

CHAPTER 3

Page 41 Practice Exercises 1. between 2700 and 3600 **3.** between 2700 and 4000 **5.** between 8000 and 15,000 **7.** between $6400 and $8100 **9.** about 240 **11.** about $4900 **13.** about $6000 **15.** about 72,000 **17.** reasonable **19.** reasonable **21.** not reasonable; about 56,000 **23.** between 1500 and 2400 **25.** about $900

Page 41 Review Exercises 1. 48 **3.** 27 **5.** 54 **7.** 24 **9.** 42

Page 43 Practice Exercises 1. 336 **3.** $57,152 **5.** 6035 **7.** 270,625 **9.** 312,052 **11.** $801,602 **13.** 958 **15.** 3852 **17.** $64,272 **19.** 55,252 **21.** 112,608 **23.** $368,186 **25.** 3366 **27.** 3384 **29.** 3116 **31.** 9072 **33.** 43,740 **35.** 230,432 **37.** a **39.** b **41.** c **43.** No **45.** No

Page 44 Review Exercises 1. 50 **3.** 80 **5.** 14 **7.** 160 **9.** 17 **11.** about 12,000 **13.** about $900 **15.** about 10,000 **17.** about $10,000 **19.** about $20 **21.** about $200 **23.** 693 **25.** 1.2092 **27.** $630 **29.** 156,672 **31.** $3600 **33.** $183.21

Pages 46–47 Practice Exercises 1. 990 **3.** 1600 **5.** 39,125 **7.** 8400 **9.** 5900 **11.** 362,304 **13.** 208 **15.** 232 **17.** 4194 **19.** 8226 **21.** 5456 **23.** 6930 **25.** 1536 **27.** 8100 **29.** 4500 **31.** 93 mi

Page 47 Review Exercises 1. 8800 **3.** 1,800,000 **5.** 67,620 **7.** 18,356 **9.** 165,508 **11.** 146,624 **13.** 25,500 **15.** 4,200,000

Page 49 Practice Exercises 1. 10,000 **3.** 49 **5.** 14 **7.** 1 **9.** 1 **11.** 135 **13.** 17,496 **15.** 2160 **17.** 8748 **19.** 81 **21.** 72 **23.** 32 **25.** 245 **27.** 21 **29.** 120 **31.** 1 **33.** 100 **37.** 99^1

Page 49 Review Exercises 1. about 900 **3.** about 7000 **5.** about 15,000 **7.** about 36,000

Page 50 Skill Review 1. c **3.** b **5.** between 300 and 600 **7.** about $400 **9.** 3638 **11.** $30,770 **13.** $1155 **15.** 136,026 **17.** 242,165 **19.** 83,239 **21.** 234 **23.** 4949 **25.** 506,968 **27.** 900 **29.** 9100 **31.** 1564 **33.** 8 **35.** 10 **37.** 2401 **39.** 144 **41.** 630 **43.** 9 **45.** 405 **47.** 2646 **49.** 1728

Page 51 Exercises 1. $99.88 **3.** $112 **5.** $1.28

Page 55 Exercises 1. balance due; $98.12 **3.** refund; $1148.15 **5.** $3800 **7.** box 2 **9.** Answers may vary. Examples: go to a bank; write to the Internal Revenue Service **11.** The penalty is 5% of the unpaid tax for the first six months. After that, penalties increase according to a set schedule.

Page 56 Exercises 1. 20,328 **3.** 1078

Page 57 Chapter Review 1. commutative **3.** a **5.** c **7.** d **9.** b **11.** $3944 **13.** 94,384 **15.** 44,400 **17.** 384 **19.** 304,584 **21.** 1372 **23.** 5391 **25.** 3600 **27.** 4800 **29.** 180 **31.** 3888 **33.** 64 **35.** 2304 **37.** 27,000 **39.** $1.09

CHAPTER 4

Page 61 Practice Exercises 1. about 40 **3.** about 70 **5.** about 500 **7.** about 900 **9.** about 300; over **11.** about 9000; under **13.** about 1200; under **15.** about 100; over **17.** about 900 **19.** about 80 **21.** about 200 **23.** about 500 **25.** about $300 **27.** about $700

Page 61 Review Exercises 1. 20 **3.** 60 **5.** 100 **7.** 400 **9.** 1000 **11.** 5000 **13.** 10,000 **15.** 20,000

Pages 63–64 Practice Exercises 1. 99 **3.** 123 **5.** 245 R3 **7.** 1275 **9.** 1891 R5 **11.** 2009 R7 **13.** 5531 **15.** 5051 R2 **17.** 320 R5 **19.** 110 R1 **21.** 353 R28 **23.** 817 R31 **25.** 2107 R17 **27.** 4121 **29.** 594 **31.** 959 R53 **33.** 846 R1 **35.** 409 R2 **37.** 665 R5 **39.** 783 **41.** 1088 R26 **43.** 820 **45.** 213 R103 **47.** 726 R299 **49.** reasonable **51.** 990 **53.** reasonable **55.** 40 min **57.** $9 each hour

Page 64 Review Exercises 1. 1877 **3.** 240,000 **5.** 738.09 **7.** 33.73 **9.** 109,934 **11.** 104.95 **13.** 67,878 **15.** 864

Pages 66–67 Practice Exercises 1. 16 **3.** 63 **5.** 333 **7.** 0 **9.** 10 **11.** 20 **13.** 14 **15.** 19 **17.** 319 **19.** 32 **21.** 10 **23.** 64 **25.** 7 **27.** +; ÷ **29.** ÷; × **31.** +; ÷ **33.** $45

Page 67 Review Exercises 1. 1600 **3.** 272 **5.** about 1200 **7.** about 14,000 **9.** about 900 **11.** about 1000

Page 69 Problems 1. a. 2796 mi **b.** 7846 mi **3.** $444 **5.** 3190 mi **7.** $65.30 **9.** $52.72

Page 69 Review Exercises 1. 192 mi

Page 70 Skill Review 1. about 60 **3.** about 400 **5.** about 90 **7.** about 500 **9.** about 11,000 **11.** about 200 **13.** about 60; under **15.** about 700; over **17.** about 1100; under **19.** 107 **21.** 1670 R2 **23.** 3121 R2 **25.** 76 R11 **27.** 201 R10 **29.** 1401 R32 **31.** 43 **33.** 6 **35.** 31 **37.** 8 **39.** 53 **41.** 4 **43.** $13,225 **45.** 127 people

Page 71 Exercises 1. 5 h **3.** 34 min **5.** 11 h 25 min **7.** 14 h 11 min **9.** 8 h 47 min **11.** 19 min **13.** 10:38 A.M.

Page 73 Exercises 1. 7:50 A.M. **3.** 12:25 P.M. **5.** 7:15 A.M. **7.** 10:52 A.M. **9.** 7:48 A.M. **11.** 10 min

Page 74 Exercises 1. 936 **3.** 744; 44,640 **5.** 14; 3 **7.** 54 **9.** about 4380 **11.** about 146

Page 75 Chapter Review 1. estimated **3.** about 90 **5.** about 70 **7.** about 4000 **9.** about 80; under **11.** about 300; over **13.** about 1000; under **15.** 359 R3 **17.** 3781 **19.** 1692 R6 **21.** 51 **23.** 171 **25.** 11 **27.** $297 **29.** 8:15 A.M.

CHAPTER 5

Pages 79–80 Practice Exercises 1. 16.1 **3.** 11.9 **5.** 145.4 **7.** 480 **9.** 59.6 **11.** 0.984 **13.** 22.44 **15.** 0.012 **17.** 0.008 **19.** 0.2516 **21.** 0.02 **23.** 49.98 **25.** $135.84 **27.** $12.36 **29.** $80.40 **31.** $.70 **33.** $86.40 **35.** $11.22 **37.** 26.898 **39.** $18.87 **41.** 0.09 **43.** 1.062 **45.** $98.10 **47.** 27.26 **49.** 57.33 **51.** $1.97 **53. a.** 3.9 **b.** 39 **c.** 390 **55.** Move the decimal point to the right the same number of places as there are zeros in 10, 100, or 1000. **57.** $38.97

Page 80 Review Exercises 1. 2 **3.** 1000 **5.** 6 **7.** 10,000

Page 82 Practice Exercises 1. about 40 **3.** about 5.6 **5.** about 0.27 **7.** about 2 **9.** about 3600; underestimate **11.** 0.24; can't tell **13.** about 9; overestimate **15.** about 30; can't tell **17.** about 630 **19.** about 0.32 **21.** about 120 **23.** about 0.36 **25.** about $1 **27.** about $90

Page 82 Review Exercises 1. 968 R4 **3.** 347

Pages 84–85 Practice Exercises 1. $1.44 **3.** 2.24 **5.** 0.255 **7.** 24.8 **9.** 0.029 **11.** 5.02 **13.** 2.05 **15.** 1.208 **17.** 2.9 **19.** 129 **21.** 6.2 **23.** 1500 **25.** 63 **27.** 27.2 **29.** 92 **31.** 2.25 **33.** 0.8 **35.** 0.34 **37.** 4.15 **39.** 0.097 **41.** 22.7 **43.** 0.48 **45.** 1.35 **47.** $1.20 **49.** 1.009 **51.** 4.07 **53.** 31.67 **55.** 0.694 **57. a.** 0.15 **b.** 0.015 **c.** 0.0015 **61.** $.80

Page 85 Review Exercises 1. about 6000 **3.** about 20 **5.** about $1000 **7.** about 0.54 **9.** about 30 **11.** about 30 **13.** about 900 **15.** about 800

Pages 86–87 Practice Exercises 1. about 7 **3.** about 10 **5.** about 500 **7.** about 0.05 **9.** about 700 **11.** about 0.6 **13.** about 2000 **15.** about 0.3 **17.** about 400 **19.** about 0.4 **21.** about 60 **23.** about 600 **25.** about $.30 **27.** about $3

Page 87 Review Exercises 1. 62 **3.** 57,238 **5.** 68.742 **7.** $16,340

Page 88 Skill Review 1. 91.2 **3.** 35.4 **5.** 28.44 **7.** 0.021 **9.** 20.4363 **11.** $.79 **13.** 68 **15.** $2500 **17.** 3499.66 **19.** 000.1537 **21.** 3.353920 **23.** 000.474 **25.** 8.46 **27.** 385 **29.** 0.056 **31.** 9.033 **33.** 1.058 **35.** $2.19 **37.** about 10 **39.** about 9000 **41.** about 20 **43.** about 70 **45.** about 300 **47.** about 20 **49.** about 0.4

Page 89 Exercises 1. $248.08 **3.** $7801.20 **5.** $13,436.80 **7.** Joshua **9.** Multiply weekly salary by 52.

Pages 90–91 Exercises **1.** \$493.92 **3.** \$213.75 **5.** \$328 **7.** \$528 **9.** \$335.40 **11.** 15 h **13.** \$28.50 **15.** \$51 **17.** \$.03

Page 93 Exercises **1.** \$182.60; \$660.40 **3.** \$277.82; \$1029.60 **5.** \$402.94; \$1581.84 **7.** \$212.05 **9.** \$339.15 **11.** \$3289.38 **13.** \$1787.76

Page 94 Exercises **1.** \$147.96 **3.** by the week; \$119.90 < \$129.95

Page 95 Chapter Review **1.** is approximately equal to **3.** 25.744 **5.** \$66.50 **7.** 0.018 **9.** \$.67 **11.** 1.7469 **13.** about \$80 **15.** about 6.3 **17.** about 0.15 **19.** about 0.6 **21.** about \$40 **23.** about 0.24 **25.** 0.809 **27.** 9.5 **29.** 1200 **31.** 189.2 **33.** about 80 **35.** about 0.4 **37.** about 300 **39.** about 0.1 **41.** about 0.2 **43.** about 0.06 **45.** \$682.50

CHAPTER 6

Pages 99–100 Practice Exercises **1.** km **3.** m **5. a.** 45 mm **b.** 4.5 cm **7.** 1.365 **9.** 360 **11.** 2320 **13.** 105 cm **15.** 1.6 m **17.** a, c **19.** a, c **25.** 91 mm

Page 100 Review Exercises **1.** 24,000 **3.** 1270 **5.** 0.016 **7.** 0.371

Page 102 Practice Exercises **1.** L **3.** mL **5.** 2850 **7.** 47 **9.** 0.789 **11.** = **13.** > **15.** < **17.** 15 mL **19.** 3.8 L **21.** 500 mL **23.** 5 L

Page 102 Review Exercises **1.** 85 **3.** 93 **5.** 19

Pages 104–105 Practice Exercises **1.** g **3.** kg **5.** 14,000 **7.** 0.18 **9.** 1200 **11.** c **13.** a **15.** 3; 0.003 **17.** 12,400; 12,400,000 **19.** 0.375; 375,000 **21.** 70 kg **23.** 76 cm **25.** 6000 mg

Page 105 Review Exercises **1.** about \$40 **3.** about 8000 **5.** about 1.8 **7.** about 50

Pages 106–107 Problems **1.** \$19 **3.** 35 s **5.** \$750 **7.** 144 **9.** Answers may vary. Example: 6, 8, 16, 18, 36, 38, 76, 78, 156, 158

Page 107 Review Exercises **1.** 17 h **3.** \$37

Page 108 Skill Review **1.** cm **3. a.** 57 mm **b.** 5.7 cm **5.** 1.45 **7.** 46,300 **9.** L **11.** 257 **13.** kg **15.** 4500 **17.** 0.422 **19.** c **21.** 17

Page 109 Exercises **1.** about \$2.50 **3.** about \$32.40

Page 111 Exercises **1. a.** 1180 mg **b.** No **3.** 20 g **5.** 108 mg **7.** about 4 glasses **9.** roast beef

Pages 112–114 Exercises **1.** \$.12 **3.** 9 **5.** 17 min **7.** Sept. 23; 8:29 A.M. **9.** \$21.71 **11.** \$28.39 **15.** 772 **17.** \$.022470 **19.** \$6.15 **21.** \$64.79 **23.** 08 8417 272650 **25.** 4/10/88 to 5/12/88 **29.** \$21.70 **31.** 7/30/88 **33.** \$9.30 **35.** 12683 **37.** \$1.75 **39.** \$.00

Page 115 Chapter Review **1.** gram **3.** milliliters **5.** 2 m **7.** 15 g **9.** 16.4 **11.** 22.413 **13.** 250 **15. a.** 39 mm **b.** 3.9 cm **17.** \$4.40 **19.** 335 mg

CHAPTER 7

Pages 123–124 Practice Exercises **1.** No; No; Yes **3.** Yes; Yes; Yes **5.** Yes; Yes; Yes **7.** Yes; No; No **9.** Yes; No **11.** Yes; Yes **13.** Yes; Yes **15.** No; No **17.** No; No **19.** Yes; No **21.** Yes; No **23.** Yes; Yes **25.** Yes **27.** No **29.** No **31.** No **33.** Yes **35.** No **37.** No **39.** odd **41.** even **43.** Yes

Page 124 Review Exercises **1.** 125 **3.** 1 **5.** 108 **7.** 294

Pages 126–127 Practice Exercises **1.** 5×7 **3.** $3^2 \times 7$ **5.** 2×5^3 **7.** $2 \times 3 \times 17$ **9.** 9 **11.** 15 **13.** 1 **15.** 2 **17.** 10 **19.** 45 **21.** 144 **23.** 600 **25.** prime **27.** composite **29.** composite **31.** $2 \times 5 \times 11$ **33.** 2×11^2 **35.** $2^3 \times 3 \times 7$ **37.** 3; 27 **39.** 5; 75 **41.** 14; 140 **43.** 17; 170 **45.** 1; 42 **47.** 18; 216 **49.** 56, 64, 72, 80, 88 **51.** Yes; 2 is an even prime number. **53.** 24 of each

Page 127 Review Exercises **1.** 3000 **3.** 250 **5.** 63,000 **7.** 0.9 **9.** 224 **11.** 7.204

Pages 129–130 Practice Exercises **1.** 1 **3.** 75 **5.** 4 **7.** 6 **9.** 4 **11.** 6 **13.** $\frac{1}{2}$ **15.** $\frac{5}{7}$ **17.** $\frac{5}{6}$ **19.** $\frac{8}{9}$ **21.** $\frac{2}{11}$ **23.** $\frac{4}{5}$ **25.** 33 **27.** 15 **29.** 64 **31.** 10 **33.** $\frac{3}{4}$ **35.** $\frac{4}{9}$ **37.** $\frac{4}{5}$ **39.** $\frac{3}{4}$ **41.** $\frac{1}{2}$ **43.** 1 **45.** Answers will vary. Examples: $\frac{2}{16}$, $\frac{3}{24}$, $\frac{4}{32}$, $\frac{5}{40}$ **47.** $\frac{11}{12}$ **49.** 2 pieces

Page 130 Review Exercises **1.** 12 **3.** 27 **5.** 42 **7.** 36 **9.** < **11.** = **13.** > **15.** > **17.** >

Pages 132–133 Practice Exercises **1.** > **3.** = **5.** > **7.** < **9.** > **11.** < **13.** < **15.** > **17.** $\frac{1}{4}, \frac{2}{5}, \frac{3}{4}$ **19.** $\frac{7}{15}, \frac{7}{10}, \frac{5}{6}$ **21.** $\frac{5}{22}, \frac{6}{11}, \frac{9}{11}$ **23.** $1\frac{2}{9}, 1\frac{1}{2}, 1\frac{3}{4}$ **25.** < **27.** > **29.** < **31.** = **33.** $\frac{2}{9}, \frac{5}{9}, \frac{7}{9}$ **35.** $\frac{1}{6}, \frac{1}{3}, \frac{2}{3}$ **37.** $\frac{1}{2}, \frac{5}{7}, \frac{3}{4}$ **39.** $1\frac{4}{5}, 2\frac{1}{3}, 2\frac{3}{5}$ **41.** checking account

Page 133 Review Exercises **1.** about 750 **3.** about 40 **5.** about 6 **7.** about 90

Page 135 Practice Exercises **1.** about $\frac{1}{3}$ **3.** about $\frac{1}{2}$ **5.** about $\frac{1}{3}$ **7.** about $\frac{3}{4}$ **9.** about $\frac{1}{5}$

11. about $\frac{1}{8}$ **13.** $\frac{1}{2}$ **15.** $\frac{1}{2}$ **17.** $\frac{1}{2}$ **19.** 1 **21.** 1 **23.** 1 **25.** b **27.** a **29.** a **31.** c **33.** c **35.** c **37.** about $\frac{4}{5}$ **39.** about $\frac{1}{8}$

Page 135 Review Exercises **1.** 61.0407 **3.** 1.052 **5.** 75,600

Page 136 Skill Review **1.** Yes **3.** Yes **5.** No **7.** Yes **9.** $2 \times 3 \times 5$ **11.** $2 \times 3 \times 5 \times 7$ **13.** $2 \times 5 \times 13$ **15.** 1; 180 **17.** 2; 60 **19.** 105th bar **21.** 11 **23.** 96 **25.** $\frac{7}{8}$ **27.** $\frac{5}{6}$ **29.** $\frac{5}{6}$ **31.** $<$ **33.** $>$ **35.** $\frac{2}{9}, \frac{5}{9}, \frac{2}{3}$ **37.** $\frac{2}{52}, \frac{3}{26}, \frac{2}{13}$ **39.** 1 **41.** 1 **43.** $\frac{1}{2}$ **45.** about $\frac{2}{3}$

Page 137 Exercises **1.** less **3.** too little

Page 139 Exercises **1.** $\frac{1}{500}$ **3.** Yes **5.** $\frac{1}{1000}$ **7.** f11

Page 140 Exercises **1.** $\frac{1}{2}$-in. **3.** $\frac{11}{16}$-in.

Page 141 Chapter Review **1.** greatest common factor **3.** Yes **5.** No **7.** 24; 48 **9.** 1; 24 **11.** 21; 126 **13.** 14; 196 **15.** $\frac{1}{2}$ **17.** $\frac{5}{11}$ **19.** $\frac{1}{3}$ **21.** $\frac{1}{5}, \frac{1}{3}, \frac{5}{7}$ **23.** $2\frac{1}{4}, 2\frac{2}{5}, 2\frac{7}{10}$ **25.** about $\frac{1}{5}$ **27.** about $\frac{2}{3}$ **29.** about $\frac{1}{3}$ **31.** $\frac{1}{500}$ **33.** less

CHAPTER 8

Pages 145–146 Practice Exercises **1.** $\frac{8}{35}$ **3.** $\frac{5}{6}$ **5.** $\frac{4}{13}$ **7.** 4 **9.** $\frac{1}{6}$ **11.** $1\frac{2}{5}$ **13.** $\frac{4}{5}$ **15.** 27 **17.** $7\frac{1}{2}$ **19.** $9\frac{1}{2}$ **21.** $1\frac{2}{3}$ **23.** 9 **25.** $\frac{5}{21}$ **27.** $2\frac{4}{5}$ **29.** $22\frac{1}{2}$ **31.** $\frac{1}{24}$ **33.** $3\frac{4}{5}$ **35.** 1 **37.** $2\frac{3}{5}$ **39.** 6 **41.** 0 **43.** $4\frac{9}{10}$ **45.** $\frac{1}{28}$ **47.** $6\frac{4}{15}$ **49. a.** 3 **b.** 6 **c.** 9 **d.** 12 **51. a.** 4 **b.** 12 **c.** 20 **d.** 28 **53.** 27 qt

Page 146 Review Exercises **1.** about $\frac{1}{6}$ **3.** about $\frac{1}{8}$ **5.** about $\frac{1}{10}$ **7.** $\frac{1}{2}$ **9.** 1 **11.** 0

Page 148 Practice Exercises **1.** about 21 **3.** about 4 **5.** about 72 **7.** about 99 **9.** about 80 **11.** about 21 **13.** about 7 **15.** about 2 **17.** c **19.** about 4 h **21.** about 260 seniors

Page 148 Review Exercises **1.** $\frac{1}{6}$ **3.** $\frac{19}{50}$

Pages 150–151 Practice Exercises **1.** $\frac{9}{2}$ **3.** $\frac{5}{12}$ **5.** $\frac{2}{7}$ **7.** $\frac{1}{4}$ **9.** no reciprocal **11.** 14 **13.** $1\frac{3}{32}$ **15.** $\frac{1}{8}$ **17.** $1\frac{2}{3}$ **19.** $1\frac{1}{9}$ **21.** $\frac{15}{64}$ **23.** $3\frac{1}{7}$ **25.** 3 **27.** $7\frac{4}{5}$ **29.** 20 **31.** 24 **33.** whole numbers **35.** 10 servings

Page 151 Review Exercises **1.** 19,630 **3.** \$76.86 **5.** 0.728 **7.** 4135

Page 153 Practice Exercises **1.** about 3 **3.** about 9 **5.** about 7 **7.** about 5 **9.** about 27 **11.** about 2 **13.** about 30 **15.** about 80 **17.** about 4 **19.** about 24 **21.** about 7 **23.** about 32 **25.** d **27.** about 21 cars

Page 153 Review Exercises **1.** Yes **3.** No **5.** Yes **7.** No

Pages 154–155 Problems **1.** $3\frac{1}{4}$ lb **3.** 2 bags **5.** 3 buses **7.** \$312.50

Page 156 Review Exercises **1.** July

Page 156 Skill Review **1.** $\frac{5}{21}$ **3.** $\frac{7}{10}$ **5.** $\frac{3}{5}$ **7.** 13 **9.** $2\frac{1}{12}$ **11.** $2\frac{2}{3}$ **13.** $2\frac{1}{4}$ c **15.** about 36 **17.** about 350 **19.** about 18 **21.** about 30 **23.** 4 **25.** $\frac{1}{7}$ **27.** $11\frac{1}{5}$ **29.** 14 **31.** $\frac{4}{77}$ **33.** $1\frac{1}{2}$ **35.** 20 times **37.** about 7 **39.** about 49 **41.** about 12 **43.** about 40 **45.** 10 packages

Page 157 Exercises **1.** about 16 ft **3.** about 36 ft **5.** about 6 times

Page 159 Exercises **1.** \$25 **3.** \$56.70 **5.** \$3045.49

Page 161 Exercises **1.** \$228.20 **3.** \$65 **5.** \$432 **7.** \$170; \$135

Page 162 Exercises **1.** \$6340 **3.** \$2930

Page 163 Chapter Review **1.** reciprocals **3.** $\frac{3}{22}$ **5.** $3\frac{3}{7}$ **7.** $6\frac{2}{3}$ **9.** 2 **11.** $\frac{1}{2}$ **13.** $\frac{2}{15}$ **15.** $\frac{11}{20}$ **17.** $1\frac{5}{7}$ **19.** about 24 **21.** about 8 **23.** about 57 **25.** about 8 **27.** 5 bicycles **29.** \$27.50 **31.** \$2970

CHAPTER 9

Pages 167–168 Practice Exercises **1.** $\frac{4}{11}$ **3.** $\frac{1}{3}$ **5.** $1\frac{1}{3}$ **7.** 0 **9.** $\frac{3}{5}$ **11.** $\frac{7}{10}$ **13.** $\frac{3}{8}$ **15.** $1\frac{7}{36}$ **17.** $\frac{3}{28}$ **19.** $\frac{11}{24}$ **21.** $\frac{11}{15}$ **23.** $\frac{3}{8}$ **25.** $1\frac{2}{9}$ **27.** $\frac{1}{4}$ **29.** $\frac{8}{15}$ **31.** $\frac{1}{5}$ **33.** $1\frac{3}{28}$ **35.** $\frac{9}{22}$ **37.** $\frac{8}{25}$ **39.** 1 **41.** $\frac{5}{8}$ **43.** $\frac{3}{8}$ mi **45.** Answers will vary. Examples: $\frac{1}{5} + \frac{1}{5}$; $\frac{3}{5} - \frac{1}{5}$

Page 168 Review Exercises **1.** about $\frac{1}{4}$ **3.** about $\frac{2}{3}$ **5.** about $\frac{1}{4}$ **7.** $\frac{1}{2}$ **9.** 0 **11.** $\frac{1}{2}$

Pages 170–171 Practice Exercises **1.** about 5 **3.** about 4 **5.** about 22 **7.** about 2 **9.** about 5 **11.** about 17 **13.** about 11 **15.** about 6 **17.** about 13 **19.** about 4 **21.** about 16 **23.** b **25.** c **27.** a **29.** a **31.** about 5 h **33.** about \$15

Page 171 Review Exercises **1.** 7 **3.** 9 **5.** 30 **7.** 6 **9.** 21 **11.** 6 **13.** 6

Pages 173–174 Practice Exercises **1.** $23\frac{4}{5}$ **3.** $4\frac{8}{9}$ **5.** $6\frac{11}{35}$ **7.** $8\frac{35}{36}$ **9.** $18\frac{19}{24}$ **11.** $14\frac{7}{45}$ **13.** 7 **15.** $12\frac{9}{16}$ **17.** $16\frac{1}{4}$ **19.** $9\frac{3}{5}$ **21.** 11 **23.** $17\frac{23}{36}$ **25.** $5\frac{5}{7}$ **27.** 10 **29.** $19\frac{11}{25}$ **31.** $19\frac{2}{15}$ **33.** $8\frac{5}{8}$ mi **35.** $25\frac{1}{2}$ h **37.** \$157

Page 174 Review Exercises **1.** $<$ **3.** $>$ **5.** $>$ **7.** $=$

Pages 176–177 Practice Exercises **1.** $7\frac{4}{7}$ **3.** 4 **5.** $5\frac{1}{6}$ **7.** $6\frac{13}{24}$ **9.** $8\frac{5}{18}$ **11.** $2\frac{1}{2}$ **13.** $7\frac{1}{2}$ **15.** $1\frac{2}{3}$ **17.** $3\frac{7}{15}$ **19.** $11\frac{23}{28}$ **21.** $5\frac{1}{3}$ **23.** $2\frac{1}{7}$ **25.** $1\frac{17}{36}$ **27.** $4\frac{4}{33}$ **29.** $14\frac{5}{12}$ **31.** $1\frac{3}{4}$ in. **33.** Answers will vary. Examples: $3\frac{2}{5} - 1\frac{1}{10}$; $4\frac{1}{10} - 1\frac{4}{5}$ **35.** 1 h 40 min

Page 177 Review Exercises **1.** about 150 **3.** about 3 **5.** about 6 **7.** about 48,000 **9.** about 49 **11.** about 5 **13.** about 9 **15.** about 15

Page 178 Skill Review **1.** $\frac{2}{5}$ **3.** $\frac{4}{7}$ **5.** $1\frac{1}{4}$ **7.** $\frac{4}{7}$ **9.** $1\frac{1}{6}$ **11.** $\frac{9}{40}$ **13.** about 15 **15.** about 9 **17.** about 7 **19.** about 3 h **21.** $4\frac{4}{7}$ **23.** $4\frac{6}{7}$ **25.** 19 **27.** $20\frac{5}{24}$ **29.** $1\frac{2}{3}$ **31.** 8 **33.** $2\frac{3}{8}$ **35.** $4\frac{1}{7}$ **37.** $\frac{13}{30}$ **39.** $5\frac{5}{36}$ **41.** $2\frac{1}{2}$ games

Page 179 Exercises **1.** $\$\frac{5}{8}$ **3.** $\$1\frac{3}{4}$ **5.** $105 **7.** A stock dividend is an amount of money that a person who owns stock in a company may receive from the company. The amount of the dividend is based on the number of shares a person owns.

Pages 180–181 Exercises **1.** $1\frac{3}{4}$ in. **3.** $\frac{3}{16}$ in. **5.** 1 in. **7.** $\frac{7}{32}$ in. **9.** 2 in. **11.** $1\frac{5}{8}$ in.

Page 182 Exercises **1.** $11\frac{5}{8}$ in. **3.** 1-in. diameter; $\frac{1}{4}$ in.

Page 183 Chapter Review **1.** equivalent fractions **3.** 1 **5.** $10\frac{3}{5}$ **7.** $1\frac{1}{6}$ **9.** $\frac{1}{2}$ **11.** $7\frac{13}{24}$ **13.** $23\frac{13}{18}$ **15.** about 9 **17.** about 8 **19.** $4\frac{11}{12}$ h **21.** $1\frac{5}{6}$ c **23.** $\frac{5}{8}$ in.

CHAPTER 10

Pages 187 Practice Exercises **1.** $\frac{7}{10}$ **3.** $\frac{1}{125}$ **5.** $1\frac{1}{5}$ **7.** $3\frac{13}{20}$ **9.** $2\frac{8}{125}$ **11.** $\frac{2}{9}$ **13.** $\frac{5}{8}$ **15.** $\frac{31}{600}$ **17.** $\frac{11}{12}$ **19.** $\frac{2}{15}$ **21.** $\frac{1}{5}$ **23.** $\frac{11}{40}$ **25.** $3\frac{17}{20}$ **27.** $\frac{5}{6}$ **29.** $\frac{19}{400}$ **31.** $1\frac{1}{6}$ **33.** $\frac{2}{9}$ **35.** $2\frac{1}{12}$ **37.** $2\frac{7}{8}$ yd

Page 187 Review Exercises **1.** $\frac{27}{128}$ **3.** 4 **5.** $\frac{29}{30}$ **7.** $3\frac{5}{6}$

Pages 189–190 Practice Exercises **1.** 0.2 **3.** 0.625 **5.** 0.125 **7.** 1.875 **9.** 9.025 **11.** $0.\overline{6}$ **13.** $0.3\overline{6}$ **15.** $0.\overline{72}$ **17.** $3.\overline{7}$ **19.** $1.1\overline{6}$ **21.** $>$ **23.** $<$ **25.** $=$ **27.** $>$ **29.** $0.\overline{2}$ **31.** 0.045 **33.** 0.95 **35.** $0.9\overline{3}$ **37.** $1.\overline{72}$ **39.** $<$ **41.** $>$ **43.** $<$ **45.** $=$ **47.** a **49.** a **51.** 5.125 oz **53.** less; $2\frac{1}{2} = 2.5$, and $2.27 < 2.5$

Page 190 Review Exercises **1.** 3.837 **3.** 69,629 **5.** 406 **7.** 100

Pages 192–193 Practice Exercises **1.** about $\frac{2}{3}$ **3.** about 1 **5.** about $\frac{1}{6}$ **7.** about $5\frac{3}{4}$ **9.** about $10\frac{1}{5}$ **11.** about $90 **13.** about 800 **15.** about 6000 **17.** about 2400 **19.** about $100 **21.** about $24 **23.** B **25.** A **27.** C **29.** B **31.** about 30 **33.** about $3600 **35.** about $500 **37.** about 14 **39.** about $4 **41.** about $1

Page 193 Review Exercises **1.** 23 **3.** 27 **5.** 8900

Pages 194–195 Problems **1.** Gino: 165 lb; Rusty: 148 lb **3.** 12 bicycles; 5 mopeds **5.** 10 at 5 for $12.95; 14 at 7 for $17.50 **7.** a ham and cheese sandwich; juice; yogurt **9.** Yes; a tuna salad sandwich; 2 cartons of milk; fruit **11.** a tuna salad sandwich; milk; fruit

Page 195 Review Exercises **1.** 13 cars **3.** You need to know how long they played tennis.

Page 196 Skill Review **1.** $\frac{9}{10}$ **3.** $\frac{18}{25}$ **5.** $\frac{1}{250}$ **7.** $3\frac{3}{8}$ **9.** $\frac{5}{6}$ **11.** $\frac{3}{8}$ **13.** $2\frac{1}{4}$ mi **15.** 0.22 **17.** 4.5 **19.** $0.3\overline{8}$ **21.** $5.\overline{6}$ **23.** $8.8\overline{3}$ **25.** $<$ **27.** $<$ **29.** about $\frac{1}{3}$ **31.** about $\frac{4}{5}$ **33.** about $2\frac{1}{2}$ **35.** about $5\frac{3}{5}$ **37.** about 200 **39.** about 100 **41.** about $4000 **43.** about $3 **45.** 15 push-ups; 32 sit-ups

Page 198 Exercises **1.** $9 **3.** $10.25 **5.** monthly pass; the pass costs $14 less than paying each day **7.** $8.10

Page 200 Exercises **1.** $28.27 **3. a.** $16.78 **b.** $13.72 **5. a.** $8.63 **b.** $9.62

Page 201 Chapter Review **1.** repeating decimal **3.** $\frac{2}{5}$ **5.** $\frac{3}{8}$ **7.** $\frac{7}{25}$ **9.** $\frac{9}{40}$ **11.** $\frac{24}{125}$ **13.** $\frac{29}{60}$ **15.** 0.32 **17.** 8.6 **19.** $0.\overline{03}$ **21.** $2.9\overline{3}$ **23.** about $\frac{1}{2}$ **25.** about $\frac{1}{4}$ **27.** about 200 **29.** about 90 **31.** 36 tapes; 9 discs **33.** $73.30

CHAPTER 11

Pages 205–206 Practice Exercises **1.** 324 **3.** 7 **5.** 88 **7.** 139 **9.** 10,647 **11.** 25; 2 **13.** $1\frac{3}{4}$ in. **15.** 4 **17.** 111 **19.** 6; 33 **21.** 459 **23.** 3; 8; 2 **25.** 14 yd **27.** 5 ft **29.** 45 ft **35.** about 7 mi **37.** about 9 in.

Page 206 Review Exercises **1.** 0.625 **3.** 0.8 **5.** $0.\overline{2}$ **7.** $0.\overline{7}$

Page 208 Practice Exercises **1.** 80 **3.** 11 **5.** 12,500 **7.** 152 **9.** 9125 **11.** 16; 14 **13.** 7 **15.** 10,800 **17.** 133 **19.** $<$ **21.** $=$ **23.** $>$ **25.** about 16 lb

Page 208 Review Exercises **1.** $\frac{1}{25}$ **3.** $\frac{17}{100}$ **5.** $\frac{1}{40}$ **7.** $\frac{7}{75}$ **9.** $\frac{7}{8}$

Pages 210–211 Practice Exercises **1.** 3 **3.** 112 **5.** 14 **7.** 13; 1 **9.** 12; 2 **11.** 37 **13.** c **15.** c **17.** 36 **19.** 22 **21.** 12; 1 **23.** 50 ft **25.** 14 lb **27.** 2 oz **29.** 3 qt **31.** 9 qt

Page 211 Review Exercises **1.** 81 **3.** 1 **5.** 64 **7.** 1024

Page 213 Practice Exercises **1.** 40 mi 799 yd **3.** 98 lb 6 oz **5.** 37 ft 3 in. **7.** 82 c 4 fl oz **9.** 2 c 1 fl oz **11.** 10 lb 11 oz **13.** 9 ft 7 in. **15.** 5 lb 3 oz **17.** 36 c 3 fl oz **19.** 11 ft 7 in. **21.** 51 mi 260 yd **23.** 5 gal 1 qt **25.** 91 gal 3 qt 1 pt **27.** 96 ft 6 in.

Page 213 Review Exercises **1.** 3000 **3.** 0.029 **5.** 90,000 **7.** 160

Page 215 Practice Exercises **1.** 3; 1 yd **3.** 4; 0.01 cm **5.** 2; 100 ft **7.** 3; 10 mi **9.** 5220 ft **11.** 145 km **13.** 6600 lb **15.** 14,500 ft **17.** 50.24 m **19.** 43 mg **21.** 37.3 m **23.** 1270 mi **25.** 6.02 kg **27.** No. Since 89,000 yd is precise to thousands, the answer can only be precise to thousands.

Page 215 Review Exercises **1.** Yes **3.** Yes **5.** No **7.** Yes

Page 216 Skill Review **1.** 8 **3.** 234 **5.** 15,934 **7.** $2\frac{3}{16}$ in. **9.** 1420 yd **11.** 10,000 **13.** 247 **15.** 4; 15 **17.** 17 **19.** 58 **21.** 16; 1 **23.** a **25.** 14 t 530 lb **27.** 70 qt 1 pt **29.** 1056 ft **31.** 4; 0.01 L **33.** 2; 0.001 km **35.** 224 kg

Page 217 Exercises **1.** 17.0 mi/gal **3.** 28.8 mi/gal **5.** 24.7 mi/gal

Page 219 Exercises **1.** 18 lb 12 oz; $199.95; $34.95; $234.90; $8.04; $242.94 **3.** $134.13 **5.** $407.99

Page 220 Exercises **1.** 7 **3.** 8 **5.** 4

Page 221 Chapter Review **1.** yard **3.** 6 **5.** 3; 120 **7.** 7; 15 **9.** 11 **11.** 11; 1 **13.** 164 **15.** $\frac{13}{16}$ in. **17.** b **19.** 27 ft 4 in. **21.** 104 t 1850 lb **23.** 3; 0.1 m **25.** 4; 1 ft **27.** 25.1 mi/gal

CHAPTER 12

Pages 229–230 Practice Exercises **1.** $\frac{2}{13}$ **3.** $\frac{5}{3}$ **5.** $\frac{3}{1}$ **7.** $\frac{5}{78}$ **9.** $\frac{1}{20}$ **11.** $\frac{1}{18}$ **13.** $\frac{45}{2}$ **15.** 50 mi/h **17.** $3.50/ticket **19.** 10.2 m/s **21.** $\frac{2}{3}$ **23.** $\frac{2}{1}$ **25.** $\frac{20}{1}$ **27.** $\frac{500}{89}$ **29.** 3 quizzes/week **31.** 19 mi/day **33.** 13 ft/s **35.** $2.40/ticket **37.** $7.50/day **39.** 5 m/s **41.** 24 to 11; 24:11; $\frac{24}{11}$ **43.** $270

Page 230 Review Exercises **1.** 15 **3.** 5 **5.** 28 **7.** 1

Pages 232–233 Practice Exercises **1.** True **3.** False **5.** False **7.** True **9.** False **11.** True **13.** 4 **15.** 180 **17.** 18 **19.** 18 **21.** 35 **23.** 6 **25.** True **27.** False **29.** True **31.** False **33.** False **35.** True **37.** 1 **39.** 63 **41.** 114 **43.** 96 **45.** $7:14 = 3:6$; $\frac{7}{14} = \frac{3}{6}$

Page 233 Review Exercises **1.** about 1000 **3.** about 8 **5.** about 14 **7.** about 7

Pages 234–235 Problems **1.** 60 gal **3.** 64 times **5.** 6 gal **7.** 12 lawns **9.** 1 c **11.** 3 gal

Page 235 Review Exercises **1.** 21 pages **3.** 269 students

Page 236 Skill Review **1.** $\frac{6}{7}$ **3.** $\frac{34}{31}$ **5.** $\frac{5}{1}$ **7.** $\frac{7}{90}$ **9.** $\frac{1}{10}$ **11.** $\frac{60}{1}$ **13.** 375 mi/week **15.** $6.60/h **17.** 39.5 mi/h **19.** $1.29 **21.** True **23.** False **25.** False **27.** True **29.** 6 **31.** 114 **33.** 28 **35.** 36 **37.** $21 **39.** 45 spruce trees

Page 237 Exercises **1.** 5 qt **3.** 3-lb bag **5.** 12 cans **7.** $.45

Page 239 Exercises **1.** about 75 mi **3.** about 188 mi **5.** about 38 mi **7.** about 80 mi

Page 241 Exercises **1.** 20 ft; 24 ft **3.** 12 ft; 16 ft **5.** 16 ft; 36 ft **7.** 12 ft **9.** 2 in.; $1\frac{1}{2}$ in. **11.** 10 ft; 14 ft

Page 242 Exercises **1.** 30 in. **3.** $7\frac{1}{2}$ in.; $4\frac{1}{2}$ in.

Page 243 Chapter Review **1.** rate **3.** $\frac{9}{17}$ **5.** $\frac{2}{1}$ **7.** $1.09/pen **9.** 3.7 km/h **11.** True **13.** False **15.** 2 **17.** 39 **19.** $48 **21.** 50 min **23.** about 400 km

CHAPTER 13

Page 247 Practice Exercises **1.** 0.32 **3.** 0.0675 **5.** 1.57 **7.** 0.007 **9.** 0.374 **11.** $0.66\frac{2}{3}$ **13.** 1% **15.** 80.6% **17.** 245% **19.** 30% **21.** 0.5% **23.** $62\frac{1}{2}\%$ **25.** 0.08 **27.** 43% **29.** 2.3 **31.** 259% **33.** 0.3% **35.** 0.8725 **37.** 63.4% **39.** Pacers **41.** 0.065

Page 247 Review Exercises **1.** 0.9 **3.** 2.8 **5.** 1.5625

Pages 249–250 Practice Exercises **1.** $\frac{7}{100}$ **3.** $\frac{9}{20}$ **5.** $1\frac{1}{4}$ **7.** $\frac{93}{1000}$ **9.** $\frac{84}{125}$ **11.** $\frac{3}{5000}$ **13.** $\frac{1}{300}$ **15.** $\frac{1}{225}$ **17.** $\frac{23}{300}$ **19.** $\frac{47}{200}$ **21.** $\frac{1}{8}$ **23.** $\frac{2}{3}$ **25.** 50% **27.** 94% **29.** 130% **31.** $22\frac{2}{9}\%$ **33.** $83\frac{1}{3}\%$ **35.** $257\frac{1}{7}\%$ **37.** $\frac{7}{20}$ **39.** 75% **41.** $\frac{1}{125}$ **43.** $\frac{3}{8}$ **45.** $\frac{21}{5000}$ **47.** 148% **49.** 15% **51.** $\frac{94}{125}$

Page 250 Review Exercises **1.** $\frac{11}{21}$ **3.** $5\frac{1}{2}$ **5.** 5 **7.** 4.494 **9.** 95.76 **11.** 10.7

Pages 252–253 Practice Exercises **1.** 17 **3.** 70 **5.** $121\frac{4}{5}$ **7.** 27 **9.** 4.14 **11.** 118.125 **13.** 0.25; $\frac{1}{4}$ **15.** 0.75; $\frac{3}{4}$ **17.** 0.2; $\frac{1}{5}$ **19.** $0.\overline{3}$; $\frac{1}{3}$ **21.** $0.8\overline{3}$; $\frac{5}{6}$ **23.** 0.625; $\frac{5}{8}$ **25.** 34 **27.** 16.34 **29.** 21.648 **31.** 33 **33.** No. 35% + 57% + 18% = 110%; the sum of the percents must be 100%. **35.** 108 games

Page 253 Review Exercises **1.** about $\frac{1}{4}$ **3.** about $1\frac{2}{3}$ **5.** about 40 **7.** about 80 **9.** about 80 **11.** about 3000

Pages 254–255 Practice Exercises **1.** about 70 **3.** about 120 **5.** about 24 **7.** about 70 **9.** about $140 **11.** about $80 **13.** about 720 **15.** about $2.40 **17.** about $40 **19.** about $1500 **21.** about $60 **23.** about 210 **25.** about 240 **27.** about $40 **29.** about 90,000

Page 255 Review Exercises **1.** $\frac{4}{5}$ **3.** $3\frac{1}{2}$ **5.** $5\frac{3}{20}$ **7.** $12\frac{4}{7}$

Page 256 Skill Review **1.** 0.04 **3.** 1.73 **5.** 0.142 **7.** 46.2% **9.** 0.5% **11.** 29.1% **13.** $1\frac{41}{50}$ **15.** $\frac{1}{300}$ **17.** 90% **19.** $43\frac{3}{4}$% **21.** $212\frac{1}{2}$% **23.** 40.9 **25.** $55\frac{1}{2}$ **27.** 112 **29.** 2.512 **31.** 74.4 **33.** 18 women **35.** about 80 **37.** about 27 **39.** about $80 **41.** about $21 **43.** about $40

Page 257 Exercises **1.** $6856.29 **3.** $48,397.76 **5.** $1754.50 **7.** $2249.52

Pages 258–259 Exercises **1.** $.12 **3.** $.01 **5.** $.26 **7.** $9.63 **9.** $.12; $2.11 **11.** $2.08; $61.58 **13.** $759.60 **15.** $12,476.34 **17.** $81.68

Pages 260–261 Exercises **1.** $954.04 **3.** $2239.15 **5.** $800.08 **7.** $1139.46 **9.** refund; $281.26 **11.** $513.20 **13.** $904.75

Page 262 Exercises **1.** $14,400; $470.88 **3.** $5850; $191.30 **5.** A mill is $.001. So a tax rate of 8 mills per $1 is the same as $8 per $1000.

Page 263 Chapter Review **1.** percent **3.** 0.01 **5.** 0.4625 **7.** 180% **9.** $1\frac{2}{5}$ **11.** $\frac{9}{125}$ **13.** $28\frac{4}{7}$% **15.** $376\frac{1}{2}$ **17.** 45 **19.** 35.25 **21.** about $14 **23.** about 18 **25.** 704 votes **27.** $231.13

CHAPTER 14

Pages 267–268 Practice Exercises **1.** 95% **3.** 20% **5.** 40% **7.** $12\frac{1}{2}$% **9.** $16\frac{2}{3}$% **11.** $187\frac{1}{2}$% **13.** $31\frac{1}{4}$% **15.** $92\frac{1}{2}$% **17.** $106\frac{2}{3}$% **19.** 32% **21.** $12\frac{1}{2}$% **23.** 40% **25.** 182% **27.** $62\frac{1}{2}$% **29.** $\frac{3}{40}$

Page 268 Review Exercises **1.** 82% **3.** 50% **5.** 233% **7.** 40% **9.** $66\frac{2}{3}$%

Pages 270–271 Practice Exercises **1.** 60% **3.** $137\frac{1}{2}$% **5.** 25% **7.** $83\frac{1}{3}$% **9.** decrease; 80% **11.** increase; $37\frac{1}{2}$% **13.** increase; 125% **15.** increase; 176% **17.** $16\frac{2}{3}$% **19.** 70 students

Page 271 Review Exercises **1.** 0.25; $\frac{1}{4}$ **3.** $0.\overline{3}$; $\frac{1}{3}$ **5.** 0.6; $\frac{3}{5}$ **7.** 0.125; $\frac{1}{8}$ **9.** $0.8\overline{3}$; $\frac{5}{6}$

Pages 273–274 Practice Exercises **1.** 25 **3.** 1225 **5.** 4000 **7.** 84 **9.** 347 **11.** 225 **13.** 1030 **15.** 456 **17.** 940 **19.** 576 **21.** 25 **23.** 51 **25.** 1120 **27.** 50 **29.** 75%

Page 274 Review Exercises **1.** $\frac{3}{26}$ **3.** $\frac{4}{1}$ **5.** $\frac{32}{3}$ **7.** 8 **9.** 21

Pages 276–277 Practice Exercises **1.** 49 **3.** 22.5 **5.** 34.1 **7.** 15.36 **9.** 55 **11.** 51% **13.** 98% **15.** 2% **17.** 5% **19.** $87\frac{1}{2}$% **21.** 630 **23.** 7 **25.** 400 **27.** 50 **29.** 55 **31.** 356.7 **33.** 68.4 **35.** 48% **37.** 230% **39.** 60 **41.** 450 movies

Page 277 Review Exercises **1.** 330 **3.** 0.1 **5.** 6.84

Pages 278–279 Problems **1.** 16 h **3.** 46 h **5.** 7:30 A.M. **7.** $1652.83

Page 279 Review Exercises **1.** $4096 **3.** 219

Page 280 Skill Review **1.** 36% **3.** 82% **5.** $62\frac{1}{2}$% **7.** $83\frac{1}{3}$% **9.** $307\frac{9}{13}$% **11.** increase; 16% **13.** increase; 120% **15.** increase; 10% **17.** decrease; 40% **19.** increase; $66\frac{2}{3}$% **21.** 6 **23.** 900 **25.** 852 **27.** 1000 **29.** 320 **31.** 62 **33.** 2.6 **35.** $6\frac{1}{4}$% **37.** 170% **39.** 760 **41.** 9:40 A.M.

Page 281 Exercises **1.** about $300 **3.** about $600 **5.** about $1400 **7.** $250; $37\frac{1}{2}$%

Pages 282–283 Exercises **1.** $450; $100 **3.** $1187.20; $758.80 **5.** $504 **7.** $140 **9.** $756.60

Pages 284–285 Exercises **1.** $16.48; $266.48 **3.** $1500; $2000 **5.** $11.03; $501.03 **7.** $4097.74; $4097.74 **9.** $16.64; $266.64 **11.** $4725 **13.** $800 **15.** $2357.70 **17.** $2025

Pages 287–288 Exercises **1.** $13; $117 **3.** $62.50; $187.50 **5.** $15; $30 **7.** $17.50; $67.50 **9.** $150; $400 **11.** $75; $150 **13.** $3.80; $8.85 **15.** $21.83; $26.67 **17.** $.15; $3.15 **19.** 90%

Page 289 Chapter Review **1.** wholesale **3.** 58% **5.** 100 **7.** 500 **9.** 200 **11.** increase; 30% **13.** decrease; $33\frac{1}{3}$% **15.** 189 **17.** 80% **19.** 26% **21.** $541.80

CHAPTER 15

Pages 293–294 Practice Exercises 1. \$100 **3.** \$16.78 **5.** \$125 **7.** \$375 **9.** \$70 **11.** \$162 **13.** \$678 **15.** \$5300 **17.** \$1085 **19.** \$9344 **21.** \$2510 **23.** \$750.38 **25.** \$35; \$735 **27.** \$312.50; \$1567.50 **29.** \$281.25; \$4781.25 **31.** \$225 at $5\frac{1}{2}\%$ **33.** \$7500 at 6.6% **35.** \$210 **37.** \$4077.50

Page 294 Review Exercises 1. \$2277 **3.** \$5004 **5.** \$1.70 **7.** 285.99 **9.** $1\frac{1}{8}$ **11.** $27\frac{3}{5}$

Pages 296–297 Practice Exercises 1. \$6615 **3.** \$2184.05 **5.** \$424.56 **7.** \$895.58 **9.** \$1155.63 **11.** \$2653.11 **13.** \$527.88 **15.** \$1406.04 **17.** 16% compounded quarterly

Page 297 Review Exercises 1. $2^3 \times 7$ **3.** 5^3 **5.** $2^4 \times 3^2$

Page 298 Skill Review 1. \$35 **3.** \$2187.50 **5.** \$195.23 **7.** \$525 **9.** \$696 **11.** \$1128.05 **13.** \$2114.51 **15.** \$32.50 **17.** \$742.63 **19.** \$9417.80 **21.** \$795.68 **23.** \$1432.92 **25.** \$2165.80 **27.** \$2613.77 **29.** \$4776.40

Page 300 Exercises 1. \$1.23; \$301.23; \$1.24; \$302.47 **3.** \$19.02; \$5819.02; \$19.09; \$5838.11 **5.** \$5.86; \$1105.86; \$5.89; \$1111.75 **7.** \$914.48 **9.** \$3.61

Pages 301–303 Exercises 1. \$22.50 **3.** \$73.25 **5.** \$142.40 **7.** \$25.20 **9.** \$56

Pages 304–305 Exercises 1. \$.47; \$85.56 **3.** \$.48; \$40.88 **5.** \$4.16; \$312.71 **7.** \$.79 **9.** \$.88; \$81.10 **11.** Global Credit Card

Pages 306–307 Exercises 1. \$10,135.20 **3.** \$10,902.44 **5.** \$11,841.92 **7.** \$1083.28 **9.** \$565.12 **11.** used car

Pages 308–309 Exercises 1. \$20,000; \$80,000 **3.** \$23,000; \$69,000 **5.** \$60,000; \$140,000 **7.** \$8000; \$32,000 **9.** \$28,000; \$52,000 **11.** \$34,500; \$80,500 **15.** \$120,000; \$30,000 **17.** \$76,000; \$1000 **19.** \$72,000

Page 310 Exercises 1. \$152 **3.** \$56.03 **5.** \$10,000

Page 311 Chapter Review 1. compound interest **3.** \$25; \$525 **5.** \$498.23; \$3053.23 **7.** \$198; \$3498 **9.** \$1166.40 **11.** \$541.20 **13.** \$3215.74 **15.** \$81.28 **17.** \$20,000

CHAPTER 16

Pages 319–320 Practice Exercises 1. about 150 lb **3.** about 700 lb **5.** horse **7.** polar bears and horses **9.** 2 **11.** Cardinals **13.** Orioles

15.

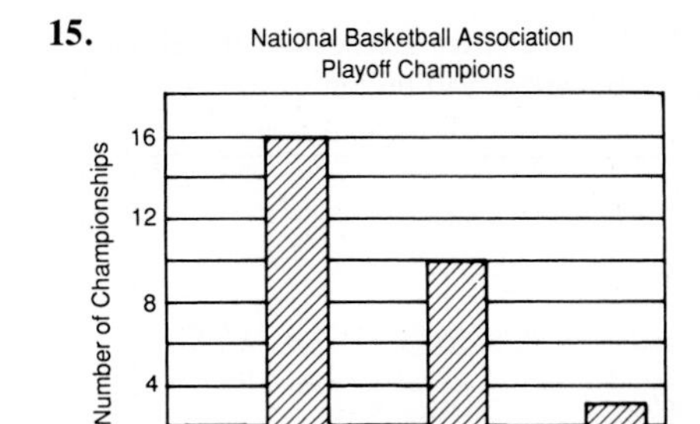

19. Los Angeles and Miami

Page 321 Review Exercises 1. 7359; 71,532; 73,521 **3.** \$859; \$958; \$985 **5.** 0.0033; 0.03; 0.033 **7.** $\frac{2}{11}$; $\frac{5}{11}$; $\frac{7}{11}$ **9.** $\frac{5}{8}$; $\frac{7}{10}$; $\frac{3}{4}$

Pages 323–325 Practice Exercises 1. 16 **3.** 7 **5.** 5 **7.** 2:00 P.M. **9.** about 500 thousand **11.** about 950 thousand **13.** 1980 and 1985

15.

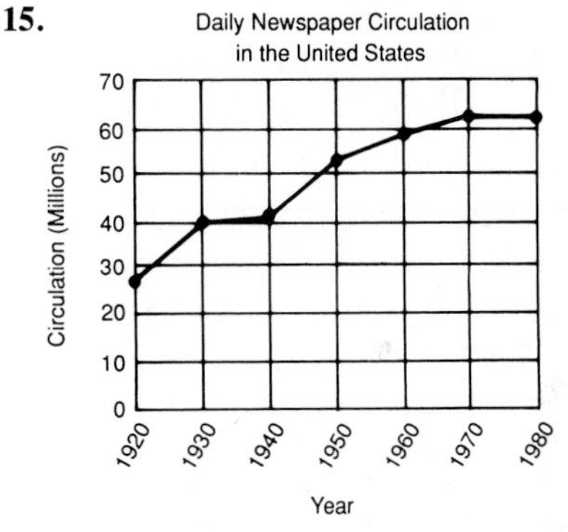

17.

Public School Enrollment in the United States

Students (Millions) — Elementary; Secondary — Year: 1965, 1970, 1975, 1980, 1985, 1990 (projected)

19. 1940 and 1960 **21.** about 70% **23.** bar graph

Page 325 Review Exercises 1. 8 thousand, 602 **3.** 13 hundredths **5.** 2 and 71 ten-thousandths

Pages 327–329 Practice Exercises **1.** about \$30 billion **3.** about \$40 billion
5. T-Shirt Sales

Freshmen
Sophomores
Juniors
Seniors
Key: = 10 T-Shirts

7. about 6000 **9.** about 3000 **11.** $7\frac{1}{2}$ **13.** about 40 **15.** Video Village **17.** TV Time **19.** $5\frac{1}{2}$
21. Radio and Television Stations

Cincinnati
Houston
Kansas City
Los Angeles
Key: = 6 radio stations = 6 television stations

23. bar graph

Page 329 Review Exercises **1.** 26 **3.** 200 **5.** 10.5 **7.** 150%

Pages 331–333 Practice Exercises **1.** about 28 million mi^2 **3.** about 32.2 million mi^2 **5.** Indian **7.** about $\frac{1}{4}$ **9.** Indian and Atlantic **11.** \$900,000 **13.** \$1,200,000 **15.** \$200,000 **17.** \$400,000 **19.** 550 **21.** 825 **23.** 4000 **25.** 6500 **27.** 320 **29.** country
31.

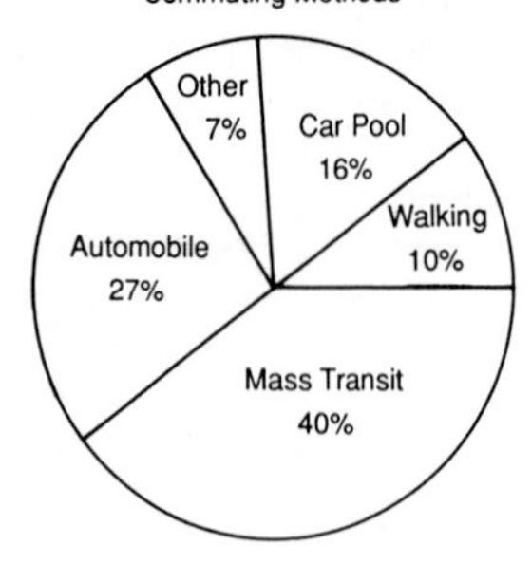

33. about 175 million **35.** about 3 to 1

Page 333 Review Exercises **1.** 128 **3.** 0.6 **5.** 22 **7.** 38,000 **9.** 3

Page 335 Problems **1.** about 15 million **3.** about 25 million **5.** Answers may vary. Example: about 35 million **7.** about 30° F **9.** about 70° F **11.** April and October **13.** Answers may vary. Example: about 42 in. **15.** Answers may vary. Example: about 57 in. **17.** 7 years

Page 335 Review Exercises **1.** 40 classes

Page 336 Skill Review **1.** 11 **3.** store sales
5.

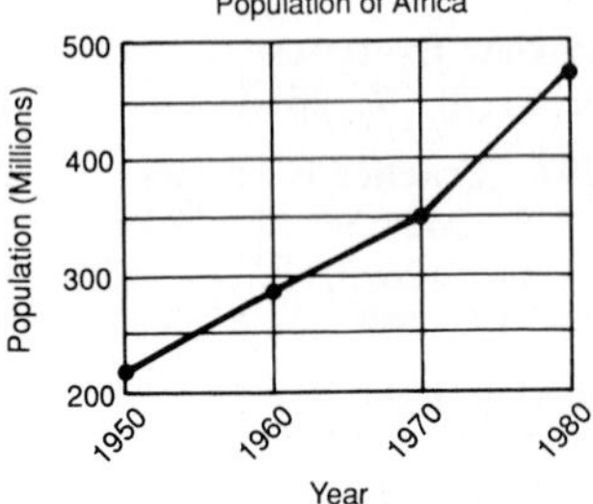

7. 1250 **9.** 165 **11.** about \$15 thousand **13.** about \$25 thousand **15.** about \$40 thousand

Page 338 Exercises **1.** Jan.: Carter—about 18,000; Hooper—about 17,000 Sept.: Carter—about 17,000; Hooper—about 18,000
5. Small Business Remains Steady in Pleasant City

Number of Small Businesses: 30, 40, 50, 60
Year: 1960, 1965, 1970, 1975, 1980, 1985

Pages 339–340 Exercises **1.** about \$50 **3.** about \$50 **5.** about \$30 **7.** \$600 **9.** \$360 **11.** \$240 **13.** \$144 **15.** about 25% **17.** about 15% **19.** about 10% **21.** about 10%

Pages 341–342 Chapter Review **1.** budget **3.** about 3 in. **5.** Center Town, Old Falls **7.** about 20 million **9.** 1970 and 1980 **11.** about 500 million **13.** about 800 million **15.** about \$26.4 billion **17.** radio and magazines **19.** about 95 million **21.** about 110 million **23.** about 135 million **25.** Dana's Diner: about \$25; The Grill: about \$25; Sandy's Kitchen: about \$20 **29.** about \$70

CHAPTER 17

Pages 347–348 Practice Exercises **1.** 68; 68; 68 and 70; 10 **3.** 61; 55; none; 57 **5.** 13; 12.6; none; 4.4 **7.** 4.6; 3.6; 3.6 and 8.4; 7.2 **9.** \$103.33; \$103; \$99 and \$107; \$36

11. \$5.79; \$5.98; \$7.29; \$3.92 **13.** 73.2; 71; 63; 24 **15.** 2.2; 1.8; none; 2.7 **17.** 5.0; 5.25; 4.7; 6.4 **19.** \$179.20; \$189; \$192; \$42 **21.** 9.5; 9.5; 9.4 and 9.6; 0.2 **23.** \$.44; \$.44; \$.35 and \$.53; \$.18 **25.** \$10,000 **27.** \$28,000 **29.** about 30 lb **31.** The weight would be greater than 7 lb.

Page 348 Review Exercises **1.** 3,500,000 **3.** 100,100,000,000 **5.** 10,000,320 **7.** 0.0038

Pages 350–351 Practice Exercises

1. Ages of Band Members

Age (Years)	Tally	Frequency
15	𝍸 II	7
16	𝍸 𝍸	10
17	𝍸 IIII	9
18	𝍸 I	6

3. Lengths of Bridges (m)

Length	Tally	Frequency
401–500	II	2
501–600	IIII	4
601–700	II	2
701–800	III	3
801–900	I	1

5. Prices of Records (\$)

Price	Tally	Frequency
7.99	𝍸	5
8.99	IIII	4
9.99	II	2
10.99	IIII	4

Page 351 Review Exercises

1. bar graph;

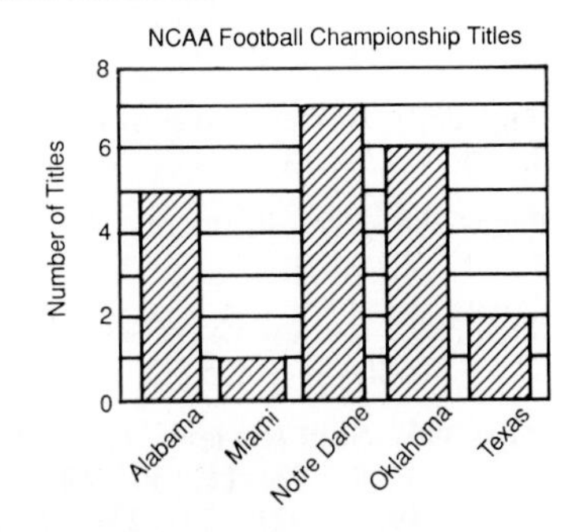

Page 354 Practice Exercises **1.** 5 **3.** 11 **5.** 11 **7.** 14 **9.** 16 **11.** 4 **13.** 3 and 4

15. Runs Scored by Teams

Number of Runs	Tally	Frequency
3	𝍸 I	6
4	𝍸 I	6
5	II	2
6	II	2

Page 354 Review Exercises **1.** $\frac{4}{5}$ **3.** $16\frac{1}{2}$ **5.** $2\frac{1}{8}$

Pages 356–357 Practice Exercises **1.** 10 gal **3.** 2 **5.** 4 gal and 5 gal **7.** about 45 min **9.** positive **11.** 24 **13.** 2 **15.** about 2 **17.** No

Page 357 Review Exercises **1.** increase; 50% **3.** increase; 100% **5.** decrease; 20%

Pages 359–360 Practice Exercises

1. Number of Home Runs Per Season

3	1, 6, 6, 6, 7, 7, 7, 8
4	0, 0, 0, 7, 7, 7, 8, 8, 8, 9, 9
5	0, 1, 4, 4, 8, 9
6	0, 1

3. 3 **5.** 0.042

7. Heights of World's Tallest Dams (m)

21	6, 9
22	0, 0, 0, 1, 6, 6
23	3, 5, 7, 7
24	2, 2, 2
25	0, 3
26	1, 5, 5
27	2
28	5
29	
30	0
31	
32	5

9. none **11.** 7 **13.** 24 **15.** They are the same.

Page 360 Review Exercises **1.** \$1440; \$7440

Page 363 Practice Exercises

1.

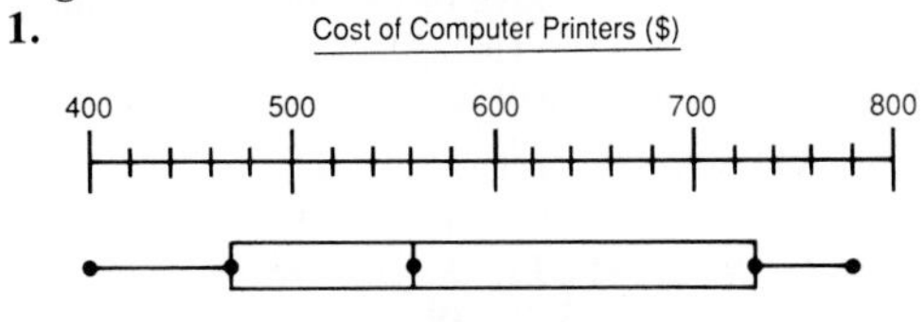

3. Eastern: 23.0; 21.0; 24.0; Western: 23.5; 21.0; 25.0

5.

Survey of Cities

20 40 60 80 100 120 140 160 180 200

Riverton

Central Falls

Page 363 Review Exercises 1. $\frac{3}{5}$ **3.** $\frac{2}{3}$ **5.** $\frac{2}{5}$

Page 364 Skill Review 1. 45; 38; 38; 25 **3.** 4.9; 5.25; 8.1; 6.8 **5. Number of Cousins**

Number of Cousins	Tally	Frequency
6	~~IIII~~ III	8
7	~~IIII~~	5
8	~~IIII~~ I	6
9	III	3

7. 26 **9.**

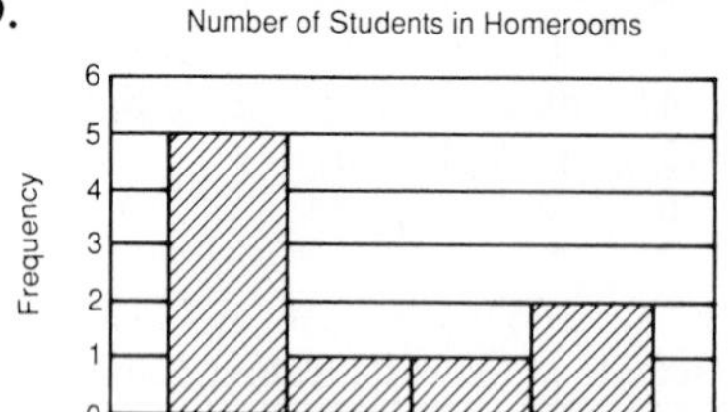

11. 15 min **13. Record Weights of Fresh-Water Fish (lb)**

1	1, 1, 2, 2, 2, 4, 8
2	1, 2, 2, 2, 6, 7
3	1, 2, 2, 6, 8
4	1
5	0, 1, 4, 8, 8

Page 365 Exercises 1. \$132.50 **3.** \$150

Pages 366–367 Exercises 1. No. Five of the seven selling prices are much greater than the mode. **3.** Yes. The median is \$100,000. Three of the seven selling prices are equal to or near the median. **5.** No. The commissions are either considerably higher or lower than \$3200. **7.** Yes; the median is \$2450. Five of the commissions are near the median. **9.** The agency is likely to quote the median, which is lower than the mean.

Pages 369–370 Exercises 1. \$512; \$7840 **3.** \$122.50; \$2870 **5.** \$499.50; \$21,285 **7.** \$609 **9.** \$440 **11.** \$1755 **13.** \$3880 **15.** \$10,916 **17.** \$27 **19.** \$76.50 **21.** \$535.50 **23.** \$238

Page 371 Chapter Review 1. mean **3.** 65; 61; none; 55 **5.** 3 **7. Plane Fares (Dollars)** **9.** 6 h **11.** \$25

16	2, 4, 5
17	0
18	0, 6, 8
19	8
20	0, 5, 7, 7

CHAPTER 18

Pages 375–376 Practice Exercises 1. $\frac{1}{3}$ **3.** $\frac{1}{2}$ **5.** 0 **7.** $\frac{5}{6}$ **9.** $\frac{5}{6}$ **11.** 1 to 7 **13.** 3 to 1 **15.** 3 to 5 **17.** 1 to 1 **19.** 1 to 1 **21.** $\frac{7}{18}$ **23.** $\frac{1}{3}$ **25.** 7 to 29 **27.** 13 to 5 **29.** 30% **31.** 80% **33.** 4 **35.** $\frac{1}{4}$; 25%

Page 376 Review Exercises 1. $\frac{4}{25}$ **3.** $\frac{16}{21}$ **5.** 7 **7.** $6\frac{2}{9}$ **9.** $\frac{5}{12}$ **11.** $8\frac{1}{10}$ **13.** 1 **15.** $\frac{1}{2}$ **17.** 1

Pages 378–379 Practice Exercises 1. 12 **3.** 6 **5.** 8 **7.** $\frac{1}{8}$ **9.** $\frac{1}{8}$ **11.** $\frac{7}{8}$ **13.** 5 **15.** denim **17.** $\frac{3}{5}$ **19.** 6 **21.** 6 **23.** $\frac{1}{4}$

Page 379 Review Exercises 1. 132 **3.** 72 **5.** 17

Page 381 Practice Exercises 1. 54 **3.** 1024 **5.** $\frac{1}{216}$ **7.** $\frac{125}{216}$ **9.** 12 **11.** 18 **13.** $\frac{1}{24}$ **15.** $\frac{1}{2}$ **17.** 17,576 **19.** 8,000,000; $\frac{1}{8{,}000{,}000}$

Page 381 Review Exercises 1. No **3.** Yes **5.** Yes

Pages 383–384 Practice Exercises 1. $\frac{1}{18}$ **3.** $\frac{1}{3}$ **5.** 0 **7.** $\frac{9}{56}$ **9.** $\frac{1}{28}$ **11.** $\frac{15}{56}$ **13.** $\frac{2}{75}$ **15.** $\frac{1}{9}$ **17.** $\frac{77}{225}$ **19.** $\frac{1}{30}$ **21.** $\frac{4}{15}$ **23.** $\frac{6}{35}$ **25.** $\frac{16}{675}$ **27.** dependent **29.** $\frac{11}{60}$ **31.** $\frac{1}{5}$; 1 to 4

Page 384 Review Exercises 1. about 300 **3.** about \$1400 **5.** about \$20 **7.** about \$28 **9.** about 5 **11.** about 5 **13.** about \$100 **15.** about 60

Pages 386–387 Practice Exercises 1. 720 **3.** 6 **5.** 6720 **7.** 840 **9.** 4 **11.** 495 **13.** 24 **15.** 5040 **17.** 60; 10 **19.** 120; 5 **21.** 210 **23.** 792; 120

Page 387 Review Exercises 1. 12; 12; none; 6 **3.** 53; 52; 59; 13

Pages 388–389 Problems 1. 10 **3.** 11 **5.** 11 **7.** 9 **9.** 5

Page 389 Review Exercises 1. Steven's: \$7.75; Shelley's: \$6.25 **3.** 7

Page 390 Skill Review 1. $\frac{3}{8}$ **3.** $\frac{1}{2}$ **5.** 1 to 7 **7.** 5 to 3 **9.** 12 **11.** 30 **13.** $\frac{1}{12}$ **15.** $\frac{1}{18}$ **17.** $\frac{1}{5}$ **19.** $\frac{1}{15}$ **21.** 24 **23.** 21 **25.** 8

Page 391 Exercises 1. $\frac{1}{5}$ **3.** $\frac{9}{10}$ **5.** side **7.** $\frac{14}{25}$

Pages 392–393 Exercises 1. about 208 **3.** about 338 **5.** The percent is incorrect. $\frac{14}{100} = 14\%$ **7. a.** about 7200 **b.** about 2880 **c.** about 6480 **9.** about 360 **11.** No; because they are friends, the ten people probably have the same favorite rock band. **13.** Yes **15.** Yes **17.** No; studying current data about the weather.

Page 394 Exercises 1. about 80 **3.** basketball

Page 395 Chapter Review 1. independent **3.** $\frac{1}{3}$ **5.** $\frac{2}{3}$ **7.** 5 to 1 **9.** 1 to 1 **11.** $\frac{5}{16}$ **13.** $\frac{1}{11}$ **15.** $\frac{1}{22}$ **17.** 90; 45 **19.** $\frac{1}{8}$ **21.** 10 **23.** about 50

CHAPTER 19

Page 403 Practice Exercises 1. $^{-}10$ **3.** 75 **5.** $^{-}4$ **7.** $\frac{15}{1}$ **9.** $^{-}\frac{44}{1}$ **11.** $\frac{83}{9}$ **13.** $^{-}\frac{116}{100}$ **15.** $^{-}\frac{2}{3}$ **17.** 23 **19.** $^{-}75$ **21.** 89 **23.** 175.45 **25.** $8\frac{1}{2}$ **27.** $\frac{1}{2}$ **29.–33.** Answers may vary. Examples are given. **29.** $\frac{5}{1}$, $\frac{10}{2}$ **31.** $\frac{24}{100}$, $\frac{6}{25}$ **33.** $\frac{2}{9}$, $\frac{4}{18}$

Page 403 Review Exercises 1. 29; 30; 33; 10

Pages 405–406 Practice Exercises 1. *C* **3.** *B* **5.** *E* **7.** $^{-}15$ **9.** 100 **11.** $\frac{3}{4}$ **13.** $7\frac{7}{8}$ **15.** $^{-}0.67$ **17.** 5 **19.** 41 **21.** 0 **23.** $\frac{3}{4}$ **25.** 4.99 **27.** *F* **29.** *E* **31.** *B* **33.** $^{-}95$; 95 **35.** $^{-}\frac{6}{17}$; $\frac{6}{17}$ **37.** 5.87; 5.87 **39.** A **41.** A

Page 406 Review Exercises 1. $<$ **3.** $>$ **5.** $<$ **7.** 15 h

Page 408 Practice Exercises 1. $<$ **3.** $>$ **5.** $<$ **7.** $>$ **9.** $^{-}3$; $^{-}2$; $^{-}1$; 3 **11.** $^{-}4.5$; $^{-}2.1$; 1.6; 8.5 **13.** $^{-}1\frac{7}{8}$; $^{-}\frac{7}{8}$; $\frac{3}{8}$; $3\frac{5}{8}$ **15.** False **17.** True **19.** True **21.** True **25.** Africa

Page 408 Review Exercises 1. 298 **3.** 15.43 **5.** $\frac{9}{10}$ **7.** $6\frac{1}{4}$

Pages 410–411 Practice Exercises
1. 9 **3.** 1 **5.** $^{-}2$ **7.** 1 **9.** 55 **11.** $^{-}35$ **13.** $^{-}6.4$ **15.** $5\frac{5}{6}$ **17.** $^{-}7$ **19.** 8 **21.** $^{-}2.6$ **23.** $^{-}2\frac{1}{3}$ **25.** 60 **27.** $^{-}54$ **29.** 0 **31.** $8\frac{6}{7}$ **33.** 1.2 **35.** $^{-}10$ **37.** 50°F above zero **39.** Frank's; 4

Page 411 Review Exercises 1. 0.16 **3.** 4% **5.** 0.014 **7.** 150% **9.** 0.002 **11.** $11\frac{1}{3}$%

Pages 413–414 Practice Exercises 1. 21 + $^{-}57$ **3.** $^{-}4$ + $^{-}98$ **5.** 49.5 + $^{-}32.7$ **7.** $^{-}1\frac{1}{7}$ + $^{-}6\frac{2}{7}$ **9.** $^{-}3$ **11.** 66 **13.** $^{-}80$ **15.** $^{-}3$ **17.** $^{-}9$ **19.** $^{-}27$ **21.** $^{-}3.3$ **23.** $^{-}\frac{1}{11}$ **25.** 6 + $^{-}15$; $^{-}9$ **27.** $^{-}5$ + 27; 22 **29.** $^{-}6\frac{1}{25}$ + $^{-}16\frac{1}{25}$; $^{-}22\frac{2}{25}$ **31.** 9 − $^{-}12$; 21 **33.** $^{-}18$°F **35.** $^{-}7$ **37.** 164 pages

Page 414 Review Exercises 1. 62 **3.** 1.65 **5.** $\frac{1}{25}$ **7.** $\frac{2}{3}$ **9.** 0.8 **11.** 35 **13.** 4; 0.1 m **15.** 1; 0.01 kg

Pages 416–417 Practice Exercises 1. 18 **3.** $^{-}30$ **5.** $^{-}184$ **7.** $^{-}9.1$ **9.** 270 **11.** $^{-}112$ **13.** 176 **15.** $^{-}\frac{1}{6}$ **17.** 9 **19.** $^{-}8$ **21.** $^{-}28$ **23.** $^{-}\frac{9}{20}$ **25.** $^{-}25$ **27.** $^{-}12$ **29.** 20.5 **31.** $\frac{5}{18}$ **33.** 0 **35.** $^{-}11.2$ **37.** 16 **39.** $^{-}8$ **41.** 1 **43.** a decrease of 20°F **45.** $10\frac{1}{2}$ min

Page 417 Review Exercises 1. $\frac{1}{20}$ **3.** $\frac{17}{500}$ **5.** $33\frac{1}{3}$%

Page 418 Skill Review 1. $^{-}30$ **3.** 20 **5.** $^{-}\frac{17}{4}$ **7.** $\frac{47}{100}$ **9.** $\frac{0}{1}$ **11.** 39; 39 **13.** $^{-}14.7$; 14.7 **15.** $\frac{1}{5}$; $\frac{1}{5}$ **17.** $^{-}15\frac{1}{4}$; $15\frac{1}{4}$ **19.** 0; 0 **21.** $<$ **23.** $>$ **25.** $^{-}6$; $^{-}5$; $^{-}3$; 4 **27.** $^{-}3.2$; $^{-}0.2$; 1.4; 4 **29.** $^{-}1\frac{1}{5}$; $^{-}\frac{4}{5}$; 0; $\frac{4}{5}$ **31.** 23 **33.** 29 **35.** $^{-}1.3$ **37.** $^{-}13\frac{6}{7}$ **39.** decrease of $\$7\frac{7}{8}$ per share **41.** 15 **43.** 60 **45.** $^{-}8.6$ **47.** $^{-}\frac{5}{9}$ **49.** 11.9 **51.** 9 **53.** 0 **55.** $^{-}\frac{3}{14}$

Page 419 Exercises 1. °C **3.** °F **5.** °F **7.** °F

Pages 420–422 Exercises 1. Yes **3.** No **5.** 1.275×10^3 **7.** 1×10^7 **9.** 600 **11.** 870,000,000,000 **13.** 2×10^7 **15.** 6.5×10^7 **17.** 4.19×10^2 **19.** 9.265×10^7 **21.** 2.19316×10^3 **23.** 500,000 **25.** 2000 **27.** Yes **29.** No **31.** 3×10^{-1} **33.** 2.7×10^{-2} **35.** 0.05 **37.** 0.000091 **39.** 3×10^{-2} **41.** 1×10^{-6} **43.** 0.002 **45.** 0.0313

Page 423 Chapter Review 1. opposites **3.** $^{-}50$ **5.** 45 **7.** $\frac{49}{1}$ **9.** $\frac{8}{9}$ **11.** $\frac{0}{1}$ **13.** 78 **15.** 0.5 **17.** $^{-}14\frac{1}{3}$ **19.** 12 **21.** 44.5 **23.** $\frac{13}{15}$ **25.** $^{-}3.1$; $^{-}1.9$; 1.8; 2.3 **27.** $^{-}14$ **29.** 54 **31.** 0 **33.** $^{-}12$ **35.** $5\frac{5}{9}$ **37.** 0 **39.** °F **41.** 8×10^4

CHAPTER 20

Pages 427–428 Practice Exercises 1. 1 **3.** $^{-}50$ **5.** 6 **7.** 49 **9.** $^{-}2$ **11.** $^{-}10$ **13.** 12 **15.** $^{-}6$ **17.** 0 **19.** 0 **21.** undefined **23.** 0 **25.** 6 **27.** 99 **29.** 55 **31.** $^{-}2$ **33.** 1 **35.** 4 **37.** 9 **39.** $^{-}40$ **41.** $^{-}5$ **43.** $^{-}12$ **45.** 200 **47.** 0 **49.** 0 **51.** 0

Page 428 Review Exercises 1. 34 **3.** $^{-}96$ **5.** $\frac{7}{30}$ **7.** $^{-}2$ **9.** 27 **11.** $16\frac{2}{3}$% **13.** 6

Pages 430–431 Practice Exercises 1. 13 **3.** 4 **5.** $^{-}13$ **7.** 14 **9.** $^{-}43$ **11.** 0.2 **13.** 14 **15.** $^{-}5$ **17.** 80 **19.** 11 **21.** $^{-}14$ **23.** $5\frac{6}{7}$ **25.** 13 **27.** 32 **29.** $^{-}11$ **31.** $^{-}1$ **33.** $^{-}11$ **35.** 16 **37.** 3 **39.** 15.2 **41.** $^{-}\frac{8}{9}$

Page 431 Review Exercises 1. $\frac{4}{3}$ **3.** $\frac{1}{16}$ **5.** 8 **7.** $8\frac{1}{2}$ min **9.** 14 months

Pages 433–434 Practice Exercises 1. $^{-}8$ **3.** $^{-}12$ **5.** 3.6 **7.** 0.03 **9.** $^{-}200$ **11.** $^{-}45$ **13.** 8.8 **15.** 2.8 **17.** 16 **19.** $^{-}20$ **21.** $^{-}12$

23. 108 **25.** 8 **27.** $^{-}42$ **29.** $^{-}20$ **31.** 3 **33.** 30 **35.** $^{-}2.8$ **37.** 0.9 **39.** $^{-}10$

Page 434 Review Exercises 1. No; 500 mL + 750 mL = 1250 mL = 1.25 L

Page 436 Practice Exercises 1. 6 **3.** $^{-}3$ **5.** $^{-}18$ **7.** $^{-}48$ **9.** 15 **11.** $^{-}54$ **13.** $^{-}1$ **15.** 1.3 **17.** 124 **19.** $^{-}12$ **21.** $^{-}120$ **23.** 16 **25.** B, D

Page 436 Review Exercises 1. \$15 **3.** 9

Pages 438–439 Practice Exercises 1. $z > {}^{-}15$ **3.** $a < 24$ **5.** $w \geq {}^{-}19$ **7.** $c \leq 5$ **9.** $z > 200$ **11.** $c \leq {}^{-}3$ **13.** $b \geq 3$ **15.** $s < {}^{-}14$

17.

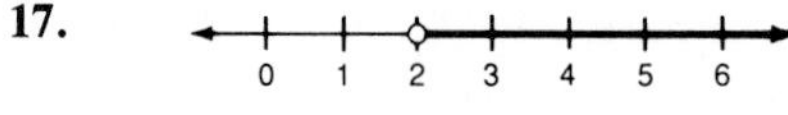

19.

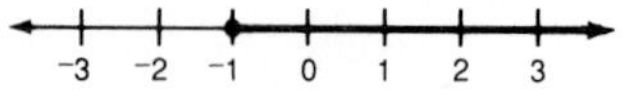

21.

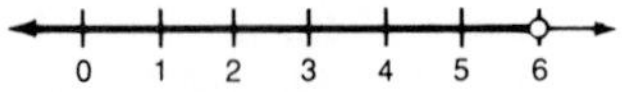

23.

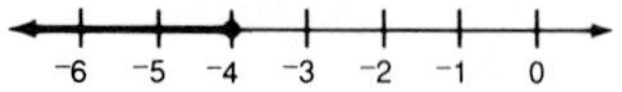

25. $a > 1$ **27.** $q \geq {}^{-}1$ **29.** $r < {}^{-}6$ **31.** $b \leq 3$ **33.** C, D, E **35.** C **37.** B **39.** $g < 18$

Page 439 Review Exercises 1. 2 **3.** 6; 2 **5.** 10,560

Page 441 Problems 1. $y - 3 = 12$; 15 tapes **3.** $\frac{y}{3} = 12$; \$36 **5.** \$1.50 **7.** \$2.40

Page 441 Review Exercises 1. about 5000 **3.** about 7500 **5.** about 20,000

Page 442 Skill Review 1. 15 **3.** 37 **5.** $^{-}4$ **7.** 44 **9.** 0 **11.** $^{-}11$ **13.** $^{-}50$ **15.** $^{-}4$ **17.** 10.4 **19.** $^{-}10$ **21.** $^{-}20$ **23.** $^{-}48$ **25.** $^{-}6$ **27.** 0.6 **29.** 81 **31.** 30 **33.** $^{-}99$ **35.** $^{-}42$ **37.** $z > 1$ **39.** $a \leq 4$ **41.** $t > {}^{-}24$

43.

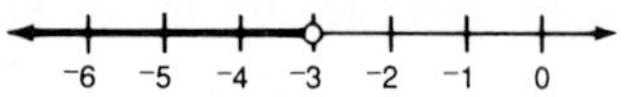

45.

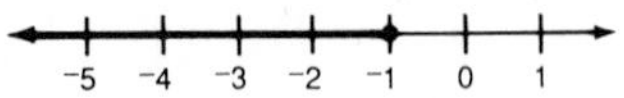

47.

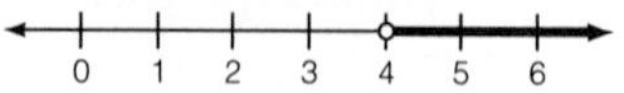

49. 73 ft

Pages 443–444 Exercises 1. 194°F **3.** $^{-}4$°F **5.** $A = \frac{h}{b}$ **7.** $p = \frac{m}{a}$ **9.** 3300 ft **11.** 580 mi/h **13.** 165 mi **15.** 1056 km/h

Pages 445–446 Exercises 1. 4 **3.** 40 **5.** 3 h **7.** $s = \frac{P}{4}$ **9.** $p = \frac{C}{n}$ **11.** $t = An$ **13.** $h = \frac{2A}{b}$ **15.** $s = \frac{D - r}{2}$ **17.** 45 **19.** 5 ft **21.** \$2000 **23.** 5 **25.** 3 **27.** 3%

Page 447 Chapter Review 1. variable **3.** 3 **5.** $^{-}128$ **7.** $^{-}7$ **9.** $^{-}18$ **11.** 6 **13.** $^{-}1.1$ **15.** $3\frac{3}{5}$ **17.** $p > 3$

19.

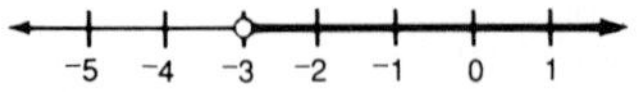

21.

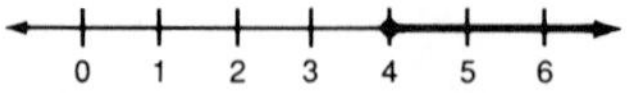

23. \$280 **25.** 4 h

CHAPTER 21

Pages 451–452 Practice Exercises 1. ($^{-}3$, $^{-}5$) **3.** (3, $^{-}3$) **5.** (5, 5) **7.** ($^{-}5$, 5) **9.** (3, 0) **11.** (0, $^{-}2$)

13.–24.

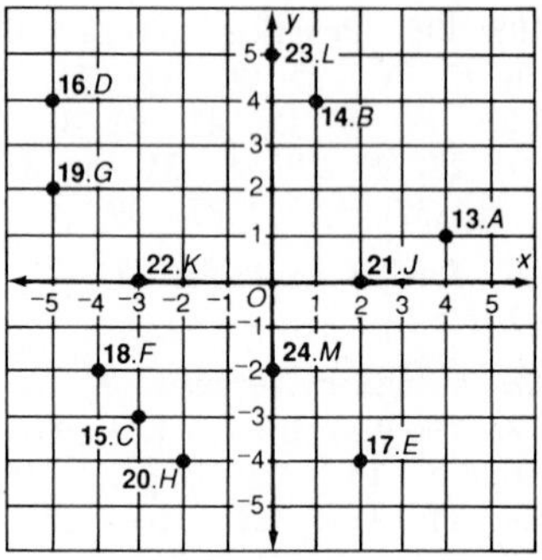

25. Q **27.** N **29.** P **31.** S **33.** Z **35.** U **47.** I **49.** None **51.** II **53.** None **55.** IV **57.** I **59.** 5 blocks due west

Page 452 Review Exercises 1. $^{-}11$ **3.** $^{-}14$ **5.** $^{-}2$ **7.** $^{-}15.6$

Pages 454–455 Practice Exercises 1. Yes **3.** No **5.** Yes **7.** No **9.** (2, 11) **11.** ($^{-}1$, $^{-}1$) **13.** ($^{-}4$, $^{-}13$) **15.** ($\frac{1}{2}$, 5)

21.

$y = 3x - 1$		
x	y	(x, y)
$^{-}2$	$^{-}7$	($^{-}2$, $^{-}7$)
$^{-}1$	$^{-}4$	($^{-}1$, $^{-}4$)
0	$^{-}1$	(0, $^{-}1$)
1	2	(1, 2)
2	5	(2, 5)

23.

$y = {}^{-}2x + 2$		
x	y	(x, y)
$^{-}2$	6	($^{-}2$, 6)
$^{-}1$	4	($^{-}1$, 4)
0	2	(0, 2)
1	0	(1, 0)
2	$^{-}2$	(2, $^{-}2$)

25. b, c **27.** a, c, d

Page 455 Review Exercises **1.** $^-24$; 24 **3.** 0; 0 **5.** 0.75; 0.75 **7.** $475

Pages 457–458 Practice Exercises

1.

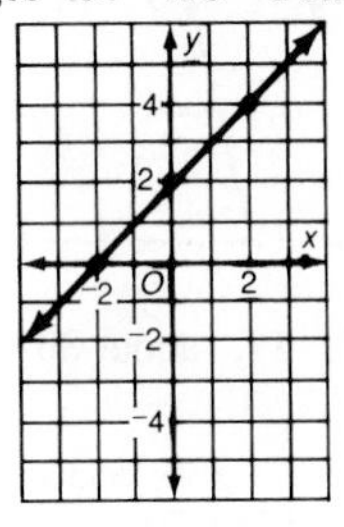

5.

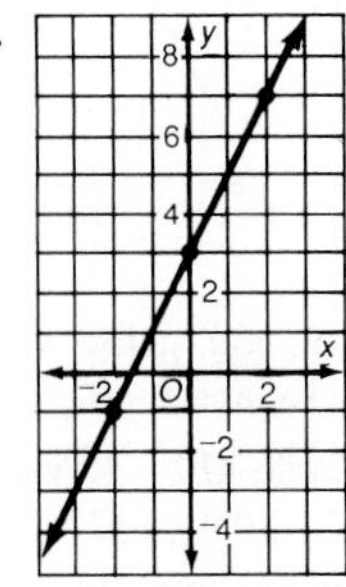

13.

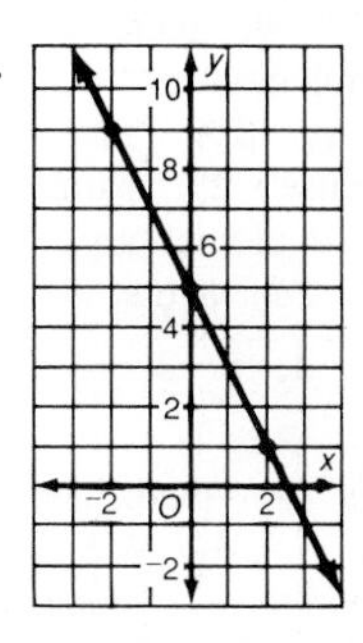

15.

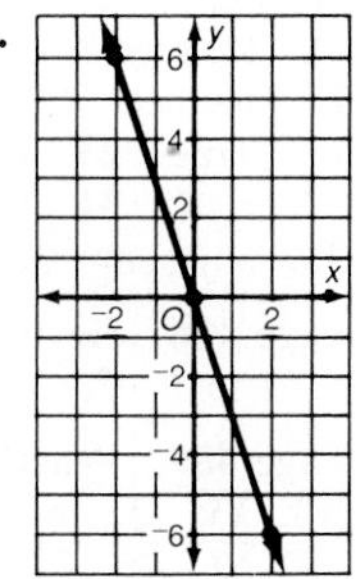

17. C **19.** B

21.

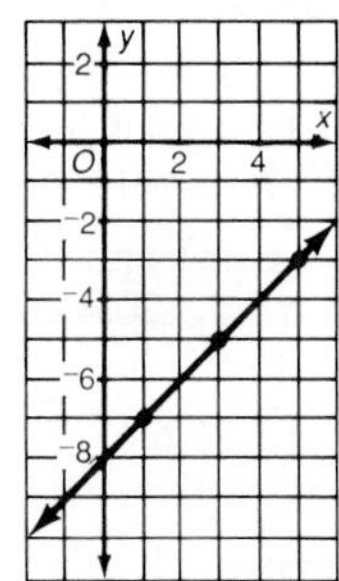

25.

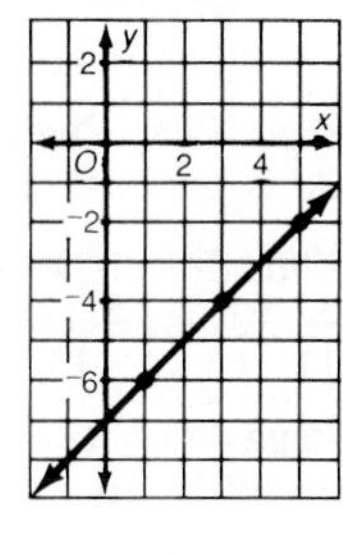

Page 458 Review Exercises **1.** 48 mi/h **3.** 4.8 m/s **5.** about 200%

Pages 460–461 Practice Exercises **1.** Yes **3.** Yes **5.** Yes **7.** No **9.** (2, 7) **11.** ($^-1$, 3) **13.** No solution **15.** (0, $^-1$) **17.** (1, 3) **19.** (4, 1) **21.** No solution

Page 461 Review Exercises **1.** about 400 **3.** about 320 **5.** about 700 **7.** about 6 **9.** about 14,000 **11.** about 7 **13.** about 4 **15.** about 5 **17.** about $3

Page 462 Skill Review **1.** *C* **3.** *B* **5.** *D* **7.** *G* **9.** *F* **19.** Yes **21.** No **23.** No **25.** No

27.

$y = x + 7$		
x	y	(x, y)
$^-2$	5	($^-2$, 5)
$^-1$	6	($^-1$, 6)
0	7	(0, 7)
1	8	(1, 8)
2	9	(2, 9)

29.

$y = 3x$		
x	y	(x, y)
$^-2$	$^-6$	($^-2$, $^-6$)
$^-1$	$^-3$	($^-1$, $^-3$)
0	0	(0, 0)
1	3	(1, 3)
2	6	(2, 6)

31.

$y = 2x + 3$		
x	y	(x, y)
$^-2$	$^-1$	($^-2$, $^-1$)
$^-1$	1	($^-1$, 1)
0	3	(0, 3)
1	5	(1, 5)
2	7	(2, 7)

33.

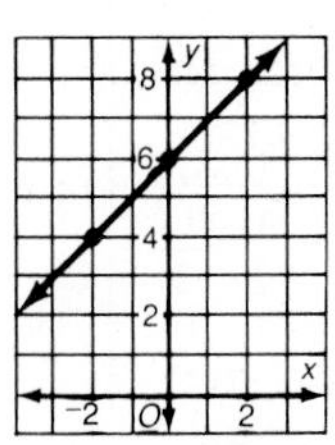

35.

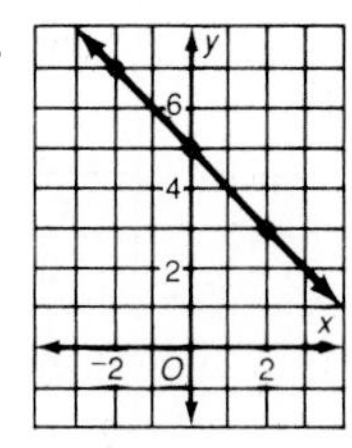

37.

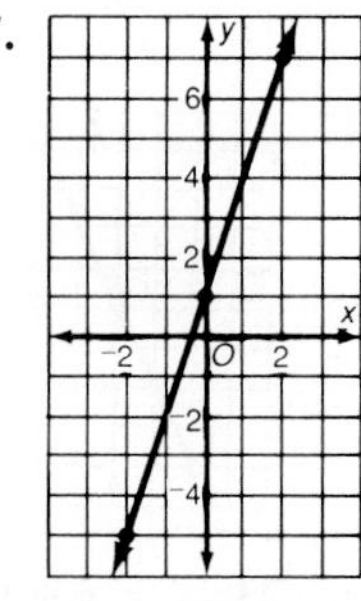

39. Yes **41.** Yes **43.** No **45.** (1, $^-1$) **47.** No solution **49.** No solution

Page 463 Exercises **1.** B2 **3.** E1 **5.** F2 and G2 **7.** Coordinates will vary. Different maps do not use the same system.

Pages 464–466 Exercises **1.** Yes **3.** Yes **5.** No **7.** $y = 20 - x$ **9.** $y = 7x$ **11.** $y = 55x$ **13.** 18; 15; 8; 14; 0 **15.** 13; 8; 12; 27; 36; 38 **17.** 23; 32; 50; 95; 100 **19.** $y = x + 5$ **21.** B **23.** A **25.** $y = x + 3$

Pages 467–469 Exercises 1. 24 mi/h **3.** 40 mi/h **5.** *Q* **7.** *P* **9.** E **11.** D **13.** A

Page 470 Exercises 1. $1.65 **3.** 4 oz

Page 471 Chapter Review 1. *x*-axis

3.–10.

11.

	$y = {}^{-}6x$	
x	y	(x, y)
$^{-}2$	12	$(^{-}2, 12)$
$^{-}1$	6	$(^{-}1, 6)$
0	0	(0, 0)
1	$^{-}6$	$(1, {}^{-}6)$
2	$^{-}12$	$(2, {}^{-}12)$

13.

	$y = {}^{-}3x + 7$	
x	y	(x, y)
$^{-}2$	13	$(^{-}2, 13)$
$^{-}1$	10	$(^{-}1, 10)$
0	7	(0, 7)
1	4	(1, 4)
2	1	(2, 1)

15.

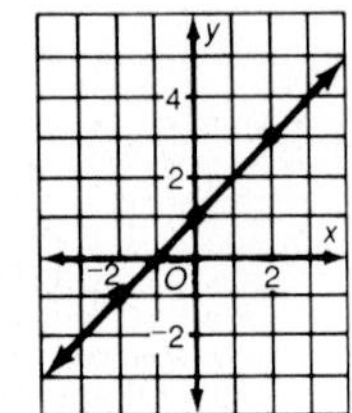

17.

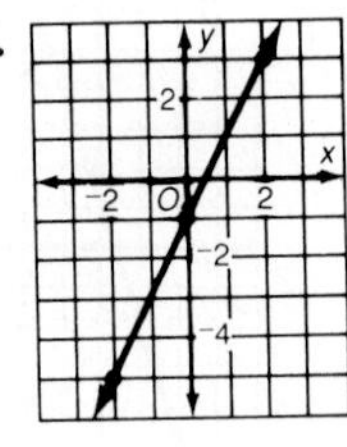

19. $(^{-}1, {}^{-}1)$ **21.** No solution **23.** $y = x + 9$ **25.** *A*

CHAPTER 22

Pages 479–480 Practice Exercises 1. *E, F, G, H, O, M, N, P, X, Y* **3.** $\overleftrightarrow{EF}$, $\overleftrightarrow{XY}$ **5.** $\overline{EF}$, $\overline{NO}$, $\overline{NM}$, $\overline{XY}$, $\overline{GH}$ **7.** adjacent **9.** vertical **11.** Answers may vary. Examples: $\overleftrightarrow{AB}$, $\overleftrightarrow{ME}$, $\overleftrightarrow{TP}$ **13.** Answers may vary. Examples: $\overrightarrow{GT}$, $\overrightarrow{FA}$, $\overrightarrow{GB}$ **21.** *A, B, C, D* **23.** $\overline{AB}$, $\overline{AC}$, $\overline{AD}$

Page 480 Review Exercises 1. $x + 2\frac{1}{2} = 8$; $5\frac{1}{2}$ min **3.** $5p + 1.37 = 3.12$; $.35

Page 482 Practice Exercises 1. 45° **3.** 30° **5.** 35° **7.** 60° **9.** 75°

13. 60° **15.** 90° **17.** 145° **19.** about 80° **21.** about 30°

Page 482 Review Exercises 1. $30; $530 **3.** $630; $12,630 **5.** $34.50; $264.50

Pages 485–486 Practice Exercises 1. acute **3.** obtuse **5.** acute **7.** right **9.** acute **11.** supplementary **13.** complementary **15.** neither **17.** supplementary **19.** False **21.** True **23.** False **25.** False **27.** parallel **29.** right **31.** obtuse **33.** supplementary **35.** adjacent **37.** acute; 50° **39.** acute; 30°

Page 486 Review Exercises 1. $\frac{1}{4}$ **3.** $\frac{19}{20}$ **5.** 1 to 1 **7.** 3 to 1

Page 489 Practice Exercises 1. quadrilateral; $\overline{WY}$, $\overline{XV}$ **3.** regular hexagon; $\overline{MP}$, $\overline{MJ}$, $\overline{MK}$, $\overline{NJ}$, $\overline{NK}$, $\overline{NL}$, $\overline{PK}$, $\overline{PL}$, $\overline{JL}$ **5.** rhombus **7.** trapezoid **9.** rectangle **11.** triangle **13.** quadrilateral **15.** hexagon **17.** rhombus, square **19.** trapezoid

Page 489 Review Exercises 1. 13 ft 1 in. **3.** 3 gal 2 qt **5.** 77 c 4 fl oz **7.** 3 gal 1 qt

Page 491 Practice Exercises 1. 41 cm **3.** 61 m **5.** 328 mm **7.** 48 ft **9.** 20.4 cm **11.** 68 yd **13.** 2

Page 491 Review Exercises 1. 23.6 **3.** 39.25 **5.** $1\frac{4}{7}$ **7.** $\frac{2}{3}$

Pages 493–494 Practice Exercises 1. circle *O* **3.** $\overline{KO}$, $\overline{OH}$, $\overline{OL}$ **5.** $\angle KOL$, $\angle HOL$, $\angle KOH$ **7.** 94.2 in. **9.** 43.96 cm **11.** 20.41 km **13.** 176 ft **15.** 154 m **17.** $1\frac{4}{7}$ yd **19.** radius **21.** diameter **23.** radius **25.** 28.26 yd **27.** 320.28 in. **29.** 4.5 ft; 28.26 ft **31.** 16 cm; 50.24 cm **33.** 6.5 mm; 13 mm **35.** 264 in. **37.** 88 ft; 18

Page 494 Review Exercises 1. 17 **3.** 64 **5.** 3

Pages 497–498 Practice Exercises 1. 196 mm^2 **3.** 608.4 m^2 **5.** 300 m^2 **7.** 29.04 in.^{2} **9.** 196 km^2 **11.** 8 ft^2 **13.** 6.76 m^2 **15.** 6 yd^2 **17.** 6000 yd^2 **19.** 3 **21.** about 8 square units **23.** about 5 square units

Page 498 Review Exercises **1.** 16 **3.** 48 **5.** 588 **7.** 12 **9.** 40

Pages 500–501 Practice Exercises
1. 254.34 m^2 **3.** 154 yd^2 **5.** 907.46 yd^2 **7.** 1386 in.^{2} **9.** 3.14 km^2 **11.** 154 yd^2 **13.** 1133.54 ft^2 **15.** 616 in.^{2} **17.** 314 m^2 **19.** 200.96 mm^2 **21.** 616 ft^2 **23.** 2464 yd^2 **25.** 153.86 ft^2 **27.** 301.44 m^2 **29.** 1256 cm^2 **31.** 2464 mi^2

Page 501 Review Exercises **1.** $\frac{1}{3}$ **3.** $\frac{3}{16}$ **5.** $\frac{20}{1}$

Page 502 Skill Review **1.** Answers may vary. Examples: A, B, C, D, E, F, G, H, J, K **3.** Answers may vary. Examples: $\overline{BC}$, $\overline{DE}$, $\overline{GF}$, $\overline{GH}$, $\overline{JK}$ **5.** 90° **7.** 130° **9.** obtuse **11.** right **13.** square, rhombus **15.** 48 ft **17.** 10.2 yd **19.** circle O **21.** $\overline{OA}$, $\overline{OB}$, $\overline{OC}$ **23.** 37.68 cm **25.** 8.1 cm^2 **27.** 616 ft^2

Pages 503–504 Exercises **1.** Yes **3.** No
5. **7.**

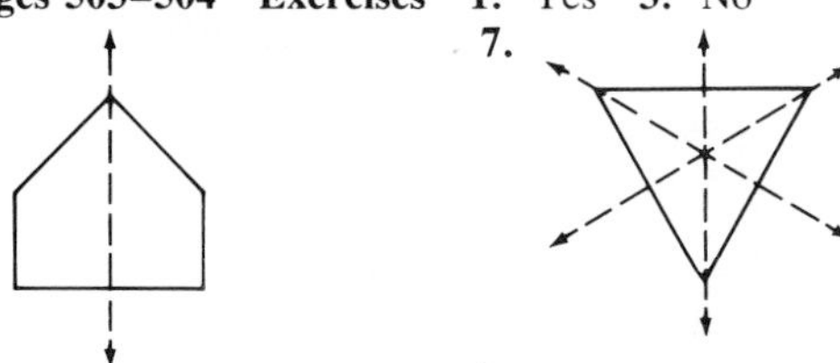

9. Yes **11.** No **13.** 2 **15.** 3 **17. a.** 3 **b.** 3 **19. a.** 5 **b.** 5

Page 506 Exercises
1.

How High School Students Use Computers in School

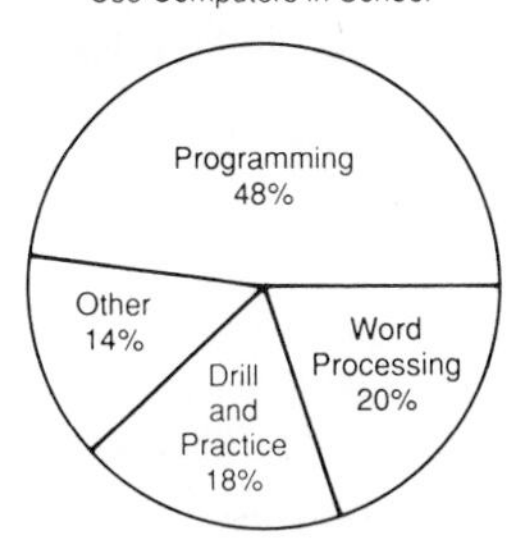

3.

Composition of Earth's Atmosphere

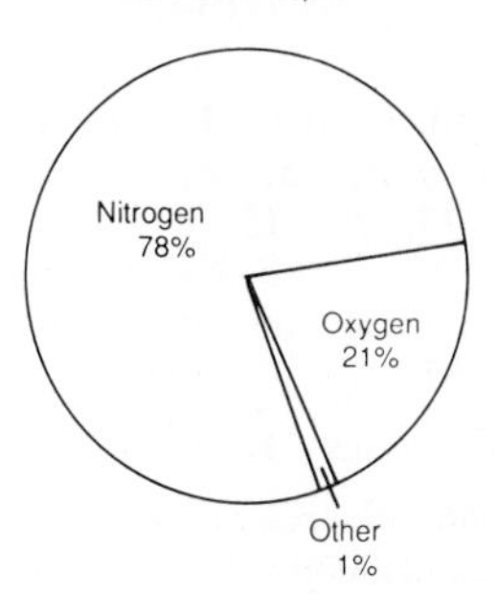

Pages 507–508 Exercises **1.** 2025 m^2 **3.** 158.24 yd^2 **5.** 80 cm^2 **7.** 246 ft^2 **9.** 35 ft^2

Pages 509–510 Chapter Review **1.** Adjacent **3.** perpendicular **5.** Answers may vary. Examples: $\angle AHC$ and $\angle DHG$; $\angle EGB$ and $\angle FGH$ **7.** Answers may vary. $\overleftrightarrow{AB}$ and $\overleftrightarrow{CD}$ or $\overleftrightarrow{AB}$ and $\overleftrightarrow{EF}$
9. **11.**

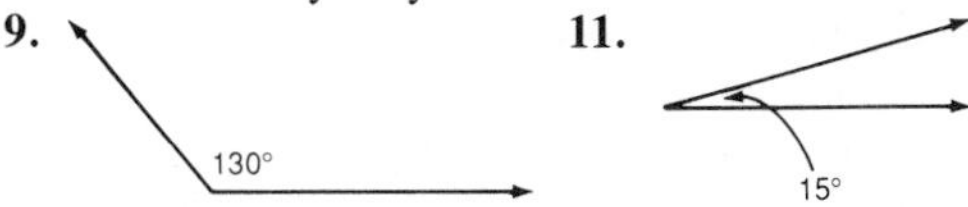

13. neither **15.** complementary **17.** neither **19.** triangle; none **21.** rectangle, square **23.** trapezoid **25.** 12 cm **27.** 22.5 m **29.** 2.56 m^2 **31.** 9 mi^2 **33.** 216 yd^2 **35.** 44 ft; 154 ft^2 **37.** 56.52 km; 254.34 km^2 **39.** 132 in.; 1386 in.^{2} **41. a.** 2 **b.** 2 **43.** 1200 ft^2

CHAPTER 23

Pages 515–516 Practice Exercises **1.** rectangular pyramid **3.** cone **5.** rectangular prism **7.** cylinder **9.** a **11.** 6; 8; 12; 2 **13.** 5; 5; 8; 2

Page 516 Review Exercises **1.** 90 in.^{2} **3.** 49 ft^2 **5.** 264 in. **7.** $\frac{1}{10}$; 1 to 9

Pages 518–519 Practice Exercises **1.** 518 m^2 **3.** 26.46 in.^{2} **5.** 600 yd^2 **7.** 1350 mm^2 **9.** 4350 mm^2 **11.** 83 in.^{2} **13.** 824 in.^{2}

Page 519 Review Exercises **1.** $^{-}17$ **3.** $^{-}9$ **5.** $^{-}\frac{7}{16}$ **7.** $\frac{1}{3}$ **9.** $^{-}0.3$ **11.** 2.6

Pages 521–522 Practice Exercises
1. 251.2 cm^2 **3.** 565.2 ft^2 **5.** 10,208 m^2 **7.** 125.6 in.^{2} **9.** 15,972 mm^2 **11.** 1540 cm^2 **13.** 1017.36 mm^2 **15.** 11,264 mm^2

Page 522 Review Exercises
1.–6.

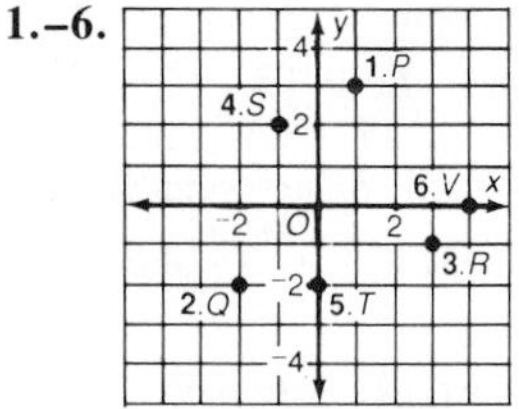

7.

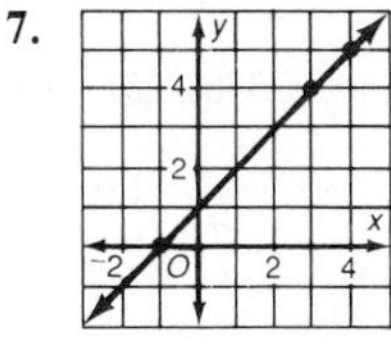

9.

11.

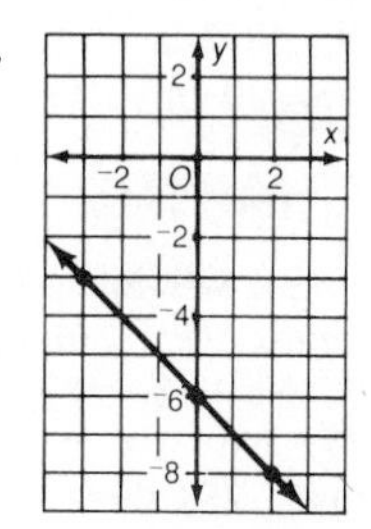

Pages 524–526 Practice Exercises **1.** 165 ft^3 **3.** 258.3 mm^3 **5.** 2730 $in.^3$ **7.** 113.04 m^3 **9.** 6269.01 cm^3 **11.** 2040 cm^3 **13.** 67,200 cm^3 **15.** 339.12 cm^3 **17.** 2,541,440 ft^3 **19.** 336 $in.^3$ **21.** 186.3 L **23.** 180 mL

Page 526 Review Exercises **1.** 0.25; $\frac{1}{4}$ **3.** 0.75; $\frac{3}{4}$ **5.** $0.\overline{3}$; $\frac{1}{3}$

Pages 528–529 Practice Exercises **1.** 9 ft^3 **3.** 72 m^3 **5.** 2376 yd^3 **7.** 270 ft^3 **9.** 3391.2 cm^3 **11.** 6 cm

Page 529 Review Exercises **1.** $b > 2$ **3.** $y \geq {}^-1$

Page 531 Practice Exercises **1.** 1519.76 cm^2 **3.** 803.84 $in.^2$ **5.** 7234.56 mm^3 **7.** 11,304 yd^2; 113,040 yd^3 **9.** 13,677.84 m^2; 150,456.24 m^3 **11.** 3052.08 mm^3

Page 531 Review Exercises **1.** $\frac{1}{5}$ **3.** $\frac{5}{8}$ **5.** $5\frac{7}{8}$

Pages 532–533 Exercises **1.** 14 cm^2 **3.** 20 cm^2 **5.** 26 cm^2 **7.** 32 cm^2 **9.** 34 cm^2 **11.** 38 cm^2 **13.** d **15.** 8 **17.** 6

Page 533 Review Exercises **1.** 60 toys **3.** 4

Page 534 Skill Review **1.** cone **3.** triangular prism **5.** 139 $in.^2$ **7.** 5652 ft^2 **9.** 64 cm^3 **11.** 105 mm^3 **13.** 1808.64 cm^2; 7234.56 cm^3 **15.** 26 cm^2

Pages 535–537 Exercises **1.** 2016 ft^3 **3.** 216 $in.^3$ **5.** 12,246 cm^3 **7.** 320.28 mm^3 **9.** 12,000 m^3 **11.** 1646.08 ft^3 **13.** 8320.32 yd^3 **15.** 452.16 cm^3 **17.** 540 ft^3 **19.** c

21.

TOP VIEW FRONT VIEW SIDE VIEW

Pages 538–539 Exercises **1.** 9 **3.** 14 **5.** 1 gal **7.** 2 gal **9.** Yes; $25 \times 13 = 325$ (ft^2); $325 < 440$ **11.** \$31.48 **13.** paint; about \$140

Page 540 Exercises **1.** 3140 ft^3 **3.** about 757 bushels

Page 541 Chapter Review **1.** volume **3.** cylinder **5.** rectangular pyramid **7.** 616 m^2 **9.** 3872 $in.^2$ **11.** 1560 yd^3 **13.** 1125 $in.^3$ **15.** 20 cm^2 **17.** 672 m^3 **19.** 2 gal

CHAPTER 24

Pages 546–547 Practice Exercises **1.** scalene; obtuse **3.** equilateral; acute **5.** Yes **7.** No **9.** No **11.** 68° **13.** 38° **15.** Yes; equilateral **17.** No **19.** No **21.** 12°; obtuse **23.** 23°; obtuse **25.** three **27.** greater **29.** 60°, 60°, 60°

Page 547 Review Exercises **1.** $^-3$ **3.** $^-98$ **5.** 0 **7.** 57 **9.** 18 **11.** 21

Page 550 Practice Exercises **1.** 3 **3.** 40 **5.** 0 **7.** 6.325 **9.** 23 **11.** 1 **13.** 9.487 **15.** 100 **17.** 11.874 **19.** 106 **21.** True **23.** True **25.** $>$ **27.** $<$ **29.** 441 cm^2 **31.** 50 ft

Page 550 Review Exercises **1.** 15 **3.** $^-37$ **5.** $^-22$

Pages 552–553 Practice Exercises **1.** No **3.** Yes **5.** Yes **7.** Yes **9.** 15 km **11.** 18 mi **13.** $c = 10$ **15.** $b = 24$ **17.** $a = 20$ **19.** 8 mi **21.** 60 ft **23.** 8.9 ft **25.** 3.2 mi

Page 553 Review Exercises **1.** 132 **3.** 4 **5.** 7 **7.** 5% **9.** 700

Pages 555–556 Practice Exercises **1.** similar; congruent **3.** not similar **5.** $x = 15$ **7.** $x = 36$ **9.** $x = 8$, $y = 10$; congruent **11.** $x = 24$; $y = 30$; not congruent

Page 556 Review Exercises

1.

$y = 3x - 5$		
x	y	(x, y)
$^-2$	$^-11$	$(^-2, {}^-11)$
$^-1$	$^-8$	$(^-1, {}^-8)$
0	$^-5$	$(0, {}^-5)$
1	$^-2$	$(1, {}^-2)$
2	1	(2, 1)

3. 32; 33; 33 and 36; 20

Pages 558–559 Practice Exercises **1.** $\frac{12}{5}$ **3.** $\frac{12}{13}$ **5.** $\frac{12}{13}$ **7.** $\frac{24}{25}$ **9.** $\frac{24}{7}$ **11.** $\frac{7}{25}$ **13.** $\frac{20}{29}$ **15.** $\frac{21}{20}$ **17.** $\frac{21}{29}$ **19.** $\frac{15}{17}$ **21.** $\frac{8}{15}$ **23.** $\frac{8}{17}$ **27.** 25 m **29.** $\frac{3}{5}$ **31.** $\frac{3}{5}$ **33.** $\frac{3}{4}$

Page 559 Review Exercises **1.** about 400 **3.** about 200 **5.** about 10 **7.** about 1200 **9.** about 15 **11.** about 5 **13.** about 54 **15.** about 80 **17.** about 700 **19.** about 20

Pages 560–561 Problems **1.** 110 **3.** 891 **5.** 2580 **7.** 14

Page 561 Review Exercises **1.** 14 cm^2 **3.** 24 cm^2

Page 562 Skill Review **1.** scalene; obtuse **3.** equilateral; acute **5.** No **7.** 7 **9.** 20 **11.** 10.100 **13.** 36 **15.** $c = 13$ **17.** $b = 20$ **19.** $a = 15$ **21.** $x = 5$; $y = 4$; congruent **23.** $x = 27$; $y = 54$; congruent **25.** $\frac{12}{13}$ **27.** $\frac{5}{13}$ **29.** $\frac{5}{13}$ **31.** 87

Page 563 Exercises **1.** 9 ft **3.** 18 ft

Pages 564–566 Exercises **1.** 0.5000 **3.** 0.1392 **5.** 1.0000 **7.** $a \approx 4.0$; $b \approx 5.7$

9. $a \approx 10.3$; $b \approx 14.7$ **11.** $a \approx 8.0$; $c \approx 17.0$ **13. a.** 60° **b.** 90° **c.** 5 **d.** 8.7 **15.** 13 ft **17.** 38 m

Pages 569–570 Exercises **13.** Yes **15.** *R* and *S* are on the circle. **17.** Fold the paper so that the sides of the angle land directly on top of each other. The crease in the paper will be the bisector of the angle.

Page 573 Chapter Review **1.** obtuse **3.** 71°; acute **5.** 37°; right **7.** 12 **9.** 60 **11.** 132 **13.** $\frac{4}{3}$ **15.** $\frac{4}{5}$ **17.** $\frac{3}{5}$ **19.** 33

Extra Practice

Chapter 1, page 580 **1.** 278 and 361 thousandths **3.** 11 billion, 320 thousand, 434 **5.** 5302 ten-thousandths **7.** 75 and 4 hundredths **9.** 24,034,000,129 **11.** 9,019,211 **13.** two hundred sixty-two thousand, seventeen **15.** 129,400 **17.** 200,600 **19.** 60,405,000 **21.** thousand **23.** million **25.** billion **27.** 700 **29.** \$25 **31.** \$14.80 **33.** 11 **35.** 100 **37.** \$87.53 **39.** 66,000 **41.** $<$ **43.** $<$ **45.** $>$ **47.** $<$ **49.** $>$ **51.** $<$

Chapter 2, page 581 **1.** 196 **3.** \$24.25 **5.** 12,041 **7.** \$82 **9.** \$7.20 **11.** 17.29 **13.** 342 **15.** \$1101 **17.** 2658 **19.** 14.4 **21.** \$65.75 **23.** 912 **25.** \$6.76 **27.** about \$2100 **29.** about 300 **31.** about \$40 **33.** about 70 **35.** about 70 **37.** about 120 **39.** You need to know how much money she gave the salesperson. **41.** 187

Chapter 3, page 582 **1.** between \$100 and \$120 **3.** between 1600 and 2700 **5.** between \$300 and \$600 **7.** between 12,000 and 20,000 **9.** between 800 and 1500 **11.** about \$540 **13.** about 2800 **15.** about 400 **17.** about 21,000 **19.** about \$30,000 **21.** about 70,000 **23.** \$2112 **25.** 1519 **27.** 3681 **29.** \$2220 **31.** 1260 **33.** 5704 **35.** 239,730 **37.** 95,190 **39.** 216,980 **41.** \$11,210 **43.** 5478 **45.** 3200 **47.** 294 **49.** 316 **51.** 344,448 **53.** 370 **55.** 333 **57.** 294 **59.** 463,524 **61.** 4300 **63.** 190 **65.** 27 **67.** 125 **69.** 64 **71.** 10,000 **73.** 192 **75.** 343 **77.** 448 **79.** 1296

Chapter 4, page 583 **1.** about 900 **3.** about 1000 **5.** about 9000 **7.** about 1000 **9.** about 2000 **11.** about 9000 **13.** about 2000 **15.** about 800 **17.** 2291 **19.** 23,509 R1 **21.** 18 R23 **23.** 1439 **25.** 532 R21 **27.** 18,790 R18 **29.** 29,356 R31 **31.** 2782 R54 **33.** 12 **35.** 95 **37.** 2 **39.** 2 **41.** 4 **43.** 57 **45.** 89 **47.** 4 **49.** \$19.75 **51.** 456

Chapter 5, page 584 **1.** 196.2 **3.** 1.905 **5.** 14.098 **7.** \$10.80 **9.** \$495.18 **11.** \$.15 **13.** \$33.60 **15.** about 50 **17.** about 600 **19.** about 400 **21.** about 3.6 **23.** about \$54 **25.** about 2000 **27.** about \$600 **29.** \$7.03 **31.** 2.957 **33.** 220 **35.** 12 **37.** 18.139 **39.** 3071 **41.** \$1.89 **43.** about 200 **45.** about 0.4 **47.** about 0.09 **49.** about 70 **51.** about 7 **53.** about 0.08

Chapter 6, page 585 **1.** cm **3.** mm **5. a.** 34 mm **b.** 3.4 cm **7.** 72 **9.** 12.3 **11.** 1900 **13.** 11 cm **15.** L **17.** L **19.** 92.371 **21.** 295 **23.** 4300 **25.** g **27.** mg **29.** 0.163 **31.** 300 **33.** 17,000 **35.** \$10.75

Chapter 7, page 586 **1.** No **3.** Yes **5.** Yes **7.** Yes **9.** $3 \times 5 \times 13$ **11.** $2^3 \times 13$ **13.** 3×7^2 **15.** 14; 28 **17.** 5; 105 **19.** 2; 132 **21.** 5; 1040 **23.** 23, 29, 31 **25.** 3 **27.** 1 **29.** $\frac{2}{5}$ **31.** $\frac{1}{2}$ **33.** $\frac{31}{35}$ **35.** $<$ **37.** $<$ **39.** $\frac{7}{18}, \frac{4}{9}, \frac{2}{3}$ **41.** $1\frac{3}{14}, 1\frac{1}{4}, 1\frac{2}{7}$ **43.** about $\frac{1}{3}$ **45.** about $\frac{1}{5}$ **47.** about $\frac{2}{5}$ **49.** 1 **51.** $\frac{1}{2}$ **53.** 1 **55.** about $\frac{3}{4}$

Chapter 8, page 587 **1.** $\frac{2}{7}$ **3.** $1\frac{7}{8}$ **5.** $\frac{3}{4}$ **7.** $4\frac{4}{9}$ **9.** about 28 **11.** about 55 **13.** about 30 **15.** about 3 **17.** about 36 h **19.** $\frac{8}{7}$ **21.** $\frac{1}{10}$ **23.** $\frac{2}{7}$ **25.** $1\frac{3}{5}$ **27.** $\frac{1}{9}$ **29.** $8\frac{2}{5}$ **31.** $\frac{1}{2}$ **33.** about 3 **35.** about 4 **37.** about 8 **39.** about 45 **41.** 3 **43.** \$1.75

Chapter 9, page 588 **1.** 1 **3.** $\frac{7}{9}$ **5.** $\frac{2}{15}$ **7.** $1\frac{11}{20}$ **9.** $\frac{5}{24}$ **11.** $\frac{11}{20}$ **13.** about 23 **15.** about 5 **17.** about 56 **19.** about 14 **21.** about 13 **23.** about 15 **25.** about 21 **27.** about 11 **29.** $11\frac{1}{2}$ **31.** $13\frac{7}{10}$ **33.** $9\frac{19}{24}$ **35.** 14 **37.** $13\frac{1}{2}$ **39.** $9\frac{5}{21}$ **41.** $12\frac{1}{4}$ mi **43.** $5\frac{1}{6}$ **45.** $2\frac{1}{24}$ **47.** $25\frac{13}{28}$ **49.** $11\frac{15}{16}$ **51.** $6\frac{1}{3}$ **53.** $7\frac{17}{35}$

Chapter 10, page 589 **1.** $\frac{4}{5}$ **3.** $\frac{13}{40}$ **5.** $\frac{1}{500}$ **7.** $2\frac{7}{50}$ **9.** $7\frac{1}{40}$ **11.** $\frac{2}{7}$ **13.** $\frac{1}{3}$ **15.** $\frac{1}{6}$ **17.** $\frac{23}{45}$ **19.** $\frac{3}{80}$ **21.** 0.6 **23.** 0.52 **25.** 13.375 **27.** $0.1\overline{6}$ **29.** $7.1\overline{3}$ **31.** $<$ **33.** $<$ **35.** $>$ **37.** $=$ **39.** $6\frac{3}{4}$ oz; $6\frac{3}{4} = 6.75$, $6.75 > 6.5$ **41.** about 1 **43.** about $\frac{3}{5}$ **45.** about $\frac{3}{4}$ **47.** about $24\frac{2}{3}$ **49.** about $18\frac{1}{2}$ **51.** about \$180 **53.** about \$6 **55.** about 90 **57.** about \$3000 **59.** 13 aprons; 22 potholders

Chapter 11, page 590 **1.** 5 **3.** 156 **5.** 69 **7.** 152 **9.** 23 **11.** $1\frac{5}{8}$ in. **13.** 112 **15.** 12,000 **17.** 14,400 **19.** 4584 **21.** 8 **23.** 11 **25.** 93 **27.** 34 **29.** 53 **31.** 39 ft 3 in. **33.** 22 lb 2 oz **35.** 4 t 1550 lb

37. 3 yd 8 in. **39.** Yes; 6×2 ft 8 in. = 12 ft 48 in. = 16 ft; 18 ft $>$ 16 ft **41.** 4; 10 mi **43.** 3; 10 lb **45.** 3280 ft **47.** 16,000 mi

Chapter 12, page 591 **1.** $\frac{9}{25}$ **3.** $\frac{13}{3}$ **5.** $\frac{2}{7}$ **7.** $\frac{16}{7}$ **9.** $\frac{40}{3}$ **11.** $\frac{36}{5}$ **13.** 55 mi/h **15.** 21 books/shelf **17.** 21.5 mi/gal **19.** \$6.35/h **21.** False **23.** False **25.** False **27.** True **29.** 3 **31.** 2 **33.** 10 **35.** 4 **37.** 48 **39.** 4 **41.** \$23.75 **43.** \$2.40

Chapter 13, page 592 **1.** 0.27 **3.** 0.09 **5.** 85% **7.** 50% **9.** $3\frac{3}{4}$% **11.** $0.15\frac{1}{3}$ **13.** 140% **15.** 0.6% **17.** 0.089 **19.** 0.004 **21.** $\frac{11}{50}$ **23.** $1\frac{3}{4}$ **25.** $\frac{23}{500}$ **27.** $\frac{3}{500}$ **29.** $\frac{113}{200}$ **31.** 5% **33.** $12\frac{1}{2}$% **35.** 16% **37.** $18\frac{3}{4}$% **39.** 250% **41.** 164 **43.** 16 **45.** 147 **47.** 21.648 **49.** 0.0671 **51.** \$728 **53.** about 60 **55.** about \$200 **57.** about \$10 **59.** about \$40 **61.** about 1000

Chapter 14, page 593 **1.** 80% **3.** $37\frac{1}{2}$% **5.** 14% **7.** $5\frac{5}{7}$% **9.** increase; $83\frac{1}{3}$% **11.** decrease; 40% **13.** decrease; $66\frac{2}{3}$% **15.** 14 **17.** 3250 **19.** 172 **21.** 37.7 **23.** 8% **25.** $83\frac{1}{3}$% **27.** 640 **29.** \$7

Chapter 15, page 594 **1.** \$42 **3.** \$1495.66 **5.** \$573.75 **7.** \$585.33 **9.** \$570 **11.** \$749.70 **13.** \$2194.20 **15.** \$150; \$2150 **17.** \$525.31 **19.** \$1623.60 **21.** \$4579.60 **23.** \$3746.16 **25.** \$889.59 **27.** \$2175.87 **29.** \$1583.63

Chapter 16, page 595 **1.** about 100 **3.** about 250 **7.** \$2500 **9.** about $\frac{1}{4}$ **11.–15.** Answers may vary. **11.** about \$13,000 **13.** about \$16,000 **15.** about \$18,000

Chapter 17, page 596 **1.** 38; 45; 52; 42 **3.** 4.9; 4.25; 2.7; 7.9 **7.** 43 **11.** 60 and 80

Chapter 18, page 597 **1.** $\frac{1}{6}$ **3.** 1 **5.** $\frac{1}{3}$ **7.** $\frac{1}{3}$ **9.** 1 to 2 **11.** 5 to 1 **13.** 1 to 5 **15.** 1 to 5 **17.** 24 **19.** $\frac{1}{80}$ **21.** $\frac{7}{80}$ **23.** $\frac{1}{16}$ **25.** $\frac{1}{18}$ **27.** $\frac{1}{18}$ **29.** $\frac{1}{10}$ **31.** $\frac{1}{5}$ **33.** 20; 10 **35.** 840; 35 **37.** 16

Chapter 19, page 598 **1.** $\frac{7}{1}$ **3.** $\frac{123}{1}$ **5.** $\frac{21}{4}$ **7.** $^{-}\frac{9}{10}$ **9.** $\frac{214}{100}$ **11.** $\frac{1}{3}$ **13.** $^{-}15$; 15 **15.** 0; 0 **17.** 23.8; 23.8 **19.** $^{-}\frac{5}{8}$; $\frac{5}{8}$ **21.** $19\frac{4}{11}$; $19\frac{4}{11}$ **23.** $<$ **25.** $>$ **27.** $<$ **29.** $>$ **31.** $^{-}8$; $^{-}7$; 6; 7 **33.** $^{-}4\frac{1}{6}$; $^{-}\frac{5}{6}$; $1\frac{1}{6}$; $2\frac{1}{6}$ **35.** 28 **37.** $^{-}30$ **39.** 22 **41.** $^{-}4.9$ **43.** $^{-}3.3$ **45.** $^{-}3\frac{1}{2}$ **47.** $^{-}7$ **49.** 53 **51.** $^{-}27$ **53.** 47.9 **55.** 2.3 **57.** 1 **59.** 24 **61.** $^{-}42$ **63.** $^{-}6$ **65.** 4 **67.** $\frac{6}{35}$ **69.** $^{-}48$ **71.** $^{-}2$°F

Chapter 20, page 599 **1.** 7 **3.** 4 **5.** $^{-}12$ **7.** 64 **9.** 0 **11.** $^{-}5$ **13.** $^{-}2$ **15.** $^{-}\frac{2}{7}$ **17.** $^{-}5$ **19.** $^{-}4$ **21.** $^{-}3.6$ **23.** 189 **25.** 3 **27.** $^{-}63$ **29.** $^{-}18$ **31.** $y > 4$ **33.** $a \leq {}^{-}5$ **35.** $m > 0$ **37.** \$21.50 **39.** 200 mi

Chapter 21, page 600 **1.** Q **3.** T **5.** V **7.** O **9.** K **11.** M **23.** No **25.** No **27.** Yes **29.** Yes

31.

$y = x - 6$		
x	y	(x, y)
$^{-}2$	$^{-}8$	$(^{-}2, {}^{-}8)$
$^{-}1$	$^{-}7$	$(^{-}1, {}^{-}7)$
0	$^{-}6$	$(0, {}^{-}6)$
1	$^{-}5$	$(1, {}^{-}5)$
2	$^{-}4$	$(2, {}^{-}4)$

33.

$y = {}^{-}5x$		
x	y	(x, y)
$^{-}2$	10	$(^{-}2, 10)$
$^{-}1$	5	$(^{-}1, 5)$
0	0	$(0, 0)$
1	$^{-}5$	$(1, {}^{-}5)$
2	$^{-}10$	$(2, {}^{-}10)$

35.

$y = {}^{-}2x + 9$		
x	y	(x, y)
$^{-}2$	13	$(^{-}2, 13)$
$^{-}1$	11	$(^{-}1, 11)$
0	9	$(0, 9)$
1	7	$(1, 7)$
2	5	$(2, 5)$

49. Yes **51.** $(^{-}2, {}^{-}1)$ **53.** No solution **55.** $(0, {}^{-}4)$

Chapter 22, page 601 **1.–5.** Answers may vary. Examples are given. **1.** P, R, S, V, T, W, U, Q **3.** $\overline{RV}$, $\overline{VS}$, $\overline{TW}$ **5.** $\angle UWQ$ and $\angle VWT$; $\angle PVR$ and $\angle SVW$ **11.** complementary **13.** supplementary **15.** parallelogram **17.** trapezoid **19.** 56 m **21.** 11.3 cm **23.** 47.1 cm **25.** 396 m **27.** 16.32 m^2 **29.** 187 yd^2 **31.** 3850 m^2

Chapter 23, page 602 **1.** sphere **3.** cube **5.** 2072 $in.^2$ **7.** 238.64 cm^2 **9.** 1728 yd^3 **11.** 1078 $in.^3$ **13.** 1808.64 mm^2; 7234.56 mm^3 **15.** 28 cm^2

Chapter 24, page 603 **1.** scalene; right **3.** scalene; obtuse **5.** Yes **7.** 5 **9.** 70 **11.** 9.747 **13.** 62 **15.** $c = 25$ **17.** $b = 12$ **19.** $a = 15$ **21.** $x = 5$, $y = 12$; not congruent **23.** $x = 9$, $y = 15$; not congruent **25.** $\frac{5}{13}$ **27.** $\frac{12}{13}$ **29.** $\frac{12}{13}$ **31.** 21